Contributions in Mathematical and Computational Sciences • Volume 7

Editors
Hans Georg Bock
Willi Jäger
Otmar Venjakob

For further volumes:
http://www.springer.com/series/8861

Thanasis Bouganis • Otmar Venjakob
Editors

Iwasawa Theory 2012

State of the Art and Recent Advances

Springer

Editors
Thanasis Bouganis
Dept. of Mathematical Sciences
Durham University
Durham
United Kingdom

Otmar Venjakob
Institute of Mathematics
University of Heidelberg
Heidelberg
Germany

ISSN 2191-303X ISSN 2191-3048 (electronic)
Contributions in Mathematical and Computational Sciences
ISBN 978-3-642-55244-1 ISBN 978-3-642-55245-8 (eBook)
DOI 10.1007/978-3-642-55245-8

Springer Heidelberg New York Dordrecht London

Library of Congress Control Number: 2014957536

Mathematics Subject Classification (2010): 11R23

Printed on acid-free paper

Springer is part of Springer Science+Business Media (www.springer.com)

Preface to the Series

Contributions to Mathematical and Computational Sciences

Mathematical theories and methods and effective computational algorithms are crucial in coping with the challenges arising in the sciences and in many areas of their application. New concepts and approaches are necessary in order to overcome the complexity barriers particularly created by nonlinearity, high-dimensionality, multiple scales and uncertainty. Combining advanced mathematical and computational methods and computer technology is an essential key to achieving progress, often even in purely theoretical research.

The term mathematical sciences refers to mathematics and its genuine sub-fields, as well as to scientific disciplines that are based on mathematical concepts and methods, including sub-fields of the natural and life sciences, the engineering and social sciences and recently also of the humanities. It is a major aim of this series to integrate the different sub-fields within mathematics and the computational sciences, and to build bridges to all academic disciplines, to industry and other fields of society, where mathematical and computational methods are necessary tools for progress. Fundamental and application-oriented research will be covered in proper balance.

The series will further offer contributions on areas at the frontier of research, providing both detailed information on topical research, as well as surveys of the state-of-the-art in a manner not usually possible in standard journal publications. Its volumes are intended to cover themes involving more than just a single "spectral line" of the rich spectrum of mathematical and computational research.

The Mathematics Center Heidelberg (MATCH) and the Interdisciplinary Center for Scientific Computing (IWR) with its Heidelberg Graduate School of Mathematical and Computational Methods for the Sciences (HGS) are in charge of providing and preparing the material for publication. A substantial part of the material will be acquired in workshops and symposia organized by these institutions in topical areas of research. The resulting volumes should be more than just proceedings collecting papers submitted in advance. The exchange of information and the discussions during the meetings should also have a substantial influence on the contributions.

This series is a venture posing challenges to all partners involved. A unique style attracting a larger audience beyond the group of experts in the subject areas of specific volumes will have to be developed.

Springer Verlag deserves our special appreciation for its most efficient support in structuring and initiating this series.

Heidelberg, Germany

Hans Georg Bock
Willi Jäger
Otmar Venjakob

Preface

Iwasawa Theory is one of the most active fields of research in modern Number Theory. The great interest in Iwasawa Theory is reflected by the highly successful bi-annual series of international conferences, starting in 2004 in Besancon and continuing in Limoges, Irsee and Toronto with a scientific committee formed by John Coates, Ralph Greenberg, Cornelius Greither, Masato Kurihara, and Thong Nguyen Quang Do. The Iwasawa 2012 Conference, organized by Otmar Venjakob and Thanasis Bouganis, took place in Heidelberg (July 30–August 3) and drew in over 120 participants. It was supported by the Mathematics Center Heidelberg (MATCH) and by the European Research Council (ERC) Starting Grant IWASAWA awarded to Otmar Venjakob. This volume, Iwasawa Theory 2012 – State of the Art and Recent Advances, presents research and overview articles contributed by conference speakers and participants, as well as lecture notes from an introductory mini-course given by Chris Wultrich and Xin Wan and held the week before the conference.

One can argue that Iwasawa Theory has its roots in the early nineteenth century and in the work of Ernst Kummer (29 January 1810–14 May 1893), who studied the class number of the cyclotomic field $Q(\zeta_p)$, in his approach to prove Fermat's Last Theorem. Kummer not only provided a solution to the theorem for a large class of prime exponents, but also discovered a link between the p-divisibility of the class number of $Q(\zeta_p)$ and the values of the Riemann zeta function at the negative integers. This link between arithmetic expressions and special values of zeta functions, which was later refined in the work of Herbrand and Ribet, lies at the heart of modern number theory. It is the earliest example of a range of highly conjectural deep relations between arithmetic expressions and L-values, the most celebrated of which is the Conjecture of Birch and Swinnerton-Dyer.

However it was Kenkichi Iwasawa (September 11, 1917–October 26, 1998) and his Main Conjecture that completely transformed our view of the arithmetic of cyclotomic fields. Indeed Iwasawa, inspired by the work of Andre Weil on the Zeta Function of varieties over finite fields, initiated the systematic investigation of the p-part of the class number in the cyclotomic extension of Q. Not only did he manage to prove his deep theorems with respect to the growth of the p-part of the class number in such extensions; he also formulated his Main Conjecture, which relates the size of a particular Galois module to the Kubota-Leopold p-adic L-function. This conjecture would go on to serve as the prototype for an array of Main Conjectures, which predict a deep relation between p-adic L-functions and arithmetic invariants of abelian varieties or, even more generally, motives.

The Main Conjecture for cyclotomic fields is now a theorem, and considerable progress has also been made on other fronts, such as the Main Conjectures for CM fields, for elliptic modular forms and the Main Conjectures for abelian varieties over function fields. The proofs of all these Main Conjectures involve an impressive combination of various strands of pure mathematics such as K-theory, automorphic forms and algebraic geometry, contributing enormously to the popularity of the subject. Iwasawa Theory has not stopped growing in terms of its complexity and generalization. Undoubtedly the work of Hida, and his investigation of what are now referred to as Hida families, has transformed the way that we view Iwasawa Theory today. There has been also great interest in extending Iwasawa Theory to a non-abelian setting, where the focus is on the arithmetic behavior of the underlying motive over a p-adic Lie extension. A vast generalization of the Main Conjectures to this non-abelian setting has now been formulated and there have already been some first results, both in the number field and in the function field case.

It is exactly these astonishing new and rapid developments that the Iwasawa Conference series seeks to address. The main aim is to bring together experts from different strands in or closely related to Iwasawa Theory to report on recent developments and exchange ideas. The series has also established a tradition of very lively and pleasant meetings, a tradition that was strengthened by the 2012 conference. Events such as the half-day-long cruise on the Neckar River undoubtedly helped to create an inviting and stimulating atmosphere for the conference participants.

The week before the 2012 conference a preparatory mini-course was offered by Chris Wultrich and Xin Wan, aimed to introduce graduate students and newcomers to the field. While Chris Wultrich offered an overview of several basic aspects of Iwasawa Theory, Xin Wan provided an introduction to the work of Skinner and Urban on the Main Conjecture for elliptic modular forms. Their lecture notes, *Overview of Some Iwasawa Theory* by Chris Wultrich and *Introduction to Skinner-Urban's Work on the Main Conjecture for GL2* by Xin Wan, are now presented as part of this volume. The organizers would like to take this opportunity to thank them again for their excellent lecture series in the summer of 2012 and their contributions to this volume.

The talks given in the Iwasawa 2012 conference covered the wide range of development in Iwasawa Theory over the past several years, and were complemented by a poster session. Not every contribution in this volume is based on a talk given during the conference; some of the contributions are survey articles, while others are original research articles appearing for first time in printed form.

Acknowledgements It is of course only the efforts of the contributors that made this volume possible, and the editors are grateful to them. Moreover the editors would like to thank the referees for their valued work, and to thank once again the speakers and the participants of the

2012 conference and the preparatory lecture series. Further the editors would like to express their gratitude to Birgit Schmoetten-Jonas for her support with organizing the 2012 conference and editing this volume, which has been nothing less than indispensable. Lastly, it is our pleasure to thank MATCH and ERC for their financial support, as well as Mrs. Allewelt and Dr. Peters from Springer Verlag for their excellent collaboration in editing this volume.

Durham, UK Thanasis Bouganis
Heidelberg, Germany Otmar Venjakob

Contents

Part I
Lecture Notes

Overview of Some Iwasawa Theory

Christian Wuthrich

1 Introduction

These are the notes to lectures given at Heidelberg in July 2012. The intention was to give an concise overview of some topics in Iwasawa theory to prepare the participants for the conference. As a consequence, they will contain a lot of definitions and results, but hardly any proofs and details. Especially I would like to emphasise that the word "proof" should be replaced by "sketch of proof" in all cases below. Also, I have no claim at making this a complete introduction to the subject, nor is the list of references at the end. For this the reader might find (Greenberg 2001) a better source.

The talks were given in four sessions, which form the four sections of these notes. We start by the classical Iwasawa theory for the class group, including the fundamental result of Iwasawa on the growth of class groups in $\mathbb{Z}_p$-extensions. We also describe Stickelberger elements, cyclotomic units and the main conjecture. This first section also contains the basic facts about Iwasawa algebras.

The second section introduces Iwasawa theory for elliptic curves by studying the growth of the Selmer group. We define Mazur-Stickelberger elements and the p-adic L-functions and state the main conjecture in this context. The third section includes the proof of the control theorem for Selmer groups (in the ordinary case) and the formula for the leading term of the characteristic series of the Selmer group. The last section shows how one generalises Selmer groups to various Galois representations. We conclude with a rough and short explanation about Kato's Euler system.

C. Wuthrich (✉)
School of Mathematical Sciences, University of Nottingham, Nottingham, UK
e-mail: christian.wuthrich@nottingham.ac.uk; christian.wuthrich@gmail.com

T. Bouganis and O. Venjakob (eds.), *Iwasawa Theory 2012*, Contributions in Mathematical and Computational Sciences 7, DOI 10.1007/978-3-642-55245-8_1,

2 Iwasawa Theory of the Class Group

Let F be a number field and let p be an **odd** prime. Suppose we are given a tower of Galois extensions $F = F^{(0)} \subset F^{(1)} \subset F^{(2)} \subset \cdots$ such that the Galois group of $F^{(n)}/F$ is cyclic of order p^n for all $n \geqslant 1$. Write $C^{(n)}$ for the p-primary part of the class group of $F^{(n)}$ and write p^{e_n} for its order.

Theorem 1 (Iwasawa 1959). *There exist integers μ, λ, ν, and n_0 such that*

$$e_n = \mu\, p^n + \lambda\, n + \nu \qquad \textit{for all } n \geqslant n_0.$$

2.1 $\mathbb{Z}_p$-Extensions

Let me first describe the tower of extensions that we are talking about. Set $F^{(\infty)} = \bigcup F^{(n)}$. The extension $F^{(\infty)}/F$ is called a *$\mathbb{Z}_p$-extension* as its Galois group Γ is isomorphic to the additive group of p-adic integers since it is the projective limit of cyclic groups of order p^n. The most important example is the *cyclotomic $\mathbb{Z}_p$-extension*: If $F = \mathbb{Q}$, then the Galois group of $\mathbb{Q}(\mu_{p^n})/\mathbb{Q}$ is $\left(\mathbb{Z}/p^n\mathbb{Z}\right)^{\times}$, which is cyclic of order $(p-1)p^{n-1}$. So there is an extension $\mathbb{Q}^{(n-1)}/\mathbb{Q}$ contained in $\mathbb{Q}(\mu_{p^n})$ such that $\mathbb{Q}^{(\infty)}$ is a $\mathbb{Z}_p$-extension of $\mathbb{Q}$. For a general F, the cyclotomic $\mathbb{Z}_p$-extension is $F^{(\infty)} = \mathbb{Q}^{(\infty)} \cdot F$.

It follows from the Kronecker-Weber theorem that $\mathbb{Q}^{(\infty)}$ is the unique $\mathbb{Z}_p$-extension of $\mathbb{Q}$. It would be a consequence of Leopoldt's conjecture that the cyclotomic $\mathbb{Z}_p$-extension is the only one for any totally real number field, see Neukirch et al. (2000, Theorem 11.1.2). For a general number field F, the compositum of all $\mathbb{Z}_p$-extension contains at least $\mathbb{Z}_p^{r_2+1}$ in its Galois group where r_2 denotes the number of complex places in F. For an imaginary quadratic number field F, for instance, the theory of elliptic curves with complex multiplication provides us with another interesting $\mathbb{Z}_p$-extension, the *anti-cyclotomic $\mathbb{Z}_p$-extension*. It can be characterised as the only $\mathbb{Z}_p$-extension $F^{(\infty)}/F$ such that $F^{(\infty)}/\mathbb{Q}$ is a non-abelian Galois extension.

Lemma 1 (Proposition 11.1.1 in Neukirch et al. 2000). *The only places that can ramify in $F^{(\infty)}/F$ divide p and at least one of them must ramify.*

In the cyclotomic $\mathbb{Z}_p$-extension of F, all places above p are ramified and there are only finitely many places above all other places.

2.2 The Iwasawa Algebra and Its Modules

Let $\mathcal{O}$ be a "coefficient ring", for us this will always be the ring of integers in a p-adic field; so $\mathcal{O} = \mathbb{Z}_p$ is typical. There is a natural morphism between the group rings $\mathcal{O}\big[\mathrm{Gal}(F^{(n)}/F)\big]$, which allows us to form the limit

$$\Lambda = \varprojlim_{n} \mathcal{O}\big[\mathrm{Gal}(F^{(n)}/F)\big].$$

This completed topological group ring, called the *Iwasawa algebra* and also denoted by $\mathcal{O}[\![\Gamma]\!]$, is far better to work with than the huge group ring $\mathcal{O}[\Gamma]$.

Proposition 1. *To a choice of a topological generator γ of Γ, there is an isomorphism from Λ to the ring of formal power series $\mathcal{O}[\![T]\!]$ sending γ to $T+1$.*

The proof is given in Theorem 5.3.5 of Neukirch et al. (2000). By the Weierstrass preparation theorem, an element $f(T) \in \mathcal{O}[\![T]\!]$ can be written as a product of a power of the uniformiser of $\mathcal{O}$ times a unit of $\mathcal{O}[\![T]\!]$ and times a distinguished polynomial, which, by definition, is a monic polynomial whose non-leading coefficients belong to the maximal ideal.

Let $X^{(n)}$ be a system of abelian groups with an action by $\mathrm{Gal}(F^{(n)}/F)$. If there is a naturally defined norm map $X^{(n+1)} \to X^{(n)}$, then we can form $X = \varprojlim X^{(n)}$ and consider it as a compact Λ-module. For instance the class groups $C^{(n)}$ above have a natural norm map between them. Also lots of naturally defined cohomology groups will have such a map, too. Suppose now $\mathcal{O}/\mathbb{Z}_p$ is unramified, otherwise the power of p below must be replaced by a power of the uniformiser of $\mathcal{O}$.

Proposition 2. *Let X be a finitely generated Λ-module. Then there exist integers $r, s, t, m_1, m_2, \ldots, m_s, n_1, n_2, \ldots n_t$, irreducible distinguished polynomials $f_1, f_2, \ldots f_t$, and a morphism of Λ-modules*

$$X \longrightarrow \Lambda^r \oplus \bigoplus_{i=1}^{s} \Lambda \big/ p^{m_i}\Lambda \oplus \bigoplus_{j=1}^{t} \Lambda \big/ f_j^{n_j}\Lambda$$

whose kernel and cokernel are finite.

Proofs can be found in Serre (1995), Neukirch et al. (2000, Theorem 5.3.8) or quite different in Washington (1997, Theorem 13.12) and Lang (1990, Theorem 5.3.1). The main reason is that Λ is a 2-dimensional local, unique factorisation domain.

As the ideals $f_j\Lambda$ and the integers $r, \ldots, n_t$ are uniquely determined by X, we can define the following invariants attached to X. The *rank* of X is $\mathrm{rank}_\Lambda(X) = r$. The *$\mu$-invariant* is $\mu(X) = \sum_{i=1}^{s} m_i$ and the *λ-invariant* is $\lambda(X) = \sum_{j=1}^{t} n_j \cdot \deg(f_j)$. Finally, if $r = 0$,

$$\mathrm{char}(X) = p^{\mu(X)} \cdot \prod_{j=1}^{t} f_j^{n_j}\,\Lambda$$

is called the *characteristic ideal* of X. If $r = 0$, $s \leqslant 1$ and all f_j are pairwise coprime, then $X \to \Lambda/\mathrm{char}(X)$ has finite kernel and cokernel.

Let us summarise the useful properties of Λ-modules in a lemma. Write $X_{\Gamma^{(n)}}$ for the largest quotient of X on which $\Gamma^{(n)} = \mathrm{Gal}(F^{(\infty)}/F^{(n)})$ acts trivially.

Lemma 2. *Let X be a Λ-module.*

1. *X is finitely generated if and only if X is compact and X_Γ is a finitely generated $\mathbb{Z}_p$-module.*
2. *Suppose X is a finitely generated Λ-module. Then X is Λ-torsion, i.e., the Λ-rank of X is 0, if and only if $X_{\Gamma^{(n)}}$ has bounded $\mathbb{Z}_p$-rank.*
3. *If $X_{\Gamma^{(n)}}$ is finite for all n, then there are constants ν and n_0 such that $|X_{\Gamma^{(n)}}| = p^{e_n}$ with $e_n = \mu(X) \cdot p^n + \lambda(X) \cdot n + \nu$ for all $n \geqslant n_0$.*

Proofs can be found in §5.3 of Neukirch et al. (2000). Note that if $X = \Lambda/f$ for an irreducible f, then $X_{\Gamma^{(n)}}$ is finite, unless f is a factor of the distinguished polynomial $\omega^{(n)} = (1+T)^{p^n} - 1$ corresponding to a topological generator of $\Gamma^{(n)}$.

2.3 *Proof of Iwasawa's Theorem*

I will sketch the proof of Theorem 1 only in the simplified case when F has a single prime $\mathfrak{p}$ above p and that this prime is totally ramified in $F^{(\infty)}/F$. Let $L^{(n)}$ be the p-Hilbert class field of $F^{(n)}$, i.e., the largest unramified extension of $F^{(n)}$ whose Galois group is abelian and a p-group. By class field theory the Galois group of $L^{(n)}/F^{(n)}$ is isomorphic to $C^{(n)}$.

Set $L^{(\infty)} = \bigcup L^{(n)}$, which is a Galois extension of $F^{(\infty)}$ with Galois group $X = \varprojlim C^{(n)}$. The action of $\Gamma^{(n)}$ on $C^{(n)}$ translates to an action of Γ on X given by the following. Let $\gamma \in \Gamma$ and $x \in X$. Choose a lift g of γ to the Galois group of $L^{(\infty)}/F$ and set $x^\gamma = gxg^{-1}$. So X is a compact Λ-module.

Define K to be the largest abelian extension of F inside $L^{(\infty)}$. Then $L^{(0)}$ and $F^{(\infty)}$ are contained in K. The maximality of K shows that the Galois group of $K/F^{(\infty)}$ must be equal to X_Γ.

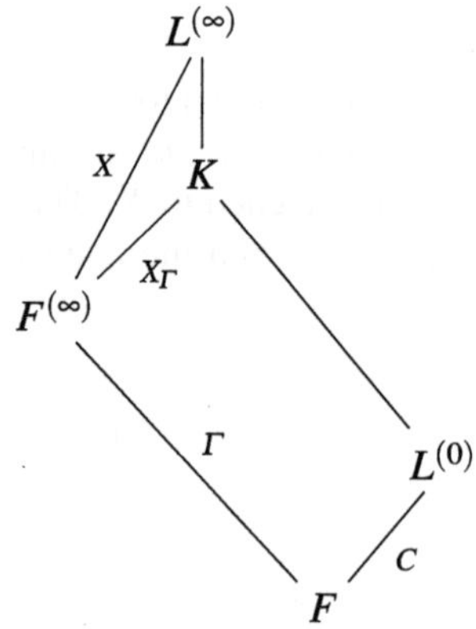

Since $K/F^{(\infty)}$ is unramified and $F^{(\infty)}/F$ is totally ramified at $\mathfrak{p}$, the inertia group I at a prime above $\mathfrak{p}$ in K gives a section of the map from $\mathrm{Gal}(K/F) \to \Gamma$.

Since $K^I = L^{(0)}$, we have $\mathrm{Gal}(K/F^{(\infty)}) = C^{(0)} =: C$ and it has a trivial action of Γ on it. Hence C is isomorphic to X_Γ. Replacing in this argument F by $F^{(n)}$, we can also conclude that $X_{\Gamma^{(n)}} \cong C^{(n)}$.

In particular, it is always finite. Hence X is a finitely generated torsion Λ-module and Lemma 2 (c) implies the theorem. □

Iwasawa has given an example in Iwasawa (1973) of a $\mathbb{Z}_p$-extension with $\mu(X) > 0$, however he conjectured that $\mu(X) = 0$ whenever the tower is the cyclotomic $\mathbb{Z}_p$-extension. This was shown to be true by Ferrero–Washington (1979) when $F/\mathbb{Q}$ is abelian.

The above proof can also be used to show that if $F = \mathbb{Q}$ then the class group of $\mathbb{Q}^{(n)}$ has no p-torsion. Conjecturally this may even be true for $F = \mathbb{Q}(\mu_p)^+$, see Sect. 2.8.

2.4 *Stickelberger Elements*

Let K be an abelian extension of $\mathbb{Q}$. By the Kronecker-Weber theorem, there is a smallest integer m such that $K \subset \mathbb{Q}(\mu_m)$ called the conductor of K. For each $a \in (\mathbb{Z}/m\mathbb{Z})^\times$ write σ_a for the image of a under the map $(\mathbb{Z}/m\mathbb{Z})^\times \cong \mathrm{Gal}(\mathbb{Q}(\mu_m)/\mathbb{Q}) \to \mathrm{Gal}(K/\mathbb{Q}) = G$. The *Stickelberger element* for K is defined to be

$$\theta_K = -\frac{1}{m} \sum_{\substack{1 \leqslant a < m \\ (a,m)=1}} a \cdot \sigma_a^{-1} \in \mathbb{Q}[G]$$

and the *Stickelberger ideal* is $I = \mathbb{Z}[G] \cap \theta_K \mathbb{Z}[G]$. It is not difficult to show that $I = I'\theta_K$ with I' being the ideal in $\mathbb{Z}[G]$ generated by all $c - \sigma_c$ with $(c, m) = 1$, see Washington (1997, Lemma 6.9).

Theorem 2 (Stickelberger). *The Stickelberger ideal I annihilates the class group of K.*

This means that for any fractional ideal $\mathfrak{a}$ and any integer c coprime to m, the ideal becomes principal after applying $(c - \sigma_c)\,\theta_K$ to it. It is important to note that this theorem does not say anything interesting when K is totally real as then θ_K is a multiple of the norm $N_{K/\mathbb{Q}}$. Hence it will not give us information about the class number of $\mathbb{Q}(\mu_p)^+$. For a quadratic imaginary field K, this is an algebraic version of the analytic class number formula for K, see the remark (b) after theorem 6.10 in Washington (1997).

Here is the idea of the proof, for details see Washington (1997, Theorem 6.10) or Lang (1990, Theorem 2.4).

Proof. We consider only the case $K = \mathbb{Q}(\mu_p)$ for some odd prime p. In each ideal class there is a prime ideal $\mathfrak{q}$ of degree 1, i.e., it is split above some prime $\ell \equiv 1 \pmod{p}$. Take the Dirichlet character χ modulo ℓ of order p such that

$\chi(a) \equiv a^{(\ell-1)/p} \pmod{\mathfrak{q}}$ for all a. Fix a primitive ℓ-th root of unity ξ. The *Gauss sum* of χ is defined to be

$$\mathrm{Ga}(\chi) = - \sum_{u \bmod \ell} \chi(u)\, \xi^u \in \mathbb{Q}(\mu_p, \mu_\ell).$$

One can show that $\mathrm{Ga}(\chi) \cdot \overline{\mathrm{Ga}(\chi)} = \ell$ and that we have $\mathrm{Ga}(\chi)^{(c-\sigma_c)} \in \mathbb{Q}(\mu_p)$ for all c coprime to p. Finally a detailed analysis of the valuation of this Gauss sum at all primes above ℓ reveals that for any $\beta \in \mathbb{Z}[G]$ such that $\beta\theta_K \in \mathbb{Z}[G]$, we have that $\mathfrak{q}^{\beta\theta_K} = \big(\mathrm{Ga}(\chi)\big)^{\beta}\, \mathcal{O}_K$ is a principal ideal in the ring of integers $\mathcal{O}_K$ of K.

2.5 p-Adic L-Functions

Consider the cyclotomic $\mathbb{Z}_p$-extension $F^{(\infty)}$ of $F = \mathbb{Q}(\mu_p)$ for some odd prime p. Write $G^{(n)}$ for the Galois group of $F^{(n)} = \mathbb{Q}(\mu_{p^{n+1}})$ over $\mathbb{Q}$ and $G = \varprojlim G^{(n)} = \mathrm{Gal}(F^{(\infty)}/\mathbb{Q})$. Then $G \cong \Delta \times \Gamma$ with $\Delta = G^{(0)}$ and $\Gamma = \mathrm{Gal}(F^{(\infty)}/F)$. We write

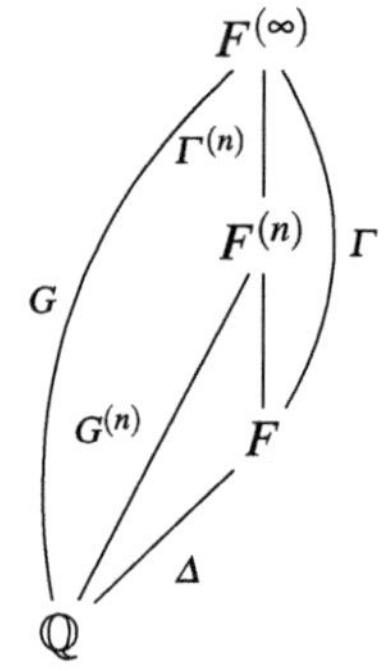

γ_a for the image of σ_a in Γ. The cyclotomic character $\chi\colon G \to \mathbb{Z}_p^\times$ splits accordingly into the Teichmüller character $\omega\colon \Delta \to \mathbb{Z}_p^\times$ and $\kappa\colon \Gamma \to 1+p\mathbb{Z}_p$. So for any $a \in \mathbb{Z}_p^\times$, the character ω sends σ_a to a $(p-1)$-st root of unity with $\omega(a) - a \in p\mathbb{Z}_p$ and $\kappa(\gamma_a) = \langle a \rangle = a/\omega(a)$.

For $i \in {}^{\mathbb{Z}}/_{(p-1)\mathbb{Z}}$, consider the projector

$$\varepsilon_i = \frac{1}{p-1} \sum_{a=1}^{p-1} \omega^i(a)\, \sigma_a^{-1} \in \mathbb{Z}_p[\Delta]$$

to the ω^i-eigenspaces. We split up the Stickelberger element $\theta^{(n)} = \theta_{F^{(n)}} \in \mathbb{Q}[G^{(n)}]$ for the field $F^{(n)}$ into $p-1$ elements $\theta_i^{(n)} \in \mathbb{Q}_p[\mathrm{Gal}(F^{(n)}/F)]$ defined by $\varepsilon_i \cdot \theta^{(n)} = \theta_i^{(n)} \cdot \varepsilon_i$; explicitly

$$\theta_i^{(n)} = -\frac{1}{p^{n+1}} \sum_{\substack{1 \leqslant a < p^{n+1} \\ p \nmid a}} a \cdot \omega^{-i}(a) \cdot \gamma_a^{-1} \in \mathbb{Q}_p[\mathrm{Gal}(F^{(n)}/F)].$$

Lemma 3. *If $i \neq 1$ then $\theta_i^{(n)} \in \mathbb{Z}_p[\mathrm{Gal}(F^{(n)}/F)]$ and if $i \neq 0, 1$, then $\theta_i = (\theta_i^{(n)})_{n\geqslant 1}$ belongs to $\varprojlim \mathbb{Z}_p[\mathrm{Gal}(F^{(n)}/F)] = \Lambda$. If $i \neq 0$ is even then $\theta_i = 0$.*

Recall that the generalised Bernoulli numbers for a Dirichlet character χ of conductor m are defined by

$$\sum_{a=1}^{m} \chi(a) \frac{t\, e^{at}}{e^{mt} - 1} = \sum_{r=0}^{\infty} B_{r,\chi} \frac{t^r}{r!}.$$

An explicit computation (Washington 1997, Theorem 7.10) links the elements θ_i to these Bernoulli numbers and the traditional Bernoulli numbers B_r. Recall that the $B_{r,\chi}$ and B_r also turn up as values of the complex L-function $L(s,\chi)$ and the Riemann zeta-function (Washington 1997, Theorem 4.2). Hence we find the interpolation property.

Theorem 3. *For any even integer $r \geqslant 1$, we have*

$$\kappa^{1-r}(\theta_{1-r}) = -(1 - p^{r-1}) \frac{B_r}{r} = (1 - p^{r-1})\, \zeta(1 - r).$$

Furthermore, for any $r \geqslant 1$ and any even $j \not\equiv r \pmod{p-1}$, we have

$$\kappa^{1-r}(\theta_{1-j}) = -\frac{B_{r,\omega^{j-r}}}{r} = L(1 - r, \omega^j).$$

For any $s \in \mathbb{Z}_p$ we can extend $\kappa^s\colon \Gamma \to 1 + p\mathbb{Z}_p$ linearly to $\kappa^s\colon \Lambda \to \mathbb{Z}_p$. The p-adic L-functions are defined to be $L_p(s,\omega^j) = \kappa^s(\theta_{1-j})$. To represent the p-adic L-function as a map $\chi \mapsto \chi(\theta_{1-j})$ is analogue to Tate's description of complex L-functions in his thesis; often these maps are written as measures on the Galois group Γ.

Now $L_p(s,\omega^j)$ is an analytic function in s and the existence of such a function satisfying the above theorem is equivalent to strong congruences between the values of $L(s,\omega^j)$ for negative integers s. For instance, one can deduce the Kummer congruences (Washington 1997, Corollary 5.14) from the theorem.

Leopoldt showed that $L_p(1,\omega^j)$ satisfies a p-adic analytic class number formula involving the p-adic regulator, see Washington (1997, Theorems 5.18 and 5.24). The p-adic L-function for $j = 0$ corresponding to θ_1 is not in Λ, instead it has a simple pole at $s = 1$.

The above L-functions are in fact the branches of the p-adic zeta-function discovered by Kubota and Leopoldt. There are generalisations to a much larger class of L-functions: Suppose K is a totally real number field and F/K an abelian extension of degree prime to p. Let χ be a character of the Galois group of F/K into the algebraic closure of $\mathbb{Q}_p$ and suppose that F is still totally real. Take $\mathcal{O}$ to be the ring $\mathbb{Z}_p[\chi]$ generated by the values of χ. Then there is a p-adic L-function $L_\chi \in \mathcal{O}[\![\Gamma]\!]$ such that $\kappa^s(L_\chi) = L_p(s,\chi)$ satisfies $L_p(1-r,\chi) = L(1-r,\chi\omega^{-r}) \cdot \prod_{\mathfrak{p}|p}(1 - \chi\omega^{-r}(\mathfrak{p})N(\mathfrak{p})^{r-1})$ for all $r \geqslant 1$. See for instance Wiles (1990).

2.6 The Main Conjecture

Let $3 \leqslant i \leqslant p-2$ be an odd integer. Consider the projective limit X of the p-primary parts of the class groups of $F^{(n)} = \mathbb{Q}(\mu_{p^{n+1}})$. Since Δ acts on this $\mathbb{Z}_p$-module, we can decompose it into eigenspaces for this action. Let $X_i = \varepsilon_i X$, which is now a finitely generated torsion $\Lambda = \mathbb{Z}_p[\![\Gamma]\!]$-module. Hence it makes sense to talk about its characteristic ideal.

Theorem 4 (Main conjecture). *The ideal* $\mathrm{char}(X_i)$ *is generated by* θ_i *for all odd* $3 \leqslant i \leqslant p-2$.

This was first proven by Mazur-Wiles (1984), then generalised to totally real fields by Wiles (1990). These proofs use crucially the arithmetic of modular forms. Later a proof was found using the Euler system of cyclotomic units, see Coates and Sujatha (2006) and the appendix in Lang (1990).

This theorem has many implications (some of which were known before the conjecture was proved). We can split up the p-primary part C of the class group of $\mathbb{Q}(\mu_p)$ into eigenspaces $C_i = \varepsilon_i C$

Theorem 5. *For every odd* $3 \leqslant i \leqslant p-2$, *the order of* C_i *is equal to the order of* $\mathbb{Z}_p/B_{1,\omega^{-i}}$.

Theorem 6 (Herbrand-Ribet (Ribet 1976)). *For any odd* $3 \leqslant i \leqslant p-2$, *the character* ω^i *appears in* C/C^p *if and only if* p *divides the numerator of* B_{p-i}.

2.7 Cyclotomic Units

The p-adic L-function can also be constructed out of the following units. For each c coprime to m, the element $(\zeta_m^c - 1)/(\zeta_m - 1)$ is a unit in $\mathbb{Z}[\zeta_m]$ where ζ_m a primitive m-th root of unity, called a *cyclotomic unit*. On the one hand they are linked to the p-adic L-function as m varies in the powers of p; in fact the p-adic L-functions can be obtained as a logarithmic derivatives of the Coleman series associated to the cyclotomic unit. See Propositions 2.6.3 and 4.2.4 in Coates and Sujatha (2006). On the other hand they are linked to the class group: When m is a power of p, the index of the group generated by the cyclotomic units and the roots of unity in $\mathbb{Q}(\mu_m)$ is equal to the class group order of $\mathbb{Q}(\mu_m)^+$ within the group of units in $\mathbb{Z}[\zeta_m]$.

The cyclotomic p-units $\zeta_m^c - 1$ form an Euler system, see §3.2 in Rubin (2000), the appendix in Lang (1990) and §5.2 in Coates and Sujatha (2006), due to the fact that they make the Euler factors of the L-function appear in their compatibility with respect to the norm map:

$$N_{\mathbb{Q}(\mu_{m\ell})/\mathbb{Q}(\mu_m)}(\zeta_{m\ell} - 1) = (1 - \sigma_\ell^{-1})(\zeta_m - 1)$$

for any prime $\ell \nmid m$. These special elements provides a powerful way of bounding the class group in terms of values of the p-adic L-function and yield a proof of the main conjecture.

2.8 Vandiver's Conjecture

The theory so far only covered the minus part of the class group, i.e., C_i for odd i. Note that $\oplus_{i \text{ even}} C_i$ is the p-primary part of the class group of $\mathbb{Q}(\mu_p)^+$.

Conjecture 1 (Vandiver). The class number of $\mathbb{Q}(\mu_p)^+$ is not divisible by p.

Although one may argue (see end of §5.4 in Washington 1997) that it is not likely to hold for all p, it is known to hold for all primes $p \leqslant 39 \cdot 2^{22}$ see Buhler and Harvey (2011). Moreover for all these 9,163,831 primes, the components C_i are always cyclic of order p and there are at most 7 non-trivial components. However, probably there are primes with C_i of order larger than p and probably the λ-invariant can get arbitrarily large.

It is known since Kummer that if p divides the class number of $\mathbb{Q}(\mu_p)^+$ then p divides $|C_i|$ for some odd i, see Corollary 8.17 in Washington (1997).

Proposition 3. *If Vandiver's conjecture holds for p, then C_i is isomorphic to $\mathbb{Z}_p/B_{1,\omega^{-i}}$ for all odd i. Moreover $C_i^{(n)}$ is a cyclic $\mathbb{Z}_p[\mathrm{Gal}(F_n/F)]$-module for all n.*

This is shown in Corollary 10.15 in Washington (1997).

Conjecture 2 (Greenberg 2001). If F is totally real, then $X = \varprojlim C^{(n)}$ is finite.

2.9 Examples

Let us first take $p = 5$ and so $i = 3$ is the only interesting value. We take γ_{1+p} to be the generator of Γ corresponding to $T + 1$. Then

$$\begin{aligned}\theta_3^{(4)} =& 2 + 2 \cdot 5 + 5^2 + 3 \cdot 5^3 + 4 \cdot 5^4 + \mathbf{O}(5^5) + \left(4 + 4 \cdot 5 + 5^2 + 4 \cdot 5^3 + \mathbf{O}(5^4)\right) \cdot T \\ &+ \left(1 + 5 + 4 \cdot 5^2 + \mathbf{O}(5^4)\right) \cdot T^2 + \mathbf{O}(T^3)\end{aligned}$$

which is congruent to θ_3 modulo $\omega^{(4)} = (1 + T)^{5^4} - 1$; in particular the above expression is the correct approximation for the 5-adic L-function θ_3. It is a unit in Λ as the leading term $-B_{1,\omega^2} = 2 + 2 \cdot 5 + \cdots$ is a 5-adic unit. Of course this is not surprising as the class group of $\mathbb{Q}(\mu_5)$ is trivial. So here $X = 0$ and $e_n = 0$ for all n.

Now to the first irregular prime $p = 37$. Here the Bernoulli number B_{32} is divisible by 37. Accordingly, we expect a non-trivial ω^5 part in the class group of $\mathbb{Q}(\mu_{37})$. Indeed the approximation to the 37-adic L-function is

$$\begin{aligned}\theta_5^{(3)} =& 14 \cdot 37 + 33 \cdot 37^2 + 13 \cdot 37^3 + \mathbf{O}(37^4) + \left(16 + 6 \cdot 37 + 32 \cdot 37^2 + \mathbf{O}(37^3)\right) \cdot T \\ &+ \left(29 + 9 \cdot 37 + 13 \cdot 37^2 + \mathbf{O}(37^3)\right) \cdot T^2 + \mathbf{O}(T^3).\end{aligned}$$

This is not a unit as $-B_{1,\omega^{-5}}$ is divisible by 37. From the fact that the second coefficient is a unit, we conclude that θ_5 is a unit times a linear factor. Hence X is a free $\mathbb{Z}_{37}$-module of rank 1 and $e_n = n + 1$ for all n. The fact which underlies the proof of Ribet's theorem is that the Eisenstein series

$$\begin{aligned}G &= -\frac{B_{32}}{2 \cdot 32} + \sum_{n \geqslant 1} \sum_{d \mid n} d^{31} q^n \\ &= \tfrac{7709321041217}{32640} + q + 2147483649\, q^2 + 617673396283948\, q^3 + 4611686020574871553\, q^4 + \cdots\end{aligned}$$

of weight 32 is congruent modulo one of the primes above 37 in $\mathbb{Q}(\mu_{12})$ to the cuspform

$$f = q + \zeta_{12}\, q^2 + (-\zeta_{12}^3 + \zeta_{12}^2 - \zeta_{12})\, q^3 - \zeta_{12}^2\, q^4 + (2\,\zeta_{12}^3 + \zeta_{12}^2 - 2\,\zeta_{12}^2 - 2)\, q^5 + \cdots$$

of weight 2 for the group $\Gamma_1(37)$ and character ω^{30}.

3 Iwasawa Theory for Elliptic Curves

3.1 Examples

Let $\mathbb{Q}^{(\infty)}/\mathbb{Q}$ be the cyclotomic $\mathbb{Z}_p$-extension of $\mathbb{Q}$ and let $E/\mathbb{Q}$ be an elliptic curve. The theorem of Mordell-Weil shows that the group $E(\mathbb{Q}^{(n)})$ is finitely generated for all n. Is this still true for $E(\mathbb{Q}^{(\infty)})$? In particular is the rank of $E(\mathbb{Q}^{(n)})$ bounded as n grows? The analogy with the case of global function fields suggests that this should be the case.

There is a second interesting group attached to E. For any elliptic curve E over a number field F, the Tate-Shafarevich group $\text{Ш}(E/F^{(n)})$ is a certain torsion abelian group whose definition we give in Sect. 3.2. We write $\text{Ш}^{(n)}$ for its p-primary part which is conjectured to be finite for all n. The first four examples were computed with the methods in Stein and Wuthrich (2013).

3.1.1 Example 1

Let E be the elliptic curve given by

$$E: \qquad y^2 + x\,y = x^3 - 6511\,x - 203353$$

which has $E(\mathbb{Q}) = Ш(E/\mathbb{Q}) = 0$ and it is labelled 174b2 in Cremona's table (Cremona 1997). It has bad reduction at 2 (additive), 3, and 29 (both split multiplicative).

If $p = 5$ then the rank of $E(\mathbb{Q}^{(n)})$ is zero for all n and the group $Ш^{(n)}$ is trivial, too. Since a p-torsion group can not act with a single fixed point on a p-primary group, we have that $E(\mathbb{Q}^{(n)})$ has no p-torsion for all n.

3.1.2 Example 2

Let us take the same curve but now with $p = 7$. Then the rank will still be zero for all n. However if $|Ш^{(n)}| = p^{e_n}$, then $e_n = p^n + 2n - 1$ for all $n \geqslant 0$. So the Tate-Shafarevich group will explode in this case. Note that this curve has a 7-isogeny defined over $\mathbb{Q}$ and one Tamagawa number is 7 and the number of points in the reduction over $\mathbb{F}_7$ has 7 points. So $p = 7$ appears in various places. In fact $Ш^{(n)}$ is formed of $p^n - 1$ copies of $\mathbb{Z}/p\mathbb{Z}$ and two copies of $\mathbb{Z}/p^n\mathbb{Z}$.

3.1.3 Example 3

Again with the same curve, but this time for the prime $p = 13$. Once more the rank remains 0 in the tower, however the p-primary part of $Ш(E/\mathbb{Q}^{(n)})$ grows with

$$e_n = \left\lfloor \frac{p}{p^2 - 1}\, p^n - \frac{n}{2} \right\rfloor$$

for all n. This formula is shown in Kurihara (2002). For instance $e_0 = e_1 = 0$, $e_2 = 12$, $e_3 = 168$, $e_4 = 2{,}208, \ldots$ Visibly the growth does not obey the same type of regularity as in the previous examples. The difference is that E has supersingular reduction at $p = 13$.

3.1.4 Example 4

Let us consider now the curve 5692a1

$$E: \qquad y^2 = x^3 + x^2 - 18\,x + 25$$

which has $E(\mathbb{Q}) = \mathbb{Z}(0,5) \oplus \mathbb{Z}(1,3)$. For $p = 3$, one can show that the rank is 6 over $\mathbb{Q}^{(1)}$ and it is 12 for all $\mathbb{Q}^{(n)}$ with $n \geqslant 2$. The 3-primary part of $Ш(E/\mathbb{Q}^{(n)})$ is trivial for all n. Note however that we do not know if $Ш(E/\mathbb{Q})$ is finite or not.

3.1.5 Example 5

Finally, consider the curve 11a3

$$E: \qquad y^2 + y = x^3 - x^2$$

and consider the anti-cyclotomic $\mathbb{Z}_3$-extension above $F = \mathbb{Q}(\sqrt{-7})$. The construction of Heegner points allows us to produce points of infinite order $P^{(n)} \in E(F^{(n)})$. The tower of points is compatible in the sense that the trace of $P^{(n)}$ to the layer below is $(-1) \cdot P^{(n-1)}$. It can be shown that these points and their Galois conjugates generate a group of rank p^n in $E(F^{(n)})$. Hence this is an example in which the rank is not bounded. See Bertolini (2001).

3.2 *Selmer Groups*

Let E/F be an elliptic curve over a number field F. Set Σ to be the finite set of places in F consisting of all places above p, all places of bad reduction for E and all infinite places.

For any field K, we write $H^i(K, \cdot)$ for the group cohomology of continuous cochains for the profinite absolute Galois group $\mathrm{Gal}(\bar{K}/K)$. The notation $H^i_\Sigma(F, \cdot)$ will stand for the cohomology for the Galois group $G_\Sigma(F)$ of the maximal extension of F that is unramified outside F, see Neukirch et al. (2000, §8.3); it can also be described as the étale cohomology group $H^i_{\text{ét}}(\mathrm{Spec}(\mathcal{O}_F)\backslash\Sigma, \cdot)$ for the corresponding étale group scheme. If the Galois module M is finite p-primary, then $H^i_\Sigma(F, M)$ is finite, see Neukirch et al. (2000, Theorem 8.3.19). If M is a finitely generated $\mathbb{Z}_p$-module then so is $H^i_\Sigma(F, M)$, see Rubin (2000, Proposition B.2.7). For any abelian group A, we will denote the Pontryagin dual $\mathrm{Hom}_{\mathbb{Z}}(A, \mathbb{Q}/\mathbb{Z})$ by $\hat{A}$.

For any finite extension K/F, we define the Tate-Shafarevich group $Ш(E/K)$ to be the kernel of the localisation map

$$H^1(K, E) \to \prod_v H^1(K_v, E)$$

where the product runs over all places v of K and K_v denotes the completion at v. The non-trivial elements in $Ш(E/K)$ have an interpretation as curves of genus 1 defined over K with Jacobian isomorphic to E and which are counter-examples to the Hasse principle, see Milne (2006, §17). It is known that $Ш(E/K)$ is a torsion abelian group such that the Pontryagin dual of the p-primary part is a finitely generated $\mathbb{Z}_p$-module for every prime p. It is conjectured that $Ш(E/K)$ is finite.

Let m be a power of p. For any extension K of F, the long exact sequence of cohomology for $E[m]$ gives the Kummer exact sequence

$$0 \longrightarrow E(K)\big/ mE(K) \xrightarrow{\kappa} H^1\big(K, E[m]\big) \longrightarrow H^1(K,E)[m] \longrightarrow 0.$$

For any finite extension K of F, we define the m-Selmer group $\mathrm{Sel}^m(E/K)$ as the elements in $H^1(K, E[m])$ that restrict to elements in the image of the local Kummer map $\kappa_v\colon E(K_v)/p^k E(K_v) \to H^1(K_v, E[m])$ for all places v of K. This contains naturally $E(K)/mE(K)$ as a subgroup whose quotient is $\text{Ш}(E/K)[m]$. Since all cocycles in the Selmer group are unramified outside Σ, we get an exact sequence

$$0 \longrightarrow \mathrm{Sel}^m(E/K) \longrightarrow H^1_\Sigma\big(K, E[m]\big) \longrightarrow \bigoplus_{v\in\Sigma} H^1(K_v, E)[m].$$

In particular, this shows that $\mathrm{Sel}^m(E/K)$ is finite. We can now form the two limits, induced by the inclusion and the multiplication by p map between $E[p^k]$ and $E[p^{k+1}]$. We set

$$\mathcal{S}(E/K) = \varinjlim_k \mathrm{Sel}^{p^k}(E/K) \subset H^1_\Sigma(K, W) \qquad \text{and}$$

$$\mathfrak{S}(E/K) = \varprojlim_k \mathrm{Sel}^{p^k}(E/K) \subset H^1_\Sigma(K, T)$$

where $T = \varprojlim_k E[p^k]$ is the (compact) p-adic Tate module and $W = \varinjlim_k E[p^k] = E[p^\infty]$ is the (discrete) p-primary torsion of E. It is true that $\varprojlim H^1_\Sigma(K, E[p^k]) = H^1_\Sigma(K, T)$ by an argument of Tate, see Neukirch et al. (2000, Corollary 2.3.5).

The corresponding limit versions of the Mordell-Weil group are $\varinjlim E(K)/p^k E(K)$ and $\varprojlim E(K)/p^k E(K)$. The first can be seen to be equal to $E(K) \otimes_{\mathbb{Z}} \mathbb{Q}_p/\mathbb{Z}_p$, which is isomorphic to a direct sum of rank($E(K)$) copies of $\mathbb{Q}_p/\mathbb{Z}_p$. The latter is equal to $E(K) \otimes_{\mathbb{Z}} \mathbb{Z}_p$, which is equal to the sum of rank($E(K)$) copies of $\mathbb{Z}_p$ plus the finite group $E(K)[p^\infty]$. By passing to the limits, we find the exact sequences

$$0 \longrightarrow E(K) \otimes \mathbb{Q}_p/\mathbb{Z}_p \longrightarrow \mathcal{S}(E/K) \longrightarrow \text{Ш}(E/K)[p^\infty] \longrightarrow 0$$

$$0 \longrightarrow E(K) \otimes \mathbb{Z}_p \longrightarrow \mathfrak{S}(E/K) \longrightarrow \varprojlim_k \text{Ш}(E/K)[p^k] \longrightarrow 0$$

where the lower sequence remains exact because we have taken projective limits of finite groups. The group on the right hand side of the second line is a free $\mathbb{Z}_p$-module which is conjecturally trivial. The first line combines nicely the rank information with the Tate-Shafarevich group. The Pontryagin duals of the first line and all the groups in the second line are finitely generated $\mathbb{Z}_p$-modules.

Later in Sect. 5, we will give another description of $\mathcal{S}(E/K)$ which does not use the Kummer map, but uses the modules W and T only.

3.3 Iwasawa Theory for the Selmer Group

Given a $\mathbb{Z}_p$-extension $F^{(\infty)}/F$, we consider the limit $\mathcal{S}(E/F^{(\infty)}) = \varinjlim_n \mathcal{S}(E/F^{(n)})$ and its dual

$$X = \widehat{\mathcal{S}(E/F^{(\infty)})} = \varprojlim_n \widehat{\mathcal{S}(E/F^{(n)})} \tag{1}$$

which is naturally a compact Λ-module. The maps are induced by the natural inclusion $E(F^{(n)}) \to E(F^{(n+1)})$ and the restriction map on the Tate-Shafarevich groups. Hence, if the Mordell-Weil group stabilises after a few steps, as in all but the last example above, then X will contain a $\mathbb{Z}_p$-module of this rank. The other natural limit $\varprojlim_n \mathfrak{S}(E/F^{(n)})$ with respect to the corestriction map is less interesting: If the Mordell-Weil group stabilises, meaning that $E(F^{(\infty)}) = E(F^{(n)})$ for some n, and the Tate-Shafarevich groups are finite, then this limit is trivial.

Lemma 4. *The Selmer group X is a finitely generated Λ-module for any $\mathbb{Z}_p$-extension.*

Proof. We should show by Lemma 2 that X_Γ is a finitely generated $\mathbb{Z}_p$-module; this is the dual of the Γ-fixed part of $\mathcal{S}(E/F^{(\infty)})$. Later in Theorem 11, we will show that this is not too far from the dual of $\mathcal{S}(E/F)$, which is a finitely generated $\mathbb{Z}_p$-module.

Conjecture 3 (Mazur 1972). If E has good ordinary reduction at all places in F above p and $F^{(\infty)}/F$ is the cyclotomic $\mathbb{Z}_p$-extension, then the Selmer group X is a torsion Λ-module.

Note that this conjecture implies, by Proposition 2, that the largest free $\mathbb{Z}_p$-module in X has finite rank $\lambda(X)$, so by the above this will imply that the rank of $E(F^{(n)})$ stabilises. Moreover we have:

Proposition 4. *If the conjecture holds then $E(F^{(\infty)})$ is a finitely generated $\mathbb{Z}$-module. Suppose that $Ш(E/F^{(n)})[p^\infty]$ is finite for all n, then there are constants μ, λ, ν, n_0 such that if $\left|Ш(E/F^{(n)})[p^\infty]\right| = p^{e_n}$, then $e_n = \mu\, p^n + \lambda\, n + \nu$ for all $n \geqslant n_0$.*

Proof. The first part is Theorem I.5 in Greenberg (1999). For the second part, we will have to show that $X_{\Gamma^{(n)}}$ is very close to the dual of $\mathcal{S}(E/F^{(n)})$ for all n. This is done in the control Theorem 12 below.

As shown by the Examples 3.1.4 and 3.1.5, none of the two assumptions in Mazur's conjecture can be removed. Here are two important result in support of the conjecture.

Theorem 7 (Mazur 1972). *If $E(F)$ and $Ш(E/F)[p^\infty]$ are finite, then the conjecture holds.*

The main result of Kato (2004) implies the following.

Theorem 8. *Let E be an elliptic curve defined over $\mathbb{Q}$ and let F be an abelian extension of $\mathbb{Q}$. Suppose that E has good ordinary reduction at p, then the Selmer group for the cyclotomic $\mathbb{Z}_p$-extension is a torsion Λ-module.*

3.4 Mazur-Stickelberger Elements

Let $E/\mathbb{Q}$ be an elliptic curve. We will suppose that E has good reduction at p. Let ω_E be a Néron differential on E; this is just $\frac{dx}{2y}$ when E is given by a global minimal model. The canonical lattice Z_E for E is the image of $\int: H_1\big(E(\mathbb{C}), \mathbb{Z}\big) \to \mathbb{C}$ sending a closed path γ on $E(\mathbb{C})$ to $\int_\gamma \omega_E$. We define Ω_E to be the smallest positive element of Z_E.

The theorem of modularity (Breuil et al. 2001) shows that there exists a morphism φ_E of curves $X_0(N) \to E$ defined over $\mathbb{Q}$. We take one of minimal degree. If f is the newform corresponding to the isogeny class of E, then there is a natural number c_E, called the Manin constant, such that $c_E \cdot \varphi_E^*(\omega_E)$ is equal to the differential $2\pi i f(z)dz$ on $X_0(N)$ corresponding to f, written here as a differential in the variable z on the upper half plane $\mathcal{H}$. For the so-called optimal curve in the isogeny class one expects $c_E = 1$.

For any rational number $r = \frac{a}{m}$, consider the ray from r to $i\,\infty$ in the upper half plane. Its image in $X_0(N)(\mathbb{C})$ is a (not necessarily closed) path $\{r, \infty\}$ between two cusps.

Proposition 5 (Manin 1972). *There is a natural number $t \geqslant 1$ such that, for all $r \in \mathbb{Q}$, the value of $\lambda_f(r) = 2\pi i \int_{i\infty}^{r} f(z)\,dz$ belongs to $\frac{1}{t} Z_E$.*

This is clear for the closed paths, i.e., when r is $\Gamma_0(N)$-equivalent to $i\infty$. The proof for general r uses the Hecke operators T_ℓ on $X_0(N)$.

We define the (plus) *modular symbol* $[r]^+$ by

$$[r]^+ = \frac{1}{\Omega_E} \cdot \mathrm{Re}\Big(2\pi i \int_{i\infty}^{r} f(z)\,dz\Big) \in \mathbb{Q},$$

see Cremona (1997) for more details. For an abelian field K of conductor m, we define the Stickelberger element for E to be

$$\Theta_{E/K} = \sum_{\substack{1 \leqslant a < m \\ (a,m)=1}} \left[\frac{a}{m}\right]^+ \sigma_a \in \mathbb{Q}\big[\mathrm{Gal}(K/\mathbb{Q})\big].$$

Let ℓ be a prime of good reduction. The ℓ-th coefficient a_ℓ of f satisfies $\ell - a_\ell + 1 = \#E(\mathbb{F}_\ell)$. If ℓ does not divide m, then

$$N_{\mathbb{Q}(\mu_{m\ell})/\mathbb{Q}(\mu_m)}\big(\Theta_{E/\mathbb{Q}(\mu_{m\ell})}\big) = (-\sigma_\ell)\big(1 - a_\ell\,\sigma_\ell^{-1} + \sigma_\ell^{-2}\big)\big(\Theta_{E/\mathbb{Q}(\mu_m)}\big),$$

which can be deduced from the action of the Hecke operator T_ℓ on $X_0(N)$.

3.5 The p-Adic L-Function

Let p be a prime of good reduction for E. Write now $\Theta_E^{(n)}$ for the Stickelberger element for the field $\mathbb{Q}^{(n)}$ and write $G^{(n)} = \mathrm{Gal}(\mathbb{Q}^{(n)}/\mathbb{Q})$. We define the map $j\colon \mathbb{Q}[G^{(n)}] \to \mathbb{Q}[G^{(n+1)}]$ to send an element of $G^{(n)}$ to the sum over all its preimages in $G^{(n+1)}$. Then the norm $N\colon \mathbb{Q}[G^{(n+1)}] \to \mathbb{Q}[G^{(n)}]$ sends $\Theta_E^{(n+1)}$ to

$$N\big(\Theta_E^{(n+1)}\big) = a_p \cdot \Theta_E^{(n)} - j\big(\Theta_E^{(n-1)}\big).$$

This is shown using the Hecke operator T_p. Let α be a root of the polynomial $X^2 - a_p\,X + p$. We set

$$\mathcal{L}_E^{(n)} = \frac{1}{\alpha^{n+1}} \cdot \Theta_E^{(n)} - \frac{1}{\alpha^{n+2}}\, j\big(\Theta_E^{(n-1)}\big) \tag{2}$$

for all $n \geqslant 1$. Then $\mathcal{L}_E = (\mathcal{L}_E^{(n)})_{n\geqslant 1}$ belongs to $\varprojlim \mathbb{Q}[\mathrm{Gal}(\mathbb{Q}^{(n)}/\mathbb{Q})]$ and it is called the *p-adic L-function* of E. Explicitly, we have

$$\mathcal{L}_E^{(n)} = \sum_{\substack{1\leqslant a<p^{n+1}\\ p\nmid a}} \left(\frac{1}{\alpha^{n+1}}\left[\frac{a}{p^{n+1}}\right]^+ - \frac{1}{\alpha^{n+2}}\left[\frac{a}{p^n}\right]^+\right)\cdot \gamma_a.$$

Suppose now that E has good ordinary reduction at p. Then a_p is a p-adic unit and hence we can find one root α which belongs to $\mathbb{Z}_p^\times$. Because the denominator of $[\frac{a}{m}]$ is uniformly bounded, $\mathcal{L}_E$ actually belongs to $\mathbb{Q}_p \otimes \Lambda$ and in many cases it is known that $\mathcal{L}_E \in \Lambda$. For the supersingular case there is no unit root α and $\mathcal{L}_E$ will never belong to Λ, see Pollack (2003).

Theorem 9. *The p-adic L-function satisfies the interpolation properties*

$$\mathbf{1}(\mathcal{L}_E) = \left(1 - \frac{1}{\alpha}\right)^2 \cdot \frac{L(E,1)}{\Omega_E} \tag{3}$$

and

$$\chi(\mathcal{L}_E) = \frac{\mathrm{Ga}(\chi)}{\alpha^{n+1}} \cdot \frac{L(E,\bar{\chi},1)}{\Omega_E} \tag{4}$$

for all character χ of conductor p^{n+1} on Γ, i.e., that factor through $\mathrm{Gal}(\mathbb{Q}^{(n)}/\mathbb{Q})$ *but not through* $\mathrm{Gal}(\mathbb{Q}^{(n-1)}/\mathbb{Q})$.

The proof connecting the corresponding finite sums of modular symbols to the Mellin transform of the modular form can be found in formula (8.6) and §14 of Mazur et al. (1986).

Theorem 10 (Rohrlich 1984). *Only finitely many of the values $\chi(\mathcal{L}_E)$ in Eq.* (4) *are zero. In particular $\mathcal{L}_E \neq 0$.*

Again, we can define the analytic function $L_p(E, s) = \kappa^{s-1}(\mathcal{L}_E)$. If $L(E, 1) \neq 0$, we know that $E(\mathbb{Q})$ is finite. As a consequence of the above theorem, we see that $L_p(E, 1) \neq 0$ as $\alpha \neq 1$. Moreover the value $L_p(E, 1)$ is then predicted by the Birch and Swinnerton-Dyer conjecture.

Conjecture 4 (p-adic version of the Birch and Swinnerton-Dyer conjecture (Mazur et al. 1986)). The order of vanishing of $L_p(E, s)$ at $s = 1$ is equal to the rank of $E(\mathbb{Q})$. The leading term of its series at $s = 1$ is equal to

$$\left(1 - \frac{1}{\alpha}\right)^2 \cdot \frac{\mathrm{Reg}_p(E/\mathbb{Q}) \cdot \#\text{Ш}(E/\mathbb{Q}) \cdot \prod_v c_v}{\left(\#E(\mathbb{Q})_{\mathrm{tors}}\right)^2}$$

where c_v are the Tamagawa numbers and $\mathrm{Reg}_p(E/\mathbb{Q})$ is the p-adic regulator, see Sect. 4.2.

It would be very interesting to know that the order of vanishing of $L_p(E, s)$ is equal to the order of vanishing of $L(E, s)$. However this is only known when the p-adic L-function vanishes to order at most 1 by Perrin-Riou (1987).

3.6 The Main Conjecture

Let $E/\mathbb{Q}$ and suppose E has good ordinary reduction at p. By Theorem 8, we can associate to E the characteristic ideal char(X) of the dual of the Selmer group in (1). Under the same hypothesis, we have constructed a p-adic L-function (2).

Conjecture 5 (main conjecture). The p-adic L-function $\mathcal{L}_E$ is a generator for the characteristic ideal char(X).

The other series of lectures will talk about the main result by Skinner and Urban on this conjecture. See Wan (2014) in these proceedings.

The generalisations to higher weight modular forms for $\Gamma_0(N)$ with $p \nmid N$ and $p \nmid a_p$ is fairly straight forward, see Mazur et al. (1986). For the extensions to $p \mid N$, but $p^2 \nmid N$, one has to be a bit careful as the case of split multiplicative reduction behaves differently due to the presence of exceptional zeroes because $\alpha = 1$. Finally the generalisation to the supersingular case is clearly much more complicated. To my knowledge the generalisation to additive reduction, i.e., when $p^2 \mid N$, is not yet fully done.

4 The Leading Term Formula

4.1 Control Theorem

As before let E/F be an elliptic curve and let $F^{(\infty)}/F$ be a $\mathbb{Z}_p$-extension. Recall that X is the dual of the limit Selmer group $\mathcal{S}(E/F^{(\infty)})$ as defined in (1) and we are interested in comparing X_Γ with the dual of $\mathcal{S}(E/F)$. For a place $v \in \Sigma$, write $J_v^{(\infty)} = \prod_{v|w} H^1(F_w^{(\infty)}, E)[p^\infty]$. We have the following diagram

$$\begin{array}{ccccccc} 0 \longrightarrow & \mathcal{S}(E/F^{(\infty)})^\Gamma & \longrightarrow & H^1_\Sigma(F^{(\infty)},W)^\Gamma & \longrightarrow & \bigoplus_{v\in\Sigma} (J_v^{(\infty)})^\Gamma \\ & \uparrow \alpha & & \uparrow \beta & & \uparrow \oplus_v \gamma_v \\ 0 \longrightarrow & \mathcal{S}(E/F) & \longrightarrow & H^1_\Sigma(F,W) & \longrightarrow & \bigoplus_{v\in\Sigma} H^1(F_v,E)[p^\infty]. \end{array} \tag{5}$$

We want to bound the kernel and cokernel of α. Even if it is clear that $E(F^{(\infty)})^\Gamma = E(F)$, it is not obvious what happens to the map

$$E(F) \otimes {}^{\mathbb{Q}_p}\!/\!_{\mathbb{Z}_p} \to \left(E(F^{(\infty)}) \otimes {}^{\mathbb{Q}_p}\!/\!_{\mathbb{Z}_p}\right)^\Gamma$$

as a non-divisible point in $E(F)$ can become divisible by p in $E(F^{(n)})$.

Theorem 11. *The kernel of α is finite and the dual of the cokernel is a finitely generated $\mathbb{Z}_p$-module.*

Proof. The inflation-restriction sequence (Neukirch et al. 2000, Proposition 1.6.6) for the H^i_Σ-cohomology of $W = E[p^\infty]$ gives that the kernel of β is $H^1\big(\Gamma, H^0_\Sigma(F^{(\infty)}, W)\big)$ and the cokernel lies in $H^2\big(\Gamma, H^0_\Sigma(F^{(\infty)}, W)\big)$. Now the dual of $D = H^0(F^{(\infty)}, W) = E(F^{(\infty)})[p^\infty]$ has $\mathbb{Z}_p$-rank at most 2. Hence the dual of the exact sequence

$$0 \longrightarrow D^\Gamma \longrightarrow D \xrightarrow{\gamma-1} D \longrightarrow D_\Gamma \longrightarrow 0$$

shows that $H^1(\Gamma, D) = D_\Gamma$ has the same corank as $D^\Gamma = E(F)[p^\infty]$, which is finite. Hence the kernel of β and α are finite. In fact, if D is finite as in almost all cases, then the kernel of β has the same order as $E(F)[p^\infty]$.

The cokernel of β is trivial, because $H^2(\Gamma, D) = \varinjlim_k H^2\big(\Gamma, E(F^{(\infty)})[p^k]\big)$ and the latter groups are trivial because Γ has cohomological dimension 1, see Neukirch et al. (2000, Proposition 1.6.13). Since β is surjective we see that the duals of $H^1_\Sigma(F^{(\infty)}, W)^\Gamma$ and $\mathcal{S}(E/F^{(\infty)})^\Gamma$ are finitely generated $\mathbb{Z}_p$-modules.

Note that this proves Lemma 4 saying that X is a finitely generated Λ-module.

Theorem 12 (Control theorem). *Suppose that E has good ordinary reduction at all primes that ramify in $F^{(\infty)}/F$. Then the map α has finite kernel and cokernel.*

Proof. From the previous proof, we see that we are left to show that the cokernel of α is finite. By the snake lemma applied to (5), it suffices to show that the kernel of $\bigoplus\gamma_v$ is finite under our hypothesis.

Let $v \in \Sigma$. By local Tate duality (Tate 1995b), the group $H^1(F_v, E)[p^\infty]$ is the Pontryagin dual of the p-adic completion $E(F_v)^\star = \varprojlim E(F_v)/p^k E(F_v)$ of the local points. The structure of elliptic curves over local fields, see chapter 7 in Silverman (2009), can be used to show that $E(F_v)^\star$ is finite if $v \nmid p$ and it is a finitely generated $\mathbb{Z}_p$-module of rank $[F_v : \mathbb{Q}_p]$ if $v \mid p$. Choose a place w above v in $F^{(\infty)}$. We wish to show that the kernel of

$$\gamma_v \colon H^1(F_v, E)[p^\infty] \longrightarrow H^1(F_w^{(\infty)}, E)[p^\infty]$$

is finite. Again by Tate duality this map is dual to the norm map

$$\widehat{\gamma}_v \colon E(F^{(\infty)})^\star \longrightarrow E(F_v)^\star.$$

The following lemmas will conclude this proof.

We will write $x \overset{\times}{=} y$ if there is a p-adic unit u such that $x = u \cdot y$.

Lemma 5. *If v splits completely in $F^{(\infty)}/F$, then $\ker \gamma_v = 0$. Otherwise, if v is unramified, then $\# \ker \gamma_v \overset{\times}{=} c_v$, the Tamagawa number of E/F_v. In particular $\ker \gamma_v$ is non-zero for only finitely many v.*

Proof. If v splits completely, then $F_w^{(\infty)} = F_v$ and the "norm" map is clearly surjective.

First assume $v \nmid p$. The local extension $F_w^{(\infty)}/F_v$ is unramified and so the Néron model of E will not change in this extension. Let Φ be its group of components and write E_v^0 for the connected component of the special fibre. Then we have that

$$0 \longrightarrow E_v^0(\mathbb{F}_v)^\star \longrightarrow E(F_v)^\star \longrightarrow \Phi(\mathbb{F}_v)^\star \longrightarrow 0$$

because the points in the formal group $\hat{E}$ are divisible by p when $v \nmid p$. Now the norm map on the left hand side is surjective because of Lang's theorem (Lang 1956). On the right hand side instead the norm map will be the zero map for sufficiently large n. Hence $\ker \gamma_v$ is dual to $\Phi(\mathbb{F}_v)^\star$.

Now assume $v \mid p$, but the extension $F^{(\infty)}/F$ is unramified at v. The argument is the same as above, except that we now have to show that the norm is surjective on the formal groups. That is done in the part (a) of the following lemma.

Lemma 6. *Let E be an elliptic curve over a p-adic field K and let L/K be a cyclic extension of degree a power of p. Let $\mathfrak{m}_L$ and $\mathfrak{m}_K$ be the maximal ideals of L and K respectively.*

1. *If L/K is unramified then the norm map on the formal group $\hat{E}(\mathfrak{m}_L) \to \hat{E}(\mathfrak{m}_K)$ is surjective.*
2. *If L/K is ramified and E has good ordinary reduction. Then the cokernel of the norm map on the formal groups is finite.*
3. *If $v \mid p$ is totally ramified and E has good ordinary reduction, then $\#\ker \gamma_v \overset{\times}{=} N_v^2$ where N_v is the number of points in the reduction $E(\mathbb{F}_v)$.*

Proof. For the proof of the first point uses the filtration by $\hat{E}(\mathfrak{m}^k)$, the formal logarithm that gives an isomorphism $\hat{E}(\mathfrak{m}_L^k) \cong 1 + \mathfrak{m}_L^k$ for large enough k, the fact that $H^1(\mathbb{F}_L/\mathbb{F}_K, \mathbb{F}_L) = 0$ for the residue fields, and the surjectivity of the norm map on units (Serre 1968, Proposition V.3).

The proof of the latter two can be found in Lubin and Rosen (1978). It relies on the fact that the formal group of E becomes isomorphic to the multiplicative formal group over the ring of integers of the completion of the maximal unramified extension of F_v.

A different and more accessible proof of item 3 in the above lemma is explained in Lemma 4.6 in Greenberg (1999). It should be noted that both item 2 and item 3 are no longer valid when the reduction is supersingular. The theory in the case of good supersingular reduction at primes above p is quite different.

One can now add a proof for Theorem 7. If $E(F)$ and $Ш(E/F)[p^\infty]$ are both finite, then so is $\mathcal{S}(E/F)$. By the control Theorem 12, this implies that X_Γ is finite. Since the Γ-coinvariants $\Lambda_\Gamma \cong \mathbb{Z}_p$ of Λ are not finite, we see that X is a torsion Λ-module. In fact, we see that this holds for all $\mathbb{Z}_p$-extensions, not just the cyclotomic. In Example 3.1.5, the rank of $E(F)$ is 1.

4.2 p-Adic Heights

We will now construct an analogue to the real-valued Néron-Tate height. We present a version inspired by Bloch (1980). Let E/F be an elliptic curve and we suppose that E has good ordinary reduction at all places above p that are ramified. To each cohomology class in $H^1(F,T)$ represented by a cocycle $\xi\colon G_F = \mathrm{Gal}(\bar{F}/F) \to T = T_p(E)$, we associate an extension

$$0 \longrightarrow T_p\mu \longrightarrow T_\xi \longrightarrow T \longrightarrow 0$$

where $T_p\mu = \varprojlim \mu[p^k]$, also denoted by $\mathbb{Z}_p(1)$, is a free $\mathbb{Z}_p$-module of rank 1 on which G_F acts via the cyclotomic character. As a $\mathbb{Z}_p$-module $T_\xi = T_p\mu \oplus T$, but the G_F-action is given by

$$\sigma(\zeta, P) = \Big(\sigma(\zeta)\cdot e\big(\xi_\sigma, \sigma(P)\big),\ \sigma(P)\Big) \qquad \text{for } \zeta \in T_p\mu \text{ and } P \in T,$$

with $e: T \times T \to T_p\mu$ denoting the Weil-pairing (Silverman 2009, §3.8). It is not hard to show that the class of the extension T_ξ does not depend on the chosen cocycle and that the boundary maps $\partial: H^i(F,T) \to H^{i+1}(F,T_p\mu)$ are given by applying the Weil-pairing to the cup-product with ξ, at least up to sign.

Consider now the commutative diagram with exact rows

$$\begin{array}{ccccccccc} 0 & \longrightarrow & H^1(F,T_p\mu) & \longrightarrow & H^1(F,T_\xi) & \longrightarrow & H^1(F,T) & \longrightarrow & H^2(F,T_p\mu) \\ & & \downarrow & & \downarrow & & \downarrow & & \downarrow \\ 0 & \longrightarrow & \prod_v H^1(F_v,T_p\mu) & \longrightarrow & \prod_v H^1(F_v,T_\xi) & \longrightarrow & \prod_v H^1(F_v,T) & \xrightarrow{\prod \partial_v} & \prod_v H^2(F_v,T_p\mu) \end{array} \tag{6}$$

with the product running over all places in F. It follows from global class field theory that the downwards arrow on the right is injective (Neukirch et al. 2000, Corollary 9.1.8.ii) if T is replaced by μ_{p^k}; that the limit is still surjective needs an additional argument (Tate 1976, Corollary to Proposition 2.2). On the left we have the map from the p-adic completion $(F^\times)^\star$ of $F^\times$ to $\prod_v (F_v^\times)^\star$.

Choose an topological generator γ. Let $l: \Gamma \to \mathbb{Z}_p$ be the map that send γ to 1. For each place v consider the composition

$$\lambda_v: F_v^\times \longrightarrow \mathbb{A}_F^\times \longrightarrow \Gamma \xrightarrow{\ l\ } \mathbb{Z}_p$$

where $\mathbb{A}_F^\times$ is the idèle group of F and the map that follows it is the reciprocity map. This map extends to the completion $\lambda_v: (F_v^\times)^\star \to \mathbb{Z}_p$. In case $F^{(\infty)}/F$ is the cyclotomic $\mathbb{Z}_p$-extension, then l is a multiple of $\log_p \circ \kappa$. For finite places v away from p, the map is simply $\lambda_v(x) = (\log_p(\kappa(\gamma)))^{-1} \cdot \log(\#\mathbb{F}_v) \cdot v(x)$ where $\mathbb{F}_v$ is the residue field. For places above p, we get $\lambda_v(x) = -(\log_p(\kappa(\gamma))^{-1} \cdot \log_p(N_{K_v/\mathbb{Q}_p}(x))$.

Suppose now ξ belongs to $\mathfrak{S}(E/F)$. Let $\mathfrak{S}(E/F)^0$ be the subgroup of $\mathfrak{S}(E/F)$ of all elements η such that $\mathrm{res}_v(\eta) \in E(F_v)^\star$ lies in the image of the norm from $E(F_w^{(\infty)})^\star$ for all places v. By Lemmas 5 and 6, this subgroup has finite index in $\mathfrak{S}(E/F)$. Let $\eta \in \mathfrak{S}(E/F)^0$. Since both $\mathrm{res}_v(\eta)$ and $\mathrm{res}_v(\xi)$ belong to the image of $E(F_v)^\star$ inside $H^1(F_v, T)$, their cup-pairing is trivial. This is again a consequence of local Tate duality (Tate 1995b). Hence $\mathrm{res}_v(\eta)$ is sent to 0 by ∂_v. By the injectivity of the right arrow in (6), we conclude that there is an element ζ in $H^1(F, T_\xi)$ that maps to η in $H^1(F, T)$.

For each place v, we will define a local lift $\zeta_v \in H^1(F_v, T_\xi)$. Since $\eta \in \mathfrak{S}(E/F)^0$, there is an element $\tilde{\eta}_v \in H^1(F_w^{(\infty)}, T)$ whose norm is $\mathrm{res}_v(\eta)$. Pick any lift of $\tilde{\eta}_v$ to $H^1(F_w^{(\infty)}, T_\xi)$ and define ζ_v to be its norm in $H^1(F_v, T_\xi)$.

By construction $\mathrm{res}_v(\zeta) - \zeta_v \in H^1(F_v, T_\xi)$ maps to 0 in $H^1(F_v, T)$ and hence we can viewed it as an element in $H^1(F_v, T_p\mu) = (F_v^\times)^\star$. We set

$$\langle \xi, \eta \rangle = \sum_v \lambda_v\big(\mathrm{res}_v(\zeta) - \zeta_v\big) \in \mathbb{Z}_p.$$

It is not hard to check that this element is independent of the choices made, because both $(F^{\times})^{\star}$ and the norms from $H^1(F_w^{(\infty)}, T_p\mu)$ lie in the kernel of $\sum_v \lambda_v$.

Since $\mathfrak{S}(E/F)^0$ has finite index, we can linearly extend this to a pairing on $\mathfrak{S}(E/F)$ with values in $\mathbb{Q}_p$. This is called the *canonical p-adic height pairing* corresponding to the $\mathbb{Z}_p$-extension and the choice of γ. Note that this construction only works under the assumption that E has good ordinary reduction at the ramified places. For the generalisation to any Galois representation, which is de Rham at places above p, see Perrin-Riou (1995, §3.1.2).

There is a variant of this construction: Let $\xi \in \mathfrak{S}(E/F)^0$ and $\eta = (\eta^{(n)})_n \in \mathcal{S}(E/F^{(\infty)})$, then one can construct in a similar way an element of ${}^{\mathbb{Q}_p}/_{\mathbb{Z}_p}$. This time one lifts $\mathrm{res}_w(\eta^{(n)}) \in H^1(F_w^{(n)}, W)$ to $H^1(F_w^{(n)}, T_\xi \otimes {}^{\mathbb{Q}_p}/_{\mathbb{Z}_p})$ etc. It turns out that the map $\hat{\delta}\colon \mathfrak{S}(E/F)^0 \to \mathrm{Hom}\big(\mathcal{S}(E/F^{(\infty)}), {}^{\mathbb{Q}_p}/_{\mathbb{Z}_p}\big) = X$ has its image in X^{Γ}. Now the p-adic height pairing can be described involving the map $\pi\colon X^{\Gamma} \to X \to X_{\Gamma}$.

Proposition 6 (Proposition 3.4.5 in Perrin-Riou 1992). *There is a commutative diagram*

$$\begin{array}{ccccc} X_\Gamma & \xleftarrow{\qquad\qquad\pi\qquad\qquad} & & & X^\Gamma \\ \downarrow{\scriptstyle\alpha} & & & & \uparrow{\scriptstyle\hat\delta} \\ \widehat{\mathcal{S}(E/F)} & \xrightarrow[\iota]{} & \mathrm{Hom}_{\mathbb{Z}_p}\big(\mathfrak{S}(E/F),\mathbb{Z}_p\big) & \xleftarrow[h_p]{} & \mathfrak{S}(E/F)^0 \end{array} \tag{7}$$

where h_p is the p-adic height pairing and ι is a naturally defined surjective $\mathbb{Z}_p$-morphism with finite kernel.

Finally one should mention that the p-adic height pairing has also an analytic description using canonical p-adic sigma-functions σ_v for all ramified places. These are well-explained in Mazur and Tate (1991) and a fast algorithm for computing them was found in Mazur et al. (2006) using Kedlaya's algorithm. For instance, if $E/\mathbb{Q}$ and $P = (x, y) \in E(\mathbb{Q})$ is a point that has good reduction everywhere and reduces to 0 at p, then $h_p(P) = \log_p(\sigma_p(P)) - \log_p(e)$ where e is the square root of the denominator of x. In general the formula allows one to compute the p-adic height with only the information of E over F together with the explicit maps λ_v.

Conjecture 6 (Schneider 1982). The canonical p-adic height pairing for the cyclotomic $\mathbb{Z}_p$-extension on an elliptic curve with good ordinary reduction at all places above p is non-degenerate.

Suppose $\text{Ш}(E/F)[p^{\infty}]$ is finite. Choose a basis of $E(F)$ modulo torsion and let $\mathrm{Reg}_\gamma(E/F) \in \mathbb{Q}_p$ be the determinant of the p-adic height pairing on this basis. The number $\mathrm{Reg}_p(E/F) = \mathrm{Reg}_\gamma(E/F) \cdot \log_p(\kappa(\gamma))^{\mathrm{rank}\, E(F)}$ is independent of the choice of γ. The above conjecture then says that $\mathrm{Reg}_\gamma(E/F) \neq 0$ in the cyclotomic case. For the anti-cyclotomic $\mathbb{Z}_p$-extension it can well be that the p-adic height is degenerate.

4.3 Leading Term

The following theorem was proved by Perrin-Riou for curves with complex multiplication then in general by Schneider (1985). See Perrin-Riou (1993) for the details to complete our sketch of proof.

Let $F^{(\infty)}/F$ be a $\mathbb{Z}_p$-extension such that all ramified places are totally ramified. Write $\Sigma(\mathrm{ram})$ for the set of all the ramified places in F and denote by S the set of all places that split completely in $F^{(\infty)}$.

As before, identify Λ with $\mathbb{Z}_p[\![T]\!]$ via the choice of a topological generator γ. Let $F_X(T) \in \mathbb{Z}_p[\![T]\!]$ be a generator of the characteristic ideal for X.

Theorem 13. *Suppose E has good ordinary reduction at all ramified places above p and suppose that the canonical p-adic height for $F^{(\infty)}/F$ is non-degenerate. Then*

1. *X is Λ-torsion;*
2. *The characteristic series $F_X(T)$ has a zero of order $\mathrm{rank}_{\mathbb{Z}_p}\big(\mathfrak{S}(E/F)\big)$ at $T = 0$;*
3. *If $\text{Ш}(E/F)[p^\infty]$ is finite then the leading term $F_X^*(0)$ of $F_X(T)$ at $T = 0$ satisfies*

$$F_X^*(0) \overset{\times}{=} \prod_{v \in \Sigma(\mathrm{ram})} N_v^2 \cdot \frac{\mathrm{Reg}_p(E/F) \cdot \#\text{Ш}(E/F)[p^\infty] \cdot \prod_{v \notin S} c_v \cdot}{\left(\#E(F)_{tors}\right)^2} .$$

If the main conjecture holds for a curve $E/\mathbb{Q}$, then the finiteness of $\text{Ш}(E/\mathbb{Q})[p^\infty]$ and the non-degeneracy of the p-adic height pairing imply the p-adic BSD conjecture, up to a p-adic unit in the leading term. This is because $1 - 1/\alpha \overset{\times}{=} N_v$. Theorem 13 together with Kato's Theorem 16 can be used to give a new efficient algorithm (Stein and Wuthrich 2013) for the determination of the rank of $E(\mathbb{Q})$ and upper bounds of $\text{Ш}(E/\mathbb{Q})[p^\infty]$ for almost all p.

The proof of this theorem follows surprisingly closely what Tate (1995a) did in the function field case to reduce BSD to the finiteness of the p-primary part of the Tate-Shafarevich group. The algebraic part relies on the following lemma which can be deduced from Proposition 2.

Lemma 7. *Let X be a finitely generated Λ-module and suppose the cokernel of $\pi\colon X^\Gamma \to X_\Gamma$ is finite. Then*

1. *X is Λ-torsion;*
2. *π has finite kernel;*
3. *The leading term of its characteristic series satisfies $F_X^*(0) \overset{\times}{=} \frac{\#\mathrm{coker}(\pi)}{\#\ker(\pi)} =: q(\pi)$.*

If X_Γ is finite, then $q(\pi)$ is the Euler-characteristic $\prod_{i=0}^{1} \#H^i\big(\Gamma, \mathcal{S}(E/F^{(\infty)})\big)^{(-1)^i}$. A proof in this case can be found in §3 of Coates and Sujatha (2000) or in §4 of the chapter of Coates et al. (1999) contributed by Greenberg.

Proof (Proof of Theorem 13). From diagram (7) and the assumption that h_p has finite kernel and cokernel, we find that π must have finite cokernel. Hence the lemma applies and we are left to determine the value of $q(\pi)$. Note that δ has now also finite kernel and cokernel.

Global duality (Neukirch et al. 2000, Theorem 8.6.13) gives us a long exact sequence.

$$0 \longrightarrow \mathcal{S}(E/F) \longrightarrow H^1_\Sigma(F,W) \longrightarrow \bigoplus_{v\in\Sigma} H^1(F_v,E)[p^\infty] \longrightarrow \widehat{\mathfrak{S}(E/F)} \longrightarrow H^2_\Sigma(F,W) \longrightarrow 0$$

Because X is Λ-torsion we have that $\varprojlim \mathfrak{S}(E/F^{(n)}) = 0$ as shown in Perrin-Riou (1992, §3.1.7). When taking the limit of these above sequence over n, the resulting exact sequence is

$$0 \longrightarrow \mathcal{S}(E/F^{(\infty)}) \longrightarrow H^1_\Sigma(F^{(\infty)},W) \longrightarrow \bigoplus_{v\in\Sigma} J_v^{(\infty)} \longrightarrow 0$$

where $J_v^{(\infty)} = \prod_{w|v} H^1(F_w^{(\infty)}, E)[p^\infty]$.

Now we can produce the following big diagram with exact rows. As it is too long we write it in two parts, the top two arrows on the right continue as the lower two arrows on the left.

$$\begin{array}{ccccccccc}
0 & \longrightarrow & \mathcal{S}(E/F^{(\infty)})^\Gamma & \longrightarrow & H^1_\Sigma(F^{(\infty)},W)^\Gamma & \longrightarrow & \bigoplus_{v\in\Sigma} (J_v^{(\infty)})^\Gamma & \longrightarrow & \cdots \\
 & & \alpha\uparrow & & \beta\uparrow & & \oplus\gamma_v\uparrow & & \\
0 & \longrightarrow & \mathcal{S}(E/F) & \longrightarrow & H^1_\Sigma(F,W) & \longrightarrow & \bigoplus_{v\in\Sigma} H^1(F_v,E)[p^\infty] & \longrightarrow & \cdots
\end{array}$$

$$\begin{array}{ccccccc}
\cdots & \longrightarrow & \mathcal{S}(E/F^{(\infty)})_\Gamma & \longrightarrow & H^1_\Sigma(F^{(\infty)},W)_\Gamma & \longrightarrow & 0 \\
 & & \delta\downarrow & & \varepsilon\downarrow & & \\
\cdots & \longrightarrow & \widehat{\mathfrak{S}(E/F)} & \longrightarrow & H^2_\Sigma(F,W) & \longrightarrow & 0
\end{array}$$

To show that the two right squares commute requires some work (Perrin-Riou 1992, §4.4 and 4.5). By the way, I am cheating slightly as in fact the term $\mathfrak{S}(E/F)$ should be replaced by $\mathfrak{S}(E/F)^0$ and other terms should be modified accordingly. Also one has to show that $(J_v^{(\infty)})_\Gamma = 0$ to get the exactness in the top right corner; this follows from the triviality of $H^2(F_v, E)[p^\infty]$, again by local Tate duality.

Next, the transgression map ε is part of the short exact sequences from the degenerating Hochschild-Serre spectral sequence. It is injective and the cokernel is equal to $H^2_\Sigma(F^{(\infty)}, W)^\Gamma$. However the group $H^2_\Sigma(F^{(\infty)}, W)$ is trivial as shown in §3.4.1 in Perrin-Riou (1992). This shows that ε is an isomorphism. So the big diagram gives the equality

$$q(\alpha) \cdot q(\beta)^{-1} \cdot q\Big(\bigoplus \gamma_v\Big) \cdot q(\delta) = 1.$$

On the other hand the diagram (7) gives the equation

$$q(\hat{\delta}) \cdot q(\pi) \cdot q(\alpha) \cdot q(\iota) = q(h_p)$$

In the proof of Theorem 11 we have seen that $q(\beta) \stackrel{\times}{=} \left(\#E(F)_{\mathrm{tors}}\right)^{-1}$ if $E(F^{(\infty)})[p^\infty]$ is finite, which follows without too much difficulty from Lubin and Rosen (1978) under our assumptions. In Lemmas 5 and 6 we found that

$$q\Big(\bigoplus \gamma_v\Big) \stackrel{\times}{=} \Big(\prod_{v \notin S} c_v \cdot \prod_{v \in \Sigma(\mathrm{ram})} N_v^2\Big)^{-1}.$$

Finally, we know that $E(F) \otimes \mathbb{Z}_p$ has index $\#E(F)[p^\infty]$ in $\mathfrak{S}(E/F)$ if the Tate-Shafarevich group is finite. Hence

$$q(h_p) \stackrel{\times}{=} \frac{\mathrm{Reg}_p(E/F)}{\#E(F)_{\mathrm{tors}}}.$$

It is not difficult to see that $q(\iota) = \#Ш(E/F)[p^\infty]$ under our hypothesis. The neglected index of $[\mathfrak{S}(E/F) : \mathfrak{S}(E/F)^0]$ would have cancelled.

5 Selmer Groups for General Galois Representations

Let T be a free $\mathbb{Z}_p$-module of finite rank with an action of $G_F = \mathrm{Gal}(\bar{F}/F)$. We will suppose that only finitely many places ramify in T; so T has an action of $G_\Sigma(K)$ for a finite set Σ containing the places above p and ∞. Let $V = T \otimes \mathbb{Q}_p$ and $W = V/T = T \otimes {}^{\mathbb{Q}_p}/_{\mathbb{Z}_p}$. So far we were dealing with the example $T_E = T_p E$ and $W_E = E[p^\infty]$ and V_E is then a 2-dimensional Galois representation. But of course there are lots of other examples, such as more general subquotients of étale cohomology groups of varieties defined over F with $\mathbb{Q}_p$-coefficients or Galois representation attached to modular forms.

We wish to define the Selmer group however we have no longer a Kummer map $\kappa :? \to H^1(F_v, W)$ or $? \to H^1(F_v, T)$. In order to understand how to define it in the general case, we look at how we could describe the image of κ in the case of an elliptic curve.

5.1 Local Conditions at Places Away from p

Let $v \in \Sigma$ be a prime not dividing p. Then the Kummer maps are $E(F_v) \otimes {}^{\mathbb{Q}_p}/_{\mathbb{Z}_p} \to H^1(F_v, W_E)$ and $E(F_v)^\star \to H^1(F_v, T_E)$. Recall that $E(F_v)$ contains with finite index a group isomorphic to $\mathfrak{m}_v$, the maximal ideal of F_v. Hence $E(F_v) \otimes {}^{\mathbb{Q}_p}/_{\mathbb{Z}_p} = 0$ and $E(F_v)^\star = E(F_v)[p^\infty] = W_E^{G_{F_v}}$ is finite.

Here is the general definition for $v \nmid p$. Define the subgroup $H^1_f(F_v, V)$ by the exact sequence

$$0 \longrightarrow H^1_f(F_v, V) \longrightarrow H^1(F_v, V) \longrightarrow H^1(I, V) \tag{8}$$

where $I = I_v$ is the inertia subgroup in $\mathrm{Gal}(\bar{F}_v/F_v)$. By the restriction-inflation sequence, $H^1_f(F_v, V)$ is isomorphic to $H^1(F_v^{\mathrm{ur}}/F_v, V^I)$ where F_v^{ur} is the maximal unramified extension of F_v. Then we define $H^1_f(F_v, W)$ as the image of $H^1_f(F_v, V)$ under the map $H^1(F_v, V) \to H^1(F_v, W)$. Also $H^1_f(F_v, T)$ is the preimage of $H^1_f(F_v, V)$ from the map $H^1(F_v, T) \to H^1(F_v, V)$.

For the case of the elliptic curve, we find $H^1_f(F_v, V_E) = 0$ for all $v \nmid p$: In fact, we have in general that $H^1_f(F_v, V) = V^I/(\mathrm{Fr}_v - 1)V^I$ by Lemma 1.3.2 in Rubin (2000). If the reduction is good then $V_E^I = V_E$ by the Néron–Ogg–Shafarevich criterion (Silverman 2009, Theorem 7.7.1) and Fr_v, the Frobenius of $\mathrm{Gal}(F_v^{\mathrm{ur}}/F_v)$, acts with eigenvalues different from 1. If the reduction is multiplicative, then $V_E^I \cong \mathbb{Q}_p(1)$ and the group $H^1_f(F_v, V_E)$ is again trivial. Finally if the reduction is additive, then $V_E^I = 0$. Since we have the exact sequence

$$0 = V_E^{G_{F_v}} \longrightarrow W_E^{G_{F_v}} \longrightarrow H^1(F_v, T_E) \longrightarrow H^1(F_v, V_E) \longrightarrow H^1(F_v, W_E)$$

we obtain that $H^1_f(F_v, W_E) = 0 = E(F_v) \otimes {}^{\mathbb{Q}_p}/_{\mathbb{Z}_p}$ and $H^1_f(F_v, T_E) = E(F_v)[p^\infty] = E(F_v)^\star$.

It would be tempting to define in general $H^1_f(F_v, W)$ without passing through V by replacing V with W in (8). However the elliptic curve example explains that this does not work. In general, we have that $H^1_f(F_v, W)$ is the divisible part of the kernel of $H^1(F_v, W) \to H^1(I_v, W)$ and $H^1_f(F_v, T)$ has finite index in the corresponding kernel for T; see Rubin (2000, Lemma 1.3.5).

5.2 Local Conditions at Places Above p

Let now v be a place above p. The definition of the group $H^1_f(F_v, V)$ is given by Bloch-Kato (1990) by asking that

$$0 \longrightarrow H^1_f(F_v, V) \longrightarrow H^1(F_v, V) \longrightarrow H^1(F_v, V \otimes B_{\mathrm{cris}}) \tag{9}$$

is an exact sequence, where B_{cris} is a certain period ring of Fontaine.

Now, if V is ordinary, one can give an easier definition. Here a general representation V is called *ordinary* if there is a decreasing filtration $\mathrm{Fil}^i\, V$ of $\mathrm{Gal}(\bar{F}_v/F_v)$-stable subspaces of V such that the inertia group I acts like the i-th power of the cyclotomic character on the quotient $\mathrm{Fil}^i\, V/\,\mathrm{Fil}^{i+1}\, V$. For an ordinary elliptic curve, we consider the kernel of the reduction on $E[p^k]$ which is the p^k-torsion $\hat{E}[p^k]$ of the formal group. Then $\mathrm{Fil}^1\, V_E = T_p\hat{E} \otimes \mathbb{Q}_p$ sits in the middle between $\mathrm{Fil}^2\, V_E = 0$ and $\mathrm{Fil}^0\, V_E = V_E$.

We set $F^+V = \mathrm{Fil}^1\, V$ and then Greenberg (1989) shows that

$$0 \longrightarrow H^1_f(F_v, V) \longrightarrow H^1(F_v, V) \longrightarrow H^1\left(I, {}^{V}/_{F^+V}\right) \tag{10}$$

is exact. The subgroups $H^1_f(F_v, W)$ and $H^1_f(F_v, T)$ are again defined as the image and preimage of $H^1_f(F_v, V)$, respectively.

5.3 The Selmer Groups

The Selmer groups are now defined as the following kernels. They are often denoted by $H^1_f(F, W)$ and $H^1_f(F, T)$.

$$0 \longrightarrow \mathcal{S}(W) \longrightarrow H^1_\Sigma(F, W) \longrightarrow \bigoplus_{v \in \Sigma} {}^{H^1(F_v, W)}\Big/_{H^1_f(F_v, W)},$$

$$0 \longrightarrow \mathfrak{S}(T) \longrightarrow H^1_\Sigma(F, T) \longrightarrow \bigoplus_{v \in \Sigma} {}^{H^1(F_v, T)}\Big/_{H^1_f(F_v, T)}$$

They are now defined only in terms of T and equal to the previously defined Selmer group for elliptic curves. If $\text{Ш}(E/F)$ is finite, then we have a way to determine the rank and the order of the Tate-Shafarevich group from the Galois representation V_E only. Note that the L-function $L(E, s)$ is also constructed from V_E only.

For instance, if $T = \mathbb{Z}_p$ has a trivial G_F-action on it, then $\mathcal{S}({}^{\mathbb{Q}_p}/_{\mathbb{Z}_p})$ is the dual of the p-primary part of the class group. If $T = \mathbb{Z}_p(1)$ is of rank 1 with the action by G_F given by the cyclotomic character, then $\mathcal{S}\big({}^{\mathbb{Q}_p}/_{\mathbb{Z}_p}(1)\big)$ sits in a short exact sequence between $\mathcal{O}_F^\times \otimes {}^{\mathbb{Q}_p}/_{\mathbb{Z}_p}$ and the p-primary part of the class group.

6 Kato's Euler System

In this section we give a quick and very imprecise overview of the work of Kato (2004) where he proves one divisibility in the main conjecture for elliptic curves (and modular forms of higher weight). See also Scholl (1998) and Colmez (2004).

Let $E/\mathbb{Q}$ be an elliptic curve and p an odd prime at which E has good reduction. Let N be the conductor of E. We assume that $E[p]$ is an irreducible $G_{\mathbb{Q}}$-module and hence all $G_{\mathbb{Q}}$-stable lattices in V_pE are equal up to a scaling factor.

6.1 Construction of the Euler System

Let M be an integer. Pick an integer $m \geqslant 5$ coprime to $6M$. For any elliptic curve A over a field $k/\mathbb{Q}$, we can construct a division polynomial $f_m \in k(A)^\times$, which is a function of divisor $\operatorname{div}(f_m) = -m^2(O) + \sum_{P\in A[m]}(P)$ normalised such that $[a]^* f_m = f_m$ for all a coprime to m. Consider the maps $g_i^{(m)}$ for $i = 1$ or 2 that sends a point in the modular curve $Y(M)$ represented by (A, Q_1, Q_2) to $f_m(Q_i)$. It is a rational function on $Y(M)$ without zero, i.e., $g_i \in \mathcal{O}(Y(M))^\times$, called a *Siegel unit.* It can be shown that the function $g_i = g_i^{(m)} \otimes \frac{1}{m^2-1}$ in $\mathcal{O}(Y(M)) \otimes \mathbb{Q}$ is independent of the choice of m. They give rise to Beilinson element in $K_2\big(\mathcal{O}(Y(M))^\times\big) \otimes \mathbb{Q}$, defined as the Steinberg symbol $\{g_1, g_2\}$.

Such pairs of modular units can now be sent through a chain of maps. For a square-free r coprime to pN, we take $M = N\,p^{n+1}\,r$ and consider the map

$$X(M) \longrightarrow X(N) \otimes \mathbb{Q}^{(n)}(\mu_r) \xrightarrow{\ \varphi_E\ } E \otimes \mathbb{Q}^{(n)}(\mu_r).$$

We chase the pair of modular units through the maps (see §8.4 in Kato 2004)

$$\begin{array}{ccc}
\mathcal{O}\big(Y(M)\big)^\times \times \mathcal{O}\big(Y(M)\big)^\times & \xrightarrow{\ \cup\ } & H^2_{\text{ét}}\Big(Y(N)/\mathbb{Q}^{(n)}(\mu_r),\ \mathbb{Z}_p(2)\Big) \\
 & & \Big\downarrow \text{twist à la Soulé} \\
 & & H^2_{\text{ét}}\Big(Y(N)/\mathbb{Q}^{(n)}(\mu_r),\ \mathbb{Z}_p(1)\Big) \\
 & & \Big\downarrow \\
 & & H^1\Big(\mathbb{Q}^{(n)}(\mu_r), H^1_{\text{ét}}\big(\overline{Y(N)},\ \mathbb{Z}_p(1)\big)\Big) \\
 & & \Big\downarrow \\
 & & H^1\big(\mathbb{Q}^{(n)}(\mu_r),\ T_E\big) \otimes \mathbb{Q}
\end{array}$$

where we used that $H^1_{\text{ét}}(\overline{E}, \mathbb{Z}_p(1)) = T_E$. Poincaré duality relates $H^1_{\text{ét}}(\overline{Y_0(N)}, \mathbb{Q}_p)$ to modular symbols as it is equal to the homology $H_1(\overline{X_0(N)}(\mathbb{C}), \{\text{cusps}\}, \mathbb{Z}) \otimes \mathbb{Q}_p$ of paths between cusps. See §4.7 and §8.3 in Kato (2004).

The image of the Siegel units produce now elements $c_r^{(n)}$ in $H^1_\Sigma(\mathbb{Q}^{(n)}(\mu_r), T)$ that form an Euler system for a certain lattice T in $T_E \otimes \mathbb{Q}_p$. See example 13.3 in Kato (2004) for details. In particular, $(c_r^{(n)})_n$ belongs to $\varprojlim_n H^1(\mathbb{Q}^{(n)}(\mu_r), T)$. If ℓ is a prime not dividing rpN then they satisfy the norm relations

$$\operatorname{cor}\bigl(c_{r\ell}^{(n)}\bigr) = \Bigl(1 - \frac{a_\ell}{\ell}\sigma_\ell^{-1} + \frac{1}{\ell}\sigma_\ell^{-2}\Bigr)(c_r^{(n)}).$$

See proposition 8.12 in Kato (2004) for the precise statement deduced from the Hecke operators on the modular curves.

6.2 *Relation to p-Adic L-Function*

Suppose E has good ordinary reduction at p and let $\alpha \in \mathbb{Z}_p$ be the unit root of Frobenius. We continue to assume that $E[p]$ is irreducible. The general "dual of exponential" à la Bloch-Kato has a very explicit description for elliptic curves. It is the map

$$\exp^*\colon E(\mathbb{Q}_p^{(n)})^\star \to \Bigl(E(\mathbb{Q}_p^{(n)}) \otimes {}^{\mathbb{Q}_p}\!/\!_{\mathbb{Z}_p}\Bigr)^\wedge \to \mathbb{Q}_p^{(n)}$$

which is a linear extension of the formal logarithm on the formal group with respect to the invariant differential ω_E. Based on the work of Coleman, Perrin-Riou has constructed an Iwasawa theoretic version which interpolates these maps:

$$\operatorname{Col}\colon \mathbb{H}^1_s := \Bigl(E(\mathbb{Q}_p^{(\infty)}) \otimes {}^{\mathbb{Q}_p}\!/\!_{\mathbb{Z}_p}\Bigr)^\wedge \to \Lambda.$$

It is an injective Λ-morphism with finite cokernel such that for all character χ of Γ of conductor p^{n+1}, we have

$$\chi\bigl(\operatorname{Col}(z)\bigr) = \frac{\operatorname{Ga}(\chi)}{\alpha^{n+1}} \sum_{\sigma \in \operatorname{Gal}(\mathbb{Q}^{(n)}/\mathbb{Q})} \bar{\chi}(\sigma)\exp^*\bigl(\sigma(z^{(n)})\bigr).$$

See the appendix of Rubin (1998) for details of the construction.

One of the main theorems of Kato is

Theorem 14. *There is an element $c^{(\infty)} \in \varprojlim H^1_\Sigma(\mathbb{Q}^{(n)}, T) \otimes \mathbb{Q}$ closely related to the ones constructed above such that $\operatorname{Col}(c^{(\infty)}) = \mathcal{L}_E$*

This is theorem 16.6 in Kato (2004) with the "good choice" of γ^+ as in 17.5. This is the technically most difficult part of Kato (2004). It implies that the Euler system is non-trivial by Theorem 10. If the representation $\bar{\rho}\colon G_{\mathbb{Q}} \to \mathrm{GL}(T_E)$ is surjective, then $c^{(\infty)}$ is integral by his theorem 12.5.4.

6.3 Euler System Method

For each i, the limit $\mathbb{H}^i = \varprojlim_n H^i_{\Sigma}(\mathbb{Q}^{(n)}, T_E)$ is a finitely generated Λ-module, which does not depend on Σ as long as it contains p. The existence of a non-trivial Euler system $c_r^{(\infty)} \in \varprojlim H^1(\mathbb{Q}^{(n)}(\mu_r), T_E)$ proves the following.

Theorem 15 (Theorem 12.4 in Kato 2004).

1. *$\mathbb{H}^2$ is Λ-torsion.*
2. *$\mathbb{H}^1$ is a torsion-free Λ-module of rank* 1.
3. *If $E[p]$ is an irreducible $G_{\mathbb{Q}}$-module, then $\mathbb{H}^1$ is free of rank* 1.

See also Rubin (2000). The statement that $\mathbb{H}^2$ is Λ-torsion is called the weak Leopoldt conjecture and it is believed to hold for many Galois representations T. Global duality together with basic results deduced from the above theorem provides us with an exact sequence

$$0 \longrightarrow \mathbb{H}^1 / c^{(\infty)}\Lambda \longrightarrow \mathbb{H}^1_s / c^{(\infty)}\Lambda \longrightarrow X \longrightarrow \mathbb{H}^2 \longrightarrow 0.$$

Here $c^{(\infty)} \in \mathbb{H}^1 \otimes \mathbb{Q}$ is the part of the Euler system that is sent to the p-adic L-function $\mathcal{L}_E$ by the Coleman map; therefore Col sends the second term into $\Lambda/\mathcal{L}_E\Lambda$ with finite cokernel. Hence the main conjecture is equivalent to

Conjecture 7. The characteristic ideal of $\mathbb{H}^2$ and of $\mathbb{H}^1/c^{(\infty)}\Lambda$ are equal.

The advantage of this formulation is that it does not involve the p-adic L-function and makes sense in the supersingular case as well.

Theorem 16 (Theorem 17.4 in Kato 2004). *Suppose E has good ordinary reduction at p. Then*

1. *X is a torsion Λ-module;*
2. *There is an integer $t \geqslant 0$ such that the characteristic ideal* $\mathrm{char}(X)$ *divides $p^t\,\mathcal{L}_E\,\Lambda$;*
3. *If the representation $\bar{\rho}\colon G_{\mathbb{Q}} \to \mathrm{GL}(T_E)$ is surjective, then* $\mathrm{char}(X)$ *divides $\mathcal{L}_E\,\Lambda$.*

Acknowledgements It is my pleasure to thank Thanasis Bouganis, Sylvia Guibert, Chern-Yang Lee, Birgit Schmoetten-Jonas and Otmar Venjakob for their efforts to make my stay in Heidelberg a very pleasant time and for their help when writing and improving these notes. All computations were done in Sage (Stein et al. 2012).

References

Bertolini, M.: Iwasawa theory for elliptic curves over imaginary quadratic fields. J. Théor. Nombres Bordeaux **13**(1), 1–25 (2001). 21st Journées Arithmétiques, Rome (2001)

Breuil, C., Conrad, B., Diamond, F., Taylor, R.: On the modularity of elliptic curves over $\mathbb{Q}$: wild 3-adic exercises. J. Am. Math. Soc. **14**(4), 843–939 (2001)

Bloch, S.: A note on height pairings, Tamagawa numbers, and the Birch and Swinnerton-Dyer conjecture. Invent. Math. **58**(1), 65–76 (1980)

Bloch, S., Kato, K.: L-functions and Tamagawa numbers of motives. In: The Grothendieck Festschrift, Vol. I. Progress in Mathematics, vol. 86, pp. 333–400. Birkhäuser, Boston (1990)

Buhler, J.P., Harvey, D.: Irregular primes to 163 million. Math. Comput. **80**(276), 2435–2444 (2011)

Coates, J., Sujatha, R.: Galois Cohomology of Elliptic Curves. Tata Institute of Fundamental Research Lectures on Mathematics, vol. 88. Narosa Publishing House, New Delhi (2000)

Coates, J., Sujatha, R.: Cyclotomic Fields and Zeta Values. Springer Monographs in Mathematics. Springer, Berlin (2006)

Coates, J., Greenberg, R., Ribet, K.A., Rubin, K., Viola, C. (ed.): Arithmetic Theory of Elliptic Curves: Lectures from the 3rd C.I.M.E. Session held in Cetraro, 12–19 July 1997. Lecture Notes in Mathematics, vol. 1716. Springer, Berlin/New York (1999).

Colmez, P.: La conjecture de Birch et Swinnerton-Dyer p-adique. Astérisque **294**(ix), 251–319 (2004)

Cremona, J.E.: Algorithms for Modular Elliptic Curves, 2nd edn. Cambridge University Press, Cambridge (1997)

Ferrero, B., Washington, L.C.: The Iwasawa invariant μ_p vanishes for abelian number fields. Ann. Math. (2) **109**(2), 377–395 (1979)

Greenberg, R.: Iwasawa theory for p-adic representations. In: Algebraic Number Theory. Advanced Studies in Pure Mathematics, vol. 17, pp. 97–137. Academic, Boston (1989)

Greenberg, R.: Introduction to Iwasawa theory for elliptic curves. In: Arithmetic Algebraic Geometry. IAS/Park City Mathamatical Series, vol. 9, pp. 407–464. Park City (1999). MathSciNet: MR1860044

Greenberg, R.: Iwasawa theory—past and present. In: Class Field Theory—Its Centenary and Prospect (Tokyo, 1998). Advanced Studies in Pure Mathematics, vol. 30, pp. 335–385. Mathematical Society of Japan, Tokyo (2001)

Iwasawa, K.: On Γ-extensions of algebraic number fields. Bull. Am. Math. Soc. **65**, 183–226 (1959)

Iwasawa, K.: On the μ-invariants of $\mathbb{Z}_l$-extensions. In: Number Theory, Algebraic Geometry and Commutative Algebra, In Honor of Yasuo Akizuki, pp. 1–11. Kinokuniya, Tokyo (1973)

Kato, K.: p-adic Hodge theory and values of zeta functions of modular forms. Astérisque **295**(ix), 117–290 (2004). Cohomologies p-adiques et applications arithmétiques. III

Kurihara, M.: On the Tate Shafarevich groups over cyclotomic fields of an elliptic curve with supersingular reduction. I. Invent. Math. **149**(1), 195–224 (2002)

Lang, S.: Algebraic groups over finite fields. Am. J. Math. **78**, 555–563 (1956)

Lang, S.: Cyclotomic Fields I and II. Graduate Texts in Mathematics, vol. 121, 2nd edn. Springer, New York (1990). With an appendix by Karl Rubin

Lubin, J., Rosen, M.I.: The norm map for ordinary abelian varieties. J. Algebra **52**(1), 236–240 (1978)

Manin, J.I.: Parabolic points and zeta functions of modular curves. Izv. Akad. Nauk SSSR Ser. Mat. **36**, 19–66 (1972)

Mazur, B.: Rational points of abelian varieties with values in towers of number fields. Invent. Math. **18**, 183–266 (1972)

Mazur, B., Tate, J.: The p-adic sigma function. Duke Math. J. **62**(3), 663–688 (1991)

Mazur, B., Wiles, A.: Class fields of abelian extensions of $\mathbb{Q}$. Invent. Math. **76**(2), 179–330 (1984)

Mazur, B., Tate, J., Teitelbaum, J.: On p-adic analogues of the conjectures of Birch and Swinnerton-Dyer. Invent. Math. **84**(1), 1–48 (1986)

Mazur, B., Stein, W., Tate, J.: Computation of p-adic heights and log convergence. Doc. Math. **Extra Vol.**, 577–614 (2006)

Milne, J.S.: Elliptic Curves. BookSurge, Charleston (2006)

Neukirch, J., Schmidt, A., Wingberg, K.: Cohomology of Number Fields. Grundlehren der Mathematischen Wissenschaften, vol. 323. Springer, Berlin/New York (2000)

Perrin-Riou, B.: Points de Heegner et dérivées de fonctions L p-adiques. Invent. Math. **89**(3), 455–510 (1987)

Perrin-Riou, B.: Théorie d'Iwasawa et hauteurs p-adiques. Invent. Math. **109**(1), 137–185 (1992)

Perrin-Riou, B.: Théorie d'Iwasawa et hauteurs p-adiques (cas des variétés abéliennes). Séminaire de Théorie des Nombres, Paris, 1990–1991. Progress in Mathematics, vol. 108, pp. 203–220. Birkhäuser (1993). MathSciNet: MR1263528

Perrin-Riou, B.: Fonctions L p-adiques des représentations p-adiques. Astérisque **229** (1995). MathSciNet: MR1327803

Pollack, R.: On the p-adic L-function of a modular form at a supersingular prime. Duke Math. J. **118**(3), 523–558 (2003)

Ribet, K.A.: A modular construction of unramified p-extensions of $\mathbb{Q}(\mu_p)$. Invent. Math. **34**(3), 151–162 (1976)

Rohrlich, D.E.: On L-functions of elliptic curves and cyclotomic towers. Invent. Math. **75**(3), 409–423 (1984)

Rubin, K.: Euler systems and modular elliptic curves. In: Galois Representations in Arithmetic Algebraic Geometry (Durham, 1996). London Mathematical Society Lecture Note Series, vol. 254, pp. 351–367. Cambridge University Press, Cambridge (1998)

Rubin, K.: Euler Systems. Annals of Mathematics Studies, vol. 147. Hermann Weyl Lectures. The Institute for Advanced Study. Princeton University Press, Princeton (2000)

Schneider, P.: p-adic height pairings. I. Invent. Math. **69**(3), 401–409 (1982)

Schneider, P.: p-adic height pairings. II. Invent. Math. **79**(2), 329–374 (1985)

Scholl, A.J.: An introduction to Kato's Euler systems. In: Galois Representations in Arithmetic Algebraic Geometry (Durham, 1996). London Mathematical Society Lecture Note Series, vol. 254, pp. 379–460. Cambridge University Press, Cambridge (1998)

Serre, J.-P.: Corps Locaux, Publications de l'Université de Nancago, No. VIII, Deuxième édition. Hermann, Paris (1968)

Serre, J.-P.: Classes des corps cyclotomiques (d'après K. Iwasawa). Séminaire Bourbaki, vol. 5, pp. Exp. No. 174, 83–93. Société Mathématique de France, Paris (1995)

Silverman, J.H.: The Arithmetic of Elliptic Curves. Graduate Texts in Mathematics, vol. 106, 2nd edn. Springer, Dordrecht (2009)

Stein, W., Wuthrich, C.: Algorithms for the arithmetic of elliptic curves using Iwasawa theory. Math. Comput. **82**(283), 1757–1792 (2013)

Stein, W.A., et al.: Sage Mathematics Software (Version 5.1). The Sage Development Team (2012). Available from http://www.sagemath.org

Tate, J.: Relations between K_2 and Galois cohomology. Invent. Math. **36**, 257–274 (1976)

Tate, J.: On the Conjectures of Birch and Swinnerton-Dyer and a Geometric Analog. Séminaire Bourbaki, vol. 9, pp. Exp. No. 306, 415–440. Société Mathématique de France, Paris (1995a)

Tate, J.: WC-groups over p-adic fields. Séminaire Bourbaki, vol. 4, pp. Exp. No. 156, 265–277. Société Mathématique de France, Paris (1995b)

Wan, X.: Introduction to Skinner-Urban's work on the Iwasawa main conjecture for GL2. Iwasawa Theory 2012: State of the Art and Recent Advances. Springer, Berlin/Heidelberg (2014)

Washington, L.C.: Introduction to Cyclotomic Fields. Graduate Texts in Mathematics, vol. 83, 2nd edn. Springer, New York (1997)

Wiles, A.: The Iwasawa conjecture for totally real fields. Ann. Math. (2) **131**(3), 493–540 (1990)

Introduction to Skinner-Urban's Work on the Iwasawa Main Conjecture for GL_2

Xin Wan

1 Introduction

These notes are an introduction to the recent work of Christopher Skinner and Eric Urban (2010) proving (one divisibility of) the Iwasawa main conjecture for $\mathrm{GL}_2/\mathbb{Q}$ (see Theorem 1). We give the necessary background materials and explain the proofs. We focus on the main ideas instead of the details and therefore will sometimes be brief and even imprecise.

These notes are organized as follows. In Sect. 2 we formulate various Iwasawa main conjectures for modular forms. We also explain an old result of Ribet to illustrate the rough idea of the strategy behind the later proofs. In Sects. 3 and 4 we introduce the notions of automorphic forms and Eisenstein series on the unitary group $GU(2,2)$. Section 5 is devoted to explaining the Galois argument. Sections 6–9 give the tools used in the computation for the Fourier and Fourier-Jacobi coefficients of various Eisenstein series, which is a crucial ingredient in the argument. In Sect. 10 we give an example of a theorem of the author generalizing the Skinner-Urban work.

2 Main Conjectures

We introduce the objects required to state the Iwasawa main conjectures for GL_2.

X. Wan (✉)
Mathematics Department, Columbia University, New York, NY, 10027, USA
e-mail: xw2295@math.columbia.edu

T. Bouganis and O. Venjakob (eds.), *Iwasawa Theory 2012*, Contributions in Mathematical and Computational Sciences 7, DOI 10.1007/978-3-642-55245-8_2,

2.1 Families of Characters

Let p be an odd prime. Choose $\iota : \mathbb{C} \simeq \mathbb{C}_p$. Let $G_{\mathbb{Q}} = \mathrm{Gal}(\bar{\mathbb{Q}}/\mathbb{Q})$ and $\mathbb{Q}_\infty \subset \mathbb{Q}(\mu_{p^\infty})$ be the cyclotomic $\mathbb{Z}_p$-extension of $\mathbb{Q}$. Let $\Gamma_{\mathbb{Q}} = \mathrm{Gal}(\mathbb{Q}_\infty/\mathbb{Q})$. Let $\Lambda_{\mathbb{Q}} := \mathbb{Z}_p[[\Gamma_{\mathbb{Q}}]]$. We also define $\Lambda_A = \Lambda_{\mathbb{Q},A} = \Lambda_{\mathbb{Q}} \otimes_{\mathbb{Z}_p} A$ for A a $\mathbb{Z}_p$-algebra. Let $\Psi = \Psi_{\mathbb{Q}} : G_{\mathbb{Q}} \to \Lambda_{\mathbb{Q}}^\times$ be the composition of $G_{\mathbb{Q}} \twoheadrightarrow \Gamma_{\mathbb{Q}}$ with $\Gamma_{\mathbb{Q}} \hookrightarrow \Lambda_{\mathbb{Q}}^\times$. Let $\varepsilon_{\mathbb{Q}}$ be a character of $\mathbb{Q}^\times\backslash\mathbb{A}_{\mathbb{Q}}^\times$ which is the composition of $\Psi_{\mathbb{Q}}$ with the reciprocity map of class field theory (normalized using geometric Frobenius elements). Take $\gamma \in \Gamma$ to be the topological generator such that $\epsilon(\gamma) = 1 + p$ where ϵ is the cyclotomic character giving the canonical isomorphism $\mathrm{Gal}(\mathbb{Q}(\mu_{p^\infty})/\mathbb{Q}) \simeq \mathbb{Z}_p^\times$. For each $\zeta \in \mu_{p^\infty}$ and integer k we let $\psi_{k,\zeta}$ be the finite order character of $\mathbb{Q}^\times\backslash\mathbb{A}_{\mathbb{Q}}^\times$ that is the composition of $\Psi_{\mathbb{Q}}$ with the map $\Lambda_{\mathbb{Q}}^\times \to \mathbb{C}_p^\times$ that maps γ to $\zeta(1+p)^k$. We also write ψ_ζ for $\psi_{0,\zeta}$. We let ω be the Teichimuller character.

2.2 Characteristic Ideals and Fitting Ideals

Let A be a Noetherian normal domain and X a finite A-module. The characteristic ideal $\mathrm{char}_A(X) \subset A$ is defined to be zero if X is not torsion and

$$\mathrm{char}_A(X) = \{x \in A | \mathrm{ord}_P x \geq \mathrm{length}_{A_P}(X_P), \text{ for all height one primes } P \subset A\}.$$

Now take any presentation

$$A^r \to A^s \to X \to 0$$

of X. The Fitting ideal is defined by the ideal of A generated by all the determinants of the $s \times s$ minors of the matrix representing the first arrow.

Remark 1. Fitting ideals respect any base change while characteristic ideals do not in general.

2.3 Selmer Groups for Modular Forms

Let $f = \sum_{n=1}^\infty a_n q^n \in S_k(N, \psi_0)$, $k \geq 2$, be a cuspidal eigenform with character ψ_0 of $(\mathbb{Z}/N\mathbb{Z})^\times$ and let $L/\mathbb{Q}_p$ be a finite extension containing all Fourier coefficients a_n of f. Let $\mathcal{O}_L$ be the ring of integers of L. Assume that f is ordinary, which means that a_p is a unit in $\mathcal{O}_L$. Let $\rho = \rho_f : G_{\mathbb{Q}} \to \mathrm{Aut}_L V_f$ be the usual two dimensional Galois representation associated to f. Then it is well known (by Wiles 1988, for example) that there is a $G_{\mathbb{Q}_p}$-stable L-line $V_f^+ \subset V_f$ such that V_f/V_f^+ is unramified. We fix a $G_{\mathbb{Q}}$-stable $\mathcal{O}_L$ lattice $T_f \subset V_f$ and let $T_f^+ = T_f \cap V_f^+$.

Definition 1 (Selmer Groups). Let Σ be a finite set of primes.

$$\mathrm{Sel}_L^{\Sigma}(T_f) := \ker\{H^1(\mathbb{Q}, T_f \otimes_{\mathcal{O}_L} \Lambda_{\mathcal{O}_L}^*(\Psi^{-1})) \to H^1(I_p, (T_f/T_f^+) \otimes_{\mathcal{O}_L} \Lambda_{\mathcal{O}_L}^*(\Psi^{-1})) \times \prod_{\ell \neq p, \ell \notin \Sigma} H^1(I_\ell, T_f \otimes \Lambda_{\mathcal{O}_L}^*(\Psi^{-1}))\}$$

where $\Lambda_A^* = \mathrm{Hom}_{\mathbb{Z}_p}(\Lambda_A, \mathbb{Q}_p/\mathbb{Z}_p)$ is the Pontryagin dual and $\Lambda_{\mathcal{O}_L}^*(\Psi^{-1})$ means that the Galois action is given by the character Ψ^{-1}. Let

$$X_L^{\Sigma}(T_f) := \mathrm{Hom}_{\mathbb{Z}_p}(\mathrm{Sel_L}^{\Sigma}(T_f), \mathbb{Q}_p/\mathbb{Z}_p)$$

and

$$\mathrm{char}_{f,\mathbb{Q}}^{\Sigma}(f) = \mathrm{char}_{\Lambda_{\mathbb{Q},\mathcal{O}_L}}(X_L^{\Sigma}(T_f)).$$

2.4 p-Adic L-Functions

Let $0 < n < k-2$ be an integer and $\zeta \neq 1$ a p^{t-1}th root of unity. Let Σ be a finite set of primes. We define the algebraic part of a special L-value for f by:

$$L_{alg}^{\Sigma}(f, \psi_\zeta^{-1}\omega^n, n+1) := a_p(f)^{-t} \frac{p^{t(n+1)} n! L^{\Sigma}(f, \psi_\zeta^{-1}\omega^n, n+1)}{(-2\pi i)^n \tau(\psi_\zeta^{-1}\omega^n) \Omega_f^{\mathrm{sgn}((-1)^m)}}$$

where $a_p(f)$ is the p-adic unit root of $x^2 - a_p x + p^{k-1}\psi_0 = 0$, $\tau(\psi)$ is a Gauss sum for ψ and $\Omega_f^{\pm}$ are Hida's canonical periods of f. (There is also a formula for $\zeta = 1$ which is more complicated which we omit here.) The p-adic L-function is a certain element $\mathcal{L}_{f,\mathbb{Q}}^{\Sigma} \in \Lambda_{\mathbb{Q},\mathcal{O}_L}$ characterized by the following interpolation property. Let $\phi_{n,\zeta} : \Lambda_{\mathcal{O}_L} \to \mathcal{O}_L(\zeta)$ be the $\mathcal{O}_L$ homomorphism sending γ to $\zeta(1+p)^n$. Then:

$$\phi_{n,\zeta}(\mathcal{L}_{f,\mathbb{Q}}^{\Sigma}) = L_{alg}^{\Sigma}(f, \psi_\zeta^{-1}\omega^n, n+1), 0 \leq n \leq k-2.$$

This was constructed in Amice and Velu (1975), and also Mazur et al. (1986).

2.5 The Main Conjecture

The Iwasawa main conjecture for f is the following

Conjecture 1. The module $X_L^{\Sigma}(T_f)$ is a finite torsion $\Lambda_{\mathbb{Q},\mathcal{O}_L}$-module and $\mathrm{char}_{f,\mathbb{Q}}^{\Sigma}$ is generated by $\mathcal{L}_{f,\mathbb{Q}}^{\Sigma}$.

The main result that we are going to prove in this lecture series is:

Theorem 1 (Kato, Skinner-Urban). *Suppose f has trivial character, weight 2 and good ordinary reduction at p. Suppose also that:*

- *The residual representation $\bar{\rho}_f$ is irreducible.*
- *For some $p \neq \ell || N$, $\bar{\rho}_f$ is ramified at ℓ.*

Then the Iwasawa main conjecture is true in $\Lambda_{\mathbb{Q},\mathcal{O}_L} \otimes_{\mathbb{Z}_p} \mathbb{Q}_p$. If, moreover, there exists an $\mathcal{O}_L$-basis of T_f with respect to which the image of ρ_f contains $SL_2(\mathbb{Z}_p)$, then the equality holds in $\Lambda_{\mathbb{Q},\mathcal{O}_L}$.

The last condition is put by Kato (2004) who proved "$\supseteq$". It is satisfied, for example, by the p-adic Tate modules of semistable elliptic curves if $p \geq 11$. We will focus on Skinner-Urban's proof for "$\subseteq$". The technical condition (ii) can be removed by working with forms over totally real fields and a base change trick.

2.6 Two and Three-Variable Main Conjectures

More p-adic characters: Let $\mathcal{K}/\mathbb{Q}$ be an imaginary quadratic extension such that p splits as $v_0\bar{v}_0$, where v_0 is determined by our chosen isomorphism $\iota : \mathbb{C} \simeq \mathbb{C}_p$. By class field theory there is a unique $\mathbb{Z}_p^2$-extension of $\mathcal{K}$ unramified outside p, which we denote by $\mathcal{K}_\infty$. Let $G_\mathcal{K} := \mathrm{Gal}(\bar{\mathcal{K}}/\mathcal{K})$ and $\Gamma_\mathcal{K} := \mathrm{Gal}(\mathcal{K}_\infty/\mathcal{K})$. There is an action of complex conjugation c on $\Gamma_\mathcal{K}$. We write $\Gamma_\mathcal{K}^\pm$ for the subgroup on which c acts by ± 1. For any $\mathbb{Z}_p$-algebra $A \subset \bar{\mathbb{Q}}_p$ we define: $\Lambda_{\mathcal{K},A} = A[[\Gamma_\mathcal{K}]]$, $\Lambda_{\mathcal{K},A}^\pm = A[[\Gamma_\mathcal{K}^\pm]]$ and generators $\gamma^\pm$ of $\Gamma_\mathcal{K}^\pm$ by requiring $\mathrm{rec}_{\mathcal{K}_p}((1+p)^{\frac{1}{2}}, (1+p)^{\pm\frac{1}{2}}) = \gamma^\pm$. Here $\mathrm{rec}_{\mathcal{K}_p}$ is the reciprocity map of class field theory. (Note that $\mathcal{K}_p \simeq \mathbb{Q}_p \times \mathbb{Q}_p$.) Let $\Psi_\mathcal{K}$ be the composition $G_\mathcal{K} \to \Gamma_\mathcal{K} \hookrightarrow \Lambda_{\mathcal{K},A}^\times$. We define $\Psi_\mathcal{K}^\pm$ similarly. We also define characters $\varepsilon_\mathcal{K}, \varepsilon_\mathcal{K}^\pm$ of $\mathcal{K}^\times\backslash\mathbb{A}_\mathcal{K}^\times$ by composing $\varepsilon_\mathcal{K}, \varepsilon_\mathcal{K}^\pm$ with the reciprocity map.

Now let f be a cuspidal eigenform of weight $k \geq 2$. We define the set of arithmetic points by:

$$\mathcal{X}_{f,\mathcal{K}}^a := \{\phi : \mathcal{O}_L \text{ homomorphism } \Lambda_{\mathcal{K},\mathcal{O}_L} \to \bar{\mathbb{Q}}_p : \phi(\gamma^+) = \zeta^+(1+p)^{k-2}, \phi(\gamma^-) = \zeta^-, \zeta^\pm \in \mu_{p^\infty}\}.$$

For $\phi \in \mathcal{X}_{f,\mathcal{K}}^a$, let $\theta_\phi := \omega^{2-k}\chi_f^{-1}\xi_\phi$ with $\xi_\phi := (\phi \circ \Psi_\mathcal{K})(\epsilon^{2-k}\varepsilon^{k-2}.\chi_f)$ and let $\mathfrak{f}_{\theta_\phi}$ be the conductor of θ_ϕ. In particular we show that for any finite set Σ of primes containing all the bad primes (all the primes where f or $\mathcal{K}$ is ramified), we have the two variable p-adic L-function $\mathcal{L}_{f,\mathcal{K}}^\Sigma \in \Lambda_{\mathcal{K},\mathcal{O}_L}$ such that:

$$\phi(\mathcal{L}_{f,\mathcal{K}}^\Sigma) = u_f a_p(f)^{ord_p(Nm(\mathfrak{f}_{\theta_\phi}))} . \frac{((k-2)!)^2 \mathfrak{g}(\theta_\phi) Nm(\mathfrak{f}_{\theta_\phi}\mathfrak{d})^{k-2} L_\mathcal{K}^\Sigma(f, \theta_\phi, k-1)}{(2\pi i)^{2k-2}\Omega_f^+\Omega_f^-}$$

for any sufficiently ramified $\phi \in \mathcal{X}^a_{f,\mathcal{K}}$, where $\mathfrak{d}$ is the different of $\mathcal{K}$ and u_f is a p-adic unit depending on f. We remark that if $\zeta^- = 1$ then our special L-value is just the product of the special L-values for f and $f \otimes \chi_{\mathcal{K}}$ twisted by some ψ such that $\psi \circ \mathrm{Nm} = \theta_\phi$. Here $\chi_{\mathcal{K}}$ is the quadratic character for $\mathcal{K}/\mathbb{Q}$.

We can also define Selmer groups $Sel^{\Sigma}_{f,\mathcal{K}}$ and $X^{\Sigma}_{f,\mathcal{K}}$, $\mathrm{char}^{\Sigma}_{f,\mathcal{K}}$ in the exact same way as in the one-variable case. We have the two-variable main conjecture:

Conjecture 2. $X^{\Sigma}_{f,\mathcal{K}}$ is a finite torsion $\Lambda_{\mathcal{K},\mathcal{O}_L}$-module. Furthermore $\mathrm{char}^{\Sigma}_{\mathcal{K},f}$ is principal and generated by $\mathcal{L}^{\Sigma}_{f,\mathcal{K}}$.

Additionally f can be embedded in a Hida family of ordinary cuspidal eigenforms $\mathbf{f}$ (we discuss these in the next section in more detail). We can form a three-variable p-adic L-function $\mathcal{L}^{\Sigma}_{\mathbf{f},\mathcal{K}}$ and formulate a 3 variable main conjecture.

2.7 Comment on the Proof

The cyclotomic main conjecture for modular forms (Conjecture 1) is deduced from a partial inclusion in three-variable main conjecture. Roughly speaking the inclusion $\mathrm{char}_{\mathbf{f},\mathcal{K}} \subseteq (\mathcal{L}^{\Sigma}_{\mathbf{f},\mathcal{K}})$ in the three-variable main conjecture, when specialized to the cyclotomic $\mathbb{Z}_p$-extension of $\mathcal{K}$, implies that the inclusion $\mathrm{char}_{f,\mathbb{Q}}.\mathrm{char}_{f\otimes\chi_{\mathcal{K}},\mathbb{Q}} \subseteq (\mathcal{L}_{f,\mathbb{Q}}.\mathcal{L}_{f\otimes\chi_{\mathcal{K}},\mathbb{Q}})$ in the cyclotomic main conjecture for f and $f \otimes \chi_{\mathcal{K}}$. By Kato's work, this in turn implies that $\mathrm{char}_{f,\mathbb{Q}} = (\mathcal{L}_{f,\mathbb{Q}})$ and $\mathrm{char}_{f\otimes\chi_{\mathcal{K}},\mathbb{Q}} = (\mathcal{L}_{f\otimes\chi_{\mathcal{K}},\mathbb{Q}})$.

We will focus on proving the inclusion $\mathrm{char}^{\Sigma}_{\mathbf{f},\mathcal{K}} \subseteq (\mathcal{L}^{\Sigma}_{\mathbf{f},\mathcal{K}})$ in the three variable main conjecture in the rest of the paper. The general strategy for proving this inclusion is: the product of $\mathcal{L}^{\Sigma}_{\mathcal{K}}$ and the Σ-imprimitive Kubota-Leopodlt p-adic L-function $\mathcal{L}^{\Sigma}_{1,\mathbb{Q}}$ attached to the trivial character gives the congruences between Eisenstein series and cusp forms on the unitary similitude group $GU(2,2) \Rightarrow$ the same congruences between reducible and irreducible Galois representations $\Rightarrow$ required extension class in $H^1(\mathcal{K}, -)$. The first arrow is by the Langlands correspondence and the second is a Galois theoretic argument, the so-called "lattice construction".

2.8 A Theorem of Ribet

In this section we review Ribet's proof of the converse to Herbrand's theorem (Ribet 1976). This illustrates the main ideas in the strategy. In this section we set $\mathcal{O}_L = \mathbb{Z}_p$, λ the maximal ideal of $\mathcal{O}_L$, $\kappa = \mathcal{O}_L/\lambda$.

Theorem 2. *Suppose $j \in [2, p-3]$ is an even number. If $p|\zeta(1-j)$ then $H^1_{ur}(G_{\mathbb{Q}}, \mathbb{F}(\omega^{1-j})) \neq 0$ (the group of everywhere unramified classes is non-zero).*

Proof. For $j \neq 2$, we make use of the level 1 weight j Eisenstein series:

$$E_j(q) = \frac{\zeta(1-j)}{2} + \sum_{n \geq 1} \sigma_{j-1}(n) q^n$$

where $\sigma_{j-1}(n) = \sum_{d|n} d^{j-1}$. If $p|\zeta(1-j)$, then E_j "looks" like a cusp form modulo p. We divide the proof into three steps:

Step1: Construct a cusp form $f' \in S_j(SL_2(\mathbb{Z}), \mathbb{Z}_p)$ such that $f' \equiv E_j(mod\ p)$ (in terms of q-expansion). This is a case by case study using the fact that the ring of modular forms of level 1 is $\mathbb{C}[E_4, E_6]$.

Step 2: Prove that f' can be replaced by an eigenform $f \in S_j(SL_2(\mathbb{Z}_p), \mathcal{O}_L)$ whose Hecke eigenvalues are the same as those of E_j modulo p. This can be proven by easy commutative algebra (essentially a lemma of Deligne and Serre).

Step 3: The lattice construction: construct the class by comparing the Galois representations of E_j and f. Note that the Galois representation for E_j is $\epsilon^{j-1} \oplus 1$. It is easy to see that there is a $\sigma_0 \in I_p$ such that $\epsilon^{j-1}(\sigma_0) \not\equiv 1 (\mathrm{mod}\ p)$. Since $a_p(f) \equiv \sigma_{j-1}(p) \equiv 1 (\mathrm{mod}\ p)$, f is ordinary. As we have noted before, $\rho_f|_{G_{\mathbb{Q}_p}} = \begin{pmatrix} \alpha\epsilon^{j-1} & * \\ & \alpha \end{pmatrix}$ for some unramified character α. Take a basis $\{v_1, v_2\}$ such that

$$\rho_f(\sigma_0) = \begin{pmatrix} \epsilon^{j-1}(\sigma_0) & \\ & 1 \end{pmatrix}.$$

Write $\rho = \rho_f$ and $\rho(\sigma) = \begin{pmatrix} a_\sigma & b_\sigma \\ c_\sigma & d_\sigma \end{pmatrix}$ for $\sigma \in \mathcal{O}[G_{\mathbb{Q}}]$.

Claim:

(a) $a_\sigma, d_\sigma, b_\sigma c_\tau \in \mathcal{O}$ for $\sigma, \tau \in \mathcal{O}_L[G_{\mathbb{Q}}]$ and $a_\sigma \equiv \omega^{j-1}(\sigma), d_\sigma \equiv 1, b_\sigma c_\tau \equiv 0 (\mathrm{mod}\ p)$;

(b) $\mathcal{C} := \{c_\sigma : \sigma \in \mathcal{O}_L[G_{\mathbb{Q}}]\}$ is a non-zero fractional ideal.

(c) $c_\sigma = 0$ if $\sigma \in I_\ell$ for all ℓ.

Proof of the claim: Let $\epsilon_1 := \frac{1}{\epsilon^{j-1}(\sigma_0)-1}(\sigma_0 - 1), \epsilon_2 := \frac{1}{1-\epsilon^{j-1}(\sigma_0)}(\sigma_0 - \epsilon^{j-1}(\sigma_0))$; one can check: $\rho(\epsilon_1) = \begin{pmatrix} 1 & 0 \\ 0 & 0 \end{pmatrix}$ and $\rho(\epsilon_2) = \begin{pmatrix} 0 & 0 \\ 0 & 1 \end{pmatrix}$. Thus $a_\sigma = \mathrm{trace}\rho(\epsilon_1\sigma) \in \mathcal{O}_L$ and $\mathrm{trace}\rho(\epsilon_1\sigma) \equiv \mathrm{trace}(\epsilon^{j-1} + 1)(\epsilon_1\sigma) = \epsilon^{j-1}(\sigma)$. The claim for d_σ is proven similarly. Also since $\rho(\sigma)\rho(\tau) = \rho(\sigma\tau)$, $b_\sigma c_\tau = a_{\sigma\tau} - a_\sigma a_\tau \equiv 0 (\mathrm{mod}\ p)$.

(b) follows from the irreducibility of ρ and (c) can be seen from the description for $\rho|_{G_{\mathbb{Q}_p}}$ above and the triviality of $\rho|_{I_\ell}$ for $\ell \neq p$.

Let $M_1 = \mathcal{O}_L v_1$, $M_2 = \mathcal{C} v_2$, $M = M_1 \oplus M_2$ (which is easily seen to be the $\mathcal{O}_L[G_{\mathbb{Q}}]$-submodule generated by v_1).

- $\bar{M}_2 := M_2/\lambda M_2 \simeq \kappa$ (note that $\mathcal{C}$ is non-zero by (b)) is a $G_{\mathbb{Q}}$ stable submodule of $\bar{M} := M/\lambda M$. This is because for any $m_2 = cv_2 \in M_2$, $\rho(\sigma)m_2 = b_\sigma cv_1 + d_\sigma cv_2 \in \lambda v_1 + \mathcal{C}v_2$ by (a);
- By (a), $G_{\mathbb{Q}}$ acts by ω^{j-1} and 1 on $\bar{M}_1 = \bar{M}/\bar{M}_2$ and $\bar{M}_2$ respectively;
- The extension: $0 \to \bar{M}_2 \to \bar{M} \to \bar{M}_1 \to 0$ is non split since M is generated by v_1 over $\mathcal{O}_L[G_{\mathbb{Q}}]$.

Thus $\bar{M}$ gives a nontrivial extension class and it actually in $H^1_{ur}(\mathbb{Q}, \kappa(\omega^{1-j}))$ by claim (c).

If $j = 2$ the Eisenstein series E_2 is not holomorphic and we use E_{p+1} in the place of E_2.

3 Hermitian Modular Forms on $GU(n,n)$

3.1 Hermitian Half Space and Automorphic Forms

Let $\mathcal{K}/\mathbb{Q}$ be an imaginary quadratic extension and let $\mathcal{O}$ be the ring of integers of $\mathcal{K}$. Let $GU(n,n)$ be the unitary similitude group associated to the pairing $\begin{pmatrix} & 1_n \\ -1_n & \end{pmatrix} = \omega_n$ on $\mathcal{K}^{2n}$:

$$G := GU(n,n)(A) = \{g \in GL_{2n}(\mathcal{O} \otimes A) : g\omega_n{}^t\bar{g} = \lambda_g \omega_n, \lambda_g \in A^\times\}.$$

Here $\mu(g) := \lambda_g$ is the similitude character, and we write $U := U(n,n) \subset G$ for the kernel of μ. We define $Q = Q_n$ to be the Siegel parabolic subgroup of G consisting of block matrices of the form $\begin{pmatrix} A & B \\ C & D \end{pmatrix}$ such that $C = 0$. Let

$$\mathbf{H}_n := \{Z \in M_n(\mathbb{C}) : -i(Z - {}^t\bar{Z}) > 0\}.$$

(Note that $\mathbf{H}_1$ is the usual upper half plane).

Let $Z \in \mathbf{H}_n$. For $\alpha = \begin{pmatrix} A & B \\ C & D \end{pmatrix} \in G(\mathbb{R})$ with A, B, C, D $n \times n$ block matrices. Let $\mu_\alpha(Z) := CZ + D, \kappa_\alpha(Z) = \bar{C}^tZ + \bar{D}$. We define the automorphy factor:

$$J(\alpha, Z) := (\mu_\alpha(Z), \kappa_\alpha(Z)).$$

Let $G(\mathbb{R})^+ = \{g \in G(\mathbb{R}), \mu(g) > 0\}$ then $G(\mathbb{R})^+$ acts on $\mathbf{H}_n$ by

$$g(Z) := (A_g Z + B_g)(C_g Z + D_g)^{-1},\ g = \begin{pmatrix} A_g & B_g \\ C_g & D_g \end{pmatrix}.$$

Let $K_\infty^+ = \{g \in U(\mathbb{R}) : g(i) = i\}$ (we write i for the matrix $i1_n \in \mathbf{H}_n$) and Z_∞ be the center of $G(\mathbb{R})$. We define $C_\infty := Z_\infty K_\infty^+$. Then $k_\infty \mapsto J(k_\infty, i)$ defines a homomorphism from C_∞ to $GL_n(\mathbb{C}) \times GL_n(\mathbb{C})$.

Definition 2. A weight $\underline{k}$ is a set of integers $(k_{n+1}, \ldots, k_{2n}; k_n, \ldots, k_1)$ such that $k_1 \geq k_2 \geq \ldots \geq k_{2n}$ and $k_n \geq k_{n+1} + 2n$.

A weight $\underline{k}$ defines an algebraic representation of $GL_n(\mathbb{C}) \times GL_n(\mathbb{C})$ by

$$\rho_{\underline{k}}(g_+, g_-) := \rho_{(k_n,\ldots,k_1)}(g_+) \otimes \rho_{(-k_{n+1},\ldots,-k_{2n})}(g_-)$$

where $\rho_{(a_1,\ldots,a_n)}$ is the dual of the usual irreducible algebraic representation of GL_n with highest weight $(a_1, \ldots, a_n)$. Let $V_{\underline{k}}(\mathbb{C})$ be the representation of C_∞ given by

$$k_\infty \to \rho_{\underline{k}} \circ J(k_\infty, i).$$

Fix K an open compact of $G(\mathbb{A}_f)$. We let

$$Sh_K(G) = G(\mathbb{Q})^+ \backslash \mathbf{H}_n \times G(\mathbb{A}_f)/KC_\infty.$$

The automorphic sheaf $\omega_{\underline{k}}$ is the sheaf of holomorphic sections of

$$G(\mathbb{Q})^+ \backslash \mathbf{H}_n \times G(\mathbb{A}_f) \times V_{\underline{k}}(\mathbb{C})/KC_\infty \to G(\mathbb{Q})^+ \backslash \mathbf{H}_n^+ \times G(\mathbb{A}_f)/KC_\infty.$$

One can also define these Shimura varieties and automorphic sheaves in terms of moduli of abelian varieties. We omit these here.

The global sections of $\omega_{\underline{k}}$ is the space of modular forms consisting of holomorphic functions:

$$f : \mathbf{H}_n \times G(\mathbb{A}_f) \to V_{\underline{k}}(\mathbb{C})$$

which are invariant by some open compact K of the second variable, and satisfy:

$$\mu(\gamma)^{\frac{k_1+\ldots+k_{2n}}{2}} \rho_{\underline{k}}(J(\gamma, Z))^{-1} f(\gamma(Z), g) = f(Z, g)$$

for all $\gamma \in gKg^{-1} \cap G^+(\mathbb{Q})$. Also, when $n = 1$ we require a moderate growth condition.

Remark 2. We will be mainly interested in the scalar-valued forms. In this case $V_{\underline{k}}(\mathbb{C})$ is 1-dimensional of weight $\underline{k} = (0, \ldots, 0; \kappa, \ldots, \kappa)$ for some integer $\kappa \geq 2$.

3.2 *Hida Theory*

Hida Theory $GL_2/\mathbb{Q}$ We choose a quick way to present Hida theory. Let M be prime to p and χ a character of $(\mathbb{Z}/pM\mathbb{Z})^\times$. The *weight space* is $\mathrm{Spec}\Lambda$ for

$\Lambda := \mathbb{Z}_p[[T]]$. Let $\mathbb{I}$ be a domain finite over Λ. A point $\phi \in \mathrm{Spec}\mathbb{I}$ is called arithmetic if the image of ϕ in $\mathrm{Spec}\Lambda$ is given by the $\mathbb{Z}_p$-homomorphism sending $(1+T) \mapsto \zeta(1+p)^{\kappa-2}$ for some $\kappa \geq 2$ and ζ a p-power root of unity. We usually write κ_ϕ for this κ, called the weight of ϕ. We also define χ_ϕ to be the character of $\mathbb{Z}_p^\times \simeq (\mathbb{Z}/p\mathbb{Z})^\times \times (1+p\mathbb{Z}_p)$ that is trivial on the first factor and is given by $(1+p) \mapsto \zeta$ on the second factor.

Definition 3. An $\mathbb{I}$-family of forms of tame level M and character χ is a formal q-expansion $\mathbf{f} = \sum_{n=0}^{\infty} a_n q^n, a_n \in \mathbb{I}$, such that for a Zariski dense set of arithmetic points ϕ the specialization $f_\phi = \sum_{n=0}^{\infty} \phi(a_n) q^n$ of $\mathbf{f}$ at ϕ is the q-expansion of a modular form of weight κ_ϕ, character $\chi\chi_\phi\omega^{2-\kappa_\phi}$ where ω is the Techimuller character, and level M times some power of p.

Definition 4. The U_p operator is defined on both the spaces of modular forms and families. It is given by:

$$U_p(\sum_{n=0}^{\infty} a_n q^n) = \sum_{n=0}^{\infty} a_{pn} q^n.$$

Hida's ordinary idempotent e_p is defined by $e_p := \lim_{n\to\infty} U_p^{n!}$. A form f or family $\mathbf{f}$ is called ordinary if $e_p f = f$ or $e_p \mathbf{f} = \mathbf{f}$.

FACT The space of ordinary families is finite and free over the ring $\mathbb{I}$.

Remark 3. For Hilbert modular forms the analogues space is still finite but not free in general. The subspace of ordinary cuspidal families is both finite and free.

Hida Theory for $GU(2,2)$ For simplicity let us restrict to the case when the prime to p part of the nebentypus is trivial. Fix some prime to p level group K of $G(\hat{\mathbb{Z}})$. Let T be the diagonal torus of $U = U(2,2)$. Let χ be a character of $T(\mathbb{Z}/p\mathbb{Z})$. The weight space is $\mathrm{Spec}\Lambda_2$ where Λ_2 is defined to be the completed group algebra of $T(1+p\mathbb{Z}) = (1+p\mathbb{Z})^4$. Let A be any domain finite over Λ_2.

Definition 5. A weight $\underline{k} = (k_1, k_2; k_3, k_4)$ is a set of integers k_i such that $k_1 \geq k_2 + 2 \geq k_3 + 4 \geq k_4 + 4$.

Definition 6. A point $\phi \in \mathrm{Spec}A$ is called arithmetic if its image in $\mathrm{Spec}\Lambda_2$ is given by the character $[\underline{k}]\chi_\phi.\chi$ where $\underline{k}$ is a weight and $[\underline{k}]$ is given by:

$$\mathrm{diag}(t_1, t_2, t_3, t_4) \mapsto t_1^{k_3} t_2^{k_4} t_3^{k_2} t_4^{k_1}$$

(We identify $U(\mathbb{Z}_p) \simeq GL_4(\mathbb{Z}_p)$ by the first projection of $\mathcal{K}_p \simeq \mathcal{K}_{v_0} \times \mathcal{K}_{\bar{v}_0}$) and χ_ϕ is a finite order character of $T(1+p\mathbb{Z})$.

We are going to define Hida families by a finite number of q-expansions: Let $K \subset G(\mathbb{A}_f)$ be a level group $X(K)$ be a finite set of representatives x of $G(\mathbb{Q})\backslash G(\mathbb{A}_f)/K$ with $x_p \in Q(\mathbb{Z}_p)$. For any $g \in GU(2,2)(\mathbb{A}_\mathbb{Q})$ let $S_{[g]}^+$ comprise those positive semi-definite Hermitian matrices h in $M_n(\mathcal{K})$ such that $\mathrm{Tr}hh' \in \mathbb{Z}$ for all Hermitian matrices h' such that $\begin{pmatrix} 1 & h' \\ & 1 \end{pmatrix} \in N_Q(\mathbb{Q}) \cap gKg^{-1}$.

Definition 7. For any ring A finite over Λ_2 we define space of A-adic forms with tame level $K \subset G(\mathbb{A}_f)$ and coefficient ring A to be the elements:

$$F := \{F_x\}_{x \in X(K)} \in \oplus_{x \in X(K)} A[[q^{S^+_{[x]}}]]$$

such that for a Zariski dense set of arithmetic points $\phi \in \mathrm{A}$ the specialization F_ϕ of F at ϕ is the q-expansion of the matrix coefficient of the highest weight vector of holomorphic modular forms of weight $\underline{k}_\phi$ and nebentypus $\chi\chi_\phi\omega(t_1^{-k_3}t_2^{-k_4}t_3^{-k_2}t_4^{-k_1})$.

(In the Skinner-Urban case the interpolated points ϕ are of scalar weights and thus do not need to take the highest weight vector.)

Definition 8. Some U_p operators: for $t^+ = \mathrm{diag}(t_1, t_2, t_3, t_4) \in T(\mathbb{Q}_p)$ such that $t_2/t_1, t_3/t_2, t_3/t_4 \in p\mathbb{Z}_p$. We define an operator U_{t^+} on the space of Hermitian modular forms by: $U_{t^+}.f = |[\underline{k}^*](t)|_p^{-1} f|_{\underline{k}} u_{t^+}$ where $[\underline{k}^*] = [\underline{k} + (2,2;-2,-2)]$ and $f|_{\underline{k}} u_{t^+}$ is the usual Hecke operator defined by double coset decomposition (with no normalization factors).

Hida proved that this u_{t^+} preserves integrality of modular forms and defined an idempotent:

$$e^{ord} := \lim_{n\to\infty} u_{t^+}^{n!}.$$

A form or family F is called nearly ordinary if $e^{ord}F = F$. Again, we have that the space of nearly ordinary Hida families with coefficient ring A is finite and free over A. This is called the *Hida's control theorem* for ordinary forms.

Remark 4. In order to prove the finiteness and freeness (both in the GL_2 and unitary group case) we need to go back to the notion of p-adic modular forms using the Igusa tower, which we omit here.

Another important input of Hida theory is the fundamental exact sequence proved (Skinner and Urban 2010, Chapter 6). We let $C_1(K)$ be the set of cusp labels of genus 2 and label K (Skinner and Urban 2010, 5.4.3). Write Λ_1 for the weight ring of $U(1,1)/\mathbb{Q}$. Then Skinner-Urban proved the following

Theorem 3. *For any Λ_2-algebra A there is a short exact sequence*

$$0 \to \mathcal{M}^0_{ord}(K^p, A) \to \mathcal{M}^1_{ord}(K^p, A) \to \oplus_{[g]\in C_1(K)} \mathcal{M}^0_{ord}(K^p_{1,g}, \Lambda_1) \otimes_{\Lambda_1} A \to 0.$$

Here $\mathcal{M}^0_{ord}(K^p, A)$ is the space of A-valued families of ordinary cusp forms on $GU(2,2)$, $\mathcal{M}^1_{ord}(K^p, A)$ is the space of ordinary forms taking 0 at all genus 0 cusps (Skinner and Urban 2010, 5.4). The $\mathcal{M}^0_{ord}(K^p_{1,g}, \Lambda_1)$ is the space of ordinary cusp *forms on* $U(1,1)$ *with tame level group* $K^p_{1,g}$ *for* $K^p_{1,g} = GU(1,1)(\mathbb{A}_f) \cap gKg^{-1}$ *and* $GU(1,1)$ *is embedded as the levi subgroup of the Klingen Parabolic subgroup of* $GU(2,2)$. *The* Φ *is the "Siegel operator" giving the restricting to boundary map. The* Λ_1*-algebra structure for* Λ_2 *is given by the embedding* $T_1 \hookrightarrow T_2 : (t_1, t_2) \to (t_1, 1, t_2, 1)$.

The proof is a careful study of the geometry of the boundary of the Igusa varieties (Skinner and Urban 2010, 6.2,6.3). This theorem is used to construct a cuspidal Hida family on $GU(2,2)$ that is congruent to the Klingen Eisenstein series modulo the p-adic L-function.

One more important property of ordinary families is that the specialization of a nearly ordinary family to a very regular weight is classical. This will be used to ensure that the $\Lambda_{\mathcal{D}}$-adic Hecke algebra of ordinary $\Lambda_{\mathcal{D}}$-adic form can not have CAP components.

4 Eisenstein Series on GU(2,2)

4.1 Klingen Eisenstein Series

Let P be the Klingen Parabolic subgroup of $GU(2,2)$ consisting of matrices of the form

$$\begin{pmatrix} \times & 0 & \times & \times \\ \times & \times & \times & \times \\ \times & 0 & \times & \times \\ 0 & 0 & 0 & \times \end{pmatrix}.$$

Let M_P be the levi subgroup of P defined by

$$M_P \simeq GU(1,1) \times \mathrm{Res}_{\mathcal{O}_{\mathcal{K}}/\mathbb{Z}}\mathbb{G}_m, (g, x) \mapsto \begin{pmatrix} A_g & & B_g & \\ & \mu(g)\bar{x}^{-1} & & \\ C_g & & D_g & \\ & & & x \end{pmatrix}.$$

Let N_P be the unipotent radical of P.

Observe that if π is an automorphic representation of GL_2 and ψ is a Hecke character of $\mathbb{A}_{\mathcal{K}}^\times$ which restricts to the central character χ_π of π on $\mathbb{A}_{\mathbb{Q}}^\times$, then these uniquely determine an automorphic representation π_ψ of $GU(1,1)$ with central character ψ. Now suppose we have a triple (π, ψ, τ) where π is an irreducible cuspidal automorphic representation of GL_2 and ψ and τ are Hecke characters of $\mathbb{A}_{\mathcal{K}}^\times$ such that $\psi|_{\mathbb{A}_{\mathbb{Q}}^\times} = \chi_\pi$. Then $\pi_\psi \boxtimes \tau$ is an automorphic representation of M_P. We extend this to a representation of P by requiring that N_P act trivially.

Then Klingen Eisenstein series are forms on $GU(2,2)$ which are induced the above representation of P. In fact we need to first work locally for each place v (say, finite) of $\mathbb{Q}$. Let (π_v, ψ_v, τ_v) be the local triple then $(\pi_\psi)_v \boxtimes \tau_v$ is a representation of $M_P(\mathbb{Q}_v)$. We extend it to a representation ρ_v of $P(\mathbb{Q}_v)$ by requiring that $N_P(\mathbb{Q}_v)$ acts trivially. Then we form the induced representation $I(\rho_v) = \mathrm{Ind}_{P(\mathbb{Q}_v)}^{G(\mathbb{Q}_v)}\rho_v$. When everything is unramified and ϕ_v is a spherical vector of π_v, there is a unique vector $f^0_{\phi_v} \in I(\rho_v)$ which is invariant under $G(\mathbb{Z}_v)$ and $f^0_{\phi_v}(1) = \phi_v$. The Archimedean picture is slightly different (see Skinner and Urban 2010, section 9.1).

Let $\phi = \otimes_v\phi_v \in \pi$ and let $I(\rho)$ be the restricted product of the $I(\rho_v)$'s with respect to the unramified vectors above. If $f \in I(\rho)$ we let $f_z(g) = \delta(m)^{\frac{3}{2}+z}\rho(m)f(k)$ for $g = mnk \in M_PN_PK$. Here we let K be $G(\hat{\mathbb{Z}})$. Note that the f_z takes values in the representation space V of π. However π can be embedded to the space of automorphic forms on $GL_2(\mathbb{A}_\mathbb{Q})$. We also write $f_z(g)$ for the function on $GU(2,2)(\mathbb{A}_\mathbb{Q})$ given by $f_z(g)(1)$.

The *Klingen Eisenstein Series* is defined by:

$$E(f;z,g) := \sum_{\gamma \in P(\mathbb{Q})\backslash G(\mathbb{Q})} f_z(\gamma g).$$

This is absolutely convergent for $\mathrm{Re}z \gg 0$ and can be meromorphically continued to all $z \in \mathbb{C}$.

4.2 p-Adic Families

Let $\mathbb{I}$ be a normal domain finite over $\mathbb{Z}_p[[W]]$. (W is a variable) and $\mathbf{f}$ is a normalized ordinary eigenform with coefficient ring $\mathbb{I}$. In Sect. 8 we are going to define the "Eisenstein Datum" $\mathcal{D}$ which contains the information of $\mathbf{f}, \mathbb{I}, \mathcal{K}$, etc. Define $\Lambda_\mathcal{D} := \mathbb{I}[[\Gamma_\mathcal{K}]][[\Gamma_\mathcal{K}^-]]$. We are going to define the set of arithmetic points $\phi \in \mathrm{Spec}\Lambda_\mathcal{D}$ and this $\Lambda_\mathcal{D}$ p-adically parameterizes triples $(f_\phi, \psi_\phi, \tau_\phi)$ to which we associate the Klingen Eisenstein series. Later we will also give $\Lambda_\mathcal{D}$ the structure of a finite Λ_2-algebra and construct a $\Lambda_\mathcal{D}$-adic nearly ordinary Klingen Eisenstein family, which we denote by $\mathbf{E}_\mathcal{D}$.

Now we consider $\Lambda_\mathcal{D}$-adic cusp forms on $GU(2,2)$. Let $h_\mathcal{D} := h^{\Sigma,0}_{ord}(K, \Lambda_\mathcal{D})$ be the Hecke algebra for the space of $\Lambda_\mathcal{D}$-coefficient nearly ordinary cuspidal forms with respect to some level group K. It is generated by Hecke operators at primes outside Σ and the prime p.

Definition 9. Let $I_\mathcal{D}$ be the ideal of $h_\mathcal{D}$ generated by $\{T - \lambda_{\mathbf{E}_\mathcal{D}}(T)\}$'s for T elements in the abstract algebra. Here $\lambda_{\mathbf{E}_\mathcal{D}}(T)$ is the Hecke eigenvalue of T on $\mathbf{E}_\mathcal{D}$. The structure map $\Lambda_\mathcal{D} \to h_\mathcal{D}/I_\mathcal{D}$ is easily seen to be surjective. Thus there is an ideal $\mathcal{E}_\mathcal{D}$ of $\Lambda_\mathcal{D}$ such that $\Lambda_\mathcal{D}/\mathcal{E}_\mathcal{D} \simeq h_\mathcal{D}/I_\mathcal{D}$. This $\mathcal{E}_\mathcal{D}$ is called the Klingen Eisenstein ideal.

The motivation to define this ideal will be more clear after we have discussed the Galois representations.

5 Galois Representations and Lattice Construction

5.1 Galois Representations

We first recall the following theorem (due to Harris-Taylor, S.W Shin, S. Morel, C. Skinner et al.) attaching Galois representations to automorphic representations on $GU(n,n)$.

Theorem 4. *Let π be an irreducible cuspidal representation of $GU(n,n)(\mathbb{A}_{\mathbb{Q}})$ and let χ_π be its central character. Let $\Sigma(\pi)$ be the finite set of primes ℓ such that either π_ℓ or $\mathcal{K}$ is ramified. Suppose π_∞ is the regular holomorphic discrete series of weight $\underline{k} := (k_{n+1},\ldots,k_{2n};k_1,\ldots,k_n)$ such that*

$$k_1 \geq \ldots \geq k_n, k_n \geq k_{n+1} + 2n, k_{n+1} \geq \ldots \geq k_{2n}.$$

Then there is a continuous semisimple representation:

$$R_p(\pi) : G_{\mathcal{K}} \to GL_n(\bar{\mathbb{Q}}_p)$$

such that:

(i) $R_p(\pi)^\vee(1-2n) \otimes \sigma_{\chi_\pi}^{1+c} \simeq R_p(\pi)^c$.
(ii) $R_p(\pi)$ *is unramified outside primes above those in* $\Sigma(\pi) \cup \{p\}$ *and for such primes* w *we have*

$$\det(1 - R_p(\pi)(\mathrm{frob}_w)q_w^{-s}) = L(BC(\pi)_w \otimes \psi_w, s + \frac{1}{2} - n)^{-1}.$$

(iii) If π is nearly ordinary at p, then:

$$R_p(\pi)_{G_{\mathcal{K}_{v_0}}} \simeq \begin{pmatrix} \xi_{2n,v_0}\epsilon^{-\kappa_{2n}} & * & * \\ 0 & \ldots & * \\ 0 & 0 & \xi_{1,v_0}\epsilon^{-\kappa_1} \end{pmatrix}$$

and

$$R_p(\pi)|_{G_{\mathcal{K},\bar{v}_0}} \simeq \begin{pmatrix} \xi_{1,\bar{v}_0}\epsilon^{\kappa_1+1-2n-|\underline{k}|} & * & * \\ 0 & \ldots & * \\ 0 & 0 & \xi_{2n,\bar{v}_0}\epsilon^{\kappa_{2n}+1-2n-|\underline{k}|} \end{pmatrix}.$$

Here ξ_{i,v_0} and $\xi_{i,\bar{v}_0}$ are unramified characters and ϵ is the cyclotomic character, $|\underline{k}| = k_1 + \ldots + k_{2n}$, $\kappa_i = k_i + n - i$ for $1 \leq i \leq n$ and $\kappa_i = k_i + 3n - i$ for $n+1 \leq i \leq 2n$.

Returning to the $GU(2,2)$ case, it is formal to patch the Galois representations attached to cuspidal nearly ordinary forms to a Galois pseudo-character $R_{\mathcal{D}}$ of $G_{\mathcal{K}}$ with values in $h_{\mathcal{D}}$. (Pseudo characters are firstly introduced by Wiles (1988). They are function on $G_{\mathcal{K}}$ satisfying the relations that should be satisfied by the trace of a representation. However it does not necessarily come from a representation. We omit the definitions.) We can associate a Galois representation $\rho_{\mathbf{E}_{\mathcal{D}}}$ to the Klingen Eisenstein family $\mathbf{E}_{\mathcal{D}}$ with coefficient ring $\Lambda_{\mathcal{D}}$ by a similar recipe. It is essentially the direct sum of the Galois representation $\rho_{\mathbf{f}}$ associated to the Hida family $\mathbf{f}$ with two $\Lambda_{\mathcal{D}}$-adic characters.

The *motivation* for the Klingen Eisenstein ideal is:

$$R_{\mathcal{D}}(\mathrm{mod} I_{\mathcal{D}}) = \mathrm{tr}\rho_{\mathbf{E}_{\mathcal{D}}}(\mathrm{mod}\mathcal{E}_{\mathcal{D}}).$$

(Recall that $h_{\mathcal{D}}/I_{\mathcal{D}} \simeq \Lambda_{\mathcal{D}}/\mathcal{E}_{\mathcal{D}}$.) This relation follows from the congruences for the corresponding Hecke eigenvalues. Also, $R_{\mathcal{D}}$ is generically "more irreducible" than $\rho_{\mathbf{E}_{\mathcal{D}}}$ in the sense that it can be written as the sum of at most two "generically irreducible" pseudo-characters while $\rho_{\mathbf{E}_{\mathcal{D}}}$ has three pieces. (This is proven in Skinner and Urban (2010, 7.3.1) using a result of M. Harris on non-existence of CAP forms of very regular weights.)

The next thing to do is use the "lattice construction" to get the Galois cohomology classes from the congruences between irreducible and reducible Galois representations.

Recall in the last section we have:

$$\Lambda_{\mathcal{D}}/\mathcal{E}_{\mathcal{D}} \xrightarrow{\sim} \mathbf{h}_{\mathcal{D}}/I_{\mathcal{D}}$$

$$\mathrm{trace}\rho_{\mathbf{E}_{\mathcal{D}}}(\mathrm{mod}\mathcal{E}_{\mathcal{D}}) = R_{\mathcal{D}}(\mathrm{mod} I_{\mathcal{D}}).$$

Our goal is to prove:

$$(\mathcal{L}_{1,\mathbb{Q}}^{\Sigma}\mathcal{L}_{\mathbf{f},\mathcal{K}}^{\Sigma}) \supset \mathcal{E}_{\mathcal{D}} \supset \mathrm{char}_{\mathbf{f},\mathcal{K}}^{\Sigma}.$$

Now we are going to prove the second inclusion using the lattice construction. The first one will be proved at the end of Sect. 9.

5.2 Galois Argument: Lattice Construction

The lattice construction in Skinner and Urban (2010) involves three irreducible pieces and is complicated. Instead we are going to give the lattice construction which involves only two pieces (the case in Wiles 1990) and briefly mention the difference at the end. We partly follow Skinner (2009). Let us axiomize the situation: let Λ be the weight algebra and $\mathbb{I}$ a reduced ring which is a finite Λ-algebra. Let ρ be a Galois representation of $G_{\mathbb{Q}}$ on $\mathbb{I}^2$. Let J and I be nonzero ideals of Λ and $\mathbb{I}$ such that the structure map induces $\Lambda/J \simeq \mathbb{I}/I$. Let P be a height one prime of Λ such that $\mathrm{ord}_P(J) = t > 0$. Then there is a unique height one prime P' of $\mathbb{I}$ containing (I, P). Since $\mathbb{I}$ is reduced we can talk about its total fraction ring $K = \prod_i F_{\mathbb{J}_i}$ where the $\mathbb{J}_i$'s are domains finite over $\mathbb{I}$ and the $F_{\mathbb{J}_i}$'s are the fraction fields of the $\mathbb{J}_i$'s.

Suppose:

1. Each representation $\rho_{\mathbb{J}_i}$ on $F_{\mathbb{J}_i}^2$ induced from ρ via projection to $F_{\mathbb{J}_i}$ is irreducible.
2. There are $\Lambda^\times$-valued characters χ_1 and χ_2 of $G_{\mathbb{Q}}$ such that:
$$\mathrm{tr}\rho(\sigma) \equiv \chi_1(\sigma) + \chi_2(\sigma)(\mathrm{mod} I)$$
for each $\sigma \in G_{\mathbb{Q}}$.
3. There are $\mathbb{I}^\times$-valued characters χ_1' and χ_2' of $G_{\mathbb{Q}_p}$ such that
$$\rho|_{G_{\mathbb{Q}_p}} \simeq \begin{pmatrix} \chi_1' & * \\ & \chi_2' \end{pmatrix}$$
and there is a $\sigma_0 \in G_{\mathbb{Q}_p}$ such that $\chi_1'(\sigma_0) \not\equiv \chi_2'(\sigma_0)(\mathrm{mod} P')$.
4. $\chi_1(\sigma) \equiv \chi_1'(\sigma)(\mathrm{mod}\ I)$, $\chi_2(\sigma) \equiv \chi_2'(\sigma)(\mathrm{mod}\ I)$ for each $\sigma \in \mathbb{I}[G_{\mathbb{Q}_p}]$.
5. ρ is unramified outside p.

We define the dual Selmer group $X := H^1_{ur}(\mathbb{Q}, \Lambda^*(\chi_1^{-1}\chi_2))^*$. Here "ur" means extensions unramified everywhere and $*$ means Pontryagin dual.

Definition 10. Let $\mathrm{char}_\Lambda(X)$ be the characteristic ideal of X as a Λ module.

We are going to prove:

Proposition 1. *Under the assumptions above,* $\mathrm{ord}_P(\mathrm{char}_\Lambda(X)) \geq \mathrm{ord}_P(J)$.

Proof. Suppose $t = \mathrm{ord}_P(J) > 0$. We take the σ_0 in assumption (3) and a basis (v_1, v_2) so that $\rho(\sigma_0)$ has the form $\begin{pmatrix} \chi_1(\sigma_0) & \\ & \chi_2(\sigma_0) \end{pmatrix}$. We write $\rho(\sigma) = \begin{pmatrix} a_\sigma & b_\sigma \\ c_\sigma & d_\sigma \end{pmatrix} \in M_2(K)$ for each $\sigma \in K[G_{\mathbb{Q}}]$ with respect to this basis. Then we claim the following. Let $r := \chi_1(\sigma_0) - \chi_2(\sigma_0)$, (so $r \notin P$)

(a) $ra_\sigma, rd_\sigma, r^2 b_\sigma c_\tau \in \mathbb{I}$ for all $\sigma, \tau \in \mathbb{I}[G_{\mathbb{Q}}]$ and $ra_\sigma \equiv r\chi_1(\sigma)(\mathrm{mod} I)$, $rd_\sigma \equiv r\chi_2(\sigma)(\mathrm{mod} I)$, $r^2 b_\sigma c_\tau \equiv 0(\mathrm{mod} I)$.
(b) $\mathcal{C} := \{c_\sigma : \sigma \in \mathbb{I}[G_{\mathbb{Q}}]\}$ is a finite faithful A-module;
(c) $c_\sigma = 0$ for $\sigma \in I_p$.

(c) is by (3) and (b) follows easily from assumption (1) and $G_{\mathbb{Q}}$ being compact. (a) is by calculation: e.g. set $\delta_1 := \sigma_0 - \chi_2(\sigma_0)$ then $ra_\sigma = \mathrm{trace}\rho(\delta_1\sigma) \equiv r\chi_1(\sigma)(\mathrm{mod} I)$ by assumption (2).

Now we deduce the proposition using these claims. We write $\Lambda_P, \mathbb{I}_P$, etc. for the localizations at P and define $\mathcal{M} := \mathbb{I}_P[G_{\mathbb{Q}}]v_1 \subset V$. Then it is easy to check that $\mathcal{M} = \mathbb{I}_P v_1 \oplus \mathcal{C}_P v_2$. Define $\mathcal{M}_2 := \mathcal{C}_P v_2$. Then (a) implies $\bar{\mathcal{M}}_2 := \mathcal{M}_2/I\mathcal{M}_2 \subset \bar{\mathcal{M}} := \mathcal{M}/I\mathcal{M}$ is a direct summand that is $G_{\mathbb{Q}}$ stable. Define $\mathcal{M}_1 = \mathbb{I}_P v_1$ and $\bar{\mathcal{M}}_1 = \mathcal{M}_1/I\mathcal{M}_1$ then we have

$$0 \to \bar{\mathcal{M}}_2 \to \bar{\mathcal{M}} \to \bar{\mathcal{M}}_1 \to 0.$$

Now we return to the lattice construction without localization at P. We will find a finite Λ-module $\mathfrak{m}_2 \subseteq \bar{\mathcal{M}}_2$ such that

(i) $\mathfrak{m}_{2,P} = \bar{\mathcal{M}}_2$;
(ii) There exists a Λ-map $X \to \mathfrak{m}_2$ that is a surjection after localizing at P.

Then $\mathrm{ord}_P(\mathrm{char}_\Lambda(X)) = \mathrm{ord}_P(\mathrm{char}_{\Lambda_P}(X_P)) = \mathrm{ord}_P(\mathrm{Fitt}_{\Lambda_P}(\Lambda_P/\mathrm{char}_{\Lambda_P}(X_P))) \geq \mathrm{ord}_P(\mathrm{Fitt}_{\Lambda_P}\mathfrak{m}_{2,P}) = \mathrm{ord}_P(\mathrm{Fitt}_{\Lambda_P}\bar{\mathcal{M}}_2)$. But $\mathrm{Fitt}_{\Lambda_P}\bar{\mathcal{M}}_2(\mathrm{mod}\ J) = \mathrm{Fitt}_{\Lambda_P/J\Lambda_P}\bar{\mathcal{M}}_2 = \mathrm{Fitt}_{\mathbb{I}_P/I\mathbb{I}_P}\bar{\mathcal{M}}_2 = \mathrm{Fitt}_{\mathbb{I}_P}\bar{\mathcal{M}}_2(\mathrm{mod} I) = \mathrm{Fitt}_{\mathbb{I}_P}\mathcal{M}_2(\mathrm{mod} I) = 0$, the last equality follows from the fact that $\mathcal{M}_2$ is a faithful submodule of $\mathrm{Frac}\mathbb{I}$ in view of the generic irreducibility of the Galois representation ρ. So $\mathrm{Fitt}_{\Lambda_P}\bar{M}_2 \subseteq J$ and $\mathrm{ord}_P(\mathrm{char}_\Lambda(X)) \geq \mathrm{ord}_P J$.

Now let $\mathfrak{m} \subset \bar{\mathcal{M}}$ be the $\mathbb{I}[G_{\mathbb{Q}}]$-module generated by $\bar{v}_1$, $\mathfrak{m}_2 := \mathfrak{m} \cap \bar{\mathcal{M}}_2$, $\mathfrak{m}_1 := \mathfrak{m}/\mathfrak{m}_2 \subset \bar{\mathcal{M}}_1$. Note that $\mathfrak{m}_{2,P} = \bar{\mathcal{M}}_2$. Then we have:

$$0 \to \mathfrak{m}_2 \to \mathfrak{m} \to \mathfrak{m}_1 \to 0. \qquad (*)$$

- By assumption (5) and (c) above this extension is everywhere unramified.
- $\bar{\mathcal{M}}_1 \simeq \Lambda/P^t\Lambda$ as Λ-module by definition. So it is easy to see $\mathfrak{m}_1 \simeq \Lambda/P^t\Lambda$ as well.
- By (a) the $G_{\mathbb{Q}}$-action on $\mathfrak{m}_2$ and $\mathfrak{m}_1$ are given by χ_2 and χ_1 respectively.

We expect the (*) in the matrix to give the desired extension. More precisely let $[\mathfrak{m}] \in H^1(\mathbb{Q}, \mathfrak{m}_2(\chi_1^{-1}\chi_2))$ be the class defined by (*). Then we get a canonical map $\theta : \mathrm{Hom}_\Lambda(\mathfrak{m}_1, \Lambda^*) \to H^1(\mathbb{Q}, \Lambda^*(\chi_1^{-1}\chi_2))$. Taking the Pontryagin dual $\theta^* : H^1(\mathbb{Q}, \Lambda^*(\chi_1^{-1}\chi_2))^* \to \mathfrak{m}_2$. We claim that θ^* becomes surjective after taking localization at P. (As in Sect. 2 this is basically because $\mathfrak{m}$ is generated by $\bar{v}_1$ over $\mathbb{I}[G_{\mathbb{Q}}]$.)

Proof of the claim Let $\mathfrak{N} = \ker(\theta)$ and let $S \subset \mathfrak{N}$ be any finite subset, $\mathfrak{m}_S := \cap_{\phi \in S}\ker\phi$. Then we have:

$$0 \to \mathfrak{m}_2/\mathfrak{m}_S \to \prod_{\phi \in S} \Lambda^* \to \prod_{\phi} \Lambda^*/(\mathfrak{m}_2/\mathfrak{m}_S) \to 0. \qquad (**)$$

Equip each module with the $G_{\mathbb{Q}}$ action $\chi_1^{-1}\chi_2$ and take the cohomology long exact sequence. By the definition of $\mathfrak{N}$ the image of $[\mathfrak{m}]$ in $H^1(\mathbb{Q}, \mathfrak{m}_2/\mathfrak{m}_S(\chi_1^{-1}\chi_2))$ is in the kernel of the map $H^1(\mathbb{Q}, \mathfrak{m}_2/\mathfrak{m}_S(\chi_1^{-1}\chi_2)) \to H^1(\mathbb{Q}, \prod_{\phi \in S} \Lambda^*(\chi_1^{-1}\chi_2))$ which is a quotient of $\prod_{\phi \in S} \Lambda^*(\chi_1^{-1}\chi_2)^{G_\mathbb{Q}}$ which is killed by $r = \chi_1(\sigma_0) - \chi_2(\sigma_0) \notin P$. Thus the exact sequence

$$0 \to (\mathfrak{m}_2/\mathfrak{m}_S)_P \to (\mathfrak{m}/\mathfrak{m}_S)_P \to \mathfrak{m}_{1,P} \to 0$$

is split. If $(\mathfrak{m}_2/\mathfrak{m}_S)_P \neq 0$ then this contradicts the fact that $\mathfrak{m}$ is generated by $\bar{v}_1$ over $\mathbb{I}[G_\mathbb{Q}]$. Thus $\mathfrak{m}_{2,P} = \mathfrak{m}_{S,P}$. By the arbitrariness of S we get $\mathfrak{R}_P = 0$. This proves the claim.

Now we compare with the Skinner and Urban's (2010) case. There we have three irreducible pieces and the matrix is like $\begin{pmatrix} \chi_1 & & \\ * & \rho_f & \\ * & & \chi_2 \end{pmatrix}$. We expect the upper $*$ in the matrix to give the required extension. However we are not able to distinguish the contribution of $(*)$ to $H^1_f(\mathcal{K}, \chi_1^{-1}\rho_f)$ and $H^1_f(\mathcal{K}, \chi_1^{-1}\chi_2)^{c=1} = H^1_f(\mathbb{Q}, \tau)$ where τ is the composition of the transfer map $V : G_\mathbb{Q}^{ab} \to G_\mathcal{K}^{ab}$ and $\chi_1^{-1}\chi_2$. But by the Iwasawa main conjecture for Hecke characters proved in Wiles (1990), this $H^1_f(\mathcal{K}, \chi^{-1}\chi_2)^{c=1}$ is controlled by the p-adic L-function for the trivial character, which is a unit.

6 Doubling Methods

6.1 Siegel Eisenstein Series on $GU(n, n)$

Let Q_n be the Siegel parabolic consists of block matrices $\begin{pmatrix} \times & \times \\ & \times \end{pmatrix}$. Let v be a finite prime of $\mathbb{Q}$, write $K_{n,v}$ for $GU(n,n)(\mathbb{Z}_v)$. Fix χ a character of $\mathcal{K}_v^\times$. Let $I_n(\chi)$ be the space of smooth and K_n-finite functions $f : K_{n,v} \to \mathbb{C}$ such that $f(qk) = \chi(\det D_q) f(k)$ for $q = \begin{pmatrix} A_q & B_q \\ & D_q \end{pmatrix} \in Q_n$ from such f we define

$$f(z, -) : G_n(\mathbb{Q}_v) \to \mathbb{C}$$

by

$$f(z, qk) := \chi(\det D_q) |\det A_q D_q^{-1}|_v^{z+\frac{n}{2}} f(k).$$

Suppose $\mathcal{K}_v$ is unramified over $\mathbb{Q}_v$ and χ is unramified, then there is a unique vector $f^0 \in I(\chi)$ which is invariant under $K_{n,v}$ and $f^0(1) = 1$. There is an Archimedean picture as well (see Skinner and Urban 2010, 11.4.1).

Now let $\chi = \otimes_v \chi_v$ be a Hecke character of $\mathbb{A}_{\mathcal{K}}^\times/\mathcal{K}^\times$. Then we define $I(\chi)$ as a restricted product of local $I(\chi_v)$'s as above with respect to the above unramified vectors. For any $f \in I(\chi)$ we define the Siegel Eisenstein series

$$E(f;z,g) := \sum_{\gamma \in Q_n(\mathbb{Q})\backslash G_n(\mathbb{Q})} f(z,\gamma g).$$

This is absolutely convergent if $\mathrm{Re} z \gg 0$ and has a meromorphic continuation to all $z \in \mathbb{C}$.

6.2 Some Embedding

The Klingen Eisenstein series are difficult to compute, while Siegel Eisenstein series are much easier. The point of doubling method is to reduce the computation of the former to the latter. We are going to introduce some important embeddings that are used in the Pullback formulas. Let (V_1, ω_1) be the Hermitian space for $U(1,1)$ and (V_1^-) another Hermitian space whose metric is $(-\omega_1)$. Elements of V_1 and V_1^- are denoted (v_1, v_2) and (u_1, u_2) for $v_i, u_i \in \mathcal{K}$. Let $V_2 = V_1 \oplus X \oplus Y$ be the Hermitian space for $U(2,2)$ where $X \oplus Y$ is a 2-dimensional Hermitian space for the metric $\begin{pmatrix} & 1 \\ -1 & \end{pmatrix}$ and elements are written as (x, y) for $x, y \in \mathcal{K}$ with respect to this basis.

Let $W = V_2 \oplus V_1^-$ be the Hermitian space for $U(3,3)$. Let $R = \begin{pmatrix} 1 &&&&& \\ & 1 &&&& \\ &&& 1 && \\ &&&& 1 & \\ &&&&& 1 \\ && 1 &&& \end{pmatrix}$ and

$S = \begin{pmatrix} 1 &&&&& \\ & 1 &&&& \\ && 1 &&& \\ && -1 & 1 && \\ &&&& 1 & \\ -1 &&&&& 1 \end{pmatrix}$ These give maps:

$$(v_1, x, v_2, y, u_1, u_2) \mapsto (v_1, x, u_2, v_2, y, u_1)$$

$$(v_1, x, u_2, v_2, y, u_1) \mapsto (v_1 - u_1, x, u_2 - v_2, v_2, y, u_1).$$

Now we define the embedding:

$$\gamma_3 : G_{2,1} := \{(g, g') \in GU(2,2,) \times GU(1,1), \mu(g) = \mu(g')\} \hookrightarrow GU(3,3)$$

by

$$\begin{pmatrix} g & \\ & g' \end{pmatrix} \mapsto S^{-1}R^{-1}\begin{pmatrix} g & \\ & g' \end{pmatrix} RS.$$

Let V^d be the image of V_1 in $V_1 \oplus V_1^-$ by the diagonal embedding. Let τ_1 be any element of $U(3,3)(\mathbb{Q})$ which maps the maximal isotrophic subspace $V^d \oplus X$ to $\mathcal{K}v_1 \oplus \mathcal{K}u_1 \oplus X$, then one can check that: $\tau_1^{-1}Q_3\tau_1 \cap \gamma_3(U(2,2) \times U(1,1)) = \gamma_3(Q_2 \times B_1)$. An important property of such embedding is:

$$\{(m(g,x)n, g) : g \in GU(1,1), x \in \mathrm{Res}_{\mathcal{K}/\mathbb{Q}}\mathbb{G}_m, n \in N_P\} \subset Q_3.$$

6.3 Pullback Formula

Let χ be a unitary Hecke character as before and $f \in I(\chi)$. Given a cusp form φ on G_1, define the pullback section by:

$$F_\varphi(f, z, g) := \int_{U(1,1)(\mathbb{A})} f(z, \gamma_3(g, g_1 h))\bar{\chi}(\det g_1 h)\varphi(g_1 h) dg_1$$

where $h \in GU(1,1)(\mathbb{A})$ is any element such that $\mu(h) = \mu(g)$. This is absolutely convergent if $\mathrm{Re} z \gg 0$. It is easy to see that F_ϕ does not depend on the choice of h. Note that this is a Klingen Eisenstein section. Then

Proposition 2. *For z in the region of absolute convergence and h as above, we have:*

$$\int_{U(1,1)(\mathbb{Q})\backslash U(1,1)(\mathbb{A})} E(f; z, \gamma_3(g, g_1 h))\bar{\chi}(\det g_1 h)\varphi(g_1, h) dg_1$$
$$= \sum_{P(\mathbb{Q})\backslash GU(2,2)(\mathbb{Q})} F_\varphi(f; z, \gamma g).$$

Remark 5. The right hand side is nothing but the expression of the Klingen Eisenstein series.

Proof. This is proven by Shimura (1997). There Shimura proved the following double coset decomposition in (2.4) and (2.7) in loc. cit.:

$$U(3,3) = Q_3\gamma_3(U(2,2) \times U(1,1)) \cup Q_3\tau_1\gamma_3(U(2,2) \times U(1,1))$$

and

$$Q_3\gamma_3(U(2,2)\times U(1,1)) = \cup_{\beta\in U(2,2),\xi\in U(1,1)} Q_3\gamma_3((\beta,\xi)),$$

$$Q_3\tau_1\gamma_3(U(2,2)\times U(1,1)) = \cup_{\beta\in Q_2\backslash U(2,2),\gamma\in B_1\backslash U(1,1)} Q_3\tau_1\gamma_3((\beta,\gamma)).$$

Thus by unfolding the Siegel Eisenstein series we write the integration into two parts. We claim that the integration for the part involving terms with τ_1 is 0. We first fix β and sum over the γ's, this equals

$$\int_{B_1(\mathbb{Q})\backslash U(1,1)(\mathbb{A})} f(z;\tau_1\gamma_3((\beta g, g_1 h)))\varphi(g_1 h)dg_1.$$

Recall we have noted that $\tau_1\gamma_3(1, B_1)\tau_1^{-1} \subseteq Q_3$. Since φ is cuspidal, $\int_{B_1(\mathbb{Q})\backslash B_1(\mathbb{A}_\mathbb{Q})}\phi(bg_1h)db = 0$ for all g_1. Thus the integration is 0. This proves the claim. The proposition then follows from our description for $Q_3\gamma_3(U(2,2)\times U(1,1))$.

7 Constant Terms

Suppose ϕ is of weight κ and let $z_\kappa = \frac{\kappa-3}{2}$. Let P be the Klingen parabolic and R any standard $\mathbb{Q}$ parabolic of $GU(2,2)$. We are going to compute the constant terms $E(f,z,g)_R$ of the Klingen Eisenstein series $E(f,z,g)$ along R. We write N_R for the unipotent radical of P. The constant term along R is given by:

$$E(f,z,g)_R = \int_{N_R(\mathbb{Q})\backslash N_R(\mathbb{A})} E(f,z,ng)dn.$$

A famous computation of Langlands tells us that: if $R \neq P$ then $E(f,z,g)_R = 0$. For $R = P$ we first define the intertwining operator:

$$A(\rho,z,f)(g) := \int_{N_P(\mathbb{A})} f_z(wng)dn.$$

This is absolutely convergent for $\mathrm{Re}z \gg 0$ and is defined for all $z \in \mathbb{C}$ by meromorphic continuation. It is a product of local integrals. This intertwines the representations $I(\rho)$ and some $I(\rho_1)$ where ρ_1 is defined similar to ρ but replacing (π,ψ,τ) by $(\pi\times(\tau\circ\det), \psi\tau\tau^c, \bar{\tau}^c)$. Then $E(f,z,g)_P = f_z(g) + A(\rho,z,f)(g)$. It turns out that under our choices $z = z_\kappa$ and $\kappa > 6$, $A(\rho,z,f)$ is absolutely convergent and the Archimedean component is 0. Thus $A(\rho,z_\kappa,f)$ equals 0. Thus

$$E(f,z_\kappa,g)_P = f_{z_\kappa}(g).$$

Let us explain how the special L-values that we are interested in show up in the constant term of the Klingen Eisenstein series. The Klingen section f is realized as the pullback section of some Siegel Eisenstein series on $GU(3,3)$. At the unramified places a computation of Lapid and Rallis (2005) tells us that if the Siegel section is f_v^0 then the pullback section is $f_{\phi_v}^0 L(\tilde{\pi}, \psi/\tau, z+1) L(\bar{\chi}_\pi(\psi/\tau)', 2z+1)$. Here the first L-factor is the local Euler factor for the base change of the dual $\tilde{\pi}$ twisted by ψ/τ and the second is a Dirichlet L-factor. Taking the product over all good primes, the special L-values we are interested in show up as the constant term of the Klingen Eisenstein series obtained by pullback.

8 p-Adic Interpolation

Definition 11. An Eisenstein datum is a sextuple $\mathcal{D} := (A, \mathbb{I}, \mathbf{f}, \psi, \xi, \Sigma)$ where

- A is a finite $\mathbb{Z}_p$-algebra and $\mathbb{I}$ is a normal domain finite over $A[[W]]$.
- $\mathbf{f}$ is a Hida family of cuspidal newforms with coefficient ring $\mathbb{I}$.
- ψ is an A-valued finite order character which restricts to the tame part of the central character of $\mathbf{f}$ on $\mathbb{A}_{\mathbb{Q}}^\times$.
- ξ is another A-valued finite order Hecke character of $\mathbb{A}_{\mathcal{K}}^\times$.
- Σ is a finite set of primes containing all the bad primes.

Recall that we have defined a ring $\Lambda_{\mathcal{D}} = \mathbb{I}[[\Gamma_{\mathcal{K}}]][[\Gamma_{\mathcal{K}}^-]]$. We use $\Lambda_{\mathcal{D}}$ to interpolate triples (f, ψ, τ) that are used to construct Klingen Eisenstein series. Recall that we have defined a weight ring $\Lambda_2 \simeq \mathbb{Z}_p[[\Gamma_2]]$ for $\Gamma_2 \simeq (1+p\mathbb{Z}_p)^4$. We first give $\Lambda_{\mathcal{D}}$ a Λ_2-algebra structure. We define homomorphisms $\alpha : A[[\Gamma_{\mathcal{K}}]] \to \mathbb{I}[[\Gamma_{\mathcal{K}}^-]]$ and $\beta : A[[\Gamma_{\mathcal{K}}]] \to \mathbb{I}[[\Gamma_{\mathcal{K}}]]$ (we omit the formulas). Then the Λ_2-algebra structure map is given by composing $\alpha \otimes \beta : A[[\Gamma_{\mathcal{K}} \times \Gamma_{\mathcal{K}}]] \to \Lambda_{\mathcal{D}}$ with the map $\Gamma_2 \to \Gamma_{\mathcal{K}} \times \Gamma_{\mathcal{K}}$ given by:

$$(t_1, t_2, t_3, t_4) \mapsto \mathrm{rec}_{\mathcal{K}}(t_3 t_4, t_1^{-1} t_2^{-1}) \times \mathrm{rec}_{\mathcal{K}}(t_4, t_2^{-1}),$$

where $\mathrm{rec}_{\mathcal{K}}$ is the reciprocity map in class field theory normalized by the geometric Frobenius. Let $\boldsymbol{\psi} := \alpha \circ \omega^{-1} \psi \Psi_{\mathcal{K}}^{-1}$ and $\boldsymbol{\xi} := \beta \circ \chi_{\mathbf{f}} \xi \Psi_{\mathcal{K}}$.

Definition 12. A point $\phi \in \mathrm{Spec}\Lambda_{\mathcal{D}}$ is called arithmetic if $\phi|_{\mathbb{I}}$ is arithmetic with some weight $\kappa_\phi \geq 2$ and there are $\zeta_\pm, \zeta'_- \in \mu_{p^\infty}$ such that $\phi(\gamma^+) = \zeta_+(1+p)^{\kappa_\phi - 2}, \phi(\gamma^-) = \zeta_-$ for $\gamma^\pm \in \Gamma_{\mathcal{K}}^\pm$ and $\phi(\gamma^{-'}) = \zeta'_-$ for $\gamma^{-'}$ the topological generator of $\Gamma_{\mathcal{K}}^-$.

For every such ϕ we define Hecke characters. Let $p = v_0 \bar{v}_0$ be the decomposition in $\mathcal{K}$ and let

$$\psi_\phi(x) := x_\infty^{-\kappa_\phi} x_{v_0}^{\kappa_\phi} (\phi \circ \boldsymbol{\psi})(x), \ \xi_\phi = \phi \circ \xi.$$

Then we can construct a $\Lambda_{\mathfrak{D}}$-coefficient formal q-expansion E_{sieg} that, when specialize to a Zariski dense set of arithmetic points ϕ, is the nearly ordinary Klingen Eisenstein series $E_{Kling,\phi}$ we constructed using the triple: $(\mathbf{f}_\phi, \psi_\phi|.|^{\frac{\kappa_\phi}{2}}, \tau_\phi = \psi_\phi\xi_\phi^{-1}|.|^{\frac{\kappa_\phi}{2}})$. This is achieved by first constructing a $\Lambda_{\mathfrak{D}}$-adic Siegel Eisenstein series on $GU(3,3)$ and using the pullback formula to construct the Klingen Eisenstein family on $GU(2,2)$. To do this we need to choose a Siegel section f_ϕ at each arithmetic point ϕ so that

1. f_ϕ depends p-adic analytically on ϕ;
2. The pulls back of f_ϕ to (a multiple of) the nearly ordinary Klingen Eisenstein section.

The hardest part is the computations at the primes dividing p (Skinner and Urban 2010, 11.4.14,15,19). It turns out that certain Siegel-Weil Eisenstein sections work well. In fact in Skinner and Urban (2010), the section is not given in terms of the Siegel-Weil section. However it indeed provided the idea of how the section given in loc. cit. is figured out. Let us briefly explain the idea.

Let Φ be the Schwartz function on $M_{(3,6)}(\mathbb{Q}_p)$ defined by:

$$\Phi(X,Y) := \Phi_1(X)\hat{\Phi}_2(Y),$$

where X and Y are 3×3 matrices and define a Siegel-Weil section by:

$$f^\Phi(g) = \chi_2^{-1}(\det g)|\det g|_p^{-s+\frac{3}{2}} \times \int_{GL_3(Q_p)} \Phi((0,X)g)\chi_1^{-1}\chi_2^{-1}(\det X)|\det X|_p^{-2s+3}d^\times X$$

for $\chi_p = (\chi_1, \chi_2)$. The $\hat{\Phi}_2$ means the Fourier transform of Φ_2. We let Φ_1 be a Schwartz function supported on the set of matrices X such that the X_{13} and X_{31} are in $\mathbb{Z}_p^\times$ and the values on it is given by the product of two characters of X_{13} and X_{31}. Choosing Φ_2 properly and unfolding the formula for the β-th local Fourier coefficients, we can make sure that it is essentially given by $\Phi_1({}^t\beta)$ (up to some easier constant depending on β). Thus the first requirement is ensured. This Siegel Weil section is explicitly given by

$$f_p(g) = \sum_{a\in(\mathcal{O}_p/x)^\times} f_z^{0,(2)}(g\begin{pmatrix} a^{-1} & \\ & \bar{a}\end{pmatrix})$$

where $(x) = \mathrm{cond}(\xi^c)$, τ is our χ defining the Siegel Eisenstein section, $\xi = \psi/\tau$ and $f_z^{0,(2)}$ in loc. cit. lemma 11.4.20.

How to interpolate the Klingen Eisenstein series? Hida proved the existence of a Hecke operator $1_f \in \mathbb{T}_\kappa^{ord}(N,\chi_f,A)\otimes_A F_A$ on the space $S_\kappa^{ord}(N,\chi_f)$ of ordinary cusp forms with weight κ level N and character χ_f, such that

$$1_f.g = \frac{< g, f^c|_\kappa \begin{pmatrix} & -1 \\ N & \end{pmatrix} >}{< f, f^c|_\kappa \begin{pmatrix} & -1 \\ N & \end{pmatrix} >} f.$$

This 1_f is not necessarily p-adically integral (Hida 1985). The congruence number η_f is defined (up to a p-adic unit) to be the minimally divisible by p number such that $\ell_f := \eta_f 1_f$ is in $\mathbb{T}_\kappa^{ord}(N, \chi_f, A)$. The candidate that we choose for the Klingen Eisenstein series $E_{Kling,\phi}$ at the arithmetic point ϕ is the one such that:

$$\ell_f^{U(1,1)} e^{U(1,1)} E_{sieg,\phi}|_{U(2,2)\times U(1,1)} = E_{Kling,\phi} \boxtimes f.$$

Here the superscript means the Hecke operators are applied to the forms considered as a form on $U(1,1)$. If we replace f by a Hida family $\mathbf{f}$ and suppose the local Hecke algebra $\mathbb{T}_{\mathfrak{m}_f}$ (the localization of the Hecke algebra at the maximal ideal $\mathfrak{m}_f$ corresponding to f) is Gorenstein, then we can similarly define $1_{\mathbf{f}}$ and $\eta_{\mathbf{f}}$, $\ell_{\mathbf{f}}$, thus interpolating everything in p-adic families.

In particular, we get the Klingen-Eisenstein series interpolating $E_{Kling,\phi}$ whose constant terms are divisible by $\mathcal{L}_{\mathbf{f},\mathcal{K}}^\Sigma.\mathcal{L}_1^\Sigma$ in light of the discussion at the end of the last section.

9 Fourier-Jacobi Coefficients

Recall that we have seen that the constant terms of the Klingen Eisenstein family are divisible by the p-adic L-function. In order to show that the Eisenstein ideal is contained in the principal ideal $(\mathcal{L}_{\mathbf{f},\mathcal{K}}^\Sigma)$, we still need to show that some Fourier coefficient is co-prime to the p-adic L-function.

9.1 Generalities

We are going to compute the Fourier-Jacobi coefficient of the Siegel Eisenstein series E_{sieg} as a function on $U(1,1)(\mathbb{A})$ via the embedding $\gamma_3 : U(2,2)\times U(1,1) \hookrightarrow U(3,3)$. The purpose is, by the pullback formula introduced in the previous section, to express the Fourier coefficients of the Klingen Eisenstein series in terms of the Petersson inner product with the cusp form we start with. For $\mathcal{Z} \in \mathbf{H}_3$

$$E_{sieg}(\mathcal{Z}) = \sum_{T\geq 0} a_{Te}(\mathrm{Tr}T\mathcal{Z}).$$

Write $S_2(\mathbb{Q})$ or $S_2(\mathbb{Q}_v)$ for the set of 2×2 Hermitian matrices over $\mathbb{Q}$ or $\mathbb{Q}_v$. For β a 2×2 Hermitian matrix the βth Fourier-Jacobi coefficient is

$$\sum_{T=\begin{pmatrix}\beta & * \\ * & *\end{pmatrix}} a_{Te}(\mathrm{Tr} T \mathcal{Z}).$$

We have an integral representation for the Fourier-Jacobi coefficients:

$$E_{sieg,\beta}(g) = \int_{N_Q(\mathbb{Q})\backslash N_Q(\mathbb{A})} E^{sieg}(\begin{pmatrix} 1_3 & \begin{matrix} S & 0 \\ 0 & 0 \end{matrix} \\ & 1_3 \end{pmatrix} g) e_{\mathbb{A}}(-\mathrm{Tr}\beta S) dS.$$

Here $e_{\mathbb{A}} = \otimes e_v$ and $e_v(x_v) = e^{-2\pi i \{x_v\}}$ for v a finite primes and $e_v(x_v) = e^{2\pi x_v}$ for $x \in \infty$. The following lemma gives a way to compute the Fourier-Jacobi coefficients of E_{sieg}.

Lemma 1. *Let $f \in I_3(\chi)$, $\beta \in S_2(\mathbb{Q})$. Suppose $\beta > 0$. Let V be the 2-dimensional $\mathcal{K}$-space of column vectors. If* $\mathrm{Re}(z) > \frac{3}{2}$. *Then:*

$$E_{sieg,\beta}(f; z, g) = \sum_{\gamma \in Q_1(\mathbb{Q})\backslash G_1(\mathbb{Q})} \sum_{x \in V} \int_{S_2(\mathbb{A})} f(w_3 \begin{pmatrix} 1_3 & \begin{matrix} S & x \\ {}^t\bar{x} & 0 \end{matrix} \\ & 1_3 \end{pmatrix} \alpha_1(1, \gamma) g) e_{\mathbb{A}}(-\mathrm{Tr}\beta S) dS.$$

Proof. We omit it here. See Skinner and Urban (2010, 11.3)

The integrand in the lemma is a product of local integrals. We are mainly interested in evaluating the Fourier Jacobi coefficients at $\alpha_1(\mathrm{diag}(y, {}^t\bar{y}^{-1}), g)$ for $y \in \mathrm{GL}_2(\mathbb{A}_{\mathcal{K}})$ and $g \in U_1(\mathbb{A}_{\mathbb{Q}})$.

Definition 13. For each prime v of $\mathbb{Q}$ and $f \in I_3(\chi_v)$, set*12pt

$$FJ_\beta(f; z, x, g, y) = \int_{S_2(\mathbb{Q}_v)} f(z, w_3 \begin{pmatrix} 1_3 & \begin{matrix} S & x \\ {}^t\bar{x} & 0 \end{matrix} \\ & 1_3 \end{pmatrix} \alpha_1(\mathrm{diag}(y, {}^t\bar{y}^{-1}), g)) e_v(-\mathrm{Tr}\beta S) dS.$$

We are going to identify the Fourier Jacobi coefficients with some forms that we are more familiar with.

9.2 *Backgrounds for Theta Functions*

Local Picture Let v be a prime of $\mathbb{Q}$ and $h \in S_2(\mathbb{Q}_v)$, $\det h \neq 0$. Then h defines a two-dimensional Hermitian space V_v. Let U_h be the corresponding unitary group.

Let λ_v be a character of $\mathcal{K}_v^\times$ whose restriction to $\mathbb{Q}_v^\times$ is trivial. One can define the Weil representation $\omega_{h,\lambda}$ of $U_h(\mathbb{Q}_v) \times U(1,1)(\mathbb{Q}_v)$ on the space $S(V_v)$ of Schwartz functions on V_v (we omit the formulas).

Global Picture Now let $h \in S_2(\mathbb{Q}), h > 0$ and a Hecke character $\lambda = \otimes\lambda_v$ of $\mathbb{A}_{\mathcal{K}}^\times/\mathcal{K}^\times$ such that $\lambda|_{\mathbb{A}_{\mathbb{Q}}^\times} = 1$. Then we define a Weil representation $\omega_{h,\lambda}$ of $U_h(\mathbb{A}_{\mathbb{Q}}) \times U(1,1)(\mathbb{A}_{\mathbb{Q}})$ on $S(V \otimes \mathbb{A})$ by tensoring the local representations.

Theta Functions Given $\Phi \in S(V \otimes \mathbb{A}_{\mathbb{Q}})$ we define

$$\Theta_h(u,g;\Phi) := \sum_{x \in V} \omega_{h,\lambda}(u,g)\Phi(x)$$

which is an automorphic form on $U_h(\mathbb{A}_{\mathbb{Q}}) \times U_1(\mathbb{A}_{\mathbb{Q}})$ and gives the theta correspondence between U_h and $U(1,1)$.

9.3 *Coprime to the p-Adic L-Function*

Now let us return to the Fourier Jacobi coefficients. It turns out that by some local computations, for each v, $FJ_\beta(f;z,x,g,y)$ has the form $f_1(g)(\omega_{\beta,\lambda_v}(y,g)\Phi_v)(0)$ where we have chosen a Hecke character λ as above, $f_1 \in I_1(\chi_v/\lambda_v)$ and Φ_v is a Schwartz function on $\mathcal{K}_v^2$, ω_{β,λ_v} is defined using the character λ_v. Thus from Lemma 1 the Fourier Jacobi coefficient is the product of an Eisenstein series $E_1(g)$ and a theta series $\Theta_\beta(y,g)$.

Now we prove that the Klingen Eisenstein series is coprime to the p-adic L-function. Let us take an auxiliary Hida family $\mathbf{g}$ of cuspidal eigenforms. Using the functorial property of the theta correspondence we can find some linear combinations of $E_\beta(f;z_\kappa,\alpha_1(\mathrm{diag}(y,{}^t\bar{y}^{-1}),g))$'s which "picks up" the $\mathbf{g}$-eigencomponent of $\Theta_\beta(y,g)$ (as a function of g). By pairing this with the original $\phi \in \pi$ we started with, we find certain linear combinations of the Fourier coefficients of the Klingen Eisenstein family which can be expressed in the form $\mathcal{A}_{\mathbf{g}}\mathcal{B}_{\mathbf{g}}$ where $\mathcal{B}_{\mathbf{g}}$ is the "multiple" of $\mathbf{g}$ showing up in $\Theta_\beta(y,g)$. By choosing $\mathbf{g}$ properly $\mathcal{B}_{\mathbf{g}}$ can be made a unit in $\Lambda_{\mathcal{D}}$ ($\mathbf{g}$ is chosen to be a Hida family of theta series from the quadratic imaginary field $\mathcal{K}$. $\mathcal{B}_{\mathbf{g}}$ interpolates a square of central critical values of Hecke L-functions of CM characters. One needs to use a result of Finis (2006) on the non-vanishing modulo p of anticyclotomic Hecke L-values to conclude $\mathcal{B}_{\mathbf{g}}$ can be chosen to be a unit). The factor $\mathcal{A}_{\mathbf{g}}$ is interpolating $< E_1(g).\mathbf{g}, \mathbf{f}_\phi >$, essentially the Rankin Selberg L-values of $\mathbf{g}$ with $\mathbf{f}$. By checking the nebentypus we find that $\mathcal{A}_{\mathbf{g}}$ only involves $\mathbb{I}[[\Gamma_{\mathcal{K}}^+]]$ and is non-zero by the temperedness of $\mathbf{f}$ and $\mathbf{g}$.

Now we make the following assumption: $N = N^+N^-$ where N^+ is a product of primes split in $\mathcal{K}$ and N^- is a square-free product of an odd number of primes inert in $\mathcal{K}$. Furthermore we assume that for each $\ell | N^-$, $\bar{\rho}_{\mathfrak{f}}$ is ramified at ℓ. Under this assumption Vatsal (2003) proved if we expand the p-adic L-function as:

$$\mathcal{L}_{\mathbf{f},\mathcal{K}}^{\Sigma} = a_0 + a_1(\gamma^- - 1) + a_2(\gamma^- - 1)^2 + \ldots\ldots$$

for $a_i \in \mathbb{I}[[\Gamma_{\mathcal{K}}^{+}]]$, then some a_i must be in $(\mathbb{I}[[\Gamma_{\mathcal{K}}^{+}]])^{\times}$. This implies (easy exercise) that $\mathcal{A}_{\mathbf{h}}$ is outside any height one prime P of $\mathbb{I}[[\Gamma_{\mathcal{K}}]]$ containing $\mathcal{L}_{\mathbf{f},\mathcal{K}}^{\Sigma}$ (since $\mathcal{A}_{\mathbf{g}}$ belongs to $\mathbb{I}[[\Gamma_{\mathcal{K}}^{+}]]$, one may assume $P = P_{+}\mathbb{I}[[\Gamma_{\mathcal{K}}^{+}]]$ for some height one prime P_{+} of $\mathbb{I}[[\Gamma_{\mathcal{K}}^{+}]]$. By Vatsal's result, $\mathrm{ord}_{P_{+}}(\mathcal{L}_{\mathbf{f},\mathcal{K}}^{\Sigma}) = 0$.)

We are ready to prove the result promised in the previous section: $\mathrm{ord}_P(\mathcal{E}_{\mathcal{D}}) \geq \mathrm{ord}_P(\mathcal{L}_{\mathbf{f},\mathcal{K}}^{\Sigma}.\mathcal{L}_{1}^{\Sigma})$ for any height one prime P (Here $\mathcal{L}_{1}^{\Sigma}$ is the p-adic L-function for the trivial character, which is co-prime to $(\mathcal{L}_{\mathbf{f},\mathcal{K}}^{\Sigma})$ by the work of Vatsal. Skinner-Urban (2010) actually worked in a more general setting by allowing non-trivial characters). First recall that all constant terms of the Klingen Eisenstein family are divisible by $\mathcal{L}_{\mathbf{f},\mathcal{K}}^{\Sigma}\mathcal{L}_{1}^{\Sigma}$. By the fundamental exact sequence one can find some family $\mathbf{F}$ of forms on $GU(2,2)$ such that $\mathbf{E}_{\mathcal{D}} - (\mathcal{L}_{\mathbf{f},\mathcal{K}}^{\Sigma})(\mathcal{L}_{1}^{\Sigma})\mathbf{F} := \mathbf{H}$ is a cuspidal family. Now we prove the desired inequality. Suppose $r = \mathrm{ord}_P(\mathcal{L}_{\mathbf{f},\mathcal{K}}^{\Sigma}) \geq 1$. By construction there is a Fourier coefficient of the above constructed cuspidal family $\mathbf{H}$ outside P. Denote it as $c(\beta, x; \mathbf{H})$ where $\beta \in S_2(\mathbb{Q})$ and $x \in GU(2,2)(\mathbb{A}_{\mathbb{Q}})$. We define a map:

$$\mu := h_{\mathcal{D}} \to \Lambda_P / P^r \Lambda_P$$

by: $\mu(h) = c(\beta, x; h\mathbf{H})/c(\beta, x; \mathbf{H})$. This is $\Lambda_{\mathcal{D}}$-linear and surjective. Moreover,

$$c(\beta, x; h\mathbf{H}) \equiv c(\beta, x; h\mathbf{E}_{\mathcal{D}}) \equiv \lambda_{\mathcal{D}}(h) c(\beta, x; \mathbf{E}_{\mathcal{D}}) \equiv \lambda_{\mathcal{D}}(h) c(\beta, x, \mathbf{H}) (\mathrm{mod}\ P^r).$$

Thus $I_{\mathcal{D}} \subseteq \ker\mu$. So we have a surjection $\mu : h_{\mathcal{D}}/I_{\mathcal{D}} \twoheadrightarrow \Lambda_P/P^r\Lambda_P$. But the right hand side is $\Lambda_{\mathcal{D}}/\mathcal{E}_{\mathcal{D}}$. This gives the inequality.

10 Generalizations of the Skinner-Urban Work

We have seen that the key ingredient of this work is a study of the p-adic properties of the Fourier coefficients of the Klingen Eisenstein series. To generalize this argument to more general unitary groups we need some non-vanishing modulo p results for special values of L-functions, which so far is only available for forms on unitary groups of rank at most 2. We are able to study the Klingen Eisenstein series for $U(1,1) \hookrightarrow U(2,2)$ and $U(2,0) \hookrightarrow U(3,1)$, proving the corresponding main conjectures for two different Rankin Selberg p-adic L-functions. Here we only mention the following by product (proved in the Wan (2014)):

Theorem 5. *Let F be a totally real field in which p splits completely. Let f be a Hilbert modular form over F with trivial character and parallel weight 2. Let ρ_f be the p-adic representation of G_F associated to f. Suppose:*

1. *f has good ordinary reduction at all primes dividing p;*
2. *$\bar{\rho}_f$ is absolutely irreducible.*
3. *If $[F = Q]$ is even and the global sign of f is -1, then the automorphic representation of f is not principal series in at least one finite place.*

If the central value $L(f, 1) = 0$, *then* $H^1_f(F, \rho_f)$*is infinite.*

In the case when the sign of $L(f, s)$ is -1 this is an early result of Nekovar (2006) and Zhang (2001). The cases when this sign is $+1$ is new. Note that even in the case when $F = \mathbb{Q}$ our result is slightly stronger than the one in Skinner and Urban (2010). The reason is that by working with general totally real fields we can use a base change trick to remove some of the technical local conditions (Skinner and Urban 2010).

References

Amice, Y., Velu, J.: Distributions p-adiques associes aux sries de Hecke. In: Asterisque, vols. 24–25, pp. 119–131. Société Mathématique de France, Paris (1975)

Finis, T.: Divisibility of anticyclotomic L-functions and theta functions with complex multiplication. Ann. Math. (2) **163**(3), 767–807 (2006)

Hida, H.: A p-adic measure attached to the zeta functions associated with two elliptic modular forms. I. Invent. Math. **79**(1), 159–195 (1985)

Kato, K.: p-adic Hodge theory and values of zeta functions of modular forms (Cohomologies p-adiques et applications arithmetiques III). Asterisque **295**(ix), 117–290 (2004)

Lapid, E., Rallis, S.: On the local factors of representations of classical groups. In: Automorphic Representations, L-Functions and Applications: Progress and Prospects. Ohio State University Mathematical Research Institute Publications, vol. 11, pp. 309–359. de Gruyter, Berlin (2005)

Mazur, B., Tate, J., Teitelbaum, J.: On p-adic analogues of the conjectures of Birch and Swinnerton-Dyer. Invent. Math. **84**(1), 1–48 (1986)

Nekovar, J.: Selmer Complexes. Asterisque, vol. 310. Sociéteé Mathématique de France, Paris (2006)

Ribet, K.A.: A modular construction of unramified p-extensions of $\mathbb{Q}(\zeta_p)$. Invent. Math. **34**(3), 151–162 (1976)

Shimura, G.: Euler Products and Eisenstein Series. CBMS Regional Conference Series in Mathematics, vol. 93. American Mathematical Society, Providence (1997)

Skinner, C.: Galois representations, Iwasawa theory and special values of L-functions. CMI Lecture Notes (2009)

Skinner, C., Urban, E.: The Iwasawa main conjecture for GL_2 (2010)

Vatsal, V.: Special values of anticyclotomic L-functions. Duke Math. J. **116**(2), 219–261 (2003)

Wan, X.: Iwasawa Main Conjecture for Hilbert Modular Forms, preprint, 2014

Wiles, A.: On ordinary Λ-adic representations associated to modular forms. Invent. Math. **94**, 529–573 (1988)

Wiles, A.: The Iwasawa conjecture for totally real fields. Ann. Math. **131**, 493–540 (1990)

Zhang, S.: Heights of Heegner points on Shimura curves. Ann. Math. **153**, 27–147 (2001)

Part II
Research and Survey Articles

On Extra Zeros of p-Adic L-Functions: The Crystalline Case

Denis Benois

Nous avons toutefois supposé pour simplifier que les opérateurs $1-\varphi$ et $1-p^{-1}\varphi^{-1}$ sont inversibles laissant les autres cas, pourtant extrêmement intéressants pour plus tard.

Introduction to Chapter III of Perrin-Riou (1995)

1 Introduction

1.1 Extra Zeros

Let M be a pure motive over $\mathbb{Q}$. Assume that the complex L-function $L(M,s)$ of M extends to a meromorphic function on the whole complex plane $\mathbb{C}$. Fix an odd prime p. It is expected that one can construct p-adic analogues of $L(M,s)$ p-adically interpolating algebraic parts of its special values. The above program has been realised and the corresponding p-adic L-functions constructed in many cases, but the general theory remains conjectural. Perrin-Riou (1995) formulated precise conjectures about the existence and arithmetic properties of p-adic L-functions in the case where the p-adic realisation V of M is crystalline at p. Let $\mathbf{D}_{\mathrm{cris}}(V)$ denote the filtered Dieudonné module associated to V by the theory of Fontaine. Let D be a subspace of $\mathbf{D}_{\mathrm{cris}}(V)$ of dimension $d_+(V)=\dim_{\mathbb{Q}_p} V^{c=1}$ stable under the action of φ. (As usual, c denotes the complex conjugation.) One says that D is regular if one can associate to D a p-adic analogue of the six-term exact sequence of Fontaine and Perrin-Riou (see Sect. 4.1.3 and Perrin-Riou (1995) for an exact definition). Fix a lattice T of V stable under the action of the Galois group and a lattice N of a regular

D. Benois (✉)
Institut de Mathématiques, Université de Bordeaux, 351 cours de la Libération, 33405, Talence, France
e-mail: denis.benois@math.u-bordeaux1.fr

T. Bouganis and O. Venjakob (eds.), *Iwasawa Theory 2012*, Contributions in Mathematical and Computational Sciences 7, DOI 10.1007/978-3-642-55245-8_3,

module D. Perrin-Riou conjectured that one can associate to this data a p-adic L-function $L_p(T, N, s)$ satisfying some explicit interpolation property. Let r denote the order of vanishing of $L(M, s)$ at $s = 0$ and let $L^*(M, 0) = \lim_{s \to 0} s^{-r} L(M, s)$. Then at $s = 0$ the interpolation property reads

$$\lim_{s \to 0} \frac{L_p(T, N, s)}{s^r} = \mathcal{E}(V, D) R_{V,D}(\omega_{V,N}) \frac{L^*(M, 0)}{R_{M,\infty}(\omega_M)}.$$

Here $R_{M,\infty}(\omega_M)$ (resp. $R_{V,D}(\omega_{V,N})$) is the determinant of the Beilinson (resp. the p-adic) regulator computed in some compatible bases ω_M and $\omega_{V,N}$ and $\mathcal{E}(V, D)$ is an Euler-like factor given by

$$\mathcal{E}(V, D) = \det(1 - p^{-1}\varphi^{-1} \mid D) \det(1 - \varphi \mid \mathbf{D}_{\mathrm{cris}}(V)/D).$$

If either $D^{\varphi = p^{-1}} \neq 0$ or $(\mathbf{D}_{\mathrm{cris}}(V)/D)^{\varphi = 1} \neq 0$ we have $\mathcal{E}(V, D) = 0$ and the order of vanishing of $L_p(N, T, s)$ should be $> r$. In this case we say that $L_p(T, N, s)$ has an extra zero at $s = 0$. The same phenomenon occurs in the case where V is semistable and non-crystalline at p. An archetypical example is provided by elliptic curves having split multiplicative reduction (Mazur et al. 1986). Assume that 0 is a *critical* point for $L(M, s)$ and that $H^0(\mathbb{Q}, V) = H^0(\mathbb{Q}, V) = 0$.[1] In Benois (2011) using the theory of (φ, Γ)-modules we associated to each regular D an invariant $\mathcal{L}(V, D) \in \mathbb{Q}_p$ generalising both Greenberg's $\mathcal{L}$-invariant (Greenberg 1994) and Fontaine–Mazur's $\mathcal{L}$-invariant (Mazur 1994). This allows one to formulate a quite general conjecture about the behavior of p-adic L-functions at extra zeros in the spirit of Greenberg (1994). To the best of our knowledge this conjecture is actually proved in the following cases:

1. Kubota–Leopoldt p-adic L-functions (Ferrero and Greenberg 1978, 1979; Gross and Koblitz 1979). Here the $\mathcal{L}$-invariant can be interpreted in terms of Gross' p-adic regulator (Greenberg and Stevens 1993; Gross 1981). We also remark that in Dasgupta et al. (2011) this result was generalized to totally real fields (assuming the Leopoldt conjecture).
2. Modular forms of even weight (Greenberg and Stevens 1993; Kato 2004; Stevens 2010). Here the $\mathcal{L}$-invariant coincides with Fontaine–Mazur's $\mathcal{L}(f)$. In Mok (2009), Spiess (to appear) and Rosso (2013) the method of Greenberg and Stevens (1993) was generalized to study trivial zeros of elliptic curves/modular forms of weight 2 over totally real fields.
3. Modular forms of odd weight (Benois 2014). The associated p-adic representation V is either crystalline or potentially crystalline at p and we do need the theory of (φ, Γ)-modules to define the $\mathcal{L}$-invariant.
4. Symmetric squares of modular forms having either split multiplicative or good reduction (Rosso 2014). In the split multiplicative and good ordinary reduction

[1] By Tate's conjecture this condition should be equivalent to the vanishing of $H^0(M)$ and $H^0(M^*(1))$.

cases the associated p-adic representation V is ordinary and the $\mathscr{L}$-invariant reduces to Greenberg's construction (Greenberg 1994). In the supersingular case again the definition of the $\mathscr{L}$-invariant involves (φ, Γ)-modules. For elliptic curves having good ordinary reduction another proof, based on factoring the several-variable Rankin–Selberg p-adic L-function along the diagonal has been suggested by Dasgupta (work in progress). The fact that a factorization would lead to the proof of the conjecture appears previously in Citro (2008).

5. Symmetric powers of CM-modular forms (Harron 2013; Harron and Lei to appear). See also (Harron and Jorza 2013).

1.2 Extra Zero Conjecture

The goal of this paper is to generalise the conjecture from Benois (2011) to the *non critical* point case. Assume that V is crystalline at p. Then a weight argument shows that $\mathcal{E}(V, D)$ can vanish only if $wt(M) = 0$ or -2. In particular, we expect that the interpolation factor does not vanish at $s = 0$ if $wt(M) = -1$ i.e. that the p-adic L-function can not have an extra zero at the central point in the good reduction case. To fix ideas assume that $wt(M) \leqslant -2$ and that M has no subquotients isomorphic to $\mathbb{Q}(1)$. [2] We denote by $H^1_f(V)$ the Bloch–Kato Selmer group of V. Then D is regular if and only if the associated p-adic regulator map

$$r_{V,D} \ : \ H^1_f(V) \to \mathbf{D}_{\mathrm{cris}}(V)/(\mathrm{Fil}^0\mathbf{D}_{\mathrm{cris}}(V) + D)$$

is an isomorphism. The semisimplicity of $\varphi \ : \ \mathbf{D}_{\mathrm{cris}}(V) \ \to \ \mathbf{D}_{\mathrm{cris}}(V)$ (which conjecturally always holds) allows one to decompose D into a direct sum

$$D = D_{-1} \oplus D^{\varphi=p^{-1}}.$$

Under some mild assumptions (see Sects. 4.1.2 and 5.1.2 below) we associate to D an $\mathscr{L}$-invariant $\mathscr{L}(V, D)$ which is a direct generalization of the main construction of Benois (2011). The Beilinson–Deligne conjecture predicts that $L(M, s)$ does not vanish at $s = 0$ and that $L(M^*(1), s)$ has a zero of order $r = \dim_{\mathbb{Q}_p} H^1_f(V)$ at $s = 0$. Let $L^*_p(T, N, 0)$ denote the first non-zero coefficient in the Taylor expansion of $L_p(T, N, s)$. We propose the following conjecture (see Conjecture 4 of Sect. 5.2.2):

Conjecture 1. Let D be a regular subspace of $\mathbf{D}_{\mathrm{cris}}(V)$ and let $e = \dim_{\mathbb{Q}_p}(D^{\varphi=p^{-1}})$. Then

[2]The last condition is not really essential and can be suppressed.

(i) The p-adic L-function $L_p(T,N,s)$ has a zero of order e at $s=0$ and

$$\frac{L_p^*(T,N,0)}{R_{V,D}(\omega_{V,N})} = -\mathscr{L}(V,D)\,\mathcal{E}^+(V,D)\,\frac{L(M,0)}{R_{M,\infty}(\omega_M)}.$$

(ii) Let $D^\perp$ denote the orthogonal complement to D under the canonical duality $\mathbf{D}_{\mathrm{cris}}(V)\times \mathbf{D}_{\mathrm{cris}}(V^*(1)) \to \mathbb{Q}_p$. The p-adic L-function $L_p(T^*(1),N^\perp,s)$ has a zero of order $e+r$ where $r=\dim_{\mathbb{Q}} H^1_f(V)$ at $s=0$ and

$$\frac{L_p^*(T^*(1),N^\perp,0)}{R_{V^*(1),D^\perp}(\omega_{V^*(1),N^\perp})} = \mathscr{L}(V,D)\,\mathcal{E}^+(V^*(1),D^\perp)\,\frac{L^*(M^*(1),0)}{R_{M^*(1),\infty}(\omega_{M^*(1)})}.$$

In the both cases

$$\begin{aligned}\mathcal{E}^+(V,D) &= \mathcal{E}^+(V^*(1),D^\perp)\\ &= \det(1-p^{-1}\varphi^{-1}\mid D_{-1})\,\det(1-p^{-1}\varphi^{-1}\mid \mathbf{D}_{\mathrm{cris}}(V^*(1))).\end{aligned}$$

Remarks. (1) $\mathcal{E}^+(V,D)$ is obtained from $\mathcal{E}(V,D)$ by excluding the zero factors. It can also be written in the form

$$\mathcal{E}^+(V,D) = E_p^*(V,1)\,\det\nolimits_{\mathbb{Q}_p}\left(\frac{1-p^{-1}\varphi^{-1}}{1-\varphi}\,|D_{-1}\right)$$

where $E_p(V,t)=\det(1-\varphi t\mid \mathbf{D}_{\mathrm{cris}}(V))$ is the Euler factor at p and

$$E_p^*(V,t) = E_p(V,t)\left(1-\frac{t}{p}\right)^{-e}.$$

(2) Assume that $H^1_f(V)=0$. Since $H^1_f(V^*(1))$ should also vanish by the weight argument, our conjecture in this case reduces to Conjecture 2.3.2 from Benois (2011).

(3) The regularity of D supposes that the localisation $H^1_f(V)\to H^1_f(\mathbb{Q}_p,V)$ is injective. Jannsen's conjecture (made more precise by Bloch and Kato) says that the p-adic realisation map $H^1_f(M)\otimes\mathbb{Q}_p \to H^1_f(V)$ is an isomorphism. The composition $H^1_f(M)\to H^1_f(\mathbb{Q}_p,V)$ of these two maps is essentially the syntomic regulator. Its injectivity seems to be a difficult open problem.

(4) Let $f=\sum\limits_{n=1}^{\infty} a_n(f)q^n$ and $g=\sum\limits_{n=1}^{\infty} b_n(f)q^n$ be two different newforms of weight 2 and nebentypus η_f and η_g on $\Gamma_1(N)$. For any prime l we denote by $\alpha_{l,1}$ and $\alpha_{l,2}$ (resp. $\beta_{l,1}$ and $\beta_{l,2}$) the roots of the Hecke polynomial $X^2-a_l(f)X+\eta_f(l)l$ (resp. $X^2-b_l(f)X+\eta_g(l)l$). The Rankin–Selberg L-function $L(f\otimes g,s)$ is defined by

$$L(f \otimes g, s) = \prod_l P_l(f \otimes g, l^{-s})^{-1}$$

where

$$P_l(f \otimes g, X) = (1-\alpha_{l,1}\beta_{l,1}X)(1-\alpha_{l,1}\beta_{l,2}X)(1-\alpha_{l,2}\beta_{l,1}X)(1-\alpha_{l,2}\beta_{l,2}X).$$

Let W_f and W_g be the p-adic representations associated to f and g respectively. Let $W_{f,g} = W_f \otimes W_g$ and $T_{f,g}$ a fixed lattice of $W_{f,g}$. The complex L-function $L(W_{f,g}, s)$ associated to $W_{f,g}$ coincides with $L(f \otimes g, s)$ up to Euler factors at $l \mid N$. The p-adic representation $W_{f,g}(2)$ can be viewed as the p-adic realisation of a motive $M_{f,g}(2)$ of weight -2. We remark that $M_{f,g}(2)$ is non-critical and the special value $L(W_{f,g}(2), 0) = L(W_{f,g}, 2)$ can be expressed in terms of the regulator map (Beilinson 1984). If $p \nmid N$, the restriction of $W_{f,g}(2)$ on the decomposition group at p is crystalline with Hodge–Tate weights $(-2, -1, -1, 0)$. The crystalline module $\mathbf{D}_{\mathrm{cris}}(W_{f,g}(2))$ is a 4-dimensional vector space generated by eigenvectors d_{ij}, $1 \leq i, j \leq 2$ such that $\varphi(d_{ij}) = \alpha_i \beta_j p^{-2} d_{ij}$. Let D_{ij} $(1 \leqslant i, j \leq 2)$ be the 3-dimensional subspace of $\mathbf{D}_{\mathrm{cris}}(W_{f,g}(2))$ generated by $d_{r,s}$, $(r, s) \neq (i, j)$ and N_{ij} a fixed lattice of D_{ij}. One expects that if f and g are not CM, then D_{ij} is regular. If $\alpha_{p,r}\beta_{p,s} = p$ for some $(r, s) \neq (i, j)$, then $D_{ij}^{\varphi=p^{-1}} \neq 0$ and our extra zero conjecture predicts the behavior of $L_p(T_{f,g}, N_{ij}, s)$ at $s = 0$. As $W_{f,g}(2)$ is non-critical, this gives an example of extra zero which is not covered by the trivial zero conjecture from Benois (2011).

1.3 Selmer Complexes and Perrin-Riou's Theory

In the last part of the paper we show that our extra zero conjecture is compatible with the Main Conjecture of Iwasawa theory as formulated in Perrin-Riou (1995). The main technical tool here is the descent theory for Selmer complexes (Nekovář 2006). We hope that the approach to Perrin-Riou's theory based on the formalism of Selmer complexes can be of independent interest.

For a profinite group G and a continuous G-module X we denote by $C_c^\bullet(G, X)$ the standard complex of continuous cochains. Let S be a finite set of primes containing p. Denote by G_S the Galois group of the maximal algebraic extension of $\mathbb{Q}$ unramified outside $S \cup \{\infty\}$. Set $\mathbf{R}\Gamma_S(X) = C_c^\bullet(G_S, X)$ and $\mathbf{R}\Gamma(\mathbb{Q}_v, X) = C_c^\bullet(G_v, X)$, where G_v is the absolute Galois group of $\mathbb{Q}_v$. Let Γ be the Galois group of the cyclotomic p-extension $\mathbb{Q}(\zeta_{p^\infty})/\mathbb{Q}$, $\Gamma_1 = \mathrm{Gal}(\mathbb{Q}(\zeta_{p^\infty})/\mathbb{Q}(\zeta_p))$ and $\Delta = \mathrm{Gal}(\mathbb{Q}(\zeta_p)/\mathbb{Q})$. Let $\Lambda(\Gamma) = \mathbb{Z}_p[[\Gamma]]$ denote the Iwasawa algebra of Γ. Each $\Lambda(\Gamma)$-module X decomposes into the direct sum of its isotypical components $X = \oplus_{\eta \in \hat{\Delta}} X^{(\eta)}$ and we denote by $X^{(\eta_0)}$ the component which corresponds to the trivial character η_0. Set $\Lambda = \Lambda(\Gamma)^{(\eta_0)}$. Let $\mathcal{H}$ denote the algebra of power series with coefficients in $\mathbb{Q}_p$ which converge on the open unit disk. We will denote again by $\mathcal{H}$ the associated large Iwasawa algebra $\mathcal{H}(\Gamma_1)$ (see Sect. 3.2.1). In this paper we consider only the trivial character component of the module of p-adic L-functions because it is sufficient for applications to trivial zeros, but in the general case the construction is exactly the same. We keep notation and assumptions of Sect. 1.2. Assume that the weak Leopoldt conjecture holds for (V, η_0) and $(V^*(1), \eta_0)$. We consider global and local Iwasawa cohomology $\mathbf{R}\Gamma_{\mathrm{Iw},S}(T) = \mathbf{R}\Gamma_S((\Lambda(\Gamma)\otimes_{\mathbb{Z}_p} T)^\iota)$ and $\mathbf{R}\Gamma_{\mathrm{Iw}}(\mathbb{Q}_v, T) = \mathbf{R}\Gamma(\mathbb{Q}_v, (\Lambda(\Gamma)\otimes_{\mathbb{Z}_p} T)^\iota)$ where ι is the canonical involution on $\Lambda(\Gamma)$. Let D be a regular submodule of $\mathbf{D}_{\mathrm{cris}}(V)$. For each non-Archimedean place v we define a local condition at v in the sense of Nekovář (2006) as follows. If $v \neq p$ we use the unramified local condition which is defined by

$$\mathbf{R}\Gamma_{\mathrm{Iw}}^{(\eta_0)}(\mathbb{Q}_v, N, T) = \mathbf{R}\Gamma_{\mathrm{Iw},f}^{(\eta_0)}(\mathbb{Q}_v, T) = \left[T^{I_v} \otimes \Lambda^\iota \xrightarrow{1-f_v} T^{I_v} \otimes \Lambda^\iota \right]$$

where I_v is the inertia subgroup at v and f_v is the geometric Frobenius. If $v = p$ we define

$$\mathbf{R}\Gamma_{\mathrm{Iw}}^{(\eta_0)}(\mathbb{Q}_v, N, T) = (N \otimes \Lambda)[-1].$$

The derived version of the large exponential map $\mathrm{Exp}_{V,h}$, for $h \gg 0$ (see Perrin-Riou 1994) gives a morphism

$$\mathbf{R}\Gamma_{\mathrm{Iw}}^{(\eta_0)}(\mathbb{Q}_p, N, T) \to \mathbf{R}\Gamma_{\mathrm{Iw}}^{(\eta_0)}(\mathbb{Q}_p, T) \otimes \mathcal{H}.$$

Therefore we have a diagram

$$\begin{array}{ccc}
\mathbf{R}\Gamma_{\mathrm{Iw},S}^{(\eta_0)}(T) \otimes_\Lambda \mathcal{H} & \longrightarrow & \bigoplus_{v\in S} \mathbf{R}\Gamma_{\mathrm{Iw}}^{(\eta_0)}(\mathbb{Q}_v, T) \otimes_\Lambda \mathcal{H} \\
 & & \uparrow \\
 & & \left(\bigoplus_{v\in S} \mathbf{R}\Gamma_{\mathrm{Iw}}^{(\eta_0)}(\mathbb{Q}_v, N, T) \right) \otimes_\Lambda \mathcal{H}.
\end{array}$$

Let $\mathbf{R}\Gamma_{\mathrm{Iw},h}^{(\eta_0)}(D, V)$ denote the Selmer complex associated to this data. By definition it sits in the distinguished triangle

$$\mathbf{R}\Gamma_{\mathrm{Iw},h}^{(\eta_0)}(D, V) \to \left(\mathbf{R}\Gamma_{\mathrm{Iw},S}^{(\eta_0)}(V) \bigoplus \left(\bigoplus_{v\in S} \mathbf{R}\Gamma_{\mathrm{Iw}}^{(\eta_0)}(\mathbb{Q}_v, D, V) \right) \right) \otimes \mathcal{H} \to \left(\bigoplus_{v\in S} \mathbf{R}\Gamma_{\mathrm{Iw}}^{(\eta_0)}(\mathbb{Q}_v, V) \right) \otimes \mathcal{H}. \quad (1)$$

Define

$$\Delta_{\mathrm{Iw},h}(N,T) =$$

$$\det\nolimits_{\Lambda}^{-1}\left(\mathbf{R}\Gamma_{\mathrm{Iw},S}^{(\eta_0)}(T)\bigoplus\left(\bigoplus_{v\in S}\mathbf{R}\Gamma_{\mathrm{Iw}}^{(\eta_0)}(\mathbb{Q}_v,N,T)\right)\right)\otimes\det\nolimits_{\Lambda}\left(\bigoplus_{v\in S}\mathbf{R}\Gamma_{\mathrm{Iw}}^{(\eta_0)}(\mathbb{Q}_v,T)\right).$$

Our results can be summarized as follows (see Theorems 4, 5 and Corollary 2).

Theorem 1. *Assume that $\mathscr{L}(V,D)\neq 0$. Then*

(i) The cohomology $\mathbf{R}^i\Gamma_{\mathrm{Iw},h}^{(\eta_0)}(D,V)$ are $\mathcal{H}$-torsion modules for all i.
(ii) $\mathbf{R}^i\Gamma_{\mathrm{Iw},h}^{(\eta_0)}(D,V)=0$ for $i\neq 2,3$ and

$$\mathbf{R}^3\Gamma_{\mathrm{Iw},h}^{(\eta_0)}(D,V)\simeq\left(H^0(\mathbb{Q}(\zeta_{p^\infty}),V^*(1))^*\right)^{(\eta_0)}\otimes_{\Lambda}\mathcal{H}.$$

(iii) The complex $\mathbf{R}\Gamma_{\mathrm{Iw},h}^{(\eta_0)}(D,V)$ is semisimple in the sense that for each i the natural map

$$\mathbf{R}^i\Gamma_{\mathrm{Iw},h}^{(\eta_0)}(D,V)^{\Gamma}\to\mathbf{R}^i\Gamma_{\mathrm{Iw},h}^{(\eta_0)}(D,V)_{\Gamma}$$

is an isomorphism.

This theorem allows us to apply to our Selmer complexes the descent machinery developed by Nekovář (2006). Assume that $\mathscr{L}(V,D)\neq 0$. Let $\mathcal{K}$ be the field of fractions of $\mathcal{H}$. Then Theorem 1 together with (1) define an injective map

$$i_{V,\mathrm{Iw},h}\;:\;\Delta_{\mathrm{Iw},h}(N,T)\to\mathcal{K}$$

and the module of p-adic L-functions is defined as

$$\mathbf{L}_{\mathrm{Iw},h}^{(\eta_0)}(N,T)=i_{V,\mathrm{Iw},h}(\Delta_{\mathrm{Iw},h}(N,T))\subset\mathcal{K}.$$

Let γ_1 be a fixed generator of Γ_1. Choose a generator $f(\gamma_1-1)$ of the free Λ-module $\mathbf{L}_{\mathrm{Iw},h}^{(\eta_0)}(N,T)$ and define a meromorphic p-adic function

$$L_{\mathrm{Iw},h}(T,N,s)=f(\chi(\gamma_1)^s-1),$$

where $\chi\;:\;\Gamma\to\mathbb{Z}_p^*$ is the cyclotomic character. For $a,b\in\mathbb{Q}_p^*$ we will write $a\sim_p b$ if a and b coincide up to a p-adic unit.

Theorem 2. *Assume that $\mathscr{L}(V,D)\neq 0$. Then*

(i) The p-adic L-function $L_{\mathrm{Iw},h}(T,N,s)$ has a zero of order $e=\dim_{\mathbb{Q}_p}(D^{\varphi=p^{-1}})$ at $s=0$.

(ii) One has

$$\frac{L^*_{\mathrm{Iw},h}(T,N,0)}{R_{V,D}(\omega_{T,N})} \sim_p \Gamma(h)^{d_+(V)}\,\mathscr{L}(V,D)\,\mathcal{E}^+(V,D)\,\frac{\#\text{Ш}(T^*(1))\,\mathrm{Tam}^0_{\omega_M}(T)}{\#H^0_S(V/T)\,\#H^0_S(V^*(1)/T^*(1))},$$

where $\text{Ш}(T^*(1))$ *is the Tate–Shafarevich group of Bloch–Kato (1990) and* $\mathrm{Tam}^0_{\omega_M}(T)$ *is the product of local Tamagawa numbers of* T.

Remarks. (1) Using the compatibility of Perrin-Riou's theory with the functional equation we obtain analogous results for the $L_p(T^*(1), N^{\perp}, s)$ (see Sect. 6.2.6).

(2) If $\mathbf{D}_{\mathrm{cris}}(V)^{\varphi=1} = \mathbf{D}_{\mathrm{cris}}(V)^{\varphi=p^{-1}} = 0$ the phenomenon of extra zeros does not appear, $\mathscr{L}(V,D) = 1$ and Theorem 2 was proved in Perrin-Riou (1995), Theorem 3.6.5. We remark that even in this case our proof is different. We compare the leading term of $L^*_{\mathrm{Iw},h}(T,N,s)$ with the trivialisation $i_{\omega_M,p} : \Delta_{\mathrm{EP}}(T) \to \mathbb{Q}_p$ of the Euler–Poincaré line $\Delta_{\mathrm{EP}}(T)$ (see Fontaine 1992) and show that in compatible bases one has

$$\frac{L^*_{\mathrm{Iw},h}(T,N,0)}{R_{V,D}(\omega_{V,N})} \sim_p \Gamma(h)^{d_+(V)}\,\mathscr{L}(V,D)\,\mathcal{E}^+(V,D)\,i_{\omega_M,p}(\Delta_{\mathrm{EP}}(T)) \qquad (2)$$

(see the proof of Theorem 5). Now Theorem 2 follows from the well known computation of $i_{\omega_M,p}(\Delta_{\mathrm{EP}}(T))$ in terms of the Tate–Shafarevich group and Tamagawa numbers (see Fontaine and Perrin-Riou 1994, Chapitre II).

(3) Let $E/\mathbb{Q}$ be an elliptic curve having good reduction at p. Consider the p-adic representation $V = \mathrm{Sym}^2(T_p(E)) \otimes \mathbb{Q}_p$, where $T_p(E)$ is the p-adic Tate module of E. It is easy to see that $D = \mathbf{D}_{\mathrm{cris}}(V)^{\varphi=p^{-1}}$ is one dimensional. In this case some versions of Theorem 2 were proved in Perrin-Riou (1998) and Delbourgo (2002) with an ad hoc definition of the $\mathscr{L}$-invariant. Remark that p-adic L-functions attached to the symmetric square of a newform were constructed by Dabrowski and Delbourgo (1997) and Rosso (2014).

(4) Another approach to Iwasawa theory in the non-ordinary case was developed by Pottharst (2012, 2013). Pottharst uses the formalism of Selmer complexes but works with local conditions coming from submodules of the (φ,Γ)-module associated to V rather than with the large exponential map. This approach has many advantages, in particular it allows to develop an interesting theory for representations which are not necessarily crystalline. Nevertheless it seems that the large exponential map is crucial for the study of extra zeros at least in the good reduction case.

(5) The Main conjecture of Iwasawa theory (Colmez 2000; Perrin-Riou 1995) says that for $h \gg 0$ the analytic p-adic L-function $L_p(N,T,s)$ multiplied by a simple explicit Γ-factor $\Gamma_{V,h}(s)$ depending on h can be written in the form $\Gamma_{V,h}(s)L_p(N,T,s) = f(\chi(\gamma_1)^s - 1)$ for an appropriate generator $f(\gamma_1 - 1)$ of $\mathbf{L}^{(\eta_0)}_{\mathrm{Iw},h}(N,T)$. Therefore the Main conjecture implies Bloch–Kato style formulas

for special values of $L_p(N,T,s)$. We remark that the Bloch–Kato conjecture predicts that

$$\frac{L^*(M,0)}{R_{M,\infty}(\omega_M)} \sim_p \frac{\#\text{Ш}(T^*(1))\,\mathrm{Tam}^0_{\omega_M}(T)}{\#H^0_S(V/T)\,\#H^0_S(V^*(1)/T^*(1))}$$

and therefore Theorem 2 implies the compatibility of our extra zero conjecture with the Main conjecture. Note that this also follows directly from (2) if we use the formalism of Fontaine and Perrin-Riou (see Fontaine 1992) to formulate Bloch–Kato conjectures.

1.4 The Plan of the Paper

The organisation of the paper is as follows. In Sect. 2 we review the theory of (φ,Γ)-modules which is the main technical tool in our definition of the $\mathscr{L}$-invariant. We also give the derived version of the computation of Galois cohomology in terms of (φ,Γ)-modules. This follows easily from the results of Herr (1998) and Liu (2007) and the proofs are placed in Appendix. Similar results can be found in Pottharst (2012, 2013). In Sect. 3 we recall preliminaries on the Bloch–Kato exponential map and review the construction of the large exponential map of Perrin-Riou given by Berger (2003) using again the basic language of derived categories. The $\mathscr{L}$-invariant is constructed in Sect. 4.1. In Sect. 4.2 we relate this construction to the derivative of the large exponential map. This result plays a key role in the proof of Theorem 2. The extra zero conjecture is formulated in Sect. 5. In Sect. 6 we interpret Perrin-Riou's theory in terms of Selmer complexes and prove Theorems 1 and 2.

This paper is a revised and extended version of the preprint Benois (2009). In Benois (2009), Theorems 1 and 2 above were proved under the additional assumption that $H^1_f(V)=0$. As we pointed out before, in this case our extra-zero conjecture reduces to the Conjecture 2.3.2 from Benois (2011).

2 Preliminaries

2.1 (φ,Γ)-Modules

2.1.1 The Robba Ring

The main references for the material of this section are (Berger 2002) and (Colmez 2003). In this section K is a finite unramified extension of $\mathbb{Q}_p$ with residue field k_K, O_K its ring of integers, and σ the absolute Frobenius of K. Let $\overline{K}$ an algebraic closure of K, $G_K=\mathrm{Gal}(\overline{K}/K)$ and C the completion of $\overline{K}$. Let $v_p\ :\ C\to$

$\mathbb{R} \cup \{\infty\}$ denote the p-adic valuation normalized so that $v_p(p) = 1$ and set $|x|_p = \left(\frac{1}{p}\right)^{v_p(x)}$. Write $B(r,1)$ for the p-adic annulus $B(r,1) = \{x \in C \mid r \leqslant |x| < 1\}$. As usually, μ_{p^n} denotes the group of p^n-th roots of unity. Fix a system of primitive roots of unity $\varepsilon = (\zeta_{p^n})_{n\geqslant 0}$, $\zeta_{p^n} \in \mu_{p^n}$ such that $\zeta_{p^n}^p = \zeta_{p^{n-1}}$ for all n. Set $K_n = K(\zeta_{p^n})$, $K_\infty = \bigcup_{n=0}^{\infty} K_n$, $H_K = \mathrm{Gal}(\overline{K}/K_\infty)$, $\Gamma = \mathrm{Gal}(K_\infty/K)$ and denote by $\chi : \Gamma \to \mathbb{Z}_p^*$ the cyclotomic character.

Set

$$\tilde{\mathbf{E}}^+ = \varprojlim_{x \mapsto x^p} O_C/\,p\,O_C = \{x = (x_0, x_1, \ldots, x_n, \ldots) \mid x_i^p = x_i \ \forall i \in \mathbb{N}\}.$$

Let $\hat{x}_n \in O_C$ be a lifting of x_n. Then for all $m \geqslant 0$ the sequence $\hat{x}_{m+n}^{p^n}$ converges to $x^{(m)} = \lim_{n\to\infty} \hat{x}_{m+n}^{p^n} \in O_C$ which does not depend on the choice of liftings. The ring $\tilde{\mathbf{E}}^+$ equipped with the valuation $v_{\mathbf{E}}(x) = v_p(x^{(0)})$ is a complete local ring of characteristic p with residue field $\bar{k}_K$. Moreover it is integrally closed in its field of fractions $\tilde{\mathbf{E}} = \mathrm{Fr}(\tilde{\mathbf{E}}^+)$.

Let $\tilde{\mathbf{A}} = W(\tilde{\mathbf{E}})$ be the ring of Witt vectors with coefficients in $\tilde{\mathbf{E}}$. Denote by $[\,] : \tilde{\mathbf{E}} \to W(\tilde{\mathbf{E}})$ the Teichmüller lift. Any $u = (u_0, u_1, \ldots) \in \tilde{\mathbf{A}}$ can be written in the form

$$u = \sum_{n=0}^{\infty} [u^{p^{-n}}]p^n.$$

Set $\pi = [\varepsilon] - 1$, $\mathbf{A}_{K_0}^+ = O_{K_0}[[\pi]]$ and denote by $\mathbf{A}_K$ the p-adic completion of $\mathbf{A}_K^+[1/\pi]$. Let $\tilde{\mathbf{B}} = \tilde{\mathbf{A}}[1/p]$, $\mathbf{B}_K = \mathbf{A}_K[1/p]$ and let $\mathbf{B}$ denote the completion of the maximal unramified extension of $\mathbf{B}_K$ in $\tilde{\mathbf{B}}$. Set $\mathbf{A} = \mathbf{B} \cap \tilde{\mathbf{A}}$, $\tilde{\mathbf{A}}^+ = W(\mathbf{E}^+)$, $\mathbf{A}^+ = \tilde{\mathbf{A}}^+ \cap \mathbf{A}$ and $\mathbf{B}^+ = \mathbf{A}^+[1/p]$. All these rings are endowed with natural actions of the Galois group G_K and Frobenius φ.

Set $\mathbf{A}_K = \mathbf{A}^{H_K}$ and $\mathbf{B}_K = \mathbf{A}_K[1/p]$. We remark that Γ and φ act on $\mathbf{B}_K$ by

$$\tau(\pi) = (1+\pi)^{\chi(\tau)} - 1, \qquad \tau \in \Gamma,$$
$$\varphi(\pi) = (1+\pi)^p - 1.$$

For any $r > 0$ define

$$\tilde{\mathbf{B}}^{\dagger,r} = \left\{ x \in \tilde{\mathbf{B}} \mid \lim_{k\to+\infty} \left(v_{\mathrm{e}}(x_k) + \frac{pr}{p-1} k \right) = +\infty \right\}.$$

Set $\mathbf{B}^{\dagger,r} = \mathbf{B} \cap \tilde{\mathbf{B}}^{\dagger,r}$, $\mathbf{B}_K^{\dagger,r} = \mathbf{B}_K \cap \mathbf{B}^{\dagger,r}$, $\mathbf{B}^{\dagger} = \bigcup_{r>0} \mathbf{B}^{\dagger,r}$ $\mathbf{A}^{\dagger} = \mathbf{A} \cap \mathbf{B}^{\dagger}$ and $\mathbf{B}_K^{\dagger} = \bigcup_{r>0} \mathbf{B}_K^{\dagger,r}$.

It can be shown that for any $r \geqslant p-1$

$$\mathbf{B}_K^{\dagger,r} = \left\{ f(\pi) = \sum_{k\in\mathbb{Z}} a_k\pi^k \mid a_k \in K \text{ and } f \text{ is holomorphic and bounded on } B(r,1) \right\}.$$

Define

$$\mathbf{B}_{\mathrm{rig},K}^{\dagger,r} = \left\{ f(\pi) = \sum_{k\in\mathbb{Z}} a_k\pi^k \mid a_k \in K \text{ and } f \text{ is holomorphic on } B(r,1) \right\}.$$

Set $\mathcal{R}(K) = \bigcup_{r\geqslant p-1} \mathbf{B}_{\mathrm{rig},K}^{\dagger,r}$ and $\mathcal{R}^+(K) = \mathcal{R}(K) \cap K[[\pi]]$. It is not difficult to check that these rings are stable under Γ and φ. To simplify notations we will write $\mathcal{R} = \mathcal{R}(\mathbb{Q}_p)$ and $\mathcal{R}^+ = \mathcal{R}^+(\mathbb{Q}_p)$. As usual, we set

$$t = \log(1+\pi) = \sum_{n=1}^{\infty}(-1)^{n+1}\frac{\pi^n}{n} \in \mathcal{R}.$$

Note that $\varphi(t) = pt$ and $\tau(t) = \chi(\tau)t$, $\tau \in \Gamma$.

2.1.2 (φ, Γ)-Modules

The main references for the material of this section are (Cherbonnier and Colmez 1998; Colmez 2008; Fontaine 1991). Let A be either $\mathbf{B}_K^{\dagger}$ or $\mathcal{R}(K)$. A (φ,Γ)-module over A is a finitely generated free A-module $\mathbf{D}$ equipped with semilinear actions of φ and Γ commuting with each other and such that the induced linear map φ : $A\otimes_{\varphi}\mathbf{D} \to \mathbf{D}$ is an isomorphism. Such a module is said to be étale if it admits a $\mathrm{A}_K^{\dagger}$-lattice N stable under φ and Γ and such that $\varphi\ :\ \mathbf{A}_K^{\dagger}\otimes_{\varphi}N \to N$ is an isomorphism. The functor $\mathbf{D} \mapsto \mathcal{R}(K)\otimes_{\mathbf{B}_K^{\dagger}}\mathbf{D}$ induces an equivalence between the category of étale (φ,Γ)-modules over $\mathbf{B}_K^{\dagger}$ and the category of (φ,Γ)-modules over $\mathcal{R}(K)$ which are of slope 0 in the sense of Kedlaya's theory (Kedlaya 2004 and Colmez 2008, Corollary 1.5). Then Fontaine's classification of p-adic representations (Fontaine 1991) together with the main result of Cherbonnier and Colmez (1998) lead to the following statement.

Proposition 1. *(i) The functor*

$$\mathbf{D}^{\dagger}\ :\ V \mapsto \mathbf{D}^{\dagger}(V) = (\mathbf{B}^{\dagger}\otimes_{\mathbb{Q}_p} V)^{H_K}$$

establishes an equivalence between the category of p-adic representations of G_K and the category of étale (φ,Γ)-modules over $\mathbf{B}_K^{\dagger}$.

(ii) *The functor* $\mathbf{D}^{\dagger}_{\mathrm{rig}}(V) = \mathcal{R}(K) \otimes_{\mathbf{B}^{\dagger}_K} \mathbf{D}^{\dagger}(V)$ *gives an equivalence between the category of* p*-adic representations of* G_K *and the category of* (φ, Γ)*-modules over* $\mathcal{R}(K)$ *of slope* 0.

Proof. See Colmez (2008), Proposition 1.7.

2.1.3 Cohomology of (φ, Γ)-Modules

We refer to (Herr 1998, 2001; Liu 2007) for the material discussed in this section. Fix a generator γ of Γ. If $\mathbf{D}$ is a (φ, Γ)-module over A, we denote by $C_{\varphi,\gamma}(\mathbf{D})$ the complex

$$C_{\varphi,\gamma}(\mathbf{D}) \ : \ 0 \longrightarrow \mathbf{D} \xrightarrow{f} \mathbf{D} \oplus \mathbf{D} \xrightarrow{g} \mathbf{D} \longrightarrow 0$$

where $f(x) = ((\varphi - 1)\, x, (\gamma - 1)\, x)$ and $g(y,z) = (\gamma - 1)\, y - (\varphi - 1)\, z$. Set $H^i(\mathbf{D}) = H^i(C_{\varphi,\gamma}(\mathbf{D}))$. A short exact sequence of (φ, Γ)-modules

$$0 \longrightarrow \mathbf{D}' \longrightarrow \mathbf{D} \longrightarrow \mathbf{D}'' \longrightarrow 0$$

gives rise to an exact cohomology sequence:

$$0 \to H^0(\mathbf{D}') \to H^0(\mathbf{D}) \to H^0(\mathbf{D}'') \to H^1(\mathbf{D}') \to \cdots \to H^2(\mathbf{D}'') \to 0.$$

Proposition 2. *Let* V *be a* p*-adic representation of* G_K. *Then the complexes* $\mathbf{R}\Gamma(K, V)$, $C_{\varphi,\gamma}(\mathbf{D}^{\dagger}(V))$ *and* $C_{\varphi,\gamma}(\mathbf{D}^{\dagger}_{\mathrm{rig}}(V))$ *are isomorphic in the derived category of* $\mathbb{Q}_p$*-vector spaces* $\mathcal{D}(\mathbb{Q}_p)$.

Proof. This is a derived version of Herr's computation of Galois cohomology (Herr 1998). The proof is given in the Appendix (see Proposition 9 and Corollary 3).

2.1.4 Iwasawa Cohomology

Recall that $\Lambda = \mathbb{Z}_p[[\Gamma_1]]$ denotes the Iwasawa algebra of Γ_1. Set $\Delta = \mathrm{Gal}(K_1/K)$ and $\Lambda(\Gamma) = \mathbb{Z}_p[\Delta] \otimes_{\mathbb{Z}_p} \Lambda$. Let $\iota \ : \ \Lambda(\Gamma) \to \Lambda(\Gamma)$ denote the involution defined by $\iota(g) = g^{-1}$, $g \in \Gamma$. If T is a $\mathbb{Z}_p$-adic representation of G_K, then the induced module $\mathrm{Ind}_{K_\infty/K}(T)$ is isomorphic to $(\Lambda(\Gamma) \otimes_{\mathbb{Z}_p} T)^{\iota}$ and we set

$$\mathbf{R}\Gamma_{\mathrm{Iw}}(K, T) \ = \ \mathbf{R}\Gamma(K, \mathrm{Ind}_{K_\infty/K}(T)).$$

Write $H^i_{\mathrm{Iw}}(K, T)$ for the Iwasawa cohomology

$$H^i_{\mathrm{Iw}}(K, T) = \varprojlim_{\mathrm{cor}_{K_n/K_{n-1}}} H^i(K_n, T).$$

Recall that there are canonical and functorial isomorphisms

$$\mathbf{R}^i\Gamma_{\mathrm{Iw}}(K,T) \simeq H^i_{\mathrm{Iw}}(K,T), \qquad i \geqslant 0,$$
$$\mathbf{R}\Gamma_{\mathrm{Iw}}(K,T) \otimes^{\mathbf{L}}_{\Lambda(\Gamma)} \mathbb{Z}_p[G_n] \simeq \mathbf{R}\Gamma(K_n,T)$$

(see Nekovář 2006, Proposition 8.4.22). The interpretation of the Iwasawa cohomology in terms of (φ,Γ)-modules was found by Fontaine (unpublished but see Cherbonnier and Colmez 1999). We give here the derived version of this result. Let $\psi : \mathbf{B} \to \mathbf{B}$ be the operator defined by the formula $\psi(x) = \frac{1}{p}\varphi^{-1}\left(\mathrm{Tr}_{\mathbf{B}/\varphi(\mathbf{B})}(x)\right)$. We see immediately that $\psi \circ \varphi = \mathrm{id}$. Moreover ψ commutes with the action of G_K and $\psi(\mathbf{A}^\dagger) = \mathbf{A}^\dagger$. Consider the complexes

$$C_{\mathrm{Iw},\psi}(T) \ : \ \mathbf{D}(T) \xrightarrow{\psi-1} \mathbf{D}(T),$$
$$C^\dagger_{\mathrm{Iw},\psi}(T) \ : \ \mathbf{D}^\dagger(T) \xrightarrow{\psi-1} \mathbf{D}^\dagger(T).$$

Proposition 3. *The complexes* $\mathbf{R}\Gamma_{\mathrm{Iw}}(K,T)$, $C_{\mathrm{Iw},\psi}(T)$ *and* $C^\dagger_{\mathrm{Iw},\psi}(T)$ *are naturally isomorphic in the derived category* $\mathcal{D}(\Lambda(\Gamma))$ *of* $\Lambda(\Gamma)$*-modules.*

Proof. See Proposition 11 and Corollary 4.

2.1.5 (φ,Γ)-Modules of Rank 1

Recall the computation of the cohomology of (φ,Γ)-modules of rank 1 following Colmez (2008). As in op. cit., we consider the case $K = \mathbb{Q}_p$ and put $\mathcal{R} = \mathbf{B}^\dagger_{\mathrm{rig},\mathbb{Q}_p}$ and $\mathcal{R}^+ = \mathbf{B}^+_{\mathrm{rig},\mathbb{Q}_p}$. The differential operator $\partial = (1+\pi)\dfrac{d}{d\pi}$ acts on $\mathcal{R}$ and $\mathcal{R}^+$. If $\delta : \mathbb{Q}_p^* \to \mathbb{Q}_p^*$ is a continuous character, we write $\mathcal{R}(\delta)$ for the (φ,Γ)-module $\mathcal{R}e_\delta$ defined by $\varphi(e_\delta) = \delta(p)e_\delta$ and $\gamma(e_\delta) = \delta(\chi(\tau))\,e_\delta$. Let x denote the character induced by the natural inclusion of $\mathbb{Q}_p$ in L and $|x|$ the character defined by $|x| = p^{-v_p(x)}$.

Proposition 4. *Let* $\delta : \mathbb{Q}_p^* \to \mathbb{Q}_p^*$ *be a continuous character. Then:*

(i)

$$H^0(\mathcal{R}(\delta)) = \begin{cases} \mathbb{Q}_p t^m & \text{if } \delta = x^{-m}, m \in \mathbb{N} \\ 0 & \text{otherwise.} \end{cases}$$

(ii)

$$\dim_{\mathbb{Q}_p}(H^1(\mathcal{R}(\delta)))$$
$$= \begin{cases} 2 & \text{if either } \delta(x) = x^{-m}, m \geqslant 0 \text{ or } \delta(x) = |x|x^m, m \geqslant 1, \\ 1 & \text{otherwise.} \end{cases}$$

(iii) *Assume that* $\delta(x) = x^{-m}$, $m \geqslant 0$. *The classes* $\mathrm{cl}(t^m, 0)\, e_\delta$ *and* $\mathrm{cl}(0, t^m)\, e_\delta$ *form a basis of* $H^1(\mathcal{R}(x^{-m}))$.
(iv) *Assume that* $\delta(x) = |x|x^m$, $m \geqslant 1$. *Then* $H^1(\mathcal{R}(|x|x^m))$, $m \geqslant 1$ *is generated by* $\mathrm{cl}(\alpha_m)$ *and* $\mathrm{cl}(\beta_m)$ *where*

$$\alpha_m = \frac{(-1)^{m-1}}{(m-1)!}\,\partial^{m-1}\left(\frac{1}{\pi}+\frac{1}{2}, a\right) e_\delta, \quad (1-\varphi)\,a = (1-\chi(\gamma)\gamma)\left(\frac{1}{\pi}+\frac{1}{2}\right),$$
$$\beta_m = \frac{(-1)^{m-1}}{(m-1)!}\,\partial^{m-1}\left(b, \frac{1}{\pi}\right) e_\delta, \quad (1-\varphi)\left(\frac{1}{\pi}\right) = (1-\chi(\gamma)\,\gamma)\,b$$

Proof. See Colmez (2008), Sects. 2.3–2.5.

2.2 Crystalline Representations

2.2.1 The Rings $\mathbf{B}_{\mathrm{cris}}$ and $\mathbf{B}_{\mathrm{dR}}$

The main references for the material of this section are (Fontaine 1982, 1994a). Let $\theta_0 : \mathbf{A}^+ \to O_C$ be the map given by the formula

$$\theta_0\left(\sum_{n=0}^{\infty}[u_n]p^n\right) = \sum_{n=0}^{\infty} u_n^{(0)} p^n.$$

It can be shown that θ_0 is a surjective ring homomorphism and that $\ker(\theta_0)$ is the principal ideal generated by $\omega = \sum_{i=0}^{p-1}[\epsilon]^{i/p}$. By linearity, θ_0 can be extended to a map $\theta : \tilde{\mathbf{B}}^+ \to C$. The ring $\mathbf{B}_{\mathrm{dR}}^+$ is defined to be the completion of $\tilde{\mathbf{B}}^+$ for the $\ker(\theta)$-adic topology:

$$\mathbf{B}_{\mathrm{dR}}^+ = \varprojlim_n \tilde{\mathbf{B}}^+/\ker(\theta)^n.$$

This is a complete discrete valuation ring with residue field C equipped with a natural action of G_K. Moreover, there exists a canonical embedding $\overline{K} \subset \mathbf{B}_{\mathrm{dR}}^+$. The series $t = \sum_{n=0}^{\infty}(-1)^{n-1}\pi^n/n$ converges in the topology of $\mathbf{B}_{\mathrm{dR}}^+$ and it is easy to see that t generates the maximal ideal of $\mathbf{B}_{\mathrm{dR}}^+$. The Galois group acts on t by the formula $g(t) = \chi(g)t$. Let $\mathbf{B}_{\mathrm{dR}} = \mathbf{B}_{\mathrm{dR}}^+[t^{-1}]$ be the field of fractions of $\mathbf{B}_{\mathrm{dR}}^+$. This is a complete discrete valuation field equipped with a G_K-action and an exhaustive separated decreasing filtration $\mathrm{Fil}^i\mathbf{B}_{\mathrm{dR}} = t^i\mathbf{B}_{\mathrm{dR}}^+$. As G_K-modules, $\mathrm{Fil}^i\mathbf{B}_{\mathrm{dR}}/\mathrm{Fil}^{i+1}\mathbf{B}_{\mathrm{dR}} \simeq C(i)$ and $\mathbf{B}_{\mathrm{dR}}^{G_K} = K$.

Consider the *PD*-envelope of $\mathbf{A}^+$ with a respect to the map θ_0

$$\mathbf{A}^{\mathrm{PD}} = \mathbf{A}^+\left[\frac{\omega^2}{2!}, \frac{\omega^3}{3!}, \dots, \frac{\omega^n}{n!}, \dots\right]$$

and denote by $\mathbf{A}_{\mathrm{cris}}^+$ its p-adic completion. Let $\mathbf{B}_{\mathrm{cris}}^+ = \mathbf{A}_{\mathrm{cris}}^+ \otimes_{\mathbb{Z}_p} \mathbb{Q}_p$ and $\mathbf{B}_{\mathrm{cris}} = \mathbf{B}_{\mathrm{cris}}^+[t^{-1}]$. Then $\mathbf{B}_{\mathrm{cris}}$ is a subring of $\mathbf{B}_{\mathrm{dR}}$ endowed with the induced filtration and Galois action. Moreover, it is equipped with a continuous Frobenius φ, extending the map $\varphi : \mathbf{A}^+ \to \mathbf{A}^+$. One has $\varphi(t) = pt$.

2.2.2 Crystalline Representations

In this section we review the notion of crystalline representation and its relationship to the theory of (φ, Γ)-modules (see Berger 2002, 2004; Fontaine 1994b). Let L be a finite extension of $\mathbb{Q}_p$. Denote by K its maximal unramified subextension. A filtered Dieudonné module over L is a finite dimensional K-vector space M equipped with the following structures:

- A σ-semilinear bijective map $\varphi : M \to M$;
- An exhaustive decreasing filtration $(\mathrm{Fil}^i M_L)_{i\in Z}$ on the L-vector space $M_L = L \otimes_K M$.

  A K-linear map $f : M \to M'$ is said to be a morphism of filtered modules if
- $f(\varphi(d)) = \varphi(f(d))$ for all $d \in M$;
- $f(\mathrm{Fil}^i M_L) \subset \mathrm{Fil}^i M'_L$ for all $i \in Z$.

The category $\mathbf{MF}_L^{\varphi}$ of filtered Dieudonné modules is additive, has kernels and cokernels but is not abelian. Denote by $\mathbf{1}$ the vector space K_0 with the natural action of σ and the filtration given by

$$\mathrm{Fil}^i\mathbf{1} = \begin{cases} K, & \text{if } i \leqslant 0, \\ 0, & \text{if} i > 0. \end{cases}$$

Then $\mathbf{1}$ is a unit object of $\mathbf{MF}_L^{\varphi}$ i.e. $M \otimes \mathbf{1} \simeq \mathbf{1} \otimes M \simeq M$ for any M.

If M is a one dimensional Dieudonné module and d is a basis vector of M, then $\varphi(d) = \alpha d$ for some $\alpha \in K$. Set $t_N(M) = v_p(\alpha)$ and denote by $t_H(M)$ the unique filtration jump of M. If M is of an arbitrary finite dimension d, set $t_N(M) = t_N(\wedge^d M)$ and $t_H(M) = t_H(\wedge^d M)$. A Dieudonné module M is said to be weakly admissible if $t_H(M) = t_N(M)$ and if $t_H(M') \leqslant t_N(M')$ for any φ-submodule $M' \subset M$ equipped with the induced filtration. Weakly admissible modules form a subcategory of $\mathbf{MF}_L^{\varphi}$ which we denote by $\mathbf{MF}_L^{\varphi,f}$.

If V is a p-adic representation of G_L, define $\mathbf{D}_{\mathrm{dR}}(V) = (\mathbf{B}_{\mathrm{dR}} \otimes V)^{G_L}$. Then $\mathbf{D}_{\mathrm{dR}}(V)$ is a L-vector space equipped with the decreasing filtration $\mathrm{Fil}^i \mathbf{D}_{\mathrm{dR}}(V) = (\mathrm{Fil}^i \mathbf{B}_{\mathrm{dR}} \otimes V)^{G_L}$. One has $\dim_L \mathbf{D}_{\mathrm{dR}}(V) \leqslant \dim_{\mathbb{Q}_p}(V)$ and V is said to be de Rham if $\dim_L \mathbf{D}_{\mathrm{dR}}(V) = \dim_{\mathbb{Q}_p}(V)$. Analogously one defines $\mathbf{D}_{\mathrm{cris}}(V) = (\mathbf{B}_{\mathrm{cris}} \otimes V)^{G_L}$. Then $\mathbf{D}_{\mathrm{cris}}(V)$ is a filtered Dieudonné module over L of dimension $\dim_K \mathbf{D}_{\mathrm{cris}}(V) \leqslant \dim_{\mathbb{Q}_p}(V)$ and V is said to be crystalline if the equality holds here. In particular, for crystalline representations one has $\mathbf{D}_{\mathrm{dR}}(V) = \mathbf{D}_{\mathrm{cris}}(V) \otimes_K L$. By the theorem of Colmez–Fontaine (2000), the functor $\mathbf{D}_{\mathrm{cris}}$ establishes an equivalence between the category of crystalline representations of G_L and $\mathbf{MF}_L^{\varphi,f}$. Its quasi-inverse $\mathbf{V}_{\mathrm{cris}}$ is given by $\mathbf{V}_{\mathrm{cris}}(D) = \mathrm{Fil}^0(D \otimes_K \mathbf{B}_{\mathrm{cris}})^{\varphi=1}$.

An important result of Berger (2002, Theorem 0.2) says that $\mathbf{D}_{\mathrm{cris}}(V)$ can be recovered from the (φ, Γ)-module $\mathbf{D}^{\dagger}_{\mathrm{rig}}(V)$. The situation is particularly simple if $L/\mathbb{Q}_p$ is unramified. In this case set $\mathbf{D}^+(V) = (V \otimes_{\mathbb{Q}_p} \mathbf{B}^+)^{H_K}$ and $\mathbf{D}^+_{\mathrm{rig}}(V) = \mathcal{R}^+(K) \otimes_{\mathbf{B}^+_K} \mathbf{D}^+(V)$. Then

$$\mathbf{D}_{\mathrm{cris}}(V) = \left(\mathbf{D}^+_{\mathrm{rig}}(V)\left[\frac{1}{t}\right]\right)^{\Gamma}$$

(see Berger 2004, Proposition 3.4).

3 The Exponential Map

3.1 The Bloch–Kato Exponential Map

3.1.1 Cohomology of Dieudonné Modules

Let L be a finite extension of $\mathbb{Q}_p$ and K its maximal unramified subextension. Recall that we denote by $\mathbf{MF}_L^{\varphi}$ the category of filtered Dieudonné modules over L. If M is an object of $\mathbf{MF}_L^{\varphi}$, define

$$H^i(L, M) = \mathrm{Ext}^i_{\mathbf{MF}_L^{\varphi}}(\mathbf{1}, M), \qquad i = 0, 1.$$

We remark that $H^*(L, M)$ can be computed explicitly as the cohomology of the complex

$$C^{\bullet}(M) \ : \ M \xrightarrow{f} (M_L/\mathrm{Fil}^0 M_L) \oplus M$$

where the modules are placed in degrees 0 and 1 and $f(d) = (d \pmod{\mathrm{Fil}^0 M_L}, (1 - \varphi)(d))$ (see Fontaine and Perrin-Riou 1994; Nekovář 1993). Note that if M is weakly admissible then each extension $0 \to M \to M' \to \mathbf{1} \to 0$ is weakly admissible too and we can write $H^i(L, M) = \mathrm{Ext}^i_{\mathbf{MF}_L^{\varphi,f}}(\mathbf{1}, M)$.

3.1.2 The Exponential Map

Let $\mathbf{Rep}_{\mathrm{cris}}(G_L)$ denote the category of crystalline representations of G_L. For any object V of $\mathbf{Rep}_{\mathrm{cris}}(G_L)$ define

$$H^i_f(L,V)=\mathrm{Ext}^i_{\mathbf{Rep}_{\mathrm{cris}}(G_L)}(\mathbb{Q}_p(0),V).$$

An easy computation shows that

$$H^i_f(L,V)=\begin{cases} H^0(L,V) & \text{if } i=0,\\ \ker\left(H^1(L,V)\to H^1(L,V\otimes \mathbf{B}_{\mathrm{cris}})\right) & \text{if } i=1,\\ 0 & \text{if } i\geqslant 2.\end{cases}$$

Let $t_V(L)=\mathbf{D}_{\mathrm{dR}}(V)/\mathrm{Fil}^0\mathbf{D}_{\mathrm{dR}}(V)$ denote the tangent space of V. The rings $\mathbf{B}_{\mathrm{dR}}$ and $\mathbf{B}_{\mathrm{cris}}$ are related to each other via the fundamental exact sequence

$$0\longrightarrow \mathbb{Q}_p\longrightarrow \mathbf{B}_{\mathrm{cris}}\xrightarrow{f}\mathbf{B}_{\mathrm{dR}}/\mathrm{Fil}^0\mathbf{B}_{\mathrm{dR}}\oplus \mathbf{B}_{\mathrm{cris}}\longrightarrow 0$$

where $f(x)=(x \pmod{\mathrm{Fil}^0\mathbf{B}_{\mathrm{dR}}},(1-\varphi)\,x)$ (see Bloch and Kato 1990, §4). Tensoring this sequence with V and taking cohomology one obtains an exact sequence

$$0\longrightarrow H^0(L,V)\longrightarrow \mathbf{D}_{\mathrm{cris}}(V)\longrightarrow t_V(L)\oplus \mathbf{D}_{\mathrm{cris}}(V)\longrightarrow H^1_f(L,V)\longrightarrow 0.$$

The last map of this sequence gives rise to the Bloch–Kato exponential map

$$\exp_{V,L}\ :\ t_V(L)\oplus \mathbf{D}_{\mathrm{cris}}(V)\longrightarrow H^1(L,V).$$

Following (Fontaine 1992) set

$$\mathbf{R}\Gamma_f(L,V)\ =\ C^\bullet(\mathbf{D}_{\mathrm{cris}}(V))\ =\ \left[\mathbf{D}_{\mathrm{cris}}(V)\xrightarrow{f}t_V(L)\oplus \mathbf{D}_{\mathrm{cris}}(V)\right].$$

From the classification of crystalline representations in terms of Dieudonné modules it follows that the functor $\mathbf{V}_{\mathrm{cris}}$ induces natural isomorphisms

$$r^i_{V,p}\ :\ \mathbf{R}^i\Gamma_f(L,V)\longrightarrow H^i_f(L,V),\qquad i=0,1.$$

The composite homomorphism

$$t_V(L)\oplus \mathbf{D}_{\mathrm{cris}}(V)\longrightarrow \mathbf{R}^1\Gamma_f(L,V)\xrightarrow{r^1_{V,p}}H^1(L,V)$$

coincides with the Bloch–Kato exponential map $\exp_{V,L}$ (Nekovář 1993, Proposition 1.21).

3.1.3 The Map $\mathbf{R}\Gamma_f(L,V) \to \mathbf{R}\Gamma(L,V)$

Let $g : B^\bullet \to C^\bullet$ be a morphism of complexes. We denote by $\mathrm{Tot}^\bullet(g)$ the complex $\mathrm{Tot}^n(g) = C^{n-1} \oplus B^n$ with differentials $d^n : \mathrm{Tot}^n(g) \to \mathrm{Tot}^{n+1}(g)$ defined by the formula $d^n(c,b) = ((-1)^n g^n(b) + d^{n-1}(c), d^n(b))$. It is well known that if $0 \longrightarrow A^\bullet \xrightarrow{f} B^\bullet \xrightarrow{g} C^\bullet \longrightarrow 0$ is an exact sequence of complexes, then f induces a quasi isomorphism $A^\bullet \xrightarrow{\sim} \mathrm{Tot}^\bullet(g)$. In particular, tensoring the fundamental exact sequence with V, we obtain an exact sequence of complexes

$$0 \to \mathbf{R}\Gamma(L,V) \to C_c^\bullet(G_L, V \otimes \mathbf{B}_{\mathrm{cris}}) \xrightarrow{f} \\ C_c^\bullet(G_L, (V \otimes (\mathbf{B}_{\mathrm{dR}}/\mathrm{Fil}^0\mathbf{B}_{\mathrm{dR}})) \oplus (V \otimes \mathbf{B}_{\mathrm{cris}})) \to 0$$

which gives a quasi isomorphism $\mathbf{R}\Gamma(L,V) \xrightarrow{\sim} \mathrm{Tot}^\bullet(f)$. Since $\mathbf{R}\Gamma_f(L,V)$ coincides tautologically with the complex

$$C_c^0(G_L, V \otimes \mathbf{B}_{\mathrm{cris}}) \xrightarrow{f} C_c^0(G_L, (V \otimes (\mathbf{B}_{\mathrm{dR}}/\mathrm{Fil}^0\mathbf{B}_{\mathrm{dR}})) \oplus (V \otimes \mathbf{B}_{\mathrm{cris}}))$$

we obtain a diagram

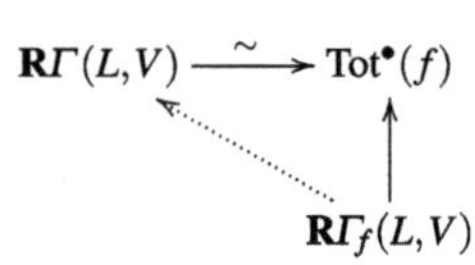

which defines a morphism $\mathbf{R}\Gamma_f(L,V) \to \mathbf{R}\Gamma(L,V)$ in $\mathcal{D}(\mathbb{Q}_p)$ (see Burns and Flach 1996, Sect. 1.2.1). We remark that the induced homomorphisms $\mathbf{R}^i\Gamma_f(L,V) \to H^i(L,V)$ $(i = 0,1)$ coincide with the composition of $r^i_{V,p}$ with natural embeddings $H^i_f(L,V) \to H^i(L,V)$.

3.1.4 Exponential Map for (φ, Γ)-Modules

In this subsection we define an analogue of the exponential map for crystalline (φ,Γ)-modules. See Nakamura (2014) for a more general setting. Let $K/\mathbb{Q}_p$ be an unramified extension. If $\mathbf{D}$ is a (φ,Γ)-module over $\mathcal{R}(K)$ define

$$\mathcal{D}_{\mathrm{cris}}(\mathbf{D}) = (\mathbf{D}[1/t])^\Gamma .$$

It can be shown that $\mathcal{D}_{\mathrm{cris}}(\mathbf{D})$ is a finite dimensional K-vector space equipped with a natural decreasing filtration $\mathrm{Fil}^i\mathcal{D}_{\mathrm{cris}}(\mathbf{D})$ and a semilinear action of φ. One says that D is crystalline if

$$\dim_K(\mathcal{D}_{\mathrm{cris}}(\mathbf{D})) = \mathrm{rank}(\mathbf{D}).$$

From Berger (2008), Théorème A it follows that the functor $\mathbf{D} \mapsto \mathcal{D}_{\mathrm{cris}}(\mathbf{D})$ is an equivalence between the category of crystalline (φ, Γ)-modules and $\mathbf{MF}_K^{\varphi}$. Note that if V is a p-adic representation of G_K then $\mathbf{D}_{\mathrm{cris}}(V) = \mathcal{D}_{\mathrm{cris}}(\mathbf{D}_{\mathrm{rig}}^{\dagger}(V))$ and V is crystalline if and only if $\mathbf{D}_{\mathrm{rig}}^{\dagger}(V)$ is.

Let $\mathbf{D}$ be a (φ, Γ)-module. To any cocycle $\alpha = (a, b) \in Z^1(C_{\varphi,\gamma}(\mathbf{D}))$ one can associate the extension

$$0 \longrightarrow \mathbf{D} \longrightarrow \mathbf{D}_{\alpha} \longrightarrow \mathcal{R}(K) \longrightarrow 0$$

defined by

$$\mathbf{D}_{\alpha} = D \oplus \mathcal{R}(K)\, e, \qquad (\varphi - 1)\, e = a, \quad (\gamma - 1)\, e = b.$$

As usual, this gives rise to an isomorphism $H^1(\mathbf{D}) \simeq \mathrm{Ext}^1_{\mathcal{R}}(\mathcal{R}(K), \mathbf{D})$. We say that $\mathrm{cl}(\alpha)$ is crystalline if $\dim_K(\mathcal{D}_{\mathrm{cris}}(\mathbf{D}_{\alpha})) = \dim_K(\mathcal{D}_{\mathrm{cris}}(\mathbf{D})) + 1$ and define

$$H^1_f(\mathbf{D}) = \{\mathrm{cl}(\alpha) \in H^1(\mathbf{D}) \mid \mathrm{cl}(\alpha) \text{ is crystalline}\}$$

(see Benois 2011, Sect. 1.4.1). If $\mathbf{D}$ is crystalline (or more generally potentially semistable) one has a natural isomorphism

$$H^1(K, \mathcal{D}_{\mathrm{cris}}(\mathbf{D})) \longrightarrow H^1_f(\mathbf{D}).$$

Set $t_{\mathbf{D}} = \mathcal{D}_{\mathrm{cris}}(\mathbf{D})/\mathrm{Fil}^0\mathcal{D}_{\mathrm{cris}}(\mathbf{D})$ and denote by

$$\exp_{\mathbf{D}} \; : \; t_{\mathbf{D}} \oplus \mathcal{D}_{\mathrm{cris}}(\mathbf{D}) \to H^1(\mathbf{D})$$

the composition of this isomorphism with the projection

$$t_{\mathbf{D}} \oplus \mathcal{D}_{\mathrm{cris}}(\mathbf{D}) \to H^1(K, \mathcal{D}_{\mathrm{cris}}(\mathbf{D}))$$

and the embedding $H^1_f(\mathbf{D}) \hookrightarrow H^1(\mathbf{D})$.

Assume that $K = \mathbb{Q}_p$. To simplify notation we will write $\mathbf{D}_m$ for $\mathcal{R}(|x|x^m)$ and e_m for its canonical basis. Then $\mathcal{D}_{\mathrm{cris}}(\mathbf{D}_m)$ is the one dimensional $\mathbb{Q}_p$-vector space generated by $t^{-m}e_m$. As in Benois (2011), we normalise the basis $(\mathrm{cl}(\alpha_m), \mathrm{cl}(\beta_m))$ of $H^1(\mathbf{D}_m)$ putting $\alpha_m^* = (1 - 1/p)\, \mathrm{cl}(\alpha_m)$ and $\beta_m^* = (1 - 1/p) \log(\chi(\gamma))\, \mathrm{cl}(\beta_m)$.

Proposition 5. *(i) $H^1_f(\mathbf{D}_m)$ is the one-dimensional $\mathbb{Q}_p$-vector space generated by α_m^*.*

(ii) The exponential map

$$\exp_{\mathbf{D}_m} \; : \; t_{\mathbf{D}_m} \longrightarrow H^1(\mathbf{D}_m)$$

sends $t^{-m} w_m$ to $-\alpha_m^*$.

Proof. This is a reformulation of Benois (2011), Proposition 1.5.8 (ii).

3.2 *The Large Exponential Map*

3.2.1 Notation

In this section p is an odd prime number, K is a finite unramified extension of $\mathbb{Q}_p$ and σ the absolute Frobenius acting on K. Recall that $K_n = K(\zeta_{p^n})$ and $K_\infty = \bigcup_{n=1}^{\infty} K_n$. We set $\Gamma = \mathrm{Gal}(K_\infty/K)$, $\Gamma_n = \mathrm{Gal}(K_\infty/K_n)$ and $\Delta = \mathrm{Gal}(K_1/K)$. Let $\Lambda = \mathbb{Z}_p[[\Gamma_1]]$ and $\Lambda(\Gamma) = \mathbb{Z}_p[\Delta] \otimes_{\mathbb{Z}_p} \Lambda$. We will consider the following operators acting on the ring $K[[X]]$ of formal power series with coefficients in K:

- The ring homomorphism $\sigma \, : \, K[[X]] \longrightarrow K[[X]]$ defined by

$$\sigma\left(\sum_{i=0}^{\infty} a_i X^i\right) = \sum_{i=0}^{\infty} \sigma(a_i) X^i.$$

- The ring homomorphism $\varphi \, : \, K[[X]] \longrightarrow K[[X]]$ defined by

$$\varphi\left(\sum_{i=0}^{\infty} a_i X^i\right) = \sum_{i=0}^{\infty} \sigma(a_i) \varphi(X)^i, \qquad \varphi(X) = (1+X)^p - 1.$$

- The differential operator $\partial = (1+X)\dfrac{d}{dX}$. One has $\partial \circ \varphi = p\varphi \circ \partial$.
- The operator $\psi \, : \, K[[X]] \longrightarrow K[[X]]$ defined by

$$\psi(f(X)) = \frac{1}{p}\varphi^{-1}\left(\sum_{\zeta^p=1} f((1+X)\zeta - 1)\right).$$

  It is easy to see that ψ is a left inverse to φ, i.e. that $\psi \circ \varphi = \mathrm{id}$.
- An action of Γ given by $\gamma\left(\sum_{i=0}^{\infty} a_i X^i\right) = \sum_{i=0}^{\infty} a_i \gamma(X)^i$, $\gamma(X) = (1+X)^{\chi(\gamma)} - 1$.

  We remark that these formulas are compatible with the definitions from Sects. 2.1.1 and 2.1.4. Fix a generator $\gamma_1 \in \Gamma_1$ and define

$$\mathcal{H} = \{f(\gamma_1 - 1) \mid f \in \mathbb{Q}_p[[X]] \text{ is holomorphic on } B(0,1)\},$$
$$\mathcal{H}(\Gamma) = \mathbb{Z}_p[\Delta] \otimes_{\mathbb{Z}_p} \mathcal{H}.$$

3.2.2 The Map $\Xi^{\varepsilon}_{V,n}$

It is well known that $\mathbb{Z}_p[[X]]^{\psi=0}$ is a free Λ-module generated by $(1+X)$ and the operator ∂ is bijective on $\mathbb{Z}_p[[X]]^{\psi=0}$. If V is a crystalline representation of G_K put $\mathcal{D}(V) = \mathbf{D}_{\mathrm{cris}}(V) \otimes_{\mathbb{Z}_p} \mathbb{Z}_p[[X]]^{\psi=0}$. Let $\Xi^{\varepsilon}_{V,n} : \mathcal{D}(V)_{\Gamma_n}[-1] \longrightarrow \mathbf{R}\Gamma_f(K_n, V)$ be the map defined by

$$\Xi^{\varepsilon}_{V,n}(\alpha) = \begin{cases} p^{-n}\left(\sum_{k=1}^{n} (\sigma \otimes \varphi)^{-k} \alpha(\zeta_{p^k} - 1),\ -\alpha(0)\right) & \text{if } n \geqslant 1, \\ \mathrm{Tr}_{K_1/K}\left(\Xi^{\varepsilon}_{V,1}(\alpha)\right) & \text{if } n = 0. \end{cases}$$

An easy computation shows that $\Xi^{\varepsilon}_{V,0} : \mathbf{D}_{\mathrm{cris}}(V)[-1] \to \mathbf{R}\Gamma_f(K,V)$ is given by the formula

$$\Xi^{\varepsilon}_{V,0}(a) = \frac{1}{p}\,(-\varphi^{-1}(a), -(p-1)\,a).$$

In particular, it is homotopic to the map $a \mapsto -(0, (1 - p^{-1}\varphi^{-1})\,a)$. Let

$$\Xi^{\varepsilon}_{V,n} : \mathcal{D}(V) \to \mathbf{R}^1\Gamma(K_n, V) = \frac{t_V(K_n) \oplus \mathbf{D}_{\mathrm{cris}}(V)}{\mathbf{D}_{\mathrm{cris}}(V)/V^{G_K}}$$

denote the homomorphism induced by $\Xi^{\varepsilon}_{V,n}$. Then

$$\Xi^{\varepsilon}_{V,0}(a) = -(0, (1 - p^{-1}\varphi^{-1})\,a) \pmod{\mathbf{D}_{\mathrm{cris}}(V)/V^{G_K}}.$$

If $\mathbf{D}_{\mathrm{cris}}(V)^{\varphi=1} = 0$ the operator $1-\varphi$ is invertible on $\mathbf{D}_{\mathrm{cris}}(V)$ and we can write

$$\Xi^{\varepsilon}_{V,0}(a) = \left(\frac{1 - p^{-1}\varphi^{-1}}{1-\varphi}\,a, 0\right) \pmod{\mathbf{D}_{\mathrm{cris}}(V)/V^{G_K}}. \qquad (3)$$

For any $i \in \mathbb{Z}$ let $\Delta_i : \mathcal{D}(V) \to \dfrac{\mathbf{D}_{\mathrm{cris}}(V)}{(1-p^i\varphi)\mathbf{D}_{\mathrm{cris}}(V)} \otimes \mathbb{Q}_p(i)$ be the map given by

$$\Delta_i(\alpha(X)) = \partial^i \alpha(0) \otimes \varepsilon^{\otimes i} \pmod{(1-p^i\varphi)\mathbf{D}_{\mathrm{cris}}(V)}.$$

Set $\Delta = \oplus_{i\in\mathbb{Z}}\Delta_i$. If $\alpha \in \mathcal{D}(V)^{\Delta=0}$, then by Perrin-Riou (1994), Proposition 2.2.1 there exists $F \in \mathbf{D}_{\mathrm{cris}}(V) \otimes_{\mathbb{Q}_p} \mathbb{Q}_p[[X]]$ which converges on the open unit disk and such that $(1-\varphi)F = \alpha$. A short computation shows that

$$\Xi^{\varepsilon}_{V,n}(\alpha) = p^{-n}((\sigma \otimes \varphi)^{-n}(F)(\zeta_{p^n}-1),0) \pmod{\mathbf{D}_{\mathrm{cris}}(V)/V^{G_K}}, \qquad \text{if } n \geqslant 1$$

(see Benois and Berger 2008, Lemme 4.9).

3.2.3 Construction of the Large Exponential Map

As $\mathbb{Z}_p[[X]][1/p]$ is a principal ideal domain and $\mathcal{H}$ is $\mathbb{Z}_p[[X]][1/p]$-torsion free, $\mathcal{H}$ is flat. Thus

$$\begin{aligned} C^{\dagger}_{\mathrm{Iw},\psi}(V) \otimes^{\mathbf{L}}_{\Lambda_{\mathbb{Q}_p}} \mathcal{H}(\Gamma) &= C^{\dagger}_{\mathrm{Iw},\psi}(V) \otimes_{\Lambda_{\mathbb{Q}_p}} \mathcal{H}(\Gamma) \\ &= \left[\mathcal{H}(\Gamma) \otimes_{\Lambda_{\mathbb{Q}_p}} \mathbf{D}^{\dagger}(V) \xrightarrow{\psi-1} \mathcal{H}(\Gamma) \otimes_{\Lambda_{\mathbb{Q}_p}} \mathbf{D}^{\dagger}(V) \right]. \end{aligned}$$

By Proposition 3 one has an isomorphism in $\mathcal{D}(\mathcal{H}(\Gamma))$

$$\mathbf{R}\Gamma_{\mathrm{Iw}}(K,V) \otimes^{\mathbf{L}}_{\Lambda_{\mathbb{Q}_p}} \mathcal{H}(\Gamma) \simeq C^{\dagger}_{\mathrm{Iw},\psi}(V) \otimes_{\Lambda_{\mathbb{Q}_p}} \mathcal{H}(\Gamma).$$

The action of $\mathcal{H}(\Gamma)$ on $\mathbf{D}^{\dagger}(V)^{\psi=1}$ induces an injection $\mathcal{H}(\Gamma) \otimes_{\Lambda_{\mathbb{Q}_p}} \mathbf{D}^{\dagger}(V)^{\psi=1} \hookrightarrow \mathbf{D}^{\dagger}_{\mathrm{rig}}(V)^{\psi=1}$. Composing this map with the canonical isomorphism $H^1_{\mathrm{Iw}}(K,V) \simeq \mathbf{D}^{\dagger}(V)^{\psi=1}$ we obtain a map $\mathcal{H}(\Gamma) \otimes_{\Lambda_{\mathbb{Q}_p}} H^1_{\mathrm{Iw}}(K,V) \hookrightarrow \mathbf{D}^{\dagger}_{\mathrm{rig}}(V)^{\psi=1}$. For any $k \in \mathbb{Z}$ set $\nabla_k = t\partial - k = t\dfrac{d}{dt} - k$. An easy induction shows that $\nabla_{k-1} \circ \nabla_{k-2} \circ \cdots \circ \nabla_0 = t^k \partial^k$.

Fix $h \geqslant 1$ such that $\mathrm{Fil}^{-h}\mathbf{D}_{\mathrm{cris}}(V) = \mathbf{D}_{\mathrm{cris}}(V)$ and $V(-h)^{G_K} = 0$. For any $\alpha \in \mathcal{D}(V)^{\Delta=0}$ define

$$\Omega^{\varepsilon}_{V,h}(\alpha) = (-1)^{h-1} \frac{\log \chi(\gamma_1)}{p} \nabla_{h-1} \circ \nabla_{h-2} \circ \cdots \nabla_0(F(\pi)),$$

where $F \in \mathcal{H}(V)$ is such that $(1-\varphi)F = \alpha$. It is easy to see that $\Omega^{\varepsilon}_{V,h}(\alpha) \in \mathbf{D}^{+}_{\mathrm{rig}}(V)^{\psi=1}$. Berger (2003) shows that $\Omega^{\varepsilon}_{V,h}(\alpha) \in \mathcal{H}(\Gamma) \otimes_{\Lambda_{\mathbb{Q}_p}} \mathbf{D}^{\dagger}(V)^{\psi=1}$ and therefore gives rise to a map

$$\mathbf{Exp}^{\varepsilon}_{V,h} : \mathcal{D}(V)^{\Delta=0}[-1] \to \mathbf{R}\Gamma_{\mathrm{Iw}}(K,V) \otimes^{\mathbf{L}}_{\Lambda_{\mathbb{Q}_p}} \mathcal{H}(\Gamma)$$

Let

$$\mathrm{Exp}^{\varepsilon}_{V,h} : \mathcal{D}(V)^{\Delta=0} \to \mathcal{H}(\Gamma) \otimes_{\Lambda_{\mathbb{Q}_p}} H^1_{\mathrm{Iw}}(K,V)$$

denote the map induced by $\mathbf{Exp}^{\varepsilon}_{V,h}$ in degree 1. The following theorem is a reformulation of the construction of the large exponential map given by Berger (2003).

Theorem 3. *Let*

$$\mathbf{Exp}^{\varepsilon}_{V,h,n} \;:\; \mathcal{D}(V)^{\Delta=0}_{\Gamma_n}[-1] \longrightarrow \mathbf{R}\Gamma_{\mathrm{Iw}}(K,V) \otimes^{\mathbf{L}}_{\Lambda_{\mathbb{Q}_p}} \mathbb{Q}_p[G_n].$$

denote the map induced by $\mathbf{Exp}^{\varepsilon}_{V,h}$*. Then for any* $n \geqslant 0$ *the following diagram in* $\mathcal{D}(\mathbb{Q}_p[G_n])$ *is commutative:*

$$\begin{array}{ccc}
\mathcal{D}(V)^{\Delta=0}_{\Gamma_n}[-1] & \xrightarrow{\mathbf{Exp}^{\varepsilon}_{V,h,n}} & \mathbf{R}\Gamma_{\mathrm{Iw}}(K,V) \otimes^{\mathbf{L}}_{\Lambda_{\mathbb{Q}_p}} \mathbb{Q}_p[G_n] \\
\Big\downarrow{\scriptstyle \Xi^{\varepsilon}_{V,n}} & & \Big\downarrow{\scriptstyle \simeq} \\
\mathbf{R}\Gamma_f(K_n,V) & \xrightarrow{(h-1)!} & \mathbf{R}\Gamma(K_n,V).
\end{array}$$

In particular, $\mathrm{Exp}^{\varepsilon}_{V,h}$ *coincides with the large exponential map of Perrin-Riou.*

Proof. Passing to cohomology in the previous diagram one obtains the diagram

$$\begin{array}{ccc}
\mathcal{D}(V)^{\Delta=0} & \xrightarrow{\mathrm{Exp}^{\varepsilon}_{V,h}} & \mathcal{H}(\Gamma) \otimes_{\Lambda_{\mathbb{Q}_p}} H^1_{\mathrm{Iw}}(K,V) \\
\Big\downarrow{\scriptstyle \Xi^{\varepsilon}_{V,n}} & & \Big\downarrow{\scriptstyle \mathrm{pr}_{V,n}} \\
\mathbf{D}_{\mathrm{dR}/K_n}(V) \oplus \mathbf{D}_{\mathrm{cris}}(V) & \xrightarrow{(h-1)!\,\exp_{V,K_n}} & H^1(K_n,V)
\end{array}$$

which is exactly the definition of the large exponential map. Its commutativity is proved in Berger (2003), Theorem II.13. Now, the theorem is an immediate consequence of the following remark. Let D be a free A-module and let $f_1, f_2 : D[-1] \to K^{\bullet}$ be two maps from $D[-1]$ to a complex of A-modules such that the induced maps $H^1(f_1)$ and $H^1(f_2) : D \to H^1(K^{\bullet})$ coincide. Then f_1 and f_2 are homotopic.

Remark. The large exponential map was first constructed in Perrin-Riou (1994). See Colmez (1998) and Benois (2000) for alternative constructions and Perrin-Riou (2001), Nakamura (2014) and Riedel (2013) for generalisations.

4 The $\mathscr{L}$-Invariant

4.1 Definition of the $\mathscr{L}$-Invariant

4.1.1 Preliminaries

Let S be a finite set of primes of $\mathbb{Q}$ containing p and G_S the Galois group of the maximal algebraic extension of $\mathbb{Q}$ unramified outside $S \cup \{\infty\}$. For each place v we denote by G_v the decomposition group at v, by $\mathbb{Q}_v^{\mathrm{ur}}$ the maximal unramified extension of $\mathbb{Q}_v$ and by I_v and f_v the inertia subgroup and Frobenius automorphism respectively. Let V be a pseudo-geometric p-adic representation of G_S. This means that the restriction of V on the decomposition group at p is a de Rham representation. Following Greenberg, for any $v \notin \{p,\infty\}$ set

$$\mathbf{R}\Gamma_f(\mathbb{Q}_v,V) = \left[V^{I_v} \xrightarrow{1-f_v} V^{I_v}\right],$$

where the terms are placed in degrees 0 and 1 (see Burns and Flach 1996; Fontaine 1992). We remark that there is a natural quasi-isomorphism $\mathbf{R}\Gamma_f(\mathbb{Q}_v,V) \simeq C_c^\bullet(G_v/I_v,V^{I_v})$. Note that $\mathbf{R}^0\Gamma_f(\mathbb{Q}_v,V) = H^0(\mathbb{Q}_v,V)$ and $\mathbf{R}^1\Gamma_f(\mathbb{Q}_v,V) = H^1_f(\mathbb{Q}_v,V)$ where

$$H^1_f(\mathbb{Q}_v,V) = \ker(H^1(\mathbb{Q}_v,V) \to H^1(\mathbb{Q}_v^{\mathrm{ur}},V)).$$

For $v = p$ the complex $\mathbf{R}\Gamma_f(\mathbb{Q}_v,V)$ was defined in Sect. 3.1.2. To simplify notation write $H^i_S(V) = H^i(G_S,V)$ for the continuous Galois cohomology of G_S with coefficients in V. The Bloch–Kato Selmer group of V is defined as

$$H^1_f(V) = \ker\left(H^1_S(V) \to \bigoplus_{v\in S} \frac{H^1(\mathbb{Q}_v,V)}{H^1_f(\mathbb{Q}_v,V)}\right).$$

We also set

$$H^1_{f,\{p\}}(V) = \ker\left(H^1_S(V) \to \bigoplus_{v\in S-\{p\}} \frac{H^1(\mathbb{Q}_v,V)}{H^1_f(\mathbb{Q}_v,V)}\right).$$

From the Poitou–Tate exact sequence one obtains the following exact sequence relating these groups (see for example Perrin-Riou 1995, Lemme 3.3.6)

$$0 \to H^1_f(V) \to H^1_{f,\{p\}}(V) \to \frac{H^1(\mathbb{Q}_p,V)}{H^1_f(\mathbb{Q}_p,V)} \to H^1_f(V^*(1))\ .$$

We also have the following formula relating dimensions of Selmer groups (see Fontaine and Perrin-Riou 1994, II, 2.2.2)

$$\dim_{\mathbb{Q}_p} H^1_f(V) - \dim_{\mathbb{Q}_p} H^1_f(V^*(1)) - \dim_{\mathbb{Q}_p} H^0_S(V) + \dim_{\mathbb{Q}_p} H^0_S(V^*(1)) = \dim_{\mathbb{Q}_p} t_V(\mathbb{Q}_p) - \dim_{\mathbb{Q}_p} H^0(\mathbb{R},V).$$

Set $d_\pm(V) = \dim_{\mathbb{Q}_p}(V^{c=\pm 1})$, where c denotes complex conjugation.

4.1.2 Basic Assumptions

Assume that V satisfies the following conditions

(C1) $H^1_f(V^*(1))=0.$
(C2) $H^0_S(V)=H^0_S(V^*(1))=0.$
(C3) V is crystalline at p and $\varphi : \mathbf{D}_{\mathrm{cris}}(V)\to\mathbf{D}_{\mathrm{cris}}(V)$ is semisimple at 1 and p^{-1}.
(C4) $\mathbf{D}_{\mathrm{cris}}(V)^{\varphi=1}=0.$
(C5) The localisation map

$$\mathrm{loc}_p\ :\ :\ H^1_f(V)\to H^1_f(\mathbb{Q}_p,V)$$

is injective.

These conditions appear naturally in the following situation. Let X be a proper smooth variety over $\mathbb{Q}$. Let $H^i_p(X)$ denote the p-adic étale cohomology of X. Consider the Galois representations $V=H^i_p(X)(m)$. By Poincaré duality together with the hard Lefschetz theorem we have

$$H^i_p(X)^*\simeq H^i_p(X)\,(i)$$

and thus $V^*(1)\simeq V(i+1-2m)$. The Beilinson conjecture (in the formulation of Bloch and Kato) predicts that

$$H^1_f(V^*(1))=0\qquad\text{if}\qquad w\leqslant -2.$$

This corresponds to the hope that there are no nontrivial extensions of $\mathbb{Q}(0)$ by motives of weight $\geqslant 0$. If X has a good reduction at p, then V is crystalline (Faltings 1989) and the semisimplicity of φ is a well known (and difficult) conjecture. By a result of Katz and Messing (1974) $\mathbf{D}_{\mathrm{cris}}(V)^{\varphi=1}\neq 0$ can occur only if $i=2m$. Therefore up to eventually replace V by $V^*(1)$ the conditions **(C1, C3–C4)** conjecturally hold except the weight -1 case $i=2m-1$.

The condition $\mathbf{D}_{\mathrm{cris}}(V)^{\varphi=1}=0$ implies that the exponential map $t_V(\mathbb{Q}_p)\to H^1_f(\mathbb{Q}_p,V)$ is an isomorphism and we denote by $\log_V$ its inverse. The composition of the localisation map loc_p with the Bloch–Kato logarithm

$$r_V\ :\ H^1_f(V)\to t_V(\mathbb{Q}_p)$$

coincides conjecturally with the p-adic (syntomic) regulator. We remark that if $H^0(\mathbb{Q}_v,V)=0$ for all $v\neq p$ (and therefore $H^1_f(\mathbb{Q}_v,V)=0$ for all $v\neq p$) then loc_p is injective for all $m\neq i/2, i/2+1$ by a result of Jannsen (1989, Lemma 4 and Theorem 3).

If $H^0_S(V)\neq 0$, then V contains a trivial subextension $V_0=\mathbb{Q}_p(0)^k$. For $\mathbb{Q}_p(0)$ our theory describes the behavior of the Kubota–Leopoldt p-adic L-function and is well known. Therefore we can assume that $H^0_S(V) = 0$. Applying the same argument to $V^*(1)$ we can also assume that $H^0_S(V^*(1)) = 0$.

From our assumptions we obtain an exact sequence

$$0 \to H^1_f(V) \to H^1_{f,\{p\}}(V) \to \frac{H^1(\mathbb{Q}_p, V)}{H^1_f(\mathbb{Q}_p, V)} \to 0. \tag{4}$$

Moreover

$$\dim_{\mathbb{Q}_p} H^1_f(V) = \dim_{\mathbb{Q}_p} t_V(\mathbb{Q}_p) - d_+(V), \tag{5}$$

$$\dim_{\mathbb{Q}_p} H^1_{f,\{p\}}(V) = d_-(V) + \dim_{\mathbb{Q}_p} H^0(\mathbb{Q}_p, V^*(1)). \tag{6}$$

4.1.3 Regular Submodules

In the remainder of this section we assume that V satisfies **(C1–C5)**.

Definition (Perrin-Riou).

(1) A φ-submodule D of $\mathbf{D}_{\mathrm{cris}}(V)$ is regular if $D \cap \mathrm{Fil}^0\mathbf{D}_{\mathrm{cris}}(V) = 0$ and the map

$$r_{V,D} : H^1_f(V) \to \mathbf{D}_{\mathrm{cris}}(V)/(\mathrm{Fil}^0\mathbf{D}_{\mathrm{cris}}(V) + D)$$

induced by r_V is an isomorphism.

(2) Dually, a φ-submodule D of $\mathbf{D}_{\mathrm{cris}}(V^*(1))$ is regular if

$$D + \mathrm{Fil}^0\mathbf{D}_{\mathrm{cris}}(V^*(1)) = \mathbf{D}_{\mathrm{cris}}(V^*(1))$$

and the map

$$D \cap \mathrm{Fil}^0\mathbf{D}_{\mathrm{cris}}(V^*(1)) \to H^1(V)^*$$

induced by the dual map $r_V^* : \mathrm{Fil}^0\mathbf{D}_{\mathrm{cris}}(V^*(1)) \to H^1(V)^*$ is an isomorphism.

It is easy to see that if D is a regular submodule of $\mathbf{D}_{\mathrm{cris}}(V)$, then

$$D^\perp = \mathrm{Hom}(\mathbf{D}_{\mathrm{cris}}(V)/D, \mathbf{D}_{\mathrm{cris}}(\mathbb{Q}_p(1)))$$

is a regular submodule of $\mathbf{D}_{\mathrm{cris}}(V^*(1))$. From (5) we also obtain that

$$\dim D = d_+(V), \qquad \dim D^\perp = d_-(V) = d_+(V^*(1)).$$

Let $D \subset \mathbf{D}_{\mathrm{cris}}(V)$ be a regular submodule. As in Benois (2011) we use the semisimplicity of φ to decompose D into the direct sum

$$D = D_{-1} \oplus D^{\varphi=p^{-1}}.$$

which gives a four step filtration

$$\{0\} \subset D_{-1} \subset D \subset \mathbf{D}_{\mathrm{cris}}(V).$$

Let $\mathbf{D}_0 = \mathbf{D}$ and $\mathbf{D}_{-1}$ denote the (φ, Γ)-submodules associated to D and D_{-1} by Berger's theory, thus

$$D = \mathscr{D}_{\mathrm{cris}}(\mathbf{D}), \qquad D_{-1} = \mathscr{D}_{\mathrm{cris}}(\mathbf{D}_{-1}).$$

Set $W = \mathrm{gr}_0 \mathbf{D}^{\dagger}_{\mathrm{rig}}(V)$. Thus we have two tautological exact sequences

$$0 \to \mathbf{D} \to \mathbf{D}^{\dagger}_{\mathrm{rig}}(V) \to \mathbf{D}' \to 0,$$
$$0 \to \mathbf{D}_{-1} \to \mathbf{D} \to W \to 0.$$

Note the following properties of cohomology of these modules:

(a) The natural maps $H^1(\mathbf{D}_{-1}) \to H^1(\mathbf{D})$ and $H^1(\mathbf{D}) \to H^1(\mathbf{D}^{\dagger}_{\mathrm{rig}}(V)) = H^1(\mathbb{Q}_p, V)$ are injective. This follows from the observation that $\mathscr{D}_{\mathrm{cris}}(\mathbf{D}')^{\varphi=1} = 0$ by **(C4)**. Since $H^0(\mathbf{D}') = \mathrm{Fil}^0 \mathscr{D}_{\mathrm{cris}}(\mathbf{D}')^{\varphi=1}$ (Benois 2011, Proposition 1.4.4) we have $H^0(\mathbf{D}') = 0$. The same argument works for W.

(b) $H^1_f(\mathbf{D}_{-1}) = H^1(\mathbf{D}_{-1})$. In particular the exponential map $\exp_{\mathbf{D}_{-1}} : D_{-1} \to H^1(\mathbf{D}_{-1})$ is an isomorphism. This follows from the computation of dimensions of $H^1(\mathbf{D}_{-1})$ and $H^1_f(\mathbf{D}_{-1})$. Namely, since $D_{-1}^{\varphi=1} = D_{-1}^{\varphi=p^{-1}} = \{0\}$ the Euler–Poincaré characteristic formula (Liu 2007) together with Poincaré duality give

$$\begin{aligned}&\dim_{\mathbb{Q}_p} H^1(\mathbf{D}_{-1})\\ &\quad = \mathrm{rank}(\mathbf{D}_{-1}) - \dim_{\mathbb{Q}_p} H^0(\mathbf{D}_{-1}) - \dim_{\mathbb{Q}_p} H^0(\mathbf{D}^*_{-1}(\chi)) = \dim_{\mathbb{Q}_p}(D_{-1}).\end{aligned}$$

On the other hand since

$$\mathrm{Fil}^0 D_{-1} = D_{-1} \cap \mathrm{Fil}^0 \mathbf{D}_{\mathrm{cris}}(V) = \{0\}$$

one has $\dim_{\mathbb{Q}_p} H^1_f(\mathbf{D}_{-1}) = \dim_{\mathbb{Q}_p}(D_{-1})$ by Benois (2011), Corollary 1.4.5.

(c) The exponential map $\exp_{\mathbf{D}} : D \to H^1_f(\mathbf{D})$ is an isomorphism. This follows from $\mathrm{Fil}^0 D = \{0\}$ and $D^{\varphi=1} = \{0\}$.

The regularity of D is equivalent to the decomposition

$$H^1_f(\mathbb{Q}_p, V) = H^1_f(V) \oplus H^1_f(\mathbf{D}). \tag{7}$$

Since loc_p is injective by **(C5)**, the localisation map $H^1_{f,\{p\}}(V) \to H^1(\mathbb{Q}_p, V)$ is also injective. Let

$$\kappa_D \ :\ H^1_{f,\{p\}}(V) \to \frac{H^1(\mathbb{Q}_p, V)}{H^1_f(\mathbf{D})}$$

denote the composition of this map with the canonical projection.

Lemma 1. *(i) One has*

$$H^1_f(\mathbb{Q}_p, V) \cap H^1(\mathbf{D}) = H^1_f(\mathbf{D}).$$

(ii) κ_D *is an isomorphism.*

Proof. (i) Since $H^0(\mathbf{D}') = 0$ we have a commutative diagram with exact rows and injective colomns

$$\begin{array}{ccccccc} 0 & \longrightarrow & H^1_f(\mathbf{D}) & \longrightarrow & H^1_f(\mathbb{Q}_p, V) & \longrightarrow & H^1_f(\mathbf{D}') \\ & & \downarrow & & \downarrow & & \downarrow \\ 0 & \longrightarrow & H^1(\mathbf{D}) & \longrightarrow & H^1(\mathbb{Q}_p, V) & \longrightarrow & H^1(\mathbf{D}'). \end{array}$$

This gives (i).

(ii) Since $H^1_f(\mathbf{D}) \subset H^1_f(\mathbb{Q}_p, V)$ one has $\ker(\kappa_D) \subset H^1_f(\mathbb{Q}_p, V)$. One the other hand (7) shows that κ_D is injective on $H^1_f(V)$. Thus $\ker(\kappa_D) = \{0\}$. On the other hand, because $\dim_{\mathbb{Q}_p} H^1_f(\mathbf{D}) = \dim_{\mathbb{Q}_p}(D)$ we have

$$\dim_{\mathbb{Q}_p}\left(\frac{H^1(\mathbb{Q}_p, V)}{H^1_f(\mathbf{D})}\right) = d_-(V) + \dim_{\mathbb{Q}_p} H^0(\mathbb{Q}_p, V^*(1)).$$

Comparing this with (5–6) we obtain that κ_D is an isomorphism.

4.1.4 The Main Construction

Set $e = \dim_{\mathbb{Q}_p}(D^{\varphi=p^{-1}})$. The (φ, Γ)-module W satisfies

$$\mathrm{Fil}^0 \mathscr{D}_{\mathrm{cris}}(W) = 0, \qquad \mathscr{D}_{\mathrm{cris}}(W)^{\varphi=p^{-1}} = \mathscr{D}_{\mathrm{cris}}(W).$$

(Recall that $\mathscr{D}_{\mathrm{cris}}(W) = D^{\varphi=p^{-1}}$.) The cohomology of such modules was studied in detail in Benois (2011), Proposition 1.5.9 and Sect. 1.5.10. Namely, $H^0(W) = 0$, $\dim_{\mathbb{Q}_p} H^1(W) = 2e$ and $\dim_{\mathbb{Q}_p}(W) = e$. There exists a canonical decomposition

$$H^1(W) = H^1_f(W) \oplus H^1_c(W)$$

of $H^1(W)$ into the direct sum of $H^1_f(W)$ and some canonical space $H^1_c(W)$. Moreover there exist canonical isomorphisms

$$i_{D,f} \;:\; \mathscr{D}_{\mathrm{cris}}(W) \simeq H^1_f(W), \qquad i_{D,c} \;:\; \mathscr{D}_{\mathrm{cris}}(W) \simeq H^1_c(W).$$

These isomorphisms can be described explicitly. By Proposition 1.5.9 of Benois (2011)

$$W \simeq \overset{e}{\underset{i=1}{\oplus}} \mathbf{D}_{m_i},$$

where $\mathbf{D}_{m_i} = \mathscr{R}(|x|x^{m_i})$, $m_i \geqslant 1$. By Proposition 5 $H^1_f(\mathbf{D}_m)$ is generated by α^*_m and $H^1_c(\mathbf{D}_m)$ is the subspace generated by β^*_m (see also Proposition 4). Then

$$i_{D,f}(x) = x\alpha^*_m, \qquad i_{D,c}(x) = x\beta^*_m.$$

Since $H^0(W) = 0$ and $H^2(\mathbf{D}_{-1}) = 0$ we have exact sequences

$$0 \to H^1(\mathbf{D}_{-1}) \to H^1(\mathbf{D}) \to H^1(W) \to 0,$$
$$0 \to H^1_f(\mathbf{D}_{-1}) \to H^1_f(\mathbf{D}) \to H^1_f(W) \to 0.$$

Since $H^1_f(\mathbf{D}_{-1}) = H^1(\mathbf{D})$ we obtain that

$$\frac{H^1(\mathbf{D})}{H^1_f(\mathbf{D})} \simeq \frac{H^1(W)}{H^1_f(W)}.$$

Let $H^1(D,V)$ denote the inverse image of $H^1(\mathbf{D})/H^1_f(\mathbf{D})$ by κ_D. Then κ_D induces an isomorphism

$$H^1(D,V) \simeq \frac{H^1(\mathbf{D})}{H^1_f(\mathbf{D})}.$$

By Lemma 1 the localisation map $H^1(D,V) \to H^1(W)$ is well defined and injective. Hence, we have a diagram

$$\begin{array}{ccc}
\mathscr{D}_{\mathrm{cris}}(W) & \xrightarrow[\sim]{i_{D,f}} & H^1_f(W) \\
\uparrow{\scriptstyle p_{D,f}} & & \uparrow{\scriptstyle p_{D,f}} \\
H^1(D,V) & \longrightarrow & H^1(W) \\
\downarrow{\scriptstyle p_{D,c}} & & \downarrow{\scriptstyle p_{D,c}} \\
\mathscr{D}_{\mathrm{cris}}(W) & \xrightarrow[\sim]{i_{D,c}} & H^1_c(W),
\end{array}$$

where $\rho_{D,f}$ and $\rho_{D,c}$ are defined as the unique maps making this diagram commute. From Lemma 1 (ii) it follows that $\rho_{D,c}$ is an isomorphism. The following definition generalises (in the crystalline case) the main construction of Benois (2011) where we assumed in addition that $H^1_f(V) = 0$.

Definition. *The determinant*

$$\mathscr{L}(V, D) = \det\left(\rho_{D,f} \circ \rho_{D,c}^{-1} \mid \mathscr{D}_{\mathrm{cris}}(W)\right)$$

will be called the $\mathscr{L}$-invariant associated to V and D.

4.2 $\mathscr{L}$-Invariant and the Large Exponential Map

4.2.1 Differentiation of the Large Exponential Map

In this section we interpret $\mathscr{L}(V, D)$ in terms of the derivative of the large exponential map. This interpretation is crucial for the proof of the main theorem of this paper. Recall that $H^1(\mathbb{Q}_p, \mathcal{H}(\Gamma) \otimes_{\mathbb{Q}_p} V) = \mathcal{H}(\Gamma) \otimes_{\Lambda(\Gamma)} H^1_{\mathrm{Iw}}(\mathbb{Q}_p, V)$ injects into $\mathbf{D}^{\dagger}_{\mathrm{rig}}(V)$. Set

$$F_0 H^1(\mathbb{Q}_p, \mathcal{H}(\Gamma) \otimes_{\mathbb{Q}_p} V) = \mathbf{D} \cap H^1(\mathbb{Q}_p, \mathcal{H}(\Gamma) \otimes_{\mathbb{Q}_p} V),$$
$$F_{-1} H^1(\mathbb{Q}_p, \mathcal{H}(\Gamma) \otimes_{\mathbb{Q}_p} V) = \mathbf{D}_{-1} \cap H^1(\mathbb{Q}_p, \mathcal{H}(\Gamma) \otimes_{\mathbb{Q}_p} V).$$

As in Sect. 3.2 we fix a generator $\gamma \in \Gamma$. The following result is a straightforward generalisation of Benois (2014), Prop. 2.2.2. For the convenience of the reader we give here the proof which is the same as in op. cit. modulo obvious modifications.

Proposition 6. *Let D be a regular subspace of $\mathbf{D}_{\mathrm{cris}}(V)$. For any $a \in D^{\varphi=p^{-1}}$ let $\alpha \in \mathcal{D}(V)$ be such that $\alpha(0) = a$. Then*

(i) There exists a unique $\beta \in F_0 H^1(\mathbb{Q}_p, \mathcal{H}(\Gamma) \otimes V)$ such that

$$(\gamma - 1)\,\beta = \mathrm{Exp}^{\varepsilon}_{V,h}(\alpha).$$

(ii) The composite map

$$\delta_{D,h} : D^{\varphi=p^{-1}} \to F_0 H^1(\mathbb{Q}_p, \mathcal{H}(\Gamma) \otimes V) \to H^1(W)$$
$$\delta_{D,h}(a) = \beta \pmod{H^1(\mathbf{D}_{-1})}$$

is given explicitly by the following formula:

$$\delta_{D,h}(a) = -(h-1)! \left(1 - \frac{1}{p}\right)^{-1} (\log \chi(\gamma))^{-1}\, i_{D,c}(\alpha).$$

Proof. Since $\mathbf{D}_{\mathrm{cris}}(V)^{\varphi=1} = 0$, the operator $1-\varphi$ is invertible on $\mathbf{D}_{\mathrm{cris}}(V)$ and we have a diagram

$$\begin{array}{ccc} \mathcal{D}(V)^{\Delta=0} & \xrightarrow{\mathrm{Exp}^{\varepsilon}_{V,h}} & H^1(\mathbb{Q}_p, \mathcal{H}(\Gamma)\otimes V) \\ \big\downarrow{\scriptstyle \Xi^{\varepsilon}_{V,0}} & & \big\downarrow{\scriptstyle \mathrm{pr}_V} \\ \mathbf{D}_{\mathrm{cris}}(V) & \xrightarrow{(h-1)!\exp_V} & H^1(\mathbb{Q}_p, V). \end{array}$$

where $\Xi^{\varepsilon}_{V,0}(\alpha) = \dfrac{1-p^{-1}\varphi^{-1}}{1-\varphi}\,\alpha(0)$ (see (3)). If $\alpha \in D^{\varphi=p^{-1}} \otimes \mathbb{Z}_p[[X]]^{\psi=0}$, then $\Xi^{\varepsilon}_{V,0}(\alpha) = 0$ and $\mathrm{pr}_V\left(\mathrm{Exp}^{\varepsilon}_{V,h}(\alpha)\right) = 0$. On the other hand, as $V^{G_K} = 0$ the map $\left(\mathcal{H}(\Gamma)\otimes_{\Lambda_{\mathbb{Q}_p}} H^1_{\mathrm{Iw}}(\mathbb{Q}_p, V)\right)_{\Gamma} \to H^1(\mathbb{Q}_p, V)$ is injective. Thus there exists a unique $\beta \in \mathcal{H}(\Gamma)\otimes_{\Lambda} H^1_{\mathrm{Iw}}(\mathbb{Q}_p, T)$ such that $\mathrm{Exp}^{\varepsilon}_{V,h}(\alpha) = (\gamma-1)\,\beta$. Now take $a \in D^{\varphi=p^{-1}}$ and set

$$f = a\otimes \ell\left(\frac{(1+X)^{\chi(\gamma)}-1}{X}\right),$$

where $\ell(g) = \dfrac{1}{p}\log\left(\dfrac{g^p}{\varphi(g)}\right)$. An easy computation shows that

$$\sum_{\zeta^p=1} \ell\left(\frac{\zeta^{\chi(\gamma)}(1+X)^{\chi(\gamma)}-1}{\zeta(1+X)-1}\right) = 0.$$

Thus $f \in D^{\varphi=p^{-1}} \otimes \mathbb{Z}_p[[X]]^{\psi=0}$. Write α in the form $\alpha = (1-\varphi)\,(1-\gamma)\,(a\otimes \log(X))$. Then

$$\Omega_{V,h}(\alpha) = (-1)^{h-1}\frac{\log\chi(\gamma_1)}{p}\,t^h\partial^h((\gamma-1)\,(a\log(\pi))) = \frac{\log\chi(\gamma_1)}{p}\,(\gamma-1)\,\beta$$

where

$$\beta = (-1)^{h-1}t^h\partial^h(a\log(\pi)) = (-1)^{h-1}at^h\partial^{h-1}\left(\frac{1+\pi}{\pi}\right).$$

This implies immediately that $\beta \in \mathbf{D}$. On the other hand $D^{\varphi=p^{-1}} = \mathcal{D}_{\mathrm{cris}}(W) = (W[1/t])^{\Gamma}$ and we will write $\tilde{a}$ for the image of a in $W\,[1/t]$. By Benois (2011), Sect. 1.5.8–1.5.10 one has $W \simeq \bigoplus_{i=1}^{e} \mathbf{D}_{m_i}$ where $\mathbf{D}_m = \mathscr{R}(|x|x^m)$ and we denote by e_m the canonical base of $\mathbf{D}_m$. Then without loss of generality we may assume that $\tilde{a} = t^{-m_i}e_{m_i}$ for some i. Let $\tilde{\beta}$ be denote the image of β in $W^{\psi=1}$ and let $h^1_0 : W^{\psi=1} \to H^1(W)$ be the canonical map furnished by Proposition 3. Recall that $h^1_0(\tilde{\beta}) = \mathrm{cl}(c,\tilde{\beta})$ where $(1-\gamma)\,c = (1-\varphi)\,\tilde{\beta}$. Then $\tilde{\beta} = (-1)^{h-1}t^{h-m_i}\partial^h\log(\pi)$.

By Lemma 1.5.1 of Cherbonnier and Colmez (1998) there exists a unique $b_0 \in \mathbf{B}_{\mathbb{Q}_p}^{\dagger,\psi=0}$ such that $(\gamma-1)\,b_0 = \ell(\pi)$. This implies that

$$(1-\gamma)\,(t^{h-m_i}\,\partial^h b_0 e_{m_i}) \;=\; (1-\varphi)\,(t^{h-m_i}\,\partial^h \log(\pi) e_{m_i}) \;=\; (-1)^{h-1}(1-\varphi)\,\tilde{\beta}.$$

Thus $c = (-1)^{h-1}t^{h-m_i}\,\partial^h b_0 e_{m_i}$ and $\mathrm{res}(ct^{m_i-1}dt) = (-1)^{h-1}\mathrm{res}(t^{h-1}\partial^h b_0 dt)\, e_{m_i} = 0$. Next from the congruence $\tilde{\beta} \equiv (h-1)!\,t^{-m_i}e_{m_i} \pmod{\mathbb{Q}_p[[\pi]]\,e_{m_i}}$ it follows that $\mathrm{res}(\tilde{\beta}t^{m_i-1}dt) = (h-1)!\,e_{m_i}$. Therefore by Benois (2011), Corollary 1.5.6 we have

$$\mathrm{cl}(c,\tilde{\beta}) \;=\; (h-1)!\,\mathrm{cl}(\beta_m) \;=\; (h-1)!\frac{p}{\log\chi(\gamma_1)}\,i_{W,c}(a).$$

On the other hand

$$\alpha(0) \;=\; a\otimes\ell\left(\frac{(1+X)^{\chi(\gamma)}-1}{X}\right)\bigg|_{X=0} \;=\; a\left(1-\frac{1}{p}\right)\log(\chi(\gamma)).$$

These formulas imply that

$$\delta_{D,h}(a) = (h-1)!\left(1-\frac{1}{p}\right)^{-1}(\log\chi(\gamma))^{-1}\,i_{W,c}(a).$$

and the proposition is proved.

4.2.2 Interpretation of the $\mathscr{L}$-Invariant

From the definition of $H^1(D,V)$ and Lemma 1 we immediately obtain that

$$\frac{H^1(\mathbb{Q}_p,V)}{H^1_{f,\{p\}}(V)+H^1(\mathbf{D}_{-1})} \simeq \frac{H^1(\mathbf{D})}{H^1(D,V)+H^1(\mathbf{D}_{-1})} \simeq \frac{H^1(W)}{H^1(D,V)}.$$

Thus, the map $\delta_{D,h}$ constructed in Proposition 6 induces a map

$$D^{\varphi=p^{-1}} \to \frac{H^1(\mathbb{Q}_p,V)}{H^1_{f,\{p\}}(V)+H^1(\mathbf{D}_{-1})}$$

which we will denote again by $\delta_{D,h}$. On the other hand, we have isomorphisms

$$D^{\varphi=p^{-1}} \xrightarrow{\exp_V} \frac{H^1_f(\mathbb{Q}_p,V)}{\exp_{V,\mathbb{Q}_p}(D_{-1})} \simeq \frac{H^1_f(\mathbb{Q}_p,V)}{H^1(\mathbf{D}_{-1})} \simeq \frac{H^1(\mathbb{Q}_p,V)}{H^1_{f,\{p\}}(V)+H^1(\mathbf{D}_{-1})}.$$

Proposition 7. *Let $\lambda_D : D^{\varphi=p^{-1}} \to D^{\varphi=p^{-1}}$ denote the homomorphism making the diagram*

$$\begin{array}{ccc} D^{\varphi=p^{-1}} & \xrightarrow{\quad\lambda_D\quad} & D^{\varphi=p^{-1}} \\ & {\scriptstyle\delta_{D,h}}\searrow \quad \swarrow{\scriptstyle (h-1)!\exp_V} & \\ & \dfrac{H^1(\mathbb{Q}_p,V)}{H^1_{f,\{p\}}(V)+H^1(\mathbf{D}_{-1})} & \end{array}$$

commute. Then

$$\det\left(\lambda_D \mid D^{\varphi=p^{-1}}\right) = (\log\chi(\gamma))^{-e}\left(1-\frac{1}{p}\right)^{-e}\mathscr{L}(V,D).$$

Proof. The proposition follows from Proposition 6 and the following elementary fact. Let $U = U_1 \oplus U_2$ be the decomposition of a vector space U of dimension $2e$ into the direct sum of two subspaces of dimension e. Let $X \subset U$ be a subspace of dimension e such that $X \cap U_1 = \{0\}$. Consider the diagrams

$$\begin{array}{ccc} X & \xrightarrow{p_1} & U_1 \\ {\scriptstyle p_2}\downarrow & \nearrow{\scriptstyle f} & \\ U_2 & & \end{array} \qquad\qquad \begin{array}{ccc} U/X & \xleftarrow{i_1} & U_1 \\ {\scriptstyle i_2}\uparrow & \nearrow{\scriptstyle g} & \\ U_2 & & \end{array}$$

where p_k and i_k are induced by natural projections and inclusions. Then $f = -g$. Applying this remark to $U = H^1(W)$, $X = H^1(D,V)$, $U_1 = H^1_f(W)$, $U_2 = H^1_c(W)$ and taking determinants we obtain the proposition.

5 Special Values of p-Adic L-Functions

5.1 *The Bloch–Kato Conjecture*

5.1.1 The Euler–Poincaré Line

The main references for the material of this section are (Burns and Flach 1996; Fontaine 1992; Fontaine and Perrin-Riou 1994). Let V be a p-adic pseudo-geometric representation of $\mathrm{Gal}(\overline{\mathbb{Q}}/\mathbb{Q})$. Thus V is a finite-dimensional $\mathbb{Q}_p$-vector space equipped with a continuous action of the Galois group G_S for a suitable finite set of places S containing p. Write $\mathbf{R}\Gamma_S(V) = C^\bullet_c(G_S, V)$ and define

$$\mathbf{R}\Gamma_{S,c}(V) = \text{cone}\left(\mathbf{R}\Gamma_S(V) \to \bigoplus_{v\in S\cup\{\infty\}} \mathbf{R}\Gamma(\mathbb{Q}_v, V)\right)[-1].$$

Fix a $\mathbb{Z}_p$-lattice T of V stable under the action of G_S and set

$$\Delta_S(V) = \det{}_{\mathbb{Q}_p}^{-1}\mathbf{R}\Gamma_{S,c}(V), \qquad \Delta_S(T) = \det{}_{\mathbb{Z}_p}^{-1}\mathbf{R}\Gamma_{S,c}(T).$$

Then $\Delta_S(T)$ is a $\mathbb{Z}_p$-lattice of the one-dimensional $\mathbb{Q}_p$-vector space $\Delta_S(V)$ which does not depend on the choice of T. Therefore it defines a p-adic norm on $\Delta_S(V)$ which we denote by $\|\cdot\|_S$. Moreover, $(\Delta_S(V), \|\cdot\|_S)$ does not depend on the choice of S. More precisely, if Σ is a finite set of places which contains S, then there exists a natural isomorphism $\Delta_S(V) \xrightarrow{\sim} \Delta_\Sigma(V)$ such that $\|\cdot\|_\Sigma = \|\cdot\|_S$. This allows one to define the Euler–Poincaré line $\Delta_{\mathrm{EP}}(V)$ as $(\Delta_S(V), \|\cdot\|_S)$ where S is sufficiently large. Recall that for any finite place $v \in S$ we defined

$$\mathbf{R}\Gamma_f(\mathbb{Q}_v, V) = \begin{cases} \left[V^{I_v} \xrightarrow{1-f_v} V^{I_v}\right] & \text{if } v \neq p \\ \left[\mathbf{D}_{\mathrm{cris}}(V) \xrightarrow{(\mathrm{pr},1-\varphi)} t_V(\mathbb{Q}_p) \oplus \mathbf{D}_{\mathrm{cris}}(V)\right] & \text{if } v = p. \end{cases}$$

At $v = \infty$ we set $\mathbf{R}\Gamma_f(R, V) = \left[V^+ \to 0\right]$, where the first term is placed in degree 0. Thus $\mathbf{R}\Gamma_f(R, V) \xrightarrow{\sim} \mathbf{R}\Gamma(R, V)$. For any v we have a canonical morphism $\mathrm{loc}_v : \mathbf{R}\Gamma_f(\mathbb{Q}_v, V) \to \mathbf{R}\Gamma(\mathbb{Q}_v, V)$ which can be viewed as a local condition in the sense of Nekovář (2006). Consider the diagram

$$\begin{array}{ccc} \mathbf{R}\Gamma_S(V) & \longrightarrow & \bigoplus_{v\in S\cup\{\infty\}} \mathbf{R}\Gamma(\mathbb{Q}_v, V) \\ & & \uparrow \\ & & \bigoplus_{v\in S\cup\{\infty\}} \mathbf{R}\Gamma_f(\mathbb{Q}_v, V) \end{array}$$

and define

$$\mathbf{R}\Gamma_f(V) = \text{cone}\left(\mathbf{R}\Gamma_S(V) \bigoplus \left(\bigoplus_{v\in S\cup\{\infty\}} \mathbf{R}\Gamma_f(\mathbb{Q}_v, V)\right) \to \bigoplus_{v\in S\cup\{\infty\}} \mathbf{R}\Gamma(\mathbb{Q}_v, V)\right)[-1].$$

Thus, we have a distinguished triangle

$$\mathbf{R}\Gamma_f(V) \to \mathbf{R}\Gamma_S(V) \bigoplus \left(\bigoplus_{v\in S\cup\{\infty\}} \mathbf{R}\Gamma_f(\mathbb{Q}_v, V)\right) \to \bigoplus_{v\in S\cup\{\infty\}} \mathbf{R}\Gamma(\mathbb{Q}_v, V). \quad (8)$$

Set

$$\Delta_f(V) = \det_{\mathbb{Q}_p}^{-1} \mathbf{R}\Gamma_f(V) \otimes \det_{\mathbb{Q}_p}^{-1} t_V(\mathbb{Q}_p) \otimes \det_{\mathbb{Q}_p} V^+.$$

It is easy to see that $\mathbf{R}\Gamma_f(V)$ and $\Delta_f(V)$ do not depend on the choice of S. Consider the distinguished triangle

$$\mathbf{R}\Gamma_{S,c}(V) \to \mathbf{R}\Gamma_f(V) \to \bigoplus_{v\in S\cup\{\infty\}} \mathbf{R}\Gamma_f(\mathbb{Q}_v, V).$$

The identity map id $: \mathbf{D}_{\mathrm{cris}}(V) \to \mathbf{D}_{\mathrm{cris}}(V)$ induces an isomorphism

$$\det_{\mathbb{Q}_p} \mathbf{R}\Gamma_f(\mathbb{Q}_p, V) \simeq \det_{\mathbb{Q}_p}^{-1} t_V(\mathbb{Q}_p). \tag{9}$$

For $v \neq p$ the identity map id $: V^{I_v} \to V^{I_v}$ gives a trivialisation

$$\det_{\mathbb{Q}_p} \mathbf{R}\Gamma_f(\mathbb{Q}_v, V) \simeq \mathbb{Q}_p. \tag{10}$$

Since $\det_{\mathbb{Q}_p} \mathbf{R}\Gamma_f(R, V) = \det_{\mathbb{Q}_p} V^+$ tautologically, we obtain canonical isomorphisms

$$\Delta_f(V) \simeq \det_{\mathbb{Q}_p}^{-1} \mathbf{R}\Gamma_{S,c}(V) \simeq \Delta_{\mathrm{EP}}(V). \tag{11}$$

The cohomology of $\mathbf{R}\Gamma_f(V)$ is as follows:

$$\begin{aligned} &\mathbf{R}^0\Gamma_f(V) = H^0_S(V), \quad \mathbf{R}^1\Gamma_f(V) = H^1_f(V), \quad \mathbf{R}^2\Gamma_f(V) \simeq H^1_f(V^*(1))^*, \\ &\mathbf{R}^3\Gamma_f(V) = \operatorname{coker}\left(H^2_S(V) \to \bigoplus_{v\in S} H^2(\mathbb{Q}_v, V) \right) \simeq H^0_S(V^*(1))^*. \end{aligned} \tag{12}$$

These groups sit in the following exact sequence:

$$\begin{aligned} 0 \to \mathbf{R}^1\Gamma_f(V) \to H^1_S(V) \to \bigoplus_{v\in S} \frac{H^1(\mathbb{Q}_v, V)}{H^1_f(\mathbb{Q}_v, V)} \to \mathbf{R}^2\Gamma_f(V) \to \\ H^2_S(V) \to \bigoplus_{v\in S} H^2(\mathbb{Q}_v, V) \to \mathbf{R}^3\Gamma_f(V) \to 0. \end{aligned}$$

The L-function of V is defined as the Euler product

$$L(V,s) = \prod_v E_v(V, (Nv)^{-s})^{-1}$$

where

$$E_v(V,t)=\begin{cases}\det\left(1-f_vt\mid V^{I_v}\right), & \text{if } v\neq p\\ \det\left(1-\varphi t\mid \mathbf{D}_{\mathrm{cris}}(V)\right) & \text{if } v=p.\end{cases}$$

5.1.2 Canonical Trivialisations

In this paper we treat motives in the formal sense and assume all conjectures about the category of mixed motives $\mathcal{MM}$ over $\mathbb{Q}$ which are necessary to state the Bloch–Kato conjecture (see Fontaine 1992; Fontaine and Perrin-Riou 1994). Let M be a pure motive over $\mathbb{Q}$ and let M_{B} and M_{dR} denote its Betti and de Rham realisations respectively. Fix an odd prime p and denote by $V=M_p$ the p-adic realisation of M. Then one has comparison isomorphisms

$$M_{\mathrm{B}}\otimes_{\mathbb{Q}}\mathbb{C}\overset{\sim}{\to} M_{\mathrm{dR}}\otimes_{\mathbb{Q}}\mathbb{C}, \tag{13}$$

$$M_{\mathrm{B}}\otimes_{\mathbb{Q}}\mathbb{Q}_p\overset{\sim}{\to} V. \tag{14}$$

The isomorphism (14) induces a trivialisation

$$\Omega_M^{(\acute{e}t,p)}\,:\,\det_{\mathbb{Q}_p}V\otimes\det_{\mathbb{Q}}^{-1}M_{\mathrm{B}}\to\mathbb{Q}_p. \tag{15}$$

The complex conjugation acts compatibly on M_{B} and V and decomposes the last isomorphism into $\pm$ parts which we denote again by $\Omega_M^{(\acute{e}t,p)}$ to simplify notation

$$\Omega_M^{(\acute{e}t,p)}\,:\,\det_{\mathbb{Q}_p}V^{\pm}\otimes\det_{\mathbb{Q}}^{-1}M_{\mathrm{B}}^{\pm}\to\mathbb{Q}_p. \tag{16}$$

The restriction of V to the decomposition group at p is a de Rham representation and $\mathbf{D}_{\mathrm{dR}}(V)\simeq M_{\mathrm{dR}}\otimes_{\mathbb{Q}}\mathbb{Q}_p$. The comparison isomorphism

$$V\otimes\mathbf{B}_{\mathrm{dR}}\overset{\sim}{\to}\mathbf{D}_{\mathrm{dR}}(V)\otimes\mathbf{B}_{\mathrm{dR}} \tag{17}$$

induces a map

$$\tilde{\Omega}_M^{(H,p)}\,:\,\det_{\mathbb{Q}_p}V\otimes\det_{\mathbb{Q}_p}^{-1}\mathbf{D}_{\mathrm{dR}}(V)\to\mathbf{B}_{\mathrm{dR}}.$$

It is not difficult to see that there exists a finite extension L of $\widehat{\mathbb{Q}_p^{\mathrm{ur}}}$ such that $\mathrm{Im}(\tilde{\Omega}_M^{(H,p)})\subset Lt^{t_H(V)}$ and we define

$$\Omega_M^{(H,p)}\,:\,\det_{\mathbb{Q}_p}V\otimes\det_{\mathbb{Q}_p}^{-1}\mathbf{D}_{\mathrm{dR}}(V)\to L \tag{18}$$

by $\Omega_M^{(H,p)} = t^{-t_H(V)}\tilde{\Omega}_M^{(H,p)}$. We remark that if V is crystalline at p then one can take $L = \widehat{\mathbb{Q}_p^{\mathrm{ur}}}$ (see Perrin-Riou 1995, Appendice C.2).

Assume that the groups $H^i(M) = \mathrm{Ext}^i_{\mathcal{MM}}(\mathbb{Q}(0), M)$ are well defined and vanish for $i \neq 0, 1$. It should be possible to define a $\mathbb{Q}$-subspace $H^1_f(M)$ of $H^1(M)$ consisting of "integral" classes of extensions which is expected to be finite dimensional. It is convenient to set $H^0_f(M) = H^0(M)$. The conjectures of Tate and Jannsen predict that the regulator map induces isomorphisms

$$H^i_f(M) \otimes_{\mathbb{Q}} \mathbb{Q}_p \simeq H^i_f(V), \qquad i = 0, 1. \tag{19}$$

In this paper M will always denote a motive satisfying the following conditions

(M1) M is pure of weight $w \leqslant -2$.
(M2) The p-adic realisation V of M is crystalline at p.
(M3) M has no subquotients isomorphic to $\mathbb{Q}(1)$.

These conditions imply that $H^0(M) = H^0(M^*(1)) = 0$ and $H^1(M^*(1)) = 0$ by the weight argument. Furthermore, (19) implies that the representation V should satisfy the conditions **(C1, C2, C4)** of Sect. 4.1.2. Also, from (12) it follows that

$$\det_{\mathbb{Q}_p} \mathbf{R}\Gamma_f(V) \xrightarrow{\sim} \det^{-1}_{\mathbb{Q}_p} H^1_f(V). \tag{20}$$

The semisimplicity of φ is a well known conjecture which is actually known for abelian varieties. Finally **(C5)** should follow from the injectivity of the syntomic regulator.

The comparison isomorphism (13) induces an injective map

$$\alpha_M \; : \; M_{\mathrm{B}}^+ \otimes_{\mathbb{Q}} \mathbb{R} \to t_M(\mathbb{R})$$

and the six-term exact sequence of Fontaine and Perrin-Riou (see Fontaine 1992, Sect. 6.10) degenerates into an isomorphism (the regulator map)

$$r_{M,\infty} \; : \; H^1_f(M) \otimes_{\mathbb{Q}} \mathbb{R} \xrightarrow{\sim} \mathrm{coker}(\alpha_M).$$

The maps α_M and $r_{M,\infty}$ define a map

$$R_{M,\infty} \; : \; \det^{-1}_{\mathbb{Q}} t_M(\mathbb{Q}) \otimes \det_{\mathbb{Q}} M_{\mathrm{B}}^+ \otimes \det_{\mathbb{Q}} H^1_f(M) \to \mathbb{R}$$

Fix bases $\omega_f \in \det_{\mathbb{Q}} H^1_f(M)$, $\omega_{t_M} \in \det_{\mathbb{Q}} t_M(\mathbb{Q})$ and $\omega^+_{M_{\mathrm{B}}} \in \det_{\mathbb{Q}} M^+_{\mathrm{B}}$. Set $\omega_M = (\omega_f, \omega_{t_M}, \omega^+_{M_{\mathrm{B}}})$ and define

$$R_{M,\infty}(\omega_M) = R_{M,\infty}(\omega^{-1}_{t_M} \otimes \omega^+_{M_{\mathrm{B}}} \otimes \omega_f).$$

Using (11), (16), (20) and the isomorphisms (19) define

$$i_{\omega_M,p} \;:\; \Delta_{\mathrm{EP}}(V) \simeq \det_{\mathbb{Q}_p}^{-1} t_V(\mathbb{Q}_p) \otimes \det_{\mathbb{Q}_p} V^+ \otimes \det_{\mathbb{Q}_p} H^1_f(V) \to \mathbb{Q}_p \qquad (21)$$

by $x = i_{\omega_M,p}(x)\,(\omega_{t_M}^{-1} \otimes \omega_{M_{\mathrm{B}}}^{+} \otimes \omega_f)$.

Consider now the case of the dual motive $M^*(1)$. Again one has $\Delta_{\mathrm{EP}}(V^*(1)) \simeq \Delta_f(V^*(1))$ where

$$\Delta_f(V^*(1)) \simeq \det_{\mathbb{Q}_p}^{-1} t_{V^*(1)}(\mathbb{Q}_p) \otimes \det_{\mathbb{Q}_p} V^*(1)^+ \otimes \det_{\mathbb{Q}_p} H^1_f(V).$$

The map $\alpha_{M^*(1)} \;:\; M^*(1)^+_{\mathrm{B}} \otimes_{\mathbb{Q}} \mathbb{R} \to t_{M^*(1)}(\mathbb{R})$ is surjective and it is related to α_M by the canonical duality $\mathrm{coker}(\alpha_M) \times \ker(\alpha_{M^*(1)}) \to \mathbb{R}$ (see Fontaine 1992, Sect. 5.4). The six-term exact sequence degenerates into an isomorphism

$$r_{M^*(1),\infty} \;:\; H^1_f(M)^* \otimes_{\mathbb{Q}} \mathbb{R} \simeq \ker(\alpha_{M^*(1)}).$$

This allows one to define a map

$$R_{M^*(1),\infty} \;:\; \det_{\mathbb{Q}}^{-1} t_{M^*(1)}(\mathbb{Q}) \otimes \det_{\mathbb{Q}} M^*(1)^+_{\mathrm{B}} \otimes \det_{\mathbb{Q}} H^1_f(M) \to \mathbb{R}.$$

We fix bases $\omega_{t_{M^*(1)}} \in \det_{\mathbb{Q}} t_{M^*(1)}(\mathbb{Q})$ and $\omega^+_{M^*(1)_{\mathrm{B}}} \in \det_{\mathbb{Q}} M^*(1)^+_{\mathrm{B}}$ and set

$$\omega_{M^*(1)} = (\omega_{t_{M^*(1)}}, \omega^+_{M^*(1)_{\mathrm{B}}}, \omega_f),$$
$$R_{M^*(1),\infty}(\omega_{M^*(1)}) = R_{M^*(1),\infty}(\omega^{-1}_{t_{M^*(1)}} \otimes \omega^+_{M^*(1)_{\mathrm{B}}} \otimes \omega_f).$$

Again this data defines a trivialisation

$$i_{\omega_{M^*(1)},p} \;:\; \Delta_{\mathrm{EP}}(V^*(1)) \to \mathbb{Q}_p. \qquad (22)$$

It is conjectured that the L-functions $L(V,s)$ and $L(V^*(1),s)$ are well defined complex functions, have meromorphic continuation to the whole of $\mathbb{C}$ and satisfy some explicit functional equation (see Fontaine and Perrin-Riou 1994, Chap. III). One expects that they do not depend on the choice of the prime p and we will denote them by $L(M,s)$ and $L(M^*(1),s)$ respectively. The conjectures about special values of these functions can be stated as follows.

Conjecture 2 (Beilinson–Deligne). The L-function $L(M,s)$ does not vanish at $s=0$ and

$$\frac{L(V,0)}{R_{M,\infty}(\omega_M)} \in \mathbb{Q}^*.$$

The L-function $L(M^*(1),s)$ has a zero of order $r = \dim_{\mathbb{Q}_p} H^1_f(M)$ at $s=0$. Let $L(M^*(1),0) = \lim_{s\to 0} s^{-r} L(M^*(1),s)$. Then

$$\frac{L(M^*(1),0)}{R_{M^*(1),\infty}(\omega_{M^*(1)})} \in \mathbb{Q}^*.$$

Conjecture 3 (Bloch–Kato). Let T be a $\mathbb{Z}_p$-lattice of V stable under the action of G_S. Then

$$i_{\omega_M,p}(\Delta_{\mathrm{EP}}(T)) = \frac{L(M,0)}{R_{M,\infty}(\omega_M)}\mathbb{Z}_p,$$

$$i_{\omega_{M^*(1)},p}(\Delta_{\mathrm{EP}}(T^*(1))) = \frac{L(M^*(1),0)}{R_{M^*(1),\infty}(\omega_{M^*(1)})}\mathbb{Z}_p.$$

5.1.3 Compatibility with Functional Equation

The compatibility of the Bloch–Kato conjecture with the functional equation follows from the conjecture $C_{EP}(V)$ of Fontaine and Perrin-Riou about local Tamagawa numbers (see Fontaine and Perrin-Riou 1994, Chap. III, Sect. 4.5.4). More precisely, define

$$\Gamma^*(V) = \prod_{i\in Z} \Gamma^*(-i)^{h_i(V)} \tag{23}$$

where $h_i(V) = \dim_{\mathbb{Q}_p}(\mathrm{gr}_i(\mathbf{D}_{\mathrm{dR}}(V)))$ and

$$\Gamma^*(i) = \begin{cases} (i-1)! & \text{if } i > 0, \\ \frac{(-1)^i}{(-i)!} & \text{if } i \leqslant 0. \end{cases}$$

The exact sequence

$$0 \to t_{V^*(1)}(\mathbb{Q}_p)^* \to \mathbf{D}_{\mathrm{dR}}(V) \to t_V(\mathbb{Q}_p) \to 0$$

allows to consider $\omega_{M_{\mathrm{dR}}} = \omega_{t_M} \otimes \omega_{t_{M^*(1)}}^{-1} \in \det_{\mathbb{Q}_p}\mathbf{D}_{\mathrm{dR}}(V)$. Choose bases $\omega_T^+ \in \det_{\mathbb{Z}_p} T^+$ and $\omega_T^- \in \det_{\mathbb{Z}_p} T^-$ and set $\omega_T = \omega_T^+ \otimes \omega_T^- \in \det_{\mathbb{Z}_p} T$ and $\omega_{T^*(1)}^+ = (\omega_T^-)^* \in \det_{\mathbb{Z}_p} T^*(1)^+$. Define p-adic periods $\Omega_M^{(\acute{e}t,p)}(\omega_T^+,\omega_{M_{\mathrm{B}}}^+)$ and $\Omega_M^{(H,p)}(\omega_T,\omega_{M_{\mathrm{dR}}})$ by $\omega_T^+ = \Omega_M^{(\acute{e}t,p)}(\omega_T^+,\omega_{M_{\mathrm{B}}}^+)\omega_{M_{\mathrm{B}}}^+$ and $\omega_T = \Omega_M^{(H,p)}(\omega_T,\omega_{M_{\mathrm{dR}}})\omega_{M_{\mathrm{dR}}}$ using isomorphisms (15) and (18). Then the conjecture $C_{EP}(V)$ implies that

$$\frac{i_{\omega_{M^*(1)},p}(\Delta_{\mathrm{EP}}(T^*(1)))}{\Omega_{M^*(1)}^{(\acute{e}t,p)}(\omega_{T^*(1)}^+,\omega_{M^*(1)_{\mathrm{B}}}^+)} = \Gamma^*(V)\,\Omega_M^{(H,p)}(\omega_T,\omega_{M_{\mathrm{dR}}})\,\frac{i_{\omega_M,p}(\Delta_{\mathrm{EP}}(T))}{\Omega_M^{(\acute{e}t,p)}(\omega_T^+,\omega_{M_{\mathrm{B}}}^+)} \tag{24}$$

(see Perrin-Riou 1995, Appendice C). We remark that for crystalline representations $C_{EP}(V)$ is proved in Benois and Berger (2008).

5.2 p-Adic L-Functions

5.2.1 p-Adic Beilinson Conjecture

We maintain previous notation and conventions. Let M be a motive which satisfies the conditions **(M1–M3)** of Sect. 5.1.2 and let V denote the p-adic realisation of M. We fix bases $\omega_{M_{\mathrm{B}}}^+ \in \det_{\mathbb{Q}} M_{\mathrm{B}}^+$, $\omega_{t_M} \in \det_{\mathbb{Q}} t_M(\mathbb{Q})$ and $\omega_f \in \det_{\mathbb{Q}} H^1_f(M)$. We also fix a lattice T in V stable under the action of G_S and a basis $\omega_T^+ \in \det_{\mathbb{Z}_p} T^+$. To simplify notation we will assume that the choices of $\omega_{M_{\mathrm{B}}}^+$ and ω_T^+ are compatible, namely that $\Omega_M^{(\acute{e}t,p)}(\omega_T^+, \omega_{M_{\mathrm{B}}}^+) = 1$. Let D be a regular subspace of $\mathbf{D}_{\mathrm{cris}}(V)$. We fix a $\mathbb{Z}_p$-lattice N of D and a basis $\omega_N \in \det_{\mathbb{Z}_p} N$. By the analogy with the Archimedean case we can consider the p-adic regulator as a map $r_{V,D} : H^1_f(V) \to \mathrm{coker}(\alpha_{V,D})$ where

$$\alpha_{V,D} : D \to t_V(\mathbb{Q}_p)$$

is the natural projection. Set $\omega_{V,N} = (\omega_{t_M}, \omega_N, \omega_f)$ and denote by $R_{V,D}(\omega_{V,N})$ the determinant of $r_{V,D}$ computed in the bases ω_f and $\omega_{t_M} \otimes \omega_N^{-1}$. Namely, $R_{V,D}(\omega_{V,N})$ is the image of $\omega_{t_M}^{-1} \otimes \omega_N \otimes \omega_f$ under the induced isomorphism

$$R_{V,D} : \det{}_{\mathbb{Q}_p}^{-1} t_V(\mathbb{Q}_p) \otimes \det{}_{\mathbb{Q}_p} D \otimes \det{}_{\mathbb{Q}_p} H^1_f(V) \to \mathbb{Q}_p.$$

Now, consider the projection

$$\alpha_{V^*(1),D^\perp} : D^\perp \to t_{V^*(1)}(\mathbb{Q}_p).$$

A standard argument from linear algebra shows that $\alpha_{V^*(1),D^\perp}$ is surjective and is related to $\alpha_{V,D}$ by the canonical duality $\mathrm{coker}(\alpha_{V,D}) \times \ker(\alpha_{V^*(1),D}) \to \mathbb{Q}_p$. This defines isomorphisms

$$\det{}_{\mathbb{Q}_p}^{-1} t_{V^*(1)}(\mathbb{Q}_p) \otimes \det{}_{\mathbb{Q}_p} D^\perp \simeq \det{}_{\mathbb{Q}_p}(\ker(\alpha_{V^*(1),D^\perp})) \simeq \det{}_{\mathbb{Q}_p}^{-1}(\mathrm{coker}(\alpha_{V,D}))$$

and composing this map with the determinant of $r_{V,D}$ we have again a trivialisation

$$R_{V^*(1),D^\perp} : \det{}_{\mathbb{Q}_p}^{-1} t_{V^*(1)}(\mathbb{Q}_p) \otimes \det{}_{\mathbb{Q}_p} D^\perp \otimes \det{}_{\mathbb{Q}_p} H^1_f(V) \to \mathbb{Q}_p.$$

Choose a lattice $N^\perp \subset D^\perp$, fix bases $\omega_{t_{M^*(1)}}$ and $\omega_{N^\perp}$ of $\det_{\mathbb{Q}_p} t_{V^*(1)}(\mathbb{Q}_p)$ and $\det_{\mathbb{Z}_p} N^\perp$ respectively and set $\omega_{V,N^\perp} = (\omega_{t_{M^*(1)}}, \omega_{N^\perp}, \omega_f)$.

Perrin-Riou conjectured (see Perrin-Riou 1995) that there exists an analytic p-adic L-function $L_p(T,N,s)$ which interpolates special values of the complex L-function $L(M,s)$. In particular one expects that if p^{-1} is not an eigenvalue of φ acting on D then $L_p(T,N,s)$ does not vanish at $s=0$ and

$$\frac{L_p(T,N,0)}{R_{V,D}(\omega_{V,N})} = \mathcal{E}(V,D)\,\frac{L(M,0)}{R_{M,\infty}(\omega_M)}$$

where

$$\begin{aligned}\mathcal{E}(V,D) &= \det(1-p^{-1}\varphi^{-1}\mid D)\,\det(1-p^{-1}\varphi^{-1}\mid D^{\perp})\\ &= \det(1-p^{-1}\varphi^{-1}\mid D)\,\det(1-\varphi\mid \mathbf{D}_{\mathrm{cris}}(V)/D).\end{aligned}$$

Dually it is conjectured that there exists a p-adic L-function $L_p(T^*(1),N^{\perp},s)$ which interpolates special values of $L(M^*(1),s)$. One expects that if 1 is not an eigenvalue of φ acting on the quotient $\mathbf{D}_{\mathrm{cris}}(V^*(1))/D^{\perp}$ then $L_p(T^*(1),N^{\perp},s)$ has a zero of order $r=\dim_Q H^1_f(M)$ at $s=0$ and

$$\frac{L_p^*(T^*(1),N^{\perp},0)}{R_{V^*(1),D^{\perp}}(\omega_{V^*(1),N^{\perp}})} = \mathcal{E}(V^*(1),D^{\perp})\,\frac{L^*(M^*(1),0)}{R_{M^*(1),\infty}(\omega_M)}.$$

These properties of p-adic L-functions can be viewed as p-adic analogues of Beilinson conjectures and we refer the reader to Perrin-Riou (1995), Chap. 4 and Colmez (2000), Sect. 2.8 for more detail. Note that from the definition it is clear that $\mathcal{E}(V,D)=\mathcal{E}(V^*(1),D^{\perp})$. One can also write $\mathcal{E}(V,D)$ in the form

$$\mathcal{E}(V,D) = E_p(V,1)\det\left(\frac{1-p^{-1}\varphi^{-1}}{1-\varphi}|D\right).$$

5.2.2 Extra Zero Conjecture

Assume now that $D^{\varphi=p^{-1}}\neq 0$. Since M is crystalline at p, this can occur only if M is of weight -2. Set

$$e=\dim_{\mathbb{Q}_p} D^{\varphi=p^{-1}} = \dim_{\mathbb{Q}_p}((D^{\perp}+\mathbf{D}_{\mathrm{cris}}(V^*(1))^{\varphi=1})/D^{\perp}).$$

Assume that the p-adic realisation V of M satisfies the conditions **(C1–C5)** of Sect. 4.1.2. Decompose D into the direct sum $D=D_{-1}\oplus D^{\varphi=p^{-1}}$ and define

$$\mathcal{E}^+(V,D)=\mathcal{E}^+(V^*(1),D^{\perp})=\det(1-p^{-1}\varphi^{-1}\mid D_{-1})\,\det(1-p^{-1}\varphi^{-1}\mid D^{\perp}). \tag{25}$$

We propose the following conjecture about the behavior of p-adic L-functions at $s=0$ (see Conjecture 1 of Introduction).

Conjecture 4. (i) The p-adic L-function $L_p(T,N,s)$ has a zero of order e at $s=0$ and

$$\frac{L_p^*(T,N,0)}{R_{V,D}(\omega_{V,N})} = -\mathscr{L}(V,D)\,\mathcal{E}^+(V,D)\,\frac{L(M,0)}{R_{M,\infty}(\omega_M)}.$$

(ii) The p-adic L-function $L_p(T^*(1),N^\perp,s)$ has a zero of order $e+r$ where $r=\dim_Q H^1_f(M)$ at $s=0$ and

$$\frac{L_p^*(T^*(1),N^\perp,0)}{R_{V^*(1),D^\perp}(\omega_{V^*(1),N^\perp})} = \mathscr{L}(V,D)\,\mathcal{E}^+(V^*(1),D^\perp)\,\frac{L^*(M^*(1),0)}{R_{M^*(1),\infty}(\omega_{M^*(1)})}.$$

Remarks. (1) If $H^1_f(M)=0$ the p-adic regulator vanishes and we recover the conjecture formulated in Benois (2011), Sect. 2.3.2.

(2) The regulators $R_{M,\infty}(\omega_M)$ and $R_{V,D}(\omega_{V,N})$ are well defined up to a sign and in order to obtain equalities in the formulation of our conjecture one should make the same choice of signs in the definitions of $R_{M,\infty}(\omega_M)$ and $R_{V,D}(\omega_{V,N})$. See Perrin-Riou (1995), Sect. 4.2 for more detail.

(3) Our conjecture is compatible with the expected functional equation for p-adic L-functions. See Sect. 2.5 of Perrin-Riou (1995) and Sects. 6.2.5–6.2.6 below.

6 The Module of p-Adic L-Functions

6.1 The Selmer Complex

6.1.1 Iwasawa Cohomology

Let Γ denote the Galois group of $\mathbf{Q}(\zeta_{p^\infty})/\mathbb{Q}$ and $\Gamma_n=\mathrm{Gal}(\mathbb{Q}(\zeta_{p^\infty})/\mathbb{Q}(\zeta_{p^n}))$. Set $\Lambda=\mathbb{Z}_p[[\Gamma_1]]$ and $\Lambda(\Gamma)=\mathbb{Z}_p[\Delta]\otimes_{\mathbb{Z}_p}\Lambda$. For any character $\eta\in\hat{\Delta}$ put

$$e_\eta = \frac{1}{|\Delta|}\sum_{g\in\Delta}\eta^{-1}(g)g.$$

Then $\Lambda(\Gamma)=\bigoplus_{\eta\in\hat{\Delta}}\Lambda(\Gamma)^{(\eta)}$ where $\Lambda(\Gamma)^{(\eta)}=\Lambda e_\eta$ and for any $\Lambda(\Gamma)$-module M one has a canonical decomposition

$$M \simeq \bigoplus_{\eta \in \hat{\Delta}} M^{(\eta)}, \qquad M^{(\eta)} = e_\eta(M).$$

We write η_0 for the trivial character of Δ and identify Λ with $\Lambda(\Gamma)e_{\eta_0}$.

Let V be a p-adic pseudo-geometric representation unramified outside S. Set $d(V) = \dim(V)$ and $d_\pm(V) = \dim(V^{c=\pm 1})$. Fix a $\mathbb{Z}_p$-lattice T of V stable under the action of G_S. Let $\iota : \Lambda(\Gamma) \to \Lambda(\Gamma)$ denote the canonical involution $g \mapsto g^{-1}$. Recall that the induced module $\mathrm{Ind}_{\mathbb{Q}(\zeta_{p^\infty})/\mathbb{Q}}(T)$ is isomorphic to $(\Lambda(\Gamma) \otimes_{\mathbb{Z}_p} T)^\iota$ (see Nekovář 2006, Sect. 8.1). Define

$$\begin{aligned} H^i_{\mathrm{Iw},S}(T) &= H^i_S((\Lambda(\Gamma) \otimes_{\mathbb{Z}_p} T)^\iota), \\ H^i_{\mathrm{Iw}}(\mathbb{Q}_v, T) &= H^i(\mathbb{Q}_v, (\Lambda(\Gamma) \otimes_{\mathbb{Z}_p} T)^\iota) \qquad \text{for any finite place } v. \end{aligned}$$

From Shapiro's lemma it follows immediately that

$$H^i_{\mathrm{Iw},S}(T) = \varprojlim_{\mathrm{cores}} H^i_S(\mathbb{Q}(\zeta_{p^n}), T), \qquad H^i_{\mathrm{Iw}}(\mathbb{Q}_p, T) = \varprojlim_{\mathrm{cores}} H^i(\mathbb{Q}_p(\zeta_{p^n}), T).$$

Set $H^i_{\mathrm{Iw},S}(V) = H^i_{\mathrm{Iw},S}(T) \otimes_{\mathbb{Z}_p} \mathbb{Q}_p$ and $H^i_{\mathrm{Iw}}(\mathbb{Q}_v, V) = H^i_{\mathrm{Iw}}(\mathbb{Q}_v, T) \otimes_{\mathbb{Z}_p} \mathbb{Q}_p$. In Perrin-Riou (1995) Perrin-Riou proved the following results about the structure of these modules.

(i) $H^i_{\mathrm{Iw},S}(V) = 0$ and $H^i_{\mathrm{Iw}}(\mathbb{Q}_v, T) = 0$ if $i \neq 1, 2$;

(ii) If $v \neq p$, then for each $\eta \in \hat{\Delta}$ the η-component $H^i_{\mathrm{Iw}}(\mathbb{Q}_v, T)^{(\eta)}$ is a finitely-generated Λ-torsion module. In particular, $H^1_{\mathrm{Iw}}(\mathbb{Q}_v, T) \simeq H^1(\mathbb{Q}_v^{\mathrm{ur}}/\mathbb{Q}_v, (\Lambda(\Gamma) \otimes_{\mathbb{Z}_p} T^{I_v})^\iota)$.

(iii) If $v = p$ then $H^2_{\mathrm{Iw}}(\mathbb{Q}_p, T)^{(\eta)}$ are finitely-generated Λ-torsion modules. Moreover, for each $\eta \in \hat{\Delta}$

$$\mathrm{rank}_\Lambda\left(H^1_{\mathrm{Iw}}(\mathbb{Q}_p, T)^{(\eta)}\right) = d, \qquad H^1_{\mathrm{Iw}}(\mathbb{Q}_p, T)^{(\eta)}_{\mathrm{tor}} \simeq H^0(\mathbb{Q}_p(\zeta_{p^\infty}), T)^{(\eta)}.$$

Remark that by local duality $H^2_{\mathrm{Iw}}(\mathbb{Q}_p, T) \simeq H^0(\mathbb{Q}_p(\zeta_{p^\infty}), V^*(1)/T^*(1))$.

(iv) If the weak Leopoldt conjecture holds for the pair (V, η), i.e. if $H^2_S(\mathbb{Q}(\zeta_{p^\infty}), V/T)^{(\eta)} = 0$, then $H^2_{\mathrm{Iw},S}(T)^{(\eta)}$ is Λ-torsion and

$$\mathrm{rank}_\Lambda\left(H^1_{\mathrm{Iw},S}(T)^{(\eta)}\right) = \begin{cases} d_-(V), & \text{if } \eta(c) = 1 \\ d_+(V), & \text{if } \eta(c) = -1. \end{cases}$$

Passing to the projective limit in the Poitou-Tate exact sequence one obtains an exact sequence

$$0 \to H^2_S(\mathbb{Q}(\zeta_{p^\infty}), V^*(1)/T^*(1))^\wedge \to H^1_{\mathrm{Iw},S}(T) \to \bigoplus_{v\in S} H^1_{\mathrm{Iw}}(\mathbb{Q}_v, T)$$
$$\to H^1_S(\mathbb{Q}(\zeta_{p^\infty}), V^*(1)/T^*(1))^\wedge \to H^2_{\mathrm{Iw},S}(T)$$
$$\to \bigoplus_{v\in S} H^2_{\mathrm{Iw}}(\mathbb{Q}_v, T) \to H^0_S(\mathbb{Q}(\zeta_{p^\infty}), V^*(1)/T^*(1))^\wedge \to 0. \quad (26)$$

Define

$$\begin{aligned}
\mathbf{R}\Gamma_{\mathrm{Iw},S}(T) &= C^\bullet_c(G_S, (\Lambda(\Gamma)\otimes_{\mathbb{Z}_p} T)^\iota),\\
\mathbf{R}\Gamma_{\mathrm{Iw}}(\mathbb{Q}_v, T) &= C^\bullet_c(G_v, (\Lambda(\Gamma)\otimes_{\mathbb{Z}_p} T)^\iota),\\
\mathbf{R}\Gamma_S(\mathbb{Q}(\zeta_{p^\infty}), V^*(1)/T^*(1)) &= C^\bullet_c(G_S, \mathrm{Hom}_{\mathbb{Z}_p}(\Lambda(\Gamma), V^*(1)/T^*(1))).
\end{aligned}$$

Then the sequence (26) is induced by the distinguished triangle

$$\mathbf{R}\Gamma_{\mathrm{Iw},S}(T) \to \bigoplus_{v\in S}\mathbf{R}\Gamma_{\mathrm{Iw}}(\mathbb{Q}_v, T) \to \left(\mathbf{R}\Gamma_S(\mathbb{Q}(\zeta_{p^\infty}), V^*(1)/T^*(1))^\iota\right)^\wedge [-2]$$

(see Nekovář 2006, Theorem 8.5.6). Finally, we have the usual descent formulas

$$\mathbf{R}\Gamma_{\mathrm{Iw},S}(T)\otimes^{\mathbf{L}}_\Lambda \mathbb{Z}_p \simeq \mathbf{R}\Gamma_S(T), \qquad \mathbf{R}\Gamma_{\mathrm{Iw}}(\mathbb{Q}_v, T)\otimes^{\mathbf{L}}_\Lambda \mathbb{Z}_p \simeq \mathbf{R}\Gamma(\mathbb{Q}_v, T)$$

(see Nekovář 2006, Proposition 8.4.22).

6.1.2 The Complex $\mathbf{R}\Gamma^{(\eta_0)}_{\mathrm{Iw},h}(D, V)$

For the remainder of this chapter we assume that V satisfies the conditions **(C1–C5)** of Sect. 4.1.2 and that the weak Leopoldt conjecture holds for (V, η_0) and $(V^*(1), \eta_0)$. We remark that these assumptions are not independent. Namely, by Perrin-Riou (1995), Proposition B.5 **(C4)** and **(C5)** imply the weak Leopoldt conjecture for $(V^*(1), \eta_0)$. From the same result it follows that the vanishing of $H^1_f(V^*(1))$ implies the weak Leopoldt conjecture for (V, η_0) if in addition we assume that $H^0(\mathbb{Q}_p, V^*(1)) = 0$.

To simplify notations we write $\mathcal{H}$ for $\mathcal{H}(\Gamma_1)$. Fix a regular subspace D of $\mathbf{D}_{\mathrm{cris}}(V)$ and a $\mathbb{Z}_p$-lattice N of D. Set $\mathcal{D}_p(N, T)^{(\eta_0)} = N\otimes_{\mathbb{Z}_p}\mathbb{Z}_p[[X]]^{\psi=0} \simeq N\otimes_{\mathbb{Z}_p}\Lambda$, and define

$$\begin{aligned}
\mathbf{R}\Gamma^{(\eta_0)}_{\mathrm{Iw}}(\mathbb{Q}_p, N, T) &= \mathcal{D}_p(N, T)^{(\eta_0)}[-1],\\
\mathbf{R}\Gamma^{(\eta_0)}_{\mathrm{Iw}}(\mathbb{Q}_p, D, V) &= \mathbf{R}\Gamma^{(\eta_0)}_{\mathrm{Iw}}(\mathbb{Q}_p, N, T)\otimes_{\mathbb{Z}_p}\mathbb{Q}_p.
\end{aligned}$$

Consider Perrin-Riou's exponential map

$$\mathbf{Exp}^{\varepsilon}_{V,h} \,:\, \mathbf{R}\Gamma^{(\eta_0)}_{\mathrm{Iw}}(\mathbb{Q}_p, N, T) \otimes_{\Lambda} \mathcal{H} \to \mathbf{R}\Gamma^{(\eta_0)}_{\mathrm{Iw}}(\mathbb{Q}_p, T) \otimes^{\mathbf{L}}_{\Lambda} \mathcal{H}$$

which will be viewed as a local condition at p. If $v \neq p$ the inertia group I_v acts trivially on Λ. Set

$$\mathbf{R}\Gamma^{(\eta_0)}_{\mathrm{Iw}}(\mathbb{Q}_v, N, T) \;=\; \left[T^{I_v} \otimes \Lambda^{\iota} \xrightarrow{1-f_v} T^{I_v} \otimes \Lambda^{\iota}\right]$$

where the first term is placed in degree 0. We have a commutative diagram

$$\begin{array}{ccc} \mathbf{R}\Gamma^{(\eta_0)}_{\mathrm{Iw},S}(T) \otimes_{\Lambda} \mathcal{H} & \longrightarrow & \bigoplus\limits_{v\in S} \mathbf{R}\Gamma^{(\eta_0)}_{\mathrm{Iw}}(\mathbb{Q}_v, T) \otimes_{\Lambda} \mathcal{H} \\ & & \uparrow \\ & & \left(\bigoplus\limits_{v\in S} \mathbf{R}\Gamma^{(\eta_0)}_{\mathrm{Iw}}(\mathbb{Q}_v, N, T)\right) \otimes_{\Lambda} \mathcal{H}\,. \end{array} \tag{27}$$

Consider the associated Selmer complex

$$\mathbf{R}\Gamma^{(\eta_0)}_{\mathrm{Iw},h}(D,V) \;=\; \mathrm{cone}\left[\left(\mathbf{R}\Gamma^{(\eta_0)}_{\mathrm{Iw},S}(T) \bigoplus \left(\bigoplus_{v\in S} \mathbf{R}\Gamma^{(\eta_0)}_{\mathrm{Iw}}(\mathbb{Q}_v, N, T)\right)\right) \otimes_{\Lambda} \mathcal{H} \to \right.$$
$$\left. \bigoplus_{v\in S} \mathbf{R}\Gamma^{(\eta_0)}_{\mathrm{Iw}}(\mathbb{Q}_v, T) \otimes_{\Lambda} \mathcal{H}\right][-1].$$

It is easy to see that it does not depend on the choice of S. Our main result about this complex is the following theorem.

Theorem 4. *Assume that V satisfies the conditions* **(C1–C5)** *and that the weak Leopoldt conjecture holds for (V, η_0) and $(V^*(1), \eta_0)$. Let D be a regular subspace of $\mathbf{D}_{\mathrm{cris}}(V)$. Assume that $\mathscr{L}(V, D) \neq 0$. Then*

(i) $\mathbf{R}^i\Gamma^{(\eta_0)}_{\mathrm{Iw},h}(D,V)$ *are $\mathcal{H}$-torsion modules for all i.*
(ii) $\mathbf{R}^i\Gamma^{(\eta_0)}_{\mathrm{Iw},h}(D,V) = 0$ *for $i \neq 2,3$ and*

$$\mathbf{R}^3\Gamma^{(\eta_0)}_{\mathrm{Iw},h}(D,V) \simeq \left(H^0(\mathbb{Q}(\zeta_{p^\infty}), V^*(1))^*\right)^{(\eta_0)} \otimes_{\Lambda} \mathcal{H}.$$

(iii) The complex $\mathbf{R}\Gamma^{(\eta_0)}_{\mathrm{Iw},h}(D,V)$ *is semisimple in the sense that for each i the natural map*

$$\mathbf{R}^i\Gamma^{(\eta_0)}_{\mathrm{Iw},h}(D,V)^{\Gamma} \to \mathbf{R}^i\Gamma^{(\eta_0)}_{\mathrm{Iw},h}(D,V)_{\Gamma}$$

is an isomorphism.

6.1.3 Proof of Theorem 4

6.1.3.1 We leave the proof of the following lemma as an easy exercise.

Lemma 2. *Let A and B be two submodules of a finitely-generated free $\mathcal{H}$-module M. Assume that the natural maps $A_{\Gamma_1} \to M_{\Gamma_1}$ and $B_{\Gamma_1} \to M_{\Gamma_1}$ are both injective. Then $A_{\Gamma_1} \cap B_{\Gamma_1} = \{0\}$ implies that $A \cap B = \{0\}$.*

6.1.3.2 Since $H^0_{\mathrm{Iw},S}(V)$ and $H^0_{\mathrm{Iw}}(\mathbb{Q}_v, V)$ are zero, we have $\mathbf{R}^0\Gamma^{(\eta_0)}_{\mathrm{Iw},h}(D,V) = 0$. Next, by definition $\mathbf{R}^1\Gamma^{(\eta_0)}_{\mathrm{Iw},h}(D,V) = \ker(f)$ where

$$f : \left(H^1_{\mathrm{Iw},S}(T)^{(\eta_0)} \bigoplus \mathcal{D}_p(N,T)^{(\eta_0)} \bigoplus_{v\in S-\{p\}} H^1_{\mathrm{Iw},f}(\mathbb{Q}_v, T)^{(\eta_0)} \right) \otimes \mathcal{H} \to \bigoplus_{v\in S} H^1_{\mathrm{Iw}}(\mathbb{Q}_v, T)^{(\eta_0)} \otimes \mathcal{H}$$

is the map induced by (27). If $v \in S - \{p\}$ one has

$$H^1_{\mathrm{Iw},f}(\mathbb{Q}_v, T)^{(\eta_0)} = H^1_{\mathrm{Iw}}(\mathbb{Q}_v, T)^{(\eta_0)} = H^1(\mathbb{Q}_v^{\mathrm{ur}}/\mathbb{Q}_v, (\Lambda \otimes T^{I_v})^{\iota}).$$

Thus

$$\mathbf{R}^1\Gamma^{(\eta_0)}_{\mathrm{Iw},h}(D,V) = \left(H^1_{\mathrm{Iw},S}(T)^{(\eta_0)} \otimes_\Lambda \mathcal{H}\right) \cap \left(\mathrm{Exp}^{\varepsilon}_{V,h}\left(\mathcal{D}_p(D,T)^{(\eta_0)}\right) \otimes_\Lambda \mathcal{H}\right)$$

in $H^1_{\mathrm{Iw}}(\mathbb{Q}_p, T)^{(\eta_0)} \otimes_\Lambda \mathcal{H}$. Put

$$A = \mathrm{Exp}^{\varepsilon}_{V,h}(D_{-1} \otimes \mathcal{H}) \oplus X^{-1}\mathrm{Exp}^{\varepsilon}_{V,h}(D^{\varphi=p^{-1}} \otimes \mathcal{H}) \subset H^1_{\mathrm{Iw}}(\mathbb{Q}_p, T)^{(\eta_0)} \otimes_\Lambda \mathcal{H},$$

where we identify X with the operator $\gamma_1 - 1$ (see Sect. 3.2.1). By Theorem 3 and Proposition 6 A_{Γ_1} injects into $H^1(\mathbb{Q}_p, V)$. Since $T^{H_{\mathbb{Q}_p}}$ is the torsion submodule of $H^1_{\mathrm{Iw}}(\mathbb{Q}_p, T)$, the $\mathcal{H}$-module $M = \left(\dfrac{H^1_{\mathrm{Iw}}(\mathbb{Q}_p, T)}{T^{H_{\mathbb{Q}_p}}}\right)^{(\eta_0)} \otimes_\Lambda \mathcal{H}$ is free and $A \hookrightarrow M$. Since $T^{G_{\mathbb{Q}_p}} = 0$ one has $M_{\Gamma_1} = H^1_{\mathrm{Iw}}(\mathbb{Q}_p, V)_\Gamma \subset H^1(\mathbb{Q}_p, V)$ and we obtain that A_{Γ_1} injects into M_{Γ_1}.

Set $B = \left(\dfrac{H^1_{\mathrm{Iw},S}(T)}{T^{H_{\mathbb{Q}}}}\right)^{(\eta_0)} \otimes_\Lambda \mathcal{H}$. The weak Leopoldt conjecture for $(V^*(1), \eta_0)$ together with the fact that $H^1_{\mathrm{Iw}}(\mathbb{Q}_v, T)$ are Λ-torsion for $v \in S - \{p\}$ imply that $B \hookrightarrow M$. Since the image of $H^1_{\mathrm{Iw}}(\mathbb{Q}_v, V)_\Gamma$ in $H^1(\mathbb{Q}_v, V)$ is contained in $H^1_f(\mathbb{Q}_v, V)$, the image of $H^1_{\mathrm{Iw},S}(V)_\Gamma$ in $H^1_S(V)$ is in fact contained in $H^1_{f,\{p\}}(V)$. From **(C5)** it follows that $H^1_{f,\{p\}}(V)$ injects into $H^1(\mathbb{Q}_p, V)$ and we have

$$H^1_{\mathrm{Iw},S}(V)^{(\eta_0)}_{\Gamma_1} = H^1_{\mathrm{Iw},S}(V)_\Gamma \hookrightarrow H^1_{f,\{p\}}(V) \hookrightarrow H^1(\mathbb{Q}_p, V).$$

Thus $B_{\Gamma_1} \subset M_{\Gamma_1}$. We shall prove that $\mathbf{R}^1\Gamma_{\mathrm{Iw},h}^{(\eta_0)}(D,V)=0$. By Lemma 2 it suffices to show that $A_{\Gamma_1}\cap B_{\Gamma_1}=\{0\}$. Now we claim that $A_{\Gamma_1}\cap H^1_{f,\{p\}}(V)=\{0\}$. First note that by Lemma 1

$$H^1_{f,\{p\}}(V)\hookrightarrow \frac{H^1(\mathbb{Q}_p,V)}{H^1(\mathbf{D}_{-1})}.$$

On the other hand, from Theorem 3 it follows that

$$\mathrm{Exp}^{\varepsilon}_{V,h}(D_{-1}\otimes\mathcal{H})_{\Gamma_1}=\exp_{V,\mathbb{Q}_p}(D_{-1})\subset H^1(\mathbf{D}_{-1}).$$

Now Proposition 6 implies that the image of A_{Γ_1} in $\dfrac{H^1(\mathbb{Q}_p,V)}{H^1(\mathbf{D}_{-1})}$ coincides with $H^1_c(W)$. But $\mathscr{L}(V,D)\neq 0$ if and only if $H^1_D(V)\cap H^1_c(W)=0$ where $H^1_D(V)$ denotes the inverse image of $H^1(W)$ in $H^1_{f,\{p\}}(V)$ (see Lemma 1 (ii)). This proves the claim and implies that $\mathbf{R}^1\Gamma_{\mathrm{Iw},h}^{(\eta_0)}(D,V)=0$.

6.1.3.3 We shall show that $\mathbf{R}^2\Gamma_{\mathrm{Iw},h}^{(\eta_0)}(D,V)$ is $\mathcal{H}$-torsion. By definition, we have an exact sequence

$$0\to \mathrm{coker}(f)\to \mathbf{R}^2\Gamma_{\mathrm{Iw},h}^{(\eta_0)}(D,V)\to Ш^2_{\mathrm{Iw},S}(V)^{(\eta_0)}\otimes_{\Lambda_{\mathbb{Q}_p}}\mathcal{H}\to 0, \qquad (28)$$

where

$$Ш^2_{\mathrm{Iw},S}(V)=\ker\left(H^2_{\mathrm{Iw},S}(V)\to\bigoplus_{v\in S}H^2_{\mathrm{Iw}}(\mathbb{Q}_v,V)\right).$$

It follows from the weak Leopoldt conjecture that $Ш^2_{\mathrm{Iw},S}(V)$ is $\Lambda_{\mathbb{Q}_p}$-torsion. On the other hand, as $\mathcal{H}$ is a Bézout ring (see Lazard 1962), the formulas

$$\begin{aligned}\mathrm{rank}_\Lambda H^1_{\mathrm{Iw},S}(T)^{(\eta_0)}&=d_-(V),\\ \mathrm{rank}_\Lambda H^1_{\mathrm{Iw}}(\mathbb{Q}_p,T)^{(\eta_0)}&=d(V),\\ \mathrm{rank}_\Lambda \mathfrak{D}_p(N,T)&=d_+(V)\end{aligned}$$

together with the fact that $\mathbf{R}^1\Gamma_{\mathrm{Iw},h}^{(\eta_0)}(D,V)=0$ imply that $\mathrm{coker}(f)$ is $\mathcal{H}$-torsion. We have therefore proved that $\mathbf{R}^2\Gamma_{\mathrm{Iw},h}(D,V)$ is $\mathcal{H}$-torsion. Finally, the Poitou–Tate exact sequence gives that

$$\mathbf{R}^3\Gamma_{\mathrm{Iw},h}^{(\eta_0)}(D,V)=\left(H^0(\mathbb{Q}(\zeta_{p^\infty}),V^*(1))^*\right)^{(\eta_0)}\otimes_{\Lambda_{\mathbb{Q}_p}}\mathcal{H}$$

is also $\mathcal{H}$-torsion.

6.1.3.4 Now we prove the semisimplicity of $\mathbf{R}\Gamma^{(\eta_0)}_{\mathrm{Iw},h}(D,V)$. First write

$$H^1_{\mathrm{Iw},S}(V)^{(\eta_0)} \simeq \Lambda^{d_-(V)}_{\mathbb{Q}_p} \oplus H^1_{\mathrm{Iw},S}(V)^{(\eta_0)}_{\mathrm{tor}}.$$

Since $H^1_{\mathrm{Iw},S}(V)_{\mathrm{tor}} \subset H^1_{\mathrm{Iw}}(\mathbb{Q}_p,V)_{\mathrm{tor}} = V^{H_{\mathbb{Q}_p}}$, $(H^1_{\mathrm{Iw},S}(V)_{\mathrm{tor}})_\Gamma = 0$ by the snake lemma. Thus $\dim_{\mathbb{Q}_p} H^1_{\mathrm{Iw},S}(V)^{(\eta_0)}_{\Gamma_1} = d_-(V)$. On the other hand $\dim_{\mathbb{Q}_p} H^1_{f,\{p\}}(V) = d_-(V) + \dim_{\mathbb{Q}_p} H^0(\mathbb{Q}_p, V^*(1))$ by (6) and the dimension argument shows that in the commutative diagram

$$\begin{array}{ccccccccc}
0 & \longrightarrow & H^1_{\mathrm{Iw},S}(V)^{(\eta_0)}_{\Gamma_1} & \longrightarrow & H^1_{f,\{p\}}(V) & \longrightarrow & H^0(\mathbb{Q}_p,V^*(1))^* & \longrightarrow & 0 \\
 & & \downarrow & & \downarrow & & \downarrow & & \\
0 & \longrightarrow & H^1_{\mathrm{Iw}}(\mathbb{Q}_p,V)^{(\eta_0)}_{\Gamma_1} & \longrightarrow & H^1(\mathbb{Q}_p,V) & \longrightarrow & H^0(\mathbb{Q}_p,V^*(1))^* & \longrightarrow & 0
\end{array} \tag{29}$$

with obviously exact upper line the bottom line is also exact. This implies immediately that the natural map

$$\frac{H^1_{\mathrm{Iw}}(\mathbb{Q}_p,V)^{(\eta_0)}_{\Gamma_1}}{H^1_{\mathrm{Iw},S}(V)^{(\eta_0)}_{\Gamma_1} + H^1(\mathbf{D}_{-1})} \longrightarrow \frac{H^1(\mathbb{Q}_p,V)}{H^1_{f,\{p\}}(V) + H^1(\mathbf{D}_{-1})}$$

is an isomorphism.

Consider the exact sequence

$$0 \to \left(H^1_{\mathrm{Iw},S}(T)^{(\eta_0)} \oplus \mathcal{D}_p(N,T)^{(\eta_0)}\right) \otimes \mathcal{H} \to H^1_{\mathrm{Iw}}(\mathbb{Q}_p,T)^{(\eta_0)} \otimes \mathcal{H} \to \operatorname{coker}(f) \to 0.$$

Recall that $\mathrm{Exp}^{\varepsilon}_{V,h,0} : D \to H^1_{\mathrm{Iw}}(\mathbb{Q}_p,V)_\Gamma$ denotes the homomorphism induced by the large exponential map. Applying the snake lemma, and taking into account that $\mathrm{Im}(\mathrm{Exp}^{\varepsilon}_{V,h,0}) = \exp_{V,\mathbb{Q}_p}(D_{-1}) = H^1(\mathbf{D}_{-1})$ and $\ker(\mathrm{Exp}^{\varepsilon}_{V,h,0}) = D^{\varphi=p^{-1}}$ (see for example Benois and Berger 2008, Propositions 4.17 and 4.18 or the proof of Proposition 6) we obtain (by regularity of D)

$$\begin{aligned}
\operatorname{coker}(f)^{\Gamma_1} &= \ker\left(H^1_{\mathrm{Iw},S}(V)^{(\eta_0)}_{\Gamma_1} \oplus D \xrightarrow{\mathrm{Exp}^{\varepsilon}_{V,h,0}} H^1(\mathbb{Q}_p,V)\right) = D^{\varphi=p^{-1}}, \\
\operatorname{coker}(f)_{\Gamma_1} &= \frac{H^1_{\mathrm{Iw}}(\mathbb{Q}_p,V)^{(\eta_0)}_{\Gamma_1}}{H^1_{\mathrm{Iw},S}(V)^{(\eta_0)}_{\Gamma_1} + H^1(\mathbf{D}_{-1})} = \frac{H^1(\mathbb{Q}_p,V)}{H^1_{f,\{p\}}(V) + H^1(\mathbf{D}_{-1})}.
\end{aligned} \tag{30}$$

Thus, one has a commutative diagram

$$\begin{array}{ccc}
\mathrm{coker}(f)^{\Gamma_1} & \longrightarrow & D^{\varphi=p^{-1}} \\
\downarrow & & \downarrow \delta_{D,h} \\
\mathrm{coker}(f)_{\Gamma_1} & \longrightarrow & \dfrac{H^1(\mathbb{Q}_p,V)}{H^1_{f,\{p\}}(V)+H^1(\mathbf{D}_{-1})},
\end{array} \tag{31}$$

where horizontal arrows are isomorphisms, the left vertical arrow is the natural projection and the right vertical row is the map defined in Sect. 4.2.2. From Proposition 7 it follows that $\mathrm{coker}(f)^{\Gamma_1} \to \mathrm{coker}(f)_{\Gamma_1}$ is an isomorphism if and only if $\mathscr{L}(V,D) \neq 0$.

On the other hand, the arguments of Perrin-Riou (1995), Sect. 3.3.4 show that $Ш^2_{\mathrm{Iw},S}(V)_\Gamma = Ш^2_{\mathrm{Iw},S}(V)^\Gamma = 0$. We remark that Perrin-Riou assumes that $\mathbf{D}_{\mathrm{cris}}(V)^{\varphi=1} = \mathbf{D}_{\mathrm{cris}}(V)^{\varphi=p^{-1}} = 0$, but her proof works in our case without modifications and we repeat it for the convenience of the reader. Consider the commutative diagram (where we write $Ш^2_{\mathrm{Iw}}(V)$ instead $Ш^2_{\mathrm{Iw},S}(V)$ to abbreviate notation)

$$\begin{array}{ccccccc}
0 & & 0 & & & & \\
\downarrow & & \downarrow & & & & \\
H^1_{\mathrm{Iw}}(V)_\Gamma & \rightarrowtail & \bigoplus\limits_{v\in S} H^1_{\mathrm{Iw}}(\mathbb{Q}_v,V)_\Gamma & \longrightarrow & H^1_S(\mathbb{Q}(\zeta_{p^\infty}),V^*(1))^*_\Gamma & \twoheadrightarrow & Ш^2_{\mathrm{Iw}}(V)_\Gamma \\
\downarrow & & \downarrow & & \downarrow = & & \\
H^1_{f,\{p\}}(V) & \rightarrowtail & \bigoplus\limits_{\substack{v\in S\\ v\neq p}} H^1_f(\mathbb{Q}_v,V) \bigoplus H^1(\mathbb{Q}_p,V) & \twoheadrightarrow & H^1_S(V^*(1))^* & & \\
\downarrow & & \downarrow & & & & \\
H^0(\mathbb{Q}_p,V^*(1))^* & \xrightarrow{=} & H^0(\mathbb{Q}_p,V^*(1))^* & & & & \\
\downarrow & & \downarrow & & & & \\
0 & & 0. & & & &
\end{array}$$

The top row of this diagram is obtained by taking coinvariants in the Poitou–Tate exact sequence. Thus it is exact. The middle row is obtained from the exact sequence

$$0 \to H^1_S(V^*(1)) \to H^1(\mathbb{Q}_p, V^*(1)) \bigoplus \bigoplus_{v\in S-\{p\}} \frac{H^1(\mathbb{Q}_v, V^*(1))}{H^1_f(\mathbb{Q}_v, V^*(1))} \to H^1_{f,\{p\}}(V)^* \to 0$$

by taking duals. In particular, in the first and second lines of the diagram the first arrows are injections. Here we use the condition $H^1_f(V^*(1)) = 0$. The exactness of the left and middle columns follows from the diagram (29). The isomorphism from the right column comes from the exact sequence

$$0 \to H^1(\Gamma, H^0_S(\mathbb{Q}(\zeta_{p^\infty}), V^*(1))) \to H^1_S(V^*(1)) \to H^1_S(\mathbb{Q}(\zeta_{p^\infty}), V^*(1))^\Gamma \to 0$$

together with the remark that $H^1(\Gamma, H^0_S(\mathbb{Q}(\zeta_{p^\infty}), V^*(1))) = 0$ because $H^0_S(\mathbb{Q}(\zeta_{p^\infty}), V^*(1))^\Gamma = H^0_S(\mathbb{Q}, V^*(1)) = 0$ by **(C2)**. Now an easy diagram chase shows that $\text{Ш}^2_{\text{Iw},S}(V)_\Gamma = 0$. Finally, from $\dim_{\mathbb{Q}_p} \text{Ш}^2_{\text{Iw},S}(V)^\Gamma \leqslant \dim_{\mathbb{Q}_p} \text{Ш}^2_{\text{Iw},S}(V)_\Gamma$ it follows that $\text{Ш}^2_{\text{Iw},S}(V)^\Gamma = 0$. Therefore, applying the snake lemma to (28) we obtain a commutative diagram

$$\begin{array}{ccc} \operatorname{coker}(f)^{\Gamma_1} & \longrightarrow & \mathbf{R}^2\Gamma^{(\eta_0)}_{\text{Iw},h}(D,V)^{\Gamma_1} \\ \downarrow & & \downarrow \\ \operatorname{coker}(f)_{\Gamma_1} & \longrightarrow & \mathbf{R}^2\Gamma^{(\eta_0)}_{\text{Iw},h}(D,V)_{\Gamma_1}, \end{array} \tag{32}$$

where the horizontal arrows are isomorphisms and the vertical arrows are natural projections. This proves that $\mathbf{R}\Gamma^{(\eta_0)}_{\text{Iw},h}(D, V)$ is semisimple in degree 2. The semisimplicity in degree 3 is obvious because by (ii) $\mathbf{R}^3\Gamma^{(\eta_0)}_{\text{Iw},h}(D, V)^{\Gamma_1} = \mathbf{R}^3\Gamma^{(\eta_0)}_{\text{Iw},h}(D, V)_{\Gamma_1} = 0$. This completes the proof of Theorem 4. □

6.1.4

This following corollary relates the projection map

$$\mathbf{R}^2\Gamma^{(\eta_0)}_{\text{Iw},h}(D, V)^{\Gamma_1} \longrightarrow \mathbf{R}^2\Gamma^{(\eta_0)}_{\text{Iw},h}(D, V)_{\Gamma_1}$$

to the map λ_D defined in Proposition 7 and therefore to the $\mathscr{L}$-invariant. This relation plays the key role in the proof of Theorem 5 below (Theorem 2 of Introduction).

Corollary 1. *(i) One has canonical isomorphisms*

$$\mathbf{R}^2\Gamma^{(\eta_0)}_{\text{Iw},h}(D, V)^{\Gamma_1} \xrightarrow{\sim} D^{\varphi=p^{-1}},$$

$$\mathbf{R}^2\Gamma^{(\eta_0)}_{\text{Iw},h}(D, V)_{\Gamma_1} \xrightarrow{\sim} \frac{H^1(\mathbb{Q}_p, V)}{H^1_{f,\{p\}}(V) + H^1(\mathbf{D}_{-1})}.$$

(ii) The exponential map induces an isomorphism of $D^{\varphi=p^{-1}}$ onto $\mathbf{R}^2\Gamma^{(\eta_0)}_{\text{Iw},h}(D, V)_{\Gamma_1}$ and the following diagram commutes

$$\begin{array}{ccc} D^{\varphi=p^{-1}} & \xrightarrow{\sim} & \mathbf{R}^2\Gamma^{(\eta_0)}_{\text{Iw},h}(D,V)^{\Gamma_1} \\ \downarrow{\scriptstyle \lambda_D} & & \downarrow \\ D^{\varphi=p^{-1}} & \xrightarrow{(h-1)!\exp_V} & \mathbf{R}^2\Gamma^{(\eta_0)}_{\text{Iw},h}(D,V)_{\Gamma_1} \end{array}$$

where the map λ_D is defined in Proposition 7.

Proof. The corollary follows from diagrams (30), (31), (32) and the definition of λ_D.

6.2 *The Module of p-Adic L-Functions*

6.2.1 The Canonical Trivialisation

We preserve the notation and conventions of Sect. 5.2. Let D be a regular subspace of $\mathbf{D}_{\mathrm{cris}}(V)$ and assume that $\mathscr{L}(V,D) \neq 0$. We review Perrin-Riou's definition of the module of p-adic L-functions using the formalism of Selmer complexes. Set

$$\Delta_{\mathrm{Iw},h}(D,V) =$$

$$\det{}^{-1}_{\Lambda_{\mathbb{Q}_p}}\left(\mathbf{R}\Gamma^{(\eta_0)}_{\mathrm{Iw},S}(V) \oplus \left(\underset{v\in S}{\oplus}\mathbf{R}\Gamma^{(\eta_0)}_{\mathrm{Iw}}(\mathbb{Q}_v,D,V)\right)\right) \otimes \det{}_{\Lambda_{\mathbb{Q}_p}}\left(\underset{v\in S}{\oplus}\mathbf{R}\Gamma^{(\eta_0)}_{\mathrm{Iw}}(\mathbb{Q}_v,V)\right).$$

The distinguished triangle

$$\mathbf{R}\Gamma^{(\eta_0)}_{\mathrm{Iw},S}(D,V) \to \left(\mathbf{R}\Gamma^{(\eta_0)}_{\mathrm{Iw},S}(V) \bigoplus \left(\bigoplus_{v\in S}\mathbf{R}\Gamma^{(\eta_0)}_{\mathrm{Iw}}(\mathbb{Q}_v,D,V)\right)\right) \otimes \mathcal{H}$$

$$\to \left(\bigoplus_{v\in S}\mathbf{R}\Gamma^{(\eta_0)}_{\mathrm{Iw}}(\mathbb{Q}_v,V)\right) \otimes \mathcal{H} \to \mathbf{R}\Gamma^{(\eta_0)}_{\mathrm{Iw},S}(D,V)[1]$$

gives an isomorphism $\Delta_{\mathrm{Iw},h}(D,V) \otimes_{\Lambda_{\mathbb{Q}_p}} \mathcal{H} \simeq \det^{-1}_{\mathcal{H}}\mathbf{R}\Gamma^{(\eta_0)}_{\mathrm{Iw},S}(D,V)$. Let $\mathcal{K}$ denote the field of fractions of $\mathcal{H}$. By Theorem 4, all $\mathbf{R}^i\Gamma^{(\eta_0)}_{\mathrm{Iw},S}(D,V)$ are $\mathcal{H}$-torsion and we have a canonical map

$$\det{}^{-1}_{\mathcal{H}}\mathbf{R}\Gamma^{(\eta_0)}_{\mathrm{Iw},S}(D,V) \simeq \underset{i\in\{2,3\}}{\otimes}\det{}_{\mathcal{H}}^{(-1)^{i+1}}\mathbf{R}^i\Gamma^{(\eta_0)}_{\mathrm{Iw},S}(D,V) \hookrightarrow \mathcal{K}.$$

The composition of these maps gives a trivialization

$$i_{V,\mathrm{Iw},h} \;:\; \Delta_{\mathrm{Iw},h}(D,V) \longrightarrow \mathcal{K}.$$

6.2.2 Local Conditions

We compare local conditions coming from Perrin-Riou's theory to those of Bloch–Kato. Roughly speaking the computations of this section explain the appearance of the Euler-like factor $\mathcal{E}^+(V,D)$ in the formula for the special value of $L_{\mathrm{Iw},h}(T,N,s)$. They will be used in the proof of Theorem 5.

Set $\mathbf{R}\Gamma(\mathbb{Q}_p, D, V) = D[-1]$ and define

$$S = \mathrm{cone}\left(\frac{1-p^{-1}\varphi^{-1}}{1-\varphi} \,:\, \mathbf{R}\Gamma(\mathbb{Q}_p, D, V) \to \mathbf{R}\Gamma_f(\mathbb{Q}_p, V)\right)[-1]. \tag{33}$$

Since $\det_{\mathbb{Q}_p} \mathbf{R}\Gamma_f(\mathbb{Q}_p, V)$ is canonically isomorphic to $\det_{\mathbb{Q}_p}^{-1} t_V(\mathbb{Q}_p)$ by (9), the distinguished triangle

$$S \to \mathbf{R}\Gamma(\mathbb{Q}_p, D, V) \to \mathbf{R}\Gamma_f(\mathbb{Q}_p, V) \to S[1]$$

induces an isomorphism

$$\alpha_S \,:\, \det_{\mathbb{Q}_p} t_V(\mathbb{Q}_p) \xrightarrow{\sim} \det_{\mathbb{Q}_p} D \otimes \det_{\mathbb{Q}_p} S. \tag{34}$$

Explicitly

$$S = \left[D \oplus \mathbf{D}_{\mathrm{cris}}(V) \to \mathbf{D}_{\mathrm{cris}}(V) \oplus t_V(\mathbb{Q}_p)\right][-1] \simeq \\ [D \oplus \mathbf{D}_{\mathrm{cris}}(V) \to \mathbf{D}_{\mathrm{cris}}(V) \oplus D]\,[-1],$$

where the unique non-trivial map is given by

$$(x, y) \mapsto \left((1-\varphi)\,y, \left(\frac{1-p^{-1}\varphi^{-1}}{1-\varphi}x + y\right) \pmod{\mathrm{Fil}^0\mathbf{D}_{\mathrm{cris}}(V)}\right).$$

Thus

$$H^1(S) = D^{\varphi=p^{-1}}, \qquad H^2(S) = \frac{t_V(\mathbb{Q}_p)}{(1-p^{-1}\varphi^{-1})D} \simeq \frac{\mathbf{D}_{\mathrm{cris}}(V)}{\mathrm{Fil}^0\mathbf{D}_{\mathrm{cris}}(V) + D_{-1}}. \tag{35}$$

From the semi-simplicity of $\dfrac{1-p^{-1}\varphi^{-1}}{1-\varphi}$ it follows that the natural projection $H^1(S) \oplus H^1_f(V) \to H^2(S)$ is an isomorphism and we have a canonical trivialisation

$$\beta_S \,:\, \det_{\mathbb{Q}_p} S \otimes \det_{\mathbb{Q}_p} \mathbf{R}\Gamma_f(V) \\ \simeq \det_{\mathbb{Q}_p}^{-1} H^1(S) \otimes \det_{\mathbb{Q}_p} H^2(S) \otimes \det_{\mathbb{Q}_p}^{-1} H^1_f(V) \simeq \mathbb{Q}_p. \tag{36}$$

The composition of β_S with α_S gives an isomorphism

$$\theta_S \,:\, \det_{\mathbb{Q}_p} t_V(\mathbb{Q}_p) \otimes \det_{\mathbb{Q}_p} \mathbf{R}\Gamma_f(V) \overset{\alpha_S}{\simeq} \\ \det_{\mathbb{Q}_p} D \otimes \det_{\mathbb{Q}_p} S \otimes \det_{\mathbb{Q}_p} \mathbf{R}\Gamma_f(V) \overset{\mathrm{id}\otimes\beta_S}{\simeq} \det_{\mathbb{Q}_p} D. \tag{37}$$

Fix bases $\omega_{t_V} \in \det_{\mathbb{Q}_p} t_V(\mathbb{Q}_p)$, $\omega_D \in \det_{\mathbb{Q}_p} D$ and $\omega_f \in \det_{\mathbb{Q}_p} H^1_f(V)$. Let $R_{V,D}(\omega_{V,D})$ denote the determinant of the regulator map

$$r_{V,D} : H^1_f(V) \to \mathbf{D}_{\mathrm{cris}}(V)/(\mathrm{Fil}^0\mathbf{D}_{\mathrm{cris}}(V) + D)$$

defined in Sect. 4.1.3 with respect to ω_f and $\omega_{t_V} \otimes \omega_D^{-1}$.

Lemma 3. *(i) Let $f : W \to W$ be a semi-simple endomorphism of a finite-dimensional k-vector space W. The canonical projection* $\ker(f) \to \mathrm{coker}(f)$ *is an isomorphism and the tautological exact sequence*

$$0 \to \ker(f) \to W \xrightarrow{f} W \to \mathrm{coker}(f) \to 0$$

induces an isomorphism

$$\det{}^* f : \det_k(W) \to \det_k(W) \otimes \det_k(\ker(f)) \otimes \det{}_k^{-1}(\mathrm{coker}(f)) \to \det_k(W).$$

Furthermore $\det^* f(x) = \det(f \mid \mathit{coker}(f))$.

(ii) The map θ_S sends $\omega_{t_V} \otimes \omega_f^{-1}$ onto

$$\det{}^* \left(\frac{1 - p^{-1}\varphi^{-1}}{1-\varphi} | D\right)^{-1} E_p(V,1)^{-1} R_{V,D}(\omega_{V,D})^{-1} \omega_D$$

Proof. The proof is straightforward and is omitted here.

6.2.3 Definition of the Module of p-Adic L-Functions

In this subsection we interpret Perrin-Riou's construction of the module of p-adic L-functions in terms of Nekovář (2006). Fix a $\mathbb{Z}_p$-lattice N of D and set

$$\Delta_{\mathrm{Iw},h}(N,T) =$$

$$\det{}_\Lambda^{-1}\left(\mathbf{R}\Gamma^{(\eta_0)}_{\mathrm{Iw},S}(T) \bigoplus \left(\bigoplus_{v \in S} \mathbf{R}\Gamma^{(\eta_0)}_{\mathrm{Iw}}(\mathbb{Q}_v, N, T)\right)\right) \otimes \det{}_\Lambda \left(\bigoplus_{v \in S} \mathbf{R}\Gamma^{(\eta_0)}_{\mathrm{Iw}}(\mathbb{Q}_v, T)\right).$$

The module of p-adic L-functions associated to (N,T) is defined as

$$\mathbf{L}^{(\eta_0)}_{\mathrm{Iw},h}(N,T) = i_{V,\mathrm{Iw},h}(\Delta_{\mathrm{Iw},h}(N,T)) \subset \mathcal{K}.$$

Fix a generator $f(\gamma_1 - 1)$ of $\mathbf{L}^{(\eta_0)}_{\mathrm{Iw},h}(N,T)$ and define a meromorphic p-adic function

$$L_{\mathrm{Iw},h}(T,N,s) = f(\chi(\gamma_1)^s - 1).$$

Let now V be the p-adic realisation of a pure motive M over $\mathbb{Q}$ which satisfies the conditions **(M1–M3)** of Sect. 5.1.2. As we saw in Sect. 5.1.2 one expects that V satisfies **(C1–C5)**. We fix bases $\omega_f \in \det_{\mathbb{Q}} H^1_f(M)$, $\omega_{t_M} \in \det_{\mathbb{Q}} t_M(\mathbb{Q})$ and use the same notation for their images in $\det_{\mathbb{Q}_p} H^1_f(V)$ and $\det_{\mathbb{Q}_p} t_V(\mathbb{Q}_p)$ respectively. As in Sect. 5.1.3 choose bases $\omega^+_{M_{\mathrm{B}}} \in \det_{\mathbb{Q}} M^+_{\mathrm{B}}$ and $\omega^+_T \in \det_{\mathbb{Z}_p} T^+$ and define the p-adic period $\Omega^{(\text{ét},p)}_M(\omega^+_T, \omega^+_{M_{\mathrm{B}}}) \in \mathbb{Q}_p$ by $\omega^+_T = \Omega^{(\text{ét},p)}_M(\omega^+_T, \omega^+_{M_{\mathrm{B}}})\omega^+_{M_{\mathrm{B}}}$ using the comparison isomorphism (13) (see also (14)). Let ω_N be a generator of $\det_{\mathbb{Z}_p} N$.

Theorem 5. *Assume that V satisfies **(C1–C5)** and that the weak Leopoldt conjecture holds for (V, η_0) and $(V^*(1), \eta_0)$. Let D be a regular submodule of $\mathbf{D}_{\mathrm{cris}}(V)$. Assume that $\mathscr{L}(V, D) \neq 0$. Then*

(i) $L_{\mathrm{Iw},h}(T, N, s)$ is a meromorphic p-adic function which has a zero at $s = 0$ of order $e = \dim_{\mathbb{Q}_p}(D^{\varphi=p^{-1}})$.
(ii) Let $L^*_{\mathrm{Iw},h}(T, N, 0) = \lim_{s\to 0} s^{-e} L_{\mathrm{Iw},h}(T, N, s)$ be the special value of $L_{\mathrm{Iw},h}(T, N, s)$ at $s = 0$. Then

$$\frac{L^*_{\mathrm{Iw},h}(T, N, 0)}{R_{V,D}(\omega_{V,N})} \sim_p \Gamma(h)^{d_+(V)}\, \mathscr{L}(V, D)\, \mathcal{E}^+(V, D)\, \frac{i_{\omega_M,p}\,(\Delta_{\mathrm{EP}}(T))}{\Omega^{(\text{ét},p)}_M(\omega^+_T, \omega^+_{M_{\mathrm{B}}})},$$

where $i_{\omega_M,p}$ and $\mathcal{E}^+(V, D)$ are defined by (21) and (25) respectively and $\Gamma(h) = (h-1)!$.

6.2.4 Proof of Theorem 5

6.2.4.1 First recall the formalism of Iwasawa descent which will be used in the proof. The result we need is proved in Burns and Greither (2003). This is a particular case of Nekovář's descent theory (Nekovář 2006). Let $C^\bullet$ be a perfect complex of $\mathcal{H}$-modules and let $C^\bullet_0 = C^\bullet \otimes^{\mathbf{L}}_{\mathcal{H}} \mathbb{Q}_p$. We have a natural distinguished triangle

$$C^\bullet \xrightarrow{X} C^\bullet \to C^\bullet_0 \to C^\bullet[1],$$

where $X = \gamma_1 - 1$. In each degree this triangle gives a short exact sequence

$$0 \to H^n(C^\bullet)_{\Gamma_1} \to H^n(C^\bullet_0) \to H^{n+1}(C^\bullet)^{\Gamma_1} \to 0.$$

One says that $C^\bullet$ is semisimple if the natural map

$$H^n(C^\bullet)^{\Gamma_1} \to H^n(C^\bullet) \to H^n(C^\bullet)_{\Gamma_1} \tag{38}$$

is an isomorphism in all degrees. If $C^\bullet$ is semisimple, there exists a canonical trivialisation of $\det_{\mathbb{Q}_p} C^\bullet_0$, namely

$$\begin{aligned}
\vartheta \,:\, \det_{\mathbb{Q}_p} C_0^\bullet &\simeq \underset{n\in\mathbb{Z}}{\otimes} \det_{\mathbb{Q}_p}^{(-1)^n} H^n(C_0) \\
&\simeq \underset{n\in\mathbb{Z}}{\otimes} \left(\det_{\mathbb{Q}_p}^{(-1)^n} H^n(C^\bullet)_{\Gamma_1} \otimes \det_{\mathbb{Q}_p}^{(-1)^n} H^{n+1}(C^\bullet)^{\Gamma_1}\right) \\
&\simeq \underset{n\in\mathbb{Z}}{\otimes} \left(\det_{\mathbb{Q}_p}^{(-1)^n} H^n(C^\bullet)_{\Gamma_1} \otimes \det_{\mathbb{Q}_p}^{(-1)^{n-1}} H^n(C^\bullet)^{\Gamma_1}\right) \simeq \mathbb{Q}_p
\end{aligned}$$

where the last map is induced by (38). We now suppose that $C^\bullet \otimes_{\mathcal{H}} \mathcal{K}$ is acyclic and write $\vartheta_\infty \,:\, \det_{\mathcal{H}} C^\bullet \to \mathcal{K}$ for the associated morphism. Then $\vartheta_\infty(\det_{\mathcal{H}} C^\bullet) = f\mathcal{H}$, where $f \in \mathcal{K}$. Let r be the unique integer such that $X^{-r} f$ is a unit of the localization $\mathcal{H}_0$ of $\mathcal{H}$ with respect to the principal ideal $X\mathcal{H}$.

Lemma 4. *Assume that $C^\bullet \otimes \mathcal{K}$ is acyclic and $C^\bullet$ is semisimple. Then*

$$r = \sum_{n\in Z} (-1)^{n+1} \dim_{\mathbb{Q}_p} H^n(C^\bullet)^{\Gamma_1}$$

and there exists a commutative diagram

$$\begin{array}{ccc}
\det_{\mathcal{H}} C^\bullet & \xrightarrow{\;X^{-r}\vartheta_\infty\;} & \mathcal{H}_0 \\
\Big\downarrow {\scriptstyle \otimes^{\mathbf{L}}_{\mathcal{H}} \mathbb{Q}_p} & & \Big\downarrow \\
\det_{\mathbb{Q}_p} C_0^\bullet & \xrightarrow{\;\vartheta\;} & \mathbb{Q}_p
\end{array}$$

in which the right vertical arrow is the augmentation map.

Proof. See Burns and Greither (2003), Lemma 8.1. Remark that Burns and Greither consider complexes over $\Lambda \otimes_{\mathbb{Z}_p} \mathbb{Q}_p$ but since $\mathcal{H}$ is a Bézout ring, all their arguments work in our case and are omitted here.

6.2.4.2 Now we can prove Theorem 5. By Theorem 4 the complex $\mathbf{R}\Gamma_{\mathrm{Iw},h}^{(\eta_0)}(D,V)$ is semisimple, its cohomology is $\mathcal{H}$-torsion and the first assertion follows from Lemma 4 together with Corollary 1.

6.2.4.3 In this subsection we prove (ii). Define

$$\begin{aligned}
\mathbf{R}\Gamma(\mathbb{Q}_v, N, T) &= \mathbf{R}\Gamma_{\mathrm{Iw}}^{(\eta_0)}(\mathbb{Q}_v, N, T) \otimes^{\mathbf{L}}_{\Lambda} \mathbb{Z}_p, \\
\mathbf{R}\Gamma(\mathbb{Q}_v, D, V) &= \mathbf{R}\Gamma(\mathbb{Q}_v, N, T) \otimes_{\mathbb{Z}_p} \mathbb{Q}_p.
\end{aligned}$$

We remark that for $v = p$ this definition coincides with the definition given in Sect. 6.2.2. Applying $\otimes^{\mathbf{L}}_{\mathcal{H}} \mathbb{Q}_p$ to the map

$$\mathbf{R}\Gamma_{\mathrm{Iw}}^{(\eta_0)}(\mathbb{Q}_v, N, T) \otimes^{\mathbf{L}}_{\Lambda} \mathcal{H} \to \mathbf{R}\Gamma_{\mathrm{Iw}}^{(\eta_0)}(\mathbb{Q}_v, T) \otimes^{\mathbf{L}}_{\Lambda} \mathcal{H}$$

we obtain a morphism

$$\mathbf{R}\Gamma_f(\mathbb{Q}_v, D, V) \to \mathbf{R}\Gamma(\mathbb{Q}_v, V).$$

If $v \neq p$, then $\mathbf{R}\Gamma_f(\mathbb{Q}_v, D, V) = \mathbf{R}\Gamma_f(\mathbb{Q}_v, V)$ and this morphism coincides with the natural map $\mathbf{R}\Gamma_f(\mathbb{Q}_v, V) \to \mathbf{R}\Gamma(\mathbb{Q}_v, V)$. If $v = p$, then $\mathbf{R}\Gamma_f(\mathbb{Q}_v, D, V) = D[-1]$ and by Theorem 3 it coincides with the composition

$$D \xrightarrow{\frac{1-p^{-1}\varphi^{-1}}{1-\varphi}} \mathbf{D}_{\mathrm{cris}}(V) \xrightarrow{(h-1)!\exp_{V,\mathbb{Q}_p}} H^1(\mathbb{Q}_p, V).$$

Since $\mathbf{R}\Gamma_S(V) = \mathbf{R}\Gamma^{(\eta_0)}_{\mathrm{Iw},S}(V) \otimes^{\mathbf{L}}_{\Lambda} \mathbb{Q}_p$ this implies that

$$\mathbf{R}\Gamma^{(\eta_0)}_{\mathrm{Iw},h}(D, V) \otimes_{\mathcal{H}} \mathbb{Q}_p = \mathbf{R}\Gamma_h(D, V) \tag{39}$$

where $\mathbf{R}\Gamma_h(D, V)$ is the Selmer complex associated to the diagram

$$\begin{array}{ccc} \mathbf{R}\Gamma_S(V) & \longrightarrow & \bigoplus_{v\in S} \mathbf{R}\Gamma(\mathbb{Q}_v, V) \\ & & \uparrow \\ & & \bigoplus_{v\in S} \mathbf{R}\Gamma(\mathbb{Q}_v, D, V). \end{array}$$

We have a distinguished triangle

$$\mathbf{R}\Gamma_h(D, V) \to \mathbf{R}\Gamma_S(V) \bigoplus \left(\bigoplus_{v\in S} \mathbf{R}\Gamma(\mathbb{Q}_v, D, V) \right) \to \bigoplus_{v\in S} \mathbf{R}\Gamma(\mathbb{Q}_v, V) \to \mathbf{R}\Gamma_h(D, V)[1]. \tag{40}$$

Passing to determinants and using the trivialisation (10) of $\mathbf{R}\Gamma_f(\mathbb{Q}_v, V)$ for $v \neq p$ we obtain isomorphisms

$$\det{}^{-1}_{\mathbb{Q}_p} \mathbf{R}\Gamma_S(V) \otimes_{\mathbb{Q}_p} \left(\underset{v\in S}{\otimes} \det{}_{\mathbb{Q}_p} \mathbf{R}\Gamma(\mathbb{Q}_v, V) \right) \otimes \det{}_{\mathbb{Q}_p} D \xrightarrow{\sim} \det{}^{-1}_{\mathbb{Q}_p} \mathbf{R}\Gamma_h(D, V),$$

$$\xi_{D,h} \,:\, \Delta_{\mathrm{EP}}(V) \otimes_{\mathbb{Q}_p} \left(\det{}_{\mathbb{Q}_p} D \otimes \det{}^{-1}_{\mathbb{Q}_p} V^+ \right) \xrightarrow{\sim} \det{}^{-1}_{\mathbb{Q}_p} \mathbf{R}\Gamma_h(D, V), \tag{41}$$

where $\Delta_{\mathrm{EP}}(V)$ is the Euler–Poincaré line defined in Sect. 5.1.1. From (39) it follows that for any i one has an exact sequence

$$0 \to \mathbf{R}^i \Gamma^{(\eta_0)}_{\mathrm{Iw},h}(D, V)_{\Gamma_1} \to \mathbf{R}^i \Gamma_h(D, V) \to \mathbf{R}^{i+1} \Gamma^{(\eta_0)}_{\mathrm{Iw},h}(D, V)^{\Gamma_1} \to 0.$$

Now by Theorem 4

$$\mathbf{R}^i\Gamma_h(D,V) = \begin{cases} \mathbf{R}^2\Gamma_{\mathrm{Iw},h}^{(\eta_0)}(D,V)^{\Gamma_1} & \text{if } i=1 \\ \mathbf{R}^2\Gamma_{\mathrm{Iw},h}^{(\eta_0)}(D,V)_{\Gamma_1} & \text{if } i=2 \\ 0 & \text{if } i\neq 1,2. \end{cases} \tag{42}$$

By Corollary 1 the projection $\mathbf{R}^2\Gamma_{\mathrm{Iw},h}^{(\eta_0)}(D,V)^{\Gamma_1} \to \mathbf{R}^2\Gamma_{\mathrm{Iw},h}^{(\eta_0)}(D,V)_{\Gamma_1}$ is an isomorphism which induces a canonical trivialisation

$$\vartheta_{D,h} \,:\, \det{}_{\mathbb{Q}_p}^{-1}\mathbf{R}\Gamma_h(D,V) \tilde{\to} \mathbb{Q}_p$$

of $\mathbf{R}\Gamma_h(D,V)$. Applying Lemma 4 to the complex $\mathbf{R}\Gamma_{\mathrm{Iw},h}^{(\eta_0)}(D,V)$ we obtain a commutative diagram

$$\begin{array}{ccc} \det{}_{\mathcal{H}}^{-1}\mathbf{R}\Gamma_{\mathrm{Iw},h}^{(\eta_0)}(D,V) & \xrightarrow{X^{-e}i_{V,\mathrm{Iw},h}} & \mathcal{H}_0 \\ \Big\downarrow{\scriptstyle \otimes^{\mathbf{L}}_{\mathcal{H}}\mathbb{Q}_p} & & \Big\downarrow \\ \det{}_{\mathbb{Q}_p}^{-1}\mathbf{R}\Gamma_h(D,V) & \xrightarrow{\vartheta_{D,h}} & \mathbb{Q}_p \end{array} \tag{43}$$

where $i_{V,\mathrm{Iw},h}$ is the canonical trivialisation constructed in Sect. 6.2.1. From the definition of the module of p-adic L-functions

$$\Delta_{\mathrm{Iw},h}(N,T)\otimes^{\mathbf{L}}_{\Lambda}\mathbb{Z}_p \tilde{\to}$$

$$\det{}_{\mathbb{Z}_p}^{-1}\mathbf{R}\Gamma_S(T)\otimes\Big(\underset{v\in S}{\otimes}\det{}_{\mathbb{Z}_p}\mathbf{R}\Gamma(\mathbb{Q}_v,T)\Big)\otimes\Big(\underset{v\in S-\{p\}}{\otimes}\det{}_{\mathbb{Z}_p}\mathbf{R}\Gamma_f(\mathbb{Q}_v,T)\Big)\otimes\det{}_{\mathbb{Z}_p}N,$$

and again the canonical trivialisation (10) of $\mathbf{R}\Gamma_f(\mathbb{Q}_v,V)$ for $v\neq p$ induces a map

$$\Delta_{\mathrm{Iw},h}(N,T)\otimes^{\mathbf{L}}_{\Lambda}\mathbb{Z}_p \hookrightarrow \det{}_{\mathbb{Q}_p}^{-1}\mathbf{R}\Gamma_h(D,V).$$

Comparing this map with the definition (41) of $\xi_{D,h}$ one has

$$\Delta_{\mathrm{Iw},h}(N,T)\otimes^{\mathbf{L}}_{\Lambda}\mathbb{Z}_p = \xi_{D,h}\left(\Delta_{\mathrm{EP}}(T)\otimes_{\mathbb{Z}_p}\omega_N\otimes_{\mathbb{Z}_p}(\omega_T^+)^{-1}\right).$$

By definition, $L_{\mathrm{Iw}}(T,N,s)$ is a generator of $i_{V,\mathrm{Iw},h}(\Delta_{\mathrm{Iw},h}(N,T))$. Therefore, the diagram (43) gives

$$\vartheta_{D,h}\circ\xi_{D,h}(\Delta_{\mathrm{EP}}(T)\otimes_{\mathbb{Z}_p}\omega_N\otimes_{\mathbb{Z}_p}(\omega_T^+)^{-1}) = \log(\chi(\gamma))^{-e}L^*_{\mathrm{Iw},h}(T,N,0)\,\mathbb{Z}_p. \tag{44}$$

We will now compute the left hand side of this equality in terms of the canonical trivialisation of the Euler–Poincaré line. Roughly speaking we should compare the trivialisation of the Euler–Poincaré line to the trivialisation $\vartheta_{D,h}$. Consider the commutative diagram

$$\begin{array}{ccccc}
\mathbf{R}\Gamma_f(V) & \longrightarrow & \mathbf{R}\Gamma_S(V) \oplus \bigoplus_{v\in S\cup\{\infty\}} \mathbf{R}\Gamma_f(\mathbb{Q}_v,V) & \longrightarrow & \bigoplus_{v\in S\cup\{\infty\}} \mathbf{R}\Gamma(\mathbb{Q}_v,V) \\
\uparrow & & \uparrow & & \uparrow \\
\mathbf{R}\Gamma_h(D,V) & \longrightarrow & \mathbf{R}\Gamma_S(V) \oplus \bigoplus_{v\in S} \mathbf{R}\Gamma(\mathbb{Q}_v,D,V) & \longrightarrow & \bigoplus_{v\in S} \mathbf{R}\Gamma(\mathbb{Q}_v,V) \\
\uparrow & & \uparrow & & \uparrow \\
L & \longrightarrow & S\oplus V^+[-1] & \longrightarrow & V^+[-1]
\end{array} \qquad (45)$$

where $L = \operatorname{cone}\big(\mathbf{R}\Gamma_h(D,V) \to \mathbf{R}\Gamma_f(V)\big)\,[-1]$ and S is defined by (33). The upper and middle rows of (45) coincide with (8) and (40) up to the following modification: the map $\mathrm{loc}_p \;:\; \mathbf{R}\Gamma_f(\mathbb{Q}_p,V) \to \mathbf{R}\Gamma(\mathbb{Q}_p,V)$ is replaced by $\Gamma(h)\,\mathrm{loc}_p$. Hence S is isomorphic to L in the derived category $\mathcal{D}^p(\mathbb{Q}_p)$ and we have an exact triangle

$$S \to \mathbf{R}\Gamma_h(D,V) \to \mathbf{R}\Gamma_f(V) \to S[1]. \qquad (46)$$

The cohomology of S is computed in (35). On the other hand, Corollary 1 together with (42) give

$$\mathbf{R}^i\Gamma_h(D,V) \xrightarrow{\sim} \begin{cases} D^{\varphi=1} & \text{if } i=1, \\ \dfrac{H^1(\mathbb{Q}_p,V)}{H^1_{f,\{p\}}(V)+H^1(\mathbf{D}_{-1})} & \text{if } i=2. \end{cases}$$

An easy diagram chase shows that the map $H^1(S) \to \mathbf{R}^1\Gamma_h(D,V)$ induced by (46) coincides with the identity map $\mathrm{id} : D^{\varphi=p^{-1}} \to D^{\varphi=p^{-1}}$ and that one has an exact sequence

$$0 \to H^1_f(V) \to H^2(S) \to \mathbf{R}^2\Gamma_h(D,V) \to 0$$

which can be identified with

$$0 \to H^1_f(V) \to \frac{\mathbf{D}_{\mathrm{cris}}(V)}{\mathrm{Fil}^0\mathbf{D}_{\mathrm{cris}}(V)+D_{-1}} \xrightarrow{\Gamma(h)\exp_V} \frac{H^1(\mathbb{Q}_p,V)}{H^1_{f,\{p\}}(V)+H^1(\mathbf{D}_{-1})} \to 0.$$

Therefore, we have a commutative diagram

$$\begin{array}{ccc}
\det_{\mathbb{Q}_p} S \otimes \det_{\mathbb{Q}_p} \mathbf{R}\Gamma_f(V) & \xrightarrow{\ \nu\ } & \det_{\mathbb{Q}_p} \mathbf{R}\Gamma_h(D,V) \\
\downarrow \beta_S & & \downarrow \vartheta^*_{D,h} \\
\mathbb{Q}_p & \xrightarrow{\ \kappa\ } & \mathbb{Q}_p
\end{array} \qquad (47)$$

where β_S was defined in (36), $\vartheta^*_{D,h}$ is the dual of the trivialisation $\vartheta_{D,h}$ and κ is the unique map which makes this diagram commute.

From Proposition 7 and Corollary 1 (ii) giving an explicit description of the trivialisation $\mathbf{R}^1\Gamma_h(D,V) \to \mathbf{R}^2\Gamma_h(D,V)$ we obtain immediately that

$$\kappa = (\log \chi(\gamma))^e \left(1-\frac{1}{p}\right)^e \mathscr{L}(V,D)^{-1}\,\mathrm{id}_{\mathbb{Q}_p}. \tag{48}$$

Passing to determinants in the diagram (45) we obtain a commutative diagram

$$\begin{array}{ccc}
\Delta_{\mathrm{EP}}(V)\otimes\left(\det_{\mathbb{Q}_p} t_V(\mathbb{Q}_p)\otimes\det^{-1}_{\mathbb{Q}_p}V^+\right)\otimes\det_{\mathbb{Q}_p}\mathbf{R}\Gamma_f(V) & \longrightarrow & \mathbb{Q}_p \\
\downarrow{\scriptstyle \alpha_S} & & \\
\Delta_{\mathrm{EP}}(V)\otimes(\det_{\mathbb{Q}_p}D\otimes\det^{-1}_{\mathbb{Q}_p}V^+)\otimes\det_{\mathbb{Q}_p}\mathbf{R}\Gamma_h(D,V)\otimes(\det_{\mathbb{Q}_p}S\otimes\det^{-1}_{\mathbb{Q}_p}L) & & \Big\| \\
\downarrow{\scriptstyle \xi_{D,h}\otimes\mu} & & \downarrow \\
\det^{-1}_{\mathbb{Q}_p}\mathbf{R}\Gamma_h(D,V)\otimes\det_{\mathbb{Q}_p}\mathbf{R}\Gamma_h(D,V) & \xrightarrow{\text{duality}} & \mathbb{Q}_p
\end{array} \tag{49}$$

where α_S is defined by (34) and $\mu\,:\,\det S\otimes\det^{-1}L\to\mathbb{Q}_p$ is induced by (46). The upper map can be easily compared with the trivialisation map $i_{\omega_M,p}$ defined by (21): it sends $\Delta_{\mathrm{EP}}(T)\otimes(\omega_{t_M}\otimes(\omega_T^+)^{-1}\otimes\omega_f)$ onto

$$\Gamma(h)^{d_+(V)}\,\frac{i_{\omega_M,p}(\Delta_{\mathrm{EP}}(T))}{\Omega_M^{(\acute{e}t,p)}(\omega_T^+,\omega_{\mathrm{B}}^+)}. \tag{50}$$

The tautological isomorphism $\det_{\mathbb{Q}_p}L \simeq \det_{\mathbb{Q}_p}\mathbf{R}\Gamma_h(D,V)\otimes\det^{-1}_{\mathbb{Q}_p}\mathbf{R}\Gamma_f(V)$ gives a commutative diagram

$$\begin{array}{ccc}
\det_{\mathbb{Q}_p}\mathbf{R}\Gamma_h(D,V)\otimes\det_{\mathbb{Q}_p}S\otimes\det^{-1}_{\mathbb{Q}_p}L & \longrightarrow & \det_{\mathbb{Q}_p}S\otimes\det_{\mathbb{Q}_p}\mathbf{R}\Gamma_f(V) \\
\downarrow{\scriptstyle \mu} & & \downarrow{\scriptstyle \nu} \\
\det_{\mathbb{Q}_p}\mathbf{R}\Gamma_h(D,V) & \xrightarrow{=} & \det_{\mathbb{Q}_p}\mathbf{R}\Gamma_h(D,V)
\end{array} \tag{51}$$

We can summarize the diagrams (49), (51) and (47) in the following commutative diagram

$$\begin{array}{ccccc}
\Delta_{\mathrm{EP}}(V)\otimes\left(\det(t_V(\mathbb{Q}_p))\otimes\det^{-1}V^+\right)\otimes\det\mathbf{R}\Gamma_f(V) & & \longrightarrow & & \mathbb{Q}_p \\
\downarrow{\scriptstyle \alpha_S} & & & & \Big\| \\
\Delta_{\mathrm{EP}}(V)\otimes(\det D\otimes\det S\otimes\det\mathbf{R}\Gamma_f(V))\otimes\det^{-1}V^+ & \xrightarrow{\xi_{D,h}\otimes\nu} & \det^{-1}\mathbf{R}\Gamma_{f,h}(D,V)\otimes\det\mathbf{R}\Gamma_{f,h}(D,V) & \xrightarrow{\text{duality}} & \mathbb{Q}_p \\
\downarrow{\scriptstyle \mathrm{id}\otimes\beta_S} & & \downarrow{\scriptstyle \mathrm{id}\otimes\vartheta^*_{D,h}} & & \Big\| \\
\Delta_{\mathrm{EP}}(V)\otimes\left(\det D\otimes\det^{-1}_{\mathbb{Q}_p}V^+\right) & \xrightarrow{\xi_{D,h}\otimes\kappa} & \det^{-1}\mathbf{R}\Gamma_{f,h}(D,V) & \xrightarrow{\vartheta_{D,h}} & \mathbb{Q}_p.
\end{array}$$

The image of $\Delta_{\mathrm{EP}}(T) \otimes (\omega_{t_M} \otimes (\omega_T^+)^{-1} \otimes \omega_f)$ under the upper map is given by (50). The composition of the left vertical maps is the map θ_S defined by (37). By Lemma 3 θ_S sends $\Delta_{\mathrm{EP}}(T) \otimes (\omega_{t_M} \otimes (\omega_T^+)^{-1} \otimes \omega_f)$ onto

$$\det{}^* \left(\frac{1 - p^{-1}\varphi^{-1}}{1-\varphi} \mid D \right)^{-1} E_p(V,1)^{-1} R_{V,D}(\omega_{V,N})\ \Delta_{\mathrm{EP}}(T) \otimes (\omega_N \otimes (\omega_T^+)^{-1}). \tag{52}$$

Next, (44) and (48) give

$$\vartheta_{D,h} \circ (\xi_{D,h} \otimes \kappa)(\Delta_{\mathrm{EP}}(T) \otimes \omega_N \otimes (\omega_T^+)^{-1}) \\ = \left(1 - \frac{1}{p}\right)^e \mathscr{L}(V,D)^{-1} L^*_{\mathrm{Iw},h}(T,N,0)\, \mathbb{Z}_p. \tag{53}$$

Putting together (50), (52) and (53) we obtain that

$$\frac{L^*_{\mathrm{Iw},h}(T,N,0)}{R_{V,D}(\omega_{V,N})} \sim_p \sim_p \Gamma(h)^{d_+(V)} \mathscr{L}(V,D)\ E_p(V,1) \\ \det{}_{\mathbb{Q}_p}\left(\frac{1 - p^{-1}\varphi^{-1}}{1-\varphi} \,|D_{-1} \right) \frac{i_{\omega_M,p}\,(\Delta_{\mathrm{EP}}(T))}{\Omega_M^{(\acute{e}t,p)}(\omega_T^+, \omega_{\mathrm{B}}^+)}$$

and the theorem is proved. □

6.2.5 Special Values of $L^*_{\mathrm{Iw},h}(T,N,s)$

Let $\tilde{H}^1_f(T)$ denote the image of $H^1_f(T)$ in $H^1_f(V)$ and let $\omega_{T,f}$ be a base of $\det_{\mathbb{Z}_p} \tilde{H}^1_f(T)$. Let $R_{V,D}(\omega_{T,N})$ denote the determinant of $r_{V,D}$ computed in the bases ω_{t_M}, ω_N and $\omega_{T,f}$.

Corollary 2. *Under the assumptions of Theorem 5 one has*

$$\frac{L^*_{\mathrm{Iw},h}(T,N,0)}{R_{V,D}(\omega_{T,N})} \sim_p \Gamma(h)^{d_+(V)} \mathscr{L}(V,D)\, \mathcal{E}^+(V,D)\ \frac{\#\text{Ш}(T^*(1))\, \mathrm{Tam}^0_{\omega_M}(T)}{\# H^0_S(V/T)\, \# H^0_S(V^*(1)/T^*(1))},$$

where $\text{Ш}(T^*(1))$ *is the Tate–Shafarevich group of Bloch–Kato (1990) and* $\mathrm{Tam}^0_{\omega_M}(T)$ *is the product of local Tamagawa numbers of* T *taken over all primes and computed with respect to a fixed base* ω_{t_M} *of* $\det_{\mathbb{Q}} t_M(\mathbb{Q})$.

Proof. The computation of the trivialisation of the Euler–Poincaré line (see for example Fontaine and Perrin-Riou 1994, Chap. II, Théorème 5.6.3) together with the definition (21) of $i_{\omega_M,p}$ give

$$i_{\omega_M,p}(\Delta_{\mathrm{EP}}(T)) = \frac{\#\text{Ш}(T^*(1))\,\mathrm{Tam}^0_{\omega_M}(T)}{\#H^0_S(V/T)\,\#H^0_S(V^*(1)/T^*(1))}\,\Omega_M^{(\acute{e}t,p)}(\omega_T^+,\omega_{M_{\mathrm{B}}}^+)\,[\omega_f:\omega_{T,f}].$$

Since $R_{V,D}(\omega_{T,N}) = R_{V,D}(\omega_{V,N})\,[\omega_f : \omega_{T,f}]$ the corollary follows from Theorem 5.

6.2.6 The Functional Equation

Recall that we set $h_i(V) = \dim_{\mathbb{Q}_p}(\mathrm{gr}_i\mathbf{D}_{\mathrm{dR}}(V))$ and $m = \sum_{i\in\mathbb{Z}} ih_i(V)$. Since V is crystalline, $\det_{\mathbb{Q}_p}(V)$ is a one dimensional crystalline representation and $\det_{\mathbb{Z}_p}(T) = T_0(m)$ where T_0 is an unramified $G_{\mathbb{Q}_p}$-module of rank 1 over $\mathbb{Z}_p$. The module $(T_0 \otimes W(\overline{\mathbb{F}}_p))^{\varphi=1}\,e_m$ where $e_m = (t^{-1}\otimes\varepsilon)^{\otimes m}$ is a $\mathbb{Z}_p$-lattice in $\det_{\mathbb{Q}_p}(\mathbf{D}_{\mathrm{cris}}(V)) = \mathbf{D}_{\mathrm{cris}}(V_0(m))$ which depends only on T and which we denote by $\mathbf{D}_{\mathrm{cris}}(T_0(m))$.

Let $D^{\perp}$ be the dual regular module. The exact sequence

$$0 \to D \to \mathbf{D}_{\mathrm{cris}}(V) \to \left(D^{\perp}\right)^* \to 0$$

gives an isomorphism

$$\det_{\mathbb{Q}_p} D \otimes \det^{-1}_{\mathbb{Q}_p} D^{\perp} \simeq \det_{\mathbb{Q}_p}\mathbf{D}_{\mathrm{cris}}(V)$$

and we fix a lattice $N^{\perp} \subset D^{\perp}$ such that

$$\det_{\mathbb{Z}_p} N \otimes \det^{-1}_{\mathbb{Z}_p} N^{\perp} \simeq \mathbf{D}_{\mathrm{cris}}(T_0(m)).$$

Set $\Gamma_{V,h}(s) = \prod_{j>-h}(j+s)^{\dim \mathrm{Fil}^j\mathbf{D}_{\mathrm{dR}}(V)}$. The conjecture $\delta_{\mathbb{Z}_p}(V)$ of Perrin-Riou (1994) proved in Benois and Berger (2008) implies that for $h \gg 0$

$$L_{\mathrm{Iw},h}(T^*(1),N^{\perp},-s) \sim_{\Lambda^*} \Gamma_{V,h}^{-1}(s)\prod_{-h<j<h}(j+s)^{d_+(V^*(1))}L_{\mathrm{Iw},h}(T,N,s)$$

(see Perrin-Riou 1995, Théorème 2.5.2). This can seen as the algebraic counterpart of the functional equation for p-adic L-functions. An elementary computation (see Benois and Berger 2008, Lemme 4.7) shows that

$$\Gamma_{V,h}^{-1}(s)\prod_{-h<j<h}(j+s)^{d_+(V^*(1))} = \Gamma^*(V)\Gamma(h)^{d_+(V^*(1))-d_+(V)}s^r + o(s^r),$$

where $r = \dim_{\mathbb{Q}_p} t_V(\mathbb{Q}_p) - d_+(V) = \dim_{\mathbb{Q}_p} H^1_f(V)$ and $\Gamma^*(V)$ is defined by (23). Therefore $L_{\mathrm{Iw},h}(T^*(1),N^{\perp},s)$ has a zero of order $\dim_{\mathbb{Q}_p} H^1_f(V)+e$ at $s=0$. Moreover one has

$$\frac{L^*_{\mathrm{Iw},h}(T^*(1),N^\perp,0)}{\Gamma(h)^{d_+(V^*(1))}} \sim_p \Gamma^*(V)\,\frac{L_{\mathrm{Iw},h}(T,N,0)}{\Gamma(h)^{d_+(V)}}.$$

From the definition of $R_{V^*(1),D^\perp}$ (see Sect. 5.2.1) one has

$$R_{V^*(1),D^\perp}(\omega_{V^*(1),N^\perp}) = \Omega_M^{(H,p)}(\omega_T,\omega_{M_{\mathrm{dR}}})^{-1}\,R_{V,D}(\omega_{V,N})$$

where $\Omega_M^{(H,p)}$ denotes the period map defined by (17) and (18) and $\omega_{M_{\mathrm{dR}}} = \omega_{t_M} \otimes \omega^{-1}_{t_{M^*(1)}}$. Taking into account (24) we obtain that

$$\frac{L^*_{\mathrm{Iw},h}(T^*(1),N^\perp,0)}{R_{V^*(1),D^\perp}(\omega_{V,N^\perp})}$$
$$\sim_p \Gamma(h)^{d_+(V^*(1))}\,\mathscr{L}(V,D)\,\mathcal{E}^+(V^*(1),D^\perp)\,\frac{i_{\omega_M,p}\,(\Delta_{\mathrm{EP}}(T^*(1)))}{\Omega_M^{(\acute{e}t,p)}(\omega^+_{T^*(1)},\omega^+_{M^*(1)_{\mathrm{B}}})},$$

which is the analogue of Theorem 5 for $L_{\mathrm{Iw},h}(T^*(1),N^\perp,s)$.

Appendix

In this Appendix we prove some results about the cohomology of p-adic representations of local fields.

Let K be a finite extension of $\mathbb{Q}_p$ and T a p-adic representation of G_K. Fix a topological generator γ of Γ. Let $\mathbf{D}(T) = (T\otimes_{\mathbb{Z}_p}\mathbf{A})^{H_K}$ be the (φ,Γ)-module associated to T by Fontaine's theory (Fontaine 1991). Consider the complex

$$C_{\varphi,\gamma}(\mathbf{D}(T)) = \left[\mathbf{D}(T) \xrightarrow{f} \mathbf{D}(T)\oplus\mathbf{D}(T) \xrightarrow{g} \mathbf{D}(T)\right]$$

where the modules are placed in degrees 0, 1 and 2 and the maps f and g are given by

$$f(x) = ((\varphi-1)\,x\,,\,(\gamma-1)\,x), \qquad g(y,z) = (\gamma-1)\,y \;-\; (\varphi-1)\,z.$$

Proposition 8. *There are canonical and functorial isomorphisms*

$$h^i\;:\;H^i(C_{\varphi,\gamma}(\mathbf{D}(T))) \xrightarrow{\sim} H^i(K,T)$$

which can be described explicitly by the following formulas:

(i) If $i=0$, then h^0 coincides with the natural isomorphism

$$\mathbf{D}(T)^{\varphi=1,\gamma=1} \;=\; H^0(K,T\otimes_{\mathbb{Z}_p}\mathbf{A}^{\varphi=1}) \;=\; H^0(K,T).$$

(ii) *Let $\alpha, \beta \in \mathbf{D}(T)$ be such that $(\gamma - 1)\,\alpha = (1-\varphi)\,\beta$. Then h^1 sends* $\mathrm{cl}(\alpha,\beta)$ *to the class of the cocycle*

$$\mu_1(g) = (g-1)\,x + \frac{g-1}{\gamma-1}\,\beta,$$

where $x \in \mathbf{D}(T) \otimes_{\mathbf{A}_K} \mathbf{A}$ is a solution of the equation $(1-\varphi)\,x = \alpha$.

(iii) *Let $\hat{\gamma} \in G_K$ be a lifting of $g \in \Gamma$ and let x be a solution of $(\varphi - 1)\,x = \alpha$. Then h^2 sends α to the class of the 2-cocycle*

$$\mu_2(g_1, g_2) = \hat{\gamma}^{k_1}(h_1 - 1)\,\frac{\hat{\gamma}^{k_2}-1}{\hat{\gamma}-1}\,x$$

where $g_i = \hat{\gamma}^{k_i} h_i$, $h_i \in H_K$.

Proof. The isomorphisms h^i were constructed in Herr (1998), Theorem 2.1. Remark that (i) follows directly from this construction (see Herr 1998, p. 573) and that (ii) is proved in Benois (2000), Proposition 1.3.2 and Cherbonnier and Colmez (1999), Proposition I.4.1. The proof of (iii) follows along exactly the same lines. Namely, it is enough to prove this formula modulo p^n for each n. Let $\alpha \in \mathbf{D}(T)/p^n\mathbf{D}(T)$. By Proposition 2.4 of Herr (1998) there exists $r \geq 0$ and $y \in \mathbf{D}(T)/p^n\mathbf{D}(T)$ such that $(\varphi - 1)\,\alpha = (\gamma - 1)^r \beta$. Let

$$N_x = (\mathbf{D}(T)/p^n\mathbf{D}(T)) \oplus (\oplus_{i=1}^r (\mathbf{A}_K/p^n\mathbf{A}_K)\,t_i),$$

where $\varphi(t_i) = t_i + (\gamma-1)^{r-i}(\alpha)$ and $\gamma(t_i) = t_i + t_{i-1}$. Then N_x is a (φ, Γ)-module and we have a short exact sequence

$$0 \to \mathbf{D} \to N_x \to X \to 0$$

where $X = N_x/M \simeq \oplus_{i=1}^r \mathbf{A}_K/p^n\mathbf{A}_K\bar{t}_i$. An easy diagram chase shows that the connecting homomorphism $\delta^1_{\mathbf{D}} : H^1(C_{\varphi,\gamma}(\mathbf{D}(X))) \to H^2(C_{\varphi,\gamma}(\mathbf{D}(T)))$ sends $\mathrm{cl}(0, \bar{t}_r)$ to $-\mathrm{cl}(\alpha)$. The functor $\mathbf{V}(D) = (D \otimes_{\mathbf{A}_K} \mathbf{A})^{\varphi=1}$ is a quasi-inverse to $\mathbf{D}$. Thus one has an exact sequence of Galois modules

$$0 \to T/p^nT \to T_x \to \mathbf{V}(X) \to 0$$

where $T_x = \mathbf{V}(N_x)$. From the definition of x it follows immediately that $t_r - x \in T_x$. By (ii), $h^1(\mathrm{cl}(0, \bar{t}_r))$ can be represented by the cocycle $c(g) = \dfrac{g-1}{\gamma-1}\,\bar{t}_r$ and we fix its lifting $\hat{c} : G_K \to N_x$ putting $\hat{c}(g) = \dfrac{g-1}{\gamma-1}\,(t_r - x)$. As $g_1\hat{c}(g_2) - \hat{c}(g_1g_2) + \hat{c}(g_1) = -\mu_2(g_1, g_2)$, the connecting map $\delta^1_T : H^1(K, \mathbf{V}(X)) \to H^2(K, T/p^nT)$ sends $\mathrm{cl}(c)$ to $-\mathrm{cl}(\mu_2)$ and (iii) follows from the commutativity of the diagram

$$\begin{array}{ccc} H^1(C_{\varphi,\gamma}(X)) & \xrightarrow{\delta^1_{\mathbf{D}}} & H^2(C_{\varphi,\gamma}(T/p^nT)) \\ \downarrow{h^1} & & \downarrow{h^2} \\ H^1(K,\mathbf{V}(X)) & \xrightarrow{\delta^1_T} & H^2(K,T/p^nT). \end{array}$$

Proposition 9. *The complexes* $\mathbf{R}\Gamma(K,T)$ *and* $C_{\varphi,\gamma}(T)$ *are isomorphic in* $\mathcal{D}(\mathbb{Z}_p)$.

Proof. The proof is standard (see for example Burns and Flach 1996). The exact sequence

$$0 \to T \to \mathbf{D}(T) \otimes_{\mathbf{A}_K} \mathbf{A} \xrightarrow{\varphi-1} \mathbf{D}(T) \otimes_{\mathbf{A}_K} \mathbf{A} \to 0$$

gives rise to an exact sequence of complexes

$$0 \to C^\bullet_{\mathrm{c}}(G_K,T) \to C^\bullet_{\mathrm{c}}(G_K,\mathbf{D}(T) \otimes_{\mathbf{A}_K} \mathbf{A}) \xrightarrow{\varphi-1} C^\bullet_{\mathrm{c}}(G_K,\mathbf{D}(T) \otimes_{\mathbf{A}_K} \mathbf{A}) \to 0$$

Thus $\mathbf{R}\Gamma(K,T)$ is quasi-isomorphic to the total complex

$$K^\bullet(T) = \mathrm{Tot}^\bullet\left(C^\bullet_{\mathrm{c}}(G_K,\mathbf{D}(T) \otimes_{\mathbf{A}_K} \mathbf{A}) \xrightarrow{\varphi-1} C^\bullet_{\mathrm{c}}(G_K,\mathbf{D}(T) \otimes_{\mathbf{A}_K} \mathbf{A})\right).$$

On the other hand

$$C_{\varphi,\gamma}(T) = \mathrm{Tot}^\bullet\left(A^\bullet(T) \xrightarrow{\varphi-1} A^\bullet(T)\right),$$

where $A^\bullet(T) = [\mathbf{D}(T) \xrightarrow{\gamma-1} \mathbf{D}(T)]$. Consider the following commutative diagram of complexes

$$\begin{array}{ccccccc} \mathbf{D}(T) & \xrightarrow{\gamma-1} & \mathbf{D}(T) & \longrightarrow & 0 & \longrightarrow & \cdots \\ \downarrow{\beta_0} & & \downarrow{\beta_1} & & \downarrow & & \\ C^0(G_K,\mathbf{D}(T)\otimes_{\mathbf{A}_K}\mathbf{A}) & \longrightarrow & C^1(G_K,\mathbf{D}(T)\otimes_{\mathbf{A}_K}\mathbf{A}) & \longrightarrow & C^2(G_K,\mathbf{D}(T)\otimes_{\mathbf{A}_K}\mathbf{A}) & \longrightarrow & \cdots \end{array}$$

in which $\beta_0(x) = x$ viewed as a constant function on G_K and $\beta_1(x)$ denotes the map $G_K \to \mathbf{D}(T) \otimes_{\mathbf{A}_K} \mathbf{A}$ defined by $(\beta_1(x))(g) = \dfrac{g-1}{\gamma-1}x$. This diagram induces a map $\mathrm{Tot}^\bullet(A^\bullet(T) \xrightarrow{\varphi-1} A^\bullet(T)) \to K^\bullet(T)$ and we obtain a diagram

$$C_{\varphi,\gamma}(T) \to K^\bullet(T) \leftarrow \mathbf{R}\Gamma(K,T)$$

where the right map is a quasi-isomorphism. Then for each i one has a map

$$H^i(C_{\varphi,\gamma}(T)) \to H^i(K^\bullet(T)) \simeq H^i(K,T)$$

and an easy diagram chase shows that it coincides with h^i. The proposition is proved.

Corollary 3. *Let V be a p-adic representation of G_K. Then the complexes $\mathbf{R}\Gamma(K,V)$, $C_{\varphi,\gamma}(\mathbf{D}^{\dagger}(V))$ and $C_{\varphi,\gamma}(\mathbf{D}^{\dagger}_{\mathrm{rig}}(V))$ are isomorphic in $\mathcal{D}(\mathbb{Q}_p)$.*

Proof. This follows from Theorem 1.1 of Liu (2007) together with Proposition 9.

Recall that K_∞/K denotes the cyclotomic extension obtained by adjoining all p^n-th roots of unity. Let $\Gamma = \mathrm{Gal}(K_\infty/K)$ and let $\Lambda(\Gamma) = \mathbb{Z}_p[[\Gamma]]$ denote the Iwasawa algebra of Γ. For any $\mathbb{Z}_p$-adic representation T of G_K the induced representation $\mathrm{Ind}_{K_\infty/K}T$ is isomorphic to $(T\otimes_{\mathbb{Z}_p}\Lambda(\Gamma))^{\iota}$ and we set $\mathbf{R}\Gamma_{\mathrm{Iw}}(K,T) = C^{\bullet}_{\mathrm{c}}(G_K, \mathrm{Ind}_{K_\infty/K}T)$. Consider the complex

$$C_{\mathrm{Iw},\psi}(T) = \left[\mathbf{D}(T)\xrightarrow{\psi-1}\mathbf{D}(T)\right]$$

in which the first term is placed in degree 1.

Proposition 10. *There are canonical and functorial isomorphisms*

$$h^i_{\mathrm{Iw}} \;:\; H^i(C_{\mathrm{Iw},\psi}(T)) \to H^i_{\mathrm{Iw}}(K,T)$$

which can be described explicitly by the following formulas:

(i) Let $\alpha\in\mathbf{D}(T)^{\psi=1}$. Then $(\varphi-1)\,\alpha\in\mathbf{D}(T)^{\psi=0}$ and for any n there exists a unique $\beta_n\in\mathbf{D}(T)$ such that $(\gamma_n-1)\,\beta_n=(\varphi-1)\,\alpha$. The map h^1_{Iw} sends $\mathrm{cl}(\alpha)$ to $(h^1_n(\mathrm{cl}(\beta_n,\alpha)))_{n\in\mathbb{N}}\in H^1_{\mathrm{Iw}}(K_n,T)$.
(ii) If $\alpha\in\mathbf{D}(T)$, then $h^2_{\mathrm{Iw}}(\mathrm{cl}(\alpha)) = -(h^2_n(\varphi(\alpha)))_{n\in\mathbb{N}}$.

Proof. The proposition follows from Theorem II.1.3 and Remark II.3.2 of Cherbonnier and Colmez (1999) together with Proposition 8.

Proposition 11. *The complexes $\mathbf{R}\Gamma_{\mathrm{Iw}}(K,T)$ and $C_{\mathrm{Iw},\psi}(T)$ are isomorphic in the derived category $\mathcal{D}(\Lambda(\Gamma))$.*

Proof. We repeat the arguments used in the proof of Proposition 9 with some modifications. For any $n\geqslant 1$ one has an exact sequence

$$0\to\mathrm{Ind}_{K_n/K}T\rightarrowtail(\mathbf{D}(T)\otimes_{\mathbb{Z}_p}\mathbb{Z}_p[G_n]^{\iota})\otimes_{\mathbf{A}_K}\mathbf{A}\xrightarrow{\varphi-1}(\mathbf{D}(T)\otimes_{\mathbb{Z}_p}\mathbb{Z}_p[G_n]^{\iota})\otimes_{\mathbf{A}_K}\mathbf{A}\to 0.$$

Set $\mathbf{D}(\mathrm{Ind}_{K_\infty/K}T)=\mathbf{D}(T)\otimes_{\mathbb{Z}_p}\Lambda(\Gamma)^{\iota}$ and

$$\mathbf{D}(\mathrm{Ind}_{K_\infty/K}(T))\hat{\otimes}_{\mathbf{A}_K}\mathbf{A} = \varprojlim_n(\mathbf{D}(T)\otimes_{\mathbb{Z}_p}\mathbb{Z}_p[G_n]^{\iota})\otimes_{\mathbf{A}_K}\mathbf{A}.$$

As $\mathrm{Ind}_{K_n/K}T$ are compact, taking projective limit one obtains an exact sequence

$$0\to\mathrm{Ind}_{K_\infty/K}T\to\mathbf{D}(\mathrm{Ind}_{K_\infty/K}(T))\hat{\otimes}_{\mathbf{A}_K}\mathbf{A}\xrightarrow{\varphi-1}\mathbf{D}(\mathrm{Ind}_{K_\infty/K}(T))\hat{\otimes}_{\mathbf{A}_K}\mathbf{A}\to 0.$$

Thus $\mathbf{R}_{\mathrm{Iw}}(K,T)$ is quasi-isomorphic to

$$K^{\bullet}_{\mathrm{Iw}}(T)=\mathrm{Tot}^{\bullet}\left(C^{\bullet}_{\mathrm{c}}(G_K,\mathbf{D}(\mathrm{Ind}_{K_\infty/K}T)\hat{\otimes}_{\mathbf{A}_K}\mathbf{A})\xrightarrow{\varphi-1}C^{\bullet}_{\mathrm{c}}(G_K,\mathbf{D}(\mathrm{Ind}_{K_\infty/K}T)\hat{\otimes}_{\mathbf{A}_K}\mathbf{A})\right).$$

We construct a quasi-isomorphism $f_\bullet : C_{\mathrm{Iw},\psi}(T)\to K^{\bullet}_{\mathrm{Iw}}(T)$. Any $x\in\mathbf{D}(T)$ can be written in the form $x=(1-\varphi\psi)\,x+\varphi\psi(x)$ where $\psi(1-\varphi\psi)\,x=0$. Then for each $n\geqslant 0$ the equation $(\gamma_n-1)\,y_n=(\varphi\psi-1)\,x$ has a unique solution $y_n\in\mathbf{D}(T)^{\psi=0}$ (Cherbonnier and Colmez 1999, Proposition I.5.1). In particular, $y_n=\dfrac{\gamma_{n+1}-1}{\gamma_n-1}\,y_{n+1}$ and we have a compatible system of elements $Y_n=\sum\limits_{k=0}^{|G_n|-1}\gamma^k\otimes\gamma^k(y_n)\in\mathbf{D}(T)\otimes_{\mathbb{Z}_p}\mathbb{Z}_p[G_n]^{\iota}$. Put $Y=(Y_n)_{n\geq 0}\in\mathbf{D}(\mathrm{Ind}_{K_\infty/K}T)$. Then

$$(\gamma_n-1)\,Y_n=(\gamma-1)\,Y\pmod{\mathbf{D}(\mathrm{Ind}_{K_n/K}T)}.$$

Let $\eta_x\in C^1_{\mathrm{c}}(G_K,\mathbf{D}(\mathrm{Ind}_{K_\infty/K}T)\hat{\otimes}_{\mathbf{A}_K}\mathbf{A})$ be the map defined by $\eta_x(g)=\dfrac{g-1}{\gamma-1}(1\otimes x)$. Define

$$f_1:\mathbf{D}(T)\to K^1_{\mathrm{Iw}}(T)=C^0_{\mathrm{c}}(G_K,\mathbf{D}(\mathrm{Ind}_{K_\infty/K}T)\hat{\otimes}_{\mathbf{A}_K}\mathbf{A})\oplus C^1_{\mathrm{c}}(G_K,\mathbf{D}(\mathrm{Ind}_{K_\infty/K}T)\hat{\otimes}_{\mathbf{A}_K}\mathbf{A})$$

by $f_1(x)=(Y,\eta_x)$ and

$$f_2:\mathbf{D}(T)\to C^1_{\mathrm{c}}(G_K,\mathbf{D}(\mathrm{Ind}_{K_\infty/K}T)\hat{\otimes}_{\mathbf{A}_K}\mathbf{A})\subset K^2_{\mathrm{Iw}}(T)$$

by $f_2(z)=-\eta_{\varphi(z)}$. It is easy to check that $f_\bullet$ is a morphism of complexes. This gives a diagram

$$C_{\mathrm{Iw},\psi}(T)\to K^{\bullet}_{\mathrm{Iw}}(T)\leftarrow\mathbf{R}\Gamma_{\mathrm{Iw}}(K,T)$$

in which the right map is a quasi-isomorphism. Using Proposition 11 it is not difficult to check that for each i the induced map

$$H^i(C_{\mathrm{Iw},\psi}(T))\to H^i(K^{\bullet}_{\mathrm{Iw}}(T))\simeq H^i_{\mathrm{Iw}}(K,T)$$

coincides with h^i_{Iw}. The proposition is proved.

Corollary 4. *The complexes* $\mathbf{R}\Gamma_{\mathrm{Iw}}(K,T)$ *and* $C^{\dagger}_{\mathrm{Iw},\psi}(T)$ *are isomorphic in* $\mathcal{D}(\Lambda(\Gamma))$.

Proof. One has $\mathbf{D}^{\dagger}(T)^{\psi=1}=\mathbf{D}(T)^{\psi=1}$ (Cherbonnier and Colmez 1998, Proposition 3.3.2) and $\mathbf{D}^{\dagger}(T)/(\psi-1)=\mathbf{D}(T)/(\psi-1)$ (Liu 2007, Lemma 3.6). This shows that the inclusion $C^{\dagger}_{\mathrm{Iw},\psi}(T)\to C_{\mathrm{Iw},\psi}(T)$ is a quasi-isomorphism.

Remark. These results can be slightly improved. Namely, set $r_n = (p-1)p^{n-1}$. The method used in the proof of Proposition III.2.1 Cherbonnier and Colmez (1999) allows to show that $\psi(\mathbf{D}^{\dagger,r_n}(T)) \subset \mathbf{D}^{\dagger,r_{n-1}}(T)$ for $n \gg 0$. Moreover, for any $a \in \mathbf{D}^{\dagger,r_n}(T)$ the solutions of the equation $(\psi - 1)\,x = a$ are in $\mathbf{D}^{\dagger,r_n}(T)$. Thus

$$C^{\dagger,r_n}_{\mathrm{Iw},\psi}(T) = \left[\mathbf{D}^{\dagger,r_n}(T) \xrightarrow{\psi-1} \mathbf{D}^{\dagger,r_n}(T)\right], \qquad n \gg 0$$

is a well-defined complex which is quasi-isomorphic to $C^{\dagger}_{\mathrm{Iw},\psi}(T)$. Further, as $\varphi(\mathbf{A}^{\dagger,r/p}) = \mathbf{A}^{\dagger,r}$ we can consider the complex

$$C^{\dagger,r_n}_{\varphi,\gamma}(T) = \left[\mathbf{D}^{\dagger,r_{n-1}}(T) \xrightarrow{f} \mathbf{D}^{\dagger,r_n}(T) \oplus \mathbf{D}^{\dagger,r_{n-1}}(T) \xrightarrow{g} \mathbf{D}^{\dagger,r_n}(T)\right], \qquad n \gg 0$$

in which f and g are defined by the same formulas as before. Then the inclusion $C^{\dagger,r_n}_{\varphi,\gamma}(T) \to C_{\varphi,\gamma}(T)$ is a quasi-isomorphism.

Acknowledgements I am very grateful to Jan Nekovář for interesting discussions and Daniel Delbourgo, Chung Pang Mok and Thong Nguyen Quang Do for useful remarks. I would especially like to thank the referee for numerous suggestions in improving the paper and for the very patient correction of many minor points.

References

Beilinson, A.: Higher regulators and values of L-functions. In: Current Problems in Mathematics, vol. 24, pp. 181–238. Akad. Nauk USSR, Moscow (1984)

Benois, D.: Iwasawa theory of crystalline representations. Duke Math. J. **104**, 211–267 (2000)

Benois, D.: On trivial zeros of Perrin-Riou's L-functions. http://arxiv.org/abs/0906.2862 (2009)

Benois, D.: A generalization of Greenberg's $\mathcal{L}$-invariant. Am. J. Math. **133**, 1573–1632 (2011)

Benois, D.: Trivial zeros of p-adic L-functions at near central points. Journal de l'Institut de Math. de Jussieu **13**, 561–598 (2014)

Benois, D., Berger, L.: Théorie d'Iwasawa des représentations cristallines II. Commentarii Math. Helvetici **83**, 603–677 (2008)

Berger, L.: Représentations p-adiques et équations différentielles. Invent. Math. **148**, 219–284 (2002)

Berger, L.: Bloch and Kato's exponential map: three explicit formulas. In: Bloch, S., Fesenko, I., Illusie, L., Kurihara, M., Saito, S., Saito, T., Schneider, P. (eds.), Documenta Mathematica, Extra Volume: Kazuya Kato's Fiftieth Birthday, pp. 99–129. American Math. Society (2003)

Berger, L.: Limites de représentations cristallines. Compos. Math. **140**, 1473–1498 (2004)

Berger, L.: Equations différentielles p-adiques et (φ, N)-modules filtrés. Astérisque **319**, 13–38 (2008)

Bloch, S., Kato, K.: L-functions and Tamagawa numbers of motives. In: Grothendieck Festschrift I. Progress in Mathematics, vol. 86, pp. 333–400. Birkhäuser, Boston (1990)

Burns, D., Flach, M.: Motivic L-functions and Galois module structures. Math. Ann. **305**, 65–102 (1996)

Burns, D., Greither, C.: On the equivariant Tamagawa number conjecture for tate motives. Invent. Math. **153**, 303–359 (2003)

Cherbonnier, F., Colmez, P.: Représentations p-adiques surconvergentes. Invent. Math. **133**, 581–611 (1998)

Cherbonnier, F., Colmez, P.: Théorie d'Iwasawa des représentations p-adiques d'un corps local. J. Am. Math. Soc. **12**, 241–268 (1999)

Citro, C.: L-invariants of adjoint square Galois representations coming from modular forms. IMRN (2008). doi:10.1093/imrn/rnn048

Colmez, P.: Théorie d'Iwasawa des représentations de de Rham d'un corps local. Ann. Math. **148**, 485–571 (1998)

Colmez, P.: Fonctions L p-adiques. Astérisque **266**, 21–58 (2000)

Colmez, P.: Les conjectures de monodromie p-adiques. In: Séminaire Bourbaki, Volume 2001/2002. Astérisque, vol. 290, pp. 53–101. Société Mathématique de France, Paris (2003)

Colmez, P.: Représentations triangulines de dimension 2. Astérisque **319**, 213–258 (2008)

Colmez, P., Fontaine, J.-M.: Construction des représentations p-adiques semi-stables. Invent. Math. **140**, 1–43 (2000)

Dabrowski, A., Delbourgo, D.: S-adic L-functions attached to the symmetric square of a newform. Proc. Lond. Math. Soc. **74**, 559–611 (1997)

Dasgupta, S., Darmon, H., Pollack, R.: Hilbert modular forms and the Gross-Stark conjecture. Ann. Math. **174**, 439–484 (2011)

Delbourgo, D.: $\mathcal{L}$-invariants arising from conjugate measures of $\mathrm{Sym}^2 E$. Glasg. Math. J. **44**, 45–64 (2002)

Faltings, G.: Crystalline cohomology and p-adic Galois representations. In: Algebraic Analysis, Geometry and Number Theory (Baltimore, 1988), pp. 25–80. John Hopkins University Press, Baltimore (1989)

Fontaine, J.-M.: Sur certains types de représentations p-adiques du groupe de Galois d'un corps local; construction d'un anneau de Barsotti-Tate. Ann. Math. **115**, 529–577 (1982)

Fontaine, J.-M.: Représentations p-adiques des corps locaux. In: The Grothendieck Festschrift II. Progress in Mathematics, vol. 87, pp. 249–309. Birkhäuser, Boston (1991)

Fontaine, J.-M.: Valeurs spéciales de fonctions L des motifs. In: Séminaire Bourbaki, exposé 751. Astérisque, vol. 206, pp. 205–249. Société Mathématique de France, Paris (1992)

Fontaine, J.-M.: Le corps des périodes p-adiques. Astérisque **223**, 59–102 (1994a)

Fontaine, J-M.: Représentations p-adiques semi-stables. Astérisque **223**, 113–184 (1994b)

Fontaine, J.-M., Perrin-Riou, B.: Autour des conjectures de Bloch et Kato; cohomologie galoisienne et valeurs de fonctions L. In: Jannsen, U., Kleiman, S., Serre, J.-P. (eds.) Motives, Proceedings of Symposia in Pure Mathematics, Seattle, vol. 55, part 1, 599–706 (1994)

Ferrero, B., Greenberg, R.: On the behavior of p-adic L-functions at $s = 0$. Invent. Math. **50**, 91–102 (1978, 1979)

Greenberg, R.: Trivial zeros of p-adic L-functions. In: Mazur, B., Stevens, G. (eds.) p-Adic Monodromy and the Birch and Swinnerton-Dyer Conjecture. Contemporary Mathematics, vol. 165, pp. 149–174. American Mathematical Society (1994)

Greenberg, R., Stevens, G.: p-adic L-functions and p-adic periods of modular forms. Invent. Math. **111**, 407–447 (1993)

Gross, B.: p-adic L-series at $s = 0$. J. Fac. Sci. Univ. Tokyo, sect. IA Math. **28**, 979–994 (1981)

Gross, B., Koblitz, N.: Gauss sums and the p-adic Γ-function. Ann. Math. **109**, 569–581 (1979)

Harron, R.: The exceptional zero conjecture for symmetric powers of CM-modular forms: the ordinary case. IMRN **2013**, 3744–3770 (2013)

Harron, R., Jorza, A.: On symmetric power $\mathscr{L}$-invariants of Iwahori level Hilbert modular forms. http://www.its.caltech.edu/~ajorza/papers/harron-jorza-non-cm.pdf (2013)

Harron, R., Lei, A.: Iwasawa theory for symmetric powers of CM modular forms at non ordinary primes. Journal de Théorie des Nombres de Bordeaux (to appear, 2014). Preprint available on http://arxiv.org/abs/1208.1278

Herr, L.: Sur la cohomologie galoisienne des corps p-adiques. Bull. Soc. Math. Fr. **126**, 563–600 (1998)

Herr, L: Une approche nouvelle de la dualité locale de Tate. Math. Ann. **320**, 307–337 (2001)

Jannsen, U.: On the ℓ-adic cohomology of varieties over number fields and its Galois cohomology. In: Ihara, Y., Ribet, K., Serre, J.-P. (eds.) Galois Groups over **Q**. Mathematical Sciences Research Institute, vol. 16, pp. 315–360. Springer-Verlag Berlin and Heidelberg GmbH & Co. K (1989)

Kato, K.: p-adic Hodge theory and values of zeta-functions of modular forms. Astérisque **295**, 117–290 (2004)

Katz, N.M., Messing, W.: Some consequences of the Riemann hypothesis for varieties over finite fields. Invent. Math. **23**, 73–77 (1974)

Kedlaya, K.: A p-adic monodromy theorem. Ann. Math. **160**, 94–184 (2004)

Lazard, M.: Les zéros des fonctions analytiques d'une variable sur un corps valué complet. Publ. Math. IHES **14**, 47–75 (1962)

Liu, R.: Cohomology and duality for (φ, Γ)-modules over the Robba ring. IMRN **2007**(Art. IDrum150), 32 (2007)

Mazur, B.: On monodromy invariants occuring in global arithmetic and Fontaine's theory. In: Mazur, B., Stevens, G. (eds.) p-Adic Monodromy and the Birch and Swinnerton-Dyer Conjecture. Contemporary Mathematics, vol. 165, pp. 1–20. American Mathematical Society (1994)

Mok, C.P.: The exceptional zero conjecture for Hilbert modular forms. Compos. Math. **145**, 1–55 (2009)

Mazur, B., Tate, J., Teitelbaum, J.: On p-adic analogues of the conjectures of Birch and Swinnerton-Dyer. Invent. Math. **84**, 1–48 (1986)

Nakamura, K.: Iwasawa theory of de Rham (φ, Γ)-modules over the Robba ring. J. de l'Institut de Math. Jussieu **13**, 65–118 (2014)

Nekovář, J.: On p-adic height pairing. In: David, S. (ed.) Séminaire de Théorie des Nombres, Paris 1990/91. Progress in Mathematics, vol. 108, pp. 127–202. Birkhäuser, Boston (1993)

Nekovář, J.: Selmer complexes. Astérisque **310**, p. 559. Société Math. de France (2006)

Perrin-Riou, B.: Théorie d'Iwasawa des représentations p-adiques sur un corps local. Invent. Math. **115**, 81–149 (1994)

Perrin-Riou, B.: Fonctions L p-adiques des représentations p-adiques. Astérisque **229**, p. 198. Société Math. de France (1995)

Perrin-Riou, B.: Zéros triviaux des fonctions L p-adiques. Compos. Math. **114**, 37–76 (1998)

Perrin-Riou, B.: Théorie d'Iwasawa des représentations p-adiques semistables. Mémoires de la SMF, vol. 84, p. 111. Société Math. de France (2001)

Pottharst, J.: Cyclotomic Iwasawa theory of motives http://math.bu.edu/people/potthars/ (2012)

Pottharst, J.: Analytic families of finite slope Selmer groups. Algebr. Number Theory **7**, 1571–1611 (2013)

Riedel, A.: On Perrin-Riou's exponential map and reciprocity laws. PhD thesis, University of Heidelberg (2013). Available on archiv.ub.uni-heidelberg.de/volltextserver/14935/1/diss?

Rosso, G.: Derivative at $s = 1$ of the p-adic L-function of the symmetric square of a Hilbert modular form. Available on http://arxiv.org/abs/1306.4935 (2013)

Rosso, G.: Derivative of symmetric square p-adic L-functions via pull-back formula. Available on http://arxiv.org/abs/1401.1431 (2014)

Spiess, M.: On special zeros of p-adic L-functions of Hilbert modular forms. Invent. Math. **196**, 69–138 (2014)

Stevens, G.: Coleman's $\mathscr{L}$-invariant and families of modular forms. Astérisque **311**, 1–12 (2010)

On Special L-Values Attached to Siegel Modular Forms

Thanasis Bouganis

1 Introduction

Special values of L-functions play a central role in Iwasawa theory since they are indispensable for the formulation of the Main Conjectures. It is precisely this information which is encoded in the interpolation properties of the p-adic L-functions. The first step to construct these p-adic L-functions is to show that the L-functions under consideration evaluated at "critical" points have particular algebraic properties. These properties are usually described by Deligne's conjectures. In this article we address this kind of questions for L-functions associated to Siegel modular forms.

This article grew out of the author's effort to read carefully the book of Shimura "Arithmeticity in the Theory of Automorphic Forms" (Shimura 2000) which means to do also the "exercises" left by Shimura to the reader. One of them is related to the algebraicity of various special values of Siegel modular forms (see page 239, Remark 28.13 in (loc. cit.)). As Shimura points out the results left as exercises should follow by using the various techniques and results obtained in his book and various papers of him. This is indeed the case since most ideas of this article can be found in the various works of Shimura, which of course in turn requires some familiarity with them. In any rate we believe that it is useful to have the results worked out in this paper documented in the literature, and for this reason we decided to write this article. In this paper we consider the special values of Siegel modular forms of integral weight. In Bouganis (in preparation, a), the continuation of this article, we consider also special L-value attached to half-integral weight Siegel modular forms.

T. Bouganis (✉)
Department of Mathematical Sciences, Durham University, South Road, DH1 3LE, Durham, UK
e-mail: athanasios.bouganis@durham.ac.uk

T. Bouganis and O. Venjakob (eds.), *Iwasawa Theory 2012*, Contributions in Mathematical and Computational Sciences 7, DOI 10.1007/978-3-642-55245-8_4,

Let us point out some results in this article that we believe deserve special mention. The first is the reciprocity law of the action of the Galois group on half-integral weight Eisenstein series. For integral weight Eisenstein series one can find the reciprocity laws in the book of Feit (1986) (if not in the form that it is needed for our purposes). However to the best of our knowledge the reciprocity for half integral Eisenstein series has not been worked out for Siegel modular forms. Another interesting result is the definition of the period $\Omega_{\mathfrak{f}}$ appearing in Theorem 12. These kind of periods have been first considered by Sturm (1981) and Harris (1981) (and later also by Panchishkin), based on an idea of Shimura. We follow the ideas of Sturm in defining them but using some new results of Shimura we are able to improve in some cases the bounds on the weight of the Siegel modular forms that the results are applicable. Also the fact that we use the more precise form of the Andrianov-Kalinin type identity proved by Shimura, we can obtain slightly finer results, since we need to remove less Euler factors of the L-function.

This paper is organized as follows. In section two we have a very brief introduction to Siegel modular forms. Then we move to section three where after presenting various results of Shimura with respect the theory of theta series and Eisenstein series for the symplectic group, we prove the various reciprocity laws of the action of the absolute Galois group on the Eisenstein series. Some of the result have already appeared in Sturm (1981) and Feit (1986), and we use ideas of these works. For the case of half integral weight Eisenstein series we prove the reciprocity inspired by an idea of Shimura. In Sect. 4 we introduce the L-functions which are considered in this paper. All the material of this section is from Shimura's book. In Sect. 5 we also present the work of Shimura on the generalization of the so-called Adrianov-Kalinin type identity. However for our purposes we use an integral expression that it is not in the book Shimura (2000) but in a paper of Shimura (1994). The use of this integral expression will lead to study slightly different L-functions than in the ones studied in the book of Shimura (we explain more later on this). Also in this section all the material is taken from works of Shimura. In Sect. 6 we define the periods that we will use to obtain the good reciprocity laws. The idea of defining this periods as values of an L-function goes back to Shimura, and have been used by Sturm (1981), Harris (1981) and Panchishkin (1991) in the case Siegel modular forms over the rationals and of even degree. We also note that we obtain a slightly different results than in these works, partly because we use some newer results of Shimura that were not available when these works were written. In the following section we present the various results on the field of rationality of the various special functions properly normalized and in some cases we provide some reciprocity results. Finally we finish this work by briefly discussing yet another method for considering the same questions as in this paper, namely the doubling method.

One last remark with respect to the notation used in this article. Since we are using as our main reference the book of Shimura (2000) we decided to keep, the anyway excellent, notation used by him. In particular if some times we use some notions not defined in this paper the reader will find the exact same notation also in the reference. This allows to keep the length of this article reasonable since we do not need to introduce all the mathematical notions used here. We finally remark that our choice to use the notation as in Shimura's book leads us to write $Z(s, \mathbf{f})$ for the L-functions attached to a Siegel modular form $\mathbf{f}$ instead of the more standard $L(s, \mathbf{f})$.

2 Siegel Modular Forms

2.1 Integral Weight Siegel Modular Forms

In this section we introduce the notion of a Siegel modular form (classicaly and adelically). We follow closely the book of Shimura (2000).

For a positive integer $n \in \mathbb{N}$ we define the matrix $\eta_n := \begin{pmatrix} 0 & -1_n \\ 1_n & 0 \end{pmatrix}$ and for any commutative ring A with an identity the group $Sp_n(A) := \{\alpha \in GL_{2n}(A) | {}^t\alpha\eta_n\alpha = \eta_n\}$. The group $Sp_n(\mathbb{R})$ acts on the Siegel upper half space $\mathbb{H}_n := \{z \in \mathbb{C}_n^n | {}^tz = z,\ Im(z) > 0\}$ by linear fractional transformations, that is for $\alpha = \begin{pmatrix} a_\alpha & b_\alpha \\ c_\alpha & d_\alpha \end{pmatrix} \in Sp_n(\mathbb{R})$ and $z \in \mathbb{H}_n$ we have $\alpha \cdot z := (a_\alpha z + b_\alpha)(c_\alpha z + d_\alpha)^{-1} \in \mathbb{H}_n$. Moreover if we define $\mu_\alpha(z) := \mu(\alpha, z) := c_\alpha z + d_\alpha$ then we have

$$\mu(\beta\alpha, z) = \mu(\beta, \alpha z)\mu(\alpha, z), \quad \alpha, \beta \in Sp_n(\mathbb{R}), z \in \mathbb{H}_n$$

Let now F be a totally real field of degree $d := [F : \mathbb{Q}]$ and write $\mathfrak{g}$ for its ring of integers. We write $\mathbf{a}$ for the set of Archimedean places of F, $\mathbf{h}$ for the finite ones and we set $G := Sp_n(F)$. We write $G_\mathbb{A}$ for the adelic group and we decompose $G_\mathbb{A} = G_\mathbf{h} G_\mathbf{a}$ where $G_\mathbf{a} := \prod_{v \in \mathbf{a}} G_v$ and $G_\mathbf{h} := \prod_{v \in \mathbf{h}} G_v$. For two fractional ideals $\mathfrak{a}$ and $\mathfrak{b}$ of F such that $\mathfrak{a}\mathfrak{b} \subseteq \mathfrak{g}$, we define the subgroup of $G_\mathbb{A}$,

$$D[\mathfrak{a}, \mathfrak{b}] := \left\{ x = \begin{pmatrix} a_x & b_x \\ c_x & d_x \end{pmatrix} \in G_\mathbb{A} | a_x \prec \mathfrak{g}_v, b_x \prec \mathfrak{a}_v, c_x \prec \mathfrak{b}_v, d_x \prec \mathfrak{g}_v, \forall v \in \mathbf{h} \right\},$$

where we use the notation "$\prec$" of Shimura, $x \prec \mathfrak{b}_v$ meaning that the v-component of x is a matrix with entries in the ideal $\mathfrak{b}_v$. We will mainly consider groups of the form $D[\mathfrak{b}^{-1}, \mathfrak{b}\mathfrak{c}]$ for a fractional ideal $\mathfrak{b}$ and an integral ideal $\mathfrak{c}$. Strong approximation for G implies that $G_\mathbb{A} = GqD[\mathfrak{b}^{-1}, \mathfrak{b}\mathfrak{c}]$ for any $\mathfrak{b}, \mathfrak{c}$ and $q \in G_\mathbf{h}$. We define $\Gamma^q(\mathfrak{b}, \mathfrak{c}) := G \cap qD[\mathfrak{b}^{-1}, \mathfrak{b}\mathfrak{c}]q^{-1}$. Given a Hecke character ψ of F with $\psi_v(a) = 1$ for all $a \in \mathfrak{g}_v^\times$, $v \in \mathbf{h}$ such that $a - 1 \in \mathfrak{c}_v$ we define a character on $D[\mathfrak{b}^{-1}, \mathfrak{b}\mathfrak{c}]$ by $\psi(x) = \prod_{v|\mathfrak{c}} \psi_v(det(d_x)_v)$ and a character which we still denote ψ on Γ^q by $\psi(\gamma) := \psi(q^{-1}\gamma q)$.

We now write $\mathbb{Z}^\mathbf{a} := \prod_{v \in \mathbf{a}} \mathbb{Z}$ and $\mathcal{H} := \prod_{v \in \mathbf{a}} \mathbb{H}_n$. For a function $f : \mathcal{H} \to \mathbb{C}$ and an element $k \in \mathbb{Z}^\mathbf{a}$ we define

$$(f|_k\alpha)(z) := j_\alpha(z)^{-k} f(\alpha z), \ \alpha \in G, \ z \in \mathcal{H}.$$

Here we write $z = (z_v)_{v\in\mathbf{a}}$ with $z_v \in \mathbb{H}_n$ and $\alpha_v \in Sp_n(\mathbb{R})$ and define $j_\alpha(z)^{-k} := \prod_v det(\mu_{\alpha_v}(z_v))^{-k_v}$. Let now Γ be group of the form Γ^q, $q \in G_{\mathbf{h}}$ as above and ψ a Hecke character. Then we define

Definition 1. A function $f : \mathcal{H} \to \mathbb{C}$ is called a Siegel modular form for the congruence subgroup Γ of weight $k \in \mathbb{Z}^{\mathbf{a}}$ and Nebentypus ψ if

1. f is holomorphic,
2. $f|_k\gamma = \psi(\gamma)f$ for all $\gamma \in \Gamma$,
3. f is holomorphic at cusps.

The last condition is needed only if $F = \mathbb{Q}$ and $n = 1$. Then it is the classical condition of elliptic modular forms being holomorphic at cusps. The above defined space we will denote it by $\mathcal{M}_k(\Gamma, \psi)$. As it is explained in Shimura (2000, page 33) for an element $f \in \mathcal{M}_k(\Gamma, \psi)$ and an element $\alpha \in G$ we have a Fourier expansion

$$(f|_k\alpha)(z) = \sum_{h\in S_+} c_\alpha(h)e^n_{\mathbf{a}}(hz),$$

where S_+ is the set of n by n symmetric matrices with entries in F which are positive semi-definite at every real place $v \in \mathbf{a}$ and $e^n_{\mathbf{a}}(x) = exp(2\pi i \sum_{v\in\mathbf{a}} tr(x_v))$. An element $f \in M_k(\Gamma, \psi)$ is called a cusp form if $c_\alpha(h) \neq 0$ for some $\alpha \in G$ implies h_v is positive definite for all $v \in \mathbf{a}$.

We now turn to the adelic Siegel modular forms. Let D be a group of the form $D[\mathfrak{b}^{-1}, \mathfrak{bc}]$ and ψ a Hecke character of F.

Definition 2. A function $\mathbf{f} : G_{\mathbb{A}} \to \mathbb{C}$ is called adelic Siegel modular form if

1. $\mathbf{f}(\alpha x w) = \psi(w)j_w^{-k}(\mathbf{i})\mathbf{f}(x)$ for $\alpha \in G$, $w \in D$ with $w(\mathbf{i}) = \mathbf{i}$,
2. For every $p \in G_{\mathbf{h}}$ there exists $f_p \in M_k(\Gamma^p, \psi_p)$, where $\Gamma^p := G \cap pDp^{-1}$ and $\psi_p(\gamma) = \psi(p\gamma p^{-1})$ such that $\mathbf{f}(py) = (f_p|_k y)(\mathbf{i})$ for every $y \in G_{\mathbf{a}}$.

We write $\mathcal{M}_k(D, \psi)$ for this space. Strong approximation theorem for Sp_n gives $\mathcal{M}_k(D, \psi) \cong M_k(\Gamma^q, \psi_q)$ for any $q \in G_{\mathbf{h}}$. We define the space of automorphic cusp form $\mathcal{S}_k(D, \psi)$ to be the subspace of $\mathcal{M}_k(D, \psi)$ that is in bijection with $S_k(\Gamma^q, \psi)$ for any $q \in G_{\mathbf{h}}$ in the above bijection. We may also sometimes write $\mathcal{M}_k(\mathfrak{b}, \mathfrak{c}, \psi)$ for $\mathcal{M}_k(D, \psi)$. Similarly we may write $M_k(\mathfrak{b}, \mathfrak{c}, \psi)$ for $M_k(\Gamma, \psi)$ where $\Gamma = G \cap D[\mathfrak{b}^{-1}, \mathfrak{bc}]$, i.e $q = 1$.

2.2 Half-Integral Weight Siegel Modular Forms

Even though we will consider only algebraicity results for integral weight Siegel modular forms, in many case we will need to use half-integral weight modular forms. We denote by $M_{\mathbb{A}}$ the adelized metaplectic group sitting in the exact sequence $0 \to \mathbb{T} \to M_{\mathbb{A}} \to G_{\mathbb{A}} \to 0$. The last projection we denote by pr. We write C^θ for the *theta* group defined for example in Shimura (1994, page 536) and $\Gamma^\theta = G \cap C^\theta$. We also define the group $\mathfrak{M} = \{x \in M_{\mathbb{A}} | pr(x) \in P_{\mathbb{A}} C^\theta\}$, where P is the standard Siegel parabolic subgroup of G. Thanks to a canonical lift we may consider G as a subgroup of $M_{\mathbb{A}}$ and hence also Γ^θ a subgroup of $\mathfrak{M}$. For an element $\sigma \in \mathfrak{M}$ and $z \in \mathcal{H}$ we write $h_\sigma(z)$ for the holomorphic function defined by Shimura. By a half integral weight $k \in \frac{1}{2}\mathbb{Z}^{\mathbf{a}}$ we mean a tuple $(k_v)_{v \in \mathbf{a}}$ so that $k_v \in \mathbb{Z} + \frac{1}{2}$ for all $v \in \mathbf{a}$. For such a k we define the factor of automorphy

$$j_\sigma(z)^k := h_\sigma(z) j_{pr(\sigma)}(z)^{[k]}.$$

Then the definition of half integral weight modular forms, with congruence subgroup $\Gamma \leq \Gamma^\theta$ is the same as in integral case but using the new factor of automorphy. One may define also adelic automorphic forms, we refer to Shimura (2000, page 166) for this.

3 Theta and Eisenstein Series

3.1 Theta Series

Following Shimura (see page 270 in Shimura 2000) we set $W = F_n^n$ and we let $\mathcal{S}(W_{\mathbf{h}})$ denote the space of Schwartz-Bruhat functions on $W_{\mathbf{h}}$. Let τ be an n by n symmetric matrix with entries in F such that $\tau_v > 0$ for all $v \in \mathbf{a}$. For an element $\lambda \in \mathcal{S}(W_{\mathbf{h}})$ and an element $\mu \in \mathbb{Z}^{\mathbf{a}}$ such that $0 \leq \mu_v \leq 1$ for all $v \in \mathbf{a}$ we define

$$\theta(z, \lambda) = \sum_{\xi \in W} \lambda(\xi_{\mathbf{h}}) det(\xi)^\mu \mathbf{e}_{\mathbf{a}}(tr({}^t\xi \tau \xi z)), \quad z \in \mathcal{H}.$$

It is shown in the appendix of Shimura (2000) that this is an element of $\mathcal{M}_l$ with $l := \mu + \frac{n}{2}\mathbf{a}$. Moreover it is also shown that if $\mu \neq 0$ then $\theta(z, \lambda)$ is actually a cusp form. We now introduce some extra notation following Shimura (2000, Appendix A3.18). We set

$$R = \prod_{v\mathbf{h}} (\mathfrak{g}_v)_n^n, \quad E_v = GL_n(\mathfrak{g}_v), \quad R^* = R W_{\mathbf{a}} \subset W_{\mathbf{A}}.$$

We let ω be now a Hecke character of F of conductor $\mathfrak{f}$ such that $\omega_{\mathbf{a}}(-1)^n = (-1)^{n \sum_v \mu_v}$. Let now r be an element of $GL_n(F)_{\mathbf{h}}$ and define

$$\theta(z) := \sum_{W \cap r R^*} \omega_{\mathbf{a}}(det(\xi)) \omega^*(det(r^{-1}\xi)\mathfrak{g}) det(\xi)^\mu \mathbf{e}_{\mathbf{a}}^n({}^t\xi \tau \xi t),$$

where for a Hecke character ψ we denote by ψ^* the corresponding ideal character. Then Shimura proves the following proposition (Shimura 2000, Proposition A3.19).

Proposition 1 (Shimura). *Let ρ_τ be the Hecke character of F corresponding to the extension $F(c^{1/2})/F$ with $c = (-1)^{[n/2]}det(2\tau)$; put $\omega' = \omega\rho_\tau$. Then there exist a fractional ideal $\mathfrak{b}$ and an integral ideal $\mathfrak{c}$, such that the conductor of ω' divides $\mathfrak{c}$, $D[\mathfrak{b}^{-1}, \mathfrak{b}\mathfrak{c}] \subset D[2\mathfrak{d}^{-1}, 2\mathfrak{d}]$ if n is odd, and*

$$\theta(\gamma z) = \omega'_{\mathfrak{c}}(det(a_\gamma)) j^l_\gamma(z)\theta(z), \ \ \gamma \in G \cap D,$$

where $D = D[\mathfrak{b}^{-1}, \mathfrak{b}\mathfrak{c}]$. Moreover, if $\beta \in G \cap diag[q, \hat{q}]C$ with $q \in GL_n(F)_{\mathbf{h}}$, then

$$j^l_\beta(\beta^{-1}z)\theta(\beta^{-1}z) = \omega'(det(q))^{-1}\omega'_{\mathfrak{c}}(det(d_\beta q))|det(q)|^{n/2}_{\mathbf{A}} \times$$

$$\sum_{\xi \in W \cap r R^* q^{-1}} \omega_{\mathbf{a}}(det(\xi))\omega^*(det(\xi r^{-1} q)\mathfrak{g}) det(\xi)^\mu \mathbf{e}^n_{\mathbf{a}}({}^t\xi\tau\xi z).$$

In particular, let $\mathfrak{x}$ and $\mathfrak{t}$ be fractional ideals of F such that ${}^t g 2\tau g \in \mathfrak{x}$ for every $g \in r\mathfrak{g}^n_1$ and ${}^t h(2\tau)^{-1}h \in 4\mathfrak{t}^{-1}$ for every $h\hat{r}\mathfrak{g}^n_1$ and write $\mathfrak{h}$ for the conductor of ρ_τ. Then we can take

$$(\mathfrak{b}, \mathfrak{c}) = \begin{cases} (2^{-1}\mathfrak{d}\mathfrak{x}, \mathfrak{h} \cap \mathfrak{f} \cap \mathfrak{x}^{-1}\mathfrak{f}^2\mathfrak{t}), & \text{if } n \text{ is even;} \\ (2^{-1}\mathfrak{d}\mathfrak{a}^{-1}, \mathfrak{h} \cap \mathfrak{f} \cap 4\mathfrak{a} \cap \mathfrak{a}\mathfrak{f}^2\mathfrak{t}), & \text{if } n \text{ is odd.} \end{cases},$$

where $\mathfrak{a} = \mathfrak{x}^{-1} \cap \mathfrak{g}$.

3.2 Eisenstein Series

We follow Shimura (2000, pages 131–132) and define various Eisenstein series of Siegel type. Let $k \in \frac{1}{2}\mathbb{Z}^{\mathbf{a}}$ be a weight, $\mathfrak{b}$ a fractional ideal of F, $\mathfrak{c}$ an integral ideal in F and a Hecke character χ of F with infinity type $\chi_{\mathbf{a}}(x) = x^\ell_{\mathbf{a}}|x_{\mathbf{a}}|^{-\ell}$, and of conductor $\mathfrak{c}$.

When k is half integral we also assume that $D[\mathfrak{b}^{-1}, \mathfrak{b}\mathfrak{c}] \subset D[2\mathfrak{d}^{-1}, 2\mathfrak{d}]$, where $\mathfrak{d}$ the different ideal of F. Following the notation of Shimura we now define in the case of $k \in \mathbb{Z}^{\mathbf{a}}$

$$\tilde{D} = D[\mathfrak{b}^{-1}, \mathfrak{b}\mathfrak{c}],$$

and otherwise

$$\tilde{D} = \{x \in M_{\mathbf{A}} | pr(x) \in D[\mathfrak{b}^{-1}, \mathfrak{b}\mathfrak{c}]\}.$$

Write $P = \{x \in G | c_x = 0\}$ for the standard Siegel parabolic. We then define a function μ on $G_{\mathbf{A}}$ or $M_{\mathbf{A}}$ by

$$\mu(x) = 0, \text{ if } x \notin P_{\mathbf{A}}\tilde{D},$$

$$\mu(x) = \chi_{\mathbf{h}}(det(d_p))^{-1}\chi_{\mathfrak{c}}(det(d_w))^{-1} j_x^k(\mathbf{i})^{-1}|j_x(\mathbf{i})|^k,$$

if $x = pw$ with $p \in P_{\mathbf{A}}$ and $w \in \tilde{D}$. Then for a pair $(x,s) \in G_{\mathbf{A}} \times \mathbb{C}$ if $k \in \mathbb{Z}^{\mathbf{a}}$ or in $M_{\mathbf{A}} \times \mathbb{C}$ otherwise, we define the Eisenstein series (for the function ϵ below we refer to Shimura's book)

$$E_{\mathbf{A}}(x,s) = E_{\mathbf{A}}(x,s;\chi,\tilde{D}) = \sum_{\alpha \in P\backslash G} \mu(\alpha x)\epsilon(\alpha x)^{-s}.$$

We will need one more type of Eisenstein series. We define the element $\zeta \in Sp(n,F)_{\mathbf{A}}$ by

$$\zeta_{\mathbf{a}} = 1, \ \zeta_{\mathbf{h}} = \begin{pmatrix} 0 & -\delta^{-1}1_n \\ \delta 1_n & 0 \end{pmatrix},$$

where $\delta \in F_{\mathbf{h}}^{\times}$ such that $\delta\mathfrak{g} = \mathfrak{d}$. We further fix an element $\tilde{\zeta} \in M_{\mathbf{A}}$ such that $pr(\tilde{\zeta}) = \zeta$ and $h(\tilde{\zeta},z) = 1$. Then we define the Eisenstein series

$$E_{\mathbf{A}}^*(x,s) = \chi(\delta)^{-n} \times \begin{cases} E_{\mathbf{A}}(x\zeta,s), \ k \in \mathbb{Z}^{\mathbf{a}}; \\ E_{\mathbf{A}}(x\tilde{\zeta},s), \text{ otherwise.} \end{cases}$$

Finally we define the Eisenstein series

$$D_{\mathbf{A}}(x,s) = E_{\mathbf{A}}^*(x,s) \times \begin{cases} L_{\mathfrak{c}}(2s,\chi)\prod_{i=1}^{[n/2]} L_{\mathfrak{c}}(4s-2i,\chi^2), \ k \in \mathbb{Z}^{\mathbf{a}}; \\ \prod_{i=1}^{[(n+1)/2]} L_{\mathfrak{c}}(4s-2i-1,\chi^2), \quad k \notin \mathbb{Z}^{\mathbf{a}}; \end{cases},$$

Write $S = \{x \in F_n^n | {}^t x = x\}$. Then the q-expansion of $E_{\mathbf{A}}^*(x,s)$ is given by

$$E_{\mathbf{A}}^*\left(\begin{pmatrix} q & \sigma\hat{q} \\ 0 & \hat{q} \end{pmatrix}, s\right) = \sum_{h \in S} c(h,q,s)\mathbf{e}_{\mathbf{A}}^n(h\sigma),$$

where $q \in GL_n(F)_{\mathbf{A}}$ and $\sigma \in S_{\mathbf{A}}$. We now define the Eisenstein series on $(z,s) \in \mathcal{H} \times \mathbb{C}$ by

$$E^*(x(\mathbf{i}),s) = j_x^k(\mathbf{i})E_{\mathbf{A}}^*(x,s),$$

and similarly we define $D(z,s)$ and $D^{\star}(z,s)$. When we want to indicate the dependence on the various input data we will write $E(z,s;k,\chi,\mathfrak{c})$ for $E(z,s)$ or in case we want also to indicate the dependency on $\mathfrak{b}$ we will write $E(z,s;k,\chi,\Gamma)$, where $\Gamma = G \cap D[\mathfrak{b}^{-1},\mathfrak{bc}]$. We now note the q-expansion

$$E^*(z,s) = \sum_{h \in S} det(y)^{-k\mathbf{a}/2} c(h,q,s) \mathbf{e}^n_{\mathbf{a}}(hx),$$

where $q_{\mathbf{h}} = q_1$ and $q_{\mathbf{a}} = y^{1/2}$. For the coefficients $c(h,q,s)$ we have the following propositions of Shimura (2000, Proposition 16.9, 16.10, 17.6), (for notation not introduced here we refer to (loc. cit.)).

Proposition 2 (Shimura). *Suppose that $\mathfrak{c} \neq \mathfrak{g}$ and $det(q_v) > 0$ for every $v \in \mathbf{a}$. Then $c(h,q,s) \neq 0$ only if $({}^tqhq)_v \in (\mathfrak{d}\mathfrak{b}^{-1}\mathfrak{c}^{-1})_v \tilde{S}_v$ for every $v \in \mathbf{h}$. In this case*

$$c(h,q,s) = C\chi_{\mathbf{h}}(det(-q))^{-1}|det(q)_{\mathbf{h}}|^{n+1-2s}_{\mathbf{A}}|D_F|^{-2ns+3n(n+1)/4}N(\mathfrak{bc})^{-n(n+1)/2}\times$$

$$det(y)^{s\mathbf{a}}\Xi(y;h;s\mathbf{a}+k/2,s\mathbf{a}-k/2)\alpha^e_{\mathfrak{c}}(\epsilon_b^{-1}\cdot {}^tqhq,2s,\chi),$$

where $C = 1$ and $e = 0$ if $k \in \mathbb{Z}^{\mathbf{a}}$, and $C = \mathbf{e}(n[F:\mathbb{Q}]/8)$ and $e = 1$ if $k \notin \mathbb{Z}^{\mathbf{a}}$; $\epsilon_b \in F^\times_{\mathbf{h}}$ such that $\epsilon_b\mathfrak{g} = \mathfrak{b}^{-1}\mathfrak{d}$ if $k \in \mathbb{Z}^{\mathbf{a}}$, and $\epsilon_b = 1$ otherwise; D_F is the discriminant of F.The function $\Xi(g;h;\alpha,\beta) = \prod_{v\in\mathbf{a}} \xi(y_v,h_v;\alpha_v,\beta_v)$ is given in Shimura (2000, page 140).

Proposition 3 (Shimura). *Consider q and h such that $c(h,q,s) \neq 0$. Set $r = rank(h)$ and let $g \in GL_n(F)$ such that $g^{-1}hg = diag[h',0]$ with $h' \in S^r$. Let ρ_h be the Hecke character corresponding to $F(c^{1/2})/F$ where $c = (-1)^{[r/2]}det(2h')$, if $r > 0$; let $\rho_h = 1$ if $r = 0$. Then*

$$\alpha^e_{\mathfrak{c}}(\epsilon_b^{-1}\cdot {}^tqhq,2s,\chi) = \Lambda_{\mathfrak{c}}(s)^{-1}\Lambda_h(s)\prod_{v\in\mathbf{c}} f_{h,q,v}\left(\chi(\pi_v)|\pi_v|^{2s+e/2}\right),$$

where

$$\Lambda_{\mathfrak{c}}(s) = \begin{cases} L_{\mathfrak{c}}(2s,\chi)\prod_{i=1}^{[n/2]} L_{\mathfrak{c}}(4s-2i,\chi^2), & \text{if } k \in \mathbb{Z}^{\mathbf{a}};\\ \prod_{i=1}^{[(n+1)/2]} L_{\mathfrak{c}}(4s-2i+1,\chi^2), & \text{otherwise.}\end{cases}$$

$$\Lambda_h(s) = \begin{cases} L_{\mathfrak{c}}(2s-n+r/2,\chi\rho_h)\prod_{i=1}^{[(n-r)/2]} L_{\mathfrak{c}}(4s-2n+r+2i-1,\chi^2), & \text{if } k \in \mathbb{Z}^{\mathbf{a}};\\ \prod_{i=1}^{[(n-r+1)/2]} L_{\mathfrak{c}}(4s-2n+r+2i-2,\chi^2), & \text{otherwise.}\end{cases}$$

Here $f_{h,q,v}$ are polynomials with coefficients in $\mathbb{Z}$, independent of χ. For the finite set $\mathbf{c}$ see Shimura (2000).

For a number field W we follow Shimura and write $\mathcal{N}^r_k(W)$ for the space of W-rational nearly holomorphic forms of weight k (see Shimura 2000, pages 103 and 110 for the definition). The theorem below is due to Shimura (2000, Theorem 17.9).

Theorem 1 (Shimura). *Let Φ be the Galois closure of F over $\mathbb{Q}$ and let $k \in \frac{1}{2}\mathbb{Z}^{\mathbf{a}}$ with $k_v \geq (n+1)/2$ for all $v \in \mathbf{a}$ and $k_v - k_{v'} \in 2\mathbb{Z}$ for every $v, v' \in \mathbf{a}$. Let $\mu \in \frac{1}{2}\mathbf{Z}$ with $n+1-k_v \leq \mu \leq k_v$ and $|\mu - \frac{n+1}{2}| + \frac{n+1}{2} - k_v \in 2\mathbb{Z}$ for all $v \in \mathbf{a}$. Exclude the cases*

1. $\mu = (n+2)/2$, $F = \mathbb{Q}$ *and* $\chi^2 = 1$,
2. $\mu = 0$, $\mathfrak{c} = \mathfrak{g}$ *and* $\chi = 1$,
3. $0 < \mu \leq n/2$, $\mathfrak{c} = \mathfrak{g}$ *and* $\chi^2 = 1$.

Then $D(z, \mu/2; k, \chi, \mathfrak{c})$ belongs to $\pi^{\beta}\mathcal{N}_k^r(\Phi\mathbb{Q}_{ab})$, where $r = (n/2)(k - |\mu - (n+1)/2|\mathbf{a} - \frac{n+1}{2}\mathbf{a})$ except in the case where $n = 1$, $\mu = 2$, $F = \mathbb{Q}$, $\chi = 1$ and $n > 1$, $\mu = (n+3)/2$, $F = \mathbb{Q}$, $\chi^2 = 1$. In these two case we have $r = n(k - \mu + 2)/2$. Moreover we have that $\beta = (n/2)\sum_{v \in \mathbf{a}}(k_v + \mu) - [F : \mathbb{Q}]e$ where

$$e = \begin{cases} [(n+1)^2/4] - \mu, & \text{if } 2\mu + n \in 2\mathbb{Z} \text{ and } \mu \geq \lambda; \\ [n^2/4], & \text{otherwise.} \end{cases}$$

For an element $p \in \mathbb{Z}^{\mathbf{a}}$ and a weight $q \in \frac{1}{2}\mathbb{Z}^{\mathbf{a}}$ we write Δ_q^p for the differential operators defined by Shimura (2000, page 146). In particular we have $\Delta_q^p \mathcal{N}_q^t(\Phi\mathbb{Q}_{ab}) \subset \pi^{n|p|}\mathcal{N}_{q+2p}^{t+np}(\Phi\mathbb{Q}_{ab})$. Moreover for any $f \in \mathcal{N}_q^t(\Phi\mathbb{Q}_{ab})$ and any $\sigma \in Gal(\Phi\mathbb{Q}_{ab}/\Phi)$ we have that

$$\left(\pi^{-n|p|}\Delta_q^p(f)\right)^{\sigma} = \pi^{-n|p|}\Delta_q^p(f^{\sigma}) \tag{1}$$

Let $\mu \in \frac{1}{2}\mathbb{Z}$ and $k \in \frac{1}{2}\mathbb{Z}^{\mathbf{a}}$ be as in the theorem above. If $\mu \geq (n+1)/2$ then Shimura shows that (Shimura 2000, page 146)

$$\Delta_{\mu\mathbf{a}}^p D(z, \mu/2; \mu\mathbf{a}, \chi, \mathfrak{c}) = c_{\mu\mathbf{a}}^p(\mu/2)(i/2)^{n|p|} D(z, \mu/2; k\mathbf{a}, \chi, \mathfrak{c}), \tag{2}$$

where $p = (k - \mu\mathbf{a})/2$. Here $c_{\mu\mathbf{a}}^p(\mu/2) \in \mathbb{Q}^{\times}$. If $\mu < (n+1)/2$ then we have

$$\Delta_{\nu\mathbf{a}}^p D(z, \mu/2; \nu\mathbf{a}, \chi, \mathfrak{c}) = c_{\nu\mathbf{a}}^p(\mu/2)(i/2)^{n|p|} D(z, \mu/2; k\mathbf{a}, \chi, \mathfrak{c}), \tag{3}$$

where $\nu = n + 1 - \mu$, $p = (k - \nu\mathbf{a})/2$ and again $c_{\nu\mathbf{a}}^p(\mu/2) \in \mathbb{Q}^{\times}$.

The following lemma is immediate from the above equations,

Lemma 1. *Assume there exists $A(\chi), B(\chi) \in \mathbb{Q}_{ab}$ and $\beta_1, \beta_2 \in \mathbb{N}$ such that for all $\sigma \in Gal(\mathbb{Q}_{ab}/\mathbb{Q})$*

$$\left(\frac{D(z, \mu/2; \mu\mathbf{a}, \chi, \mathfrak{c})}{\pi^{\beta_1}A(\chi)}\right)^{\sigma} = \frac{D(z, \mu/2; \mu\mathbf{a}, \chi^{\sigma}, \mathfrak{c})}{\pi^{\beta_1}A(\chi^{\sigma})}, \quad \mu \geq (n+1)/2$$

and

$$\left(\frac{D(z,\mu/2;\nu\mathbf{a},\chi,\mathfrak{c})}{\pi^{\beta_2}B(\chi)}\right)^{\sigma}=\frac{D(z,\mu/2;\nu\mathbf{a},\chi^{\sigma},\mathfrak{c})}{\pi^{\beta_2}B(\chi^{\sigma})},\quad \mu\le(n+1)/2.$$

Then we have for $\mu \geq (n+1)/2$ that

$$\left(\frac{D(z,\mu/2;k,\chi,\mathfrak{c})}{\pi^{\beta_1+n|p|}i^{n|p|}A(\chi)}\right)^{\sigma}=\frac{D(z,\mu/2;k,\chi^{\sigma},\mathfrak{c})}{\pi^{\beta_1+n|p|}i^{n|p|}A(\chi^{\sigma})},\quad p=(k-\mu\mathbf{a})/2\in\mathbb{Z}^{\mathbf{a}},$$

and for $\mu \leq (n+1)/2$ that

$$\left(\frac{D(z,\mu/2;k,\chi,\mathfrak{c})}{\pi^{\beta_2+n|p|}i^{n|p|}B(\chi)}\right)^{\sigma}=\frac{D(z,\mu/2;k,\chi^{\sigma},\mathfrak{c})}{\pi^{\beta_2+n|p|}i^{n|p|}B(\chi^{\sigma})},\quad \nu=n+1-\mu\ \ p=(k-\nu\mathbf{a})/2\in\mathbb{Z}^{\mathbf{a}},$$

We will be interested in algebraicity statements of the Eisenstein series of weight sufficient large it is enough to study the effect of the action of the Galois group of the full rank coefficients. More precisely we have the following lemma.

Lemma 2. *Let $f(z)=\sum_{h\in S}c(h)\mathbf{e}^n_{\mathbf{a}}(hz)\in\mathcal{M}_{k\mathbf{a}}(\mathbb{Q}^{ab})$ with $k\geq n/2$. Assume that for an element $\sigma\in Gal(\mathbb{Q}_{ab}/\mathbb{Q})$ we have $c(h)^{\sigma}=ac(h)$ for all h with $det(h)\neq 0$ for some $a\in\mathbb{C}$. Then $c(h)^{\sigma}=ac(h)$ for all $h\in S$. In particular $f^{\sigma}=af$.*

Proof. We obviously have $f^{\sigma}\in\mathcal{M}_{k\mathbf{a}}(\mathbb{Q}_{ab})$. We consider $g:=af-f^{\sigma}\in\mathcal{M}_{k\mathbf{a}}(\mathbb{Q}_{ab})$. We note that the form g has non-zero Fourier coefficients only for $h\in S$ with $det(h)=0$. But then by Shimura (2000, Proposition 6.16) we have that $g=0$. □

We now want to consider the action of $Gal(\mathbb{Q}_{ab}/\mathbb{Q})$ on the Eisenstein series. We first consider the holomorphic ones. That is, we consider the following two Eisenstein series

1. $D(z,k/2;k\mathbf{a},\chi,\mathfrak{c})\in\pi^{\beta}\mathcal{M}_{k\mathbf{a}}(\mathbb{Q}_{ab})$ for $k\geq\frac{n+1}{2}$,
2. $D(z,\mu/2;k\mathbf{a},\chi,\mathfrak{c})\in\pi^{\beta}\mathcal{M}_{k\mathbf{a}}(\mathbb{Q}_{ab})$ for $k:=n+1-\mu$ and $\mu\leq\frac{n+1}{2}$,

where β is determined by Theorem 1. Note here that we take the field of definition to be $\mathbb{Q}_{ab}$, i.e. the extension Φ does not appear. For this we refer to Shimura (2000, Theorem 17.7).

In the following lemma we collect some properties that we will need concerning the functions $\Xi(y,h;\alpha,\beta)=\prod_{v\in\mathbf{a}}\xi(y,h;\alpha,\beta)$.

Lemma 3. *Let $h\in S$ with $det(h)\neq 0$ and $y\in S^{\mathbf{a}}_{+}(\mathbb{R})$. Then we have for $k\in\frac{1}{2}\mathbb{Z}$ we have*

$$\Xi(y,h;k,0)=2^{d(1-(n+1)/2)}i^{-dnk}(2\pi)^{dnk}\Gamma_n(k)^{-d}N(det(h))^{k-(n+1)/2}\mathbf{e}^n(iyh)$$

and for $\mu:=n+1-k$ we have

$$\Xi(y,h;(n+1)/2,(\mu-k)/2) = i^{-nk}2^{-(dn(\mu-k))/2}\pi^{dn(n+1)/2}\Gamma_n(\frac{n+1}{2})^{-d}\times$$

$$\prod_{v\in\mathbf{a}} det(y_v)^{-(\frac{\mu-k}{2})}\mathbf{e}^n(iyh)$$

Proof. The first statement is in Shimura (2000, Equation 17.12). For the second we have $\Xi(y,h;(n+1)/2,\mu/2-k/2) = \prod_{v\in\mathbf{a}}\xi(y_v,h_v;(n+1)/2,\mu/2-k/2)$, where the function $\xi(\cdot)$ is given in Shimura (2000, page 140). By Shimura (1982, Equation 4.35K) we have that $\omega(2\pi y_v,h_v;(n+1)/2,\mu/2-k/2) = 2^{-n(n+1)/2}\mathbf{e}_v(iy_vh_v)$. We conclude that

$$\xi(y_v,h_v;(n+1)/2,\mu/2-k/2) = i^{-nk}2^{-(n(\mu-k))/2}\pi^{n(n+1)/2}\Gamma_n(\frac{n+1}{2})^{-1}\times$$

$$det(y_v)^{-(\frac{\mu-k}{2})}\mathbf{e}_v(iy_vh_v),$$

where we have used the fact that $\delta_-(h_vy_v) = 1$ (the product of the negative eigenvalues of h_vy_v). Indeed we have that $\delta_-(h_vy_v) = \delta_-(y_v^{1/2}h_v{}^ty_v^{1/2})$. But the last quantity has the same number of negative eigenvalues as the matrix h_v, but $h_v > 0$. □

We will need the following Theorem (for a proof see Shimura 1997, Theorem A6.5).

Theorem 2. *Let F be a totally real field, and let ψ be a Hecke character of F with $\psi_{\mathbf{a}}(b) = \prod_{v\in\mathbf{a}}\left(\frac{b_v}{|b_v|}\right)^k$, with $0 < k \in \mathbb{Z}$. For any integral ideal $\mathfrak{c}$ of F put*

$$P_{\mathfrak{c}}(k,\psi) := \mathbf{g}(\psi)^{-1}(2\pi i)^{-kd}|D_F|^{1/2}L_{\mathfrak{c}}(k,\psi),$$

where $d = [F:\mathbb{Q}]$ and $\mathbf{g}(\psi)$ is a Gauss sum (defined Shimura 1997, page 240). Then $P_{\mathfrak{c}}(k,\psi) \in \mathbf{Q}(\psi)$ and for every $\sigma \in Gal(\mathbb{Q}(\psi)/\mathbb{Q})$ we have

$$P_{\mathfrak{c}}(k,\psi)^{\sigma} = P_{\mathfrak{c}}(k,\psi^{\sigma}).$$

We also summarize in the following lemma some more properties of Gauss sums

Lemma 4. *Let χ and ψ be two finite order Hecke characters of F and $\sigma \in Gal(\mathbb{Q}_{ab}/\mathbb{Q})$ we have*

1. $\mathbf{g}(\chi)^{\sigma} = \chi^*(q\mathfrak{g})^{-1}\mathbf{g}(\chi^{\sigma})$ *where $0 < q \in \mathbb{Z}$ so that $e(1/N(\mathfrak{f}))^{\sigma} = e(q/N(\mathfrak{f}))$, where $\mathfrak{f}$ denotes the conductor of χ.*
2. $\left(\frac{\mathbf{g}(\chi\psi)}{\mathbf{g}(\chi)\mathbf{g}(\psi)}\right)^{\sigma} = \frac{\mathbf{g}(\chi^{\sigma}\psi^{\sigma})}{\mathbf{g}(\chi^{\sigma})\mathbf{g}(\psi^{\sigma})}$.
3. *If χ is a quadratic character then $\mathbf{g}(\chi) = i^mN(\mathfrak{f})^{1/2}$ where m is the number of Archimedean primes where $\chi_v \neq 1$.*

We remark here that if we pick an element $t \in \mathbb{Z}_{\mathbf{h}}^{\times}$ so that $\mathbf{e}_{\mathbf{h}}^{[t,\mathbb{Q}]} = \mathbf{e}_{\mathbf{h}}(t^{-1}x)$ for $x \in \mathbb{Q}/\mathbb{Z}$ then we have that we can pick the $q \in \mathbb{Z}$ above so that $rt_p - 1 \in N(\mathfrak{f})\mathbb{Z}_p$ for every prime p. Then we also obtain that $\chi^*(q\mathfrak{q}) = \chi_{\mathfrak{f}}(t)$.

3.3 Eisenstein Series of Integral Weight

We first consider the integral weight case. As we mentioned in the introduction these results can be found in a slightly different form in the book of Feit (1986). We also mention that results of this kind have been obtained by Siegel, Harris, and Sturm, at least in the absolute convergence case. For completeness, and because of some notation and normalization issues, we partly reproduce these results here.
We start with the following proposition.

Proposition 4. *For the Eisenstein series*

$$D(z,k/2;k\mathbf{a},\chi,\mathfrak{c}) = L_{\mathfrak{c}}(k,\chi)\prod_{i=1}^{[n/2]} L_{\mathfrak{c}}(2k-2i,\chi^2)E(z,k/2;k\mathbf{a},\chi,\mathfrak{c})$$

with $k \geq \frac{n+1}{2}$ *we have that* $\pi^{-\beta}D(z,k/2;k\mathbf{a},\chi,\mathfrak{c}) \in \mathcal{M}_{k\mathbf{a}}(\mathbb{Q}_{ab})$ *and for all* $\sigma \in Gal(\mathbb{Q}_{ab}/\mathbb{Q})$ *we have that*

$$\left(\frac{D(z,k/2;k\mathbf{a},\chi,\mathfrak{c})}{\pi^{\beta}P(\chi)}\right)^{\sigma} = \frac{D(z,k/2;k\mathbf{a},\chi^{\sigma},\mathfrak{c})}{\pi^{\beta}P(\chi^{\sigma})}, \quad \sigma \in Gal(\mathbb{Q}_{ab}/\mathbb{Q}),$$

where $\beta = kd + \sum_{i=1}^{[n/2]}(2k-2i)$ *and* $P(\chi) := \frac{\mathbf{g}(\chi)(i)^{kd}}{|D_F|^{1/2}}\frac{\left(\prod_{i=1}^{[n/2]}(i)^{(2k-2i)d}\right)\mathbf{g}(\chi^{2[n/2]})}{|D_F|^{b(n)}}$, *with* $b(n) = 1/2$ *if* $[n/2]$ *odd and* 1 *otherwise.*

Proof. We observe that we have that $2k - 2i > 0$ for all $i = 1\ldots[n/2]$. By definition we have that $\chi_{\mathbf{a}}(b) = \prod_{v\in\mathbf{a}}\left(\frac{b_v}{|b_v|}\right)^k$. By Theorem 2 above we have for

$$A(\chi) := \frac{|D_F|^{1/2}L_{\mathfrak{c}}(k,\chi)}{\mathbf{g}(\chi)(2\pi i)^{kd}}\prod_{i=1}^{[n/2]}\frac{|D_F|^{1/2}L_{\mathfrak{c}}(2k-2i,\chi^2)}{\mathbf{g}(\chi^2)(2\pi i)^{(2k-2i)d}} \in \mathbb{Q}_{ab}$$

and for all $\sigma \in Gal(\mathbb{Q}_{ab}/\mathbb{Q})$ we have $A(\chi)^{\sigma} = A(\chi^{\sigma})$. Using Lemma 4 we may define the quantity

$$B(\chi) = \frac{|D_F|^{1/2}L_{\mathfrak{c}}(k,\chi)}{\mathbf{g}(\chi)(2\pi i)^{kd}}\left(\prod_{i=1}^{[n/2]}\frac{L_{\mathfrak{c}}(2k-2i,\chi^2)}{(2\pi i)^{(2k-2i)d}}\right)\frac{|D_F|^{b(n)}}{\mathbf{g}(\chi^{2[n/2]})},$$

where $b(n) = 1/2$ if $[n/2]$ is odd and 1 otherwise. Then we have $B(\chi)^{\sigma} = B(\chi^{\sigma})$. By Feit (1986, Theorem 15.1) we have $E(z, k/2; k\mathbf{a}, \chi, \mathfrak{c})^{\sigma} = E(z, k/2; k\mathbf{a}, \chi^{\sigma}, \mathfrak{c})$ for all $\sigma \in Gal(\mathbb{Q}(\chi)/\mathbb{Q})$. In particular we conclude that

$$\left(\frac{D(z, k/2; k\mathbf{a}, \chi, \mathfrak{c})}{\pi^{\beta} P(\chi)}\right)^{\sigma} = \frac{D(z, k/2; k\mathbf{a}, \chi^{\sigma}, \mathfrak{c})}{\pi^{\beta} P(\chi^{\sigma})}, \quad \sigma \in Gal(\mathbb{Q}_{ab}/\mathbb{Q}),$$

where $\beta = kd + \sum_{i=1}^{[n/2]}(2k - 2i)$ and $P(\chi) := \frac{\mathbf{g}(\chi)(i)^{kd}}{|D_F|^{1/2}} \frac{\left(\prod_{i=1}^{[n/2]}(i)^{(2k-2i)d}\right)\mathbf{g}(\chi^{2[n/2]})}{|D_F|^{b(n)}}$. □

Now we turn to the Eisenstein series

$$D(z, \mu/2; k\mathbf{a}, \chi, \mathfrak{c}) = L_{\mathfrak{c}}(\mu, \chi) \prod_{i=1}^{[n/2]} L_{\mathfrak{c}}(2\mu - 2i, \chi^2) E(z, \mu/2; k\mathbf{a}, \chi, \mathfrak{c}),$$

and

$$D^*(z, \mu/2; k\mathbf{a}, \chi, \mathfrak{c}) = L_{\mathfrak{c}}(\mu, \chi) \prod_{i=1}^{[n/2]} L_{\mathfrak{c}}(2\mu - 2i, \chi^2) E^*(z, \mu/2; k\mathbf{a}, \chi, \mathfrak{c}),$$

where we take $\mu \leq \frac{n+1}{2}$, and $k = n + 1 - \mu$.

We now prove

Lemma 5. *Let $\beta \in \mathbb{N}$ as in Theorem 1 so that $\pi^{-\beta} D(z, \mu/2; k\mathbf{a}, \chi, \mathfrak{c}) \in \mathcal{M}_{k\mathbf{a}}(\mathbb{Q}_{ab})$. Then we have that also $\pi^{-\beta} D^*(z, \mu/2; k\mathbf{a}, \chi, \mathfrak{c}) \in \mathcal{M}_{k\mathbf{a}}(\mathbb{Q}_{ab})$. Moreover for every $\sigma \in Gal(\mathbb{Q}_{ab}/\mathbb{Q})$ we have the reciprocity law*

$$\left(\frac{D^*(z, \mu/2; k\mathbf{a}, \chi, \mathfrak{c})}{\pi^{\beta} i^{-dnk} |D_F|^{-n\mu+3n(n+1)/4}}\right)^{\sigma} = \frac{D^*(z, \mu/2; k\mathbf{a}, \chi^{\sigma}, \mathfrak{c})}{\pi^{\beta} i^{-dnk} |D_F|^{-n\mu+3n(n+1)/4}}.$$

Proof. The first statement i.e. that $\pi^{-\beta} D^*(z, \mu/2; k\mathbf{a}, \chi, \mathfrak{c}) \in \mathcal{M}_{k\mathbf{a}}(\mathbb{Q}_{ab})$ follows from Shimura (2000, Lemma 10.10). Moreover by Lemma 2 it is enough to establish the action of $Gal(\mathbb{Q}_{ab}/\mathbb{Q})$ on the full rank coefficients. By Proposition 2 and Lemma 3 we have that the hth Fourier coefficient $c(h, \chi)$ of $\pi^{-\beta} D^*(z, \mu/2; k\mathbf{a}, \chi, \mathfrak{c})$ with $det(h) \neq 0$ is equal to

$$i^{-dnk} 2^{-(dn(\mu-k))/2} \prod_{j=0}^{n} \Gamma(\frac{n+1}{2} - j/2)^{-d} |D_F|^{-n\mu+3n(n+1)/4} N(\mathfrak{b}\mathfrak{c})^{-n(n+1)/2} \times$$

$$\prod_{v \in \mathbf{c}} f_{h,v}(\chi(\pi_v)|\pi_v|^{\mu}) \times \begin{cases} L_{\mathfrak{c}}(\mu - n/2, \chi\rho_h), & n \text{ even}; \\ 1, & n \text{ odd}. \end{cases}$$

If n is odd we have

$$\left(\frac{c(h,\chi)}{i^{-dnk}|D_F|^{-n\mu+3n(n+1)/4}}\right)^{\sigma} = \frac{c(h,\chi^{\sigma})}{i^{-dnk}|D_F|^{-n\mu+3n(n+1)/4}}.$$

Now we take $n = 2m$ even. The character $\chi\rho_h$ has infinity type $(\chi\rho_h)_{\mathbf{a}}(b) = \prod_{v\in\mathbf{a}}\left(\frac{b_v}{|b_v|}\right)^{1-\mu+m}$ since the character ρ_h is the non-trivial character of the extension $F(c^{1/2})/F$ with $c := (-1)^m det(2h)$ and $det(h) >> 0$ as h is positive definite for all real embeddings of F. Since $1-\mu+m > 0$ we have by Shimura (2000, Theorem 18.12) that $L(1-(1-\mu+m),(\chi\rho_h)^{\sigma}) = L(1-(1-\mu+m),(\chi\rho_h))^{\sigma}$ for all $\sigma \in Gal(\mathbb{Q}_{ab}/\mathbb{Q})$. Hence we conclude also in the case of n even that

$$\left(\frac{c(h,\chi)}{i^{-dnk}|D_F|^{-n\mu+3n(n+1)/4}}\right)^{\sigma} = \frac{c(h,\chi^{\sigma})}{i^{-dnk}|D_F|^{-n\mu+3n(n+1)/4}}.\square$$

We now prove the following lemma

Lemma 6. *Assume that* $(\pi^{-\beta}D^*(z,\mu/2;k\mathbf{a},\chi,\mathfrak{c}))^{\sigma} = a\pi^{-\beta}D^*(z,\mu/2;k\mathbf{a},\chi^{\sigma},\mathfrak{c})$ *for* $\sigma \in Gal(\mathbb{Q}_{ab}/\mathbb{Q})$ $a \in \mathbb{Q}_{ab}^{\times}$*. Then*

$$(\pi^{-\beta}D(z,\mu/2;k\mathbf{a},\chi,\mathfrak{c}))^{\sigma} = b\pi^{-\beta}D(z,\mu/2;k\mathbf{a},\chi^{\sigma},\mathfrak{c})$$

where $b = \chi(q\mathfrak{g})^{-n}a$*, where* $0 < q \in \mathbb{Z}$ *such that* $e(1/N(\mathfrak{c}))^{\sigma} = e(q/N(\mathfrak{c}))$.

Proof. We use an argument due to Feit (1986) and Sturm (1981, Lemma 5) first introduced by Shimura in the case of $n = 1$. We will need the reciprocity law of the action of the group $\mathcal{G}_+ \times Gal(\overline{\mathbb{Q}}/\mathbb{Q})$ defined by Shimura (2000, Theorem 10.2). We use the notation of Shimura in this theorem. Let t be an idele of F and as in Shimura we define $\iota(t) := \begin{pmatrix} 1 & 0 \\ 0 & t^{-1}\end{pmatrix}$. For a $\sigma \in Gal(\mathbb{Q}_{ab}/\mathbb{Q})$ we define the element $(\iota(t),\sigma) \in \mathcal{G}_+ \times Gal(\overline{\mathbb{Q}}/\mathbb{Q})$ where $t \in \mathbb{Z}_{\mathbf{h}}^{\times}$ corresponds to σ by class field theory and we extend σ to an element of the absolute Galois group. Moreover we may consider also $\zeta_{\mathbf{h}} \in Sp_{\mathbb{A}}$ as an element of $\mathcal{G}_+ \times Gal(\overline{\mathbb{Q}}/\mathbb{Q})$ by taking $(\zeta_{\mathbf{h}},1)$. Then we have that

$$(\iota(t),\sigma)(\zeta_{\mathbf{h}},1)(\iota(t^{-1}),\sigma^{-1})(\zeta_{\mathbf{h}}^{-1},1) = \left(\begin{pmatrix} t & 0 \\ 0 & t^{-1}\end{pmatrix},1\right)$$

In particular we have

$$\pi^{-\beta}D(z,\mu/2;k\mathbf{a},\chi,\mathfrak{c})^{(\iota(t),\sigma)(\zeta_{\mathbf{h}},1)(\iota(t^{-1}),\sigma^{-1})(\zeta_{\mathbf{h}}^{-1},1)} =$$

$$\pi^{-\beta}D(z,\mu/2;k\mathbf{a},\chi,\mathfrak{c})|_k\begin{pmatrix} t & 0 \\ 0 & t^{-1}\end{pmatrix} = \chi_{\mathfrak{c}}(t)^n\pi^{-\beta}D(z,\mu/2;k\mathbf{a},\chi,\mathfrak{c})$$

But then

$$(\pi^{-\beta} D(z, \mu/2; k\mathbf{a}, \chi, \mathfrak{c}))^{\sigma} = \pi^{-\beta} D(z, \mu/2; k\mathbf{a}, \chi, \mathfrak{c})^{(\iota(t),\sigma)} =$$

$$\pi^{-\beta} D(z, \mu/2; k\mathbf{a}, \chi, \mathfrak{c})^{(\iota(t),\sigma)(\zeta_{\mathbf{h}},1)(\iota(t^{-1}),\sigma^{-1})(\zeta_{\mathbf{h}}^{-1},1)(\zeta_{\mathbf{h}},1)(\iota(t),\sigma)(\zeta_{\mathbf{h}}^{-1},1)} =$$

$$\chi_{\mathfrak{c}}(t)^n \left(\left(\pi^{-\beta} D(z, \mu/2; k\mathbf{a}, \chi, \mathfrak{c})|_k (\zeta_{\mathbf{h}})^{\sigma}\right)|_k \zeta_{\mathbf{h}}^{-1}\right) = \chi_{\mathfrak{c}}(t)^n a \pi^{-\beta} D(z, \mu/2; k\mathbf{a}, \chi^{\sigma}, \mathfrak{c}). \tag{4}$$

□

We can now establish the following corollary

Corollary 1. *For the Eisenstein series $D(z, \mu/2; k\mathbf{a}, \chi, D)$ we have*

$$\left(\frac{D(z, \mu/2; k\mathbf{a}, \chi, \mathfrak{c})}{\pi^{\beta}\mathbf{g}(\chi^n) i^{-dnk} |D_F|^{-n\mu+3n(n+1)/4}}\right)^{\sigma} = \frac{D(z, \mu/2; k\mathbf{a}, \chi^{\sigma}, \mathfrak{c})}{\pi^{\beta}\mathbf{g}((\chi^n)^{\sigma}) i^{-dnk} |D_F|^{-n\mu+3n(n+1)/4}},$$

for all $\sigma \in Gal(\mathbb{Q}_{ab}/\mathbb{Q})$.

Proof. This follows immediately by combining Lemma 4 ((i) and (ii)), and the last two lemmas. □

3.4 Eisenstein Series of Half Integral Weight

Now we consider the case of half-integral weight. We will need the theta series $\theta(z) := \sum_{a \in \mathfrak{g}^n} \mathbf{e}_{\mathbf{a}}({}^t aza/2) \in \mathcal{M}_{\frac{1}{2}\mathbf{a}}(\mathbb{Q}, \phi)$, where the quadratic character ϕ of Γ^{θ} is defined by $h_{\gamma}(z) = \phi(\gamma) j_{\gamma}^{\mathbf{a}}(z)$ for $\gamma \in \Gamma^{\theta}$. Note that this is the series θ_F defined in Shimura (2000, page 39, Equation 6.16) by taking in the equation there, using Shimura's notation, $u = 0$ and λ the characteristic function of $\mathfrak{g}^n \subset F^n$. Note in particular that since we are taking $u = 0$ we have that $\phi_F = \theta_F$. In particular Theorem 6.8 in (loc. cit.) gives the properties of the series θ. We now prove the following lemma.

Lemma 7. *For the theta series $\theta(z)$ and for $\sigma \in Gal(\overline{\mathbb{Q}}/\mathbb{Q})$ we have that*

$$\left(\theta|_{\frac{1}{2}\mathbf{a}}\zeta_{\mathbf{h}}\right)^{\sigma}|_{\frac{1}{2}\mathbf{a}}\zeta_{\mathbf{h}}^{-1} = \theta$$

Proof. This follows immediately after observing that $\zeta_{\mathbf{h}} \in C^{\theta}$ and from Theorem 6.8 (4) in Shimura (2000). Indeed since θ is invariant under $\Gamma^{\theta} = G \cap C^{\theta}$ we have that $\theta|_{\frac{1}{2}\mathbf{a}}\zeta_{\mathbf{h}} = \theta_{\frac{1}{2}\mathbf{a}}\zeta_{\mathbf{h}}^{-1} = \theta$. Since $\theta \in \mathcal{M}_{\frac{1}{2}\mathbf{a}}(\mathbb{Q})$, we conclude the proof. □

Proposition 5. *Let λ be equal to k or μ. Let $\beta(\lambda) \in \mathbb{N}$ so that*

$$\pi^{-\beta(\lambda)} D^*(z, \lambda/2; k\mathbf{a}, \chi, \mathfrak{c}) \in \mathcal{M}_{k\mathbf{a}}(\mathbb{Q}_{ab}).$$

Let $\sigma \in Gal(\mathbb{Q}_{ab}/\mathbb{Q})$ *and assume*

$$\left(\pi^{-\beta(\lambda)} D^*(z,\lambda/2; k\mathbf{a},\chi,\mathfrak{c})\right)^{\sigma} = \alpha(\lambda)\pi^{-\beta(\lambda)} D^*(z,\lambda/2; k\mathbf{a},\chi^{\sigma},\mathfrak{c}),\ \ k = n+1-\mu,$$

for some $\alpha(\lambda) \in \overline{\mathbb{Q}}^{\times}$. *Then we have* $\pi^{-\beta(\lambda)} D(z,\lambda/2; k\mathbf{a},\chi,\mathfrak{c}) \in \mathcal{M}_{k\mathbf{a}}(\mathbb{Q}_{ab})$ *and*

$$\left(\pi^{-\beta(\lambda)} D(z,\lambda/2; k\mathbf{a},\chi,\mathfrak{c})\right)^{\sigma} = \beta\pi^{-\beta(\lambda)} D(z,\lambda/2; k\mathbf{a},\chi^{\sigma},\mathfrak{c})$$

where $\beta = (\chi\phi)_{\mathfrak{c}}(t)^n\alpha(\lambda)$.

Proof. The fact that $\pi^{-\beta(\lambda)} D(z,\lambda/2; k\mathbf{a},\chi,\mathfrak{c}) \in \mathcal{M}_k(\mathbb{Q}_{ab})$ follows from Shimura (2000, Lemma 10.10). The rest of the proof was inspired by the proof of Theorem 10.7 in Shimura (2000). We write $D(\chi,\lambda)$ for $\pi^{-\beta(\lambda)} D(z,\lambda/2; k\mathbf{a},\chi,\mathfrak{c})$. Let $k' = k + \frac{1}{2} \in \mathbb{Z}$. Then we note that $\theta D(\chi,\lambda) \in \mathcal{M}_{k'\mathbf{a}}(\mathbb{Q}_{ab})$ and for a $\sigma \in Gal(\mathbb{Q}_{ab}/\mathbb{Q})$ we have $\theta D(\chi,\lambda)^{\sigma} = (\theta D(\chi,\lambda))^{\sigma}$. Since $\theta D(\chi)$ is of integral weight we can apply the reciprocity-laws as before. Writing $t \in \mathbb{Z}_{\mathbf{h}}^{\times}$ corresponding to σ we have

$$\begin{aligned}
(\theta D(\chi,\lambda))^{\sigma} &= (\theta D(\chi,\lambda))^{((\iota(t),\sigma)(\zeta_{\mathbf{h}},1)(\iota(t^{-1}),\sigma^{-1})(\zeta_{\mathbf{h}}^{-1},1))((\zeta_{\mathbf{h}},1)(\iota(t)\sigma)(\zeta_{\mathbf{h}}^{-1},1))}\\
&= \left((\theta D(\chi,\lambda))|_{k'}\begin{pmatrix} t & 0\\ 0 & t^{-1}\end{pmatrix}\right)^{(\zeta_{\mathbf{h}},1)(\iota(t),\sigma)(\zeta_{\mathbf{h}}^{-1},1)}\\
&= (\phi\chi)_{\mathfrak{c}}(t)^n \left((\theta D(\chi)|_{k'\mathbf{a}}\zeta_{\mathbf{h}})^{\sigma}\right)_{k'\mathbf{a}}\zeta_{\mathbf{h}}^{-1} =
\end{aligned}$$

$$(\phi\chi)_{\mathfrak{c}}(t)^n\left(\phi(\zeta_0)\left(\theta|_{\frac{1}{2}\mathbf{a}}\zeta_{\mathbf{h}}\right)^{\sigma}(D(\chi,\lambda)|_{k\mathbf{a}}\zeta_{\mathbf{h}})^{\sigma}\right)|_{k'\mathbf{a}}\zeta_{\mathbf{h}}^{-1} =$$

$$\phi(\zeta_0)\phi(\zeta_0)^{-1}\left(\left(\theta|_{\frac{1}{2}\mathbf{a}}\zeta_{\mathbf{h}}\right)^{\sigma}|_{\frac{1}{2}\mathbf{a}}\zeta_{\mathbf{h}}^{-1}\right)\left((D(\chi,\lambda)|_{k\mathbf{a}}\zeta_{\mathbf{h}})^{\sigma}\right)|_{k\mathbf{a}}\zeta_{\mathbf{h}}^{-1} =$$

$$(\phi\chi)_{\mathfrak{c}}(t)^n\alpha(\lambda)\theta D(\chi^{\sigma},\lambda).$$

For the element ζ_0 we refer to Shimura (2000, page 132). The last equation follows from the last Lemma. However the previous equations deserve a comment. Note that for $f_1, f_2 \in \mathcal{M}_{\frac{1}{2}\mathbf{a}}$ and $\gamma \in \Gamma^{\theta}$ we have that $(f_1 f_2)|_{\mathbf{a}}\gamma = \phi(\gamma)(f_1|_{\frac{1}{2}\mathbf{a}}\gamma)(f_2|_{\frac{1}{2}\mathbf{a}}\gamma)$ since $h_{\gamma}(z)^2 = \phi(\gamma) j_{\gamma}^{\mathbf{a}}(z)$.

So we obtain that $\theta D(\chi,\lambda)^{\sigma} = (\phi\chi)_{\mathfrak{c}}(t)^{-n}\alpha\theta D(\chi^{\sigma})$. Since θ is not a zero divisor in the formal ring of the Fourier-expansion (see Shimura 2000, page 74) we conclude the proof. □

We now establish also in the case of half-integral weight that

Proposition 6. *Let* $\beta_1 \in \mathbb{N}$ *so that* $\pi^{-\beta_1} D^*(z,k/2; k\mathbf{a},\chi,\mathfrak{c}) \in \mathcal{M}_{k\mathbf{a}}(\mathbb{Q}_{ab})$. *Let* $\sigma \in Gal(\mathbb{Q}_{ab}/\mathbb{Q})$ *Then for n even we have*

$$\left(\frac{\pi^{-\beta_1} D^*(z, k/2; k\mathbf{a}, \chi, \mathfrak{c})}{i^{-dnk} C |D_F|^{nk/2+3n(n+1)/4}}\right)^{\sigma} = \frac{\pi^{-\beta_1} D^*(z, k/2; k\mathbf{a}, \chi^{\sigma})}{i^{-dnk} C |D_F|^{nk/2+3n(n+1)/4}}$$

and for n odd

$$\left(\frac{\pi^{-\beta_1} D^*(z, k/2; k\mathbf{a}, \chi, \mathfrak{c})}{i^{-dnk} C |D_F|^{nk/2+3n(n+1)/4} \mathbf{g}(\chi) |D_F|^{1/2} (2i)^{-(k-n)d} b([n/2])}\right)^{\sigma} =$$

$$\frac{\pi^{-\beta_1} D^*(z, k/2; k\mathbf{a}, \chi^{\sigma})}{i^{-dnk} C |D_F|^{nk/2+3n(n+1)/4} \mathbf{g}(\chi^{\sigma}) |D_F|^{1/2} (2i)^{-(k-n)d} b([n/2])},$$

where $b(m) = i^d$ if m is m is odd and 1 otherwise.

Let now $\beta_2 \in \mathbb{N}$ so that $\pi^{-\beta_2} D^*(z, \mu/2; k\mathbf{a}, \chi, \mathfrak{c}) \in \mathcal{M}_{k\mathbf{a}}(\mathbb{Q}_{ab})$. Then we have

$$\left(\frac{\pi^{-\beta_2} D^*(z, \mu/2; k\mathbf{a}, \chi, \mathfrak{c})}{i^{-dnk} C |D_F|^{-n(n+1-k)+3n(n+1)/4}}\right)^{\sigma} = \frac{\pi^{-\beta_2} D^*(z, \mu/2; k\mathbf{a}, \chi^{\sigma}, \mathfrak{c})}{i^{-dnk} C |D_F|^{-n(n+1-k)+3n(n+1)/4}},$$

where $k = n + 1 - \mu$.

Proof. Arguing as before, it is enough to consider the action of σ on the full rank coefficients. We consider an h with $det(h) \neq 0$. Then we have that the hth Fourier coefficient $c(h, \chi)$ of $\pi^{-\beta_1} D^*(z, k/2; k\mathbf{a}, \chi, \mathfrak{c})$ is equal to

$$2^{d(nk+1-(n+1)/2)} i^{-dnk} \left(\prod_{j=0}^{n-1} \Gamma(k - j/2)\right)^{-d} N(det(h))^{k-(n+1)/2} C |D_F|^{nk/2+3n(n+1)/4} \times$$

$$N(\mathfrak{b}\mathfrak{c})^{-n(n+1)/2} \prod_{v \in \mathfrak{c}} f_{h,v}\left(\chi(\pi_v)|\pi_v|^{k+1/2}\right) \times \begin{cases} \pi^{-d(k-n/2)} L_{\mathfrak{c}}(k - n/2, \chi\rho_h), & n \text{ odd}; \\ 1, & n \text{ even}. \end{cases}$$

We now note that if n is even we have that $k - (n+1)/2 \in \mathbb{Z}$ and hence $N(det(h))^{k-(n+1)/2} \in \mathbb{Q}^{\times}$. Then we conclude that

$$\left(\frac{c(h, \chi)}{i^{-dnk} C |D_F|^{nk/2+3n(n+1)/4}}\right)^{\sigma} = \frac{c(h, \chi^{\sigma})}{i^{-dnk} C |D_F|^{nk/2+3n(n+1)/4}}.$$

In the case where n is odd we have that

$$P_{\mathfrak{c}}(k - n/2, \chi\rho_h)^{\sigma} = P_{\mathfrak{c}}(k - n/2, \chi^{\sigma}\rho_h), \quad \forall \sigma \in Gal(\mathbb{Q}_{ab}/\mathbb{Q}),$$

with

$$P_{\mathfrak{c}}(k - n/2, \chi\rho_h) := \mathbf{g}(\chi\rho_h)^{-1} (2\pi i)^{-(k-n/2)d} |D_F|^{1/2} L_{\mathfrak{c}}(k - n/2, \chi\rho_h)$$

We have $\frac{\mathbf{g}(\chi\rho_h)^\sigma}{\mathbf{g}(\chi^\sigma\rho_h)} = \frac{\mathbf{g}(\chi)^\sigma\mathbf{g}(\rho_h)^\sigma}{\mathbf{g}(\chi^\sigma)\mathbf{g}(\rho_h)}$. Moreover we have that

$$\frac{\mathbf{g}(\rho_h)^\sigma}{\mathbf{g}(\rho_h)} = \begin{cases} \frac{\sqrt{N(2det(h))}^\sigma}{\sqrt{N(2det(h))}}, & \text{if } [n/2] \text{ even;} \\ \left(\frac{i^\sigma}{i}\right)^d \frac{\sqrt{N(2det(h))}^\sigma}{\sqrt{N(2det(h))}}, & \text{otherwise.} \end{cases}$$

In particular since $det(h) \in F_+$ we have

$$\frac{\left(\sqrt{N(2det(h))^{-1}}\mathbf{g}(\rho_h)\right)^\sigma}{\sqrt{N(2det(h))^{-1}}\mathbf{g}(\rho_h)} = \begin{cases} 1, & \text{if } [n/2] \text{ even;} \\ \left(\frac{i^\sigma}{i}\right)^d, & \text{otherwise.} \end{cases}$$

For n odd we have that $k-(n+1)/2$ is half integral. Hence we conclude that

$$\left(\frac{c(h,\chi)}{i^{-dnk}C|D_F|^{nk/2+3n(n+1)/4}\mathbf{g}(\chi)|D_F|^{1/2}(2i)^{-(k-n)d}b([n/2])}\right)^\sigma =$$

$$\frac{c(h,\chi^\sigma)}{i^{-dnk}C|D_F|^{nk/2+3n(n+1)/4}\mathbf{g}(\chi^\sigma)|D_F|^{1/2}(2i)^{-(k-n)d}b([n/2])},$$

where $b(i) = i^d$ if $[n/2]$ odd and 1 otherwise.

Now we turn to the Eisenstein series $D^*(z,\mu/2;k\mathbf{a},\chi,\mathfrak{c})$. The Fourier coefficient $c(h,\chi)$ of $\pi^{-\beta_2}D^*(z,\mu/2;k\mathbf{a},\chi,\mathfrak{c})$ for $det(h)\neq 0$ is equal to

$$i^{-dnk}2^{-dn(\mu-k)/2}C|D_F|^{-n(n+1-k)+3n(n+1)/4}\prod_{j=0}^{n-1}\Gamma(\frac{n+1}{2}-j/2)^{-d}N(\mathfrak{bc})^{-n(n+1)/2}\times$$

$$\prod_{v\in\mathfrak{c}} f_{h,v}\left(\chi(\pi_v)|\pi_v|^{n+1-k+1/2}\right) \times \begin{cases} L_\mathfrak{c}(n/2+1-k,\chi\rho_h), & n \text{ odd};\\ 1, & n \text{ even}.\end{cases}$$

Since we are taking $k \geq \frac{n+1}{2}$ we have that $L_\mathfrak{c}(n/2+1-k,\chi\rho_h) \in \mathbb{Q}$. Hence after observing that $n+1-k+1/2 \in \mathbb{Z}$ we conclude that

$$\left(\frac{c(h,\chi)}{i^{-dnk}C|D_F|^{-n(n+1-k)+3n(n+1)/4}}\right)^\sigma = \frac{c(h,\chi^\sigma)}{i^{-dnk}C|D_F|^{-n(n+1-k)+3n(n+1)/4}} \quad \square$$

We can now conclude

Proposition 7. *Let $\beta_1 \in \mathbb{N}$ so that $\pi^{-\beta_1}D(z,k/2;k\mathbf{a},\chi,\mathfrak{c}) \in \mathcal{M}_{k\mathbf{a}}(\mathbb{Q}_{ab})$. Let $\sigma \in Gal(\mathbb{Q}_{ab}/\mathbb{Q})$ Then for n even we have*

$$\left(\frac{\pi^{-\beta_1}D(z,k/2;k\mathbf{a},\chi,\mathfrak{c})}{\mathbf{g}(\chi\phi)^n i^{-dnk}C|D_F|^{nk/2+3n(n+1)/4}}\right)^\sigma = \frac{\pi^{-\beta_1}D(z,k/2;k\mathbf{a},\chi^\sigma)}{\mathbf{g}(\chi^\sigma\phi)^n i^{-dnk}C|D_F|^{nk/2+3n(n+1)/4}}$$

and for n odd

$$\left(\frac{\pi^{-\beta_1} D(z, k/2; k\mathbf{a}, \chi, \mathfrak{c})}{\mathbf{g}((\chi\phi)^n) i^{-dnk} C |D_F|^{nk/2+3n(n+1)/4} \mathbf{g}(\chi) |D_F|^{1/2} (2i)^{-(k-n)d} b([n/2])}\right)^{\sigma} =$$

$$\frac{\pi^{-\beta_1} D(z, k/2; k\mathbf{a}, \chi^{\sigma})}{\mathbf{g}((\chi^n)^{\sigma}\phi) i^{-dnk} C |D_F|^{nk/2+3n(n+1)/4} \mathbf{g}(\chi^{\sigma}) |D_F|^{1/2} (2i)^{-(k-n)d} b([n/2])},$$

where $b(m) = i^d$ if m is m is odd and 1 otherwise.

Let now $\beta_2 \in \mathbb{N}$ so that $\pi^{-\beta_2} D^*(z, \mu/2; k\mathbf{a}, \chi, \mathfrak{c}) \in \mathcal{M}_{k\mathbf{a}}(\mathbb{Q}_{ab})$. Then we have

$$\left(\frac{\pi^{-\beta_2} D(z, \mu/2; k\mathbf{a}, \chi, \mathfrak{c})}{\mathbf{g}(\chi\phi^n) i^{-dnk} C |D_F|^{-n(n+1-k)+3n(n+1)/4}}\right)^{\sigma} =$$

$$\frac{\pi^{-\beta_2} D(z, \mu/2; k\mathbf{a}, \chi^{\sigma}, \mathfrak{c})}{\mathbf{g}(((\chi\phi)^n)^{\sigma}) i^{-dnk} C |D_F|^{-n(n+1-k)+3n(n+1)/4}},$$

We now remark that the above proposition and Lemma 1 give a complete description of the reciprocity laws of the Eisenstein series which we are considering. We summarize all the above in the following Theorem.

Theorem 3. *Let $k \in \frac{1}{2}\mathbb{Z}^{\mathbf{a}}$ with $k_v \geq (n+1)/2$ for every $v \in \mathbf{a}$. Let $\mu \in \frac{1}{2}\mathbb{Z}$ such that $n+1-k_v \leq \mu \leq k_v$ and $|\mu - (n+1)/2| + (n+1)/2 - k_v \in 2\mathbb{Z}$ for all $v \in \mathbf{a}$. Then with a $\beta \in \mathbb{N}$ as in Theorem 1 we have*

$$\pi^{-\beta} D(z, \mu/2; k, \chi, \mathfrak{c}) \in \mathcal{N}_k^r(\Phi\mathbb{Q}_{ab}),$$

and for every $\sigma \in Gal(\Phi\mathbb{Q}_{ab}/\Phi)$ we have

$$\left(\frac{\pi^{-\beta} D(z, \mu/2; k, \chi, \mathfrak{c})}{\omega(\chi)}\right)^{\sigma} = \frac{\pi^{-\beta} D(z, \mu/2; k, \chi^{\sigma}, \mathfrak{c})}{\omega(\chi^{\sigma})},$$

where $\omega(\chi)$ is given as follows:

1. *If $k \in \mathbb{Z}^{\mathbf{a}}$, $\mu \geq (n+1)/2$:*

$$\omega(\chi) = i^{n|p|} \mathbf{g}(\chi) i^{\mu d + 2\mu[n/2] - [n/2]([n/2]+1)d} |D_F|^{-b(n)} \mathbf{g}(\chi^{2[n/2]}),$$

   where $p := \frac{(k-\mu\mathbf{a})}{2}$ and $b(n) = 0$ if $[n/2]$ odd and $1/2$ otherwise.

2. *If $k \in \mathbb{Z}^{\mathbf{a}}$, $\mu < (n+1)/2$:*

$$\omega(\chi) = i^{n|p|} \mathbf{g}(\chi^n) i^{-dn\nu} |D_F|^{-n\mu + 3n(n+1)/4},$$

   where $\nu := n+1-\mu$ and $p := \frac{k-\nu\mathbf{a}}{2}$.

3. *If $k \notin \mathbb{Z}\mathbf{a}$ and $\mu \geq (n+1)/2$:*

   a. *If n is even*

$$\omega(\chi) = i^{n|p|}\mathbf{g}(\chi^n) i^{-dnk} C |D_F|^{nk/2+3n(n+1)/4},$$

   b. *If n is odd*

$$\omega(\chi) = i^{n|p|}\mathbf{g}(\chi^n \phi) i^{-dnk} C |D_F|^{n\mu/2+3n(n+1)/4}\mathbf{g}(\chi)|D_F|^{1/2}(2i)^{-(\mu-n)d} b([n/2]),$$

   where $p := \frac{(k-\mu\mathbf{a})}{2}$ and $b(m) = i^d$ if m is m is odd and 1 otherwise and

4. *If $k \notin \mathbb{Z}\mathbf{a}$ and $\mu < (n+1)/2$:*

$$\omega(\chi) = i^{n|p|}\mathbf{g}(\chi\phi)^n i^{-dn\nu} C |D_F|^{-n(n+1-\nu)+3n(n+1)/4},$$

   where $\nu := n+1-\mu$ and $p := \frac{k-\nu\mathbf{a}}{2}$.

   In particular we have that

$$\frac{\pi^{-\beta} D(z, \mu/2; k, \chi, \mathfrak{c})}{\omega(\chi)} \in \mathcal{N}_k^r(\Phi(\chi)),$$

where $\Phi(\chi)$ is the finite extension of Φ obtained by adjoining the values of the character χ.

4 The L-Function Attached to a Siegel Modular Form

We start by discussing the Hecke algebras that we consider in this work. We follow closely Chapter V in Shimura (2000). As before we fix a fractional ideal $\mathfrak{b}$ of F and an integral ideal $\mathfrak{c}$. We write C for $D[\mathfrak{b}^{-1}, \mathfrak{b}\mathfrak{c}]$ Moreover we define

$$E = \prod_{v \in \mathbf{h}} GL_n(\mathfrak{g}_v),\ B = \{x \in GL_n(F)_\mathbf{h} | x \prec \mathfrak{g}\},\ \mathfrak{X} = CQC,\ Q = \{diag[\hat{r}, r] | r \in B\}.$$

We write $\mathfrak{R}(C, \mathfrak{X})$ for the Hecke algebra corresponding to the pair $(C, \mathfrak{X})$ and for every place $v \in \mathbf{h}$ we write $\mathfrak{R}(C_v, \mathfrak{X}_v)$ for the local Hecke algebra at v and hence $\mathfrak{R}(C, \mathfrak{X}) = \bigotimes_v \mathfrak{R}(C_v, \mathfrak{X}_v)$. We now consider the formal Dirichlet series with coefficients in the global Hecke algebra defined by $\mathfrak{T} = \sum_{C\backslash\mathfrak{X}/C} C\xi C[\nu_\mathfrak{b}(\xi)]$ and its local version at $v \in \mathbf{h}$ defined as $\mathfrak{T}_v = \sum_{C_v\backslash\mathfrak{X}_v/C_v} C_v\xi C_v[\nu_\mathfrak{b}(\xi)]$. Here $\nu_\mathfrak{b}(\xi)$ is defined by $det(q)\mathfrak{g}$ where $q \in B$ such that $\xi \in D[\mathfrak{b}^{-1}, \mathfrak{b}]diag[q^{-1}, q^*]D[\mathfrak{b}^{-1}, \mathfrak{b}]$. We have that $\mathfrak{T} = \prod_v \mathfrak{T}_v$. Moreover if we define for an integral $\mathfrak{g}$-ideal $\mathfrak{a}$ the elements $T(\mathfrak{a}) \in \mathfrak{R}(C, \mathfrak{X})$ and $T_v(\mathfrak{a}) \in \mathfrak{R}(C_v, \mathfrak{X}_v)$ for $v \in \mathbf{h}$ by

$$T(\mathfrak{a}) = \sum_{\xi \in \mathfrak{X}, \nu_{\mathfrak{b}}(\xi) = \mathfrak{a}} C\xi C, \quad T_v(\mathfrak{a}) = \sum_{\xi \in \mathfrak{X}_v, \nu_{\mathfrak{b}}(\xi) = \mathfrak{a}} C_v \xi C_v$$

then we have that $\mathfrak{T} = \sum_{\mathfrak{a}} T(\mathfrak{a})[\mathfrak{a}]$. For an element $\mathbf{f} \in \mathcal{M}_k(C, \psi)$ we have an action of the Hecke algebra $\mathfrak{R}(C, \psi)$ (see Shimura 1997). We denote this action by $\mathbf{f}|C\xi C$ for an element $C\xi C \in \mathfrak{R}(C, \psi)$. Assume now that for such an $\mathbf{f} \neq 0$ we have $f|T(\mathfrak{a}) = \lambda(\mathfrak{a})\mathbf{f}$ with $\lambda(\mathfrak{a}) \in \mathbb{C}$ for all integral $\mathfrak{g}$-ideals. Then Shimura shows see (Shimura 2000, page 171) that there exists $\lambda_{v,i} \in \mathbb{C}$ such that

$$\mathfrak{L} \cdot \sum_{\mathfrak{a}} \lambda(\mathfrak{a})[\mathfrak{a}] = \prod_{v \in \mathbf{h}} Z_v,$$

where the factors Z_v are given by

$$Z_v = \begin{cases} (1 - N(\mathfrak{p})^n [\mathfrak{p}])^{-1} \prod_{i=1}^n \left((1 - N(\mathfrak{p})^n \lambda_{v,i} [\mathfrak{p}])(1 - N(\mathfrak{p})^n \lambda_{v,i}^{-1} [\mathfrak{p}])\right)^{-1}, & \text{if } v \nmid \mathfrak{c}; \\ \prod_{i=1}^n (1 - N(\mathfrak{p})^n \lambda_{v,i} [\mathfrak{p}])^{-1}, & \text{otherwise.} \end{cases}$$

and $\mathfrak{L} := \prod_{\mathfrak{p} \nmid \mathfrak{c}} (1 - [\mathfrak{p}]) \prod_{i=1}^n \left(1 - N(\mathfrak{p})^{2i} [\mathfrak{p}]^2\right)^{-1}$, where the product is over the prime $\mathfrak{g}$-ideals prime to $\mathfrak{c}$. For a Hecke character χ of F of conductor $\mathfrak{f}$ we put

$$Z(s, \mathbf{f}, \chi) := \prod_{v \in \mathbf{h}} Z_v \left(\chi^*(\mathfrak{q}) N(\mathfrak{q})^{-s}\right), \tag{5}$$

where $Z_v\left(\chi^*(\mathfrak{q}) N(\mathfrak{q})^{-s}\right)$ is obtained from Z_v by substituting $\chi^*(\mathfrak{p}) N(\mathfrak{p})^{-s}$ for $[\mathfrak{p}]$. We will need another L-function which we will denote by $Z'(s, \mathbf{f}, \chi)$ and we define by

$$Z'(s, \mathbf{f}, \chi) := \prod_{v \in \mathbf{h}} Z_v \left(\chi^*(\mathfrak{q})(\psi/\psi_{\mathfrak{c}})(\pi_{\mathfrak{q}}) N(\mathfrak{q})^{-s}\right), \tag{6}$$

where $\pi_{\mathfrak{q}}$ a uniformizer of $F_{\mathfrak{q}}$. We note here that we may obtain the first from the second up to a finite number of Euler factors by setting $\chi\psi^{-1}$ for χ.

5 The Rankin-Selberg Method

We now explain the integral representation of the zeta function introduced above due to Shimura. Everything in this section is taken from Shimura (2000, paragraph 20 and 22) as well as Shimura (1994).

We write $\mathcal{L}$ for the set of all $\mathfrak{g}$-lattices in F_1^n. We set $L_0 := \mathfrak{g}_1^n$ and we remark that for an element $L \in \mathcal{L}$ we can find an element $y \in GL_n(F)_{\mathbf{h}}$ such that $L = yL_0$. For an element $\tau \in S$ we define

$$L_\tau := \{L \in \mathcal{L} | \ell^* \tau \ell \in \mathfrak{b}\mathfrak{d}^{-1}, \ \forall \ell \in L\}.$$

Let $\mathbf{f} \in \mathcal{M}_k(C, \psi)$, $\tau \in S_+$ and $q \in GL_n(F)_\mathbf{h}$. Following Shimura we define the following two formal Dirichlet series

$$D(\tau, q; \mathbf{f}) := \sum_{x \in B/E} \psi_\mathbf{c}(det(qx))|det(x)|_F^{-n-1} c(\tau, qx; \mathbf{f})[det(x)\mathfrak{g}], \tag{7}$$

and

$$D'(\tau, q; \mathbf{f}) := \sum_{x \in B/E} \psi(det(qx))|det(x)|_F^{-n-1} c(\tau, qx; \mathbf{f})[det(x)\mathfrak{g}]. \tag{8}$$

We note that the second is obtained from the first one by setting $(\psi/\psi_\mathbf{c})(t)[t\mathfrak{g}]$ for $[t\mathfrak{g}]$, $t \in F_\mathbf{h}^\times$ in $D(\tau, q; \mathbf{f})$ and multiplying by $(\psi/\psi_\mathbf{c})(det(q))$. We define the series

$$\mathfrak{L}_0 = \prod_{v \nmid \mathfrak{c}} \left(\prod_{i=1}^{[(n+1)/2]} (1 - N(\mathfrak{p})^{2n+2-2i}[\mathfrak{p}]^2) \right)^{-1}.$$

Then we have see (Shimura 2000, Theorem 20.9).

Theorem 4 (Shimura). *Let $0 \neq \mathbf{f} \in \mathcal{M}_k(C, \psi)$ and such that $\mathbf{f}|T(\mathfrak{a}) = \lambda(\mathfrak{a})\mathbf{f}$ for every $\mathfrak{a}$. Then for $\tau \in S_+ \cap GL_n(F)$ and $L = qL_0$ with $q \in GL_n(F)_\mathbf{h}$ we have*

$$D(\tau, q; \mathbf{f})\mathfrak{L}_0 \prod_{v \in \mathbf{b}} g_v[\mathfrak{p}] \prod_{v \nmid \mathfrak{c}} h_v([\mathfrak{p}])^{-1} =$$

$$\prod_{v \in \mathbf{h}} Z_v \sum_{L < M \in \mathcal{L}_\tau} \mu(M/L)\psi_\mathfrak{c}(det(y))[det(q^* \hat{y}\mathfrak{g}]c(\tau, y; \mathbf{f}).$$

Assume now that $k_v \geq n/2$ for some $v \in \mathbf{a}$. Then there exists $\tau \in S_+ \cap GL_n(F)$ and $r \in GL_n(F)_\mathbf{h}$ such that

$$0 \neq \psi_\mathfrak{c}(det(r))c_\mathbf{f}(\tau, r) \prod_{v \in \mathbf{h}} Z_v = D(\tau, r; \mathbf{f}) \cdot \mathfrak{L}_0 \prod_{v \nmid \mathfrak{c}} h_v([\mathfrak{p}])^{-1} \cdot \prod_{v \in \mathbf{b}} g_v([\mathfrak{p}]).$$

For the definitions of $g_v(\cdot)$ and $h_v(\cdot)$ we refer to Shimura (2000). Now given a Hecke character χ of F, $\tau \in S^+$ and $r \in GL_n(F)_\mathbf{h}$ we define a Dirichlet series as follows:

$$D'_{r,\tau}(s, \mathbf{f}, \chi) := \sum_{B/E} \psi(det(rx))\chi^*(det(x)\mathfrak{g})c_\mathbf{f}(\tau, rx)|det(x)|_F^{s-n-1}. \tag{9}$$

This series is obtained from the series in Eq. 8 by putting $\chi^*(t\mathfrak{g})|t\mathfrak{g}|_F^s$ for $[t\mathfrak{g}]$. In particular we have the equation

$$D'_{r,\tau}(s,\mathbf{f},\chi)\Lambda_{\mathfrak{c}}\left(\frac{2s-n}{4}\right)\prod_{v\in\mathbf{b}} g_v(\chi(\psi/\psi_{\mathfrak{c}})(\pi_v)|\pi_v|^s) = \tag{10}$$

$$Z'(s,\mathbf{f},\chi)(\psi/\psi_{\mathfrak{c}})^2(det(r))\times$$

$$\sum_{L<M\in\mathcal{L}_\tau} \mu(M/L)(\psi_{\mathfrak{c}}^2/\psi)(det(y))\chi^*(det(r^*\hat{y})\mathfrak{g})|det(r^*\hat{y}|_F^s c(\tau,y;\mathbf{f}),$$

where for an integral ideal $\mathfrak{a}$ we write

$$\Lambda_{\mathfrak{a}}(s) = \begin{cases} L_{\mathfrak{a}}(2s,\rho_\tau\psi\chi)\prod_{i=1}^{n/2} L_{\mathfrak{a}}(4s-2i,\psi^2\chi^2), & \text{if } n \text{ is even;} \\ \prod_{i=1}^{(n+1)/2} L_{\mathfrak{a}}(4s-2i+1,\psi^2\chi^2), & n \text{ is odd.}\end{cases},$$

Given χ as above we write $\mathfrak{f}$ for the conductor of χ. We define $t' \in \mathbf{Z}^{\mathbf{a}}$ by

$$(\psi\chi)_{\mathbf{a}}(x) = x_{\mathbf{a}}^{-t'}|x_{\mathbf{a}}|^{t'}.$$

and $\mu \in \mathbf{Z}^{\mathbf{a}}$ by the conditions $0 \leq \mu_v \leq 1$ for all $v \in \mathbf{a}$ and $\mu - [k] - t' \in 2\mathbf{Z}^{\mathbf{a}}$.

We now define a weight l and a Hecke character ψ' of F by $l = \mu + (n/2)\mathbf{a}$ and $\psi' = \chi^{-1}\rho_\tau$, where ρ_τ is the Hecke character of F corresponding to the extension $F(c^{\frac{1}{2}})/F$ with $c := (-1)^{[n/2]}det(2\tau)$. Let us write $\theta_\chi \in \mathcal{M}_l(C',\psi')$ for the theta series associated to the data (χ^{-1},μ,τ,r) in Sect. 2. Write $C' = D[\mathfrak{b}'^{-1},\mathfrak{b}'\mathfrak{c}']\}$ and define $\mathfrak{e} := \mathfrak{b} + \mathfrak{b}'$. Then we have (see Shimura 1994, page 572),

Theorem 5 (Shimura).

$$(4\pi)^{-n(su+(k+l)/2)}(\sqrt{D_F}N(\mathfrak{e})^{-1})^{n(n+1)/2}\prod_{v\in\mathbf{a}}\Gamma_n(s+(k_v+l_v)/2)D'_{r,\tau}$$

$$\times(2s+3n/2+1;\mathbf{f},\chi) =$$

$$|det(r)|_F^{-2s-n/2}det(\tau)^{+(k+\mu+nu/2)/2+su}\times$$

$$\int_\Phi f(z)\theta_\chi(z)\overline{E(z,\bar{s}+(n+1)/2,k-l,\epsilon\overline{\psi\chi\rho_\tau},\Gamma')}\delta(z)^k dz,$$

where $\Phi := \mathcal{H}/\Gamma'$ and $\Gamma' := G\cap D[\mathfrak{e}^{-1},\mathfrak{e}\mathfrak{h}]$, where $\mathfrak{h} = \mathfrak{e}^{-1}(\mathfrak{b}\mathfrak{c}\cap\mathfrak{b}'\mathfrak{c}')$. here $\epsilon = 1$ if n is even and it is the non-trivial character of $F((-1)^{\frac{n}{2}})/F$ otherwise.

In particular using the Eq. 10 we obtain

Theorem 6 (Shimura).

$$Z'(s,\mathbf{f},\chi)\prod_{v\in\mathbf{a}}\Gamma_n\left(\frac{s-n-1+k_v+\mu_v}{2}\right)\times$$

$$(\psi/\psi_{\mathfrak{c}})^2(det(r))\sum_{L<M\in\mathcal{L}_\tau}\mu(M/L)(\psi_{\mathfrak{c}}^2/\psi)(det(y))\chi^*$$

$$\times(det(r^*\hat{y})\mathfrak{g})|det(r^*\hat{y})|_F^s c(\tau,y;\mathbf{f}) =$$

$$\left(D_F^{-1/2}N(\mathfrak{e})\right)^{n(n+1)/2}(4\pi)^{n||s'u+\lambda||}det(\tau)^{s'u+\lambda}|det(r)|_{\mathbf{A}}^{n+1-s}\times$$

$$\prod_{v\in\mathbf{b}}g_v((\psi/\psi_{\mathbf{c}})(\pi_v)\chi^*(\pi_v\mathfrak{g})|\pi_v|^s)(\Lambda_{\mathfrak{c}}/\Lambda_{\mathfrak{h}})$$

$$\times((2s-n)/4)vol(\Phi) < f,\theta_\chi D((2s-n)/4) >_{\Gamma'},$$

where $s' = (2s-3n-2)/4$ *and for an integral ideal* $\mathfrak{a}$ *of* F,

$$\Lambda_{\mathfrak{a}}(s) = \begin{cases} L_{\mathfrak{a}}(2s,\rho_\tau\psi\chi)\prod_{i=1}^{n/2}L_{\mathfrak{a}}(4s-2i,\psi^2\chi^2), & \text{if } n \text{ is even;} \\ \prod_{i=1}^{(n+1)/2}L_{\mathfrak{a}}(4s-2i+1,\psi^2\chi^2), & n \text{ is odd.}\end{cases},$$

and

$$D(s) = \overline{\Lambda_{\mathfrak{h}}(s)}E(z,\bar{s};k-l,\epsilon,\rho_\tau\overline{\psi\chi},\Gamma').$$

We have normalized the Petersson inner product as follows

$$< f,\theta_\chi D((2s-n)/4) >_{\Gamma'} = \frac{1}{vol(\Phi)}\int_\Phi f(z)\overline{\theta_\chi(z)D(z,(2s-n)/4)}\delta(z)^k dz.$$

In particular there exists (τ, r) *with* $c(\tau,r;\mathbf{f}) \neq 0$ *such that*

$$Z'(s,\mathbf{f},\chi)\prod_{v\in\mathbf{a}}\Gamma_n\left(\frac{s-n-1+k_v+\mu_v}{2}\right)\psi_{\mathfrak{c}}(det(r))c(\tau,r;\mathbf{f}) = \tag{11}$$

$$\left(D_F^{-1/2}N(\mathfrak{e})\right)^{n(n+1)/2}(4\pi)^{n||s'u+\lambda||}det(\tau)^{s'u+\lambda}|det(r)|_{\mathbf{A}}^{n+1-s}\times$$

$$\prod_{v\in\mathbf{b}}g_v((\psi/\psi_{\mathbf{c}})(\pi_v)\chi^*(\pi_v\mathfrak{g})|\pi_v|^s)(\Lambda_{\mathfrak{c}}/\Lambda_{\mathfrak{h}})$$

$$\times((2s-n)/4)vol(\Phi) < f,\theta_\chi D((2s-n)/4) >_{\Gamma'}.$$

We note here that $vol(\Phi) \in \pi^{n(n+1)/2}\mathbb{Q}^\times$.

6 Petersson Inner Products and Periods

In this section we define some Archimedean periods that we will use to normalize the special values of the function $Z'(s, \mathbf{f}, \psi)$. The idea of defining these periods is due to Sturm (1981) (building on previous work of Shimura), who considered the case of n even and $F = \mathbb{Q}$. However one should notice also the difference on the bounds of the weights that we impose. In what it follows we will call a Hecke operator $T(\mathfrak{a})$, relative to the group $C = D[\mathfrak{b}^{-1}, \mathfrak{bc}]$, as "good" if $\mathfrak{a}$ is prime to $\mathfrak{c}$

Theorem 7. *Let $\mathbf{f} \in \mathcal{S}_k(\mathfrak{c}, \psi)$ be an eigenform for all the "good" Hecke operators of C. Let Φ be the Galois closure of F over $\mathbb{Q}$ and write Ψ for extension of Φ generated by the eigenvalues of $\mathbf{f}$ and their complex conjugation . Assume $m_0 := min_v(k_v) > [3n/2 + 1] + 2$. Then there exists a period $\Omega_{\mathbf{f}}$ such that for any $\mathbf{g} \in \mathcal{S}_k(\overline{\mathbb{Q}})$ we have*

$$\left(\frac{<\mathbf{f}, \mathbf{g}>}{\Omega_{\mathbf{f}}}\right)^{\sigma} = \frac{<\mathbf{f}^{\sigma}, \mathbf{g}^{\sigma'}>}{\Omega_{\mathbf{f}^{\sigma}}},$$

for all $\sigma \in Gal(\overline{\mathbb{Q}}/\Phi)$, where $\sigma' = \rho\sigma\rho$. Moreover $\Omega_{\mathbf{f}}$ depends only on the eigenvalues of $\mathbf{f}$ and we have $\frac{<\mathbf{f},\mathbf{f}>}{\Omega_{\mathbf{f}}} \in \Psi^{\times}$.

Remark 1. As we remarked above, a theorem of this form has been firstly proved by Sturm (1981), when $F = \mathbb{Q}$ and n is even. A similar theorem appears also in the work of Panchishkin (1991). It is also important to notice that in Panchishkin's theorem one can take also $\mathbf{g}$ not cuspidal. However for this he has to take the weight big enough in order to be in the range of absolute convergent for the Eisenstein series (see the Theorems after the proof). Our proof is modelled on that of Sturm (1981, Theorem 3) and of Shimura (2000, Theorem 28.5). Maybe one should here remark that one of the differences with the proof here in comparison with the one of Sturm is that we use the identity (10) and not the Andrianov-Kalinin identity used by Sturm. Finally since we are using a stronger theorem of Shimura with respect to the absolute convergence of the function $Z(s, \mathbf{f}, \chi)$ we also obtain better bounds for the weights. Finally we remark the slightly larger bound on m_0 than in Shimura (2000, Theorem 28.5). The reason for this is the above mentioned problem with the Eisenstein spectrum (i.e. separate it rationally from the cuspidal part).

Proof. We write $\{\lambda(\mathfrak{a})\}$ for the system of the eigenvalues of $\mathbf{f}$ (with respect to the "good" Hecke operators) and we define $\mathcal{V} := \{\mathbf{h} \in \mathcal{S}_k(\mathfrak{c}, \psi) | \mathbf{h}|T(\mathfrak{a}) = \lambda(\mathfrak{a})\mathbf{h}\}$. Then as in Shimura we define $\mathcal{V}(\Psi) = \mathcal{V} \cap \mathcal{S}_k(\mathfrak{c}, \psi; \Psi)$. By Garrett (1992) we have that the space $\mathcal{V}(\Psi)$ is preserved by the operators $T(\mathfrak{a})$. Moreover the "good" Hecke operators generate a ring of semi-simple Ψ-linear transformations hence we have $\mathcal{V} = \mathcal{V}(\Psi) \otimes_{\Psi} \mathbb{C}$ and $\mathcal{S}_k(C, \Psi) = \mathcal{V}(\Psi) \oplus \mathcal{U}$, with $\mathcal{U}$ a vector space over Ψ which is stable under the action of the "good" Hecke operators. Since an eigenform in $\mathcal{U} \otimes_{\Psi} \mathbb{C}$ which is not contained in $\mathcal{V}$ must be orthogonal to it we have that the above decomposition is orthogonal with respect to the Petersson inner product.

We now pick an integer σ_0 so that $3n/2+1<\sigma_0<m_0$ and $m_0-\sigma_0\notin 2\mathbb{Z}$. Note that this is always possible thanks to our assumption $m_0>[3n/2+1]+2$. Then we define $\mu\in\mathbb{Z}^{\mathbf{a}}$ by the conditions $0\leq\mu_v\leq 1$ and $\sigma_0-k_v+\mu_v\in 2\mathbb{Z}$ for all $v\in\mathbf{a}$. Our choice of σ_0 implies in particular that there exists an $v\in\mathbf{a}$ so that $\mu_v\neq 0$. We put $t':=\mu-k$. We now pick a quadratic character χ of F so that $(\psi\chi)_{\mathbf{a}}(x)=x_{\mathbf{a}}^{t'}|x_{\mathbf{a}}|^{-t'}$ and of conductor $\mathfrak{f}$ such that $\mathfrak{c}|\mathfrak{f}$. Note that such a character can be obtained as the non trivial character of the quadratic extension $F(\sqrt{\Delta})$ by picking the sign of $\Delta\in F$ properly at $v\in\mathbf{a}$ and Δ with non trivial valuation at all primes that divide $\mathfrak{c}$. The existence of such a Δ follows from the approximation theorem for F. As in Shimura (2000, page 236) we define $l:=\mu+(n/2)\mathbf{a}$ and $\nu=\sigma_0-(n/2)$. Then $\nu\geq(n+1)/2$ and $0\leq k-l-\nu\mathbf{a}\in 2\mathbb{Z}^{\mathbf{a}}$. We consider the theta series θ_χ with respect to our choices of χ and μ. By Theorem 6, after evaluating at $s=\sigma_0$ we obtain

$$Z'(\sigma_0,\mathbf{f},\chi)\prod_{v\in\mathbf{a}}\Gamma_n\left(\frac{\sigma_0-n-1+k_v+\mu_v}{2}\right)(\psi/\psi_{\mathfrak{c}})^2(det(r))\times$$

$$\sum_{L<M\in\mathcal{L}_\tau}\mu(M/L)(\psi_{\mathfrak{c}}^2/\psi)(det(y))\chi^*(det(r^*\hat{y})\mathfrak{g})|det(r^*\hat{y})|_F^{\sigma_0}c(\tau,y;\mathbf{f})=$$

$$\left(D_F^{-1/2}N(\mathfrak{e})\right)^{n(n+1)/2}(4\pi)^{n||s_0'u+\lambda||}det(\tau)^{s_0'u+\lambda}|det(r)|_{\mathbf{A}}^{n+1-\sigma_0}(\Lambda_{\mathfrak{c}}/\Lambda_{\mathfrak{h}})(\nu/2)\times$$

$$\prod_{v\in\mathbf{b}}g_v((\psi/\psi_{\mathfrak{c}})(\pi_v)\chi^*(\pi_v\mathfrak{g})|\pi_v|^{\sigma_0})vol(\Phi)<f,\theta_\chi D(\nu/2,\epsilon\rho_\tau\overline{\psi\chi})>_{\Gamma'},$$

where $s_0'=(2\sigma_0-3n-2)/4$. We now note (see Shimura 2000, page 237) that

$$\frac{\prod_{v\in\mathbf{a}}\Gamma_n\left(\frac{\sigma_0-n-1+k_v+\mu_v}{2}\right)}{vol(\Phi)}\in\pi^{n||\frac{k-l-\nu\mathbf{a}}{2}||-n||k||+d\epsilon}\mathbb{Q}^\times$$

where $\epsilon=n^2/4$ if n even and $(n^2-1)/4$ otherwise. We now write δ for the rational part of $\frac{\prod_{v\in\mathbf{a}}\Gamma_n\left(\frac{\sigma_0-n-1+k_v+\mu_v}{2}\right)}{vol(\Phi)}$. We now take $\beta\in\mathbb{N}$ so that $\pi^{-\beta}D(\nu/2)\in\mathcal{N}_{k-l}^p(\Phi\mathbb{Q}_{ab})$ with $p=\frac{k-l-\nu\mathbf{a}}{2}$ and we set $\gamma:=n||\frac{k-l-\nu\mathbf{a}}{2}||-n||k||+d\epsilon-n||s_0'u+\lambda||-\beta$. We further set

$$B(\chi,\psi,\tau,r,\mathbf{f}):=\delta(\psi/\psi_{\mathfrak{c}})^2(det(r))\times$$

$$\sum_{L<M\in\mathcal{L}_\tau}\mu(M/L)(\psi_{\mathfrak{c}}^2/\psi)(det(y))\chi^*(det(r^*\hat{y})\mathfrak{g})|det(r^*\hat{y})|_F^{\sigma_0}c(\tau,y;\mathbf{f}),$$

and

$$C(\chi,\psi,\tau,r) := (N(\mathfrak{e}))^{n(n+1)/2}\,|det(r)|_{\mathbf{A}}^{n+1-\sigma_0}\times$$

$$\prod_{v\in\mathbf{b}} g_v((\psi/\psi_{\mathfrak{c}})(\pi_v)\chi^*(\pi_v\mathfrak{g})|\pi_v|^{\sigma_0})(\Lambda_{\mathfrak{c}}/\Lambda_{\mathfrak{h}})(\nu/2).$$

We then have for every $\sigma \in Gal(\overline{\mathbb{Q}}/\Phi)$ that

$$B(\chi,\psi,\tau,r,\mathbf{f})^{\sigma} = B(\chi^{\sigma},\psi^{\sigma},\tau,r,\mathbf{f}^{\sigma}) \text{ and } C(\chi,\psi,\tau,r)^{\sigma} = C(\chi^{\sigma},\psi^{\sigma},\tau,r).$$

We now note the equalities

$$< f, \theta_{\chi} D(\nu/2, \epsilon\rho_{\tau}\overline{\psi\chi}) >_{\Gamma'}=$$

$$< f, \mathfrak{p}(\theta_{\chi} D(\nu/2, \epsilon\rho_{\tau}\overline{\psi\chi})) >_{\Gamma'}=< f, Tr_{\Gamma'}^{\Gamma}(\mathfrak{p}(\theta_{\chi} D(\nu/2, \epsilon\rho_{\tau}\overline{\psi\chi}))) >_{\Gamma},$$

where $\mathfrak{p} : \mathcal{R}_k^p \rightarrow \mathcal{S}_k$ is Shimura's holomorphic projection operators (Shimura 2000, Proposition 15.6) (note that $\theta_{\chi} D(\nu/2) \in \mathcal{R}_k^p$ since θ_{χ} is a cusp form) and $Tr_{\Gamma'}^{\Gamma}$: $\mathcal{S}_k(\Gamma',\psi) \rightarrow S_k(\Gamma,\psi)$ is the usual trace operator attached to the groups $\Gamma' \leq \Gamma$. Moreover, since $\theta_{\chi}\pi^{-\beta} D(\nu/2) \in \mathcal{N}_k^p(\Phi\mathbb{Q}_{ab})$, we may consider the action of $\sigma \in Gal(\Phi\mathbb{Q}_{ab}/\Phi)$. Then

$$\mathfrak{p}(\theta_{\chi}\pi^{-\beta} D(\nu/2, \epsilon\rho_{\tau}\overline{\psi\chi}))^{\sigma} = \mathfrak{p}(\theta_{\chi}^{\sigma}(\pi^{-\beta} D(\nu/2, \epsilon\rho_{\tau}\overline{\psi\chi}))^{\sigma}),$$

and,

$$Tr_{\Gamma'}^{\Gamma}(\theta_{\chi}\pi^{-\beta} D(\nu/2, \epsilon\rho_{\tau}\overline{\psi\chi}))^{\sigma} = Tr_{\Gamma'}^{\Gamma}(\theta_{\chi}^{\sigma} D(\nu/2, \epsilon\rho_{\tau}\overline{\psi\chi})^{\sigma}),$$

where in the last equation the last trace is from the space $\mathcal{S}_k(\Gamma',\psi^{\sigma})$ to $\mathcal{S}_k(\Gamma',\psi^{\sigma})$. The equivariant property of the holomorphic projection operator is shown in Proposition 15.6 of (loc. cit.) and the one of the trace is exactly as in Sturm where he considers the case of $F = \mathbb{Q}$, but the arguments is valid also for general F since the strong approximation theorem also hold for the group $Sp_n(F)$, the essential argument in his proof. We make this more formal in the lemma following this proof.

Keeping now the character χ fixed we know that for any given $\mathbf{f} \in \mathcal{V}$ there exists (τ, r) such that

$$B(\chi,\psi,\tau,r,\mathbf{f}) = \delta\psi(det(r))c(\tau,r;\mathbf{f}) \neq 0.$$

We note here that the same pair (τ, r) can be used for the form $\mathbf{f}^{\sigma}$, as it follows from the proof of Theorem 20.9 in Shimura (2000). As in Shimura we write $\mathfrak{G}$ for the set of pairs (τ, r) for which such an $\mathbf{f}$ exists. From the observation above the set $\mathfrak{G}$ is the same also for the system of eigenvalues $\lambda(\mathfrak{a})^{\sigma}$, for all $\sigma \in Gal(\overline{\mathbb{Q}}/\Phi)$. In particular for such an (τ, r)

$$0 \neq \pi^{\gamma} Z'(\sigma_0, \mathbf{f}, \chi)\delta\psi(det(r))c(\tau, r; \mathbf{f}) = \tag{12}$$

$$\left(D_F^{-1/2} N(\mathfrak{e})\right)^{n(n+1)/2} (4)^{n||s_0'u+\lambda||} det(\tau)^{s_0'u+\lambda} |det(r)|_{\mathbf{A}}^{n+1-\sigma_0} \times$$

$$\prod_{v \in \mathbf{b}} g_v((\psi/\psi_{\mathfrak{c}})(\pi_v)\chi^*(\pi_v \mathfrak{g})|\pi_v|^{\sigma_0})(\Lambda_{\mathfrak{c}}/\Lambda_{\mathfrak{h}})(\nu/2) < f, \theta_\chi \pi^{-\beta} D(\nu/2, \epsilon\rho_\tau \overline{\psi\chi}) >_{\Gamma'} .$$

The fact that $Z'(\sigma_0, \mathbf{f}, \chi) \neq 0$ is in principle (Shimura 2000, Theorem 20.13). Indeed in page 183 of (loc. cit.) Shimura first proves the non-vanishing of $Z'(\sigma_0, \mathbf{f}, \chi)$ for any character χ with $\mu \neq 0$, as it is the case that we consider. Further we note that this in particular implies also that $C(\chi, \psi, \tau, r) \neq 0$ for all $(\tau, r) \in \mathfrak{G}$.

We now define an element $\mathbf{g}_{\tau,r,\psi} \in \mathcal{S}_k(\Gamma, \overline{\psi}; \Phi\mathbb{Q}_{ab})$ by

$$\mathbf{g}_{\tau,r,\psi} = \pi^{-\beta} Tr_{\Gamma'}^{\Gamma} \left(\mathfrak{p}(\theta_\chi \pi^{-\beta} D(\nu/2, \epsilon\rho_\tau \overline{\psi\chi}))\right),$$

and define the space $\mathcal{W}$ to be the space generated by $g_{\tau,r,\psi}$ for $(\tau, r) \in \mathfrak{G}$. We now consider the case n even or odd separately.

The case of n even: In this case we have that ϵ is the trivial character. We now claim that there exists an $\Omega_{\mathbf{f}} \in \mathbb{C}^\times$ such that any $\mathbf{f} \in \mathcal{V}$ and any $\mathbf{g}_{\tau,r,\psi}$

$$\left(\frac{< \mathbf{f}, \mathbf{g}_{\tau,r,\psi} >}{\Omega_{\mathbf{f}}}\right)^{\sigma} = \frac{< \mathbf{f}^{\sigma}, \mathbf{g}_{\tau,r,\psi}^{\sigma'} >}{\Omega_{\mathbf{f}^{\sigma}}},$$

where $\sigma' = \rho\sigma\rho$. First we observe that

$$\mathbf{g}_{\tau,r,\psi}^{\sigma'} = Tr_{\Gamma'}^{\Gamma} \left(\mathfrak{p}(\theta_\chi \pi^{-\beta} D(\nu/2, \epsilon\rho_\tau \overline{\psi\chi}))\right)^{\sigma'} = Tr_{\Gamma'}^{\Gamma} \left(\mathfrak{p}(\theta_\chi^{\sigma'} (\pi^{-\beta} D(\nu/2, \epsilon\rho_\tau \overline{\psi\chi}))^{\sigma'})\right) =$$

$$Tr_{\Gamma'}^{\Gamma} \left(\mathfrak{p}(\theta_\chi (\pi^{-\beta} D(\nu/2, \epsilon\rho_\tau \overline{\psi\chi}))^{\sigma'})\right),$$

where the last equality follows from the fact that χ is a quadratic character. We now recall that $D(\nu/2, \rho_\tau \overline{\psi\chi}) = D(z, \nu/2; k-l, \rho_\tau \overline{\psi\chi}, \Gamma)$ and we have seen that

$$\left(\frac{\pi^{-\beta} D(z, \nu/2; k-l, \rho_\tau \overline{\psi\chi}, \Gamma)}{P(\rho_\tau \overline{\psi\chi})}\right)^{\sigma'} = \frac{\pi^{-\beta} D(z, \nu/2; k-l, \rho_\tau \overline{\psi\chi}^{\sigma'}, \Gamma)}{P(\rho_\tau \overline{\psi\chi}^{\sigma'})},$$

where $P(\rho_\tau \overline{\psi}\chi) = \frac{i^{n|p|} \mathbf{g}(\rho_\tau \overline{\psi}\chi)(i)^{\nu d}}{|D_F|^{1/2}} \frac{\left(\prod_{i=1}^{[n/2]} (i)^{(2\nu-2i)d}\right) \mathbf{g}(\overline{\psi}^{2[n/2]})}{|D_F|^{b(n)}}$, with $p = \frac{k-l-\nu\mathbf{a}}{2}$. We conclude that

$$\mathbf{g}_{\tau,r,\psi}^{\sigma'} = \frac{P(\rho_\tau \overline{\psi\chi})^{\sigma'}}{P(\rho_\tau \overline{\psi}^{\sigma'} \chi)} Tr_{\Gamma'}^{\Gamma} \left(\mathfrak{p}(\theta_\chi \pi^{-\beta} D(\nu/2, \epsilon\rho_\tau \overline{\psi}^{\sigma'} \chi))\right) = \frac{P(\rho_\tau \overline{\psi\chi})^{\sigma'}}{P(\rho_\tau \overline{\psi}^{\sigma'} \chi)} \mathbf{g}_{\tau,r,\psi^{\sigma}}.$$

We set $R(\overline{\psi}) := \frac{i^{n|p|}(i)^{\nu d}}{|D_F|^{1/2}} \frac{\left(\prod_{i=1}^{[n/2]}(i)^{(2\nu-2i)d}\right)\mathbf{g}(\overline{\psi}^{2[n/2]})}{|D_F|^{b(n)}}$. We now consider the ratio

$$\frac{\mathbf{g}(\rho_\tau \overline{\psi} \chi)^{\sigma'}}{\mathbf{g}(\rho_\tau \overline{\psi}^{\sigma'} \chi)} = \frac{\mathbf{g}(\rho_\tau)^{\sigma'}}{\mathbf{g}(\rho_\tau^{\sigma'})} \frac{\mathbf{g}(\overline{\psi})^{\sigma'}}{\mathbf{g}(\overline{\psi}^{\sigma'})} \frac{\mathbf{g}(\chi)^{\sigma'}}{\mathbf{g}(\chi^{\sigma'})}.$$

We recall that ρ_τ is the non-trivial character of the quadratic extension $F(\sqrt{c})/F$ with $c = (-1)^{[n/2]} det(2\tau)$. Since we are considering $\tau > 0$ we have that

$$\frac{\mathbf{g}(\rho_\tau)^{\sigma'}}{\mathbf{g}(\rho_\tau^{\sigma'})} = \begin{cases} \frac{\sqrt{2det(\tau)}^{\sigma'}}{\sqrt{2det(\tau)}}, & \text{if } [n/2] \text{ even;} \\ \left(\frac{i^{\sigma'}}{i}\right)^d \frac{\sqrt{N(2det(\tau))}^{\sigma'}}{\sqrt{N(2det(\tau))}}, & \text{otherwise.} \end{cases}$$

Putting all these together we conclude that

$$\mathbf{g}_{\tau,r,\psi}^{\sigma'} = \frac{\mathbf{g}(\rho_\tau)^{\sigma'}}{\mathbf{g}(\rho_\tau^{\sigma'})} \frac{\mathbf{g}(\overline{\psi})^{\sigma'}}{\mathbf{g}(\overline{\psi}^{\sigma'})} \frac{\mathbf{g}(\chi)^{\sigma'}}{\mathbf{g}(\chi^{\sigma'})} \frac{R(\overline{\psi})^{\sigma'}}{R(\overline{\psi}^{\sigma'})} \mathbf{g}_{\tau,r,\psi^\sigma}$$

For any $\mathbf{g}_{\tau,r}$ we have

$$(4)^{-n||s_0'u+\lambda||} D_F^{n(n+1)/4} \pi^\gamma Z'(\sigma_0, \mathbf{f}, \chi) B(\chi, \psi, \tau, r, \mathbf{f}) =$$

$$det(\tau)^{s_0'u+\lambda} C(\chi, \psi, \tau, r) < \mathbf{f}, \mathbf{g}_{\tau,r} >_\Gamma .$$

For any $(\tau, r) \in \mathfrak{G}$ we have seen that $C(\chi, \psi, \tau, r) \neq 0$. We obtain

$$\frac{< \mathbf{f}, \mathbf{g}_{\tau,r} >_\Gamma}{(4\pi)^{-n||s_0'u+\lambda||} D_F^{n(n+1)/4} Z'(\sigma_0, \mathbf{f}, \chi)} = det(\tau)^{-(s_0'u+\lambda)} \frac{B(\chi, \psi, \tau, r, \mathbf{f})}{C(\chi, \psi, \tau, r)}.$$

For any $\sigma \in Gal(\overline{\mathbb{Q}}/\mathbb{Q})$ we have then

$$\left(\frac{< \mathbf{f}, \mathbf{g}_{\tau,r,\psi} >_\Gamma}{(4)^{-n||s_0'u+\lambda||} D_F^{n(n+1)/4} \pi^\gamma Z'(\sigma_0, \mathbf{f}, \chi)}\right)^\sigma =$$

$$\left(det(\tau)^{-(s_0'u+\lambda)} \frac{B(\chi, \psi, \tau, r, \mathbf{f})}{C(\chi, \psi, \tau, r)}\right)^\sigma =$$

$$(det(\tau)^{-(s_0'u+\lambda)})^\sigma \frac{B(\chi^\sigma, \psi^\sigma, \tau, r, \mathbf{f}^\sigma)}{C(\chi^\sigma, \psi^\sigma, \tau, r)} =$$

$$\frac{(det(\tau)^{-(s_0'u+\lambda)})^\sigma}{det(\tau)^{-(s_0'u+\lambda)}} \frac{< \mathbf{f}^\sigma, \mathbf{g}_{\tau,r,\psi^\sigma} >_\Gamma}{(4\pi)^{-n||s_0'u+\lambda||} D_F^{n(n+1)/4} Z'(\sigma_0, \mathbf{f}^\sigma, \chi)}.$$

We remark that $s_0'u+\lambda = \frac{2\sigma_0-3n-2}{4}u + \frac{k+\mu+\frac{n}{2}u}{2} = \frac{\sigma_0 u+k+\mu}{2} - \frac{n+1}{2}u$. By our choice of σ_0 we have that $\sigma_0 u + k + \mu \in 2\mathbb{Z}^{\mathbf{a}}$. We obtain that $\frac{(det(\tau)^{-(s_0'u+\lambda)})^{\sigma}}{det(\tau)^{-(s_0'u+\lambda)}} = \frac{(det(\tau)^{\frac{1}{2}\mathbf{a}})^{\sigma}}{det(\tau)^{(\frac{1}{2}\mathbf{a})}}$ Now we note that since $det(\tau)$ is totally positive we have

$$\overline{\left(\frac{\mathbf{g}(\rho_\tau)^{\sigma'}}{\mathbf{g}(\rho_\tau^{\sigma'})}\right)}^{-1} \frac{(det(\tau)^{\frac{1}{2}\mathbf{a}})^{\sigma}}{det(\tau)^{(\frac{1}{2}\mathbf{a})}} = \begin{cases} 1, & \text{if } [n/2] \text{ even;} \\ \left(\frac{i^{\sigma'}}{i}\right)^d, & \text{otherwise.} \end{cases}$$

We have seen that

$$\mathbf{g}_{\tau,r,\psi^{\sigma}} = \left(\frac{\mathbf{g}(\rho_\tau)^{\sigma'}}{\mathbf{g}(\rho_\tau^{\sigma'})}\frac{\mathbf{g}(\overline{\psi})^{\sigma'}}{\mathbf{g}(\overline{\psi}^{\sigma'})}\frac{\mathbf{g}(\chi)^{\sigma'}}{\mathbf{g}(\chi^{\sigma'})}\frac{R(\overline{\psi})^{\sigma'}}{R(\overline{\psi}^{\sigma'})}\right)^{-1} \mathbf{g}_{\tau,r,\psi}^{\sigma'}.$$

and hence

$$\left(\frac{<\mathbf{f}, \mathbf{g}_{\tau,r,\psi}>_{\Gamma}}{(4)^{-n||s_0'u+\lambda||} D_F^{n(n+1)/4} \pi^{\gamma} Z(\sigma_0, \mathbf{f}, \chi)}\right)^{\sigma} =$$

$$\overline{\left(\frac{\mathbf{g}(\rho_\tau)^{\sigma'}}{\mathbf{g}(\rho_\tau^{\sigma'})}\frac{\mathbf{g}(\overline{\psi})^{\sigma'}}{\mathbf{g}(\overline{\psi}^{\sigma'})}\frac{\mathbf{g}(\chi)^{\sigma'}}{\mathbf{g}(\chi^{\sigma'})}\frac{R(\overline{\psi})^{\sigma'}}{R(\overline{\psi}^{\sigma'})}\right)}^{-1} \frac{(det(\tau)^{\frac{1}{2}\mathbf{a}})^{\sigma}}{det(\tau)^{(\frac{1}{2}\mathbf{a})}} \times$$

$$\frac{<\mathbf{f}^{\sigma}, \mathbf{g}_{\tau,r,\psi}^{\sigma'}>_{\Gamma}}{(4)^{-n||s_0'u+\lambda||} D_F^{n(n+1)/4} \pi^{\gamma} Z'(\sigma_0, \mathbf{f}^{\sigma}, \chi)},$$

or equivalently

$$\left(\frac{<\mathbf{f}, \mathbf{g}_{\tau,r,\psi}>_{\Gamma}}{\overline{(\mathbf{g}(\overline{\psi})\mathbf{g}(\chi)R(\overline{\psi}))}^{-1} B(n) 4^{-n||s_0'u+\lambda||} D_F^{n(n+1)/4} \pi^{\gamma} Z'(\sigma_0, \mathbf{f}, \chi)}\right)^{\sigma} =$$

$$\frac{<\mathbf{f}^{\sigma}, \mathbf{g}_{\tau,r,\psi}^{\sigma'}>_{\Gamma}}{\overline{\left(\mathbf{g}(\overline{\psi}^{\sigma'})\mathbf{g}(\chi)R(\overline{\psi}^{\sigma'})B(n)\right)}^{-1} 4^{-n||s_0'u+\lambda||} D_F^{n(n+1)/4} \pi^{\gamma} Z'(\sigma_0, \mathbf{f}^{\sigma}, \chi)}.$$

where $B(n) = i^d$ if $[n/2]$ is odd and 1 otherwise. Hence we define

$$\Omega_{\mathbf{f}} := \overline{(\mathbf{g}(\overline{\psi})\mathbf{g}(\chi)R(\overline{\psi}))}^{-1} B(n) 4^{-n||s_0'u+\lambda||} D_F^{n(n+1)/4} \pi^{\gamma} Z'(\sigma_0, \mathbf{f}, \chi)$$

The case of *n* odd: We now repeat the considerations above but with the half-integral weight Eisenstein series.

$$\left(\frac{\pi^{-\beta} D(z,\nu/2;k-l,\epsilon\chi\overline{\psi}\rho_\tau,\mathfrak{c})}{\mathbf{g}(\epsilon\rho_\tau\overline{\psi}\chi\phi)^n i^{n|p|} i^{-dn\nu} C|D_F|^{n\nu/2+3n(n+1)/4}\mathbf{g}(\epsilon\rho_\tau\overline{\psi}\chi)|D_F|^{1/2}(2i)^{-(\nu-n)d} b([n/2])}\right)^{\sigma'} =$$

$$\frac{\pi^{-\beta} D(z,\nu/2;k-l,(\epsilon\chi\overline{\psi}\rho_\tau)^{\sigma})}{\mathbf{g}((\epsilon\chi\overline{\psi}\rho_\tau)^{\sigma'}\phi)^n i^{n|p|} i^{-dn\nu} C|D_F|^{n\nu/2+3n(n+1)/4}\mathbf{g}((\epsilon\chi\overline{\psi}\rho_\tau)^{\sigma'})|D_F|^{1/2}(2i)^{-(\nu-n)d} b([n/2])},$$

where $b(m) = i^d$ if m is m is odd and 1 otherwise. We set $P(\epsilon\chi\overline{\psi}\rho_\tau)$ equal to

$$\mathbf{g}(\epsilon\rho_\tau\overline{\psi}\chi\phi)^n i^{n|p|} i^{-dn\nu} C|D_F|^{n\nu/2+3n(n+1)/4}\mathbf{g}(\epsilon\rho_\tau\overline{\psi}\chi)|D_F|^{1/2}(2i)^{-(\nu-n)d} b([n/2])$$

and as before we have

$$\mathbf{g}_{\tau,r,\psi}^{\sigma'} = \frac{P(\epsilon\rho_\tau\overline{\psi}\chi)^{\sigma'}}{P(\epsilon\rho_\tau\overline{\psi}^{\sigma'}\chi)}\mathbf{g}_{\tau,r,\psi^\sigma}.$$

We consider the ratio

$$\frac{\left(\mathbf{g}(\epsilon\rho_\tau\overline{\psi}\chi\phi)^n\mathbf{g}(\epsilon\rho_\tau\overline{\psi}\chi)\right)^{\sigma'}}{\mathbf{g}(\epsilon\rho_\tau\overline{\psi}^{\sigma'}\chi\phi)^n\mathbf{g}(\epsilon\rho_\tau\overline{\psi}^{\sigma'}\chi)} =$$

$$\left(\frac{\mathbf{g}(\epsilon)^{\sigma'}}{\mathbf{g}(\epsilon)}\right)^{n+1}\left(\frac{\mathbf{g}(\rho_\tau)^{\sigma'}}{\mathbf{g}(\rho_\tau)}\right)^{n+1}\left(\frac{\mathbf{g}(\overline{\psi})^{\sigma'}}{\mathbf{g}(\overline{\psi}^{\sigma'})}\right)^{n+1}\left(\frac{\mathbf{g}(\chi)^{\sigma'}}{\mathbf{g}(\chi)}\right)^{n+1}\left(\frac{\mathbf{g}(\phi)^{\sigma'}}{\mathbf{g}(\phi)}\right)^{n}.$$

Since $n+1$ is even and $\rho_\tau, \chi, \epsilon$ are quadratic characters we get that

$$\left(\frac{\mathbf{g}(\epsilon)^{\sigma'}}{\mathbf{g}(\epsilon)}\right)^{n+1} = \left(\frac{\mathbf{g}(\rho_\tau)^{\sigma'}}{\mathbf{g}(\rho_\tau)}\right)^{n+1} = \left(\frac{\mathbf{g}(\chi)^{\sigma'}}{\mathbf{g}(\chi)}\right)^{n+1} = 1.$$

We set $R := i^{n|p|} i^{-dn\nu} C|D_F|^{n\nu/2+3n(n+1)/4}|D_F|^{1/2}(2i)^{-(\nu-n)d} b([n/2])$, and then we have

$$\mathbf{g}_{\tau,r,\psi}^{\sigma'} = \left(\frac{\mathbf{g}(\overline{\psi})^{\sigma'}}{\mathbf{g}(\overline{\psi}^{\sigma'})}\right)^{n+1}\left(\frac{\mathbf{g}(\phi)^{\sigma'}}{\mathbf{g}(\phi)}\right)^{n}\frac{R^{\sigma'}}{R}\mathbf{g}_{\tau,r,\psi^\sigma}.$$

By the same calculations as in the case of n even, by no noticing that $s_0'u+\lambda \in \mathbb{Z}^{\mathbf{a}}$ we obtain For any $\sigma \in Gal(\overline{\mathbb{Q}}/\mathbb{Q})$ we have then

$$\left(\frac{<\mathbf{f}, \mathbf{g}_{\tau,r,\psi}>_\Gamma}{4^{-n||s_0'u+\lambda||}D_F^{n(n+1)/4}\pi^\gamma Z'(\sigma_0,\mathbf{f},\chi)}\right)^{\sigma} = \frac{<\mathbf{f}^\sigma, \mathbf{g}_{\tau,r,\psi^\sigma}>_\Gamma}{4^{-n||s_0'u+\lambda||}D_F^{n(n+1)/4}\pi^\gamma Z'(\sigma_0,\mathbf{f}^\sigma,\chi)}.$$

Hence we conclude

$$\left(\frac{<\mathbf{f},\mathbf{g}_{\tau,r,\psi}>_\Gamma}{\left(\overline{\mathbf{g}(\overline{\psi})}^{n+1}\overline{\mathbf{g}(\overline{\phi})}^{n}\bar{R}\right)^{-1}4^{-n||s_0'u+\lambda||}D_F^{n(n+1)/4}\pi^\gamma Z'(\sigma_0,\mathbf{f},\chi)}\right)^\sigma =$$

$$\frac{<\mathbf{f}^\sigma,\mathbf{g}_{\tau,r,\psi^\sigma}>_\Gamma}{\left(\overline{\mathbf{g}(\overline{\psi}^{\sigma'})}^{n+1}\overline{\mathbf{g}(\overline{\phi})}^{n}\bar{R}\right)^{-1}4^{-n||s_0'u+\lambda||}D_F^{n(n+1)/4}\pi^\gamma Z'(\sigma_0,\mathbf{f}^\sigma,\chi)}.$$

So for n odd we define

$$\Omega_{\mathbf{f}} := \left(\overline{\mathbf{g}(\overline{\psi})}^{n+1}\overline{\mathbf{g}(\overline{\phi})}^{n}\bar{R}\right)^{-1}4^{-n||s_0'u+\lambda||}D_F^{n(n+1)/4}\pi^\gamma Z'(\sigma_0,\mathbf{f},\chi).$$

By $\mathcal{W}'$ we define the space generated by the projection of $\mathcal{W}$ on $\mathcal{V}$. By definition $\mathcal{W}'=\mathcal{V}$. Indeed for any element $\mathbf{g}\in\mathcal{V}$ there exists $\mathbf{h}\in\mathcal{W}'$ such that $<\mathbf{g},\mathbf{h}>_\Gamma\neq 0$, simply by taking the projection of the corresponding $\mathbf{g}_{\tau,r}$ to $\mathcal{W}'$. So the $\mathbb{C}$ span of $\mathbf{g}_{\tau,r}$ with $\tau,r\in\mathfrak{G}$ is equal to $\mathcal{V}$. Since $\mathbf{g}_{\tau,r}$ have algebraic coefficients we have that the $\overline{\mathbb{Q}}$-span is equal to $\mathcal{V}(\overline{\mathbb{Q}})$. We can now establish the theorem for any $\mathbf{g}\in\mathcal{V}(\overline{\mathbb{Q}})$ since after writing $\mathbf{g}=\sum_j c_j\mathbf{g}_{\tau_j,r_j,\mathcal{V}}\in\mathcal{V}(\overline{\mathbb{Q}})$, where $\mathbf{g}_{\tau_j,r_j,\mathcal{V}}$ is the projection of $\mathbf{g}_{\tau_j,r_j}$ to $\mathcal{V}$, we have

$$\left(\frac{<\mathbf{f},\mathbf{g}>}{\Omega_{\mathbf{f}}}\right)^\sigma=\sum_j\overline{c_j}^\sigma\left(\frac{<\mathbf{f}^\sigma,\mathbf{g}^{\sigma'}_{\tau_j,r_j,\mathcal{V}}>}{\Omega_{\mathbf{f}^\sigma}}\right)=\frac{<\mathbf{f}^\sigma,\mathbf{g}^{\sigma'}>}{\Omega_{\mathbf{f}^\sigma}}$$

We now take any $\mathbf{g}\in\mathcal{S}_k(\Gamma,\psi;\overline{\mathbb{Q}})$. We write $\mathbf{g}=\mathbf{g}_1+\mathbf{g}_2$ with $\mathbf{g}_1\in\mathcal{V}$ and $\mathfrak{h}_2\in\mathcal{V}^\perp$. Then we have that

$$\left(\frac{<\mathbf{f},\mathbf{g}>}{\Omega_{\mathbf{f}}}\right)^\sigma=\left(\frac{<\mathbf{f},\mathbf{g}_1>}{\Omega_{\mathbf{f}}}\right)^\sigma=\frac{<\mathbf{f}^\sigma,\mathbf{g}_1^{\sigma'}>}{\Omega_{\mathbf{f}^\sigma}}=\frac{<\mathbf{f}^\sigma,\mathbf{g}^{\sigma'}>}{\Omega_{\mathbf{f}^\sigma}}$$

where the last equality follows from the fact that $<\mathbf{f},\mathbf{g}>=0$ implies that $<\mathbf{f}^\sigma,\mathbf{g}^{\sigma'}>=0$. It is enough to show this for $\mathbf{g}$ an eigenform for all the good Hecke operators in an L-packet different from that of **f**'s. That is, there exists an ideal $\mathfrak{a}$ with $(\mathfrak{a},\mathfrak{c})=1$ so that $T(\mathfrak{a})\mathbf{f}=\lambda_{\mathbf{f}}\mathbf{f}$ and $T(\mathfrak{a})\mathbf{g}=\lambda_{\mathbf{g}}\mathbf{g}$ such that $\lambda_{\mathbf{f}}\neq\lambda_{\mathbf{g}}$. But then we have

$$\lambda_{\mathbf{f}}^\sigma<\mathbf{f}^\sigma,\mathbf{g}^{\sigma'}>=<T(\mathfrak{a})\mathbf{f}^\sigma,\mathbf{g}^{\sigma'}>=$$

$$<\mathbf{f}^\sigma,T(\mathfrak{a})\mathbf{g}^{\sigma'}>=<\mathbf{f}^\sigma,\lambda_{\mathbf{g}}^{\sigma'}\mathbf{g}^{\sigma'}>=<\mathbf{f}^\sigma,\mathbf{g}^{\sigma'}>\lambda_{\mathbf{g}}^\sigma$$

and hence we conclude that $< \mathbf{f}^{\sigma}, \mathbf{g}^{\sigma'} >= 0$. Here we have used the facts that the good Hecke operators are self adjoint with respect to the Petersson inner product, and that their Hecke eigenvalues are totally real (for both facts see Shimura 2000, Lemma 23.15).

Finally taking $\mathbf{g}$ equal to $\mathbf{f}$ we obtain that $\Omega_{\mathbf{f}}$ is equal to $< \mathbf{f}, \mathbf{f} >$ up to a non-zero element in the Galois closure of the field generated by the Fourier coefficients of $\mathbf{f}$ (note that it also contains the eigenvalues). □

We now give a proof of the equivariant property of the trace that we used in the proof of the theorem. The proof follows the proof given by Sturm (1981, Lemma 11) extended to the totally real field situation.

Lemma 8. *With notation as in the proof of the above theorem we have for any* $f \in \mathcal{S}_k(\Gamma', \psi; \mathbb{Q}_{ab})$

$$Tr^{\Gamma}_{\Gamma'}(f)^{\sigma} = Tr^{\Gamma}_{\Gamma'}(f^{\sigma}),\ \ \sigma \in Gal(\Phi\mathbb{Q}_{ab}/\Phi).$$

Proof. Thanks to the strong approximation for $Sp_n(F)$ we may work adelically. We write D and D' for the corresponding to Γ and Γ' adelic groups (i.e. $\Gamma = G \cap D$). We fix elements $\{g_i\} \subset D_{\mathbf{h}}$ so that $D = \bigcup D' g_i$. For $t \in \mathbb{Z}_{\mathbf{h}}^{\times}$ corresponding to $\sigma|_{\mathbb{Q}_{ab}}$ we note that

$$\begin{pmatrix} 1_n & 0 \\ 0 & t^{-1}1_n \end{pmatrix} g_i \begin{pmatrix} 1_n & 0 \\ 0 & t1_n \end{pmatrix} \in Sp_n(\mathbb{A})_{\mathbf{h}}$$

and hence by strong approximation we can find elements $u_i \in D'$ with $f|u_i = f$ (i.e. $\psi(u_i) = 1$) and $w_i \in Sp_n(F)$ so that

$$\begin{pmatrix} 1_n & 0 \\ 0 & t^{-1}1_n \end{pmatrix} g_i \begin{pmatrix} 1_n & 0 \\ 0 & t1_n \end{pmatrix} = u_i w_i.$$

We moreover note that $w_{i\mathbf{a}} = u_{i\mathbf{a}}^{-1}$. Now we claim that since the g_i's form a set of representatives of the classes of D' in D, the same holds for $\begin{pmatrix} 1_n & 0 \\ 0 & t^{-1}1_n \end{pmatrix} g_i \begin{pmatrix} 1_n & 0 \\ 0 & t1_n \end{pmatrix}$, and hence also for w_i since $u_i \in D'$. Indeed since $t \in \mathbb{Z}_{\mathbf{h}}^{\times} \hookrightarrow F_{\mathbf{h}}^{\times}$ we have that

$$\begin{pmatrix} 1_n & 0 \\ 0 & t^{-1}1_n \end{pmatrix} \begin{pmatrix} a & b \\ c & d \end{pmatrix} \begin{pmatrix} 1_n & 0 \\ 0 & t1_n \end{pmatrix} = \begin{pmatrix} a & tb \\ t^{-1}c & d \end{pmatrix} \in D[\mathfrak{a}, \mathfrak{b}]$$

if $\begin{pmatrix} a & b \\ c & d \end{pmatrix} \in D[\mathfrak{a}, \mathfrak{b}]$, for some fractional ideals $\mathfrak{a}, \mathfrak{b}$ with $\mathfrak{a}\mathfrak{b} \subseteq \mathfrak{g}$. In particular we have that $\iota(t) g_i \iota(t^{-1}) \in D$. We claim that $D = \dot{\bigcup}_i D' \iota(t) g_i \iota(t^{-1})$. Indeed let $d \in D$. Then $\iota(t^{-1}) d \iota(t) \in D$ and hence there exists $d' \in D'$ such that $\iota(t^{-1})d\iota(t) = d'g_j$ for some j. Or equivalently $d = \iota(t)d'g_j\iota(t^{-1}) = \iota(t)d'\iota(t^{-1})\iota(t)g_j\iota(t^{-1})$, which establishes our claim since $\iota(t)d'\iota(t^{-1})\iota(t) \in D'$.

We now consider the elements $(\iota(t), \sigma), (w_i, id), (g_i, id) \in \mathcal{G}_+ \times Gal(\overline{\mathbb{Q}}/\mathbb{Q})$. Then we have

$$(Tr_{\Gamma'}^{\Gamma}(f^\sigma))^{\sigma^{-1}} = (\sum_i \psi(g_i)^\sigma f^\sigma|_k g_i)^{\sigma^{-1}} =$$

$$\sum_i (\psi(g_i)f)^{((\iota(t),\sigma)(g_i,1)(\iota(t^{-1}),\sigma^{-1}))} = \sum_i \psi(g_i) f^{(u_i w_i,1)} = \sum_i \psi(g_i) f|_k w_i.$$

The proof of the lemma is now completed after observing that $\psi(g_i) = \psi(w_i)$. □

We also mention here the following theorem of Garrett (1992).

Theorem 8 (Garrett). *Let $k > 2n + 1$ and $\mathbf{f}, \mathbf{g} \in \mathcal{S}_{k\mathbf{a}}$. Take $\mathbf{f}$ an eigenform for almost all Hecke operators. Then for all $\sigma \in Aut(\mathbb{C}/\mathbb{Q})$, we have*

$$\left(\frac{< \mathbf{f}^\rho, \mathbf{g} >}{< \mathbf{f}^\rho, \mathbf{f} >}\right)^\sigma = \frac{< \mathbf{f}^{\sigma\rho}, \mathbf{g}^\sigma >}{< \mathbf{f}^{\sigma\rho}, \mathbf{f}^\sigma >}$$

In particular if we take $\mathbf{f}, \mathbf{g} \in \mathcal{S}_{k\mathbf{a}}(\overline{\mathbb{Q}})$, and take $\mathbf{f}$ with totally real Fourier coefficients then we have that $\frac{<\mathbf{f}^\rho,\mathbf{g}>}{<\mathbf{f}^\rho,\mathbf{f}>} \in \overline{\mathbb{Q}}$ and

$$\left(\frac{< \mathbf{f}, \mathbf{g} >}{< \mathbf{f}, \mathbf{f} >}\right)^\sigma = \frac{< \mathbf{f}^\sigma, \mathbf{g}^\sigma >}{< \mathbf{f}^\sigma, \mathbf{f}^\sigma >}, \quad \sigma \in Gal(\overline{\mathbb{Q}}/\mathbb{Q}).$$

We note that if we combine the above result of Garrett with the following result of Harris on the Eisenstein spectrum

Theorem 9 (Harris). *Let $k > 2n+1$ and write $\mathcal{E}_{k\mathbf{a}}$ for the orthogonal complement of $\mathcal{S}_{k\mathbf{a}}$ in $\mathcal{M}_{k\mathbf{a}}$ (the Eisenstein series). Define $\mathcal{E}_{k\mathbf{a}}(\mathbb{Q}) := \mathcal{M}_{k\mathbf{a}}(\mathbb{Q}) \cap \mathcal{E}_{k\mathbf{a}}$. Then we have*

$$\mathcal{M}_{k\mathbf{a}}(\mathbb{Q}) = \mathcal{E}_{k\mathbf{a}}(\mathbb{Q}) \oplus \mathcal{S}_{k\mathbf{a}}(\mathbb{Q}).$$

Proof. This follows from the work of Harris (1984). Indeed in general we have that (see Shimura 2000, Theorems 27.14, and 27.16)

$$\mathcal{M}_{k\mathbf{a}}(\overline{\mathbb{Q}}) = \mathcal{E}_{k\mathbf{a}}(\overline{\mathbb{Q}}) \oplus \mathcal{S}_{k\mathbf{a}}(\overline{\mathbb{Q}})$$

and $\mathcal{E}_{k\mathbf{a}}(\overline{\mathbb{Q}}) = \oplus_{r=0}^n \mathcal{E}_{k\mathbf{a}}^r(\overline{\mathbb{Q}})$ where $\mathcal{E}_{k\mathbf{a}}^r$ the space of Klingen type Eisenstein series associated to a parabolic group P_r stabilizing an isotropic space of dimension r. Harris has shown that in the case of weight as above (i.e. the absolute convergence situation) we have that $\mathcal{E}_{k\mathbf{a}}^r(\overline{\mathbb{Q}}) = \mathcal{E}_{k\mathbf{a}}^r(\mathbb{Q}) \otimes_{\mathbb{Q}} \overline{\mathbb{Q}}$. □

Now this theorem allows us to take $\mathbf{g} \in \mathcal{M}_{k\mathbf{a}}$ in Theorem 8.

7 Algebraicity Results for Siegel Modular Forms over Totally Real Fields

In this section we present various results regarding special values of the function $Z'(s, \mathbf{f}, \chi)$, with $\mathbf{f} \in \mathcal{S}_k(\mathfrak{b}, \mathfrak{c}, \psi)$, an eigenform for all Hecke operators. We remind that we have also considered the function $Z(s, \mathbf{f}, \chi)$. The two coincide when the Nebentypus of $\mathbf{f}$ is trivial. Indeed if we write $Z_v(\chi^*(\pi_v \mathfrak{g})|\pi_v|^s)$ for the Euler factor of $Z(s, \mathbf{f}, \chi)$ at some prime $v \in \mathbf{h}$ then the corresponding Euler factor of $Z'(s, \mathbf{f}, \chi)$ is equal to $Z_v((\psi/\psi_{\mathfrak{c}})\chi^*(\pi_v \mathfrak{g})|\pi_v|^s)$. We note the equation

$$Z'(s, \mathbf{f}, \chi\psi^{-1}) = Z_{\mathfrak{c}}(s, \mathbf{f}, \chi),$$

where the sub-index on the right hand side indicates that we have removed the Euler factors all primes in the support of $\mathfrak{c}$. In particular if we take the character χ trivial (may not primitive) at the primes dividing $\mathfrak{c}$ then we have that the two functions are the same.

We start by stating a result of Shimura (2000, Theorem 28.8). We take an $\mathbf{f} \in \mathcal{S}_k(C; \overline{\mathbb{Q}})$, where

$$C = \{x \in D[\mathfrak{b}^{-1}\mathfrak{c}, \mathfrak{b}\mathfrak{c}] | a_x - 1 \prec \mathfrak{c}\}$$

We moreover take $\mathbf{f}$ of trivial Nebentypus and assume that it is an eigenform for all Hecke operators away from the primes in the support of $\mathfrak{c}$. In the notation of Shimura in Chapter V of his book, we take $\mathfrak{e} = \mathfrak{c}$, and not $\mathfrak{e} = \mathfrak{g}$. In particular here we take the Euler factors Z_v trivial for v in the support of $\mathfrak{c}$. The theorem below is stated only for $k \in \mathbb{Z}^{\mathbf{a}}$.

Theorem 10 (Shimura). *With notation as above define $m_0 := min\{k_v | v \in \mathbf{a}\}$ and assume $m_0 > (3n/2) + 1$. Let χ be a character of F such that $\chi_{\mathbf{a}}(x) = x_{\mathbf{a}}^t |x_{\mathbf{a}}|^{-t}$ with $t \in \mathbb{Z}^{\mathbf{a}}$. Set $\mu_v := 0$ if $k_v - t_v \in 2\mathbb{Z}$ and $\mu_v = 1$ if $k_v - t_v \notin 2\mathbb{Z}$. Let $\sigma_0 \in \mathbb{Z}$ such that*

1. $2n + 1 - k_v + \mu_v \leq \sigma_0 \leq k_v - \mu_v$,
2. $\sigma_0 - k_v + \mu_v \in 2\mathbb{Z}$ *for every* $v \in \mathbf{a}$ *if* $\sigma_0 > n$,
3. $\sigma_0 - 1 + k_v - \mu_v \in 2\mathbb{Z}$ *for every* $v \in \mathbf{a}$ *if* $\sigma_0 \leq n$.

We exclude the cases

1. $\sigma_0 = n + 1$, $F = \mathbb{Q}$ *and* $\chi^2 = 1$,
2. $\sigma_0 = 0$, $\mathfrak{c} = \mathfrak{g}$ *and* $\chi = 1$,
3. $0 < \sigma_0 \leq n$, $\mathfrak{c} = \mathfrak{g}$, $\chi^2 = 1$ *and the conductor of* χ *is* $\mathfrak{g}$.

Then we have

$$\frac{Z(\sigma_0, \mathbf{f}, \chi)}{< \mathbf{f}, \mathbf{f} >} \in \pi^{n(\sum_v k_v) + de} \overline{\mathbb{Q}}$$

where $d = [F : \mathbb{Q}]$ *and*

$$e := \begin{cases} (n+1)\sigma_0 - n^2 - n, \ \sigma_0 > n; \\ n\sigma_0 - n^2, \quad\quad \text{otherwise.} \end{cases}$$

We now take $\mathbf{f} \in \mathcal{S}_k(C, \psi; \overline{\mathbb{Q}})$ with C of the form $D[\mathfrak{b}^{-1}, \mathfrak{bc}]$ (i.e. the standard setting in this paper). We are interested in special values of $Z'(s, \mathbf{f}, \chi)$ for a Hecke character χ of F of conductor $\mathfrak{f}$.

Theorem 11. *Let* $\mathbf{f} \in \mathcal{S}_k(\mathfrak{b}, \mathfrak{c}, \psi; \overline{\mathbb{Q}})$ *be an eigenform for all Hecke operators. Assume that either*

1. *There exists* $v, v' \in \mathbf{a}$ *such that* $k_v \neq k_{v'}$*, and* $m_0 = min\{k_v | v \in \mathbf{a}\} > [3n/2 + 1] + 2$ *or*
2. k *is a parallel weight with* $k > 2n + 1$.

Let χ *be a character of* F *such that* $\chi_{\mathbf{a}}(x) = x_{\mathbf{a}}^t |x_{\mathbf{a}}|^{-t}$ *with* $t \in \mathbb{Z}^{\mathbf{a}}$*. Define* $t' \in \mathbb{Z}^{\mathbf{a}}$ *by* $(\psi\chi)_{\mathbf{a}}(x) = x_{\mathbf{a}}^{t'} |x_{\mathbf{a}}|^{t'}$*. Set* $\mu_v := 0$ *if* $k_v - t'_v \in 2\mathbb{Z}$ *and* $\mu_v = 1$ *if* $k_v - t'_v \notin 2\mathbb{Z}$*. Let* $\sigma_0 \in \mathbb{Z}$ *such that*

1. $2n + 1 - k_v + \mu_v \leq \sigma_0 \leq k_v - \mu_v$ *for all* $v \in \mathbf{a}$,
2. $|\sigma_0 - n - \frac{1}{2}| + n + \frac{1}{2} - k + \mu \in 2\mathbb{Z}^{\mathbf{a}}$.
3. *If* n *is even, and* $\sigma_0 = n/2 + i$ *for* $i = 0, \dots n/2$, $i \in \mathbb{N}$ *or if* n *is odd and* $\sigma_0 = n/2 - 1 + i$, $i = 1, \dots, (n+1)/2$*, then we assume the* ***Assumption*** *below.*

We exclude the cases

1. $\sigma_0 = n + 1$, $F = \mathbb{Q}$ *and* $(\chi\psi)^2 = 1$,
2. $\sigma_0 = \frac{n}{2}$, $\mathfrak{c} = \mathfrak{g}$, n *is even and there is no* (τ, r) *that satisfy our assumption such that* $\rho_\tau \neq 1$ *and* $\chi\psi = 1$,
3. $n/2 < \sigma_0 \leq n$, $\mathfrak{c} = \mathfrak{g}$ *and* $(\psi\chi)^2 = 1$.

Then with notation as in the previous theorem we have

$$\frac{Z'(\sigma_0, \mathbf{f}, \chi)}{< \mathbf{f}, \mathbf{f} >} \in \pi^{n(\sum_v k_v) + de} \overline{\mathbb{Q}}$$

Moreover, if we take a number field W *so that* $\mathbf{f}, \mathbf{f}^\rho \in \mathcal{S}_k(W)$ *and* $\Phi \subset W$*, where* Φ *is the Galois closure of* F *in* $\overline{\mathbb{Q}}$*, then*

$$\frac{Z'(\sigma_0, \mathbf{f}, \chi)}{\pi^\beta (\sqrt{D_F}^{n(n+1)/4}) i^m \omega(\epsilon \overline{\chi\psi})^\rho < \mathbf{f}, \mathbf{f} >} \in \mathcal{W} := W(\chi\psi),$$

where $\omega(\cdot)$ *is defined by using the Theorem 3 as follows*

1. *For* $\sigma_0 > n$ *and* n *even then* $\omega(\cdot)$ *is as in Theorem 3 (i),*
2. *For* $\sigma_0 > n$ *and* n *odd then* $\omega(\cdot)$ *is as in Theorem 3 (iii) (b),*

3. *For $\sigma_0 \leq n$ and n even then $\omega(\cdot)$ is as in Theorem 3 (ii),*
4. *For $\sigma_0 \leq n$ and n odd then $\omega(\cdot)$ is as in Theorem 3 (iv).*

and $m = d$ if $[n/2]$ is odd and 0 otherwise.

Assumption. Let $\theta \in F_{\mathbf{h}}^{\times}$ so that $\theta\mathfrak{g} = \mathfrak{b}^{-1}\mathfrak{d}$. Write $\mathfrak{f}'$ for the conductor of χ^2. We assume that we can find $\tau \in S_+ \cap GL_n(F)$ and $r \in GL_n(F)_{\mathbf{h}}$ so that $c(\tau, r; \mathbf{f}) \neq 0$, Eq. 11 in Theorem 6 holds and

1. If n is even and $v \nmid \mathfrak{c}\mathfrak{f}'$ then $(\theta {}^t r \tau r)_v$ is regular and $v \nmid \mathfrak{f}$,
2. If n is odd and $v \nmid \mathfrak{c}\mathfrak{f}'$ then $(\theta {}^t r \tau r)_v$ is regular and $v \nmid 2\mathfrak{f} \cap \mathfrak{b}^{-1}\mathfrak{d}$.

We note that this assumption implies that in Theorem 6 we have that $\Lambda_{\mathfrak{c}}(s)/\Lambda_{\mathfrak{h}}(s) = 1$ (see Shimura 1994, Proposition 8.3).

Proof (of Theorem 11). We first consider the Gamma factors that appear in Theorem 6. We first recall that

$$\Gamma_n(s) = \pi^{n(n-1)/4} \prod_{j=0}^{n-1} \Gamma(s - \frac{j}{2}).$$

Hence for $\prod_{v \in \mathbf{a}} \Gamma_n(\frac{\sigma_0 - n - 1 + k_v + \mu_v}{2})$ we need the condition that $\sigma_0 > 2n - k_v + \mu_v$ for all $v \in \mathbf{a}$, which is the lower bound appearing in the theorem. Moreover the Eisenstein series $D(\frac{\nu}{2})$ of weight $k - \mu - \frac{n}{2}$ for $\nu = \sigma_0 - \frac{n}{2}$ is nearly holomorphic if and only if $n + 1 - (k_v - \mu_v - \frac{n}{2}) \leq \sigma_0 - \frac{n}{2} \leq k_v - \mu_v - \frac{n}{2}$ and $|\nu - \frac{n+1}{2}| + \frac{n+1}{2} - k_v + \mu_v + \frac{n}{2} \in 2\mathbb{Z}$ for every $v \in \mathbf{a}$. These inequalities give the upper bound in the (i) condition for σ_0 and (ii). The third condition for σ_0 is imposed so that in the range where the fraction $\Lambda_{\mathfrak{c}}(s)/\Lambda_{\mathfrak{h}}(s)$ (a finite product of Euler factors associated to finite order characters) could have a pole it is equal to 1. Finally the various exclusion follows from various cases where the Eisenstein series $D(\frac{\nu}{2})$ is not nearly holomorphic.

We take $\beta \in \mathbb{N}$ so that $\pi^{-\beta} D(\frac{\nu}{2}) \in \mathcal{N}_{k-l}(\Phi\mathbb{Q}_{ab})$. Now using Theorem 6 after a proper choice of (τ, r) we have

$$\pi^{\gamma} Z'(\sigma_0, \mathbf{f}, \chi)\psi_{\mathfrak{c}}(det(r))c(\tau, r; \mathbf{f}) =$$

$$\alpha \left(D_F^{-1/2}\right)^{n(n+1)/2} det(\tau)^{s_0' u + \lambda} |det(r)|_{\mathbf{A}}^{n+1-\sigma_0} \times$$

$$\prod_{v \in \mathbf{b}} g_v((\psi/\psi_{\mathfrak{c}})(\pi_v)\chi^*(\pi_v\mathfrak{g})|\pi_v|^{\sigma_0})(\Lambda_{\mathfrak{c}}/\Lambda_{\mathfrak{h}})((2\sigma_0 - n)/4) < f, \theta_\chi(\pi^{-\beta} D(\nu/2)) >,$$

where $\alpha \in \mathbb{Q}^{\times}$, and $\gamma := n||\frac{k-l-\nu\mathbf{a}}{2}|| - n||k|| + d\epsilon - n||s_0' u + \lambda|| - \beta$ where we recall $\epsilon = n^2/4$ if n even and $(n^2 - 1)/4$ otherwise.

We now note that $\theta_\chi \in \mathcal{M}_l(\mathcal{W})$ and $\pi^{-\beta}D(\nu/2) \in \mathcal{N}^r_{k-l}(W\mathbb{Q}_{ab})$ where $r = (k-l-\nu\mathbf{a})/2$ if $\nu > (n+1)/2$ and $r = (k-l-(n+1-\nu)\mathbf{a})/2$ otherwise.

Moreover we have $s_0'u + \lambda = \frac{\sigma_0 u+k+\mu}{2} - \frac{n+1}{2}u$. In particular

1. For $\sigma_0 > n$ and n even we have that $s_0'u + \lambda \notin 2\mathbb{Z}$,
2. For $\sigma_0 > n$ and n odd we have that $s_0'u + \lambda \in 2\mathbb{Z}$,
3. For $\sigma_0 \leq n$ and n even we have that $s_0'u + \lambda \in 2\mathbb{Z}$,
4. For $\sigma_0 \leq n$ and n odd we have that $s_0'u + \lambda \notin 2\mathbb{Z}$.

We now note that $\mathbf{g}(\rho_\tau) = i^m\sqrt{N_{F/\mathbb{Q}}det(\tau)}$, with $m = d$ if $[n/2]$ is odd and 0 otherwise. Now we set $P := \sqrt{D_F}^{n(n+1)/4} i^m \omega(\epsilon\overline{\chi\psi})$ where $\omega(\cdot)$ is defined as in the statement of the theorem. Then by Theorem 3 we conclude that

$$\left(D_F^{-1/2}\right)^{n(n+1)/2} det(\tau)^{s_0'u+\lambda}\pi^{-\beta}P^{-1}D(\nu/2) \in \mathcal{N}^r_{k-l}(\mathcal{W}).$$

We set $a := \left(D_F^{-1/2}\right)^{n(n+1)/2} det(\tau)^{s_0'u+\lambda}\pi^{-\beta}P^{-1}$. By Lemma 15.8 in Shimura (2000) we have that there exists a $q \in \mathcal{M}_k(\mathcal{W})$ so that $< f, \theta_\chi a D(\nu/2) >=< f, q >$. If k is not a parallel weight, then we have that actually $q \in \mathcal{S}_k(\mathcal{W})$ since in this case $\mathcal{M}_k = \mathcal{S}_k$. Then by Theorem 7 we have that $\frac{<f,q>}{<f,f>} \in \mathcal{W}$. In the other case, that is of k being a parallel weight we can use Theorem 8 combined with the Theorem 9 to conclude again $\frac{<f,q>}{<f,f>} \in \mathcal{W}$ and hence conclude the proof. □

We now obtain also some results with reciprocity laws.

Theorem 12. *Let* $\mathbf{f} \in \mathcal{S}_k(\mathfrak{b}, \mathfrak{c}, \psi; \overline{\mathbb{Q}})$ *be an eigenform for all Hecke operators. With notation as before we take* $m_0 > [3n/2+1]+2$. *Let* χ *be a character of* F *such that* $\chi_\mathbf{a}(x) = x_\mathbf{a}^t|x_\mathbf{a}|^{-t}$ *with* $t \in \mathbb{Z}^\mathbf{a}$. *Define* $t' \in \mathbb{Z}^\mathbf{a}$ *by* $(\psi\chi)_\mathbf{a}(x) = x_\mathbf{a}^{t'}|x_\mathbf{a}|^{t'}$. *Set* $\mu_v := 0$ *if* $k_v - t_v' \in 2\mathbb{Z}$ *and* $\mu_v = 1$ *if* $k_v - t_v' \notin 2\mathbb{Z}$. *Assume that* $\mu \neq 0$.

Let $\sigma_0 \in \mathbb{Z}$ *be as in the previous Theorem. Then with* $\Omega_\mathbf{f} \in \mathbb{C}^\times$ *as defined in the previous section in Theorem 7 we have for all* $\sigma \in Gal(\overline{\mathbb{Q}}/\Phi)$ *that*

$$\left(\frac{Z'(\sigma_0, \mathbf{f}, \chi)}{\pi^{n(\sum_v k_v)+de}\sqrt{D_F^{n(n+1)/4}} i^m \omega(\epsilon\overline{\chi\psi})^\rho \Omega_\mathbf{f}}\right)^\sigma =$$

$$\frac{Z'(\sigma_0, \mathbf{f}^\sigma, \chi^\sigma)}{\pi^{n(\sum_v k_v)+de} i^m \sqrt{D_F}^{n(n+1)/4} \omega(\epsilon\overline{\chi^\sigma\psi^\sigma})^\rho \Omega_{\mathbf{f}^\sigma}}.$$

Proof. We first observe that thanks to the assumption that $\mu \neq 0$ we have that $\theta_\chi \in \mathcal{S}_l$. Moreover for $\sigma \in Gal(\overline{\mathbb{Q}}/\Phi)$ we have $\theta_\chi^{\sigma'} = \theta_{\chi^\sigma}$, as it follows from the explicit Fourier expansion of θ_χ. Moreover arguing as in the theorem above and using the reciprocity laws for Eisenstein series in Theorem 3 we have that

$$\left(\frac{\pi^{-\beta}\sqrt{D_F}^{n(n+1)/2}det(\tau)^{s_0'u+\lambda}D(\nu/2,\epsilon\overline{\psi\chi\rho_\tau})}{\omega(\epsilon\overline{\psi\chi})}\right)^{\sigma'} =$$

$$\frac{\pi^{-\beta}\sqrt{D_F}^{n(n+1)/2}det(\tau)^{s_0'u+\lambda}D(\nu/2,\epsilon\overline{\psi^\sigma\chi^\sigma})}{\omega(\epsilon\overline{\psi^\sigma\chi^\sigma})}, \quad \sigma\in Gal(\overline{\mathbb{Q}}/\Phi).$$

Moreover we have that $\theta_\chi D(\nu/2,\epsilon\overline{\psi^\sigma\chi^\sigma\rho_\tau})\in\mathcal{R}_k$. By Proposition 15.6 in Shimura (2000) we have that there exists $q=\mathfrak{p}\left(\theta_\chi D(\nu/2,\epsilon\overline{\psi^\sigma\chi^\sigma\rho_\tau})\right)\in\mathcal{S}_k$ so that $< f,\theta_\chi D(\nu/2,\epsilon\overline{\psi\chi\rho_\tau}) >=< f,q >$ and $q^\sigma=\mathfrak{p}\left(\theta_\chi^\sigma D(\nu/2,\epsilon\overline{\psi\chi\rho_\tau})^\sigma\right)$ for all $\sigma\in Aut(\mathbb{C}/\Phi)$. In particular we have that

$$\left(\frac{\sqrt{D_F}^{n(n+1)/2}det(\tau)^{s_0'u+\lambda} < f,\theta_\chi\pi^{-\beta}D(\nu/2,\epsilon\overline{\psi\chi\rho_\tau}) >}{\omega(\epsilon\overline{\psi\chi})^\rho\Omega_{\mathfrak{f}}}\right)^\sigma =$$

$$\frac{\sqrt{D_F}^{n(n+1)/2}det(\tau)^{s_0'u+\lambda} < f^\sigma,\theta_{\chi^\sigma}\pi^{-\beta}D(\nu/2,\epsilon\overline{\psi^\sigma\chi^\sigma\rho_\tau}) >}{\omega(\epsilon\overline{\psi^\sigma\chi^\sigma})^\rho\Omega_{\mathfrak{f}^\sigma}},$$

from which we conclude the proof of the theorem. □

As we have remarked in the introduction results similar to the ones proved in this paper have been obtained by Sturm (1981), Harris (1984) and Panchishkin (1991) in the case of $F=\mathbb{Q}$ and n even. We also remark that Sturm has also considered the case $n=1$ and $F=\mathbb{Q}$ in Sturm (1980). Our proofs are just generalizations of theirs building on some new results of Shimura. We close this section by mentioning that the perhaps strongest result concerning the special values of Siegel modular forms, at least when $F=\mathbb{Q}$ and under some other technical assumptions, is due to Böcherer and Schmidt (2000). Using the doubling method (see also the next section) and some holomorphic differential operators of Böcherer they obtained algebraicity results but assuming only that the weight of the Siegel modular form is larger than n rather than $\frac{3n}{2}+1$. It is of course very interesting to extend their results to the totally real field case, however the generalization of their result seems to be a quite challenging task. We comment a bit more on this in the next section.

8 Some Remarks on the Doubling Method

In this paper our main tool for the study of the special L-values of Siegel Modular Forms was the Rankin-Selberg method. However, as we also briefly mentioned above, there is yet another powerful method for the study of these special values, namely the so-called doubling method. In this paper we considered mainly the Rankin-Selberg Method, since this article, already quite long, would have increased considerably in size if the doubling method was also to be considered here. So we have decided to defer the consideration of the doubling method with respect to the same questions addressed here for a future paper. In this section we wish to very briefly discuss various aspects that are closely related to the doubling method and the questions considered in this paper.

It is perhaps fair to say that the doubling method was initiated by Garrett (1984) and extended further by Böcherer (1983, 1985a,b), Böcherer and Schmidt (2000), Shimura (1995) and in the automorphic language by Piatetski-Shapiro and Rallis (1987). Of course the list of contributors here is not meant to be complete.

Concerning the algebraicity results addressed in this paper, it seems that the two methods (Rankin-Selberg and the Doubling Method) provide in many cases the same results, but there are indeed case where one method is better than the other. Indeed, Shimura in his books (Shimura 2000, Theorem 28.8) concludes his algebraicity results by using both methods (1st method and 2nd method in Shimura's notation). However one should at this point remark the following. Shimura writes at the beginning of his proof of his Theorem 28.8 "There are two ways to prove this: the first one (i.e. doubling method) applies to the whole critical strip, and the second one (i.e. Rankin-Selberg Method) only to the right half of the strip". However this is so, because Shimura is taking $\mathfrak{e} = \mathfrak{c}$ in his book Shimura (2000, page 231) (see also our discussion just before Theorem 7.1 in this paper). Indeed in this situation the doubling method seems to be able to tackle critical points also to the left half of the critical strip, something that the Rankin-Selberg method cannot. However for $\mathfrak{e} = \mathfrak{g}$ this is not the case and this is the situation that we consider here. The main reason being that the integral expression in Theorem 6 it is available in this form (in particular this particular Siegel type Eisenstein series for which we know quite explicitly) only in the case $\mathfrak{e} = \mathfrak{g}$. We also note here that $\mathfrak{e} = \mathfrak{c}$ corresponds to $\Gamma_1(N)$-case and $\mathfrak{c} = \mathfrak{g}$ corresponds to $\Gamma_0(N)$-case in the elliptic modular form situation.

In this paper we have considered only Siegel modular forms. Of course the same questions can be addressed for other groups, as for example unitary groups. Actually Shimura in his book provides similar results (always over an algebraic closure of $\mathbb{Q}$) for hermitian modular forms, that is modular forms associated to unitary groups. For hermitian modular forms the two methods are not at all equivalent, and in particular one cannot use the Rankin-Selberg method to study special L-values for hermitian forms of unitary groups of the form $U(n, m)$ for $n \neq m$ (this is part of the case **UB** in Shimura's notation in his book). For example one cannot consider the case of hermitian modular forms for definite unitary groups. But the doubling method still does apply. Here we should say that the field of definition for hermitian modular forms has been worked out by Harris (1997) using the doubling method. However we would like further to remark here that Harris considers special L-values only in the strip of absolute convergence. In the **UT** case (i.e. $n = m$) we have obtained in some cases results (Bouganis in preparation, b) using the Rankin-Selberg method that improves the ones of Harris (i.e. beyond the absolute convergence).

As we explained at the beginning of this article one of the main motivations of our investigations is the construction of p-adic measures for Siegel modular forms. We briefly describe here what is known with respect to this, even though the reader should keep in mind that we do not wish to give here a complete and detailed picture of the situation. Historically the first results in this direction were obtained by Panchishkin (1991) (see also the joint work of Courtieu and Panchishkin 2004), who used the Rankin-Selberg method to construct these measures. However he considered the case of even degree (or genus), the main reason being that in this case the Rankin-Selberg method does not involve Eisenstein series of half-integral weight. Later Böcherer and Schmidt (2000) constructed these p-adic measures for any degree using the doubling method. One should remark here that there is a very delicate difference in the way that Böcherer and Schmidt applied the doubling method and in the way Shimura developed it in his work (Shimura 1995). Very briefly the main difference seems to be in the decomposition that is proved in Proposition 2.1 of Böcherer and Schmidt (2000) as well as the use of the holomorphic operators of Böcherer (opposite to the non-holomorphic ones in the work of Shimura). Of course one should add here that the work (Böcherer and Schmidt 2000) is restricted to Siegel modular forms over the rationals, opposite to the work of Shimura who applies to any totally real field. We simply say here that in an ongoing project we extend the work of Panchishkin (i.e. p-adic measures using the Rankin-Selberg method) in two directions. We consider also odd genus and to the totally real field case. Note, as we already said, that both the work of Panchishkin and of Böcherer and Schmidt are over the rationals. It seems to be a big challenge to obtain the analogue of Proposition 2.1 of Böcherer and Schmidt (2000) in the totally real case in the situation of strict class number bigger than one, and in particular extend the work of Böcherer and Schmidt to totally real fields. We are currently working on this. At this point it is worth mentioning that in this article we considered scalar valued Siegel modular forms. Many of the above questions can be stated also for the vector valued ones. For a first step in this direction the reader can see Ichikawa (2012). Finally we close this article by mention that of course it is very interesting to construct p-adic measures for hermitian modular forms. The doubling method has been already used in that context, as for example in Harris et al. (2005, 2006) and Skinner and Urban (2014). In Bouganis (in preparation, b) we are considering the Rankin-Selberg method for constructing these p-adic measures.

References

Böcherer, S.: Über die Fourier-jacobi-Entwicklung der Siegelschen Eisensteinreihen. Math. Z. **183**, 21–43 (1983)

Böcherer, S.: Über die Fourier-jacobi-Entwicklung der Siegelschen Eisensteinreihen II. Math. Z. **189**, 81–100 (1985a)

Böcherer, S.: Über die Funktionalgleichung automorpher L-Funktionen zur Siegelschen Modulgruppe. J. R. Ang. Math. **362**, 146–168 (1985b)

Böcherer, S., Schmidt, C.-G.: p-adic measures attached to Siegel modular forms. Annales de l' institut Fourier **50**(5), 1375–1443 (2000)

Bouganis, Th.: On special L-values attached to metaplectic modular forms (in preparation, a)

Bouganis, Th.: p-adic measures for hermitian modular forms and the Rankin-Selberg method (in preparation, b)

Courtieu, M., Panchishkin, A.A.: Non-Archimedean L-Functions and Arithmetical Siegel Modular Forms. Lecture Notes in Mathematics, vol. 1471, 2nd augmented ed. Springer, Berlin/New York (2004)

Feit, P.: Poles and Residues of Eisenstein Series for Symplectic and Unitary Groups. Volume 346 of Memoirs of the American Mathematical Society, vol. 61. American Mathematical Society, Providence (1986)

Garrett, P.B.: Pullbacks of Eisenstein series; applications. In: Satake, I., Morita, Y. (eds.) Automorphic Forms in Several Variables, pp. 114–137. Birkhäuser, Boston (1984)

Garrett, P.B.: On the arithmetic of Siegel-Hilbert cuspforms: Petersson inner products and Fourier coefficients. Invent. Math. **107**, 453–481 (1992)

Harris, M.: Special values of zeta functions attached to Siegel modular forms. Ann. Sci. Éc Norm. Super. **14**, 77–120 (1981)

Harris, M.: Eisenstein Series on Shimura varieties. Ann. Math. **119**, 59–94 (1984)

Harris, M.: L functions and periods of polarized regular motives. Journal für die reine und angewandte Mathematik **483**, 75–161 (1997)

Harris, M., Li, J.-S., Skinner, C.: The Rallis inner product formula and p-adic L-functions. In: J. Cogdell et al. (eds) Automorphic Representations, L-Functions and Applications: Progress and Prospects. Volume in Honor of S. Rallis, pp. 225–255. de Gruyter, Berlin (2005)

Harris, M., Li, J.-S., Skinner, C.: p-Adic L-Functions for Unitary Shimura Varieties, I: Construction of the Eisenstein Measure. Documenta Mathematica, Extra Volume: John H.Coates' Sixtieth Birthday, pp. 393–464 (2006)

Ichikawa, T.: Vector-valued p-adic Siegel modular forms. J. reine angew. Math. doi:10.1515/crelle-2012-0066

Panchishkin, A.A.: Non-Archimedean L-Functions of Siegel and Hilbert Modular Forms. Lecture Notes in Mathematics, vol. 1471, 166p. Springer, Berlin/New York (1991)

Piatetski-Shapiro, I., Rallis, S.: Explicit Constructions of Automorphic L-Functions. Lecture Notes in Mathematics, vol. 1254, pp 1–52 1987

Shimura, G.: Confluent hypergeometric functions on tube domains. Math. Ann. **260**, 269–302 (1982)

Shimura, G.: Euler products and Fourier coefficients of automorphic forms on symplectic groups. Invent. Math. **116**, 531–576 (1994)

Shimura, G.: Eisenstein series and zeta functions on symplectic groups. Invent. Math. **119**, 539–584 (1995)

Shimura, G.: Euler Products and Eisenstein Series. CBMS Regional Conference Series in Mathematics, No. 93. American Mathematical Society, Providence (1997)

Shimura, G.: Arithmeticity in the Theory of Automorphic Forms. Mathematical Surveys and Monographs, vol. 82. American Mathematical Society, Providence (2000)

Skinner, C., Urban, E.: The Iwasawa main conjectures for GL_2. Invent. Math. (2014). doi:10.1007/s00222-013-0448-1

Sturm, J.: Special values of zeta functions, and Eisenstein series of half integral weight. Am. J. Math. **102**(2), 219–240 (1980)

Sturm, J.: The critical values of zeta functions associated to the symplectic group. Duke Math. J. **48**(2), 219–240 (1981)

Modular Symbols in Iwasawa Theory

Takako Fukaya, Kazuya Kato, and Romyar Sharifi

1 Introduction

1.1. The starting point of this paper is the fascinatingly simple and explicit map

$$[u : v] \mapsto \{1 - \zeta_N^u, 1 - \zeta_N^v\}$$

that relates the worlds of geometry/topology and arithmetic (Busuioc 2008; Sharifi 2011). Here,

- $[u : v]$ is a Manin symbol in the relative homology group $H_1(X_1(N), \{\text{cusps}\}, \mathbb{Z})$,
- $\{1 - \zeta_N^u, 1 - \zeta_N^v\}$ is a Steinberg symbol in the algebraic K-group $K_2(\mathbb{Z}[\zeta_N, \frac{1}{N}])$,

for $N \geq 1$, where $u, v \in \mathbb{Z}/N\mathbb{Z}$ are nonzero numbers with $(u, v) = (1)$, and ζ_N is a primitive Nth root of unity.

1.2. The above map connects two different worlds in the following manner:

$$\text{geometric theory of } \mathrm{GL}_2 \Longrightarrow \text{arithmetic theory of } \mathrm{GL}_1$$

over the field $\mathbb{Q}$. Here, if we consider the geometry of the modular curve $X_1(N)$ on the left, then we consider the arithmetic of the cyclotomic field $\mathbb{Q}(\zeta_N)$ on the

---

The original version of this chapter was revised.
An erratum to this chapter can be found at DOI 10.1007/978-3-642-55245-8_17

T. Fukaya • K. Kato
Department of Mathematics, University of Chicago, 5734 S. University Ave., Chicago, IL 60637, USA
e-mail: takako@math.uchicago.edu; kkato@math.uchicago.edu

R. Sharifi (✉)
Department of Mathematics, University of Arizona, 617 N. Santa Rita Ave., Tucson, AZ 85711, USA
e-mail: sharifi@math.arizona.edu

T. Bouganis and O. Venjakob (eds.), *Iwasawa Theory 2012*, Contributions in Mathematical and Computational Sciences 7, DOI 10.1007/978-3-642-55245-8_5,

right. This connection is conjectured to be a correspondence if we work modulo the Eisenstein ideal that is defined in 2.1.6:

$$\text{geometric theory of } \mathrm{GL}_2 \text{ modulo the Eisenstein ideal} \Longleftrightarrow \text{arithmetic theory of } \mathrm{GL}_1 .$$

More generally, we are dreaming that there is a strong relationship

$$\text{geometric theory of } \mathrm{GL}_d \text{ modulo the Eisenstein ideal} \Longleftrightarrow \text{arithmetic theory of } \mathrm{GL}_{d-1}$$

over global fields. Our goals are to survey what is known and to explain this dream.

1.3. The connection with the Eisenstein ideal for GL_2 over $\mathbb{Q}$ appears as follows. The homology group we consider has the action of a Hecke algebra which contains an Eisenstein ideal, and the map of 1.1 factors through the quotient of the homology by this ideal (Fukaya and Kato Preprint). The truth of this is deep and mysterious; it is the idea of specializing at the cusp at ∞. This is the key to the connection between GL_2 and GL_1.

1.4. We note that there exist two technical issues with our simple presentation of the "map" in 1.1. We left out those Manin symbols in which one of u or v is 0, which are needed to generate the relative homology group. Also, the map is only well-defined as stated if we first invert 2 and then project to the fixed part under complex conjugation (see Sect. 2.1).

1.5. Let us consider the case that N is a power of an odd prime p and work only with p-parts. Consider the quotient

$$P_r = H_1(X_1(p^r), \mathbb{Z}_p)^+ / I_r H_1(X_1(p^r), \mathbb{Z}_p)^+$$

of the fixed part of homology under complex conjugation by the action of the Eisenstein ideal I_r in the cuspidal Hecke algebra of weight 2 and level p^r. By the well-known relationship between K_2 and $H^2_{\text{ét}}$ of $\mathbb{Z}[\zeta_{p^r}, \frac{1}{p}]$, the map of 1.1 yields a well-defined map

$$\varpi_r \colon P_r \to H^2_{\text{ét}}(\mathbb{Z}[\zeta_{p^r}, \tfrac{1}{p}], \mathbb{Z}_p(2))^+$$

that sends the image of $[u : v]$ in P_r to the cup product $(1 - \zeta^u_{p^r}) \cup (1 - \zeta^v_{p^r})$.

1.6. Let us connect this with Iwasawa theory for GL_1. As we increase r, the maps ϖ_r are compatible. The group $H^2_{\text{ét}}(\mathbb{Z}[\zeta_{p^r}, \frac{1}{p}], \mathbb{Z}_p(2))^+$ is related to the p-part A_r of the class group of $\mathbb{Q}(\zeta_{p^r})$ in the sense that its reduction modulo p^r is isomorphic to the Tate twist of $A_r^- / p^r A_r^-$. So, if we let $P = \varprojlim_r P_r$ and $X = \varprojlim_r A_r$, then we obtain a map

$$\varpi \colon P \to X^-(1)$$

that relates geometry of the tower of curves $X_1(p^r)$ modulo the Eisenstein ideal to Iwasawa theory over the union of cyclotomic fields $\mathbb{Q}(\zeta_{p^r})$. It is a map of Iwasawa modules under the action of inverses of diamond operators on the left and of Galois elements on the right.

1.7. In (Sharifi 2011), the map ϖ is conjectured to be an isomorphism. If this conjecture is true, we can understand the arithmetic object X^- by using the geometric object P. The Iwasawa main conjecture states that the characteristic ideal of X^- is the equivariant p-adic L-function. On the other hand, the characteristic ideal of P under the inverse diamond action can be computed to be a multiple of the Tate twist ξ of this L-function. If the characteristic ideals of $X^-(1)$ and P are equal, then the main conjecture follows as a consequence of the analytic class number formula. Therefore, the conjecture that ϖ is an isomorphism is an explicit refinement of the Iwasawa main conjecture.

1.8. In their proof of Iwasawa main conjecture (Mazur and Wiles 1984), Mazur and Wiles, expanding upon the work of Ribet (1976), considered the relationship between the geometric theory of GL_2 and the arithmetic theory of GL_1. Using roughly their methods, we can define a map

$$\Upsilon\colon X^-(1) \to P.$$

More precisely, Υ is constructed out of the Galois action on the projective limit of the reduction of étale homology groups $H_1^{\text{ét}}(X_1(p^r)_{/\overline{\mathbb{Q}}}, \mathbb{Z}_p)$ modulo the Eisenstein ideal. The expectation in (Sharifi 2011) is that the maps $\varpi\colon P \to X^-(1)$ and $\Upsilon\colon X^-(1) \to P$ are inverse to each other. The best evidence we have for this is the equality $\xi'\Upsilon \circ \varpi = \xi'$ after multiplication by the derivative ξ' of ξ, which is proven in (Fukaya and Kato Preprint). If the p-adic L-function ξ has no multiple zeros, this yields the conjecture up to p-torsion in P.

1.9. An analogous result for the rational function field $\mathbb{F}_q(t)$ over a finite field can be proven by following the analogy between $\mathbb{F}_q(t)$ and $\mathbb{Q}$. In both cases, the key point of the proof is that $(1-\zeta_N^u, 1-\zeta_N^v)$ and its analogue for $\mathbb{F}_q(t)$ are values at the infinity cusp of the "zeta elements," which is to say Beilinson elements and their analogues for $\mathbb{F}_q(t)$, which live in K_2 of the modular curve $X_1(N)$ and its Drinfeld analogue for $\mathbb{F}_q(t)$.

1.10. For both $\mathbb{Q}$ and $\mathbb{F}_q(t)$, the philosophy is that $\xi'\Upsilon \circ \varpi$ is the reduction modulo the Eisenstein ideal I of a map involving zeta elements. Roughly speaking, the proof consists firstly of the demonstration of the existence of a commutative diagram

$$\begin{array}{ccccc}
\mathcal{S} & \xrightarrow{z} & K & \xrightarrow{\mathrm{reg}} & \mathfrak{S} \\
{\scriptstyle \mathrm{mod}\, I}\downarrow & & \downarrow{\scriptstyle \infty} & & \downarrow{\scriptstyle \mathrm{mod}\, I} \\
P & \xrightarrow{\varpi} & Y & \xrightarrow{\xi'\Upsilon} & P.
\end{array}$$

Here, $\mathcal{S}$ is the space of modular symbols, the map z takes modular symbols to zeta elements in the K_2-group K of a modular curve, $\mathfrak{S}$ is a space of p-adic cusp forms, reg is the p-adic regulator map, and Y is either $X^-(1)$ or its analogue for $\mathbb{F}_q(t)$. The vertical arrows denoted by "mod I" are obtained by reduction modulo the Eisenstein ideal I (see Sect. 2.7 for details), and ∞ is given by specialization at a cusp at infinity. Secondly, it entails a computation of the regulator map on zeta elements that tells us that the composition $\mathcal{S} \to K \to \mathfrak{S} \to P$ coincides with ξ' times the projection $\mathcal{S} \to P$.

1.11. In this survey paper, we explain the key ideas and concepts of our work, putting aside many of the technical details that must arise in a careful treatment. While we do our best to strike a balance, the reader should be aware that some of the statements we make require minor modifications in order that they hold.

The structure of the paper is as follows. In Sect. 2, we describe the original case of the conjectures and outline the proof of the above result. In Sect. 3, we discuss and outline the proof of the analogue for $\mathbb{F}_q(t)$. In Sect. 4, we describe what might be expected for GL_d.

2 The Case of GL_2 over $\mathbb{Q}$

Fix an odd prime number p and an integer $N \geq 1$ which is not divisible by p. Let $r \geq 1$, which will vary. Let $\overline{\mathbb{Q}}$ be the algebraic closure of $\mathbb{Q}$ in $\mathbb{C}$, and let us fix an embedding of $\overline{\mathbb{Q}}$ in $\overline{\mathbb{Q}}_p$.

Recall that we want to understand the picture:

$$\text{geometric theory of } \mathrm{GL}_2 \text{ modulo the Eisenstein ideal } \underset{\Upsilon}{\overset{\varpi}{\rightleftarrows}} \text{ arithmetic theory of } \mathrm{GL}_1.$$

In Sects. 2.1–2.3, we study the map ϖ. In Sects. 2.4 and 2.5, we study the map Υ. In Sects. 2.6 and 2.7, we state the conjecture and the main result on it.

2.1 *From Modular Symbols to Cup Products*

We construct the map ϖ_r that relates modular symbols in the homology of $X_1(Np^r)$ to cup products in the cohomology of the maximal unramified outside Np extension of $\mathbb{Q}(\zeta_{Np^r})$.

2.1.1. We introduce homology groups $\mathcal{S}_r$ and $\mathcal{M}_r$ of modular curves.

Let $\mathbb{H}$ denote the complex upper half-plane and $\Gamma_1(Np^r)$ the usual congruence subgroup of matrices in $\mathrm{SL}_2(\mathbb{Z})$ that are upper-triangular and unipotent modulo Np^r. We consider the complex points $Y_1(Np^r) = \Gamma_1(Np^r)\backslash\mathbb{H}$ of the open modular curve over $\mathbb{C}$. It is traditional to use {cusps} to denote the cusps $\Gamma_1(Np^r)\backslash\mathbb{P}^1(\mathbb{Q})$, but let us instead set $C_r = \{\text{cusps}\}$. We let

$$X_1(Np^r) = Y_1(Np^r) \cup C_r = \Gamma_1(Np^r)\backslash\mathbb{H}^*,$$

be the closed modular curve, where $\mathbb{H}^* = \mathbb{H} \cup \mathbb{P}^1(\mathbb{Q})$ is the extended upper half-plane.

The usual modular symbols lie in the first homology group of the space $X_1(Np^r)$ relative to the cusps. However, $H_1(X_1(Np^r), C_r, \mathbb{Z})$ is not exactly the natural object for our study. Rather, we are interested in its quotient by the action of complex conjugation, the plus quotient.[1] We consider the plus quotients of homology and homology relative to the cusps:

$$\mathcal{S}_r = H_1(X_1(Np^r), \mathbb{Z})_+ \quad \text{and} \quad \mathcal{M}_r = H_1(X_1(Np^r), C_r, \mathbb{Z})_+,$$

where $(\ )_+$ denotes the plus quotient.

2.1.2. We introduce Manin symbols $[u : v]_r \in \mathcal{M}_r$.

Let $u, v \in \mathbb{Z}/Np^r\mathbb{Z}$ be such that $(u, v) = (1)$. For such u and v, we can find $\gamma = \left(\begin{smallmatrix} a & b \\ c & d \end{smallmatrix}\right) \in \mathrm{SL}_2(\mathbb{Z})$ with $u = c \bmod Np^r$ and $v = d \bmod Np^r$. Define

$$[u : v]_r = \left\{ \frac{d}{bNp^r} \to \frac{c}{aNp^r} \right\}_r,$$

where $\{\alpha \to \beta\}_r$ for $\alpha, \beta \in \mathbb{P}^1(\mathbb{Q})$ denotes the class in $\mathcal{M}_r$ of the hyperbolic geodesic on $\mathbb{H}^*$ from α to β. Then $[u : v]_r$ is independent of the choice of γ.

By the work of Manin (1972), we have that the group $\mathcal{M}_r$ of modular symbols is explicitly presented as an abelian group by generators $[u : v]_r$ and relations

$$[u : v]_r = [-u : v]_r = -[v : u]_r \quad \text{and} \quad [u : v]_r = [u : u + v]_r + [u + v : v]_r.$$

2.1.3. We define an intermediate relative homology group $\mathcal{M}_r^0$ used in constructing ϖ_r.

We do not use all modular symbols to connect with GL_1. Rather, we use those modular symbols with boundaries in cusps that do not lie over the cusp at 0 in $X_0(Np^r) = \Gamma_0(Np^r)\backslash\mathbb{H}^*$. Let us denote the set of cusps of $X_1(Np^r)$ that do not lie over the 0-cusp of $X_0(Np^r)$ by C_r^0. The intermediate space

$$\mathcal{M}_r^0 = H_1(X_1(Np^r), C_r^0, \mathbb{Z})_+$$

[1] This is still not quite the right object unless we invert 2. In Sect. 4, we take the point of view that the right object is the relative homology of the quotient of the space $X_1(Np^r)$ by the action of complex conjugation.

is the largest space on which we may define ϖ_r and have it factor through the Eisenstein quotient (see 2.1.7). We have $\mathcal{S}_r \subset \mathcal{M}_r^0 \subset \mathcal{M}_r$.

Our intermediate space also has a simple presentation: it is generated by the $[u : v]$ for nonzero $u, v \in \mathbb{Z}/Np^r\mathbb{Z}$ with $(u, v) = (1)$, together with the relations of 2.1.2, again for nonzero u and v, and excluding the last relation when $u + v = 0$.

2.1.4. We define the map ϖ_r, which gives our first connection between GL_2 and GL_1.

We start with the primitive Np^rth root of unity $\zeta_{Np^r} = e^{2\pi i/Np^r}$. It generates the cyclotomic field $E_r = \mathbb{Q}(\zeta_{Np^r})$ and its integer ring $\mathbb{Z}[\zeta_{Np^r}]$. Inside E_r, we have the maximal totally real subfield $F_r = \mathbb{Q}(\zeta_{Np^r})^+$ and its integer ring $\mathcal{O}_r = \mathbb{Z}[\zeta_{Np^r}]^+$.

For $a, b \in \mathbb{Z}[\zeta_{Np^r}, \frac{1}{Np}]^\times$, we let $\{a, b\}_r$ denote the norm of the Steinberg symbol of a and b to $K_2(\mathcal{O}_r[\frac{1}{Np}])$. There is a well-defined homomorphism

$$\varpi_r \colon \mathcal{M}_r^0 \otimes \mathbb{Z}[\tfrac{1}{2}] \to K_2(\mathcal{O}_r[\tfrac{1}{Np}]) \otimes \mathbb{Z}[\tfrac{1}{2}], \qquad [u : v]_r \mapsto \{1 - \zeta_{Np^r}^u, 1 - \zeta_{Np^r}^v\}_r$$

for $u, v \neq 0$. Using the Steinberg relation $\{x, 1 - x\}_r = 0$ in K_2 for $x, 1 - x \in \mathbb{Z}[\zeta_{Np^r}, \frac{1}{Np}]^\times$, one may easily check that the $\{1 - \zeta_{Np^r}^u, 1 - \zeta_{Np^r}^v\}_r$ satisfy the same relations as the $[u : v]_r$ (see Busuioc 2008; Sharifi 2011 for instance). It is necessary to invert 2 for these relations to hold.

2.1.5. We interpret ϖ_r on p-completions in terms of cup products in Galois cohomology.

For a commutative ring R in which p is invertible, the Kummer exact sequence

$$0 \to \mathbb{Z}/p^n\mathbb{Z}(1) \to \mathbb{G}_m \xrightarrow{p^n} \mathbb{G}_m \to 0$$

on $\mathrm{Spec}(R)_{\text{ét}}$ induces the connecting map $R^\times \to H^1_{\text{ét}}(R, \mathbb{Z}/p^n\mathbb{Z}(1))$. We have also the Chern class map $K_2(R) \to H^2_{\text{ét}}(R, \mathbb{Z}/p^n\mathbb{Z}(2))$. The value of this map on a product (i.e., Steinberg symbol) in $K_2(R)$ of a pair of elements of $R^\times$ is equal to the cup product of the images in $H^1_{\text{ét}}(R, \mathbb{Z}/p^n\mathbb{Z}(1))$ of the two elements.

We may apply this discussion with R equal to $\mathbb{Z}[\zeta_{Np^r}, \frac{1}{Np}]$ or $\mathcal{O}_r[\frac{1}{Np}]$, in which cases the Chern class map $K_2(R) \otimes \mathbb{Z}_p \to H^2_{\text{ét}}(R, \mathbb{Z}_p(2))$ is an isomorphism (Tate 1976). Moreover, the diagram

$$\begin{array}{ccc}
K_2(\mathbb{Z}[\zeta_{Np^r}, \frac{1}{Np}])_+ \otimes \mathbb{Z}_p & \xrightarrow{\sim} & H^2_{\text{ét}}(\mathbb{Z}[\zeta_{Np^r}, \frac{1}{Np}], \mathbb{Z}_p(2))_+ \\
{\scriptstyle N}\downarrow\wr & & {\scriptstyle \mathrm{cor}}\downarrow\wr \\
K_2(\mathcal{O}_r[\frac{1}{Np}]) \otimes \mathbb{Z}_p & \xrightarrow{\sim} & H^2_{\text{ét}}(\mathcal{O}_r[\frac{1}{Np}], \mathbb{Z}_p(2))
\end{array}$$

commutes, where N is induced by the norm and cor is induced by corestriction. The map cor is an isomorphism as $\mathcal{O}_r[\frac{1}{Np}]$ has p-cohomological dimension 2. Let

$(1-\zeta^u_{Np^r}, 1-\zeta^v_{Np^r})_r$ denote the corestriction of the cup product of the elements $1-\zeta^u_{Np^r}$ and $1-\zeta^v_{Np^r}$ of

$$\mathbb{Z}[\zeta_{Np^r}, \tfrac{1}{Np}]^\times \otimes \mathbb{Z}_p \xrightarrow{\sim} H^1_{\text{ét}}(\mathbb{Z}[\zeta_{Np^r}, \tfrac{1}{Np}], \mathbb{Z}_p(1)).$$

By definition of the symbols, the Chern class map in the lower row of the diagram satisfies

$$\{1-\zeta^u_{Np^r}, 1-\zeta^v_{Np^r}\}_r \mapsto (1-\zeta^u_{Np^r}, 1-\zeta^v_{Np^r})_r.$$

We will study the homomorphism to Galois cohomology

$$\varpi_r\colon \mathcal{M}^0_r \otimes \mathbb{Z}_p \to H^2_{\text{ét}}(\mathcal{O}_r[\tfrac{1}{Np}], \mathbb{Z}_p(2)), \qquad [u:v]_r \mapsto (1-\zeta^u_{Np^r}, 1-\zeta^v_{Np^r})_r,$$

which is identified with our original ϖ_r on p-completions.

2.1.6. We define Hecke algebras $\mathbb{T}_r$ and $\tilde{\mathbb{T}}_r$ and their Eisenstein ideals I_r and $\mathfrak{I}_r$.

The Hecke operators $T(n)$ for $n \geq 1$ generate a subalgebra $\tilde{\mathbb{T}}_r$ of $\mathrm{End}_{\mathbb{Z}_p}(\mathcal{M}_r \otimes \mathbb{Z}_p)$, the modular Hecke algebra. We will be interested in this section only in its action on $\mathcal{M}^0_r$. We also have a cuspidal Hecke algebra $\mathbb{T}_r$ of $\mathrm{End}_{\mathbb{Z}_p}(\mathcal{S}_r \otimes \mathbb{Z}_p)$ and a canonical surjection $\tilde{\mathbb{T}}_r \twoheadrightarrow \mathbb{T}_r$. These Hecke algebras contain diamond operators $\langle d\rangle$ for $d \in \mathbb{Z}$, which we take to be 0 if $(d, Np) \neq 1$.

The Hecke algebra $\tilde{\mathbb{T}}_r$ contains the Eisenstein ideal $\mathfrak{I}_r$ generated by the $T(n) - \sum_{d|n} d\langle d\rangle$ for $n \geq 1$. It is also generated by $T(\ell) - 1 - \ell\langle\ell\rangle$ for primes ℓ. The image I_r of $\mathfrak{I}_r$ in $\mathbb{T}_r$ is an Eisenstein ideal with the same generators.

2.1.7. We connect our study of ϖ_r with the Eisenstein ideal.

The third author conjectured (Sharifi 2011) (on $\mathcal{S}_r$, see also (Busuioc 2008) for $N = 1$), and the first two authors proved (Fukaya and Kato Preprint, Theorem 5.2.3) that ϖ_r is "Eisenstein." By this, we mean that ϖ_r factors through the quotient of $\mathcal{M}^0_r \otimes \mathbb{Z}_p$ by the Eisenstein ideal, that is, through a map

$$(\mathcal{M}^0_r/\mathfrak{I}_r\mathcal{M}^0_r) \otimes \mathbb{Z}_p \to H^2_{\text{ét}}(\mathcal{O}_r[\tfrac{1}{Np}], \mathbb{Z}_p(2)).$$

We can show that this follows from the fact that ϖ_r is the specialization of a map in the GL_2-setting: see Sect. 2.3.

2.1.8. Let $G_r = (\mathbb{Z}/Np^r\mathbb{Z})^\times/\{\pm 1\}$, and set $\Lambda_r = \mathbb{Z}_p[G_r]$. The algebra Λ_r appears in two different contexts in our story:

(1) On the GL_2-side, Λ_r is a $\mathbb{Z}_p$-algebra of diamond operators in $\mathbb{T}_r$ (or $\tilde{\mathbb{T}}_r$): we define a $\mathbb{Z}_p$-linear injection $\iota_r\colon \Lambda_r \hookrightarrow \mathbb{T}_r$ that sends the group element in Λ_r corresponding to $a \in (\mathbb{Z}/Np^r\mathbb{Z})^\times/\langle -1\rangle$ to the <u>inverse</u> $\langle a\rangle^{-1}$ of the diamond operator for a (i.e., for any lift of a to an integer).

(2) On the GL_1-side, Λ_r is the $\mathbb{Z}_p$-group ring of $\mathrm{Gal}(F_r/\mathbb{Q})$: we have an isomorphism

$$(\mathbb{Z}/Np^r\mathbb{Z})^\times \xrightarrow{\sim} \mathrm{Gal}(E_r/\mathbb{Q}), \qquad a \mapsto \sigma_a,$$

where $\sigma_a(\zeta_{Np^r}) = \zeta_{Np^r}^a$. This gives rise to an isomorphism $G_r \xrightarrow{\sim} \mathrm{Gal}(F_r/\mathbb{Q})$ that is the map on group elements defining $\Lambda_r \xrightarrow{\sim} \mathbb{Z}_p[\mathrm{Gal}(F_r/\mathbb{Q})]$.

These actions are compatible with ϖ_r in the sense that for any $x \in \mathcal{M}_r^0 \otimes \mathbb{Z}_p$ and $a \in (\mathbb{Z}/Np^r\mathbb{Z})^\times$, we have

$$\varpi_r(\langle a\rangle^{-1}x) = \sigma_a\varpi_r(x).$$

This is easily seen: taking $x = [u : v]_r$ for some nonzero u and v, we have

$$\langle a\rangle^{-1}[u : v]_r = [au : av]_r \quad \text{and} \quad \sigma_a(1 - \zeta_{Np^r}^u, 1 - \zeta_{Np^r}^v)_r = (1 - \zeta_{Np^r}^{au}, 1 - \zeta_{Np^r}^{av})_r.$$

So, to say that ϖ_r is Eisenstein is to say that $\varpi_r(T(\ell)x) = (1 + \ell\sigma_\ell^{-1})\varpi_r(x)$ for primes $\ell \nmid Np$ and $\varpi_r(T(\ell)x) = \varpi_r(x)$ for $\ell \mid Np$.

2.2 *Passing Up the Modular and Cyclotomic Towers: The Map* ϖ

We pass up the modular tower on the GL_2-side and the cyclotomic tower on the GL_1-side to define the map $\varpi = \varprojlim_r \varpi_r$.

2.2.1. Let $G = \varprojlim_r G_r$. Then the completed group ring

$$\Lambda = \mathbb{Z}_p[\![G]\!] = \varprojlim_r \Lambda_r$$

is the Iwasawa algebra for G. As with Λ_r, let us emphasize its dual nature:

(1) Set $\mathbb{T} = \varprojlim_r \mathbb{T}_r$ and $\tilde{\mathbb{T}} = \varprojlim_r \tilde{\mathbb{T}}_r$. The projective limit of the injections ι_r defines a map $\iota\colon \Lambda \hookrightarrow \mathbb{T}$ of profinite $\mathbb{Z}_p$-modules that takes $a \in G$ to the projective system of inverses $\langle a\rangle^{-1}$ of diamond operators corresponding to a.
(2) Set $K = \cup_{r\geq 1} F_r$, the maximal totally real subfield of $L = \cup_{r\geq 1} E_r$. Then our identifications $\Lambda_r \xrightarrow{\sim} \mathbb{Z}_p[\mathrm{Gal}(F_r/\mathbb{Q})]$ for $r \geq 1$ induce an isomorphism $\Lambda \xrightarrow{\sim} \mathbb{Z}_p[\![\mathrm{Gal}(K/\mathbb{Q})]\!]$ of completed group rings in the projective limit.

2.2.2. We have the following projective limits of spaces of modular symbols:

$$\mathcal{S} = \varprojlim_r (\mathcal{S}_r \otimes \mathbb{Z}_p) \quad \text{and} \quad \mathcal{M}^0 = \varprojlim_r (\mathcal{M}_r^0 \otimes \mathbb{Z}_p).$$

Let $\mathfrak{I} \subset \tilde{\mathbb{T}}$ and $I \subset \mathbb{T}$ be the Eisenstein ideals, defined by the same set of generators as I_r, but now viewed as compatible sequences of operators in the Hecke algebras.

Our maps ϖ_r are compatible with change of r and induce in the projective limit a map

$$\varpi\colon \mathcal{M}^0 \to \varprojlim_r H^2_{\text{ét}}(\mathcal{O}_r[\tfrac{1}{Np}], \mathbb{Z}_p(2))$$

that factors through $\mathcal{M}^0/\mathcal{I}\mathcal{M}^0$ by the result of (Fukaya and Kato Preprint). This map ϖ is a homomorphism of Λ-modules, the actions arising from part (1) of 2.2.1 on the left and part (2) of 2.2.1 on the right.

2.2.3. We recall the unramified Iwasawa module X, study its difference from Galois cohomology, and consider a related Λ-module Y.

Let X be the projective limit of the p-parts A_r of the ideal class groups of the fields E_r. Class field theory allows us to identify X with the Galois group of the maximal unramified abelian pro-p extension of L.

For R as in 2.1.5, the Kummer exact sequence induces

$$\text{Pic}(R) = H^1_{\text{ét}}(R, \mathbb{G}_m) \to H^2_{\text{ét}}(R, \mathbb{Z}/p^n\mathbb{Z}(1)).$$

Taking a projective limit of such maps for the rings $R = \mathcal{O}_r[\frac{1}{Np}]$, we obtain

$$X = \varprojlim_r A_r \to \varprojlim_r H^2_{\text{ét}}(\mathcal{O}_r[\tfrac{1}{Np}], \mathbb{Z}_p(1)).$$

In general, this map is neither injective nor surjective. Its kernel and cokernel can be explicitly described as contributions of classes of primes and Brauer groups at places dividing Np, respectively. We will deal with a part of cohomology on which this subtle difference disappears.

The Iwasawa algebra $\mathbb{Z}_p[\![\text{Gal}(L/\mathbb{Q})]\!]$ acts on X, but this action does not in general factor through Λ. We want to consider the (-1)-eigenspace X^- of X under complex conjugation. To do so, we take the Tate twist $Y = X^-(1)$, or equivalently, the fixed part $X(1)^+$. Then σ_{-1} acts trivially on Y, so Y is a Λ-module. The map from X to cohomology induces a Λ-module homomorphism

$$Y \to \varprojlim_r H^2_{\text{ét}}(\mathcal{O}_r[\tfrac{1}{Np}], \mathbb{Z}_p(2)).$$

Together with ϖ, this will allow us to relate $\mathcal{S}$ with Y.

2.2.4. We have two objects of study:

- The geometric object $P = \mathcal{S}/I\mathcal{S}$ for GL_2,
- The Iwasawa-theoretic object $Y = X^-(1)$ for GL_1.

We can relate these on θ-parts for suitable even characters θ of $(\mathbb{Z}/Np\mathbb{Z})^\times$.

For a primitive, even character $\theta\colon (\mathbb{Z}/Np\mathbb{Z})^\times \to \overline{\mathbb{Q}_p}^\times$, we may consider the quotient $\Lambda_\theta = \Lambda \otimes_{\mathbb{Z}_p[(\mathbb{Z}/Np\mathbb{Z})^\times]} \mathbb{Z}_p[\theta]$ of Λ, where $\mathbb{Z}_p[\theta]$ is the $\mathbb{Z}_p$-algebra generated by the values of θ. For a Λ-module M, we then let $M_\theta = M \otimes_\Lambda \Lambda_\theta$ denote its θ-part.

We need a technical assumption to insure that the maps

$$P_\theta \to \mathcal{M}^0_\theta / \mathfrak{I}_\theta \mathcal{M}^0_\theta \quad \text{and} \quad Y_\theta \to \varprojlim_r H^2_{\text{ét}}(\mathcal{O}_r[\tfrac{1}{Np}], \mathbb{Z}_p(2))_\theta$$

are isomorphisms. Together with primitivity, the assumption is as follows:

- $\theta\omega^{-1}|_{(\mathbb{Z}/p\mathbb{Z})^\times} \neq 1$ or $\theta\omega^{-1}|_{(\mathbb{Z}/N\mathbb{Z})^\times}(p) \neq 1$,

where $\omega \colon (\mathbb{Z}/Np\mathbb{Z})^\times \to \mathbb{Z}_p^\times$ denotes the Teichmüller character (i.e., projection to $(\mathbb{Z}/p\mathbb{Z})^\times \subset \mathbb{Z}_p^\times$). For such a θ, our ϖ induces a map on θ-parts $\varpi \colon \mathcal{S}_\theta \to Y_\theta$ that will factor through P_θ.

2.3 Zeta Elements: ϖ Is "Eisenstein"

We sketch the proof that ϖ factors through the quotient of $\mathcal{M}^0$ by the Eisenstein ideal $\mathfrak{I}$.

2.3.1. Let $Y_1(Np^r)$ be the moduli space of pairs (E, e) where E is an elliptic curve and e is a point of order Np^r on E, and let $Y(Np^r)$ be the moduli space of elliptic curves endowed with a full Np^r-level structure. We view these moduli spaces as schemes over $\mathbb{Z}[\frac{1}{Np}]$. For any nonzero $(\alpha, \beta) \in \frac{1}{Np^r}\mathbb{Z}^2/\mathbb{Z}^2$, there is a Siegel unit $g_{\alpha,\beta} \in \mathcal{O}(Y(Np^r))^\times$.[2] It has the q-expansion

$$g_{\alpha,\beta} = q^{\frac{1}{12} - \frac{\alpha}{2} + \frac{\alpha^2}{2}} \prod_{n=0}^{\infty} (1 - q^{n+\alpha} e^{2\pi i \beta}) \prod_{n=1}^{\infty} (1 - q^{n-\alpha} e^{-2\pi i \beta}) \in \mathcal{O}_r[\tfrac{1}{Np}][\![q^{1/12Np^r}]\!][q^{-1}]^\times.$$

If $\alpha = 0$, then we may view $g_{0,\beta}$ as an element of $\mathcal{O}(Y_1(Np^r))^\times$. The crucial point is that the specialization of a Siegel unit of the form $g_{0,\frac{u}{Np^r}}$ at the ∞-cusp is the cyclotomic Np-unit $1 - \zeta^u_{Np^r}$. Specifically, this specialization is given by projecting its q-expansion to $\mathcal{O}_r[\frac{1}{Np}][\![q^{1/Np^r}]\!]^\times$ and then evaluating at $q = 0$ (Fukaya and Kato Preprint, Section 5.1).

2.3.2. We have a homomorphism

$$z_r \colon \mathcal{M}^0_r \to H^2_{\text{ét}}(Y_1(Np^r), \mathbb{Z}_p(2)), \qquad z_r([u:v]_r) = g_{0,\frac{u}{Np^r}} \cup g_{0,\frac{v}{Np^r}}$$

[2]Actually, $g_{\alpha,\beta}$ is a root of a unit, but the difficulties this causes are resolvable by passing to the projective limit and descending, so we ignore this for simplicity of presentation. We will be very careless about denominators in several places, omitting them where they occur for simplicity of the discussion that follows.

of $\tilde{\mathbb{T}}_r$-modules that takes a Manin symbol to a Beilinson element given by a cup product of two Siegel units (Fukaya and Kato Preprint, Proposition 3.3.15). Related elements were studied in (Kato 2004).

2.3.3. There is again a specialization-at-∞ map

$$\infty_r\colon H^2_{\text{ét}}(Y_1(Np^r),\mathbb{Z}_p(2)) \to H^2_{\text{ét}}(\mathcal{O}_r[\tfrac{1}{Np}],\mathbb{Z}_p(2))$$

that takes $g_{0,\frac{u}{Np^r}} \cup g_{0,\frac{v}{Np^r}}$ to $(1-\zeta^u_{Np^r}, 1-\zeta^v_{Np^r})_r$. So, cup products of cyclotomic units are specializations at cusps of Beilinson elements. We have

$$\varpi_r = \infty_r \circ z_r\colon \mathcal{M}^0_r \to H^2_{\text{ét}}(\mathcal{O}_r[\tfrac{1}{Np}],\mathbb{Z}_p(2)).$$

It can be shown that specialization at ∞ is Eisenstein. Hence so is ϖ_r (Fukaya and Kato Preprint, Sections 5.1–5.2).

2.3.4. By passing the projective limit over r, we see that ϖ is Eisenstein. The identity $\varpi = \infty \circ z$ is the commutativity of the left-hand square in the diagram of 1.10.

2.4 Ordinary Homology Groups of Modular Curves

Homology groups of the modular curves are useful for us in two different ways. They contain modular symbols, allowing us to define ϖ. They also have Galois actions, allowing us to define Υ, which is our next goal. We use two different groups derived from homology, $\mathcal{S}$ as above and $\mathcal{T}$ defined below, to construct the two maps. For the modular symbols, we require only the plus part of homology. On the other hand, to have Galois actions, we cannot restrict to plus parts. Instead, we take ordinary parts to control the growth of homology groups in the modular tower and to specify the form of the local Galois action at p. The fact that we use different groups should be kept in mind in the GL_d-setting, in which we will not consider Υ.

2.4.1. We introduce Hida's ordinary p-adic cuspidal and modular Hecke algebras $\mathfrak{h}$ and $\mathfrak{H}$.

Recall our cuspidal Hecke algebra $\mathbb{T}$ from 2.2.1, which acts Λ-linearly on $\mathcal{S}$. The action of $T(p)$ breaks it into a direct product of two rings: an ordinary part in which the image of $T(p)$ is invertible and another part in which $T(p)$ is topologically nilpotent. The ordinary cuspidal p-adic Hecke algebra $\mathfrak{h} = \mathbb{T}^{\text{ord}}$ of Hida (1986) is this ordinary part. This is a Λ-subalgebra that is projective of finite Λ-rank. We may speak of Hecke operators $T(n) \in \mathfrak{h}$ by taking the images of the $T(n) \in \mathbb{T}$.

The Hecke algebra $\mathfrak{h}$ is remarkable in that it simply encapsulates information about the ordinary Hecke algebras of all weights ≥ 2 and all levels dividing some Np^r. For instance, its quotient for the action of the kernel of $G \to G_r$ is the Hecke algebra $\mathfrak{h}_r = \mathbb{T}^{\text{ord}}_r$. This highly regular behavior is the subject of Hida theory.

We also have the ordinary modular Hecke algebra $\mathfrak{H} = \tilde{\mathbb{T}}^{\text{ord}}$, of which $\mathfrak{h}$ is a quotient. In general, if M is a $\tilde{\mathbb{T}}$-module (resp., $\mathbb{T}$-module), then we use M^{ord} to denote its ordinary part, the maximal summand on which $T(p)$ acts invertibly, which is an $\mathfrak{H}$-module (resp., $\mathfrak{h}$-module).

2.4.2. We introduce the ordinary homology groups $\mathcal{T}$ and $\tilde{\mathcal{T}}$. These have commuting actions of Hecke algebras and the absolute Galois group $G_{\mathbb{Q}} = \text{Gal}(\overline{\mathbb{Q}}/\mathbb{Q})$. The study of these actions on $\mathcal{T}$ will allow us to define the map Υ in Sect. 2.5.

The Hecke operators $T(n)$ with $n \geq 1$ act on the homology of $X_1(Np^r)(\mathbb{C})$ and the homology relative to the cusps and are compatible with projective limits. We consider the ordinary parts $\mathcal{T}$ and $\tilde{\mathcal{T}}$ of the projective limits

$$\mathcal{T} = \varprojlim_r H_1(X_1(Np^r), \mathbb{Z}_p)^{\text{ord}} \subset \tilde{\mathcal{T}} = \varprojlim_r H_1(X_1(Np^r), C_r, \mathbb{Z}_p)^{\text{ord}}.$$

We are primarily interested in $\mathcal{T}$. The $\mathbb{T}$-action on $\mathcal{T}$ factors through $\mathfrak{h}$. As an $\mathfrak{h}$-module, $\mathcal{T}$ is finitely generated and torsion-free, and $\mathcal{T}$ is projective of finite rank over Λ. If we denote by $Q(\Lambda)$ the total quotient ring of Λ, then $Q(\Lambda) \otimes_\Lambda \mathcal{T}$ is a free $Q(\Lambda) \otimes_\Lambda \mathfrak{h}$-module of rank 2.

The absolute Galois group $G_{\mathbb{Q}} = \text{Gal}(\overline{\mathbb{Q}}/\mathbb{Q})$ acts on the homology of $X_1(Np^r)$ by its duality with cohomology and the identification of Betti cohomology with étale cohomology of the scheme $X_1(Np^r)_{/\overline{\mathbb{Q}}}$ over $\overline{\mathbb{Q}}$. This describes the first of the two isomorphisms

$$H_1(X_1(Np^r), \mathbb{Z}_p) \cong \text{Hom}(H^1_{\text{ét}}(X_1(Np^r)_{/\overline{\mathbb{Q}}}, \mathbb{Z}_p), \mathbb{Z}_p),$$
$$H_1(X_1(Np^r), C_r, \mathbb{Z}_p) \cong \text{Hom}(H^1_{\text{ét},c}(Y_1(Np^r)_{/\overline{\mathbb{Q}}}, \mathbb{Z}_p), \mathbb{Z}_p),$$

where in the second, the duality of the relative cohomology group is with the compactly supported cohomology of the open modular curve. This Galois action commutes with the action of the Hecke operators, so passes to ordinary parts, and it is compatible in the towers. Therefore, $\mathfrak{H}[G_{\mathbb{Q}}]$ acts compatibly on $\mathcal{T}$ and $\tilde{\mathcal{T}}$.

2.4.3. We introduce the Eisenstein ideals I and $\mathfrak{I}$ of the ordinary Hecke algebras.

Let us reuse the notation I, allowing it to denote the Eisenstein ideal of $\mathfrak{h}$, which is the image of the Eisenstein ideal I of $\mathbb{T}$ in $\mathfrak{h}$. We remark that, since $T(p) - 1 \in I$ and 1 is a unit, the quotient map $\mathbb{T}/I \to \mathfrak{h}/I$ is an isomorphism. We will also reuse the notation $\mathfrak{I}$ for the Eisenstein ideal of $\mathfrak{H}$, the image of $\mathfrak{I} \subset \tilde{\mathbb{T}}$.

2.4.4. In the GL_2-setting over $\mathbb{Q}$, there are two places which play important roles: the place at p and the real place. We study the actions of the corresponding local Galois groups.

We first study the local action at p: here we have an interesting quotient $\mathcal{T}_{\text{quo}}$. The fact that $\mathcal{T}$ is ordinary for $T(p)$ tells us about the action of $G_{\mathbb{Q}_p}$, which is to say that it is ordinary in the sense of p-adic Hodge theory. More specifically to our case, we have an exact sequence

$$0 \to \mathcal{T}_{\text{sub}} \to \mathcal{T} \to \mathcal{T}_{\text{quo}} \to 0$$

of $\mathfrak{h}[G_{\mathbb{Q}_p}]$-modules, with $\mathcal{T}_{\text{sub}}$ and $\mathcal{T}_{\text{quo}}$ defined as follows. First, $\mathcal{T}_{\text{sub}}$ is the largest submodule of $\mathcal{T}$ such that $G_{\mathbb{Q}_p}$ acts on $\mathcal{T}_{\text{sub}}(-1)$ by inverse diamond operators, and $\mathcal{T}_{\text{quo}}$ is the quotient. Put more simply, $\mathcal{T}_{\text{quo}}$ is the maximal unramified, $\mathfrak{h}$-torsion-free quotient of $\mathcal{T}$.

At the real place, we have $\mathcal{T}^+$, which is isomorphic to $\mathcal{S}^{\text{ord}}$. It fits in an exact sequence

$$0 \to \mathcal{T}^+ \to \mathcal{T} \to \mathcal{T}/\mathcal{T}^+ \to 0$$

of $\mathfrak{h}[G_{\mathbb{R}}]$-modules, and both $Q(\Lambda) \otimes_\Lambda \mathcal{T}^+$ and $Q(\Lambda) \otimes_\Lambda \mathcal{T}/\mathcal{T}^+$ are free of rank 1 over $Q(\Lambda) \otimes_\Lambda \mathfrak{h}$.

The compositions $\mathcal{T}^+ \to \mathcal{T} \to \mathcal{T}_{\text{quo}}$ and $\mathcal{T}_{\text{sub}} \to \mathcal{T} \to \mathcal{T}/\mathcal{T}^+$ relate the two exact sequences. We study these maps on Eisenstein components in 2.5.5. The interplay between the reductions modulo I of the two exact sequences allows us to construct the map Υ.

2.4.5. We discuss Λ-adic cusp forms and modular forms and their ordinary parts $\mathfrak{S}$ and $\mathfrak{M}$.

Let $S_2(Np^r)_{\mathbb{Z}}$ denote the space of cusp forms of weight 2 and level Np^r with integer coefficients. For a ring R, we then set $S_2(Np^r)_R = S_2(Np^r)_{\mathbb{Z}} \otimes R$. If $\epsilon\colon G_r \to R^\times$ is a homomorphism, then we may speak of $S_2(Np^r, \epsilon)_R$, those cusp forms in $S_2(Np^r)_R$ with nebentypus ϵ.

Any finite order character $\epsilon\colon G \to \overline{\mathbb{Q}_p}^\times$ induces a ring homomorphism $\Lambda \to \overline{\mathbb{Q}_p}$. We let $\tilde{\epsilon}\colon \Lambda[\![q]\!] \to \overline{\mathbb{Q}_p}[\![q]\!]$ be the induced map on coefficients. An element $f \in \Lambda[\![q]\!]$ is said to be a Λ-adic cusp form of weight 2 and level Np^∞ if for every ϵ, one has $\tilde{\epsilon}(f) \in S_2(Np^r, \epsilon)_{\overline{\mathbb{Q}_p}}$ with $r \geq 0$ such that ϵ factors through G_r (Ohta 1995; Wiles 1988). We denote the set of such Λ-adic cusp forms by S_Λ.

The Hecke operators $T(n)$ for $n \geq 1$ act on S_Λ via the usual formal action of Hecke operators on q-expansions. We define $\mathfrak{S}$ to be the ordinary part S_Λ^{ord} of S_Λ.[3] The ordinary Λ-adic cusp forms and the ordinary Hecke algebra are dual in the usual sense. That is, we have a perfect pairing of Λ-modules,

$$\mathfrak{h} \times \mathfrak{S} \to \Lambda, \qquad (T, f) \mapsto a_1(Tf),$$

where $a_1(g)$ denotes the q-coefficient in the q-expansion of $g \in S_\Lambda$. As a consequence, $Q(\Lambda) \otimes_\Lambda \mathfrak{S}$ is free of rank 1 over $Q(\Lambda) \otimes_\Lambda \mathfrak{h}$.

Similarly, we have a space $\mathfrak{M}$ of ordinary Λ-adic modular forms with q-expansions that are integral outside of the constant term, which sits inside $Q(\Lambda) +$

[3]There is one potentially confusing aspect: the action of $\Lambda \hookrightarrow \mathfrak{h}$ on $\mathfrak{S} \subset \Lambda[\![q]\!]$ is not given by multiplication of the coefficients of q-expansions by the element of Λ. It is instead this multiplication after first applying the inversion map $\lambda \mapsto \lambda^*$ on Λ that takes group elements to their inverses.

$\Lambda[\![q]\!]$. There is a perfect pairing $\mathfrak{H} \times \mathfrak{M} \to \Lambda$ that restricts to the pairing for cusp forms.

2.4.6. As we shall explain in a more canonical fashion in 2.7.6, there is an isomorphism $\mathfrak{T}_{\text{quo}} \cong \mathfrak{S}$ of $\mathfrak{h}$-modules given by Ohta's Λ-adic Eichler-Shimura isomorphism (Ohta 1995, 2000). Moreover, Ohta showed that $\mathfrak{T}_{\text{sub}} \cong \mathfrak{h}$ via a Λ-duality with $\mathfrak{T}_{\text{quo}}$.

2.5 *Refining the Method of Ribet and Mazur-Wiles: The Map Υ*

We define the map Υ of (Sharifi 2011) and consider the relationship with the work of Mazur and Wiles (1984). Our description is heavily influenced by the approaches of Wiles (1988) and Ohta (2000).

We suppose that $p \geq 5$ and $p \nmid \varphi(N)$.[4] We will work mostly in the θ-part (as in 2.2.4) for a fixed primitive, even character $\theta\colon (\mathbb{Z}/Np\mathbb{Z})^\times \to \overline{\mathbb{Q}}_p^\times$ such that the condition $\theta\omega^{-1}|_{(\mathbb{Z}/p\mathbb{Z})^\times} \neq 1$ or $\theta\omega^{-1}|_{(\mathbb{Z}/N\mathbb{Z})^\times}(p) \neq 1$ of 2.2.4 holds. We also suppose that $\theta \neq \omega^2$ in the case that $N = 1$.

2.5.1. We briefly outline the construction of $\Upsilon\colon Y_\theta \to P_\theta$ that will appear in this section.

We analyze the $\mathfrak{h}[G_{\mathbb{Q}}]$-action on $\mathfrak{T}_\theta/I_\theta\mathfrak{T}_\theta$, showing that it fits in an exact sequence

$$0 \to P_\theta \to \mathfrak{T}_\theta/I_\theta\mathfrak{T}_\theta \to Q_\theta \to 0$$

of $\mathfrak{h}[G_{\mathbb{Q}}]$-modules. Any such exact sequence provides a cocycle $G_{\mathbb{Q}} \to \mathrm{Hom}_{\mathfrak{h}}(Q_\theta, P_\theta)$ that defines its extension class in Galois cohomology. Our exact sequence has three key properties: the G_L-action on $\mathfrak{T}_\theta/I_\theta\mathfrak{T}_\theta$ is unramified, the G_L-actions on P_θ and Q_θ are trivial, and the $\mathfrak{h}_\theta/I_\theta$-module Q_θ is free of rank 1 with a canonical generator. We may therefore modify our cocycle as follows. First, we compose it with evaluation at the generator of Q_θ to obtain a map $G_{\mathbb{Q}} \to P_\theta$. Since G_L acts trivially on P_θ and Q_θ, this map restricts to a homomorphism $G_L \to P_\theta$. Since the G_L-action on $\mathfrak{T}_\theta/I_\theta\mathfrak{T}_\theta$ is unramified, this homomorphism in turn factors through a homomorphism $X \to P_\theta$. After a twist, it further factors through Y_θ and provides the desired map $\Upsilon\colon Y_\theta \to P_\theta$, which we can show to be of Λ-modules.

We first explain that $\mathfrak{h}_\theta/I_\theta$ is the quotient of Λ_θ by a p-adic L-function ξ_θ. This will provide the connection between the map Υ and the Iwasawa main conjecture.

2.5.2. We define the p-adic L-function ξ_θ.

[4] It should actually be possible to allow either or both of $p = 3$ and $p \mid \varphi(N)$ in what follows.

Note that any homomorphism $G \to \overline{\mathbb{Q}}^\times$ factors through some G_r and so induces an even Dirichlet character. Note also that $G_1 = (\mathbb{Z}/Np\mathbb{Z})^\times/\langle -1\rangle$ and $G \cong G_1 \times (1 + p\mathbb{Z}_p)$.

The p-adic L-function ξ_θ is the unique element of Λ_θ that interpolates Dirichlet L-values at -1 in the sense that for each character $\epsilon\colon G \to \overline{\mathbb{Q}}^\times$ such that $\epsilon|_{G_1} = \theta$, the ring homomorphism $\Lambda_\theta \to \overline{\mathbb{Q}}_p$ induced by ϵ sends ξ_θ to the value $L(\epsilon^{-1}, -1) \in \overline{\mathbb{Q}}$ of the Dirichlet L-function.

We can also describe ξ_θ in terms of Kubota-Leopoldt p-adic L-functions. We make the identification $\Lambda_\theta = \mathbb{Z}_p[\theta][\![T]\!]$ with $T = [u] - 1$, where $[u]$ is the group element of $u \in 1 + p\mathbb{Z}_p$ with p-adic logarithm $(1 - p^{-1})^{-1}$. We then have the following equality of functions of $s \in \mathbb{Z}_p$:

$$\xi_\theta(u^s - 1) = L_p(\omega^2\theta^{-1}, s - 1).$$

2.5.3. We construct a canonical isomorphism $\mathfrak{h}_\theta/I_\theta \xrightarrow{\sim} \Lambda_\theta/(\xi_\theta)$.

Consider the ordinary Λ-adic Eisenstein series

$$\mathcal{E}_\theta = \frac{1}{2}(\xi_\theta)^* + \sum_{n=1}^{\infty}\left(\sum_{\substack{d|n \\ (d,Np)=1}} d[d]\right)q^n \in \mathfrak{M}_\theta,$$

where $[d]$ is the image in $\Lambda_{\theta^{-1}}$ of the group element in G for d, and $\lambda \mapsto \lambda^*$ is the involution defined in the footnote of 2.4.5.[5] By duality with the Hecke algebra, it provides a surjective homomorphism $\mathfrak{H}_\theta \to \Lambda_\theta$, the kernel of which is $\mathfrak{I}_\theta$ by definition.

Let M_{Eis} denote the component of an $\mathfrak{H}$-module M for the unique maximal ideal $\mathfrak{m}$ containing the Eisenstein ideal $\mathfrak{I}_\theta$. By our choice of θ, the Eisenstein series $\mathcal{E}_\theta$ is not congruent modulo $\mathfrak{m}$ to any other Eisenstein series (Ohta 2003, Lemma 1.4.9). It follows from this that the injection of $\mathfrak{S}$ in $\mathfrak{M}$ induces an exact sequence

$$0 \to \mathfrak{S}_{\mathrm{Eis}} \to \mathfrak{M}_{\mathrm{Eis}} \to \Lambda_\theta \to 0,$$

where the latter map takes a modular form to the (involution of the) constant term in its q-expansion. Our map $\pi_\theta\colon \mathfrak{h}_\theta/I_\theta \to \Lambda_\theta/(\xi_\theta)$ may then be constructed from the reduction of $\mathcal{E}_\theta$ modulo ξ_θ. That is, $\mathcal{E}_\theta$ is a cusp form modulo $(\xi_\theta) \subseteq \Lambda_\theta$ by the exact sequence, and this cusp form provides the surjective map π_θ by duality with the Hecke algebra $\mathfrak{h}$. Once we know that π_θ is an isomorphism, it is inverse to the map induced by ι_θ, where ι_θ is the θ-part of the map ι defined in 2.2.1.

We explain the idea behind the injectivity of π_θ. We have an evident surjection $\Lambda_\theta \to \mathfrak{h}_\theta/I_\theta$ given by the fact that every Hecke operator $T(n)$ is identified modulo I_θ with an element of Λ_θ. So, $\mathfrak{h}_\theta/I_\theta$ is some quotient of Λ_θ. The Λ-adic forms

[5] The reader may wish to ignore the involutions in order to focus on the idea of the argument.

in $\mathfrak{M}_\theta$ have integral constant coefficients, which can be seen by the method of (Emerton 1999, Proposition 1). Given this, the existence of π_θ is equivalent to the fact $\mathcal{E}_\theta$ modulo (ξ_θ) is a Λ-adic cusp form. As the constant coefficient of $\mathcal{E}_\theta$ equals ξ_θ^* times a unit, no surjective homomorphism to a larger quotient of Λ_θ can exist.[6] Thus, π_θ is an isomorphism.

2.5.4. We define Q_θ and construct a canonical surjection $\mathcal{T}_\theta/I_\theta\mathcal{T}_\theta \to Q_\theta$ of $\mathfrak{h}[G_\mathbb{Q}]$-modules.

For a module M over a Λ-algebra h, let $M^\sharp$ denote the $h[G_\mathbb{Q}]$-module that is M as an h-module and on which $\sigma \in G_\mathbb{Q}$ acts through multiplication by the inverse of the image of σ in G. We then define $Q_\theta = (\mathfrak{h}_\theta/I_\theta)^\sharp(1)$. Consider the Jacobian variety J_r of the curve $X_1(Np^r)$. Let $J_{r,\mathrm{tor}} \subset J_r(\overline{\mathbb{Q}})$ be its torsion subgroup, and take the contravariant (i.e., dual) action of $\mathbb{T}_r$ on $J_{r,\mathrm{tor}}$. Consider the class $\alpha_r \in J_r(\mathbb{C})$ of the divisor $(0) - (\infty)$, where 0 and ∞ are viewed as cusps on $X_1(Np^r)(\mathbb{C})$. It is torsion by the theorem of Drinfeld (1973) and Manin (1972). Moreover, α_r is easily seen to be annihilated by I_r.

Let $\beta_{r,\theta}$ be the image of α_r in the θ-part of $J_r[p^\infty] = J_{r,\mathrm{tor}} \otimes \mathbb{Z}_p$. The $\mathbb{T}_{r,\theta}$-span $B_{r,\theta}$ of $\beta_{r,\theta}$ is a quotient of $\mathfrak{h}_{r,\theta}/I_{r,\theta}$ by definition. Moreover, $B_{r,\theta}$ is isomorphic to $\Lambda_{r,\theta}/(\xi_{r,\theta})$ by a computation of divisors of Siegel units that says in particular that the θ-part of the divisor of $g_{0,\frac{1}{Np^r}}$ is $\xi_{r,\theta}$ times $(0) - (\infty)$, up to a unit (see Mazur and Wiles 1984, Section 4.2).[7] Here, $\xi_{r,\theta}$ denotes the image of ξ_θ in $\Lambda_{r,\theta}$. The $G_\mathbb{Q}$-action on $B_{r,\theta}$ factors through $\mathrm{Gal}(F_r/\mathbb{Q})$, and we have $\sigma_a\beta_{r,\theta} = \langle a\rangle^{-1}\beta_{r,\theta}$ for any $a \in (\mathbb{Z}/Np^r\mathbb{Z})^\times$.

Poincaré duality allows us to identify the first étale homology group of $X_1(Np^r)_{/\overline{\mathbb{Q}}}$ with the Tate twist of the first étale cohomology group. Taking this together with the canonical pairing of cohomology and the torsion in J_r, we obtain a Galois-equivariant, perfect pairing

$$(\ ,\ )\colon H_1^{\text{ét}}(X_1(Np^r)_{/\overline{\mathbb{Q}}}, \mathbb{Z}_p) \times J_r[p^\infty] \to \mathbb{Q}_p/\mathbb{Z}_p(1)$$

with respect to which the Hecke operators are self-adjoint. Let $(\ ,\ )_\theta$ denote the induced pairing on θ-parts. Define a map ϕ by

$$\phi\colon H_1^{\text{ét}}(X_1(Np^r)_{/\overline{\mathbb{Q}}}, \mathbb{Z}_p)_\theta \to \Lambda_{r,\theta} \otimes \mathbb{Q}_p/\mathbb{Z}_p(1),\quad x \mapsto \sum_{a\in G_r} [a]_r \otimes (x, \langle a\rangle\beta_{r,\theta})_\theta,$$

where $[a]_r \in \Lambda_{r,\theta}$ denotes the group element for a.[8] Let $\xi_{r,\theta}$ be the image of ξ_θ in $\Lambda_{r,\theta}$. As $\xi_{r,\theta}\beta_{r,\theta} = 0$, the image of the map ϕ is contained in the group $(\Lambda_{r,\theta} \otimes \mathbb{Q}_p/\mathbb{Z}_p(1))[\xi_{r,\theta}]$ of $\xi_{r,\theta}$-torsion, and ϕ factors through the quotient $\mathcal{T}_{r,\theta}/I_{r,\theta}\mathcal{T}_{r,\theta}$.

---

[6] Another, more usual, way to approach injectivity is to use $I_\theta + \xi_\theta\mathfrak{h}_\theta$ in place of I_θ until one recovers the equality of these ideals through a proof of the main conjecture, as in 2.5.7 below.

[7] In the projective limit, this gives another way of defining the isomorphism $\mathfrak{h}_\theta/I_\theta \xrightarrow{\sim} \Lambda_\theta/(\xi_\theta)$.

[8] To make sense of this, note that the tensor product in the sum is taken over $\mathbb{Z}_p[\theta]$.

Consider the composition

$$\mathfrak{T}_{r,\theta}/I_{r,\theta}\mathfrak{T}_{r,\theta} \xrightarrow{\phi} (\Lambda_{r,\theta} \otimes \mathbb{Q}_p/\mathbb{Z}_p(1))[\xi_{r,\theta}] \to (\Lambda_{r,\theta}/\xi_{r,\theta})(1) \xrightarrow{\iota_{r,\theta}} (\mathfrak{h}_{r,\theta}/I_{r,\theta})(1),$$

where the second map is given by $x \mapsto \xi_{r,\theta}\tilde{x}$ for any lifting $\tilde{x}$ of x to $\mathbb{Q}_p[G_r]_\theta(1)$. It is surjective by our description of $B_{r,\theta}$ and the perfectness of $(\ ,\ )_\theta$. As seen from the Galois action on $\beta_{r,\theta}$, it is moreover an $\mathfrak{h}_r[G_\mathbb{Q}]$-module homomorphism $\mathfrak{T}_{r,\theta}/I_{r,\theta}\mathfrak{T}_{r,\theta} \to (\mathfrak{h}_{r,\theta}/I_{r,\theta})^\sharp(1)$. The maps are compatible with r, and their projective limit is the desired surjective $\mathfrak{h}[G_\mathbb{Q}]$-module homomorphism $\mathfrak{T}_\theta/I_\theta\mathfrak{T}_\theta \to Q_\theta$.

2.5.5. We explain how the surjection of 2.5.4 fits in an exact sequence

$$0 \to P_\theta \to \mathfrak{T}_\theta/I_\theta\mathfrak{T}_\theta \to Q_\theta \to 0$$

of $\mathfrak{h}[G_\mathbb{Q}]$-modules that is canonically locally split over $G_{\mathbb{Q}_p}$.

We use the fact that the Eisenstein part $\mathfrak{T}^+_{\mathrm{Eis}} \to \mathfrak{T}_{\mathrm{quo,Eis}}$ of the canonical map of 2.4.4 is an isomorphism, or equivalently, that $\mathfrak{T}_{\mathrm{sub,Eis}} \to \mathfrak{T}_{\mathrm{Eis}}/\mathfrak{T}^+_{\mathrm{Eis}}$ is an isomorphism. To see this, one uses an $\mathfrak{h}$-module splitting of the local exact sequence for $\mathfrak{T}_\theta$ (see Ohta 2000) and the method of Kurihara (1993) and Harder and Pink (1992). We refer the reader to (Fukaya and Kato Preprint, Section 6.3) for the argument.

Let us explain the use of this fact: by definition, complex conjugation acts on Q_θ by multiplication by -1. Thus, Q_θ is a quotient of $\mathfrak{T}_\theta/\mathfrak{T}^+_\theta$. By our isomorphism on Eisenstein components, it is a quotient of $\mathfrak{T}_{\mathrm{sub},\theta}/I_\theta\mathfrak{T}_{\mathrm{sub},\theta}$, which by 2.4.6 is isomorphic to $\mathfrak{h}_\theta/I_\theta$ as an $\mathfrak{h}$-module. This forces the quotient map to be an injection, so we have $Q_\theta \cong \mathfrak{T}_{\mathrm{sub},\theta}/I_\theta\mathfrak{T}_{\mathrm{sub},\theta}$. But now, this tells us that Q_θ is an $\mathfrak{h}[G_{\mathbb{Q}_p}]$-submodule of $\mathfrak{T}_\theta/I_\theta\mathfrak{T}_\theta$. In other words, the surjection $\mathfrak{T}_\theta/I_\theta\mathfrak{T}_\theta \to Q_\theta$ is canonically locally split on $G_{\mathbb{Q}_p}$. We then have necessarily that the kernel of the latter surjection is $\mathfrak{T}_{\mathrm{quo},\theta}/I_\theta\mathfrak{T}_{\mathrm{quo},\theta} \cong P_\theta$. This yields the exact sequence.

It is perhaps worth observing that this sequence is also identified with the reduction modulo I_θ of the exact sequence of $\mathfrak{h}[G_\mathbb{R}]$-modules

$$0 \to \mathfrak{T}^+_\theta \to \mathfrak{T}_\theta \to \mathfrak{T}_\theta/\mathfrak{T}^+_\theta \to 0.$$

Finally, the determinant of the $G_\mathbb{Q}$-action on $\mathfrak{T}_\theta$ is known (e.g., from the determinants of modular Galois representations) and agrees with the $G_\mathbb{Q}$-action on Q_θ, so the $G_\mathbb{Q}$-action on P_θ is trivial.

2.5.6. We have that P_θ and Q_θ have trivial actions of G_L. Hence, we have a homomorphism

$$G_L \to \mathrm{Hom}_\mathfrak{h}(Q_\theta, P_\theta), \qquad \sigma \mapsto (x \mapsto \sigma\tilde{x} - \tilde{x}),$$

where $\tilde{x}$ is a lifting of x to $\mathfrak{T}_\theta/I_\theta\mathfrak{T}_\theta$. By 2.5.5, this homomorphism factors through the unramified quotient X of G_L. Thus, we have a homomorphism $X \to$

$\mathrm{Hom}_{\mathfrak{h}}(Q_\theta, P_\theta)$ that is compatible with the action of $\mathrm{Gal}(L/\mathbb{Q})$. This gives a homomorphism of $\mathrm{Gal}(K/\mathbb{Q})$-modules

$$X^-(1) \to \mathrm{Hom}_{\mathfrak{h}}(Q_\theta(-1), P_\theta) \cong \mathrm{Hom}_{\mathfrak{h}}((\mathfrak{h}_\theta/I_\theta)^\sharp, P_\theta) \cong P_\theta^\flat,$$

where $P_\theta^\flat$ is P_θ on which $\sigma_a \in \mathrm{Gal}(L/\mathbb{Q})$ acts as multiplication by $\langle a\rangle^{-1}$. In other words, we have a Λ_θ-module homomorphism

$$\Upsilon: Y_\theta = X^-(1)_\theta \to P_\theta,$$

with the Galois action of G on the left and inverse diamond action of G on the right.

2.5.7. We describe the heart of the Mazur-Wiles proof of the Iwasawa main conjecture.

The Iwasawa main conjecture is the equality of ideals

$$\mathrm{char}_{\Lambda_\theta}(Y_\theta) = (\xi_\theta).$$

By the analytic class number formula, this conjecture is reduced to $\mathrm{char}_{\Lambda_\theta}(Y_\theta) \subseteq (\xi_\theta)$.

Let $\mathcal{L}$ be the $\mathfrak{h}[G_\mathbb{Q}]$-submodule of $\mathcal{T}_\theta$ generated by $\mathcal{T}_{\mathrm{sub},\theta}$. It follows as in 2.5.5 that we have an equality $\mathcal{L}_\mathfrak{m} = \mathcal{T}_{\mathrm{sub},\mathfrak{m}} \oplus \mathcal{L}_\mathfrak{m}^+$ of Eisenstein components. Moreover, $P'_\theta = \mathcal{L}^+/I_\theta\mathcal{L}^+$ is $G_\mathbb{Q}$-stable in $\mathcal{L}/I_\theta\mathcal{L}$. In other words, we have an exact sequence of $\mathfrak{h}[G_\mathbb{Q}]$-modules

$$0 \to P'_\theta \to \mathcal{L}/I_\theta\mathcal{L} \to Q_\theta \to 0.$$

In the same way as Υ, we may define $\Upsilon': Y_\theta \to P'_\theta$, which is now surjective by construction.

The Iwasawa main conjecture can be deduced from this surjectivity of Υ'. More precisely, we use the following facts:

(1) The map $\Upsilon': Y_\theta \to P'_\theta$ is surjective.
(2) We have that $P'_\theta = \mathcal{L}^+/I_\theta\mathcal{L}^+$ with $\mathcal{L}^+$ a finitely generated, faithful $\mathfrak{h}_\theta$-module.
(3) The kernel of the canonical surjection $\Lambda_\theta \to \mathfrak{h}_\theta/I_\theta$ is contained in (ξ_θ).[9]

From (1), we obtain

$$\mathrm{char}_{\Lambda_\theta}(Y_\theta) \subseteq \mathrm{char}_{\Lambda_\theta}(P'_\theta).$$

From (2) and (3), we can deduce that

$$\mathrm{char}_{\Lambda_\theta}(P'_\theta) \subseteq \mathrm{char}_{\Lambda_\theta}(\Lambda_\theta/(\xi_\theta)) = (\xi_\theta).$$

---

[9] Actually, we know that the kernel coincides with (ξ_θ), but this weaker statement is enough.

Hence $\mathrm{char}_{\Lambda_\theta}(Y_\theta) \subseteq (\xi_\theta)$.

2.5.8. The conjecture stated in the following section implies that Υ is surjective. This tells us that the inexplicit lattice $\mathcal{L}$ required for the Mazur-Wiles proof in 2.5.7 is precisely the canonical lattice $\mathcal{T}_\theta$. In this sense, it suggests a refinement of the method of Ribet and Mazur-Wiles.

2.6 The Conjecture: ϖ and Υ Are Inverse Maps

We state the conjecture of the third author (Sharifi 2011) and the result of the first two authors (Fukaya and Kato Preprint).

2.6.1. In 2.2.4 and 2.5.6, we defined Λ-module homomorphisms

$$\varpi\colon P_\theta \to X^-(1)_\theta \quad \text{and} \quad \Upsilon\colon X^-(1)_\theta \to P_\theta.$$

We have the conjecture of the third author. See (Sharifi 2011, Conjecture 4.12), where the conjecture is given up to a canonical unit; this stronger version was a stated hope of the third author.

Conjecture. The maps ϖ and Υ are inverse to each other.

This conjecture provides an explicit description of $X^-(1)_\theta$ in terms of modular symbols. In this sense, it may be viewed as a refinement of the main conjecture.

2.6.2. We state the result (Fukaya and Kato Preprint, Theorem 7.2.3(1)) of the first two authors. Let $\xi'_\theta \in \Lambda_\theta$ denote the derivative of the p-adic L-function ξ_θ in the s-variable (see 2.5.2).

Theorem. *We have $\xi'_\theta \Upsilon \circ \varpi = \xi'_\theta$ modulo p-torsion in P_θ.*

If ξ_θ has no multiple roots, the theorem implies the conjecture up to p-torsion in P_θ. In fact, it leads to proofs of the conjecture under various hypotheses: see (Fukaya and Kato Preprint, Section 7.2).

2.6.3. McCallum and the third author conjectured that the image of the cup product

$$H^1(\mathbb{Z}[\zeta_{Np^r}, \tfrac{1}{Np}], \mathbb{Z}_p(1)) \otimes_{\mathbb{Z}_p} H^1(\mathbb{Z}[\zeta_{Np^r}, \tfrac{1}{Np}], \mathbb{Z}_p(1)) \xrightarrow{\cup} H^2_{\text{ét}}(\mathbb{Z}[\zeta_{Np^r}, \tfrac{1}{Np}], \mathbb{Z}_p(2))$$

projects onto $H^2_{\text{ét}}(\mathbb{Z}[\zeta_{Np^r}, \frac{1}{Np}], \mathbb{Z}_p(2))^+_\theta$ (McCallum and Sharifi 2003 for $N = 1$), which implies that ϖ is surjective. This generation conjecture follows if we know that $\varpi \circ \Upsilon = 1$. In particular, it holds if ξ_θ has no multiple roots, and it also holds if $P_\theta \otimes_{\mathbb{Z}_p} \mathbb{Q}_p$ is generated by one element over $\Lambda_\theta \otimes_{\mathbb{Z}_p} \mathbb{Q}_p$ (Fukaya and Kato Preprint, Theorem 7.2.8).

2.7 The Proof that $\xi'\Upsilon \circ \varpi = \xi'$

We explain some of the important aspects of the proof of the main theorem, referring to the relevant sections of (Fukaya and Kato Preprint) for details.

2.7.1. We consider a refinement of the diagram in 1.10 in which we divide the right-hand square of that diagram into two squares:

$$\begin{array}{ccccccc}
\mathcal{S}_\theta & \xrightarrow{z} & \varprojlim_r H^2_{\text{ét}}(Y_1(Np^r), \mathbb{Z}_p(2))_\theta & \overset{\text{HS}}{\dashrightarrow} & H^1_{\text{ét}}(\mathbb{Z}[\frac{1}{Np}], \mathcal{T}_\theta(1)) & \xrightarrow{\text{reg}} & \mathfrak{S}_\theta \\
\downarrow{\scriptstyle \bmod I} & & \downarrow{\scriptstyle \infty} & & \downarrow & & \downarrow{\scriptstyle \bmod I} \\
P_\theta & \xrightarrow{\varpi} & Y_\theta & \xrightarrow{\xi'_\theta} & Y_\theta & \xrightarrow{\Upsilon} & P_\theta.
\end{array}$$

Here, the maps z and ∞ are the projective limits of the θ-components of the maps z_r and ∞_r. The commutativity of the left square of the diagram in 2.7.1 is seen in Sect. 2.3. The discussion of the rest of this diagram, and the fact that the bottom row is also multiplication by ξ'_θ, compose the rest of this subsection.

It is remarkable that ξ'_θ appears here in two very different contexts. The ξ'_θ that appears in the diagram and contributes to $\xi'_\theta\Upsilon \circ \varpi$ is related to cup product with the logarithm of the cyclotomic character. The other ξ'_θ is the constant term modulo ξ_θ of a Λ-adic modular form that appears in a computation of the regulators of zeta elements.

2.7.2. The map HS arises from the Hochschild-Serre spectral sequences

$$E_2^{i,j} = H^i_{\text{ét}}(\mathbb{Z}[\tfrac{1}{Np}], H^j_{\text{ét}}(Y_1(Np^r)_{/\overline{\mathbb{Q}}}, \mathbb{Z}_p(2))) \Rightarrow E^{i+j} = H^{i+j}_{\text{ét}}(Y_1(Np^r), \mathbb{Z}_p(2)).$$

as the projective limit over r of maps $E^2 \to E_2^{1,1}$, followed by projection to the ordinary θ-part. We remark that

$$H^1_{\text{ét}}(\mathbb{Z}[\tfrac{1}{Np}], \mathcal{T}_\theta(1)) \subset H^1_{\text{ét}}(\mathbb{Z}[\tfrac{1}{Np}], \tilde{\mathcal{T}}_\theta(1)),$$

and the image of HS is actually contained in the larger group, hence the dotted arrow. However, elements of $\mathcal{S}_\theta$ are carried to the smaller group under HS $\circ z$ (Fukaya and Kato Preprint, Proposition 3.3.14), so we can still make sense of the diagram.

2.7.3. The third vertical arrow in the diagram of 2.7.1 is a composition of maps as follows:

$$H^1_{\text{ét}}(\mathbb{Z}[\tfrac{1}{Np}], \mathcal{T}_\theta(1)) \to H^1_{\text{ét}}(\mathbb{Z}[\tfrac{1}{Np}], Q_\theta(1)) \xrightarrow{\cup(1-p^{-1})\log(\kappa)} H^2_{\text{ét}}(\mathbb{Z}[\tfrac{1}{Np}], Q_\theta(1)) \xrightarrow{\sim} Y_\theta.$$

The first map is induced by the surjection $\mathcal{T}_\theta \to Q_\theta$. The map $\kappa\colon G_{\mathbb{Q}} \to \mathbb{Z}_p^\times$ is the p-adic cyclotomic character, and log is the p-adic logarithm. The second map is the cup product, where we regard $\log(\kappa) = \log \circ \kappa$ as an element of $H^1_{\text{ét}}(\mathbb{Z}[\frac{1}{p}], \mathbb{Z}_p)$.

For the third map, note that $Q_\theta \cong (\Lambda_\theta/\xi_\theta)^\sharp(1)$ by 2.5.3 and 2.5.4. As the p-cohomological dimension of $\mathbb{Z}[\frac{1}{Np}]$ is 2, the group $H^2_{\text{ét}}(\mathbb{Z}[\frac{1}{Np}], (\Lambda_\theta/\xi_\theta)^\sharp(2))$ is isomorphic to the quotient of

$$H^2_{\text{ét}}(\mathbb{Z}[\tfrac{1}{Np}], \Lambda_\theta^\sharp(2))_\theta \xrightarrow{\sim} \varprojlim_r H^2_{\text{ét}}(\mathcal{O}_r[\tfrac{1}{Np}], \mathbb{Z}_p(2))_\theta \xrightarrow{\sim} Y_\theta$$

by the $\mathrm{Gal}(K/\mathbb{Q})$-action of ξ_θ. Here, the first isomorphism is by Shapiro's lemma, and the second is from 2.2.4. By the main conjecture and the fact that Y_θ has no finite Λ-submodules, Y_θ is ξ_θ-torsion, so $Y_\theta/\xi_\theta Y_\theta = Y_\theta$. Putting this all together, we have the map.

2.7.4. We define a functor D on pro-p $G_{\mathbb{Q}_p}$-modules.

Let $T = \varprojlim_\lambda T_\lambda$ for a projective system of finite abelian p-groups T_λ. For any abelian group M, set $T \mathbin{\hat{\otimes}} M = \varprojlim_\lambda (T_\lambda \otimes M)$. Let W denote the Witt vectors of $\overline{\mathbb{F}}_p$, and suppose that the T_λ are endowed with compatible actions of $h[G_{\mathbb{Q}_p}]$ for a pro-p ring h. We may then consider the h-module $D(T)$ that is the fixed part

$$D(T) = (T \mathbin{\hat{\otimes}} W)^{G_{\mathbb{Q}_p}}$$

for the diagonal action of $G_{\mathbb{Q}_p}$ on $T \mathbin{\hat{\otimes}} W$. If the $G_{\mathbb{Q}_p}$-actions on the T_λ are unamified, then $D(T)$ and T are isomorphic h-modules. If T has trivial $G_{\mathbb{Q}_p}$-action, then $D(T) \cong T \mathbin{\hat{\otimes}} W^{G_{\mathbb{Q}_p}} = T \mathbin{\hat{\otimes}} \mathbb{Z}_p \cong T$, and this isomorphism is canonical. See (Fukaya and Kato Preprint, Section 1.7).

2.7.5. We define p-adic regulator maps for unramified, pro-p $G_{\mathbb{Q}_p}$-modules.

Let T be as in 2.7.4, and suppose that the action of $G_{\mathbb{Q}_p}$ on T is unramified. Let $E = Q(W)$ be the maximal unramified extension of $\mathbb{Q}_p$. The p-adic regulator map (Fukaya and Kato Preprint, Section 4.2)

$$\mathrm{reg}_T \colon H^1_{\text{ét}}(\mathbb{Q}_p, T(1)) \to D(T)$$

for T is the h-module homomorphism defined as the composition

$$H^1_{\text{ét}}(\mathbb{Q}_p, T(1)) \xrightarrow{\text{inf}} H^1_{\text{ét}}(E, T(1))^{\mathrm{Fr}_p=1} \xrightarrow{\sim} (T \mathbin{\hat{\otimes}} E^\times)^{\mathrm{Fr}_p=1} \to D(T).$$

Here, the first map is inflation, the second is Kummer theory, and the final map is induced by

$$E^\times \to W(\overline{\mathbb{F}}_p), \quad x \mapsto p^{-1} \log\left(\frac{x^p}{\mathrm{Fr}_p(x)}\right),$$

where the p-adic logarithm log is defined to take p to 0.

Note that if $G_{\mathbb{Q}_p}$ acts trivially on T, then reg_T is induced by the map $(1 - p^{-1}) \log \colon \mathbb{Q}_p^\times \to \mathbb{Z}_p$ in a similar fashion.

2.7.6. We define the p-adic regulator map reg in the diagram of 2.7.1.

Note that $\mathfrak{T}_{\mathrm{quo}}$ has by definition an unramified $G_{\mathbb{Q}_p}$-action. We have a refinement (Fukaya and Kato Preprint, Section 1.7) of Ohta's Λ-adic Eichler-Shimura isomorphism (Ohta 1995). That is, there is a canonical isomorphism of $\mathfrak{h}$-modules $D(\mathfrak{T}_{\mathrm{quo}}) \xrightarrow{\sim} \mathfrak{S}$, and in particular $\mathfrak{T}_{\mathrm{quo}}$ and $\mathfrak{S}$ are noncanonically isomorphic. The map reg is then defined as the composition

$$\mathrm{reg}\colon H^1_{\text{ét}}(\mathbb{Z}[\tfrac{1}{Np}], \mathfrak{T}_\theta(1)) \to H^1_{\text{ét}}(\mathbb{Q}_p, \mathfrak{T}_{\mathrm{quo},\theta}(1)) \xrightarrow{\mathrm{reg}_{\mathfrak{T}_{\mathrm{quo}}}} D(\mathfrak{T}_{\mathrm{quo},\theta}) \xrightarrow{\sim} \mathfrak{S}_\theta.$$

2.7.7. We explain the right-hand vertical map "mod I" in the diagram of 2.7.1.

The $G_{\mathbb{Q}_p}$-action on $\mathfrak{T}_{\mathrm{quo}}/I\mathfrak{T}_{\mathrm{quo}}$ is trivial, so the canonical isomorphism $D(\mathfrak{T}_{\mathrm{quo}}) \xrightarrow{\sim} \mathfrak{S}$ provides an isomorphism $\mathfrak{T}_{\mathrm{quo}}/I\mathfrak{T}_{\mathrm{quo}} \xrightarrow{\sim} \mathfrak{S}/I\mathfrak{S}$. In particular, we obtain "mod I" as the composition of projection followed by a string of canonical isomorphisms:

$$\mathfrak{S}_\theta \twoheadrightarrow \mathfrak{S}_\theta/I_\theta\mathfrak{S}_\theta \xrightarrow{\sim} \mathfrak{T}_{\mathrm{quo},\theta}/I_\theta\mathfrak{T}_{\mathrm{quo},\theta} \xrightarrow{\sim} \mathfrak{T}^+_\theta/I_\theta\mathfrak{T}^+_\theta \xrightarrow{\sim} \mathcal{S}_\theta/I_\theta\mathcal{S}_\theta = P_\theta.$$

2.7.8. The commutativity of the two right-hand squares in the diagram of 2.7.1 are nontrivial cohomological exercises. We mention only which calculations must in the end be performed.

(1) The commutativity of the middle square is reduced to that (see Fukaya and Kato Preprint, Section 9.4) of

$$\begin{array}{ccc} H^1_{\text{ét}}(\mathbb{Z}[\tfrac{1}{Np}], (\Lambda_\theta/\xi_\theta)^\sharp(2)) & \xrightarrow{\cup(1-p^{-1})\log(\kappa)} & H^2_{\text{ét}}(\mathbb{Z}[\tfrac{1}{Np}], (\Lambda_\theta/\xi_\theta)^\sharp(2)) \\ \downarrow \wr & & \uparrow \wr \\ H^2_{\text{ét}}(\mathbb{Z}[\tfrac{1}{Np}], \Lambda^\sharp_\theta(2)) & \xrightarrow{\xi'_\theta} & H^2_{\text{ét}}(\mathbb{Z}[\tfrac{1}{Np}], \Lambda^\sharp_\theta(2)), \end{array}$$

the vertical arrows occurring in the long exact sequence in the $\mathbb{Z}[\frac{1}{Np}]$-cohomology of

$$0 \to \Lambda^\sharp_\theta(2) \to \Lambda^\sharp_\theta(2) \to (\Lambda_\theta/\xi_\theta)^\sharp(2) \to 0.$$

Thus, the ξ'_θ that appears in the diagram is found in Galois cohomology.

(2) The commutativity of the right-hand square is reduced to verifying that the map

$$Y_\theta \cong H^2_{\text{ét}}(\mathbb{Z}[\tfrac{1}{Np}], Q_\theta(1)) \leftarrow H^2_{\text{ét}}(\mathbb{Z}[\tfrac{1}{Np}], \mathfrak{T}_\theta/I_\theta\mathfrak{T}_\theta(1)) \to H^2_{\text{ét}}(\mathbb{Q}_p, P_\theta(1)) \cong P_\theta$$

given by lifting and then projecting is well-defined and agrees with Υ (Fukaya and Kato Preprint, Section 9.5). Here, the first isomorphism was discussed in 2.7.3 and the last is the invariant map of local class field theory, recalling from 2.5.5 that P_θ has trivial Galois action. This description is closer to the construction of Υ that will appear for $\mathbb{F}_q(t)$ in Sect. 3.

2.7.9. It remains to prove that the composition $\mathcal{S}_\theta/I_\theta\mathcal{S}_\theta \to \mathfrak{S}_\theta/I_\theta\mathfrak{S}_\theta \to P_\theta$, where the first arrow is the composition of the upper horizontal arrows modulo I_θ in the diagram of 2.7.1, coincides with multiplication by ξ'_θ on P_θ. This is deduced in (Fukaya and Kato Preprint, Sections 4.3 and 8.1) from the computation of the p-adic regulators of zeta elements given in (Fukaya 2003; Ochiai 2006). This is a very delicate analysis: we explain only the rough idea of how ξ'_θ appears at its end.

The map $P_\theta \to P_\theta$ is shown to be given (modulo ξ_θ) by multiplication by the constant term at $t = 1$ of a p-adic L-function in a variable t that takes values in $\mathfrak{M}_\theta$. This p-adic L-function is a product of two Λ-adic Eisenstein series which vary with t. The constant term in the q-expansion of this product is itself a product of two zeta functions $\zeta_p(t)\xi_\theta(s+t-1)$, where $\zeta_p(t)$ is the p-adic Riemann zeta function and s is the variable for $\Lambda_\theta \subset \mathfrak{h}_\theta$. Note that $\zeta_p(t)$ has a simple pole at $t = 1$ with residue 1. To evaluate $\zeta_p(t)\xi_\theta(s+t-1)$ modulo $\xi_\theta(s)$ at $t = 1$, we can first subtract $\zeta_p(t)\xi_\theta(s)$ from the product and then take the resulting limit

$$\lim_{t\to 1}\frac{\xi_\theta(s+t-1)-\xi_\theta(s)}{t-1} = \xi'_\theta(s).$$

In this manner, the map is shown to be multiplication by ξ'_θ.

3 The Case of GL_2 over $\mathbb{F}_q(t)$

We now consider the field $F = \mathbb{F}_q(t)$ for some prime power q. In this section, we provide F-analogues of the constructions, conjecture, and theorem of Sect. 2. We require the following objects:

- The ring $\mathcal{O} = \mathbb{F}_q[t]$,
- The completion $F_\infty = F((t^{-1}))$ of F at the place ∞,
- The valuation ring $\mathcal{O}_\infty = \mathbb{F}_q[\![t^{-1}]\!]$ of F_∞, which does not contain $\mathcal{O}$,
- A prime number p different from the characteristic of $\mathbb{F}_q$,
- A non-constant polynomial $N \in \mathcal{O}$.

Let us also fix an embedding $\overline{F} \hookrightarrow \overline{F_\infty}$ of separable closures. To avoid technical complications, we assume in this section that p does not divide $(q+1)|(\mathcal{O}/N\mathcal{O})^\times|$.

The organization of this section follows closely that of Sect. 2. We hope to make clear that most constructions are remarkably similar to the case of $\mathbb{Q}$, though we also highlight differences. We work with congruence subgroups of $\mathrm{GL}_2(\mathcal{O})$, rather than of $\mathrm{SL}_2(\mathbb{Z})$. Modular symbols, used to construct ϖ, are now found in the homology $\mathcal{S}$ of the compactification of the quotient of the Bruhat-Tits tree by a congruence subgroup. This $\mathcal{S}$ is a quotient of an étale homology group $\mathcal{T}$ of the Drinfeld modular curve, used in constructing Υ. As most constructions are so similar, we provide less detail than in Sect. 2. We intend for full details to appear in a forthcoming paper.

3.1 From Modular Symbols to Cup Products: The Map ϖ

3.1.1. We introduce homology groups $\mathcal{S}$ and $\mathcal{M}$ of the Bruhat-Tits tree.

Consider the Bruhat-Tits tree B for $\mathrm{PGL}_2(F_\infty)$. Its vertices are homothety classes of $\mathcal{O}_\infty$-lattices $\mathcal{L}$ of rank 2 in F_∞^2, or equivalently, elements of $\mathrm{PGL}_2(F_\infty)/\mathrm{PGL}_2(\mathcal{O}_\infty)$. This tree is $(q+1)$-valent, and two lattices $\mathcal{L} \subset \mathcal{L}'$ connected by an edge if $[\mathcal{L}' : \mathcal{L}] = q$.[10] The oriented edges then correspond to elements of $\mathrm{PGL}_2(F_\infty)/\mathcal{I}_\infty$, where $\mathcal{I}_\infty$ is the Iwahori subgroup of matrices in $\mathrm{PGL}_2(\mathcal{O}_\infty)$ that are upper-triangular modulo the maximal ideal of $\mathcal{O}_\infty$. The group $\mathrm{PGL}_2(F)$ acts on the left on B in the evident manner.

Let $\tilde{\Gamma}_1(N)$ be the congruence subgroup of $\mathrm{GL}_2(\mathcal{O})$ given by

$$\tilde{\Gamma}_1(N) = \left\{ \begin{pmatrix} a & b \\ c & d \end{pmatrix} \in \mathrm{GL}_2(\mathcal{O}) \;\middle|\; (c,d) \equiv (0,1) \bmod N \right\}.$$

We may complete the Bruhat-Tits tree to a space B^* by adding in the (rational) ends, which correspond to elements of $\mathbb{P}^1(F)$. We define

$$U(N) = \tilde{\Gamma}_1(N) \backslash B \quad \text{and} \quad \overline{U}(N) = \tilde{\Gamma}_1(N) \backslash B^*.$$

The elements of $\tilde{\Gamma}_1(N) \backslash \mathbb{P}^1(F)$ are the ends of $U(N)$. Our homology groups, or spaces of modular symbols, are then

$$\mathcal{S} = H_1(\overline{U}(N), \mathbb{Z}_p) \subset \mathcal{M} = H_1(\overline{U}(N), \{\text{ends}\}, \mathbb{Z}_p).$$

3.1.2. We introduce Manin-Teitelbaum symbols $[u : v] \in \mathcal{M}$.

Modular symbols in $\mathcal{M}$ were defined by Teitelbaum (1992) analogously to the case of $\mathbb{Q}$. In particular, given $\alpha, \beta \in \mathbb{P}^1(F)$, we have a modular symbol that is the class $\{\alpha \to \beta\}$ of any non-backtracking path in the Bruhat-Tits tree that connects the two corresponding ends of B.

Analogues of Manin symbols are defined as before. That is, for $u, v \in \mathcal{O}/N\mathcal{O}$ with $(u,v) = (1)$, we choose $\gamma = \left(\begin{smallmatrix} a & b \\ c & d \end{smallmatrix}\right) \in \mathrm{GL}_2(\mathcal{O})$ with $u = c \bmod N$ and $v = d \bmod N$, and then

$$[u : v] = \left\{ \frac{d}{bN} \to \frac{c}{aN} \right\}.$$

[10] Note that q appears in this sentence as the order of the residue field of $\mathcal{O}_\infty$.

These symbols generate $\tilde{H}$ and yield a presentation with identical relations to those of 2.1.2.

3.1.3. We introduce the intermediate space $\mathcal{M}^0$ on which we define ϖ.

Let $\mathcal{M}^0$ denote the $\mathbb{Z}_p$-submodule of $\mathcal{M}$ generated by the Manin symbols $[u : v]$ with $u, v \neq 0$. As in the case of GL_2 over $\mathbb{Q}$, we have $\mathcal{S} \subset \mathcal{M}^0 \subset \mathcal{M}$.

3.1.4. We introduce cyclotomic N-units $\lambda_{\frac{u}{N}}$ in abelian extensions $F_N \subset E_N$ of $\mathbb{F}_q(t)$. The reader may find a powerful analogy with objects in the theory of cyclotomic fields over $\mathbb{Q}$.

We consider the cyclotomic N-units $\lambda_{\frac{u}{N}}$ for $u \in \mathcal{O} - (N)$. These are the roots of the Carlitz polynomials (Carlitz 1938) for divisors of N, or are equivalently the N-torsion points of the Carlitz module. As $\lambda_{\frac{u}{N}}$ depends only on u modulo N, we abuse notation and consider it for nonzero $u \in \mathcal{O}/N\mathcal{O}$. We can visualize $\lambda_{\frac{u}{N}}$ in the completion $\mathbb{C}_\infty$ of $\overline{F_\infty}$ by

$$\lambda_{\frac{u}{N}} = \exp\left(\frac{u\pi}{N}\right) = \frac{u}{N} \prod_{a \in \mathcal{O}-\{0\}} \left(1 - \frac{u}{Na}\right),$$

where exp is the Carlitz exponential and $\pi \in \mathbb{C}_\infty$ is transcendental over F.

Let $E_N = F(\lambda_{\frac{1}{N}})$, which is an abelian extension of F of conductor $N\infty$ containing no constant field extension of F. There is an isomorphism $\mathrm{Gal}(E_N/F) \xrightarrow{\sim} (\mathcal{O}/N\mathcal{O})^\times$ such that $a \in (\mathcal{O}/N\mathcal{O})^\times$ is the image of an element $\sigma_a \in \mathrm{Gal}(E_N/F)$ that satisfies $\sigma_a(\lambda_{\frac{1}{N}}) = \lambda_{\frac{a}{N}}$. Let F_N be the largest subfield of E_N in which ∞ splits completely over F, which we might call the ray class field of modulus N. Under the above isomorphism, $\mathrm{Gal}(E_N/F_N)$ is identified with $\mathbb{F}_q^\times$. In fact, we have $\sigma_c(\lambda_{\frac{u}{N}}) = c\lambda_{\frac{u}{N}}$ for $c \in \mathbb{F}_q^\times$. These facts are found in the work of Hayes (1974).

Let $\mathcal{O}_N$ denote the integral closure of $\mathcal{O}$ in F_N. Since $p \nmid (q-1)$ by assumption, the image of $\lambda_{\frac{u}{N}}$ in the p-completion of the N-units of E_N is fixed by the action of $\mathbb{F}_q^\times$. This allows us to view $\lambda_{\frac{u}{N}}$ as an element of $H^1_{\text{ét}}(\mathcal{O}_N[\frac{1}{N}], \mathbb{Z}_p(1))$. For nonzero $u, v \in \mathcal{O}/N\mathcal{O}$, we may consider the cup product

$$\lambda_{\frac{u}{N}} \cup \lambda_{\frac{v}{N}} \in H^2_{\text{ét}}(\mathcal{O}_N[\tfrac{1}{N}], \mathbb{Z}_p(2)).$$

3.1.5. We define the map ϖ. Here, we work directly with étale cohomology, rather than K_2.

There is a homomorphism

$$\varpi : \mathcal{M}^0 \to H^2_{\text{ét}}(\mathcal{O}_N[\tfrac{1}{N}], \mathbb{Z}_p(2)), \quad [u : v] \mapsto \lambda_{\frac{u}{N}} \cup \lambda_{\frac{v}{N}}.$$

In the current setting, we can no longer quickly verify from the presentation of $\mathcal{M}^0$ that ϖ is well-defined. Rather, we see this as a consequence of the argument that ϖ is "Eisenstein" in Sect. 3.3.

3.1.6. We introduce the cuspidal Hecke algebra $\mathfrak{h}$ and its Eisenstein ideal I.

Let $\mathfrak{n}$ denote a nonzero ideal of $\mathcal{O}$. Through the action of $\mathrm{GL}_2(F)$ on B, we have a Hecke operator $T(\mathfrak{n})$ acting on $\mathcal{S}$ as the correspondence associated to $\tilde{\Gamma}_1(N)\begin{pmatrix}1&0\\0&n\end{pmatrix}\tilde{\Gamma}_1(N)$, where $\mathfrak{n}=(n)$. Let $\mathfrak{h}$ be the subring of $\mathrm{End}_{\mathbb{Z}_p}(\mathcal{S})$ generated over $\mathbb{Z}_p$ by the Hecke operators $T(\mathfrak{n})$.

We also have diamond operators $\langle\mathfrak{a}\rangle$ in $\mathfrak{h}$ for nonzero ideals $\mathfrak{a}$ of $\mathcal{O}$ prime to (N). This $\langle\mathfrak{a}\rangle$ depends only on the reduction modulo N of the monic generator of $\mathfrak{a}$.

The Eisenstein ideal I is the ideal of $\mathfrak{h}$ generated by $T(\mathfrak{n})-\sum_{\mathfrak{d}|\mathfrak{n}}\mathfrak{N}(\mathfrak{d})\langle\mathfrak{d}\rangle$ for all nonzero ideals $\mathfrak{n}$ of $\mathcal{O}$, taking $\langle\mathfrak{d}\rangle=0$ if $\mathfrak{d}+(N)\neq(1)$. Here, $\mathfrak{N}(\mathfrak{n})=[\mathcal{O}:\mathfrak{n}]$ is the absolute norm of $\mathfrak{n}$.

Similarly, we have the Eisenstein ideal $\mathfrak{I}$ of the Hecke algebra $\mathfrak{H}\subset\mathrm{End}_{\mathbb{Z}_p}(\mathcal{M})$.

3.1.7. To say that ϖ is "Eisenstein" is to say that ϖ factors through a map

$$\varpi\colon\mathcal{M}^0/\mathfrak{I}\mathcal{M}^0\to H^2_{\text{ét}}(\mathcal{O}_N[\tfrac{1}{N}],\mathbb{Z}_p(2)).$$

We explain this result in Sect. 3.3.

3.1.8. Let $G=(\mathcal{O}/N\mathcal{O})^\times/\mathbb{F}_q^\times$, and set $\Lambda=\mathbb{Z}_p[G]$.

(**1**) We have a ring homomorphism $\iota\colon\Lambda\to\mathfrak{h}$ which sends the group element $[a]$ in $\mathbb{Z}_p[G]$ for $a\in G$ to the inverse $\langle a\rangle^{-1}$ of the diamond operator corresponding to a.

(**2**) We have the isomorphism $\mathrm{Gal}(F_N/F)\xrightarrow{\sim}G$ of class field theory (see 3.1.4).

Modules over $\mathfrak{h}$ and $\mathbb{Z}_p[\mathrm{Gal}(F_N/F)]$ become Λ-modules through these identifications.

3.2 *Working with Fixed Level*

We explain why we work with fixed level in Sect. 3, and we define our two objects of study.

3.2.1. We do not pass up a tower for the following reason on the GL_1-side. By assumption on p, the field $\mathbb{F}_q$ has no nontrivial pth roots of unity. Since F_N/F contains no constant field extension, F_N also contains no nontrivial pth roots of unity. So, even if we "increase" N, we are unable to employ the Iwasawa-theoretic trick of passing Tate twists through projective limits of Galois cohomology groups. In particular, since we deal with cohomology with $\mathbb{Z}_p(2)$-coefficients, we do not work with class groups.

3.2.2. We again have two objects of study:

- The geometric object $P=\mathcal{S}/I\mathcal{S}$ for GL_2,
- The arithmetic object $Y=H^2_{\text{ét}}(\mathcal{O}_N,\mathbb{Z}_p(2))$ for GL_1.

Given a character $\theta: G \to \overline{\mathbb{Q}}_p^{\times}$, we set $\Lambda_\theta = \mathbb{Z}_p[\theta]$ and view it as a quotient of Λ through θ. For a Λ-module M, we let $M_\theta = M \otimes_\Lambda \Lambda_\theta$ denote the θ-part of M. If θ is primitive, then our assumption that p does not divide $|G|$ implies that the canonical maps

$$P_\theta \to \mathcal{M}_\theta^0/\mathfrak{I}_\theta\mathcal{M}_\theta^0 \quad \text{and} \quad Y_\theta \to H^2_{\text{ét}}(\mathcal{O}_N[\tfrac{1}{N}], \mathbb{Z}_p(2))_\theta$$

are isomorphisms.

3.3 Zeta Elements: ϖ Is "Eisenstein"

We explain that ϖ factors through the quotient of $\mathcal{M}^0$ by the Eisenstein ideal $\mathfrak{I}$.

3.3.1. We define Siegel units on Drinfeld modular curves.

Let $Y(N)$ denote the Drinfeld modular curve that is the moduli scheme for pairs consisting of a rank 2 Drinfeld module over an $\mathcal{O}[\frac{1}{N}]$-scheme and a full N-level structure (or, basis of the N-torsion) on it. Over $Y(N)$, we have a universal Drinfeld module, equipped with a full N-level structure, which locally looks like $(N^{-1}\mathcal{O}/\mathcal{O})^2 \times Y(N)$. On the universal Drinfeld module is a certain theta function Θ. Given an element of $(\frac{u_1}{N}, \frac{u_2}{N}) \in (N^{-1}\mathcal{O}/\mathcal{O})^2$, we may pull Θ back to a unit on the Drinfeld modular curve using the second coordinate of the level structure. This unit $g_{\alpha,\beta} \in \mathcal{O}_{Y(N)}^{\times}$ is the analogue of a Siegel unit.[11]

Let $Y_1(N)$ be the moduli scheme for pairs consisting of a rank 2 Drinfeld module over an $\mathcal{O}[\frac{1}{N}]$-scheme and a point of order N on it. If we take $\alpha = 0$, then the Siegel unit $g_{0,\beta}$ may again be viewed as an element of $\mathcal{O}_{Y_1(N)}^{\times}$.

3.3.2. If we take a K-theoretic product of two Siegel-type units, we obtain the Beilinson-type elements considered by Kondo and Yasuda (2012). See also the work of Kondo (2002) and Pal (2010). Much as in the case of $\mathbb{Q}$, we have a map

$$z: \mathcal{M}^0 \to H^2_{\text{ét}}(Y_1(N), \mathbb{Z}_p(2)), \qquad [u:v] \mapsto g_{0,\frac{u}{N}} \cup g_{0,\frac{v}{N}}$$

of $\mathfrak{H}$-modules. We can specialize this at the cusp corresponding to $\infty \in \mathbb{P}^1(F)$ to obtain $\lambda_{\frac{u}{N}} \cup \lambda_{\frac{v}{N}}$. This specialization map ∞ is Eisenstein. Hence, we see that

$$\varpi = \infty \circ z: \mathcal{M}^0 \to H^1_{\text{ét}}(\mathcal{O}_N[\tfrac{1}{N}], \mathbb{Z}_p(2))$$

is well-defined and Eisenstein.

[11] Actually, $g_{\alpha,\beta}$ as we have described it is not well-defined until we take its $q^2 - 1$ power. The assumption that $p \nmid (q^2 - 1)$ is used to avoid this issue when we work with étale cohomology.

3.4 Homology of Drinfeld Modular Curves

In this subsection, we study the étale homology groups of Drinfeld modular curves. Unlike in Sect. 2.4, we do not take ordinary parts. That is, the Galois representations found in the homology of Drinfeld modular curves are already "special at ∞," the required analogue of "ordinary at p." Moreover, the resulting unramfied-at-∞ quotient may in the present setting be identified with the space $\mathcal{S}$ of cuspidal symbols, which is the analogue of the plus quotient of homology of 2.2.2. In other words, the place ∞ of $\mathbb{F}_q(t)$ plays both the roles that p and the real place do in the GL_2-setting over $\mathbb{Q}$.

The statements in this subsection are consequences of the work of Drinfeld (1974).

3.4.1. We first introduce the étale homology group $\mathcal{T}$.

Over F, the Drinfeld modular curve $Y_1(N)_{/F}$ has a smooth compactification $X_1(N)_{/F}$. Over $\mathbb{C}_\infty$ (or $\overline{F}$), it is given by adding in the set of cusps $\tilde{\Gamma}_1(N)\backslash\mathbb{P}^1(F)$ of the Drinfeld upper half-plane. We define our étale homology group

$$\mathcal{T} = H_1^{\text{ét}}(X_1(N)_{/\overline{F}}, \mathbb{Z}_p)$$

as the $\mathbb{Z}_p$-dual of $H^1_{\text{ét}}(X_1(N)_{/\overline{F}}, \mathbb{Z}_p)$.

The Hecke algebra generated by the $T(\mathfrak{n})$ in $\mathrm{End}_{\mathbb{Z}_p}(\mathcal{T})$ is in fact equal to $\mathfrak{h}$. The module $\mathbb{Q}_p \otimes_{\mathbb{Z}_p} \mathcal{T}$ over the total quotient ring $\mathbb{Q}_p \otimes_{\mathbb{Z}_p} \mathfrak{h}$ of $\mathfrak{h}$ is free of rank 2.

3.4.2. We study the action of G_{F_∞} on $\mathcal{T}$.

We have an exact sequence

$$0 \to \mathcal{T}_{\text{sub}} \to \mathcal{T} \to \mathcal{T}_{\text{quo}} \to 0$$

of $\mathfrak{h}[G_{F_\infty}]$-modules, with $\mathcal{T}_{\text{sub}}$ and $\mathcal{T}_{\text{quo}}$ defined as follows. First, $\mathcal{T}_{\text{sub}}$ is the largest submodule of $\mathcal{T}$ such that G_{F_∞} acts on $\mathcal{T}_{\text{sub}}(-1)$ trivially, and $\mathcal{T}_{\text{quo}}$ is the quotient. Then $\mathcal{T}_{\text{quo}}$ is equal to the maximal unramified, $\mathfrak{h}$-torsion-free quotient of $\mathcal{T}$. In this way, the place ∞ plays the role that the place at p does in 2.4.4. In fact, G_{F_∞} acts trivially on $\mathcal{T}_{\text{quo}}$, and both $\mathbb{Q}_p \otimes_{\mathbb{Z}_p} \mathcal{T}_{\text{sub}}$ and $\mathbb{Q}_p \otimes_{\mathbb{Z}_p} \mathcal{T}_{\text{quo}}$ are free of rank 1 over $\mathbb{Q}_p \otimes_{\mathbb{Z}_p} \mathfrak{h}$.

The above short exact sequence is split as a sequence of $\mathfrak{h}$-modules: $\mathcal{T}_{\text{quo}}$ is the isomorphic image of the $\mathfrak{h}$-submodule of $\mathcal{T}$ on which a choice of Frobenius element acts trivially. This will be used in constructing Υ below.

3.4.3. Since $\overline{U}(N)$ is essentially the graph of the special fiber of a model of $X_1(N)$ over $\mathcal{O}_\infty$, we have a surjective homomorphism

$$\mathcal{T} \to \mathcal{S} = H_1(\overline{U}(N), \mathbb{Z}_p).$$

Via this map, $\mathcal{S}$ is identified with the quotient $\mathcal{T}_{\text{quo}}$ of $\mathcal{T}$ with trivial G_{F_∞}-action. In this way, $\mathcal{T}_{\text{quo}}$ is also analogous to the plus quotient of homology in 2.2.2. That is, the place ∞ also plays the role that the real place does over $\mathbb{Q}$.

3.4.4. Let $\mathfrak{S}$ be the space of those $\mathbb{Z}_p$-valued, special-at-∞ cuspidal automorphic forms

$$\phi\colon \mathrm{PGL}_2(F)\backslash \mathrm{PGL}_2(\mathbb{A}_F)/(K_f \times \mathcal{I}_\infty) \to \mathbb{Z}_p,$$

where $\mathbb{A}_F = \mathbb{A}_F^f \times F_\infty$ is the adele ring, K_f is the closure of the image of $\tilde{\Gamma}_1(N)$ in $\mathrm{PGL}_2(\mathbb{A}_F^f)$ (see 4.1.3), and $\mathcal{I}_\infty$ is the Iwahori subgroup of $\mathrm{PGL}_2(F_\infty)$. For ϕ to be special at ∞ means that its right $\mathbb{Q}_p[\mathrm{GL}_2(F_\infty)]$-span is a direct sum of copies of the "special representation." (The latter is the quotient of the locally constant functions $\mathbb{P}^1(F_\infty) \to \mathbb{Q}_p$ by the constant functions.)

The property of being special at ∞ tells us the local behavior at the prime ∞ of the 2-dimensional $\mathbb{Q}_p$-Galois representation attached to the cusp form. This is a replacement for the condition of ordinarity at p: it is what tells us the G_{F_∞}-action on $\mathcal{T}$ used in 3.4.2.

3.4.5. We explain how the groups $\mathcal{S}$ and $\mathfrak{S}$ may be identified.

The identification passes through the harmonic cocycles on $U(N)$. These are the functions on the oriented edges of $U(N)$ that change sign if we switch its orientation of an edge and which sum to zero on the edges leading into a vertex (i.e., are harmonic). The cuspidal harmonic cocycles are those supported on finitely many edges. The space of $\mathbb{Z}_p$-valued cuspidal harmonic cocycles may be directly identified with $\mathcal{S}$. It also provides a combinatorial description of $\mathfrak{S}$. To see this, one starts with the observation that the double coset space on which forms in $\mathfrak{S}$ are defined is none other than the set of oriented edges of $U(N)$. The property of being special at ∞ gives the harmonic condition, and the two notions of cuspidality coincide. Thus, the spaces $\mathfrak{S}$ and $\mathcal{S}$ that appear in the diagram of 1.10 are canonically identified in the case of $\mathbb{F}_q(t)$.

3.5 *The Map Υ*

We define the map $\Upsilon\colon Y_\theta \to P_\theta$ on θ-parts for a fixed primitive character $\theta\colon G \to \overline{\mathbb{Q}}_p^\times$.

3.5.1. We briefly outline the construction of $\Upsilon\colon Y_\theta \to P_\theta$ that will appear in this section.

As in 2.5.1, we analyze the $\mathfrak{h}[G_F]$-action on $\mathcal{T}_\theta/I_\theta\mathcal{T}_\theta$, showing that it fits in an exact sequence

$$0 \to P_\theta \to \mathcal{T}_\theta/I_\theta\mathcal{T}_\theta \to Q_\theta \to 0$$

of $\mathfrak{h}[G_F]$-modules. Similarly to the setting of GL_2 over $\mathbb{Q}$, the G_F-actions on P_θ and Q_θ are understood, and Q_θ is free of rank 1 over $\mathfrak{h}_\theta/I_\theta$ with a canonical generator. However, the domain of our map Υ is not a Galois group, so our approach to constructing Υ is different. We employ compactly supported cohomology, which is dual to Galois cohomology by Poitou-Tate duality. Instead of directly using the cocycle attached to the exact sequence, we construct Υ in 3.5.7 from a connecting homomorphism ∂ on compactly supported étale cohomology that appears as the second map in a composition of Λ_θ-module homomorphisms

$$\Upsilon : Y_\theta \xrightarrow{\sim} H^2_{\text{ét},c}(\mathcal{O}[\tfrac{1}{N}], Q_\theta(1)) \xrightarrow{\partial} H^3_{\text{ét},c}(\mathcal{O}[\tfrac{1}{N}], P_\theta(1)) \xrightarrow{\sim} P_\theta.$$

The isomorphisms are seen using the $\mathfrak{h}_\theta[G_F]$-module structure of Q_θ and the triviality of the G_F-action on P_θ, respectively.

3.5.2. We define the L-function for θ by

$$L(\theta, s) = \prod_{\mathfrak{p} \nmid N} (1 - \theta(\mathrm{Fr}_\mathfrak{p})^{-1} \mathfrak{N}(\mathfrak{p})^{-s})^{-1},$$

where the product is taken over the prime ideals $\mathfrak{p}$ of $\mathcal{O}$ not dividing N, and $\mathrm{Fr}_\mathfrak{p}$ denotes an arithmetic Frobenius at $\mathfrak{p}$. We then take $\xi_\theta \in \Lambda_\theta$ to be the nonzero value $L(\theta, -1)$.

3.5.3. We have an isomorphism $\mathfrak{h}_\theta/I_\theta \xrightarrow{\sim} \Lambda_\theta/(\xi_\theta)$. We indicate one construction of the map.

Consider the Jacobian variety J of $X_1(N)$ and the class $\alpha \in J(\mathbb{C}_\infty)$ of the divisor $(0) - (\infty)$, where 0 and ∞ are cusps on the Drinfeld modular curve. Gekeler showed that α has finite order (Gekeler 2000), and it is annihilated by I. The $\mathfrak{h}_\theta$-module generated by the θ-part of α is $\Lambda_\theta/(\xi_\theta)$ by a computation of the divisors of Siegel units, providing the desired map.

3.5.4. We define $Q_\theta = (\mathfrak{h}_\theta/I_\theta)^\sharp(1)$, where $(\ )^\sharp$ indicates a G_F-action under which any element that maps to $a \in G$ acts by multiplication by $\theta^{-1}(a)$. Much as in 2.5.4, pairing with the θ-part of α gives rise to a canonical surjection of $\mathfrak{h}_\theta[G_F]$-modules $\mathcal{T}_\theta/I\mathcal{T}_\theta \to Q_\theta$.

3.5.5. The exact sequence

$$0 \to P_\theta \to \mathcal{T}_\theta/I_\theta\mathcal{T}_\theta \to Q_\theta \to 0$$

of $\mathfrak{h}[G_F]$-modules is constructed as in 3.5.1. Here, we observe that Q_θ has a nontrivial action of the Frobenius element chosen in 3.4.2, so Q_θ is a quotient of $\mathcal{T}_{\mathrm{sub}}$. As before, $\mathcal{T}_{\mathrm{sub}}$ is $\mathbb{Z}_p$-dual to $\mathcal{T}_{\mathrm{quo}}$ and thereby isomorphic to $\mathfrak{h}$, so we have an isomorphism $\mathcal{T}_{\mathrm{sub},\theta}/I_\theta\mathcal{T}_{\mathrm{sub},\theta} \xrightarrow{\sim} Q_\theta$ that provides a G_{F_∞}-splitting of the exact sequence. The known G_F-action on Q_θ and the known determinant of the G_F-action on $\mathcal{T}_\theta$ tell us that G_F acts trivially on P_θ.

3.5.6. The analogue of the Iwasawa main conjecture over F_N is the equality

$$|Y_\theta| = [\Lambda_\theta : (\xi_\theta)]$$

of orders. This equality is a consequence of Grothendieck trace formula, so we do not require the method of Mazur-Wiles to prove it.

3.5.7. We define our map Υ.

Let $H^i_{\text{ét},c}(\mathcal{O}[\frac{1}{N}], M)$ denote the ith compactly supported étale cohomology group of a compact $\mathbb{Z}_p[\![G_F]\!]$-module M that is unramified outside $N\infty$. These groups fit in a long exact sequence

$$\cdots \to H^i_{\text{ét},c}(\mathcal{O}[\tfrac{1}{N}], M) \to H^i_{\text{ét}}(\mathcal{O}[\tfrac{1}{N}], M) \to \bigoplus_{v|N\infty} H^i_{\text{ét}}(F_v, M) \to H^{i+1}_{\text{ét},c}(\mathcal{O}[\tfrac{1}{N}], M) \to \cdots.$$

The exact sequence in 3.5.1 yields a connecting homomorphism

$$H^2_{\text{ét},c}(\mathcal{O}[\tfrac{1}{N}], Q_\theta(1)) \to H^3_{\text{ét},c}(\mathcal{O}[\tfrac{1}{N}], P_\theta(1)) = P_\theta,$$

the latter identification as P_θ has trivial G_F-action, and we can prove that the canonical map

$$H^2_{\text{ét},c}(\mathcal{O}[\tfrac{1}{N}], Q_\theta(1)) \to H^2_{\text{ét}}(\mathcal{O}[\tfrac{1}{N}], Q_\theta(1))$$

is an isomorphism. Hence, the above homomorphism is identified with $H^2_{\text{ét}}(\mathcal{O}[\frac{1}{N}], Q_\theta(1)) \to P_\theta$. Our Υ is defined as the following composition:

$$\begin{aligned}\Upsilon : Y_\theta &\xrightarrow{\sim} H^2_{\text{ét}}(\mathcal{O}_N[\tfrac{1}{N}], \mathbb{Z}_p(2))_\theta \\ &\xrightarrow{\sim} H^2_{\text{ét}}(\mathcal{O}[\tfrac{1}{N}], \Lambda^\sharp_\theta(2)) \\ &\xrightarrow{\sim} H^2_{\text{ét}}(\mathcal{O}[\tfrac{1}{N}], (\Lambda_\theta/\xi_\theta)^\sharp(2)) \\ &\xrightarrow{\sim} H^2_{\text{ét}}(\mathcal{O}[\tfrac{1}{N}], Q_\theta(1)) \to P_\theta.\end{aligned}$$

The isomorphism in the first line is by 3.2.2, the isomorphism in the second line is by Shapiro's lemma, the isomorphism in the third line follows from the fact that ξ_θ kills Y_θ by 3.5.6, and the isomorphism in the fourth line is by definition of Q_θ in 3.5.4.

3.6 The Conjecture: ϖ and Υ Are Inverse Maps

We state the conjecture and our main result in the case of GL_2 and GL_1 over $\mathbb{F}_q(t)$.

3.6.1. We state the conjecture.

Conjecture. The maps $\varpi\colon P_\theta \to Y_\theta$ and $\Upsilon\colon Y_\theta \to P_\theta$ are inverse to each other.

3.6.2. We state the theorem.

Theorem. *We have that* $\xi'_\theta \Upsilon \circ \varpi = \xi'_\theta$, *where*

$$\xi'_\theta = \frac{d}{dq^{-s}} L(\theta^{-1}, s)|_{s=-1} \in \Lambda_\theta.$$

3.6.3. We can prove the order of P_θ is divisible by the order of $\Lambda_\theta/(\xi_\theta)$ and hence by the order of Y_θ. Thus, in the case that ξ'_θ is a unit in Λ_θ, our conjecture is implied by the above theorem.

3.7 The Proof that $\xi'\Upsilon \circ \varpi = \xi'$

The method of the proof of our Theorem 3.6.2 is parallel to the proof in the $\mathbb{Q}$-case. We give only its bare outline.

3.7.1. As in 2.7.1, we consider a refinement of the diagram in 1.10 in which we divide the right-hand square of that diagram into two squares:

$$\begin{array}{ccccccc}
\mathcal{S}_\theta & \xrightarrow{z} & H^2_{\text{ét}}(Y_1(N), \mathbb{Z}_p(2))_\theta & \overset{\text{HS}}{\dashrightarrow} & H^1_{\text{ét}}(\mathcal{O}_N[\tfrac{1}{N}], \mathcal{T}_\theta(1)) & \xrightarrow{\text{reg}} & \mathfrak{S}_\theta \\
\downarrow{\scriptstyle \bmod I} & & \downarrow{\scriptstyle \infty} & & \downarrow & & \downarrow{\scriptstyle \bmod I} \\
P_\theta & \xrightarrow{\varpi} & Y_\theta & \xrightarrow{\xi'_\theta} & Y_\theta & \xrightarrow{\Upsilon} & P_\theta .
\end{array}$$

The commutativity of the leftmost square of the diagram was discussed in Sect. 3.3. We discuss the maps in the other two squares of the diagram below.

3.7.2. The map HS in the diagram arises in a Hochschild-Serre spectral sequence. An analogue of the discussion of 2.7.2 applies.

3.7.3. Let κ be the canonical generator of $H^1_{\text{ét}}(\mathbb{F}_q, \mathbb{Z}_p)$. The third vertical arrow in the diagram is the composition

$$H^1_{\text{ét}}(\mathcal{O}[\tfrac{1}{N}], \mathcal{T}_\theta(1)) \to H^1_{\text{ét}}(\mathcal{O}[\tfrac{1}{N}], \mathcal{Q}_\theta(1)) \xrightarrow{\cup\kappa} H^2_{\text{ét}}(\mathcal{O}[\tfrac{1}{N}], \mathcal{Q}_\theta(1)) \xrightarrow{\sim} Y_\theta,$$

where the last isomorphism is given in 3.5.7.

3.7.4. The map reg in the diagram is the θ-part of the p-adic regulator map

$$H^1_{\text{ét}}(F_\infty, \mathcal{S}(1)) \xrightarrow{\cup\kappa} H^2_{\text{ét}}(F_\infty, \mathcal{S}(1)) \xrightarrow{\sim} \mathcal{S} = \mathfrak{S},$$

where the second map is the invariant map of local class field theory.

3.7.5. Since $\mathcal{S}$ and $\mathfrak{S}$ are canonically identified, both "mod I" maps in the diagram are just reduction modulo I_θ.

3.7.6. The proofs of the commutativity of the other two squares are once again nontrivial, though slightly different, exercises in étale and Galois cohomology.

3.7.7. It remains to prove that the composition $\mathcal{S}_\theta \to \mathfrak{S}_\theta \to P_\theta$, where the first arrow is the composition of the upper horizontal rows, coincides with ξ'_θ times the reduction modulo I_θ map $\mathcal{S}_\theta \to P_\theta$. By the computation of Kondo and Yasuda (2012) of the values of a regulator map on the analogues of Beilinson elements, this is reduced to a comparison of their regulator map with the above p-adic regulator map.

4 What Happens for GL_d?

In this section, we discuss three settings for the study of generalizations of the conjectures in Sects. 2 and 3 for GL_d over a field F, for a fixed integer $d \geq 1$. The fields F and, thereby, the cases we consider here are:

(i) The rational numbers,
(ii) An imaginary quadratic field,
(iii) A function field in one variable over a finite field.

We have results only in the cases (i) and (iii) for $d = 2$ discussed above, but we wish to speculate and pose questions in a more general setting. Rather than formulating precise conjectures, we aim for the more modest goals of pointing in their direction and inspiring the reader to investigate further.

4.1 The Space of Modular Symbols

4.1.1. By an infinite place, we mean the unique Archimedean place in cases (i) and (ii) and a fixed place ∞ in case (iii). The remaining places are called finite places. We have the following objects:

- The subring $\mathcal{O}$ of F of elements that are integral at all finite places,
- The completion F_v of F at a place v,
- The valuation ring $\mathcal{O}_v$ of F_v at a non-Archimedean place v,
- The adele ring $\mathbb{A}_F$ of F and the adele ring $\mathbb{A}_F^f$ of finite places,
- The subring $\mathcal{O}_{\mathbb{A}}^f = \prod_{v \text{ finite}} \mathcal{O}_v$ of $\mathbb{A}_F^f$.

In the discussion below, we will use the notation $(\ )^{(d)}$ when defining an object in the GL_d-setting and then omit the notation in many instances in which d is clear.

4.1.2. We define a topological space D_d by using the standard maximal compact subgroup of $\mathrm{PGL}_d(F_\infty)$: in the respective cases, it is

(i) $\mathrm{PGL}_d(\mathbb{R})/\mathrm{PO}_d(\mathbb{R})$, so that $\mathrm{SL}_d(\mathbb{R})/\mathrm{SO}_d(\mathbb{R}) \xrightarrow{\sim} D_d$,
(ii) $\mathrm{PGL}_d(\mathbb{C})/\mathrm{PU}_d$, so that $\mathrm{SL}_d(\mathbb{C})/\mathrm{SU}_d \xrightarrow{\sim} D_d$,
(iii) The Bruhat-Tits building associated to $\mathrm{PGL}_d(F_\infty)$.

For example, in case (i) the space D_2 is the complex upper half-plane $\mathbb{H}$. In case (ii), the space D_2 is the three-dimensional hyperbolic upper-half space $\mathbb{H}_3$. Note that in case (iii), the Bruhat-Tits building has the set $\mathrm{PGL}_d(F_\infty)/\mathrm{PGL}_d(\mathcal{O}_\infty)$ of homothety classes of $\mathcal{O}_\infty$-lattices of rank d in F_∞^d as its 0-simplices.

4.1.3. Let N be a nonzero ideal of $\mathcal{O}$. Let $K_1^{(d)}(N)$ be the open compact subgroup of $\mathrm{GL}_d(\mathcal{O}_{\mathbb{A}}^f)$ given by

$$K_1^{(d)}(N) = \left\{ g \in \mathrm{GL}_d(\mathcal{O}_{\mathbb{A}}^f) \mid (g_{d,1}, \ldots, g_{d,d-1}, g_{d,d}) \equiv (0, \ldots, 0, 1) \bmod N \right\}.$$

Let

$$U^{(d)}(N) = \mathrm{GL}_d(F) \backslash (\mathrm{GL}_d(\mathbb{A}_F^f)/K_1^{(d)}(N) \times D_d).$$

The space $U^{(1)}(N)$ is the relative Picard group $\mathrm{Pic}(\mathcal{O}, N)$, viewed as a discrete space. For $d \geq 2$, the space $U^{(d)}(N)$ is homeomorphic to the disjoint union of $|\mathrm{Pic}(\mathcal{O})|$ copies of $\tilde{\Gamma}_1^{(d)}(N) \backslash D_d$, where

$$\tilde{\Gamma}_1^{(d)}(N) = \mathrm{GL}_d(\mathcal{O}) \cap K_1^{(d)}(N).$$

4.1.4. Consider cases (i) and (ii). Let $\epsilon \in \mathrm{GL}_d(\mathcal{O})$ be a diagonal matrix with entries $a_1, a_2, \ldots, a_d$ such that the product $a_1 a_2 \cdots a_d$ generates the roots of unity $\mu_F = \mathcal{O}^\times$ in F. Let

$$\Gamma_1^{(d)}(N) = \tilde{\Gamma}_1^{(d)}(N) \cap \mathrm{SL}_d(\mathcal{O}).$$

Then $\tilde{\Gamma}_1(N) \backslash D_d$ is identified with the quotient of $\Gamma_1(N) \backslash D_d$ by the action of the operator

$$\mathrm{class}(g) \mapsto \mathrm{class}(\epsilon g \epsilon^{-1})$$

for $g \in \mathrm{SL}_d(\mathbb{R})$ in case (i) and for $g \in \mathrm{SL}_d(\mathbb{C})$ in case (ii).

4.1.5. In case (i), the space $U^{(2)}(N)$ is identified with the quotient of $Y_1(N)(\mathbb{C}) = \Gamma_1(N) \backslash \mathbb{H}$ by the action of complex conjugation on $Y_1(N)(\mathbb{C})$. In fact, the description in 4.1.4 shows that it arises from the quotient of $\mathbb{H}$ by the action

$$\mathbb{H} \ni x + iy = \text{class}\begin{pmatrix} \sqrt{y} & x \\ 0 & 1/\sqrt{y} \end{pmatrix}$$

$$\mapsto \text{class}\left(\begin{pmatrix} -1 & 0 \\ 0 & 1 \end{pmatrix}\begin{pmatrix} \sqrt{y} & x \\ 0 & 1/\sqrt{y} \end{pmatrix}\begin{pmatrix} -1 & 0 \\ 0 & 1 \end{pmatrix}\right) = \text{class}\begin{pmatrix} \sqrt{y} & -x \\ 0 & 1/\sqrt{y} \end{pmatrix} = -x + iy,$$

which coincides with the action of complex conjugation on $Y_1(N)(\mathbb{C})$.

4.1.6. The space $\mathcal{S}^{(d)}(N)$ of (cuspidal) modular symbols for GL_d is defined as

$$\mathcal{S}^{(d)}(N) = \mathrm{image}(H_{d-1}(U^{(d)}(N), \mathbb{Z}) \to H^{\mathrm{BM}}_{d-1}(U^{(d)}(N), \mathbb{Z})),$$

where H^{BM}_* denotes Borel-Moore homology. Recall that if $\overline{U}(N)$ is a compactification of $U(N)$, then $H^{\mathrm{BM}}_i(U(N), \mathbb{Z})$ is canonically isomorphic to the relative homology group $H_i(\overline{U}(N), \overline{U}(N)\backslash U(N), \mathbb{Z})$. The space $\mathcal{S}(N)$ may be the homology group $H_{d-1}(\overline{U}(N), \mathbb{Z})$ for some good choice of compactification.

4.1.7. In case (i), we have by 4.1.5 a canonical map

$$H_1(X_1(N), \mathbb{Z})_+ \to \mathcal{S}^{(2)}(N) = H_1(\overline{U}(N), \mathbb{Z}),$$

where $\overline{U}(N)$ is the quotient of $X_1(N)$ by the action of complex conjugation. This map is a surjection with 2-torsion kernel.

4.1.8. For a nonzero ideal $\mathfrak{n}$ of $\mathcal{O}$, let $T(\mathfrak{n})$ denote the Hecke operator on $\mathcal{S}^{(d)}(N)$ corresponding to the sum of $K_1^{(d)}(N)$-double cosets of elements of $M_d(\mathcal{O}_{\mathbb{A}}) \cap \mathrm{GL}_d(\mathbb{A}_F^f)$ with determinant generating $\mathfrak{n}\mathcal{O}_v$ in the v-component. (For $d = 1$, we make the convention that $T(\mathfrak{n}) = 0$ if $\mathfrak{n}$ and N are not coprime.) These operators satisfy $T(\mathfrak{a}\mathfrak{b}) = T(\mathfrak{a})T(\mathfrak{b})$ for coprime $\mathfrak{a}$ and $\mathfrak{b}$.

Let $\mathbb{T}^{(d)}(N)$ denote the commutative subring of $\mathrm{End}_{\mathbb{Z}}(\mathcal{S}^{(d)}(N))$ generated by the $T(\mathfrak{n})$ with $\mathfrak{n}$ a nonzero ideal of $\mathcal{O}$.

4.1.9. For $d = 1$, the group $\mathcal{S}(N)$ of modular symbols is $H_0(U(N), \mathbb{Z}) = \mathbb{Z}[\mathrm{Pic}(\mathcal{O}, N)]$. The Hecke algebra $\mathbb{T}(N)$ is the ring $\mathbb{Z}[\mathrm{Pic}(\mathcal{O}, N)]$, with $T(\mathfrak{n})$ for $\mathfrak{n}$ coprime to N equal to the group element for $\mathfrak{n}$. Under these identifications, $\mathbb{T}(N)$ acts by left multiplication on $\mathcal{S}(N)$.

4.1.10. The modular symbol $\{0 \to \infty\}$ in Sect. 2.1 is generalized to the following element of $H^{BM}_{d-1}(U(N), \mathbb{Z})$. It is the class of the image in the identity component of $U(N)$ of the following standard subset of D_d, with a suitable orientation:

- **(i-ii)** The set of classes of diagonal matrices in $\mathrm{GL}_d(F_\infty)$ with positive real entries,
- **(iii)** The union of all $(d-1)$-simplices with 0-vertices in the set of classes in D_d of diagonal matrices in $\mathrm{GL}_d(F_\infty)$.

The modular symbols $\{\alpha \to \beta\}$ for $\alpha, \beta \in \mathbb{P}^1(\mathbb{Q})$ are generalized to the classes in $H^{BM}_{d-1}(U(N), \mathbb{Z})$ of the images in $U(N)$ of the translations by $\mathrm{GL}_d(F)$ of the above standard subset of D_d.

4.2 Questions for the General Case

We suspect that our results in Sects. 2 and 3 are special cases of a relationship

Modular symbols for GL_d modulo the Eisenstein ideal $\Longleftrightarrow$ Iwasawa theory for GL_{d-1}

that holds for $d \geq 2$. In this subsection, we describe what we expect to be true.

4.2.1. We lay out some basic objects, starting with:

- A prime number $p \neq \operatorname{char} F$,
- A nonzero ideal N of $\mathcal{O}$ that is coprime to p,
- A commutative pro-p ring R and its total ring of quotients $Q(R)$,
- A profinite R-module T with a continuous R-linear action of G_F that is unramified at every finite place not dividing Np,

Recalling from 4.1.8 the Hecke algebra $\mathbb{T}^{(d)}(N)$ and modular symbols $\mathcal{S}^{(d)}(N)$, we define

$$\mathbb{T}_R = \varprojlim_r (R \otimes \mathbb{T}^{(d)}(Np^r)) \quad \text{and} \quad \mathcal{S}_R = \varprojlim_r (R \otimes \mathcal{S}^{(d)}(Np^r)).$$

We shall often use the fact that $(p) = \mathcal{O}$ in case (iii). For instance, in this case $Np^r = N$, so we have quite simply that $\mathbb{T}_R = R \otimes \mathbb{T}^{(d)}(N)$ and $\mathcal{S}_R = R \otimes \mathcal{S}^{(d)}(N)$.

We also let $\mathbb{T}^{(d)}(Np^r)'$ be the subring of $\mathbb{T}^{(d)}(Np^r)$ generated by the $T(\mathfrak{n})$ with $\mathfrak{n}$ coprime to (p). Note that $\mathbb{T}^{(d)}(Np^r)' = \mathbb{T}^{(d)}(N)$ in case (iii) and $\mathbb{T}^{(1)}(Np^r)' = \mathbb{T}^{(1)}(Np^r)$ in all cases.

4.2.2. We place some conditions on the pair (R, T):

(**1**) The $Q(R)$-module $V = Q(R) \otimes_R T$ is free of rank $d-1$.
(**2**) For every prime ideal $\mathfrak{p}$ of $\mathcal{O}$ that does not divide Np, the characteristic polynomial $P_{\mathfrak{p}}(u) = \det_{Q(R)}(1 - \mathrm{Fr}_{\mathfrak{p}}^{-1}u \mid V)$ of an arithmetic Frobenius $\mathrm{Fr}_{\mathfrak{p}}$ lies in $R[u]$.

For a prime ideal $\mathfrak{p}$ of $\mathcal{O}$ that does not divide Np, we define $a(\mathfrak{p}^n)$ for $n \geq 0$ by

$$P_{\mathfrak{p}}(u)^{-1} = \sum_{n=0}^{\infty} a(\mathfrak{p}^n)u^n \in R[\![u]\!].$$

We then suppose:

(**3**) There exists a ring homomorphism

$$\phi_T \colon \varprojlim_r (\mathbb{Z}_p \otimes \mathbb{T}^{(d-1)}(Np^r)') \to R$$

that sends $T(\mathfrak{p}^k)$ to $a(\mathfrak{p}^k)$ for all prime ideals $\mathfrak{p}$ of $\mathcal{O}$ not dividing Np and all $k \geq 1$.

We extend a to a function on all nonzero ideals $\mathfrak{n}$ of $\mathcal{O}$ by setting $a(\mathfrak{n}) = \phi_T(T(\mathfrak{n}))$ if $\mathfrak{n}$ is coprime to p and $a(\mathfrak{n}) = 0$ otherwise. In the case $d = 2$, our definition forces $a(\mathfrak{n}) = 0$ for any $\mathfrak{n}$ not coprime to Np, while in general, these values of a may not be uniquely determined by T, so ϕ_T should be considered as part of the data.

4.2.3. We define the Eisenstein ideal I_T of $\mathbb{T}_R$ to be the ideal of the GL_d-Hecke algebra $\mathbb{T}_R$ generated by the elements

$$T(\mathfrak{n}) - \sum_{\mathfrak{d}|\mathfrak{n}} a(\mathfrak{d})\mathfrak{N}(\mathfrak{d})$$

for the nonzero ideals $\mathfrak{n}$ of $\mathcal{O}$. Note that I_T depends only on V and the choice of ϕ_T, rather than T itself. In case (i), the ideal I_T is generated by the coefficients of the formal expression

$$\sum_{n=1}^{\infty} T(n)n^{-s} - \zeta(s) \sum_{\substack{n=1 \\ (n,p)=1}}^{\infty} a(n)n^{-(s-1)}.$$

4.2.4. For any compact $R[\![G_F]\!]$-module M that is unramified outside of $S \cup \{\infty\}$ for some finite set S of finite places of F including those dividing p, we denote more simply by $H^2_{\text{ét}}(\mathcal{O}[\frac{1}{p}], M)$ the R-module $H^2_{\text{ét}}(\mathcal{O}[\frac{1}{p}], j_*M)$, where $j\colon \mathrm{Spec}(\mathcal{O}) \setminus S \hookrightarrow \mathrm{Spec}(\mathcal{O}[\frac{1}{p}])$ is the inclusion morphism. It is independent of the choice of S. We will also use a similar notation with $\mathcal{O}$ replaced by its integral closure in a finite extension of F.

4.2.5. Our two objects of study are the R-modules:

- The geometric object $P = \mathcal{S}_R/I_T\mathcal{S}_R$ on the GL_d-side,
- The arithmetic object $Y = H^2_{\text{ét}}(\mathcal{O}[\frac{1}{p}], T(d))$ on the GL_{d-1}-side.

We ask a vague question.

Question. Under what conditions does there exist a canonical isomorphism $\varpi\colon P \xrightarrow{\sim} Y$ of R-modules?

We remark that there certainly must be some conditions, as different lattices T in V may have Y that are nonisomorphic. In what follows, we introduce three settings for further study.

4.2.6. We fix some notation for abelian extensions of F and their Galois groups.

For $r \geq 0$, let H_r be the ray class field of F of modulus (p^r), and let O_r be the integral closure of $\mathcal{O}$ in H_r. Let $\Gamma_r = \mathrm{Pic}(\mathcal{O}, (p^r))$, which is canonically isomorphic to $\mathrm{Gal}(H_r/F)$ by class field theory. Let $\Gamma = \varprojlim_r \Gamma_r$.

- In case (i), we have that $H_r = \mathbb{Q}(\mu_{p^r})^+$, $O_r = \mathbb{Z}[\zeta_{p^r}]^+$, and $\Gamma = \mathbb{Z}_p^\times/\langle -1\rangle$.
- In case (ii), the field H_r is generated over F by the j-invariant $j(E)$ and x-coordinates of the p^r-torsion points of an elliptic curve E over $F(j(E))$ with CM by $\mathcal{O}$. There is an exact sequence

$$0 \to (\mathbb{Z}_p \otimes \mathcal{O})^\times/\mu_F \to \Gamma \to \mathrm{Pic}(\mathcal{O}) \to 0.$$

  Note that $\Gamma/\Gamma_{\mathrm{tor}} \cong \mathbb{Z}_p^2$, where Γ_{tor} is the torsion subgroup of Γ.
- In case (iii), we have that $H_r = H_0$, $O_r = O_0$, and $\Gamma = \Gamma_r = \mathrm{Pic}(\mathcal{O})$.

Let $[a] \in \mathbb{Z}_p[\![\Gamma]\!]$ be the group element corresponding to $a \in \Gamma$. We may also speak of $[\mathfrak{a}]$ for $\mathfrak{a}$ an ideal of $\mathcal{O}$ coprime to p by taking the sequence of classes of $\mathfrak{a}$ in the groups Γ_r. We use $(\ )^\sharp$ below to denote the (additional) G_F-action on a module over a $\mathbb{Z}_p[\![\Gamma]\!]$-algebra under which an element that restricts to $a \in \Gamma$ acts by multiplication by $[a]^{-1}$.

4.2.7. We describe setting (A_d) for $d \geq 2$.

Let R_0 be the valuation ring of a finite extension K_0 of $\mathbb{Q}_p$. Let T_0 be a free R_0-module of rank $d-1$ endowed with a continuous R-linear action of G_F. We assume that the G_F-action on T_0 is unramified at all finite places not dividing Np. We suppose that condition (3) of 4.2.2 is satisfied for (R_0, T_0), and we use $a_0(\mathfrak{n})$ to denote $a(\mathfrak{n})$ of 4.2.2 for this pair.

Let $R = R_0[\![\Gamma]\!]$ and $T = R^\sharp \otimes_{R_0} T_0$. Then the pair (R, T) satisfies conditions (1) and (2) of 4.2.2, and we suppose that it satisfies (3). It follows directly that $a(\mathfrak{n}) = [\mathfrak{n}]^{-1} \otimes a_0(\mathfrak{n})$ for any nonzero ideal $\mathfrak{n}$ of $\mathcal{O}$ that is coprime to Np. By definition of T, we also have an R-module isomorphism

$$Y = H^2_{\text{ét}}(\mathcal{O}[\tfrac{1}{p}], T(d)) \cong \varprojlim_r H^2_{\text{ét}}(O_r[\tfrac{1}{p}], T_0(d)).$$

In that the G_F-stable R_0-lattice T_0 has not been chosen with any special properties inside $V_0 = K_0 \otimes_{R_0} T_0$, we consider an additional condition.

(**4**) The G_F-representation $k_0 \otimes_{R_0} T_0$ is irreducible over the residue field k_0 of K_0.

It follows from (4) that the isomorphism class of T_0 as an $R_0[G_F]$-module depends only on the K_0-representation V_0 of G_F. That is, all G_F-stable R_0-lattices in V_0 have the same isomorphism class. Hence, the isomorphism class of the R-module Y depends only on V_0.

Finally, to avoid known exceptions in case (i), we consider a primitivity condition.

(**5**) The map ϕ_{T_0} does not factor through $\varprojlim_r (\mathbb{Z}_p \otimes \mathbb{T}^{(d-1)}(Mp^r)')$ for any ideal M of $\mathcal{O}$ properly containing N.

4.2.8. We may now ask our question for setting (A_d) under conditions (1)–(5).

Question. Does there exist a canonical isomorphism $\varpi\colon P \xrightarrow{\sim} Y$ of R-modules?

We are also interested in what happens if conditions (4) and (5) are removed. For instance, we wonder if (5) might be removed for good choices of N, p, and d, or if (4) might be removed in the presence of a good, canonical lattice T_0. In any case, we can ask the following question.

Question. If we do not suppose conditions (4) and (5), does there still exist a canonical isomorphism $\varpi_{\mathbb{Q}_p}\colon \mathbb{Q}_p \otimes_{\mathbb{Z}_p} P \xrightarrow{\sim} \mathbb{Q}_p \otimes_{\mathbb{Z}_p} Y$?

4.2.9. The p-adic Galois representations V_0 attached to the following objects of modulus or level Np^r for some $r \geq 0$ all have (R_0, T_0) and (R, T) satisfying (1)–(3):

- In case (i) for $d = 2$, an even Dirichlet character,
- In case (i) for $d = 3$, a holomorphic cupsidal eigenform,
- In case (ii) for $d = 2$, an algebraic Hecke character on $\mathbb{A}_F^\times$,
- In case (iii) for $d \geq 2$, a cuspidal eigenform of GL_{d-1} that is special at ∞.

The examples for $d = 2$ obviously satisfy (4), and in the remaining cases, (4) may be assumed. By taking each of the objects to be primitive, we may assume (5).

4.2.10. We explain how the setting (A_2) for $F = \mathbb{Q}$ and $\mathbb{F}_q(t)$ was studied in Sects. 2 and 3.

Let θ be a primitive character of $\mathrm{Pic}(\mathcal{O}, Np)$, and impose all the assumptions on p, N, and θ of Sects. 2 and 3. Take $R_0 = \mathbb{Z}_p[\theta]$, and let $T_0 = \mathbb{Z}_p[\theta]$ with G_F acting through θ^{-1}. Let $\Delta = \mathrm{Pic}(\mathcal{O}, (p))$, which we may view as a subgroup of Γ. For the objects P_θ and Y_θ of Sect. 2 in case (i) and of Sect. 3 in case (iii), we claim that

$$P_\theta = \mathbb{Z}_p \otimes_{\mathbb{Z}_p[\Delta]} P \quad \text{and} \quad Y_\theta = \mathbb{Z}_p \otimes_{\mathbb{Z}_p[\Delta]} Y,$$

with P and Y as in 4.2.5. This claim is immediate for $\mathbb{F}_q(t)$ as Δ is trivial, and it is not hard to see for Y in case (i). However, the claim for P is not evident in case (i), so we prove it.

Proof of the claim. Note that $R = \mathbb{Z}_p[\theta][\![\Gamma]\!]$ and $T = \mathbb{Z}_p[\theta][\![\Gamma]\!]^\sharp$, and note that $\mathbb{T}_R = \mathbb{T}_R^{(2)}$ of this section is $\mathbb{T}[\theta][\![\Gamma]\!]$, where $\mathbb{T}$ is as in 2.2.1. The claim for P follows if we can show that the map $T(n) \mapsto T(n)$ on Hecke operators induces an isomorphism

$$\mathbb{T}_\theta / I_\theta \xrightarrow{\sim} \mathbb{Z}_p \otimes_{\mathbb{Z}_p[\Delta]} (\mathbb{T}_R / I_T),$$

where I is the Eisenstein ideal of 2.2.1.

For a prime ℓ not dividing Np, the action of Fr_ℓ^{-1} on V is multiplication by $\theta(\ell)[\ell]^{-1}$, from which it follows that $a(\ell^k) = \theta(\ell)^k[\ell]^{-k}$ if $\ell \nmid Np$. On the other hand, condition (3) forces $a(\ell^k) = 0$ for all $k \geq 1$ for primes ℓ dividing Np. The algebra $\mathbb{T}_R$ contains diamond operators $\langle a \rangle$ for $a \in \Gamma$. This follows from the identity $\langle \ell \rangle = \ell^{-1}(T(\ell)^2 - T(\ell^2))$ for $\ell \nmid Np$, which also allows us to compute that

$\langle \ell \rangle \equiv \theta(\ell)[\ell]^{-1} \bmod I_T$. Thus, I_T is generated by $T(\ell)-1-\ell\langle \ell \rangle$ and $\langle \ell \rangle - \theta(\ell)[\ell]^{-1}$ for primes $\ell \nmid Np$ and $T(\ell) - 1$ for primes $\ell \mid Np$.

Noting that the image in $\mathbb{T}_R/I_T$ of every group element is also the image of an element of $\mathbb{T}[\theta]$, we now see that the map $T(n) \mapsto T(n)$ induces an isomorphism $(\mathbb{T}/I)[\theta] \xrightarrow{\sim} \mathbb{T}_R/I_T$ of $\mathbb{Z}[\Delta]$-modules, where $a \in \Delta$ acts by $\theta(a)\langle a \rangle^{-1}$ on the left and $\theta(a)\langle a \rangle^{-1} \equiv [a] \bmod I_T$ on the right. The induced map on Δ-coinvariants is the desired isomorphism. □

4.2.11. In setting (A_d), we have considered Galois cohomology groups of families of $(d-1)$-dimensional Galois representations in the variables given by Iwasawa theory. In case (i) of (A_d), for instance, V is a family of Galois representations in the cyclotomic variable. Of course, there are other families of Galois representations, such as Hida families, and we would like to consider them. Therefore, we introduce two additional settings (B_3) and (C_d) of study. We do not exclude any representations that are new at N from our families. Perhaps we should, but we prefer a simpler presentation.

4.2.12. We describe setting (B_3), in which we work in case (i) for $d = 3$.

Let $\mathfrak{h}$ and $\mathfrak{T}$ be as in Sect. 2.4, and consider the pair $(\mathfrak{h}^\circ, \mathfrak{T}^\circ(-1))$, where $\circ$ denotes the new-at-N part. Condition (1) holds for this pair (see 2.4.2). As a consequence of Poincaré duality, the ordinary étale homology group $\mathfrak{T}$ may be identified with the Tate twist of the ordinary étale cohomology group as $\mathfrak{h}[G_F]$-modules. The characteristic polynomials of Fr_ℓ and $T(\ell) \in \mathfrak{h}$ agree on the cohomology $\mathfrak{T}(-1)$ for any prime $\ell \neq p$. Thus, condition (2) is satisfied as well, and the map $\phi_{\mathfrak{T}^\circ(-1)}$ may be taken to be the identity map on Hecke operators.

Similarly to setting (A_d), we consider $R = \mathfrak{h}^\circ[\![\Gamma]\!]$ and $T = \mathfrak{h}[\![\Gamma]\!]^\sharp \otimes_{\mathfrak{h}} \mathfrak{T}^\circ(-1)$. The conditions (1)–(3) are again satisfied for (R, T), and we see that we have ϕ_T as in (3) such that $a(n) = T(n)[n]^{-1}$ for n prime to p. The Eisenstein ideal I_T of $\mathbb{T}_R$ is then generated by

$$1 \otimes T(n) - \sum_{\substack{m \mid n \\ (m,p)=1}} mT(m)[m]^{-1} \otimes 1 \in \varprojlim_r \mathfrak{h}[\![\Gamma]\!] \otimes \mathbb{T}^{(3)}(Np^r)$$

for all $n \geq 1$. Note also that we have an R-module isomorphism

$$Y = H^2_{\text{ét}}(\mathbb{Z}[\tfrac{1}{p}], T(3)) \cong \varprojlim_r H^2_{\text{ét}}(\mathbb{Z}[\zeta_{p^r}, \tfrac{1}{p}]^+, \mathfrak{T}^\circ(2)).$$

4.2.13. We describe setting (C_d), in which we work in case (iii) for $d \geq 2$.

Let us denote by $Y_1^{(d-1)}(N)$ the Drinfeld modular variety of dimension $d-2$ for $\tilde{K}_1^{(d-1)}(N)$ over F. We define T by

$$T = \mathrm{image}(H^{d-2}_{c,\text{ét}}(Y_1^{(d-1)}(N)_{/\overline{F}}, \mathbb{Z}_p)^\circ \to H^{d-2}_{\text{ét}}(Y_1^{(d-1)}(N)_{/\overline{F}}, \mathbb{Z}_p)^\circ),$$

where $\circ$ denotes the new part (in an appropriate sense). We then let R be the $\mathbb{Z}_p$-submodule of $\mathrm{End}_{\mathbb{Z}_p}(T)$ generated by the Hecke operators $T(\mathfrak{n})$ for nonzero ideals $\mathfrak{n}$ of $\mathcal{O}$.

We imagine but, for $d \geq 4$, are not certain that conditions (1)–(3) hold in this case and that we have ϕ_T such that $a(\mathfrak{n}) = T(\mathfrak{n})$ for all $\mathfrak{n}$. In any case, we may define the Eisenstein ideal I_T of $\mathbb{T}_R$ to be generated by

$$1 \otimes T(\mathfrak{n}) - \sum_{\mathfrak{d}|\mathfrak{n}} \mathfrak{N}(\mathfrak{d})T(\mathfrak{d}) \otimes 1 \in R \otimes \mathbb{T}^{(d)}(N),$$

for the nonzero ideals $\mathfrak{n}$ of $\mathcal{O}$. This Eisenstein ideal is all that we need to consider our question.

4.2.14. Our question for (B_3) and (C_d) is the same as it was for (A_d), so we can ask it for all:

Question. Is there a canonical isomorphism $\varpi\colon P \xrightarrow{\sim} Y$ of R-modules in any of the settings (A_d), (B_3), or (C_d)?

This question, which has been formulated rather carelessly, is still not fine enough to be a conjecture. We have more questions than answers: for instance, are the hypotheses that we have made sufficient, and to what extent are they necessary? What happens for the prime $p = 2$? We do not wish to exclude it from consideration. We have made many subtle choices that influence the story in profound yet inapparent ways: e.g., of congruence subgroups, Hecke algebras, Eisenstein ideals, and étale cohomology groups. Have we made the right choices for a correspondence? We are glad if the reader is inspired to answer these questions.

4.2.15. We end with our hope that it is possible to explicitly define the maps ϖ that are the desired isomorphisms in the settings (A_d), (B_3), and (C_d).

The groups $\mathcal{S}(N)$ often have explicit presentations very similar to those of Sect. 2.1. These are found in the work of Cremona (1984), Ash (1980), Kondo and Yasuda (2012), and others. So, explicit definitions of ϖ and affirmative answers to our questions would give explicit presentations of the arithmetic object Y.

The map ϖ should take a modular symbol to a cup product of d special units. As explained above, this has been done in cases (i) and (iii) for $d = 2$. Beyond these, the settings in which we hope to do this are:

- (A_2) in case (ii), using cup products of two elliptic units,
- (B_3) using cup products of three Siegel units,
- (C_d) using cup products of d of the Siegel units in (Kondo and Yasuda 2012).

Goncharov has made closely related investigations into the first two of these settings (Goncharov 2008).

Acknowledgements The work of the first two authors (resp., third author) was supported in part by the National Science Foundation under Grant Nos. DMS-1001729 (resp., DMS-0901526).

References

Ash, A.: Cohomology of congruence subgroups of $SL_n(\mathbb{Z})$. Math. Ann. **249**, 55–73 (1980)

Busuioc, C.: The Steinberg symbol and special values of L-functions. Trans. Am. Math. Soc. **360**, 5999–6015 (2008)

Carlitz, L.: A class of polynomials. Trans. Am. Math. Soc. **43**, 167–182 (1938)

Cremona, J.: Hyperbolic tessellations, modular symbols, and elliptic curves over complex quadratic fields. Compos. Math. **51**, 275–324 (1984)

Drinfeld, V.: Parabolic points and zeta functions of modular curves. Funkcional. Anal. i Priložen. **7**, 83–84 (1973)

Drinfeld, V.: Elliptic modules. Mat. Sb. **94**, 594–627, 656 (1974)

Emerton, E.: The Eisenstein ideal in Hida's ordinary Hecke algebra. IMRN Int. Math. Res. Not. **1999** (15), 793–802 (1999)

Fukaya, T.: Coleman power series for K_2 and p-adic zeta functions of modular forms. In: Kazuya Kato's fiftieth birthday. Doc. Math. **Extra Vol.**, 387–442 (2003)

Fukaya, T., Kato, K.: On conjectures of Sharifi, 121pp (Preprint)

Gekeler, E.: A note on the finiteness of certain cuspidal divisor class groups. Isr. J. Math. **118**, 357–368 (2000)

Goncharov, A.: Euler complexes and geometry of modular varieties. Geom. Funct. Anal. **17**, 1872–1914 (2008)

Harder, G., Pink, R.: Modular konstruierte unverzweigte abelsche p-Erweiterungen von $\mathbb{Q}(\zeta_p)$ und die Struktur ihrer Galoisgruppen. Math. Nachr. **159**, 83–99 (1992)

Hayes, D.: Explicit class field theory for rational function fields. Trans. Am. Math. Soc. **189**, 77–91 (1974)

Hida, H.: Iwasawa modules attached to congruences of cusp forms. Ann. Sci. École Norm. Sup. **19**, 231–273 (1986)

Kato, K.: p-adic Hodge theory and values of zeta functions of modular forms. In: Cohomologies p-adiques et Applications Arithmétique, III. Astérisque, vol. 295, pp. 117–290. Société Mathématique de France, Paris (2004)

Kondo, S.: Euler systems on Drinfeld modular curves and zeta values. Dissertation, University of Tokyo (2002)

Kondo, S.,Yasuda, S.: Zeta elements in the K-theory of Drinfeld modular varieties. Math. Ann. **354**, 529–587 (2012)

Kurihara, M.: Ideal class groups of cyclotomic fields and modular forms of level 1. J. Number Theory **45**, 281–294 (1993)

Manin, J.: Parabolic points and zeta-functions of modular curves. Math. USSR Izvsetija **6**, 19–64 (1972)

Mazur, B., Wiles, A.: Class fields of abelian extensions of $\mathbb{Q}$. Invent. Math. **76**, 179–330 (1984)

McCallum, W., Sharifi, R.: A cup product in the Galois cohomology of number fields. Duke Math. J. **120**, 269–310 (2003)

Ochiai, T.: On the two-variable Iwasawa main conjecture. Compos. Math. **142**, 1157–1200 (2006)

Ohta, M.: On the p-adic Eichler-Shimura isomorphism for Λ-adic cusp forms. J. reine angew. Math. **463**, 49–98 (1995)

Ohta, M.: Ordinary p-adic étale cohomology groups attached to towers of elliptic modular curves. II. Math. Ann. **318**, 557–583 (2000)

Ohta, M.: Congruence modules related to Eisenstein series. Ann. Éc. Norm. Sup. **36**, 225–269 (2003)

Pal, A.: The rigid analytical regulator and K_2 of Drinfeld modular curves. Publ. Res. Inst. Math. Sci. **46**, 289–334 (2010)

Ribet, K.: A modular construction of unramified p-extensions of $\mathbb{Q}(\mu_p)$. Invent. Math. **34**, 151–162 (1976)

Sharifi, R.: A reciprocity map and the two variable p-adic L-function. Ann. Math. **173**, 251–300 (2011)

Tate, J.: Relations between K_2 and Galois cohomology. Invent. Math. **36**, 257–274 (1976)
Teitelbaum, J.: Modular symbols for $F_q(T)$. Duke Math. J. **69**, 271–295 (1992)
Wiles, A.: On ordinary λ-adic representations associated to modular forms. Invent. Math. **94**, 529–573 (1988)

Weber's Class Number One Problem

Takashi Fukuda, Keiichi Komatsu, and Takayuki Morisawa

1 Weber's Class Number One Problem

Two hundred years ago, Gauss made the following conjecture.

Conjecture 1. There are infinitely many real quadratic fields with class number one.

This conjecture is still open not only for quadratic fields but also for algebraic number fields with arbitrary finite degree over $\mathbf{Q}$.

Conjecture 2. There are infinitely many algebraic number fields with class number one.

To study this problem, we focus our attention on the cyclotomic $\mathbf{Z}_p$-extension of the rational number field $\mathbf{Q}$.

Let p be a prime number. We denote by $\mathbf{B}(p^\infty)$ and $\mathbf{B}(p^n)$ the cyclotomic $\mathbf{Z}_p$-extension of $\mathbf{Q}$ and its n-th layer, respectively. Then, for example, we know that $\mathbf{B}(2^n) = \mathbf{Q}(\cos(2\pi/2^{n+2}))$ and $\mathbf{B}(3^n) = \mathbf{Q}(\cos(2\pi/3^{n+1}))$. We denote by $h(p^n)$ the class number of $\mathbf{B}(p^n)$. Now, we consider the following problem.

T. Fukuda
Department of Mathematics, College of Industrial Technology, Nihon University,
2-11-1 Shin-ei, Narashino, Chiba 275-0005, Japan
e-mail: fukuda.takashi@nihon-u.ac.jp

K. Komatsu
Department of Mathematics, School of Fundamental Science and Engineering,
Waseda University,3-4-1 Okubo,Shinjuku,Tokyo 169-8555,Japan
e-mail: kkomatsu@waseda.jp

T. Morisawa (✉)
Department of Mathematics, Faculty of Science and Technology, Keio University,
3-14-1 Hiyoshi, Kohoku, Yokohama, Kanagawa, 223-8522, Japan
e-mail: takayuki.morisawa@gmail.com

T. Bouganis and O. Venjakob (eds.), *Iwasawa Theory 2012*, Contributions in Mathematical and Computational Sciences 7, DOI 10.1007/978-3-642-55245-8_6,

Problem 1 (Weber's Class Number Problem). Is the class number $h(p^n)$ equal to one for any positive integer n?

Weber proved that $h(2^1) = h(2^2) = h(2^3) = 1$. Later, it was shown that $h(2^4) = 1$ by Bauer (1969), Cohn (1960) and Masley (1978) and $h(2^5) = 1$ by van der Linden (1982). In Masley (1978) and van der Linden (1982), we know that $h(3^1) = h(3^2) = h(3^3) = h(5^1) = h(7^1) = 1$. Linden also showed that $h(2^6) = h(3^4) = h(5^2) = h(11^1) = h(13^1) = 1$ under the generalized Riemann hypothesis.

However, it is very hard to compute the whole class number. So we focus on the ℓ-indivisibility of $h(p^n)$.

Problem 2. Is the class number $h(p^n)$ coprime to a prime number ℓ for any non-negative integer n?

In the case $\ell = p$, it was shown that p does not divide $h(p^n)$ by Weber (1886) ($p = 2$) and Iwasawa (1956) (in general). Thus, we study the non-p-part of $h(p^n)$. In this case, that is, $\ell \neq p$, there is Washington's result (Washington 1978) which says that the ℓ-part of $h(p^n)$ is bounded as n tends to ∞. However, we can not get any information on indivisibility from it. On the ℓ-indivisibility, there is an approach of Horie (2002, 2005a,b, 2007) and Horie-Horie (2008; 2009a; 2009b; 2010) which tries to attack $h(p^n)$ by using the cyclotomic units. The following theorem is a part of their results.

Theorem 1 (Horie, Horie-Horie).

(1) *Let $p = 2$. If a prime number ℓ satisfies $\ell \not\equiv \pm 1 \pmod 8$, then $h(2^n)$ is coprime to ℓ for any non-negative integer n.*
(2) *Let p be a prime number with $3 \leq p \leq 23$. If a prime number ℓ is a primitive root modulo p^2, then $h(p^n)$ is coprime to ℓ for any non-negative integer n.*

Although these results were very striking and very effective, there were many small prime numbers p and ℓ for which we did not know whether ℓ divides $h(p^n)$ or not. For example, it was not known whether ℓ divides $h(2^n)$ for $\ell = 7, 17, 23, 31, \ldots$. In order to consider the ℓ-indivisibility of $h(p^n)$ for such prime numbers p and ℓ, we showed the following theorem.

Theorem 2. (1) *Let $p = 2$, ℓ an odd prime number and 2^s the exact power of 2 dividing $\ell - 1$ or $\ell^2 - 1$ according as $\ell \equiv 1 \pmod 4$ or not. Put*

$$m(2,\ell) = 2s + \left\lfloor \frac{1}{2} \log_2(\ell - 1) \right\rfloor - 2,$$

where $\lfloor x \rfloor$ is the greatest integer not exceeding a real number x. If ℓ does not divide $h(2^{m(2,\ell)})$, then $h(2^n)$ is coprime to ℓ for any non-negative integer n.
(2) *Let $p = 3$, ℓ a prime number different from 3 and 3^s the exact power of 3 dividing $\ell^2 - 1$. Put*

$$m(3,\ell) = 2s + \left\lfloor \frac{1}{2}\log_3(\ell - 1) + \frac{1}{2} \right\rfloor - 1.$$

If ℓ does not divide $h(3^{m(3,\ell)})$, then $h(3^n)$ is coprime to ℓ for any non-negative integer n.

We prove the above theorem by using the method of Sinnott and Washington (see Washington 1997, section 16) and the Iwasawa main conjecture proved by Mazur-Wiles (1984). Theorem 2, together with numerical calculations based on Fukuda and Komatsu (2009) and Morisawa (2009), allows us to obtain the following theorem.

Theorem 3 (see Fukuda and Komatsu 2009, 2010, 2011 and Morisawa 2009, 2012). *If ℓ is a prime number with $\ell < 10^9$, then $h(2^n)$ and $h(3^n)$ are coprime to ℓ for any non-negative integer n.*

We also know the following theorem.

Theorem 4 (see Morisawa 2012, Morisawa and Okazaki 2013, and Okazaki preprint). *Let p and ℓ be different prime numbers. We denote by f the inertia degree at ℓ in $\mathbf{Q}(\mu_{2p})$ over $\mathbf{Q}$, where μ_m is the group of all m-th roots of unity. Let p^s be the exact power of p dividing $\ell^f - 1$ and $c = (p-1)p^{s-1}$. We put*

$$B(p,s,f) = \begin{cases} (A^c \cdot c!)^{1/f}, & \text{if } p = 2 \\ \left(\sqrt{2}^c \cdot c!\right)^{1/f}, & \text{if } p = 3 \\ \left(\left(\frac{\sqrt{6p}}{2}\right)^c \cdot c!\right)^{1/f}, & \text{if } p \geq 5 \end{cases}$$

where A is the constant defined by $A = 0.80785\cdots$. Then $h(p^n)$ is coprime to ℓ for any non-negative integer n, if $\ell > B(p,s,f)$.

By combining Theorems 3 and 4, we get the following corollary.

Corollary 1. (1) *If ℓ is a prime number with $\ell \not\equiv \pm 1 \pmod{32}$, then $h(2^n)$ is coprime to ℓ for any non-negative integer n.*
(2) *If ℓ is a prime number with $\ell \not\equiv \pm 1 \pmod{27}$, then $h(3^n)$ is coprime to ℓ for any non-negative integer n.*

2 Composites of $\mathbf{Z}_p$-Extensions of $\mathbf{Q}$

In this section, we consider the class number of the intermediate field in the$\hat{\mathbf{Z}}$-extension of $\mathbf{Q}$.

We denote by $\mathbf{B}(\infty)$ the composite of $\mathbf{B}(p^\infty)$ for all prime numbers p. Then the Galois group of $\mathbf{B}(\infty)$ over $\mathbf{Q}$ is isomorphic to $\hat{\mathbf{Z}}$ as a topological group. For any positive integer $N \in \mathbf{N}$, we denote by $\mathbf{B}(N)$ and $h(N)$ the unique subfield of $\mathbf{B}(\infty)$

with degree N over $\mathbf{Q}$ and its class number, respectively. On this extension, Coates (2012) conjectured the following.

Conjecture 3. There exists a number $C_{\mathbf{Q}}$ not depending on N, such that$h(N) \le C_{\mathbf{Q}}$ for all $N \in \mathbf{N}$.

In this direction, Coates-Liang-Mihailescu verified that $h(N) = 1$ for $1 \le N \le 28$ (see Coates 2012).

Originally, we wanted to attack the above conjecture. However, the $\hat{\mathbf{Z}}$-extension is too large to study. Therefore we restrict to the case of $\mathbf{Z}_\Sigma$-extensions.

Let Σ be a non-empty finite set of prime numbers. Put $\mathbf{Z}_\Sigma = \prod_{p\in\Sigma} \mathbf{Z}_p$ and

$$\mathbf{N}_\Sigma = \left\{ \prod_{p\in\Sigma} p^{n_p} \;\middle|\; n_p \in \mathbf{Z}_{\ge 0} \right\}.$$

Then the family $\{\mathbf{B}(N) \mid N \in \mathbf{N}_\Sigma\}$ is the family of all intermediate fields of $\mathbf{Z}_\Sigma$-extension of $\mathbf{Q}$ with finite degree over $\mathbf{Q}$. We consider the following problem.

Problem 3. Let Σ be a non-empty finite set of prime numbers and ℓ a prime number. Is the class number $h(N)$ coprime to ℓ for all $N \in \mathbf{N}_\Sigma$ or not?

Let $\Omega_\Sigma = \mathbf{Q}(\mu_N \mid N \in \mathbf{N}_\Sigma)$ and denote by $\Omega_\Sigma(\ell)$ the decomposition field of a prime number ℓ in Ω_Σ over $\mathbf{Q}$. For an intermediate field F of Ω_Σ with finite degree over $\mathbf{Q}$, we denote by $cond(F)$ the conductor of F and put $d = \varphi(cond(F))$ and $f = d[F : \mathbf{Q}]^{-1}$ where φ is the Euler function and $[F : \mathbf{Q}]$ is the degree of F over $\mathbf{Q}$. We also put

$$D(\Sigma, F) = \{\ell : \text{prime number} \notin \Sigma \mid \Omega_\Sigma(\ell) = F\}$$

and

$$B(\Sigma, F) = \left(\left(\prod_{p\in\Sigma} 2p \right)^d \cdot d! \right)^{1/f}.$$

Then, on the ℓ-indivisibility, we have the following theorem.

Theorem 5 (see Morisawa 2013). *If a prime number $\ell \in D(\Sigma, F)$ satisfies $\ell > B(\Sigma, F)$, then $h(N)$ is coprime to ℓ for all $N \in \mathbf{N}_\Sigma$.*

The above theorem says that the class numbers are coprime to ℓ if ℓ is sufficiently large in comparison with the conductor of its decomposition field in Ω_Σ.

Example 1. Put $\Sigma = \{3, 5\}$ and $F = \mathbf{Q}(\sqrt{-15})$. Then we have $cond(F) = 15$, $d = 8$ and $f = 4$. We obtain

$$B(\{3, 5\}, \mathbf{Q}(\sqrt{-15})) = 51{,}013.2075\cdots.$$

For example, $\ell = 51197, 51{,}203, 51{,}287, 51{,}347$ and $51{,}383$ are prime numbers and their decomposition field is $\mathbf{Q}(\sqrt{-15})$. Therefore, Theorem 5 implies that $h(3^n 5^m)$ is coprime to ℓ for all n and m.

On the other hand, there are small prime divisors of $h(N)$. The following prime divisors are all we know.

Theorem 6. (1) *We have* $31 \mid h(2 \cdot 31)$, $1{,}546{,}463 \mid h(2 \cdot 1{,}546(,)463)$, $73 \mid h(3 \cdot 73)$, $18{,}433 \mid h(2^8 \cdot 18{,}433)$, $114{,}689 \mid h(2^{10} \cdot 114{,}689)$, $73 \mid h(3 \cdot 73)$, $487 \mid h(3^4 \cdot 487)$, $238{,}627 \mid h(3^4 \cdot 238{,}627)$ *and* $2{,}251 \mid h(5^2 \cdot 2{,}251)$.
(2) *We have* $107 \mid h(2 \cdot 53)$.

The case (1) is a consequence of Proposition 2 in Inatomi (1986) and the case (2) was found by using Theorem 2 in Aoki and Fukuda (2006).

Acknowledgements The authors thank the organizers of Iwasawa 2012 for giving us the opportunity to talk. The authors also thank the referee for reading this paper carefully and offering several invaluable suggestions.

References

Aoki, M., Fukuda, T.: An algorithm for computing p-class groups of abelian number fields.In: Hess, F., Pauli, S., Pohst, M. (eds.) Algorithmic Number Theory. Lecture Notes in Computer Science, vol. 4076, pp. 56–71. Springer, Berlin/Heidelberg (2006)

Bauer, H.: Numerische Bestimmung von Klassenzahlen reeller zyklischer Zahlkörper. J. Number Theory **1**, 161–162 (1969)

Coates, J.: The enigmatic Tate-Shafarevich group. In: Fifth International Congress of Chinese Mathematicians. Part 1, 2, AMS/IP Studies in Advanced Mathematics, vol. 51, Part. 1, 2, pp. 43–50. American Mathematical Society, Providence (2012)

Cohn, H.: A numerical study of Weber's real class number calculation I. Numer. Math. **2**, 347–362 (1960)

Fukuda,T., Komatsu, K.: Weber's class number problem in the cyclotomic $\mathbf{Z}_2$-extension of $\mathbf{Q}$. Exp. Math. **18**(2), 213–222 (2009)

Fukuda, T., Komatsu, K.: Weber's class number problem in the cyclotomic $\mathbf{Z}_2$-extension of $\mathbf{Q}$, II. J. Théor. Nombres Bordeaux **22**(2), 359–368 (2010)

Fukuda, T., Komatsu, K.: Weber's class number problem in the cyclotomic $\mathbf{Z}_2$-extension of $\mathbf{Q}$, III. Int. J. Number Theory **7**(6), 1627–1635 (2011)

Horie, K.: Ideal class groups of Iwasawa-theoretical abelian extensions over the rational field. J. Lond. Math. Soc. **66**, 257–275 (2002)

Horie, K.: Primary components of the ideal class group of the $\mathbf{Z}_p$-extension over $\mathbf{Q}$ for Typical Inert Primes. Proc. Jpn. Acad. Ser. A Math. Sci. **81**(3), 40–43 (2005a)

Horie, K: The ideal class group of the basic $\mathbf{Z}_p$-extension over an imaginary quadratic field. Tohoku Math. J. (2) **57**, 375–394 (2005b)

Horie, K.: Certain primary components of the ideal class group of the $\mathbf{Z}_p$-extension over the rationals. Tohoku Math. J. **59**, 259–291 (2007)

Horie, K., Horie, M.: The narrow class groups of some $\mathbf{Z}_p$-extensions over the rationals. Acta Arith. **135**(2), 159–180 (2008)

Horie, K., Horie, M.: The ideal class group of the $\mathbf{Z}_p$-extension over the rationals. Tohoku Math. J. **61**, 551–570 (2009a)

Horie, K., Horie, M.: The ideal class group of the $\mathbf{Z}_{23}$-extension over the rational field. Proc. Jpn. Acad. Ser. A **85**, 155–159 (2009b)

Horie, K., Horie, M.: The narrow class groups of the $\mathbf{Z}_{17}$- and $\mathbf{Z}_{19}$-extensions over the rational field. Abh. Math. Semin. Univ. Hambg. **80**(1), 47–57 (2010)

Inatomi, A.: On $\mathbf{Z}_p$-extensions of real abelian fields. Kodai Math. J. **12**, 420–422 (1986)

Iwasawa, K.: A note on class numbers of algebraic number fields. Abh. Math. Sem. Univ. Hambg. **20**, 257–258 (1956)

Masley, J.M.: Class numbers of real cyclic number fields with small conductor. Compos. Math. **37**, 297–319 (1978)

Mazur, B., Wiles, A.: Class fields of abelian extensions of $\mathbf{Q}$. Invent. Math. **76**, 179–330 (1984)

Morisawa, T.: A class number problem in the cyclotomic $\mathbf{Z}_3$-extension of $\mathbf{Q}$. Tokyo J. Math. **32**, 549–558 (2009)

Morisawa, T.: Mahler measure of the Horie unit and Weber's class number problem in the cyclotomic $\mathbf{Z}_3$-extension of $\mathbf{Q}$. Acta Arith. **153**(1), 35–49 (2012)

Morisawa, T.: On the ℓ-part of the $\mathbf{Z}_{p_1} \times \cdots \times \mathbf{Z}_{p_s}$-extension of $\mathbf{Q}$. J. Number Theory **133**(6), 1814–1826 (2013)

Morisawa, T., Okazaki, R.: Mahler measure and Weber's class number problem in the cyclotomic $\mathbf{Z}_p$-extension of $\mathbf{Q}$ for odd prime number p. Tohoku Math. J. **65**(2), 253–272 (2013)

Okazaki, R.: On a lower bound for relative units, Schinzel's lower bound and Weber's class number problem (preprint)

van der Linden, F.J.: Class number computations of real abelian number fields. Math. Comput. **39**, 693–707 (1982)

Washington, L.C.: The non-p-part of the class number in a cyclotomic $\mathbf{Z}_p$-extension. Invent. Math. **49**, 87–97 (1978)

Washington, L.C.: Introduction to Cyclotomic Fields. Graduate Texts in Mathematics, vol. 83, 2nd edn. Springer, New York/Heidelberg/Berlin (1997)

Weber, H.: Theorie der Abel'schen Zahlkörper. Acta Math. **8**, 193–263 (1886)

On p-Adic Artin L-Functions II

Ralph Greenberg

1 Introduction

Let p be a prime. Iwasawa's famous conjecture relating Kubota-Leopoldt p-adic L-functions to the structure of certain Galois groups has been proven by Mazur and Wiles in Mazur and Wiles (1984). Wiles later proved a far-reaching generalization involving p-adic L-functions for Hecke characters of finite order for a totally real number field in Wiles (1990). As we discussed in Greenberg (1983), an analogue of Iwasawa's conjecture for p-adic Artin L-functions can then be deduced. The formulation again involves certain Galois groups. However, one can reformulate this result in terms of Selmer groups for the Artin representations. There are several advantages to such a reformulation. First of all, it fits perfectly into the much broader framework described in Greenberg (1989) which relates the p-adic L-function for a motive to the corresponding Selmer group. The crucial assumption in Greenberg (1989) that the motive be ordinary at p (or at least potentially ordinary) is satisfied by an Artin motive and all of its Tate twists.

A second advantage of a reformulation involving Selmer groups is that the issue of how to define the μ-invariant becomes resolved in a natural and transparent way. Thirdly, the arguments in Greenberg (1983) can be simplified. In particular, there is no need for singling out the class of Artin representations which are called type S in Greenberg (1983). The purpose of this paper is to explain these advantages.

Suppose that F is a totally real number field. Consider an Artin representation

$$\rho : G_F \longrightarrow \mathrm{Aut}_{\mathcal{E}}(V) ,$$

R. Greenberg (✉)
Department of Mathematics, University of Washington, Seattle, WA 98195, USA
e-mail: ralph.greenber@gmail.com

T. Bouganis and O. Venjakob (eds.), *Iwasawa Theory 2012*, Contributions in Mathematical and Computational Sciences 7, DOI 10.1007/978-3-642-55245-8_7,

where G_F is the absolute Galois group of F and V is a finite dimensional vector space over a finite extension $\mathcal{E}$ of $\mathbf{Q}_p$. We will assume that ρ is totally even. This means that ρ factors through $\Delta = \mathrm{Gal}(K/F)$, where K is a finite extension of F which is also totally real. Let $\mathcal{O}$ be the ring of integers of $\mathcal{E}$. Let T be an $\mathcal{O}$-lattice in V which is G_F-invariant. Furthermore, let $D = V/T$, a discrete $\mathcal{O}$-module.

Let F_∞ denote the cyclotomic $\mathbf{Z}_p$-extension of F. The Selmer group associated to D over F_∞ is defined by

$$\mathrm{Sel}_D(F_\infty) \;=\; \ker\Big(H^1(F_\infty, D) \longrightarrow \prod_{\eta \nmid p} H^1(F_{\infty,\eta}, D)\Big)\ .$$

Here η runs over all the primes of F_∞ except for the finitely many primes lying over p. The Archimedean primes are included in the product, although this is only important when $p = 2$. One defines the field $F_{\infty,\eta}$ to be the union of the η-adic completions of the finite extensions of F contained in F_∞.

To relate the above definition to the way Selmer groups are defined in Greenberg (1989), note that if ρ is a totally even Artin representation of G_F over $\mathbf{C}$, then the Artin L-function $L(s,\rho)$ does not have a critical value at $s = 1$ in the sense of Deligne. However, its value at $s = 1-n$ is critical in that sense when n is even and positive. One can write

$$L(1-n,\rho) = L(1,\rho(n))$$

where $\rho(n)$ is the n-th Tate twist. The underlying representation space for $\rho(n)$ over $\mathcal{E}$ is $V(n) = V \otimes \chi_F^n$, where $\chi_F : G_F \to \mathbf{Z}_p^\times$ is the p-power cyclotomic character. In the notation of Greenberg (1989), we have $F^+V(n) = V(n)$ when $n \geq 1$. (This is so for all the primes above p.) Let $T(n) = T \otimes \chi^n$ and $D(n) = V(n)/T(n)$. The corresponding Selmer group $\mathrm{Sel}_{D(n)}(F_\infty)$, as it is defined in Greenberg (1989), is just as above, but with $D(n)$ replacing D. Let $d = [F(\mu_q) : F]$, where $q = p$ when p is odd and $q = 4$ when $p = 2$. If we take $n \equiv 0 \pmod{d}$, then $D(n) \cong D$ for the action of G_{F_∞}. Thus, the two Selmer groups are then the same, although the action of $\mathrm{Gal}(F_\infty/F)$ on those groups is somewhat different. (See Remark 6.)

Since $\mathrm{Sel}_D(F_\infty)$ is a discrete $\mathcal{O}$-module and $\Gamma_F = \mathrm{Gal}(F_\infty/F)$ acts naturally and continuously on it, we can regard $\mathrm{Sel}_D(F_\infty)$ as a discrete $\Lambda_{(\mathcal{O},F)}$-module, where $\Lambda_{(\mathcal{O},F)} = \mathcal{O}[[\Gamma_F]]$. It is not difficult to show that the Pontryagin dual $X_D(F_\infty)$ of $\mathrm{Sel}_D(F_\infty)$ is a finitely generated, torsion $\Lambda_{(\mathcal{O},F)}$-module. (See Proposition 1) We denote the characteristic ideal of that $\Lambda_{(\mathcal{O},F)}$-module by I_ρ. It is a principal ideal in the ring $\Lambda_{(\mathcal{O},F)}$. As the notation suggests, this ideal depends only on ρ, and not on the choice of the Galois-invariant $\mathcal{O}$-lattice T, as we show in Proposition 2.

Another discrete $\Lambda_{(\mathcal{O},F)}$-module to be considered is $H^0(F_\infty, D)$. Its Pontryagin dual $Y_D(F_\infty)$ is clearly a finitely-generated $\mathcal{O}$-module and hence a torsion $\Lambda_{(\mathcal{O},F)}$-module. Let J_ρ denote the characteristic ideal of $Y_D(F_\infty)$. This ideal is nontrivial if and only if ρ has at least one irreducible constituent which factors through Γ_F.

The p-adic L-function associated to ρ will be denoted by $L_p(s,\rho)$. It is characterized by a certain interpolation property. In case ρ is 1-dimensional, these functions have been constructed by Deligne and Ribet (1980), by Cassou-Noguès (1979), and by Barsky (1977–1978). (See also Ribet (1979).) One can then define $L_p(s,\rho)$ if ρ has arbitrary dimension by using a classical theorem of Brauer from group theory.

One can associate to $L_p(s,\rho)$ a certain element θ_ρ in the fraction field of $\Lambda_{(\mathcal{O},F)}$. For an odd prime p, the Main Conjecture is the assertion that the fractional ideals $\Lambda_{(\mathcal{O},F)}\theta_\rho$ and $I_\rho J_\rho^{-1}$ are the same. This is proved in Sect. 4 as a consequence of theorems of Wiles proved in Wiles (1990). For $p = 2$, there is an extra power of 2 in the formulation, but this case appears to still be open.

2 Basic Results Concerning the Selmer Group

We will prove several useful propositions. We continue to make the same assumptions as in the introduction. In particular, F is a totally real number field and ρ is a totally even Artin representation of G_F defined over a field $\mathcal{E}$, a finite extension of $\mathbf{Q}_p$. The ring of integers in $\mathcal{E}$ is denoted by $\mathcal{O}$. Let $\mathfrak{m}$ be the maximal ideal of $\mathcal{O}$.

We will use the traditional terminology for modules over a topological ring Λ. If S is a discrete Λ-module, and X is its Pontryagin dual, then we say that S is a cofinitely-generated Λ-module if X is finitely-generated. If X is a torsion Λ-module, we say that S is cotorsion.

Suppose that V, T, and $D = V/T$ are as in the introduction. Let $d = \dim_{\mathcal{E}}(V)$. As an $\mathcal{O}$-module, we have $D \cong (\mathcal{E}/\mathcal{O})^d$. The Selmer group $\mathrm{Sel}_D(F_\infty)$ is a discrete $\Lambda_{(\mathcal{O},F)}$-module.

Proposition 1. *The $\Lambda_{(\mathcal{O},F)}$-module $S_D(F_\infty)$ is cofinitely-generated and cotorsion.*

Proof. Suppose that ρ factor through $\mathrm{Gal}(K/F)$, where K is a totally real, finite Galois extension of F. Let $\Delta = \mathrm{Gal}(K_\infty/F_\infty)$ and let M_∞ be the maximal abelian pro-p extension of K_∞ which is unramified at the primes of K_∞ not lying over p (including the Archimedean primes). One can consider $X(K_\infty) = \mathrm{Gal}(M_\infty/K_\infty)$ as a module over $\Lambda_{(\mathbf{Z}_p,K)} = \mathbf{Z}_p[[\Gamma_K]]$. A well-known theorem of Iwasawa asserts that $X(K_\infty)$ is finitely-generated and torsion as a $\Lambda_{(\mathbf{Z}_p,K)}$-module. The fact that it is torsion is equivalent to the fact that the so-called weak Leopoldt conjecture is valid for K_∞/K.

We have $H^0(K_\infty, D) = D$. Also, $H^1(\Delta, D)$ is finite. Hence the restriction map

$$H^1(F_\infty, D) \longrightarrow H^1(K_\infty, D)^\Delta \tag{1}$$

has finite kernel. We can identify Γ_K with $\mathrm{Gal}(F_\infty/K \cap F_\infty)$, a subgroup of Γ_F. The map (1) is then Γ_K-equivariant. Now $H^1(K_\infty, D) = \mathrm{Hom}(\mathrm{Gal}(K_\infty^{ab}/K_\infty), D)$, where K_∞^{ab} is the maximal abelian extension of K_∞. Since the inertia subgroups of $\mathrm{Gal}(K_\infty^{ab}/K_\infty)$ for all primes $\eta \nmid p$ generate $\mathrm{Gal}(K_\infty^{ab}/M_\infty)$, it is clear that the image of $\mathrm{Sel}_D(F_\infty)$ under the map (1) is contained in $\mathrm{Hom}(X(K_\infty), D)$, which is a cofinitely-generated, cotorsion $\Lambda_{(\mathbf{Z}_p,K)}$-module according to Iwasawa's theorem. Since (1) has finite kernel, it follows that $\mathrm{Sel}_D(F_\infty)$ is cofinitely-generated and cotorsion as a $\Lambda_{(\mathbf{Z}_p,K)}$-module, and therefore as a $\Lambda_{(\mathcal{O},F)}$-module. □

Remark 1. With the notation of the above proof, the cokernel of the map (1) is also finite. This follows from the fact that $H^2(\Delta, D)$ is finite. Assume that the order of $\mathrm{im}(\rho)$ is not divisible by p. We can then assume that $p \nmid |\Delta|$. In particular, $K \cap F_\infty = F$. Hence the map $\Gamma_K \to \Gamma_F$ is an isomorphism. We then have $\mathrm{Gal}(K_\infty/F) \cong \Delta \times \Gamma_F$. Furthermore, (1) is an isomorphism. The induced map

$$\mathrm{Sel}_D(F_\infty) \longrightarrow \mathrm{Hom}_\Delta(X(K_\infty), D) \tag{2}$$

is also easily verified to be an isomorphism. In addition to assuming $p \nmid |\Delta|$, assume that ρ is absolutely irreducible. Let e_ρ be the idempotent for ρ in $\mathcal{O}[\Delta]$. Then

$$\begin{aligned}\mathrm{Hom}_\Delta(X(K_\infty),\ D) &\cong \mathrm{Hom}_{\mathcal{O}[\Delta]}\big(X(K_\infty)\otimes_{\mathbf{Z}_p}\mathcal{O},\ D\big)\\ &\cong \mathrm{Hom}_{\mathcal{O}[\Delta]}\big(e_\rho X(K_\infty)\otimes_{\mathbf{Z}_p}\mathcal{O},\ D\big)\ .\end{aligned}$$

Thus, the $\Lambda_{(\mathcal{O},F)}$-modules $X_D(F_\infty)$ and $e_\rho X(K_\infty) \otimes_{\mathbf{Z}_p} \mathcal{O}$ are closely related. In fact, the characteristic ideal of the second module is I_ρ^d.

Remark 2. Suppose that X is any $\Lambda_{(\mathcal{O},F)}$-module and that I is the characteristic ideal of X. Let π be a generator of $\mathfrak{m}$, the maximal ideal of $\mathcal{O}$. The μ-invariant of X will be denoted by $\mu_F(X)$. It is the integer μ characterized by $I \subseteq \pi^\mu \Lambda_{(\mathcal{O},F)}$, $I \not\subseteq \pi^{\mu+1}\Lambda_{(\mathcal{O},F)}$. A conjecture of Iwasawa (at least for odd primes p) asserts that the μ-invariant of the $\Lambda_{(\mathcal{O},K)}$-module X_{K_∞} should vanish. This should be true even for $p = 2$. If this is so, then the proof of Proposition 1 would show that $\mu_F\big(X_D(F_\infty)\big)$ also vanishes.

It will be useful to have an alternative definition of $\mathrm{Sel}_D(F_\infty)$. Let Σ be a finite set of primes of F containing the Archimedean primes, the primes lying over p, and the ramified primes for ρ. For each $v \in \Sigma$, define

$$\mathcal{H}_v^1(F_\infty, D) = \varinjlim_n \bigoplus_{\nu|v} H^1(F_{n,\nu}, D)$$

where, for each n, ν runs over the primes of F_n lying over v. The maps defining the direct limit are induced by the local restriction maps. If v is a finite prime, then

$$\mathcal{H}_v^1(F_\infty, D) = \bigoplus_{\nu|v} H^1(F_{\infty,\nu}, D)$$

where ν runs over the finite set of primes of F_∞ lying over v. The p-cohomological dimension of $G_{F_{\infty,\nu}}$ is 1, and so $H^2(F_{\infty,\nu}, D[\pi]) = 0$. It follows that $H^1(F_{\infty,\nu}, D)$ is $\mathcal{O}$-divisible. Now assume that $v \nmid p$. Then, according to proposition 2 in Greenberg (1989), the $\mathcal{O}$-corank of $H^1(F_{\infty,\nu}, D)$ is finite. It therefore follows that the Pontryagin dual of $\mathcal{H}_v^1(F_\infty, D)$ is a torsion-free $\mathcal{O}$-module of finite rank. Hence it is a free $\mathcal{O}$-module. It is also a $\Lambda_{(\mathcal{O},F)}$-module. Since its $\mathcal{O}$-rank is finite, it must be a torsion $\Lambda_{(\mathcal{O},F)}$-module whose μ-invariant vanishes.

If p is odd and v is an Archimedean prime, then $\mathcal{H}_v^1(F_\infty, D) = 0$. However, if $p = 2$, then $\mathcal{H}_v^1(F_\infty, D)$ is nontrivial. More precisely, since all the F_n's are totally real, and ρ is totally even (so that $G_{F_{n,\nu}}$ acts trivially on D), we have

$$H^1(F_{n,\nu}, D) = H^1(\mathbf{R}, D) = D[2] \cong (\mathcal{O}/2\mathcal{O})^d$$

if v is Archimedean. One sees that $\mathcal{H}_v^1(F_\infty, D)$ is a direct limit of modules isomorphic to $(\mathcal{O}/2\mathcal{O})[\mathrm{Gal}(F_n/F)]^d$. The Pontryagin dual of $\mathcal{H}_v^1(F_\infty, D)$ is isomorphic to $\left(\Lambda_{(\mathcal{O},F)}/2\Lambda_{(\mathcal{O},F)}\right)^d$ as a $\Lambda_{(\mathcal{O},F)}$-module. It is a torsion $\Lambda_{(\mathcal{O},F)}$-module, but its μ-invariant is positive and is determined by $d = \dim_{\mathcal{E}}(V)$.

One can similarly define $\mathcal{H}_v^2(F_\infty, D)$ as a direct limit by replacing the H^1's by H^2's. However, for any finite prime v, the p-cohomological dimension of $G_{F_{\infty,\nu}}$ is 1 and so $\mathcal{H}_v^2(F_\infty, D)$ vanishes. This is also true if $v|\infty$ because $H^2(\mathbf{R}, D) = 0$.

The following definition is equivalent to the one given in the introduction:

$$\mathrm{Sel}_D(F_\infty) = \ker\Big(H^1(F_\Sigma/F_\infty, D) \longrightarrow \prod_{v\in\Sigma, v\nmid p} \mathcal{H}_v^1(F_\infty, D)\Big). \tag{3}$$

To verify the equivalence, we first point out that if v is a non-Archimedean prime of F and $v \nmid p$, and if ν is a prime of F_∞ lying over v, then $F_{\infty,\nu}$ is the unramified $\mathbf{Z}_p$-extension of F_v. It follows that the restriction map $H^1(F_{\infty,\nu}, D) \to H^1(F_v^{unr}, D)$ is injective. Here F_v^{unr} is the maximal unramified extension of F_v and contains $F_{\infty,\nu}$. Therefore, requiring a cocycle class to be trivial in $H^1(F_{\infty,\nu}, D)$ is equivalent to requiring it to be trivial in $H^1(F_v^{unr}, D)$.

Suppose now that ϕ is a 1-cocycle for G_{F_∞} with values in D. Note that we have $H^1(F_\Sigma, D) = \mathrm{Hom}(G_{F_\Sigma}, D)$. Also, G_{F_Σ} is generated topologically by the inertia subgroups I_η of G_F for all primes η of F_Σ lying over some $v \notin \Sigma$. Thus, the class $[\phi]$ in $H^1(F_\infty, D)$ has a trivial restriction to all those I_η's if and only if $[\phi]$ is in

$$\ker\big(H^1(F_\infty, D) \to H^1(F_\Sigma, D)\big) = \mathrm{im}\big(H^1(F_\Sigma/F_\infty, D) \to H^1(F_\infty, D)\big).$$

Thus, the cocycle classes in $H^1(F_\Sigma/F_\infty, D)$ can be identified under the inflation map with the cocyle classes in $H^1(F_\infty, D)$ which are unramified at all primes of F_∞ not lying over primes in Σ. The equivalence of (3) and the earlier definition follows.

It will also be useful to point out that the global-to-local map in (3) is surjective. This follows from proposition 2.1 in Greenberg and Vatsal (2000). It is only proved there for $F = \mathbf{Q}$ and odd p, but the argument works if F is totally real and for any p. One assumption is that $\mathrm{Sel}_D(F_\infty)$ is $\Lambda_{\mathcal{O},F}$-cotorsion, which is satisfied by Proposition 1 above. The other assumption is that $H^0(F_\infty, D^*)$ is finite. Here $D^* = \mathrm{Hom}(T, \mu_{p^\infty})$ and the finiteness is clear since F_∞ is totally real and $H^0(\mathbf{R}, \mu_{p^\infty})$ is finite (and even trivial if p is odd).

Let T' be another G_F-invariant $\mathcal{O}$-lattice in V. Let $D' = V/T'$. We consider $X_D(F_\infty)$ and $X_{D'}(F_\infty)$ as $\Lambda_{(\mathcal{O},F)}$-modules.

Proposition 2. *The $\Lambda_{(\mathcal{O},F)}$-modules $X_D(F_\infty)$ and $X_{D'}(F_\infty)$ have the same characteristic ideal.*

Proof. As above, let π be a generator of the maximal ideal of $\mathcal{O}$. Scaling by a power of π, we may assume that $T \subseteq T'$. We then have a G_F-equivariant map $\varphi : D \to D'$ with finite kernel. Such a map φ is called a G_F-isogeny. It is surjective. Let $\Phi = \ker(\varphi)$. Then $\Phi \subseteq D[\pi^t]$ for some $t \geq 0$. There is also a G_F-isogeny $\psi : D' \to D$ such that $\psi \circ \varphi$ is the map $D \to D$ given by multiplication by π^t.

The map φ induces a map from $\mathrm{Sel}_D(F_\infty)$ to $\mathrm{Sel}_{D'}(F_\infty)$ whose kernel is killed by π^t. Similarly, ψ induces such a map from $\mathrm{Sel}_{D'}(F_\infty)$ to $\mathrm{Sel}_D(F_\infty)$ and the compositum is multiplication by π^t. It follows that the characteristic ideals I and I' of $X_D(F_\infty)$ and $X_{D'}(F_\infty)$, respectively, are related as follows: $I' = \pi^s I$ for some $s \in \mathbf{Z}$. Thus, the proposition is equivalent to showing that the μ-invariants for the two modules are equal, and so $s = 0$.

Assume first that p is odd. In the definition (3), $\mathcal{H}_v^1(F_\infty, D) = 0$ when $v|\infty$ and $\mathcal{H}_v^1(F_\infty, D)$ has finite $\mathcal{O}$-corank when $v \in \Sigma, v \nmid p$. Thus, the μ-invariants for the Pontryagin duals of $\mathrm{Sel}_D(F_\infty)$ and $H^1(F_\Sigma/F, D)$ are equal. The same statement is true for $\mathrm{Sel}_{D'}(F_\infty)$ and $H^1(F_\Sigma/F, D')$. And so it suffices to prove that the Pontryagin duals of $H^1(F_\Sigma/F, D)$ and $H^1(F_\Sigma/F, D')$ have the same μ-invariants. This is sufficient even for $p = 2$. This follows because the map (3) and the corresponding global-to-local map for D' are both surjective. Furthermore, for any Archimedean prime v, the μ-invariants of $\mathcal{H}_v^1(F_\infty, D)$ and $\mathcal{H}_v^1(F_\infty, D')$ are equal.

Using the notation in the proof of Proposition 1, we have an exact sequence

$$\begin{aligned} H^0(F_\Sigma/F_\infty, D') &\longrightarrow H^1(F_\Sigma/F_\infty, \Phi) \\ &\longrightarrow H^1(F_\Sigma/F_\infty, D) \longrightarrow H^1(F_\Sigma/F_\infty, D') \\ &\longrightarrow H^2(F_\Sigma/F_\infty, \Phi) \longrightarrow H^2(F_\Sigma/F_\infty, D) \ . \end{aligned}$$

Now the μ-invariant of $H^0(F_\Sigma/F_\infty, D')$ certainly vanishes. Also, $H^2(F_\Sigma/F_\infty, D) = 0$. One can verify this for odd p by using propositions 3 and 4 in Greenberg (1989). For $H^2(F_\Sigma/F, D)$ is $\Lambda_{(\mathcal{O},F)}$-cotorsion by Proposition 3, and $\Lambda_{(\mathcal{O},F)}$-cofree by Proposition 4. For $p = 2$, $H^2(F_\Sigma/F, D)$ is still $\Lambda_{(\mathcal{O},F)}$-cotorsion. The analogue of Proposition 4 is that

$$\ker\big(H^2(F_\Sigma/F_\infty, D) \to \prod_{v|\infty} \mathcal{H}_v^2(F_\infty, D)\big)$$

is $\Lambda_{(\mathcal{O},F)}$-cofree, and hence must vanish. However, since $\mathcal{H}_v^2(F_\infty, D)$ vanishes, it follows that $H^2(\mathbf{R}, D) = 0$. Consequently, we indeed have $H^2(F_\Sigma/F_\infty, D) = 0$.

To complete the proof, we must show that $H^1(F_\Sigma/F_\infty, \Phi)$ and $H^2(F_\Sigma/F_\infty, \Phi)$ have the same μ-invariants as $\Lambda_{(\mathcal{O},F)}$-modules. In the statement of the proposition, we can reduce to the case where Φ is killed by π. Let $\tilde{\Lambda}_{(\mathcal{O},F)} = \Lambda_{(\mathcal{O},F)}/\pi\Lambda_{(\mathcal{O},F)}$. It then suffices to show that

$$\mathrm{corank}_{\tilde{\Lambda}_{(\mathcal{O},F)}}\big(H^1(F_\Sigma/F_\infty, \Phi)\big) = \mathrm{corank}_{\tilde{\Lambda}_{(\mathcal{O},F)}}\big(H^2(F_\Sigma/F_\infty, \Phi)\big)\ .$$

Here Φ is a representation space for G_F over $\mathcal{O}/(\pi)$ and is totally even. The Euler-Poincaré characteristic over F_∞, which is the alternating sum of the $\tilde{\Lambda}_{(\mathcal{O},F)}$-coranks of $H^i(F_\Sigma/F_\infty, \Phi)$ for $0 \le i \le 2$, turns out to be 0. The above equality follows.

The above assertion about the Euler-Poincaré characteristic for the $\mathrm{Gal}(F_\Sigma/F_\infty)$-module Φ is a consequence of the fact that if F_n denotes the n-th layer in the $\mathbf{Z}_p$-extension F_∞/F, then the Euler-Poincaré characteristic for the finite $\mathrm{Gal}(F_\Sigma/F_n)$-module Φ is equal to 1. (See Neukirch et al. 2000, (8.6.14), page 427.) Considering Φ as a vector space over $\mathfrak{f} = \mathcal{O}/(\pi)$, this means that the alternating sum of the $\mathfrak{f}$-dimensions of $H^i(F_\Sigma/F_n, \Phi)$ for $0 \le i \le 2$ is equal to 0. For the argument relating this fact to the above assertion about the $\tilde{\Lambda}_{(\mathcal{O},F)}$-coranks, we refer the reader to the proof of Proposition 4.1.1 in Greenberg (2011). The argument there is for a Galois module $E[p] \otimes \alpha$ (which is also a $\mathfrak{f}$-vector space), but applies to any such finite Galois module Φ, the only difference being the values of the Euler-Poincaré characteristics over the F_n's as n varies. □

Remark 3. As mentioned in Remark 2, $\mu\big(X_D(F_\infty)\big)$ should always vanish. The above proof would then show that $X_D(F_\infty)$ and $X_{D'}(F_\infty)$ are pseudo-isomorphic as $\Lambda_{(\mathcal{O},F)}$-modules. This is also true if $\mu\big(X_D(F_\infty)\big) = 1$. In contrast, for non-Artin motives, the μ-invariant of the Pontryagin dual of a Selmer group can be nonzero and can change under isogeny. This phenomenon was first pointed out by Mazur in Mazur (1972). The exact change in the μ-invariant under an isogeny is studied in Schneider (1987) and Perrin-Riou (1989). In fact, Proposition 2 is just a special case of the main theorem in Perrin-Riou (1989) when p is an odd prime.

Remark 4. If $\varphi : D \to D'$ is a G_F-isogeny and $\ker(\phi) = D[\mathfrak{m}^t]$ for some $t \ge 0$, then $D \cong D'$ as G_F-modules. This follows because the maximal ideal $\mathfrak{m}$ of $\mathcal{O}$ is a principal ideal. Any other G_F-isogeny will be called nontrivial. Such G_F-isogenies φ exist if and only if $T/\pi T$ is reducible as a G_F-representation space over the residue field $\mathcal{O}/(\pi)$. Now, if ρ is irreducible over $\mathcal{E}$ and $\mathrm{im}(\rho)$ has order prime to p, then it is well-known that $T/\pi T$ is also irreducible. In contrast, if $\mathrm{im}(\rho)$ has order divisible by p, then $T/\pi T$ may be reducible even if ρ is irreducible.

Proposition 3. *Suppose that ρ_1 and ρ_2 are totally even Artin representations of G_F. Let $\rho = \rho_1 \oplus \rho_2$. Then $I_\rho = I_{\rho_1} I_{\rho_2}$ and $J_\rho = J_{\rho_1} J_{\rho_2}$.*

Proof. We assume that $\mathcal{E}$ is a field of definition for ρ_1, ρ_2, and ρ. Let V_1 and V_2 be the underlying representations spaces for ρ_1 and ρ_2 over $\mathcal{E}$, and let $V = V_1 \oplus V_2$. Let T_1 and T_2 be Galois invariant $\mathcal{O}$-lattices in V_1 and V_2. Let $T = T_1 \oplus T_2$. Then $D = D_1 \oplus D_2$. With this choice of $\mathcal{O}$-lattices, it is clear that $\mathrm{Sel}_D(F_\infty) \cong \mathrm{Sel}_{D_1}(F_\infty) \oplus \mathrm{Sel}_{D_2}(F_\infty)$. The first equality in the proposition follows. We also have $H^0(F_\infty, D) \cong H^0(F_\infty, D_1) \oplus H^0(F_\infty, D_2)$, giving the second equality. □

Suppose that F' is a totally real, finite extension of F and that ρ' is a totally even Artin representation of $G_{F'}$. Thus, ρ' factors through $\mathrm{Gal}(K'/F')$, where K' is totally real. We can then define $\rho = \mathrm{Ind}_{G_{F'}}^{G_F}(\rho')$. Then ρ is an Artin representation of G_F and factors through $\mathrm{Gal}(K/F)$, where K is the Galois closure of K' over F. Note that K is totally real, and hence ρ is also totally even. Furthermore, there is an injective homomorphism $\Gamma_{F'} \to \Gamma_F$. Therefore, we can regard $\Lambda_{(\mathcal{O},F')}$ as a subring of $\Lambda_{(\mathcal{O},F)}$. One sees easily that $\Lambda_{(\mathcal{O},F)}$ is a finite integral extension of $\Lambda_{(\mathcal{O},F')}$ and that the degree is $[\Gamma_F : \Gamma_{F'}] = [F' \cap F_\infty : F]$. The characteristic ideals I and J are essentially unchanged by induction. To be precise, we have

Proposition 4. *With the above notation, we have* $I_\rho = I_{\rho'}\Lambda_{(\mathcal{O},F)}$ *and* $J_\rho = J_{\rho'}\Lambda_{(\mathcal{O},F)}$.

To simplify notation in the following proof, we will write $\mathrm{Ind}_{F'}^{F}(\rho')$ in place of $\mathrm{Ind}_{G_{F'}}^{G_F}(\rho')$. We will also write ρ_∞ and ρ'_∞ for the restrictions of ρ and ρ' to F_∞ and F'_∞, respectively.

Proof. We consider separately the two cases where $F' \cap F_\infty = F$ and $F' \subset F_\infty$. That will suffice because if $E = F' \cap F_\infty$, then $E_\infty = F_\infty$ and $\mathrm{Ind}_E^F\big(\mathrm{Ind}_{F'}^E(\rho')\big) = \rho$.

Suppose first that $F' \cap F_\infty = F$. In this case, we can identify $\Gamma_{F'}$ with Γ_F and hence $\Lambda_{(\mathcal{O},F')}$ with $\Lambda_{(\mathcal{O},F)}$. For brevity, let $G = G_F, G' = G_{F'}$, and $N = G_{F_\infty}$, a normal supgroup of G. Then $N \cap G' = G_{F'_\infty}$. Now suppose that K is a finite, totally real Galois extension of F which contains F' and such that ρ' factors through $\mathrm{Gal}(K/F')$. Then ρ factors through $\mathrm{Gal}(K/F)$. Furthermore, since $NG' = G$, we have $[G : G'] = [N : N \cap G']$, and it follows that

$$\rho|_N = \mathrm{Ind}_{G'}^{G}(\rho')\big|_N = \mathrm{Ind}_{N\cap G'}^{N}(\rho'|_N) \ .$$

Consequently, $\rho_\infty \cong \mathrm{Ind}_{F'_\infty}^{F_\infty}(\rho'_\infty)$.

Choose the Galois invariant $\mathcal{O}$-lattices for ρ and ρ' so that $\mathrm{Ind}_{G'}^{G}(D') = D$. Here we can replace G and G' by N and $N \cap G'$. Then $H^0(F_\infty, D) \cong H^0(F'_\infty D')$, and the isomorphism is equivariant for the action of $\Gamma_F = \Gamma_{F'}$. It follows that $J_\rho = J_{\rho'}$.

Note that $K_\infty = KF_\infty$ contains F'_∞. Let M_∞ be defined exactly as in the proof of Proposition 1. Then, for the reason given in that proof, we have injective maps

$$\mathrm{Sel}_D(F_\infty) \to H^1(M_\infty/F_\infty, D), \qquad \mathrm{Sel}_{D'}(F'_\infty) \to H^1(M_\infty/F'_\infty, D') \ . \quad (4)$$

The cokernels of these maps are finite. The proof is the same for both maps, and so we just discuss the first map. If η is any non-Archimedean prime of M_∞ not dividing p, let A_η denote the corresponding inertia subgroup of $\mathrm{Gal}(M_\infty/F_\infty)$. Note that the image of the first map in (4) consists of 1-cocycle classes which have a trivial restriction to $H^1(A_\eta, D)$ for all those η's. It is enough to consider the A_η's for $\eta|\nu$, where ν is a prime of F_∞, $\nu \nmid p$, and ν is ramified in K_∞/F_∞. Otherwise, A_η is itself trivial. Only finitely many such primes ν exist. It suffices to just consider one η for each ν.

Since M_∞/K_∞ is unramified at any $\eta \nmid p$, it follows that A_η is isomorphic to the corresponding inertia subgroup of $\mathrm{Gal}(K_\infty/F_\infty)$ and hence is finite. Therefore, the finiteness of the above cokernel follows from the fact that if A is any finite subgroup of $\mathrm{Gal}(M_\infty/F_\infty)$, then $H^1(A, D)$ is finite.

Let $U = G_{M_\infty}$. Then we can identify $\mathrm{Ind}_{G'/U}^{G/U}(\rho')$ with ρ, viewed as a representation of $G/U = \mathrm{Gal}(M_\infty/F)$, and also of the subgroup $\mathrm{Gal}(M_\infty/F_\infty)$. According to Shapiro's Lemma, we then have a canonical isomorphism

$$H^1(M_\infty/F_\infty, D) \longrightarrow H^1(M_\infty/F'_\infty, D') \tag{5}$$

The map is Γ_F-equivariant and so the isomorphism is as discrete $\Lambda_{(\mathcal{O},F)}$-modules. Their Pontryagin duals are isomorphic as $\Lambda_{(\mathcal{O},F)}$-modules. It follows that the Pontryagin duals of $\mathrm{Sel}_D(F_\infty)$ and $\mathrm{Sel}_{D'}(F'_\infty)$ are pseudo-isomorphic and therefore that $I_\rho = I_{\rho'}$, as stated. We remark in passing that, with a little more care, one can verify that (5) actually defines an isomorphism of $\mathrm{Sel}_D(F_\infty)$ to $\mathrm{Sel}_{D'}(F'_\infty)$.

Suppose now that $F' \subset F_\infty$. Then $F'_\infty = F_\infty$ and $\Gamma_{F'}$ is a subgroup of Γ_F of finite index t. We use the previous notation, but now we have $N \subset G' \subset G$. Thus, ρ_∞ and ρ'_∞ are the restrictions of ρ and ρ' to N, respectively. In this case, ρ_∞ is a direct sum of representations obtained from ρ'_∞ by composing with certain automorphisms of N. The automorphism are just the restriction of certain inner automorphisms of G, namely the inner automorphisms defined by some set of coset representatives $g_1, \ldots, g_t$ for G' in G. We take g_1 to be the identity. Denote these representations of G' by $\rho'_1, \ldots, \rho'_t$. (They are not necessarily distinct.) Define the corresponding discrete G'-modules $D'_1, \ldots, D'_t$ obtained from D' by composing with the above specified automorphisms of N. We have $D'_1 = D'$. For each i, $1 \le i \le t$, conjugating by g_i also defines an isomorphism of $H^1(N, D')$ to $H^1(N, D'_i)$. Since Γ_F is commutative, this isomorphism is $\Gamma_{F'}$-equivariant. Also, the isomorphism induces an isomorphism of $\mathrm{Sel}_{D'}(F_\infty)$ to $\mathrm{Sel}_{D'_i}(F_\infty)$. Thus, the $\mathrm{Sel}_{D'_i}(F_\infty)$'s are all isomorphic to $\mathrm{Sel}_{D'}(F_\infty)$ as $\Lambda_{(\mathcal{O},F')}$-modules.

As a G_{F_∞}-module, $D \cong \oplus_{1\le i\le t} D'_i$. Furthermore, $D'_i = g_i(D')$. Thus,

$$\mathrm{Sel}_D(F_\infty) \cong \bigoplus_{1\le i\le t} \mathrm{Sel}_{D'_i}(F_\infty)$$

as $\Gamma_{F'}$-modules and the action of Γ_F permutes the summands in the corresponding way. The same thing is true for the Pontryagin duals. It follows that

$$X_D(F_\infty) \cong X_{D'}(F_\infty) \otimes_{\Lambda_{(\mathcal{O},F)'}} \Lambda_{(\mathcal{O},F)}$$

and therefore $I_\rho = I_{\rho'}\Lambda_{(\mathcal{O},F)}$ as stated. □

Propositions 3 and 4 show that it is enough to consider I_τ and J_τ where τ is a totally even, absolutely irreducible Artin representation of $G_{\mathbf{Q}}$. For if ρ is a totally even Artin representation over F, then $\mathrm{Ind}_F^{\mathbf{Q}}(\rho)$ is a direct sum of absolutely irreducible Artin representations which must also be totally even. The field $\mathcal{E}$ must be chosen to be sufficiently large. Both ρ and all of the absolutely irreducible constituents of $\mathrm{Ind}_F^{\mathbf{Q}}(\rho)$ must be realizable over $\mathcal{E}$. The next remark shows that the choice of $\mathcal{E}$ is otherwise not too significant.

Remark 5. Suppose that $\mathcal{E}'$ and $\mathcal{E}$ are finite extensions of $\mathbf{Q}_p$ with rings of integers $\mathcal{O}'$ and $\mathcal{O}$, respectively. Assume that $\mathcal{E}' \subseteq \mathcal{E}$. Let $\Gamma = \mathbf{Z}_p$. Let $\Lambda' = \mathcal{O}'[[\Gamma]]$ and $\Lambda = \mathcal{O}[[\Gamma]]$, and let $\mathcal{L}'$ and $\mathcal{L}$ be their fraction fields. Thus, Λ is the integral closure of Λ' in $\mathcal{L}$. Since Λ' is integrally closed in $\mathcal{L}'$, one has $\Lambda' = \Lambda \cap \mathcal{L}$. It follows that if I' is a principal ideal in Λ', and $I = I'\Lambda$, then one can recover I' from I by $I' = I \cap \Lambda'$.

In particular, suppose that ρ' is a totally even Artin representation of G_F over $\mathcal{E}'$, and D' is the corresponding discrete Galois module. Extending scalars to $\mathcal{E}$, one obtains an Artin representation ρ over $\mathcal{E}$, and one can take $D = D' \otimes_{\mathcal{O}'} \mathcal{O}$ as the corresponding discrete Galois module. One sees easily that $\mathrm{Sel}_D(F_\infty) \cong \mathrm{Sel}_{D'}(F_\infty) \otimes_{\mathcal{O}'} \mathcal{O}$. The characteristic ideals are related by $I_\rho = I_{\rho'}\Lambda_{(\mathcal{O},F)}$. Hence $I_{\rho'} = I_\rho \cap \Lambda_{(\mathcal{O}',F)}$. Similar statements hold for the ideals $J_{\rho'}$ and J_ρ.

Our final result in this section concerns the effect of twisting the Galois representation ρ. Suppose that ξ is a 1-dimensional Artin representation of G_F which factors through Γ_F. We must choose $\mathcal{E}$ sufficiently large so that ξ has values in $\mathcal{E}$. Thus, $\xi : \Gamma_F \to \mathcal{O}^\times$ is a continuous homomorphism. We denote the twist $\rho \otimes \xi$ simply by $\rho\xi$. The corresponding discrete G_F-module is $D \otimes \xi = D \otimes_{\mathcal{O}} \mathcal{O}(\xi)$, where $\mathcal{O}(\xi)$ is the free $\mathcal{O}$-module of rank 1 on which G_F acts by ξ. We use a similar notation below for other discrete and compact $\mathcal{O}$-modules. For brevity, we denote $D \otimes \xi$ by D_ξ. Of course, as $\mathcal{O}$-modules and G_{F_∞}-modules, we can identify D_ξ and D. The actions of Γ_F on the corresponding Galois cohomology groups are related by

$$H^1(F_\Sigma/F_\infty, D_\xi) \cong H^1(F_\Sigma/F_\infty, D) \otimes \xi$$

and therefore we have the following Γ_F-equivariant isomorphism of discrete $\mathcal{O}$-modules:

$$\mathrm{Sel}_{D_\xi}(F_\infty) \cong \mathrm{Sel}_D(F_\infty) \otimes \xi \ . \tag{6}$$

Both of these $\mathcal{O}$-modules are $\Lambda_{(\mathcal{O},F)}$-modules and the isomorphism is a $\Lambda_{(\mathcal{O},F)}$-module isomorphism.

It follows from (6) that $X_{D_\xi}(F_\infty) \cong X_D(F_\infty) \otimes \xi^{-1}$ as $\Lambda_{(\mathcal{O},F)}$-modules. Furthermore, noting that

$$H^0(F_\Sigma/F_\infty, D_\xi) = H^0(F_\Sigma/F_\infty, D) \otimes \xi \ ,$$

it follows that $Y_{D_\xi}(F_\infty) \cong Y_D(F_\infty) \otimes \xi^{-1}$ as $\Lambda_{(\mathcal{O},F)}$-modules. These isomorphisms give a simple relationship between I_{ρ_ξ} and I_ρ and between J_{ρ_ξ} and J_ρ, as we now discuss.

In general, suppose that Γ is a commutative pro-p group. Let $\Lambda_\mathcal{O} = \mathcal{O}[[\Gamma]]$. We have the natural inclusion map $\varepsilon : \Gamma \to \Lambda^\times$. Let $\xi : \Gamma_F \to \mathcal{O}^\times$ be a continuous homomorphism. Since $\mathcal{O}^\times \subset \Lambda_\mathcal{O}^\times$, we obtain a continuous homomorphism $\xi\varepsilon : \Gamma \to \Lambda_\mathcal{O}^\times$. This is the map $\gamma \mapsto \xi(\gamma)\gamma$ for all $\gamma \in \Gamma$. We can then extend $\xi\varepsilon$ to a continuous $\mathcal{O}$-algebra homomorphism $tw_\xi : \Lambda_\mathcal{O} \to \Lambda_\mathcal{O}$. This is an automorphism of $\Lambda_\mathcal{O}$. The inverse map is $tw_{\xi^{-1}}$.

Now suppose that $\Gamma \cong \mathbf{Z}_p$ and that $\xi : \Gamma \to \mathcal{O}^\times$ is a continuous homomorphism. Let X be a finitely-generated, torsion $\Lambda_\mathcal{O}$-module. For any $\lambda \in \Lambda_\mathcal{O}$, $X[\lambda]$ denotes $\{x \in X \mid \lambda x = 0\}$. The characteristic ideal I_X is determined by the invariants $\mu(X)$ and $\mathrm{rank}_\mathcal{O}\big(X[\theta^t]\big)$, where θ varies over the irreducible elements in $\Lambda_\mathcal{O}$ and $t \geq 1$. We write $X_{\xi^{-1}}$ for $X \otimes \xi^{-1}$. It is easily seen that $\mu(X) = \mu(X_{\xi^{-1}})$. We regard X and $X_{\xi^{-1}}$ as the same $\mathcal{O}$-modules, but with different $\mathcal{O}$-linear actions of Γ. If $\gamma \in \Gamma$, and $x \in X = X_{\xi^{-1}}$, we denote the first action by $\gamma \cdot x$ and the second by $\gamma \cdot\cdot x$. Thus, $\gamma \cdot\cdot x = \xi(\gamma)^{-1}\gamma \cdot x$. That is, we have $(\xi(\gamma)\gamma) \cdot\cdot x = \gamma \cdot x$ for all $\gamma \in \Gamma, x \in X$. It follows that $tw_\xi(\theta) \cdot\cdot x = \theta \cdot x$ for all $\theta \in \Lambda_\mathcal{O}$ and $x \in X$. In particular, for any irreducible element $\theta \in \Lambda_\mathcal{O}$ and $t \geq 1$, we have

$$X_{\xi^{-1}}[tw_\xi(\theta^t)] = X[\theta^t] .$$

Consequently, we have the following result.

Proposition 5. *Let ξ be a character of finite order of Γ_F. Then $I_{\rho\xi} = tw_\xi(I_\rho)$ and $J_{\rho\xi} = tw_\xi(J_\rho)$.*

Remark 6. In the introduction, we mentioned the Selmer group $\mathrm{Sel}_{D(n)}(F_\infty)$ associated to the Tate twist $D(n)$ when $n \geq 2$ and $n \equiv 0 \pmod{d}$, where $d = [F(\mu_{2p}) : F]$. Note that $\chi_F^n = \kappa_F^n$ factors through Γ_F for any such n. The above discussion shows how the actions of Γ_F on $\mathrm{Sel}_D(F_\infty)$ and $\mathrm{Sel}_{D(n)}(F_\infty)$ differ. In fact, if θ generates the characteristic ideal of $X_D(F_\infty)$, then $tw_{\kappa_F^n}(\theta)$ generates the characteristic ideal of $X_{D(n)}(F_\infty)$.

3 The Definition of p-Adic Artin L-Functions

If F is totally real and $\rho : G_F \to GL_d(\mathbf{C})$ is a totally even Artin representation, then the corresponding Artin L-function $L(z, \rho)$ is a meromorphic function on $\mathbf{C}$. We will let $L^*(z, \rho)$ denote the function given by the same Euler product as $L(z, \rho)$, but with the Euler factors for the primes of F lying above p omitted. A theorem of Siegel implies that $L(1-n, \rho) \in \mathbf{Q}(\psi)$ for all integers $n \geq 1$. Here ψ is the character of ρ and $\mathbf{Q}(\psi)$ is the field generated by its values. Furthermore, $L(1 - n, \rho) \neq 0$

when n is even. For our purpose, we will consider the values $L^*(1-n,\rho)$. They are also algebraic numbers and are nonzero for even $n \geq 2$.

The above L-values behave well under conjugacy in the following sense. If $g \in \mathrm{Gal}(\overline{\mathbf{Q}}/\mathbf{Q})$, then $\psi' = g \circ \psi$ is the character of another totally even Artin representation ρ' of G_F. We then have $L(1-n,\rho') = g\big(L(1-n,\rho)\big)$ for all $n \geq 1$. The Euler factors for primes above p behave similarly, and so we have the same conjugacy properties for the values $L^*(1-n,\rho)$. Therefore, if we arbitrarily choose embeddings of $\overline{\mathbf{Q}}$ into $\mathbf{C}$ and into $\overline{\mathbf{Q}}_p$, then the complex algebraic numbers $L^*(1-n,\rho)$ and the values of ψ can all be regarded as elements of $\overline{\mathbf{Q}}_p$. The character ψ is then the character of an Artin representation $\rho : G_F \to GL_d(\overline{\mathbf{Q}}_p)$. Of course, ψ determines ρ up to equivalence. The values $L^*(1-n,\rho)$ are also determined by the $\overline{\mathbf{Q}}_p$-valued character ψ, and do not depend on the choice of embeddings. In fact, if ψ has values in a finite extension $\mathcal{E}$ of $\mathbf{Q}_p$, then the values $L^*(1-n,\rho)$ for $n \geq 1$ are also in $\mathcal{E}$, and are nonzero when n is even.

These L-values also behave well under induction. Suppose that F' is a finite, totally real extension of F. Suppose that $\rho = \mathrm{Ind}_{F'}^{F}(\rho')$, where ρ' is a totally even Artin representation of $G_{F'}$. We have $L(z,\rho) = L(z,\rho')$. The same identity is also true even if we delete the Euler factors for primes above p. Thus,

$$L^*(1-n,\rho) = L^*(1-n,\rho')$$

for all $n \geq 1$.

The p-adic L-function $L_p(s,\rho)$ satisfies the following interpolation property:

$$L_p(1-n,\rho) \;=\; L^*(1-n,\rho)$$

for all $n \equiv 0 \pmod{[F(\mu_p):F]}$ if p is odd, or all $n \equiv 0 \pmod 2$ if $p=2$. It is a meromorphic function defined on a certain disc $\mathcal{D}$ in $\overline{\mathbf{Q}}_p$. The existence of such a function was proved by Deligne and Ribet when ρ is of dimension 1. In this case, it is holomorphic on $\mathcal{D}$, except possibly at $s=1$.

Suppose now that ρ factors through $\Delta = \mathrm{Gal}(K/F)$, where K is a finite, totally real, Galois extension of F. The existence of $L_p(s,\rho)$ then follows if ρ is induced from a 1-dimensional representation ρ' of a subgroup Δ' of Δ. Then ρ is a so-called *monomial* representation. If $\Delta' = \mathrm{Gal}(K/F')$, then $\rho = \mathrm{Ind}_{F'}^{F}(\rho')$ and we have $L_p(s,\rho) = L_p(s,\rho')$. Thus, $L_p(s,\rho)$ is again holomorphic on $\mathcal{D}$, except possibly at $s=1$.

In general, a theorem of Brauer states that there exist monomial representations $\rho_1,\ldots,\rho_s\ \sigma_1,\ldots,\sigma_t$ of Δ, where $s,t \geq 0$, such that

$$\rho \oplus \Big(\bigoplus_{j=1}^{t} \sigma_j\Big) \cong \bigoplus_{i=1}^{s} \rho_i \tag{7}$$

and so we can define $L_p(s,\rho)$ as the quotient $\prod_{i=1}^{s} L_p(s,\rho_i) \big/ \prod_{j=1}^{t} L_p(s,\sigma_j)$. The above interpolation property is indeed satisfied by this function.

Let $\Gamma = \Gamma_{\mathbf{Q}} = \mathrm{Gal}(\mathbf{Q}_\infty/\mathbf{Q})$. There is a canonical isomorphism $\kappa : \Gamma \to 1 + q\mathbf{Z}_p$, where $q = p$ for odd p and $q = 4$ for $p = 2$. It is defined as the composite map

$$\mathrm{Gal}(\mathbf{Q}(\mu_{p^\infty})/\mathbf{Q}) \xrightarrow{\chi} \mathbf{Z}_p^\times \longrightarrow 1 + q\mathbf{Z}_p$$

which indeed factors through Γ. Here χ is the p-power cyclotomic character. The second map is just the projection map for the decomposition $\mathbf{Z}_p^\times = W \times (1 + q\mathbf{Z}_p)$, where W is the group of roots of unity in $\mathbf{Q}_p$. For any $s \in \mathbf{Z}_p$, one can define $\kappa^s : \Gamma \to 1 + q\mathbf{Z}_p$, which is a continuous group homomorphism. It extends to a continuous $\mathcal{O}$-algebra homomorphism $\Lambda_{(\mathcal{O},\mathbf{Q})} \to \mathcal{O}$ which we also denote by κ^s. Furthermore, for any F, the restriction map $\Gamma_F \to \Gamma$ defines an injective homomorphism $\Lambda_{(\mathcal{O},F)} \to \Lambda_{(\mathcal{O},\mathbf{Q})}$. We identify $\Lambda_{(\mathcal{O},F)}$ with its image and define κ_F^s to be the restriction of κ^s to that subring.

Let $\mathcal{L}_{(\mathcal{O},F)}$ denote the fraction field of $\Lambda_{(\mathcal{O},F)}$. Suppose that $\theta \in \mathcal{L}_{(\mathcal{O},F)}$. Write $\theta = \alpha\beta^{-1}$, where $\alpha, \beta \in \Lambda_{(\mathcal{O},F)}$ and $\beta \neq 0$. The Weierstrass preparation theorem implies that $\kappa_F^s(\beta) \neq 0$ for all but finitely many $s \in \mathbf{Z}_p$. Thus, excluding a finite set of values of s, one can make the definition $\kappa_F^s(\theta) = \kappa_F^s(\alpha)\kappa_F^s(\beta)^{-1}$. Furthermore, one has the following property:
If $\theta_1, \theta_2 \in \mathcal{L}_{(\mathcal{O},F)}$ and $\kappa_F^s(\theta_1) = \kappa_F^s(\theta_2)$ for infinitely many $s \in \mathbf{Z}_p$, then $\theta_1 = \theta_2$.
One verifies this by writing $\theta_1 = \alpha_1\beta_1^{-1}$, $\theta_2 = \alpha_2\beta_2^{-1}$, and applying the Weierstrass preparation theorem to $\alpha_1\beta_2 - \alpha_2\beta_1$.

One can associate to $L_p(s,\rho)$ a nonzero element θ_ρ of $\mathcal{L}_{(\mathcal{O},F)}$. It is characterized as follows:

$$L_p(1-s,\rho) = \kappa_F^s(\theta_\rho) \qquad \text{for all but finitely many } s \in \mathbf{Z}_p. \tag{8}$$

for all but finitely many $s \in \mathbf{Z}_p$. If ρ is 1-dimensional, then Deligne and Ribet's construction of $L_p(1-s,\rho)$ proves the existence of such a θ_ρ. Furthermore, they show that

$$J_\rho\theta_\rho \subseteq 2^{[F:\mathbf{Q}]}\Lambda_{(\mathcal{O},F)} \;,$$

where J_ρ is the ideal in $\Lambda_{(\mathcal{O},F)}$ defined in the introduction. Note that $J_\rho = \Lambda_{(\mathcal{O},F)}$ unless ρ factors through Γ_F. If ρ is monomial, one can use Proposition 4 to prove that θ_ρ exists. Then θ_ρ has the integrality property

$$J_\rho\theta_\rho \subseteq 2^{[F:\mathbf{Q}]\deg(\rho)}\Lambda_{(\mathcal{O},F)} \tag{9}$$

since if ρ is induced from a 1-dimensional Artin representation ρ' of $G_{F'}$, where F' is a finite extension of F, then $[F':\mathbf{Q}] = [F:\mathbf{Q}]\deg(\rho)$.

If ρ has arbitrary dimension, then the existence of θ_ρ satisfying (8) follows from (7). One assumes at first that $\mathcal{E}$ is large enough so that all of the monomial representations ρ_i and σ_j are realizable over $\mathcal{E}$. The θ_{ρ_i}'s and θ_{σ_j}'s are nonzero elements in the fraction fields of various subrings of $\Lambda_{(\mathcal{O},F)}$. One can then define

$$\theta_\rho = \prod_{i=1}^{s} \theta_{\rho_i} \Big/ \prod_{j=1}^{t} \theta_{\sigma_j} \tag{10}$$

With this definition, we can only say that θ_ρ is an element in the fraction field of $\Lambda_{(\mathcal{O},F)}$.

The behavior of the values $L^*(1-n,\rho)$ under conjugacy implies a similar behavior for the elements θ_ρ. To be precise, suppose that $\gamma \in G_{\mathbf{Q}_p}$. Let $\mathcal{O}' = \gamma(\mathcal{O})$. Let $\rho' = \gamma \circ \rho$. Note that γ induces a continuous isomorphism from $\Lambda_{(\mathcal{O},F)}$ to $\Lambda_{(\mathcal{O}',F)}$. This isomorphism extends to an isomorphism of the fraction fields, which we also denote by γ. We then have $\theta_{\rho'} = \gamma(\theta_\rho)$.

Concerning the choice of $\mathcal{O}$, the above conjugacy property and a straightforward Galois theory argument show that one can even take $\mathcal{O}$ to be the extension of $\mathbf{Z}_p$ generated by the values of the character of ρ. In the next section, we will see that the integrality property (9) still holds when p is odd.

The above remarks give us the following properties of the θ_ρ's which are parallel to the assertions in Propositions 3 and 4.

Proposition 6. *With the same notation as in Proposition 3, we have* $\theta_\rho = \theta_{\rho_1}\theta_{\rho_2}$.

Proposition 7. *With the same notation as in Proposition 4, we have* $\theta_\rho = \theta_{\rho'}$.

We will also need the analogue of Proposition 5. It relies on another property of the p-adic L-functions constructed by Deligne and Ribet. The interpolation property for θ_ρ stated before can be expressed as follows:

$$\kappa_F^n(\theta_\rho) \;=\; L^*(1-n,\rho) \;=\; L^*(1,\rho\kappa_F^n)$$

for all $n \geq 2$ satisfying $n \equiv 0 \pmod{p-1}$ if p is odd (or $n \equiv 0 \pmod 2$ if $p=2$). The underlying $\mathcal{E}$ representation space for $\rho\kappa_F^n$ is the Tate twist $V(n)$. However, if ξ is a character of Γ_F of finite order, and $\mathcal{O}$ contains the values of ξ, then Deligne and Ribet also show that

$$\kappa_F^n\xi(\theta_\rho) \;=\; L^*(1-n,\rho\xi) \;=\; L^*(1,\rho\xi\kappa_F^n) \;=\; L^*(1,\rho\kappa_F^n\xi).$$

Furthermore, one has $\kappa_F^n(\theta_{\rho\xi}) = L^*(1,\rho\xi\kappa_F^n)$. Thus, we have $\kappa_F^n\xi(\theta_\rho) = \kappa_F^n(\theta_{\rho\xi})$ for the above values of n.

Suppose that $\varphi : \Gamma_F \to \mathcal{O}^\times$ is any continuous homomorphism. Let ξ be as above. Then both φ and $\varphi\xi : \Gamma_F \to \mathcal{O}^\times$ extend to continuous $\mathcal{O}$-algebra homomorphisms φ and $\varphi\xi$ from $\Lambda_{(\mathcal{O},F)}$ to $\mathcal{O}$. We also have the continuous $\mathcal{O}$-algebra homomorphism $\varphi \circ tw_\xi : \Lambda_{(\mathcal{O},F)} \to \mathcal{O}$. Such $\mathcal{O}$-algebra homomorphisms are determined uniquely by their restrictions to Γ_F. Note that

$$(\varphi \circ tw_\xi)(\gamma) \;=\; \varphi(\xi(\gamma)\gamma) = (\varphi\xi)(\gamma).$$

Therefore, we also have $(\varphi \circ tw_\xi)(\theta) = (\varphi\xi)(\theta)$ for all $\theta \in \Lambda_{(\mathcal{O},F)}$. Applying this to $\varphi = \kappa_F^n$, where $n \geq 2$ and $n \equiv 0 \pmod{p-1}$ (or $n \equiv 2 \pmod 2$ if $p = 2$), we obtain

$$\kappa_F^n\big(tw_\xi(\theta_\rho)\big) = (\kappa_F^n \circ tw_\xi)(\theta_\rho) = \kappa_F^n(\theta_{\rho\xi})$$

for all such n and therefore it follows that $tw_\xi(\theta_\rho) = \theta_{\rho\xi}$.

Proposition 8. *If ξ is a 1-dimension Artin representations of G_F which factors through Γ_F, then $\theta_{\rho\xi} = tw_\xi(\theta_\rho)$.*

4 Relationship of Selmer Groups to p-Adic L-Functions

We can state the relationship quite succinctly in terms of the notation of the preceding sections. We refer to this statement as the Iwasawa Main Conjecture for ρ. As before, we assume that ρ is realizable over a finite extension $\mathcal{E}$ of $\mathbf{Q}_p$ with ring of integers $\mathcal{O}$. Let $m(\rho) = [F : \mathbf{Q}]\dim(\rho)$, which is just the degree of the representation $\mathrm{Ind}_F^{\mathbf{Q}}(\rho)$.

IMC(ρ). *Suppose that F is a totally real number field and that ρ is a totally even Artin representation of G_F. Then $I_\rho = J_\rho\theta_\rho 2^{-m(\rho)}$.*

Note that I_ρ is an ideal in $\Lambda_{(\mathcal{O},F)}$ by definition, but the assertion that $J_\rho\theta_\rho 2^{-m(\rho)}$ is an ideal in that ring, and not just a *"fractional ideal"*, is not at all clear from the definitions.

It is interesting to note that when $p = 2$, one can omit the extra power of 2 appearing in the formulation of **IMC**(ρ) by merely omitting the local conditions at the Archimedean primes in the definition of the Selmer group. That is, one can define a larger Selmer group

$$\mathrm{Sel}_D^{\#}(F_\infty) = \ker\Big(H^1(F_\infty, D) \longrightarrow \prod_{\eta \nmid p,\infty} H^1(F_{\infty,\eta}, D)\Big) .$$

It turns out that the characteristic ideal of the Pontryagin dual of the quotient group $\mathrm{Sel}_D^{\#}(F_\infty)/\mathrm{Sel}_D(F_\infty)$ is precisely $2^{m(\rho)}\Lambda_{(\mathcal{O},F)}$. One sees this by using the surjectivity of the map (3) together with the structure of $\mathcal{H}_v^1(F_\infty, D)$ for Archimedean v as described in Sect. 2. If $I_\rho^{\#}$ denotes the characteristic ideal of the Pontryagin dual of $\mathrm{Sel}_D^{\#}(F_\infty)$, then **IMC**($\rho$) is equivalent to the assertion that $I_\rho^{\#} = J_\rho\theta_\rho$.

We also consider a weaker form of **IMC**(ρ). It amounts to the above equality up to multiplication by a power of a uniformizing parameter π of $\mathcal{O}$. We will denote the ring $\Lambda_{(\mathcal{O},F)}[\frac{1}{\pi}]$ by $\Lambda_{(\mathcal{O},F)}^*$. It is a subring of the fraction field $\mathcal{L}_{(\mathcal{O},F)}$ of $\Lambda_{(\mathcal{O},F)}$.

IMC(ρ)*. *We have $I_\rho\Lambda_{(\mathcal{O},F)}^* = J_\rho\theta_\rho\Lambda_{(\mathcal{O},F)}^*$.*

The main purpose of this section is to point out that the results proved by Wiles in Wiles (1990) are sufficient to prove **IMC**(ρ) for all $p \geq 3$. Let

$$A(\rho) = I_\rho^{-1} J_\rho \theta_\rho 2^{-m(\rho)}$$

which is a principal fractional ideal of $\Lambda_{(\mathcal{O},F)}$, i.e., a nonzero free $\Lambda_{(\mathcal{O},F)}$-submodule of the fraction field $\mathcal{L}_{(\mathcal{O},F)}$. Such fractional ideals form a group.

If I is a nonzero principal ideal of $\Lambda_{(\mathcal{O},F)}$, and θ is a generator of I, then we will refer to $\mu(\Lambda_{(\mathcal{O},F)}/I)$ as the μ-invariant associated to I, or to θ. We denote it by μ_I. For a principal fractional ideal $I = I_1 I_2^{-1}$, we define the associated μ-invariant by $\mu_I = \mu_{I_1} - \mu_{I_2}$. A simple direct argument, or the fact that $\Lambda_{(\mathcal{O},F)}$ is a UFD, shows that μ_I is well-defined.

Suppose that $\rho = \rho_1 \oplus \rho_2$, where ρ_1 and ρ_2 are totally even Artin representations of G_F. It is clear that $m(\rho) = m(\rho_1) + m(\rho_2)$. It therefore follows from Propositions 3 and 6 that

$$A(\rho) = A(\rho_1)A(\rho_2) \ . \tag{11}$$

Suppose that F' is a finite, totally real extension of F, that ρ' is a totally even Artin representation of $G_{F'}$, and that $\rho = \mathrm{Ind}_{F'}^F(\rho')$. It is clear that $m(\rho) = m(\rho')$, and so Propositions 4 and 7 imply that

$$A(\rho) = A(\rho') \ . \tag{12}$$

The conjecture **IMC**(ρ) asserts that $A(\rho) = \Lambda_{(\mathcal{O},F)}$. The conjecture **IMC**(ρ)* asserts that $A(\rho)$ is generated by a power of π. Suppose that ρ factors through $\Delta = \mathrm{Gal}(K/F)$, where K is totally real. It is clear from (11) and (12) that if one proves **IMC**(ρ') (respectively, **IMC**(ρ')*) for all 1-dimensional representation ρ' of all subgroups Δ' of Δ, then **IMC**(ρ) (respectively, **IMC**(ρ)*) follows.

Now suppose that ξ is a 1-dimensional Artin representation which factors through Γ_F. Obviously, $m(\rho\xi) = m(\rho)$. It therefore follows from Propositions 5 and 8 that

$$A(\rho\xi) = tw_\xi\big(A(\rho)\big) \tag{13}$$

When p is an odd prime, Wiles proves **IMC**(ρ)* for a certain class of totally even, 1-dimensional Artin representations ρ over any totally real number field F, namely the representations which factor through $\mathrm{Gal}(K/F)$ where $K \cap F_\infty = F$. (These are the representations of type S.) This result is Theorem 1.3 in Wiles (1990). Therefore, $A(\rho)$ is generated by a power of π for all those ρ's. However, one sees easily that if ρ is 1-dimensional, but not of type S, then there exists a 1-dimensional Artin representation ξ factoring through Γ_F such that $\rho\xi$ is of type S. Since $tw_\xi(\pi) = \pi$, it follows that if $A(\rho\xi) = (\pi^t)$, then $A(\rho) = (\pi^t)$. Therefore, **IMC**(ρ)* is established when p is an odd prime, F is any totally real number field, and ρ is any 1-dimensional, even Artin representation. It then follows from (11) and

(12) that **IMC**$(\rho)^*$ holds for an arbitrary totally even Artin representation of G_F when p is odd. For $p = 2$, Wiles does prove partial results, but not enough to prove **IMC**$(\rho)^*$

Now consider **IMC**(ρ). Wiles proves this assertion when p is odd and ρ is 1-dimensional and has order prime to p. It follows from Theorem 1.3 and 1.4 in Wiles (1990). The above remarks and the lemma below allow one to establish **IMC**(ρ) for all ρ.

An alternative way to deal with the μ-invariants when ρ is 1-dimensional, but of order divisible by p, is described in Coates and Kim (2011), pages 9 and 10. It is in terms of Galois groups instead of Selmer groups. We should also add that theorem 1.4 in Wiles (1990) involves odd 1-dimensional representations instead of even. However, the equivalence of that theorem with what we need here is a consequence of the so-called *"Reflection Principle"*. It is also a consequence of theorem 2 in Greenberg (1989).

Lemma 1. *Suppose that G is a finite group and that ψ is the character of a representation of G over $\overline{\mathbf{Q}}_p$. Assume that the values of ψ are in $\mathbf{Q}_p^{unr}$. Then, there exists subgroups H_i of G and a 1-dimensional character ψ_i of each H_i, where $1 \leq i \leq t$ for some $t \geq 1$, such that*

1. $m\psi = \sum_{i=1}^{t} m_i \mathrm{Ind}_{H_i}^{G}(\psi_i)$, *where* $m, m_1, \ldots, m_t \in \mathbf{Z}$, $m \geq 1$
2. *Each ψ_i has order prime to p*

Proof. Brauer's theorem asserts that $\psi = \sum_{i=1}^{s} a_i \mathrm{Ind}_{S_i}^{G}(\varphi_i)$, where the a_i's are integers, the S_i's are subgroups of G, and, for each i, φ_i is a 1-dimensional character of S_i. If $\sigma \in G_{\mathbf{Q}_p^{unr}}$, then σ fixes the values of ψ and one also has $\psi = \sum_{i=1}^{t} n_i \mathrm{Ind}_{S_i}^{G}(\sigma \circ \varphi_i)$. The extension of $\mathbf{Q}_p^{unr}$ generated by the values of all the φ_i's is finite. If m is its degree, then, by taking the trace, one obtains

$$m\psi = \sum_{i=1}^{s} b_i \mathrm{Ind}_{S_i}^{G}(\tau_i)$$

where τ_i is the sum of the conjugates of φ_i over $\mathbf{Q}_p^{unr}$ and the b_i's are integers. Since induction is transitive, it is enough to prove the lemma for each of the characters τ_i of S_i. The values of τ_i are in $\mathbf{Q}_p^{unr}$.

Let S be a subgroup of G and let φ be a 1-dimensional character of S. Then $\varphi = \alpha\beta$, where α has order prime to p and β has p-power order. Let p^k be the order of β. We can assume that $k \geq 1$. The values of α are in $\mathbf{Q}_p^{unr}$. Let τ_φ and τ_β denote the sums of the conjugates of φ and of β over $\mathbf{Q}_p^{unr}$, respectively. Note that $\tau_\varphi = \alpha\tau_\beta$.

Let $T = \ker(\beta)$. Thus, β factors through S/T, which is cyclic of order p^k. The conjugates of β over $\mathbf{Q}_p^{unr}$ are the characters of S/T of order p^k. Let T_1 be the subgroup of S such that $T \subset T_1 \subseteq S$ and T_1/T is cyclic of order p. Let ϵ_T and ϵ_{T_1} denote the trivial characters of T and T_1, respectively. Then $\mathrm{Ind}_T^S(\epsilon_T)$ is the sum of all the irreducible characters of S/T and $\mathrm{Ind}_{T_1}^S(\epsilon_{T_1})$ is the sum of the irreducible characters factoring through S/T_1. Hence the difference is τ_β. Thus, $\tau_\beta = \mathrm{Ind}_T^S(\epsilon_T) - \mathrm{Ind}_{T_1}^S(\epsilon_{T_1})$. It follows that

$$\tau_\varphi = \mathrm{Ind}_T^S(\alpha_T) - \mathrm{Ind}_{T_1}^S(\alpha_{T_1}),$$

where α_T and α_{T_1} denote the restrictions of α to T and to T_1, respectively. Their orders are prime to p. □

Assume now that p is odd and that ρ is an arbitrary totally real Artinrepresentation over F. We assume that ρ is realizable over a finite extension $\mathcal{E}$ of $\mathbf{Q}_p$. Assume further that $\mathcal{E}$ is Galois over $\mathbf{Q}_p$. Since **IMC**$(\rho)^*$ is established, it suffices to just consider the μ-invariants to prove **IMC**(ρ). Now if ρ and ρ' are conjugate over $\mathbf{Q}_p$ under the action of Gal$(\mathcal{E}/\mathbf{Q}_p)$, then one sees easily that I_ρ and $I_{\rho'}$ are conjugate under the natural action of Gal$(\mathcal{E}/\mathbf{Q}_p)$ on $\Lambda_{(\mathcal{O},F)}$. In particular, the μ-invariants associated to those ideals are equal. The μ-invariants associated to J_ρ and $J_{\rho'}$ are 0. In addition, θ_ρ and $\theta_{\rho'}$ are conjugate too, and so the μ-invariants associated to those elements of $\mathcal{L}_{(\mathcal{O},F)}$ are equal. It follows that the μ-invariants associated to $A(\rho)$ and $A(\rho')$ are equal.

Let $\tilde{\rho} = \oplus_\sigma \sigma$, where σ runs over the conjugates of ρ over $\mathbf{Q}_p$. Then the character of $\tilde{\rho}$ has values in $\mathbf{Q}_p$, and hence in $\mathbf{Q}_p^{unr}$. Furthermore, the above remarks show that it suffices to prove that the μ-invariant for $A(\tilde{\rho})$ vanishes. The μ-invariant for $A(\rho)$ will then also vanish. Lemma 1, the behavior of $A(\cdot)$ under induction and direct sums, and Theorem 1.4 in Wiles (1990) imply that the μ-invariant for $A(\tilde{\rho})$ is indeed zero.

Thus, we have proved the following result based on Wiles (1990).

Proposition 9. *Suppose that p is an odd prime. If F is any totally real number field and ρ is any totally even Artin representation of G_F defined over a finite extension of $\mathbf{Q}_p$, then **IMC**(ρ) is true.*

References

Barsky, D.: Fonctions zêta p-adiques d'une classe de rayon des corps totalement réels, Groupe d'étude d'analyse ultramétrique (1977–1978); errata 1978–1979.

Cassou-Noguès, P.: Valeurs aux entiers négatifs des fonctions zêta et fonctions zêta p-adiques. Invent. Math. **51**, 29–59 (1979)

Coates, J., Kim, D.: Introduction to the work of M. Kakde on the non-commutative main conjectures for totally real fields. In: J. Coates, P. Schneider, R. Sujatha, O. Venjakob (eds.) Noncommutative Iwasawa Main Conjectures over Totally Real Fields, pp. 1–22. Springer, Berlin (2011)

Deligne, P., Ribet, K.: Values of abelian L-functions at negative integers over totally real fields. Invent. Math. **59**, 227–286 (1980)

Greenberg, R.: On p-adic Artin L-functions. Nagoya Math. J. **89**, 77–87 (1983)

Greenberg, R.: Iwasawa theory for p-adic representations. Adv. Stud. Pure Math. **17**, 97–137 (1989)

Greenberg, R.: Iwasawa theory, projective modules, and modular representations. Mem. AMS **992**, 1–185 (2011)
Greenberg, R., Vatsal, V.: On the Iwasawa invariants of elliptic curves. Invent. Math. **142**, 17–63 (2000)
Greenberg, R., Vatsal, V.: Iwasawa theory for Artin representations (in preparation)
Mazur, B.: Rational points of abelian varieties with values in towers of number fields. Invent. Math. **18**, 183–266 (1972)
Mazur, B., Wiles, A.: Class fields for abelian extensions of **Q**. Invent. Math. **76**, 179–330 (1984)
Neukirch, J., Schmidt, A., Wingberg, K.: Cohomology of Number Fields. Grundlehren der mathematischen Wissenschaften, vol. 323. Springer, Brlin/New York (2000)
Perrin-Riou, B.: Variation de la fonction L p-adique par isogénie. Adv. Stud. Pure Math. **17**, 347–358 (1989)
Ribet, K.: Report on p-adic L-functions over totally real fields. Astérisque **61**, 177–192 (1979)
Schneider, P.: The μ-invariant of isogenies. J. Indian Math. Soc. **52**, 159–170 (1987)
Wiles, A.: The Iwasawa conjecture for totally real fields. Ann. Math. **131**, 493–540 (1990)

Iwasawa μ-Invariants of p-Adic Hecke L-Functions

Ming-Lun Hsieh

1 Introduction

1.1 p-Adic Hecke L-Function for CM Fields

This aim of this paper is to report recent results on the vanishing of μ-invariants of p-adic Hecke L-functions for CM fields built upon fundamental works of Hida (2010). The significance of these results stems from their applications to Iwasawa main conjectures for CM fields and CM elliptic curves over totally real fields. We begin with some notation to introduce p-adic L-functions. Let $p > 2$ be an odd rational prime. Let F be a totally real field of degree d over $\mathbf{Q}$ and K be a totally imaginary quadratic extension of F. Let D_F be the absolute discriminant of F. Fix two embeddings $\iota_\infty\colon\bar{\mathbf{Q}} \to \mathbf{C}$ and $\iota_p\colon\bar{\mathbf{Q}} \to \mathbf{C}_p$ once and for all. Let c denote the complex conjugation on $\mathbf{C}$ which induces the unique non-trivial element of $\mathrm{Gal}(K/F)$. Let $\tilde{K}$ be the composite of all $\mathbf{Z}_p$-extensions of K. Then $\tilde{K}$ is Galois over F, and the complex conjugation acts on $\mathrm{Gal}(\tilde{K}/K)$ by the usual conjugation. Put

$$G^+ := \left\{\sigma \in \mathrm{Gal}(\tilde{K}/K) \mid c\sigma c = \sigma\right\}.$$

Denote by K_∞^- the fixed subfield of G^+ in $\tilde{K}$. Then the Galois group $\mathrm{Gal}(K_\infty^-/K)$ is a free $\mathbf{Z}_p$-module of rank d by class field theory. We call K_∞^- the anticyclotomic $\mathbf{Z}_p^d$-extension of K. On the other hand, let $\mathbf{Q}_\infty$ be the cyclotomic $\mathbf{Z}_p$-extension of $\mathbf{Q}$, and denote by $K_\infty^+ := K\mathbf{Q}_\infty$ the cyclotomic $\mathbf{Z}_p$-extension of K. Let $K_\infty = K_\infty^+ K_\infty^-$ be the composite of K_∞^+ and K_∞^-. If Leopoldt conjecture holds for F, then $K_\infty = \tilde{K}$.

M.-L. Hsieh (✉)
Department of Mathematics, National Taiwan University, No. 1, Sec. 4, Roosevelt Road, Taipei 10617, Taiwan
e-mail: mlhsieh@math.ntu.edu.tw

T. Bouganis and O. Venjakob (eds.), *Iwasawa Theory 2012*, Contributions in Mathematical and Computational Sciences 7, DOI 10.1007/978-3-642-55245-8_8,

Let W be the ring of integers of a finite extension of the p-adic completion of the maximal unramified extension of $\mathbf{Q}_p$. Define

$$\Lambda_\bullet := W[[\mathrm{Gal}(K_\infty^\bullet/K)]] \quad (\bullet = \emptyset, +, -).$$

Then $\Lambda = W[[\mathrm{Gal}(K_\infty/K)]]$ is an Iwasawa algebra over W of $(d+1)$-variables. Fix a set of the generators $\{\gamma_-^{(i)}\}_{i=1,\cdots,d}$ (resp. γ_+) of $\mathrm{Gal}(K_\infty^-/K)$ (resp. $\mathrm{Gal}(K_\infty^+/K)$). The restriction maps $\mathrm{Gal}(K_\infty/K) \to \mathrm{Gal}(K_\infty^\pm/K)$ induce the isomorphism

$$\Lambda = \Lambda_+ \hat{\otimes}_W \Lambda_- = W[[T_+, S_1, \cdots, S_d]]$$

with the cyclotomic variable $T_+ = \gamma_+ - 1$ and the anticyclotomic variables $\{S_i = \gamma_-^{(i)} - 1\}_{i=1,\cdots,d}$.

We assume the following ordinary hypothesis:

$$\text{Every prime of } F \text{ above } p \text{ splits in } K. \tag{ord}$$

Let Σ be a p-adic CM type of K. In other words, Σ is a set of p-adic places of K such that Σ and its complex conjugation Σc are disjoint and $\Sigma \bigsqcup \Sigma c$ is the set of all places of K above p. The existence of a p-adic CM type Σ is assured by the ordinary assumption (ord). Let Σ_∞ be the CM type corresponding to Σ, i.e. Σ is the set of embeddings $\sigma : K \hookrightarrow \bar{\mathbf{Q}}$ such that $\iota_p \circ \sigma$ induces a p-adic palace in Σ. Let χ be a Hecke character of K of infinity type $k\Sigma_\infty$ with $k \geq 1$. Suppose that χ takes value in W. According to Katz (1978) and Hida and Tilouine (1993), there exists a unique element $L_{\chi,\Sigma} \in W[[\mathrm{Gal}(K_\infty/K)]]$ characterized by the following interpolation property: for every finite order characters $\nu : \mathrm{Gal}(K_\infty/K) \to \mu_{p^\infty}$ of conductor $\prod_{\mathfrak{P}|p} \mathfrak{P}^{n_\mathfrak{P}}$,

$$\nu(L_{\chi,\Sigma}) = \frac{\Gamma(k)^d L(0,\chi\nu)}{\Omega^k} \cdot \prod_{\mathfrak{P}\in\Sigma_p} e_\mathfrak{P}(\chi\nu) \cdot \frac{[\mathcal{O}_K^\times : \mathcal{O}_F^\times]}{\sqrt{D_F}}. \tag{1}$$

where (i) $e_\mathfrak{P}(\chi\nu)$ is the p-adic multiplier defined by

$$e_\mathfrak{P}(\chi\nu) = \begin{cases} (1-\chi\nu(\mathfrak{P}))(1-\chi^{-1}\nu^{-1}(\overline{\mathfrak{P}})) & \text{if } n_\mathfrak{P} = 0, \\ \tau(\chi_\mathfrak{P}\nu_\mathfrak{P}) & \text{if } n_\mathfrak{P} > 0. \end{cases}$$

Here $\nu_\mathfrak{P}$ is the character of ν restricted on the completion $K_\mathfrak{P}^\times$ via Artin reciprocity law, and $\tau(\chi_\mathfrak{P}\nu_\mathfrak{P})$ is the Gauss sum of $\chi_\mathfrak{P}\nu_\mathfrak{P}$ (Lang 1994, p.286) (ii) $\Omega \in \mathbf{C}^\times$ is the complex CM period of K (Katz 1978, (5.1.45)). We call $L_{\chi,\Sigma}$ the (primitive) p-adic Hecke L-function attached to the character χ. Define the cyclotomic and anticyclotomic p-adic L-functions attached to χ by

$$L_{\lambda,\Sigma}^+ := L_{\chi,\Sigma}|_{S_1=\ldots=S_d=0} \in \Lambda_+, \quad L_{\chi,\Sigma}^- := L_{\chi,\Sigma}|_{T_+=0} \in \Lambda_-.$$

1.2 The μ-Invariants

We are interested in the μ-invariants of these p-adic L-functions. Recall that the μ-invariant $\mu(L^\bullet_{\chi,\Sigma})$ is defined by

$$\mu(L^\bullet_{\chi,\Sigma}) = \max\left\{r \in \mathbf{Q}_{\geq 0} \mid L^\bullet_{\chi,\Sigma} \in p^r \Lambda_\bullet\right\}.$$

Hida (2010) invented a method to compute the μ-invariants of $L_{\chi,\Sigma}$ and $L^-_{\chi,\Sigma}$ at least when p is unramified in F. For example, it is proved in Burungale and Hsieh (2013) and Hida (2011) that

Theorem 1. *If $p \nmid D_F$, then $\mu(L_{\chi,\Sigma}) = 0$.*

In other words, $L_{\chi,\Sigma} \not\equiv 0 \pmod{\mathfrak{m}_W \Lambda}$, where $\mathfrak{m}_W$ is the maximal ideal of W. However, the determination of the μ-invariant of the cyclotomic p-adic L-function $L^+_{\chi,\Sigma}$ unfortunately seems out of reach with the current techniques.

We describe the result for anticyclotomic p-adic L-functions after introducing some notation. Let $\mathfrak{C}$ be the prime-to-p conductor of χ and let $\mathfrak{C}^-$ be the product of non-split prime divisors of $\mathfrak{C}$. For each prime divisor $\mathfrak{q}|\mathfrak{C}^-$, we define the local invariant $\mu_p(\chi_\mathfrak{q}) \geq 0$ by

$$\mu_p(\chi_\mathfrak{q}) := \inf_{x \in K_\mathfrak{q}^\times} \mathrm{ord}_p(\chi_\mathfrak{q}(x) - 1),$$

where $\mathrm{ord}_p : \mathbf{C}_p^\times \to \mathbf{Q}$ is the p-adic valuation normalized so that $\mathrm{ord}_p(p) = 1$. Put

$$\mu_p(\chi) := \sum_{\mathfrak{q}|\mathfrak{C}^-} \mu_p(\chi_\mathfrak{q}).$$

We say χ is *self-dual* if $\chi|_{\mathbf{A}_F^\times} = \tau_{K/F}|\cdot|_{\mathbf{A}_F}$, where $\tau_{K/F}$ is the quadratic character associated to K/F. It is known that the complex L-function $L(s,\chi)$ of a self-dual character satisfy the function equation $L(s,\chi) = N^s W(\chi) L(-s,\chi)$, where N is a positive integer and $W(\chi) \in \{\pm 1\}$ is called the global root number of χ. It follows from the functional equation that $L^-_{\chi,\Sigma} = 0$ if $W(\chi) = -1$. On the other hand, we say χ is *residually self-dual* if $\chi(\mathfrak{a})\mathrm{N}_{F/\mathbf{Q}}(\mathfrak{a}) \equiv \tau_{K/F}(\mathfrak{a}) \pmod{\mathfrak{m}_W}$ for all prime-to-$p\mathfrak{C}$ ideals $\mathfrak{a}$ of $\mathcal{O}_F$. The following theorems are proved in Hida (2010) if $\mathfrak{C}^- = (1)$ and Hsieh (2013) for the general case.

Theorem 2. *Suppose that $p \nmid D_F$. Let χ be a self-dual Hecke character such that the global root number $W(\chi) = 1$. Then we have*

$$\mu(L^-_{\chi,\Sigma}) = \mu_p(\chi).$$

Theorem 3. *Suppose that $p \nmid D_F$ and that χ is not residually self-dual. Then*

$$\mu(L^-_{\chi,\Sigma}) = 0 \iff \mu_p(\chi) = 0.$$

Remark 1. When $F = \mathbf{Q}$, Theorem 2 was also proved by Finis (2006) by a different approach.

If χ is self-dual with the root number $W(\chi) = -1$, then $L_{\chi,\Sigma}$ has at least a simple zero at the cyclotomic variable $T_+ = 0$ by the functional equation of L-functions for CM fields. Thus the anticyclotomic p-adic L-function $L^-_{\chi,\Sigma}$ vanishes. In this case, we can consider the cyclotomic derivative

$$L^{(1),-}_{\chi,\Sigma} := \frac{\partial L_{\chi,\Sigma}}{\partial T_+}|_{T_+=0} \in \Lambda_-.$$

Let h^-_K be the relative class number of K/F. A. Burungale (2014) uses Hida's method and some inputs from Hsieh (2013) to prove the following result.

Theorem 4. *If $p \nmid h^-_K \cdot D_F$ and χ is self-dual with $W(\chi) = -1$, then*

$$\mu(L^{(1),-}_{\chi,\Sigma}) = \mu_p(\chi) + \min_{\mathfrak{q}|\mathfrak{C}^-} \left\{0, \mu'_p(\chi_\mathfrak{q})\right\},$$

where

$$\mu'_p(\chi_\mathfrak{q}) := \mathrm{ord}_p(\frac{\log_p \mathrm{N}_{K/\mathbf{Q}}(\mathfrak{q})}{\log_p(1+p)}) - \mu_p(\chi_\mathfrak{q})$$

and $\log_p : \mathbf{Z}_p^\times \to \mathbf{Z}_p$ is the p-adic logarithm, which is zero on the roots of unity and defined by the usual power series on $1 + p\mathbf{Z}_p$.

In particular, if $\mu_p(\chi) = 0$, then $\mu(L^{(1),-}_{\chi,\Sigma}) = 0$.

1.3 p-Adic L-Functions and Selmer Groups for CM Elliptic Curves

The p-adic L-functions for CM fields allow us to construct p-adic L-functions for CM elliptic curves over F. Let E be an elliptic curve defined over F. Suppose that E admits complex multiplication by the ring of integers of an imaginary quadratic field $\mathbf{Q}(\sqrt{-D})$ with the absolute discriminant $D > 0$ and that E has good ordinary reduction at all places above p. In particular, this implies that $p = \mathfrak{p}\bar{\mathfrak{p}}$ is split in $\mathbf{Q}(\sqrt{-D})$. Let K be the CM field $F(\sqrt{-D})$. Let $\Sigma_\mathfrak{p}$ be the set of primes of K above $\mathfrak{p}$. Then $\Sigma_\mathfrak{p}$ is a p-adic CM type of K. For each $\mathfrak{P} \in \Sigma_\mathfrak{p}$, let $k_\mathfrak{P}$ be the residue field and $\mathrm{N}(\mathfrak{P}) = \#(k_\mathfrak{P})$. Let $\alpha_\mathfrak{P}$ be the p-adic unit root of the Hecke polynomial $X^2 - a_\mathfrak{P}(E)X + \mathrm{N}(\mathfrak{P})$, where $a_\mathfrak{P}(E) = 1 + \mathrm{N}(\mathfrak{P}) - \#(E(k_\mathfrak{P}))$. By the theory of complex multiplication, we can associate a Hecke character χ_E of K to the CM elliptic curve $E_{/F}$ so that χ_E^{-1} is self-dual, and the Hecke L-function $L(s, \chi_E)$ of χ_E is equal to the Hasse-Weil L-function $L(s, E_{/F})$ of $E_{/F}$. We define

$$L_{E,\mathfrak{p}} := L_{\chi_E^{-1},\Sigma_\mathfrak{p}} \in \Lambda$$

to be the p-adic Hecke L-function attached to χ_E^{-1}. The p-adic L-function for E/K_∞ is defined by

$$L_p(E/K_\infty) := (L_{E,\mathfrak{p}})^2.$$

It follows from (1) that for every finite order character $\nu : \mathrm{Gal}(K_\infty/K) \to \mu_{p^\infty}$, we have

$$\nu(L_p(E/K_\infty)) = \frac{L(1, E_{/K} \otimes \nu)}{\Omega_E^2} \cdot \prod_{\mathfrak{P} \in \Sigma_\mathfrak{p}} e_\mathfrak{P}(E, \nu)^2 \cdot \frac{[\mathcal{O}_K^\times : \mathcal{O}_F^\times]^2}{D_F},$$

where $\Omega_E = \prod_{\sigma \in \Sigma} \Omega_{E^\sigma}$ is a product of the periods Ω_{E^σ} of a Néron differential form of E^σ over $\mathcal{O}_{F^\sigma,(p)}$ and $e_\mathfrak{P}(E, \nu) = e_\mathfrak{P}(\chi_E^{-1}\nu)$.

Iwasawa main conjecture for E/K_∞ asserts that the *p-adic L-function* $L_p(E/K_\infty)$ is given by the characteristic power series of the p-Selmer group of E/K_∞. We recall that for each algebraic extension L of F, the p-Selmer group $\mathrm{Sel}(E/L)$ of E over L is defined by

$$\mathrm{Sel}(E/L) = \ker\left\{ H^1(L, E[p^\infty]) \to \prod_v H^1(L_v, E) \right\},$$

where v runs over all places of L. We denote by $X_p(E/L)$ the Pontryagin dual of $\mathrm{Sel}(E/L)$. If L is a $\mathbf{Z}_p^r$-extension of K, then the Iwasawa algebra $\mathbf{Z}_p[[\mathrm{Gal}(L/K)]]$ is a formal power series ring of r variables, and $X_p(E/L)$ is a finitely generated $\mathbf{Z}_p[[\mathrm{Gal}(L/K)]]$-module. Let

$$C_p(E/L) \in \mathbf{Z}_p[[\mathrm{Gal}(L/K)]]$$

be the characteristic power series of the module $X_p(E/L)$ over $\mathbf{Z}_p[[\mathrm{Gal}(L/K)]]$. Recall that the characteristic power series of a finitely generated $\mathbf{Z}_p[[\mathrm{Gal}(L/K)]]$-module M is a generator of the intersection of all principal ideals containing the Fitting ideal of M. Now we can state Iwasawa main conjecture for E/K_∞ as follows:

Conjecture 1. Suppose that p is unramified in F. Then the principal ideal $C_p(E/K_\infty)\Lambda$ is generated by $L(E/K_\infty)$.[1]

Define the cyclotomic and anticyclotomic p-adic L-functions by $L_{E,\mathfrak{p}}^+ = L_{\chi_E^{-1},\Sigma_\mathfrak{p}}^+ \in \Lambda_+$ and $L_{E,\mathfrak{p}}^- = L_{\chi_E^{-1},\Sigma_\mathfrak{p}}^- \in \Lambda_-$. Similarly, define $L(E/K_\infty^\pm) = (L_{E,\mathfrak{p}}^\pm)^2$, and we can formula the main conjectures for these p-adic L-functions in the

[1]Note that the p-adic L-function $L_p(E/K_\infty)$ only belongs to Λ instead of $\mathbf{Z}_p[[\mathrm{Gal}(K_\infty/K)]]$. In addition, p is assumed to be unramified in F, since it is not clear to the author that the Katz p-adic L-function $L_p(E/K_\infty)$ is the right one for the main conjecture if p is ramified in F.

same way. When $F = \mathbf{Q}$, there is a completely different construction of $L_{E,\mathfrak{p}}$ by Yager (1982), combined with which the main conjectures has been settled down by Rubin (1991).

We discuss the connection between the main conjectures for CM elliptic curves and the μ-invariants of p-adic L-functions.

Theorem 5. *Suppose that $p \nmid D_F$. We have*

1. $\mu(L_{E,\mathfrak{p}}) = 0$,
2. *If the root number $W(E/F) = +1$, then $\mu(L^-_{E,\mathfrak{p}}) = 0$.*

Proof. Part (1) is an immediate consequence of Theorem 1. To see part (2), we note that $W(E/F) = W(\chi_E^{-1})$ and χ_E takes value in the imaginary quadratic field $\mathbf{Q}(\sqrt{-D})$. Therefore, $\chi_{E,\mathfrak{q}}|_{\mathcal{O}_{K_v}^\times}$ takes value in the group of the roots of unity in $\mathbf{Q}(\sqrt{-D})$ for every finite place $\mathfrak{q}$ of K. In particular, this implies $\mu_p(\chi_{E,\mathfrak{q}}) = 0$ for every $\mathfrak{q}|\mathfrak{C}^-$ whenever $p \nmid 2D$, so $\mu_p(\chi_E) = 0$, and (2) follows from Theorem 2. □

Remark 2. When $F = \mathbf{Q}$, the μ-invariants of $L_{E,\mathfrak{p}}$ along the $\mathbf{Z}_p$-extension unramified outside $\mathfrak{p}$ were proved by Gillard (1987) and Schneps (1987). In particular, these results also imply $\mu(L_{E,\mathfrak{p}}) = 0$.

If the root number $W(E/F) = -1$, then $L_{E,\mathfrak{p}}$ has at least a simple zero at the cyclotomic variable $T_+ = 0$, and $\mathrm{Sel}(E/K^-_\infty)$ is expected to have corank two over Λ_-. In this case, we consider the derivative

$$L^{(1),-}_{E,\mathfrak{p}} := \frac{\partial L_{E,\mathfrak{p}}}{\partial T_+}|_{T_+=0} \in \Lambda_-.$$

We have seen $\mu_p(\chi_E) = 0$, so the following is an immediate consequence of Theorem 4.

Theorem 6. *If $p \nmid D_F \cdot h^-_K$ and $W(E/F) = -1$, then $\mu(L^{(1),-}_{E,\mathfrak{p}}) = 0$.*

Theorem 6 implies that $L^{(1),-}_{E,\mathfrak{p}}$ is non-zero and hence $L_{E,\mathfrak{p}}$ does have a simple zero at $T_+ = 0$. When $F = \mathbf{Q}$, the non-vanishing of $L^{(1),-}_{E,\mathfrak{p}}$ was proved in Agboola and Howard (2006) by an algebraic method.

The vanishing of μ-invariants of the *anticyclotomic* p-adic L-functions $L^-_{E,\mathfrak{p}}$ and $L^{(1),-}_{E,\mathfrak{p}}$ plays a crucial role in the Eisenstein ideal approach to a one-sided divisibility towards Iwasawa main conjecture for CM fields. It has the following application to the *cyclotomic* main conjecture for CM elliptic curves over totally real fields (Hsieh 2014, Thm. 1).

Theorem 7. *Suppose that $p \nmid 3D_F \cdot h^-_K$. Then*

$$L_p(E/K^+_\infty) \text{ divides } C_p(E/K^+_\infty) \text{ in } \Lambda_+.$$

We explain very briefly how the vanishing of μ-invariants enters into the proof of the above theorem. In Hsieh (2014), we study the main conjecture by the Eisenstein congruence method for the quasi-split unitary group $U(2,1)$ of degree three over F. The general philosophy is that some p-adic L-functions can be realized as the constant term of a particular p-adic family of Eisenstein series, and the congruence between cusp forms and this special Eisenstein series modulo its constat term yields elements in the Selmer groups corresponding to p-adic L-functions. In our setting, we construct a Λ-adic Eisenstein series on $U(2,1)_{/F}$ whose constant term is a product of $L_{E,\mathfrak{p}}$ and a Deligne-Ribet p-adic L-function L_{DR}, which should contribute to two distinct Selmer groups respectively. However, it turns out that our method does not allow us to distinguish their contributions to individual Selmer group unless we know the common zeros of $L_{E,\mathfrak{p}}$ and L_{DR} are at most simple zeros. Now if $W(E/F) = +1$, $\mu(L^{-}_{E,\mathfrak{p}}) = 0$ implies $L_{E,\mathfrak{p}}$ and L_{DR} have no common zeros, while if $W(E/F) = -1$, $\mu(L^{(1),-}_{E,\mathfrak{p}}) = 0$ implies that $L_{E,\mathfrak{p}}$ and L_{DR} have at most one simple zero! Therefore, we can complete the argument to show one-sided divisibility $L_p(E/K_\infty)|C_p(E/K_\infty)$, from which Theorem 7 follows by Iwasawa descent for CM elliptic curves due to Perrin-Riou.

In the remainder of this article, we will first review Sinnott's proof of Ferrero-Washington theorem, and in Sect. 3 we explain Hida' method of proving the part 2 of Theorem 5 with emphasis on the automorphic side of the proof. In Sect. 4, we discuss the proof of Theorem 6 and the generalization of Hida's ideas to a class of p-adic Rankin-Selberg L-functions.

2 Ferrero-Washington Theorem

While the focus of the research reported here is about the μ-invariants of p-adic Hecke L-functions for CM fields, it is instructive to first review the case of p-adic L-functions for Dirichlet characters. Let $\chi : (\mathbf{Z}/N\mathbf{Z})^\times \to \mathbf{C}^\times$ be an *odd* Dirichlet character of conductor N and let $c > 1$ be an integer prime to p. Let $\mathcal{O}$ be a finite extension of $\mathbf{Z}_p$ containing values of χ and let $\mathfrak{m}_\mathcal{O}$ be the maximal ideal of $\mathcal{O}$. Let L_χ be the Kubota-Leopoldt p-adic L-function associated to χ and c. Namely, L_χ is a $\mathcal{O}$-valued p-adic measure on $1 + p\mathbf{Z}_p$ such that for all odd positive integers k,

$$\int_{1+p\mathbf{Z}_p} x^k \mathrm{d}L_\chi(x) = (1 - \chi\omega^{-k}(c)c^{k+1})L(-k, \chi\omega^{-k}). \tag{1}$$

Recall that the μ-invariant $\mu(\varphi)$ of a $\overline{\mathbf{Z}}_p$-valued p-adic measure φ on a p-adic compact group H is defined to be

$$\mu(\varphi) = \inf_{U \subset H \text{ open}} \mathrm{ord}_p(\varphi(U)).$$

The vanishing of the μ-invariant of L_χ was proved by Ferrero and Washington.

Theorem 8 (Ferrero and Washington 1979). $\mu(L_\chi) = 0$.

Sinnott (1984) gave a beautiful new proof of Ferrero-Washington theorem, which is the origin of Hida's method. Consider the torus $\mathbb{G}_{m/\mathcal{O}} = \operatorname{Spec} \mathcal{O}[t, t^{-1}]$ over $\mathcal{O}$ and its formal completion $\hat{\mathbb{G}}_{m/\mathcal{O}} = \operatorname{Spf} \mathcal{O}[[t-1]]$ at the maximal ideal $\mathfrak{m} := (p, t-1)$. Each $a \in \mathbf{Z}_p^\times$ induces an automorphism on $\hat{\mathbb{G}}_{m/\mathcal{O}}$ by sending $t \mapsto t^a$. Let $\mathbb{F} = \bar{\mathbb{F}}_p$. The first ingredient in Sinnott's proof is the following linear independence result on rational functions on $\mathbb{G}_{m/\mathbb{F}}$.

Proposition 1 (Prop. 3.1, Sinnott 1984). *Let $a_1, \ldots, a_n \in \mathbf{Z}_p^\times$ such that $a_i a_j^{-1} \notin \mathbf{Z}_{(p)}^\times$. If $f_1, \ldots, f_n \in \mathbb{F}(t)$ are non-constant functions, then $f_1(t^{a_1}), \ldots, f_n(t^{a_n}) \in \mathbb{F}((t-1))$ are linearly independent over $\mathbb{F}$.*

We can associate a formal function $G_\chi(t) \in \mathcal{O}[[t-1]]$ on $\hat{\mathbb{G}}_{m/\mathcal{O}}$ to L_χ defined by

$$G_\chi(t) := \int_{1+p\mathbf{Z}_p} t^x \mathrm{d}L_\chi(x) = \sum_{n \geq 0} \left(\int_{1+p\mathbf{Z}_p} \binom{x}{n} \mathrm{d}L_\chi(x) \right) (t-1)^n.$$

Thus, $G_\chi(t)$ is the usual power series attached to the measure on $\mathbf{Z}_p$ obtained by the extension of L_χ by zero outside $1 + p\mathbf{Z}_p$. Then it is well known that

$$\mu(L_\chi) = 0 \iff G_\chi(t) \pmod{\mathfrak{m}_\mathcal{O}} \neq 0 \in \mathbb{F}[[t-1]].$$

The second ingredient is the expression of $G_\chi(t)$ in terms of a linear combination of a rational functions on $\mathbb{G}_m$ acted by $\mathbf{Z}_p^\times$. To be precise, let $\Delta = \mu_{p-1}$ be the torsion subgroup of $\mathbf{Z}_p^\times$ and let $\Delta^{\mathrm{alg}} := \Delta \cap \mathbf{Z}_{(p)}^\times = \{\pm 1\}$. Let $\mathbf{D}$ be a set of representative of $\Delta / \Delta^{\mathrm{alg}} = \mu_{p-1} / \{\pm 1\}$ in μ_{p-1}. Then we have

Proposition 2. *Denote by $\mathcal{O}[t, t^{-1}]_{(\mathfrak{m})}$ the localization of $\mathcal{O}[t, t^{-1}]$ at $\mathfrak{m}$.*

(i) There exist a set $\{F_a\}_{a \in \mathbf{D}}$ of rational functions in $W[t, t^{-1}]_{(\mathfrak{m})}$ indexed by $\mathbf{D}$ such that

$$G_\chi(t) = \sum_{a \in \mathbf{D}} \chi(a^{-1}) F_a(t^a); \tag{2}$$

(ii) $F_a(t) \pmod{\mathfrak{m}_\mathcal{O}}$ is not a constant function for some $a \in \mathbf{D}$.

In view of Proposition 1, this proposition implies that $G_\chi(t) \pmod{\mathfrak{m}_\mathcal{O}} \neq 0$, and hence $\mu(L_\chi) = 0$.

Equation (2) indeed is a consequence of the following classical formula of Dirichlet L-values:

$$\theta^k \left(\frac{\sum_{a=1}^N \chi(a) t^a}{1 - t^N} \right) |_{t=1} = L(-k, \chi), \tag{3}$$

where $\theta := t\frac{d}{dt}$ is the invariant differential operator on $\mathbb{G}_m$. To illustrate this, we sketch a proof of Proposition 2 for the special case $\chi = \omega^i$ with an odd integer $0 < i < p-2$, where $\omega : (\mathbf{Z}/p\mathbf{Z})^\times \to \mu_{p-1}$ is the Teichmüller character. For each $a \in \mathbf{Z}_p^\times$, define a rational function

$$f_a(t) = \frac{t^r}{1-t^p}, \quad 0 \le r < p\,, r \equiv a^{-1} \,(\mathrm{mod}\ \mathfrak{m}_{\mathcal{O}})$$

and put

$$F_a(t) = 2f_a(t) - 2c \cdot f_{ca}(t^c) \in \mathcal{O}[t, t^{-1}]_{(\mathfrak{m})}.$$

Then it is straightforward to verify that

$$\theta^k(\sum_{a \in \mathbf{D}} \chi(a^{-1}) F_a(t^a))|_{t=1} = (1 - \chi\omega^{-k}(c)c^{k+1}) L(-k, \chi\omega^{-k}).$$

Hence, (2) follows from the interpolation formula (1), and the non-constancy of $F_a(t)$ (mod $\mathfrak{m}_{\mathcal{O}}$) is clear from the definition.

3 Hida's Theorem on the μ-Invariants of Anticyclotomic p-Adic Hecke L-Functions

In this section, we briefly review the method developed by Hida (2010) to compute the μ-invariant of the anticylotomic p-adic Hecke L-function $L^-_{\chi,\Sigma}$ attached to a Hecke character χ. This method is a far-reaching generalization of Sinnott's proof of Ferrero-Washington theorem. We assume

$$p > 2 \text{ is unramified in } F. \qquad \text{(unr)}$$

For simplicity, we further assume that $\mu_p(\chi) = 0$ and p does not divide the relative class number h^-_K of K/F. Let $O = \mathcal{O}_F$ and $O_p = \mathcal{O}_F \otimes_{\mathbf{Z}} \mathbf{Z}_p$. Fix a basis $\{\xi_1, \ldots, \xi_d\}$ of O over $\mathbf{Z}$ ($d = [F : \mathbf{Q}]$). The starting point is that, regarding the p-adic L-function $L^-_{\chi,\Sigma}$ as a p-adic measure $\mathrm{d}L^-_{\chi,\Sigma}$ on $\mathrm{Gal}(K^-_\infty/K_\infty) \simeq 1 + pO_p$, we can associate to $L^-_{\chi,\Sigma}$ a power series $G_\chi \in W[[S_1, \ldots, S_d]]$ given by

$$G_\chi := \int_{1+pO_p} t^x \mathrm{d}L^-_{\chi,\Sigma}(x) = \sum_{(k_1,\ldots,k_d) \in \mathbf{Z}^d_{\ge 0}} \int_{1+pO_p} \binom{x}{k_1, \ldots, k_d} \mathrm{d}L^-_{\chi,\Sigma}(x) S_1^{k_1} \cdots S_d^{k_d}$$

$$(S_1 = t^{\xi_1} - 1, \cdots, S_d = t^{\xi_d - 1}),$$

where $\binom{x}{k_1,\ldots,k_d} : O_p \to \mathbf{Z}_p$ is the function defined by

$$\binom{x}{k_1,\ldots,k_d} := \prod_{i=1}^{d} \binom{m_i}{k_i} \text{ for } x = m_1\xi_1 + \cdots + m_d\xi_d.$$

Then we have

$$\mu(L^-_{\chi,\Sigma}) = 0 \iff G_\chi \pmod{\mathfrak{m}_W} \neq 0 \in \mathbb{F}[[S_1, \cdots, S_d]].$$

To show $G_\chi \pmod{\mathfrak{m}_W} \neq 0$, Hida constructs a family of p-adic Hilbert Eisenstein series $\{\mathcal{E}^\circ_a\}_{a\in\mathbf{D}}$ indexed by a suitable finite subset $\mathbf{D}$ of transcendental automorphisms of the deformation space of an ordinary abelian variety $\mathcal{A}_{/\mathbb{F}}$ with CM by $\mathcal{O}_K$ such that G_χ is a linear combination of the *t-expansions* of $\{\mathcal{E}^\circ_a\}_{a\in\mathbf{D}}$ at the CM point $\mathcal{A}_{/\mathbb{F}}$. Proving a deep result on the linear independence of p-adic modular forms modulo p (an analogue of Proposition 1), Hida reduces the problem to showing the non-vanishing of individual p-adic Eisenstein series $\mathcal{E}^\circ_a$, which in turns is equivalent to the non-vanishing modulo p of some Fourier coefficients of $\mathcal{E}^\circ_a$ by the q-expansion principle for p-adic modular forms due to Ribet. If χ is only ramified at primes split in K (this implies K/F must be unramified and $W(\chi) = +1$ if χ is self-dual), Hida computes the Fourier coefficients of $\mathcal{E}^\circ_a$, from which he is able to easily deduce the non-vanishing of G_χ. In the general case, the computation of the Fourier coefficients of $\mathcal{E}^\circ_a$ is rather complicated. To resolve this difficulty, it seems theory of automorphic representation must be brought into play. In Hsieh (2013), we introduce a new set of p-adic Eisenstein series $\{\mathcal{E}_a\}_{a\in\mathbf{D}}$, which arises from *toric forms*, eigenforms of Hecke operators coming from $K^\times$. It turns out that the connection of these Eisenstein series to p-adic L-functions is through the toric period integral of Eisenstein series, and the non-vanishing modulo p of their Fourier coefficients are intimately related to epsilon dichotomy of the local theta correspondence for $(U(1), U(1))$ in Harris et al. (1996).

3.1 Linear Independence for p-Adic Modular Forms Modulo p

Let $N \geq 4$ be a prime-to-p integer and let $\mathfrak{c}$ be a prime-to-p integral ideal of O. Let $L_\mathfrak{c} = \mathfrak{c}^* \oplus O$ be an O-module, where $\mathfrak{c}^* = \{\xi \in F \mid \mathrm{Tr}_{F/\mathbf{Q}}(\xi\mathfrak{c}) \subset \mathbf{Z}\}$ is the dual ideal of $\mathfrak{c}$. Let

$$\mathfrak{M}(\mathfrak{c}, N) \to \operatorname{Spec} W$$

the moduli scheme of $\mathfrak{c}$-polarized abelian varieties with real multiplication by $\mathcal{O}_F$ with full N-level structure. Denote by $\mathcal{A}$ the universal abelian scheme over $\mathfrak{M}(\mathfrak{c}, N)$. The Igusa scheme $I(\mathfrak{c}, N)$ is the moduli scheme over $\mathfrak{M}(\mathfrak{c}, N)$ classifying monomorphisms $\eta_p : \mathfrak{c}^* \otimes_{\mathbf{Z}} \mu_{p^\infty/\mathfrak{M}(\mathfrak{c},N)} \hookrightarrow \mathcal{A}[p^\infty]$ of O-group schemes. Hence, each functorial point in $I(\mathfrak{c}, N)$ can be written as the isomorphism class of a quintuple $(\underline{A}, \eta_p)$, where $\underline{A} = [(A, \lambda, \iota, \eta_N)] \in \mathfrak{M}(\mathfrak{c}, N)$ is the datum consisting of an abelian scheme A, a $\mathfrak{c}$-polarization λ of A, a ring homomorphism $\iota : O \to \operatorname{End} A$ and a full N-level structure $\eta_N : L_{\mathfrak{c}} \otimes_{\mathbf{Z}} \mathbf{Z}/N\mathbf{Z} \simeq A[N]$, and $\eta_p : \mathfrak{c}^* \otimes \mu_{p^\infty} \hookrightarrow A[p^\infty]$. Put $I_m = I(\mathfrak{c}, N) \times_{\operatorname{Spec} W} \operatorname{Spec} W/p^m W$ for a positive integer m and let $\hat{I} := \varinjlim_m I_m$ be the formal completion. Let

$$V(\mathfrak{c}, N) = \varprojlim_m H^0(I_m, \mathcal{O}_{I_m})$$

be the space of p-adic modular forms, consisting of global sections of the formal scheme $\hat{I}$.

We introduce the t-expansion of p-adic modular forms around a CM point over $\mathbb{F}$. Fix a p-ordinary CM point $\mathbf{x} = [\underline{\mathcal{A}}] \in \mathfrak{M}(\mathfrak{c}, N)(W)$, where $\mathcal{A}_{/W}$ is an abelian scheme with ordinary reduction together with the complex multiplication $\iota : \mathcal{O}_K \to \operatorname{End} \mathcal{A}$. The reduction $x_0 := \mathbf{x} \otimes \mathbb{F}$ lie in the ordinary locus $\mathfrak{M}(\mathfrak{c}, N)^{\mathrm{ord}}(\mathbb{F})$. The p-adic CM-type Σ induces a decomposition $\mathcal{O}_K \otimes_{\mathbf{Z}} \mathbf{Z}_p = O_p \oplus O_p$, $a \mapsto (i_{\mathfrak{p}}(a), i_{\overline{\mathfrak{p}}}(a))$, so we can choose a level structure $\eta_{p,\mathbf{x}} : \mu_{p^\infty} \otimes \mathfrak{c}^* \hookrightarrow \mathcal{A}[p^\infty]$ at p for the CM point $\mathbf{x}$ such that $\eta_{p,\mathbf{x}}(i_{\mathfrak{p}}(a)x) = \iota(a)\eta_{p,\mathbf{x}}(x)$ for all $a \in \mathcal{O}_K$. This gives rise to a W-point $(\mathbf{x}, \eta_{p,\mathbf{x}}) \in I(\mathfrak{c}, N)(W)$. Consider $\hat{S}_{x_0}$ the local deformation space of x_0. Namely, $\hat{S}_{x_0}$ is a functor on the category of local Artinian rings with residue field $\mathbb{F}$ and $\hat{S}_{x_0}(R)$ is the set of the isomorphism classes $[\underline{A}]$ over R with $\underline{A} \otimes_R \mathbb{F} \simeq \underline{A}_0$. By the theory of Serre-Tate coordinate (Katz 1981, Theorem 2.1), there is a canonical isomorphism

$$\hat{S}_{x_0} \simeq \operatorname{Hom}(T_p A_0 \otimes_{O_p} T_p A_0, \hat{\mathbb{G}}_m), \tag{1}$$

where $T_p A_0$ is the Tate module of the étale p-divisible group $A_0[p^\infty](\mathbb{F})$. Let $T = \operatorname{Hom}(O, \mathbb{G}_{m/W})$ be the algebraic torus over W with the character group $X^*(T) = O$. Let $\hat{T} = \operatorname{Hom}(O_p, \hat{\mathbb{G}}_{m/W})$ be the associated formal torus. Denote by $t \in X^*(T)$ the character $1 \in O$. Then

$$\mathcal{O}_{\hat{T}} = \operatorname{Spf} W[O_p] = W[[S_1, \cdots, S_d]] \quad (S_i = t^{\xi_i} - 1).$$

The units group $a \in O_p^\times$ acts on $\mathcal{O}_{\hat{T}}$ by $t \mapsto t^a$ (so $S_i \mapsto (1 + S_i)^a - 1$). The level structure $\eta_{p,0}$ at p of A_0 induces an isomorphism $T_p A_0[p^\infty]^{\acute{e}t} \simeq O_p$. By (1), it induces an isomorphism $\varphi_{x_0} : \hat{S}_{x_0} \simeq \hat{T} := \operatorname{Hom}(O_p, \hat{\mathbb{G}}_m)$, and we obtain

$$\varphi_{x_0 *} : \mathcal{O}_{\hat{S}_y} \simeq \mathcal{O}_{\hat{T}} = W[[S_1, \cdots, S_d]]$$

(Hida 2010, Corollary 2.5). For $f \in V(\mathfrak{c}, N)$, we define

$$f(t) := \varphi_{x_0*}(f|_{\hat{S}_{x_0}}) \in \mathcal{O}_{\hat{T}} = W[[S_1, \cdots S_d]].$$

The formal power series $f(t)$ will be called the *t-expansion* around x_0 of f.

Define the subgroup O_p^{alg} of $O_p^\times$ by

$$O_p^{\mathrm{alg}} = \left\{ i_{\mathfrak{p}}(a/\overline{a}) \mid a \in \mathcal{O}_{K,(p)}^\times \right\}.$$

Hida proved the following deep theorem on the linear independence of modular forms modulo p.

Theorem 9 (Corollary 3.21, Hida 2010). *Let $a_1, \cdots, a_m \in O_p^\times$ such that $a_i a_j^{-1} \notin O_p^{\mathrm{alg}}$. Let $f_1, \ldots, f_m \in V(\mathfrak{c}, N) \otimes_W \mathbb{F}$ be non-constant functions. Then $f_1(t^{a_1}), \ldots, f_m(t^{a_m})$ are linearly independent over $\mathbb{F}$*

This theorem is an analogue of Proposition 1 in the setting of rational functions on the deformation space of ordinary CM abelan varieties. We are not able to talk about Hida's difficult proof here, but only mention that two of the key ingredients: (i) a rigidity result of Ching-Li Chai on formal p-divisible groups (Chai 2008, Theorem 4.3), which implies that every closed irreducible closed subvariety Z containing $z_0 := (x_0, \ldots, x_0)$ in the product of ordinary locus $\mathfrak{M}(\mathfrak{c}, N)^{\mathrm{ord}}_{/\mathbb{F}} \times \ldots \times \mathfrak{M}(\mathfrak{c}, N)^{\mathrm{ord}}_{/\mathbb{F}}$ is Tate linear at z_0 provided that Z is stable by the diagonal Hecke action of a finite index subgroup $\mathbb{T}$ of $\mathcal{O}_{K,(p)}^\times$ (see Hida 2010, Proposition 3.11 for the precise statement); (ii) the use of Zarhin's theorem on the Tate conjecture (Zarhin 1975) in the proof of Hida (2010, Corollary 3.16).

3.2 Period Integrals and Fourier Coefficients of Toric Forms

In this subsection, we explain how to combine Theorem 9 with some inputs from representation theory to show the vanishing of $\mu(L^-_{\chi,\Sigma})$ if $\mu_p(\chi) = 0$. Let $\mathcal{U}_p$ be the torsion subgroup of $O_p^\times$ and let $\mathcal{U}_p^{\mathrm{alg}} = \mathcal{U}_p \cap O_p^{\mathrm{alg}}$. Define a subgroup $\mathcal{T} \subset \mathbf{A}_K^\times$ by

$$\mathcal{T} := \left\{ a \in \mathbf{A}_K^\times \mid a/\overline{a} \in K_\infty^\times \hat{\mathcal{O}}_K^\times \right\}.$$

Let $Cl_- := \hat{F}^\times K^\times \backslash \hat{K}^\times / \hat{\mathcal{O}}_K^\times$ be the relative class group and let Cl_-^{alg} be the subgroup of Cl_- generated by the image of $\mathcal{T}$ in Cl_-. It is easy to see that Cl_-^{alg} is in fact generated by primes ramified over F. In particular, $\#Cl_-^{\mathrm{alg}}$ is a power of 2. Since we assume $p \nmid h_K^-$, the geometrically normalized reciprocity law $\mathrm{rec}_K : \hat{K}^\times \to \mathrm{Gal}(K^{ab}/K)$ induces an isomorphism

$$\mathrm{rec}_{\mathfrak{p}} : 1 + pO_p \hookrightarrow (\mathcal{O}_K \otimes \mathbf{Z}_p)^\times \to \mathrm{Gal}(K_\infty^-/K), \quad a \mapsto \mathrm{rec}_K((a, 1))|_{K_\infty^-}.$$

We define $\langle\cdot\rangle_{\mathfrak{p}} : \hat{K}^\times \to 1 + pO_p$ by $\langle a\rangle_{\mathfrak{p}} = \mathrm{rec}_{\mathfrak{p}}^{-1}(\mathrm{rec}_K(a)|_{K_\infty^-})$. Let $\mathcal{D}_0$ (resp. $\mathcal{D}_1$) be a set of representatives of $\mathcal{U}_p/\mathcal{U}_p^{\mathrm{alg}}$ in $\mathcal{U}_p$ (resp. Cl_-/Cl_-^{alg} in $\hat{K}^{(p),\times}$). Set

$$\mathbf{D} := \mathcal{D}_0 \times \mathcal{D}_1.$$

We rephrase the main result in Hsieh (2013) as follows.

Theorem 10. *(i) There exists a set of p-adic modular forms* $\{\mathcal{E}_{u,a}\}_{(u,a)\in\mathbf{D}}$ *such that*

$$G_\chi(t) = \sum_{(u,a)\in\mathbf{D}} \chi(a)\mathcal{E}_{u,a}(t^{u\langle a\rangle_{\mathfrak{p}}}).$$

(ii) If χ is not residually self-dual or χ is self-dual with the root number $W(\chi) = +1$, then $\mathcal{E}_{u,a}$ (mod $\mathfrak{m}_W$) is not a constant function for some (u,a).

We have the following simple observation.

Lemma 1. *The map* $\mathbf{D} \to O_p^\times/O_p^{\mathrm{alg}}$, $(u,a) \mapsto u\langle a\rangle_{\mathfrak{p}}$ *is injective.*

Combining these with Theorem 9, we conclude that $\mu(L^-_{\chi,\Sigma}) = 0$ if $W(\chi) = +1$.

We give a few words about how the proof of Theorem 10 (i) is related to the toric period integral of Eisenstein series. Fix an embedding $K \hookrightarrow M_2(F)$ which optimally embeds $\mathcal{O}_K$ to $M_2(\mathcal{O}_F)$. An automorphic form $f : \mathrm{GL}_2(F)\backslash\mathrm{GL}_2(\mathbf{A}_F) \to \mathbf{C}$ is called a *toric form* of character χ if

$$f(gt) = \chi(t)f(g) \text{ for all } t \in \mathcal{T}. \tag{2}$$

In Hsieh (2013, §4), we construct some special weight one toric holomorphic Eisenstein series $\mathbb{E}^h_{\chi,u} : \mathrm{GL}_2(F)\backslash\mathrm{GL}_2(\mathbf{A}_F) \to \mathbf{C}$ for each $u \in \mathcal{D}_0$. These Eisenstein series $\mathbb{E}^h_{\chi,u}$ are shown to be defined over W by q-expansion principle. Let $\mathbb{E}_{u,a} := \mathbb{E}^h_{\chi,u}|[a]$ for $a \in \mathcal{D}_1$, where $|[a]$ means the Hecke translation by a. We define the p-adic modular form $\mathcal{E}_{u,a}$ to be the p-adic avatar of $\mathbb{E}_{u,a}$. Put

$$\mathbb{E} = \sum_{u\in\mathcal{D}_0} \mathbb{E}^h_{\chi,u} \quad ; \quad \mathcal{E}(t) = \sum_{(u,a)\in\mathbf{D}} \chi(a)\mathcal{E}_{u,a}(t^{u\langle a\rangle_{\mathfrak{p}}}).$$

For $\kappa \in \mathbf{Z}_{\geq 0}[\Sigma]$, let θ^κ be the Dwork-Katz p-adic differential operator on p-adic modular forms (Katz 1978, Cor. (2.6.25)) and let δ_1^κ be the Mass-Shimura differential operator on modular forms of weight one (Hida and Tilouine 1993, (1.21)). Roughly speaking, by the interpolation property of p-adic L-functions, it is not difficult to see that

$$\theta^\kappa G_\chi(t)|_{t=1} \approx L(0, \chi\nu_\kappa),$$

where ν_κ is some anticyclotomic Hecke character of infinity type $\kappa(1-c)$, and the symbol $\approx$ means an equality up to periods and modified Euler factors at p and Γ-factors. The toric property (2) of these Eisenstein series allows us to write

$$\theta^\kappa G_\chi(t)|_{t=1} \approx \int_{\mathbf{A}_F^\times K^\times \backslash \mathbf{A}_K^\times} \delta_1^\kappa \mathbb{E}(t)\nu_\kappa(y)dt.$$

Thus, the proof of Theorem 10 boils down to the period integral formula:

$$\int_{\mathbf{A}_F^\times K^\times \backslash \mathbf{A}_K^\times} \delta_1^\kappa \mathbb{E}(t)\nu_\kappa(t)dt \approx L(0, \chi\nu_\kappa),$$

which is proved in Hsieh (2013, Prop. 4.8).

To show Theorem 10 (ii), we must prove the non-vanishing modulo p of some Fourier coefficient of $\{\mathbb{E}_{u,a}\}_{(u,a)\in\mathbf{D}}$. The Fourier coefficients $A(\psi)$ of Eisenstein series are parameterized by additive characters $\psi = \prod \psi_v : F\backslash\mathbf{A}_F \to \mathbf{C}^\times$, and it is a standard fact that $A(\psi) = \prod_v A(\psi_v)$ can be decomposed into a product of the local Whittaker integrals $A(\psi_v)$, which can be determined explicitly by a straightforward computation at all places v of F unless v is non-split in K and χ_v is ramified, which we refer to as *bad* places. Take $\delta \in K$ with $c(\delta) = -\delta$. The local Whittaker integrals $A(\psi_v)$ at places those are inert or ramified in K and divides the conductor of χ are certain partial Gauss sum defined by

$$A(\psi_v) = \int_{F_v} \chi_v(x+\delta)\psi_v(x)dx.$$

When χ is not residually self-dual, it is rather easy to show there are many $A(\psi)$ non-vanishing modulo p when $\mu_p(\chi) = 0$. The case χ is residually self-dual is subtle. For simplicity, we will assume that χ is self-dual with the root number $W(\chi) = +1$. The general case can be treated with obvious modification. To see the role of the root number assumption, we introduce the local root number

$$W(\chi_v) := \epsilon(\frac{1}{2}, \chi_v, \psi_v) \cdot \chi_v^*(-\delta) \in \{\pm 1\},$$

where $\chi_v^* := \chi_v|\cdot|_{K_v}^{-\frac{1}{2}}$ and $\epsilon(s, \chi, \psi_v)$ is the ϵ-factor (Tate 1979). It is known that $W(\chi_v) = +1$ except for finitely many non-split places and that

$$W(\chi) = \prod_v W(\chi_v).$$

In light of epsilon dichotomy of local theta correspondence for unitary groups, the non-vanishing of these $A(\psi_v)$ is connected with the sign of $W(\chi_v)$. For example, using results in (Harris et al. 1996, §8), we show that

$$W(\chi_v) = -1 \Rightarrow A(\psi_v) = 0$$

(Hsieh 2013, Lemma 6.1). Therefore, we find $A(\psi_v) = 0$ for some non-split prime v if $W(\chi) = -1$. On the other had, if $W(\chi) = +1$, we show by a *patching* argument (Hsieh 2013, Theorem. 6.5) that the non-vanishing modulo p of $A(\psi)$ for some ψ.

4 Generalizations

4.1 The Cyclotomic Derivative of $L_{\chi,\Sigma}$

We consider the case where χ is self-dual with the root number $W(\chi) = -1$ and $\mu_p(\chi) = 0$. Burungale (2014) shows the vanishing of the μ-invariant of $L^{(1),-}_{\chi,\Sigma}$ the anticyclotomic projection of the cyclotomic derivative of $L_{\chi,\Sigma}$. The main point is that we can actually families of p-adic Eisenstein series $\{\mathcal{E}_{u,a}(T_+)\}_{(u,a)\in\mathbf{D}}$ over the cyclotomic variable T_+ such that the t-expansion

$$G_\chi(T_+)(t) := \sum_{(u,a)\in\mathbf{D}} \chi(a)\mathcal{E}_{u,a}(T_+)(t^{u\langle a\rangle_{\mathfrak{p}}})$$

is the power series associated to the full p-adic L-function $L_{E,\mathfrak{p}}$, and the cyclotomic derivative

$$\sum_{(u,a)\in\mathbf{D}} \chi(a)\mathcal{E}^{(1),-}_{u,a}(t^{u\langle a\rangle_{\mathfrak{p}}}) \quad (\mathcal{E}^{(1),-}_{u,a} := \frac{\partial\mathcal{E}_{u,a}(T_+)}{\partial T_+}|_{T_+=0})$$

gives rise to the power series of $L^{(1),-}_{\chi,\Sigma}$. We can show the non-vanishing modulo p of $\mathcal{E}^{(1),-}_{u,a}$ in a similar way as in the case $W(\chi) = +1$. Applying Theorem 9, we get $\mu(L^{(1),-}_{\chi,\Sigma}) = 0$.

4.2 Anticyclotomic p-Adic Rankin-Selberg L-Functions

We can also apply Hida's theorem to compute μ-invariants of a class of anticyclotomic p-adic L-functions of Rankin-Selberg convolution of Hilbert modular forms and theta series. Let $\pi = \otimes'\pi_v$ be an irreducible unitary cuspidal automorphic representation of $\mathrm{GL}_2(\mathbf{A}_F)$ with central character ω and let π_K the base change to $\mathrm{GL}_2(\mathbf{A}_K)$. Suppose that π_v is a discrete series or limit of discrete series for each Archimedean place v. Let χ be a Hecke character of $\mathbf{A}_K^\times$ with

$$\chi|_{\mathbf{A}_F^\times} = \omega^{-1}.$$

Define a finite subset $S(\pi, \chi)$ of all places of F by

$$S(\pi, \chi) := \left\{ \text{places } v \text{ of } F \mid \epsilon^*(\pi_v, \chi_v) := \epsilon(\frac{1}{2}, \pi_{K_v} \otimes \chi_v) \cdot \omega_v(-1) = -1 \right\}.$$

We impose the following local root number condition:

$$\#(S(\pi, \chi)) = 0. \tag{ST}$$

This condition holds for example when every prime divisor of the conductor of π is split in K. Then under a technical assumption on the conductor of π, in Hsieh (2012) we construct an anticyclotomic p-adic L-functions $L_p(\pi, \chi) \in \Lambda_-$ which interpolates the algebraic part of the Rankin-Selberg central values $L(\frac{1}{2}, \pi_K \otimes \chi\nu)$ for anticyclotomic Hecke characters ν of p-power conductor. The construction is similar to that of $L_{E,\mathfrak{p}}$, but this time we have to replace p-adic toric Eisenstein series by p-adic toric cusp forms $\{\mathcal{F}_{(u,a)}\}_{(u,a)\in\mathbf{D}}$. The corresponding toric period integral is computed by an explicit version of Waldspurger formula (Waldspurger 1985), and as a consequence, we obtain an expression of the power series attached to $L_p(\pi, \chi)$ in terms of linear combination of $\mathcal{F}_{(u,a)}$ exactly as in Theorem 10. Using Hida's theorem again, we can compute the μ-invariant of $L_p(\pi, \chi)$ by an explicit calculation of Fourier coefficients of $\mathcal{F}_{(u,a)}$. We remark that the local root number hypothesis (ST) is fundamental in this method since the failure of (ST) makes the toric period integral always vanish by a theorem of Saito-Tunnell. The details can be found in Hsieh (2012).

Remark 3. Suppose that $S(\pi, \chi)$ contains all Archimedean places and $\#(S(\pi, \chi))$ is even (so (ST) does not hold). One can still construct the anticyclotomic p-adic L-function $L_p(\pi, \chi) \in \Lambda_-$ that interpolates the algebraic part of the central L-values $L(\frac{1}{2}, \pi_K \otimes \chi\nu)$ (See Bertolini and Darmon 1996; Chida and Hsieh 2012 for the elliptic case). By the work of Vatsal (2003), which depends on entirely different ideas from Hida's method, and its generalization (Chida and Hsieh 2013), the μ-invariant of $L_p(\pi, \chi)$ can be proved to vanish in many cases.

Acknowledgements The author would like to thank the referee for many suggestions on the improvement of the manuscript.

References

Agboola, A., Howard, B.: Anticyclotomic Iwasawa theory of CM elliptic curves. Ann. Inst. Fourier (Grenoble) **56**(4), 1001–1048 (2006)

Bertolini, M., Darmon, H.: Heegner points on Mumford-Tate curves. Invent. Math. **126**(3), 413–456 (1996)

Burungale, A.A.: On the μ-invariant of the cyclotomic derivative of a Katz p-adic L-function. J. Inst. Math. Jussieu (2014). doi:10.1017/S1474748013000388

Burungale, A., Hsieh, M.-L.: The vanishing of μ-invariant of p-adic Hecke L-functions for CM fields. Int. Math. Res. Not. IMRN no. 5, 1014–1027 (2013)
Chai, C.-L.: A rigidity result for p-divisible formal groups. Asian J. Math. **12**(2), 193–202 (2008)
Chida, M., Hsieh, M.-L.: Special values of anticyclotomic L-functions for modular forms (2012, preprint). ArXiv:1204.2427
Chida, M., Hsieh, M.-L.: On the anticyclotomic Iwasawa main conjecture for modular forms (2013, preprint). ArXiv:1304.3311
Ferrero, B., Washington, L.C.: The Iwasawa invariant μ_p vanishes for abelian number fields. Ann. Math. (2) **109**(2), 377–395 (1979)
Finis, T.: The μ-invariant of anticyclotomic L-functions of imaginary quadratic fields. J. Reine Angew. Math. **596**, 131–152 (2006)
Gillard, R.: Transformation de Mellin-Leopoldt des fonctions elliptiques. J. Number Theory **25**(3), 379–393 (1987)
Harris, M., Kudla, S., Sweet, W.: Theta dichotomy for unitary groups. J. Am. Math. Soc. **9**(4), 941–1004 (1996)
Hida, H.: The Iwasawa μ-invariant of p-adic Hecke L-functions. Ann. Math. (2) **172**(1), 41–137 (2010)
Hida, H.: Vanishing of the μ-invariant of p-adic Hecke L-functions. Compos. Math. **147**(4), 1151–1178 (2011)
Hida, H., Tilouine, J.: Anti-cyclotomic Katz p-adic L-functions and congruence modules. Ann. Sci. École Norm. Sup. (4) **26**(2), 189–259 (1993)
Hsieh, M.-L.: Special values of anticyclcotomic Rankin-Selberg L-functions (2012, preprint). ArXiv:1112.1580
Hsieh, M.-L.: On the μ-invariant of anticyclotomic p-adic L-functions for CM fields. J. Reine Angew. Math. (2013). doi:10.1515/crelle-2012-0056
Hsieh, M.-L.: Eisenstein congruence on unitary groups and Iwasawa main conjecture for CM fields. J. Am. Math. Soc. (2014, to appear)
Katz, N.: p-adic L-functions for CM fields. Invent. Math. **49**(3), 199–297 (1978)
Katz, N.: Serre-Tate local moduli. In: Algebraic Surfaces (Orsay, 1976–78). Lecture Notes in Mathematics, vol. 868, pp. 138–202. Springer, Berlin (1981)
Lang, S.: Algebraic Number Theory. Graduate Texts in Mathematics, vol. 110, 2nd edn. Springer, New York (1994)
Rubin, K.: The "main conjectures" of Iwasawa theory for imaginary quadratic fields. Invent. Math. **103**(1), 25–68 (1991)
Schneps, L.: On the μ-invariant of p-adic L-functions attached to elliptic curves with complex multiplication. J. Number Theory **25**(1), 20–33 (1987)
Sinnott, W.: On the μ-invariant of the Γ-transform of a rational function. Invent. Math. **75**(2), 273–282 (1984)
Tate, J.: Number theoretic background. In: Borel, A., Casselman, W., Automorphic Forms, Representations and L-Functions. Proceedings of the Symposia in Pure Mathematics, Oregon State University, Corvallis, Oregon, 1977. Part 2, Proceedings of the Symposia in Pure Mathematics, vol. XXXIII, pp. 3–26. American Mathematical Society, Providence (1979)
Vatsal, V.: Special values of anticyclotomic L-functions. Duke Math. J. **116**(2), 219–261 (2003)
Waldspurger, J.-L.: Sur les valeurs de certaines fonctions L automorphes en leur centre de symétrie. Compositio Math. **54**(2), 173–242 (1985)
Yager, R.: On two variable p-adic L-functions. Ann. Math. (2) **115**(2), 411–449 (1982)
Zarhin, J.G.: Endomorphisms of Abelian varieties over fields of finite characteristic. Izv. Akad. Nauk SSSR Ser. Mat. **39**(2), 272–277, 471 (1975)

The p-Adic Height Pairing on Abelian Varieties at Non-ordinary Primes

Shinichi Kobayashi

1 Introduction

The theory of the p-adic valued height pairing on abelian varieties was developed in the 1980s by Néron, Zarhin, Schneider, Mazur-Tate, etc. Compared with the real valued Néron-Tate height pairing, one important aspect in the p-adic valued case is that the pairing depends on several choices and in this sense there is no *canonical* p-adic height pairing. (One reason of this phenomenon is that there exist non-trivial compact subgroups in $\mathbb{Q}_p$ but not in $\mathbb{R}$.) Let A be an abelian variety over a number field F. Firstly, the p-adic pairing depends on a choice of the p-adic logarithm on the idele class group $\mathbb{A}_F^\times/F^\times$, which determines a $\mathbb{Z}_p$-extension of F. Secondly, it depends on a choice of a splitting as $\mathbb{Q}_p$-vector spaces of the Hodge filtration of the de Rham cohomology of A over the completion F_v of F at each $v|p$. This information is used to define the p-adic valued local height pairing at $v|p$. When A has ordinary reduction at $v|p$, we have a natural choice of the splitting at v obtained by the unit root subspace, and we may say there is a *canonical* p-adic local height pairing at v for a fixed $\mathbb{Z}_p$-extension. Therefore the choice of the splitting is often made implicitly in the ordinary case but one has to pay special attention for it in the non-ordinary case.

The canonical p-adic local height pairing at ordinary primes has a characterization by some integral property along the direction of the $\mathbb{Z}_p$-extension determined by the fixed p-adic logarithm, which is used crucially in the proof of the p-adic Gross-Zagier formula in Perrin-Riou (1987). This important integral property is shown by the construction of the p-adic height by using the universal norm subgroup. It is known that the universal norm subgroup does not have sufficient information at non-ordinary primes and this makes it impossible to construct the p-adic height via

S. Kobayashi (✉)
Mathematical Institute of Tohoku University, 6-3, Aoba, Aramaki, Aoba-ku, Sendai, Japan
e-mail: shinichi@math.tohoku.ac.jp

T. Bouganis and O. Venjakob (eds.), *Iwasawa Theory 2012*, Contributions in Mathematical and Computational Sciences 7, DOI 10.1007/978-3-642-55245-8_9,

the universal norm subgroup and prevents us from investigating integral properties of the p-adic height at non-ordinary primes. However, in Perrin-Riou (1990) and Kobayashi (2013), we constructed the p-adic height on elliptic curves at supersingular primes by using certain norm systems instead of the universal norm subgroup. This enables us to control integral properties of the p-adic height and we could prove the p-adic Gross-Zagier formula for elliptic curves at supersingular primes in Kobayashi (2013).

The aim of this paper is to generalize the *norm construction* of the p-adic height on elliptic curves in Perrin-Riou (1990) and Kobayashi (2013) to abelian varieties, and as an application we prove the p-adic Gross-Zagier formula for newforms with *general coefficients* (not necessarily in $\mathbb{Q}$) of weight 2 for $\Gamma_0(N)$ at good non-ordinary primes.

2 Norm Subgroup of Formal Groups

In this section, we review the universal norm subgroup and its variant in the non-ordinary case by recalling a brief history and their roles in Iwasawa theory.

Let $\mathbb{C}_p$ be the p-adic completion of the algebraic closure of $\mathbb{Q}_p$ and let K be a finite extension of $\mathbb{Q}_p$ in $\mathbb{C}_p$. We consider a ramified $\mathbb{Z}_p$-extension K_∞ of K in $\mathbb{C}_p$ (e.g. the cyclotomic $\mathbb{Z}_p$-extension) with the Galois group $\Gamma \cong \mathbb{Z}_p$. Let K_n be the n-th layer in K_∞, namely, the unique extension of K in K_∞ of degree p^n. We denote the integer ring of K_n (resp. $\mathbb{C}_p$) by $\mathcal{O}_{K_n}$ (resp. $\mathcal{O}_{\mathbb{C}_p}$). For an abelian variety A over K, the universal norm subgroup of $A(K)$ is defined by

$$\mathrm{N}A(K) := \bigcap_n \mathrm{N}_{K_n/K}\, A(K_n)$$

where $\mathrm{N}_{K_n/K}$ is the norm from $A(K_n)$ to $A(K)$. The importance of the universal norm subgroup in Iwasawa theory is already observed by Mazur (1972), and in fact, the behavior of this local object determines the (conjectural) $\mathbb{Z}_p[[\Gamma]]$-corank of the global Selmer group over the $\mathbb{Z}_p$-extension (if A and the $\mathbb{Z}_p$-extension are defined over a global number field). Schneider (1987) showed that if A has good reduction, then

$$\mathrm{rank}_{\mathbb{Z}_p} A(K)/\mathrm{N}A(K) = (\dim A - r)[K : \mathbb{Q}_p]$$

where r is the p-rank of the reduction of A. In particular, $\mathrm{N}A(K)$ is of finite index in $A(K)$ if and only if A has ordinary reduction. This fact is crucial for the norm construction of the p-adic height pairing by Schneider.

Now we slightly change the viewpoint. Instead of considering the universal norm subgroup, we consider the group of norm compatible systems

$$\varprojlim_n A(K_n) := \{(P_n)_n \in \prod_{n=0}^{\infty} A(K_n) \mid \mathrm{Tr}_{K_{n+1}/K_n} P_{n+1} = P_n\}.$$

Here we used the additive notation $\mathrm{Tr}_{K_{n+1}/K_n}$ for the norm instead of N_{K_{n+1}/K_n}. The Iwasawa algebra $\Lambda = \mathbb{Z}_p[[\Gamma]]$ acts naturally on $\varprojlim_n A(K_n)$ and by the result of Schneider, $\varprojlim_n A(K_n)$ is a finitely generated Λ-module of free rank $r[K : \mathbb{Q}_p]$. In particular, $\varprojlim_n E(K_n) = 0$ for an elliptic curve E with good supersingular reduction.

Nasybullin and Perrin-Riou showed that there is a variant of norm compatible systems which has sufficient information also in the supersingular case. Now we assume $K = \mathbb{Q}_p$ and K_∞ is the cyclotomic $\mathbb{Z}_p$-extension. We fix a system of p-power roots of unity $(\zeta_{p^n})_n$ such that $\zeta_{p^{n+1}}^p = \zeta_{p^n}$ and ζ_p is a primitive p-th root of unity. We denote the maximal ideal of $\mathbb{Z}_p[\zeta_{p^{n+1}}]$ by $\mathfrak{m}_n$ and the maximal ideal of $\mathbb{Z}_p[[T]]$ by $\mathfrak{m}_T$. Let Δ be the torsion part of $\mathrm{Gal}(\mathbb{Q}(\zeta_{p^\infty})/\mathbb{Q})$ and let κ be the cyclotomic character which canonically identifies Γ with $1 + 2p\mathbb{Z}_p$. The group $\mathrm{Gal}(\mathbb{Q}(\zeta_{p^\infty})/\mathbb{Q})$ acts naturally on $\mathbb{Z}_p[[T]]$ by $\gamma \cdot T = (T+1)^{\kappa(\gamma)} - 1$. Let L_p be the p-Euler factor polynomial $X^2 - a_p X + p$ with $a_p = 1 + p - \sharp E(\mathbb{F}_p)$. Perrin-Riou (1990) showed that there are sufficiently many "L_p-norm systems"

$$\varprojlim_{L_p} E(K_n) := \{(P_n)_n \in \prod_n E(K_n) \mid \mathrm{Tr}_{K_{n+2}/K_n} P_{n+2} - a_p \mathrm{Tr}_{K_{n+1}/K_n} P_{n+1} + pP_n = 0\},$$

which also has a non-trivial subgroup

$$\{(P_n)_n \in \prod_n E(K_n) \mid \mathrm{Tr}_{K_{n+2}/K_{n+1}} P_{n+2} - a_p P_{n+1} + P_n = 0\}. \tag{1}$$

These norm systems play important roles in the Iwasawa theory at supersingular primes, for example, see Perrin-Riou (1990) and Kobayashi (2003). (Actually, the Russian mathematician Nasybullin had already noticed the importance of such norm systems more than 10 years before Perrin-Riou and obtained the growth formula of the Tate-Shafarevich group at supersingular primes. See Nasybullin (1977).) We recall that the norm compatible system for the formal multiplicative group $\hat{\mathbb{G}}_m$ can be obtained as follows. For $x \in \hat{\mathbb{G}}_m(\mathbb{F}_p[[T]])^\Delta$, there is a canonical lift $\hat{x} \in \hat{\mathbb{G}}_m(\mathbb{Z}_p[[T]])^\Delta$ of x fixed by the Coleman norm operator. Then $\hat{x}(\zeta_{p^{n+1}} - 1)$ for n gives a norm compatible system. Let $\hat{E}$ be the formal group of E. Perrin-Riou showed that for $x \in \hat{E}(\mathbb{F}_p[[T]])^\Delta$, there is also a canonical lift $\hat{x} \in \hat{E}(\mathbb{Z}_p[[T]])^\Delta$ which satisfies "an L_p-relation" for a certain norm operator ψ related to the Dieudonné module of $\hat{E}$. Then $\hat{x}(\zeta_{p^{n+1}} - 1)$ gives an L_p-norm system. We denote by $Z_{\hat{E}}$ the image of $\hat{E}(\mathbb{F}_p[[T]])^\Delta$ in $\hat{E}(\mathbb{Z}_p[[T]])^\Delta$ by the lift. Then $Z_{\hat{E}}$ is a Λ-module of rank 2 and we also have a Λ-submodule $Z_{L_{\hat{E}}} \subset Z_{\hat{E}}$ of rank 1 related to the norm relation in (1). Perrin-Riou (1990) gave a new construction of the p-adic height pairing on E by using these modules. ($Z_{L_{\hat{E}}}$ plays a role of the splitting of the Hodge filtration.) For the calculation of the p-adic height of Heegner points, we need to characterize $Z_{\hat{E}}$ and $Z_{L_{\hat{E}}}$ purely in terms of norm systems. In other words, we want to know which L_p-norm systems can be obtained from $Z_{\hat{E}}$. This is also a question proposed by Perrin-Riou before Lemme 4.8 of Perrin-Riou (1990). It is easy to see that the question is equivalent to the problem of the interpolation of L_p-norm systems by power series, and the answer is given as a generalization of the theory of Coleman power series for $\hat{E}$ in Kobayashi (2013).

The main theorem of the theory of Coleman power series Coleman (1979) says that any norm compatible system of $\mathbb{G}_m$ can be interpolated by a power series, but in our case, there is a rather artificial element in $\varprojlim_{L_p} E(K_n)$ which cannot be obtained by elements in $Z_{\hat{E}}$ (cf. Kobayashi 2013, Remark 3.7). Therefore we need to define a good class of norm systems in $\varprojlim_{L_p} E(K_n)$, called admissible norm systems.

Definition 1. An L_p-norm compatible system $(P_n)_n \in \varprojlim_{L_p} \hat{E}(\mathfrak{m}_n)$ is called admissible if the p-th power Frobenius map on $\hat{E}(\mathcal{O}_{K_n}/p)$ sends $P_n \mod p$ to $P_{n-1} \mod p$ for all $n \geq 1$. In other words, by the reduction mod p, the system $(P_n)_n$ defines a point in $\hat{E}(R)$ with Fontaine's

$$R := \{(x_n)_n \in \prod(\mathcal{O}_{\mathbb{C}_p}/p) \,|\, x_{n+1}^p = x_n\}.$$

Then the Coleman power series theory can be generalized as follows.

Theorem 1 (Kobayashi 2013). *Let $(P_n)_n$ be an L_p-norm compatible system. Then there exists a power series $f \in \hat{E}(\mathfrak{m}_T)^{\Delta}$ such that*

$$f(\zeta_{p^{n+1}} - 1) = P_n$$

if and only if $(P_n)_n$ is admissible. Furthermore, such f satisfies

$$(\psi^2 - a_p\psi + p)\log_{\hat{E}} f = 0 \tag{2}$$

where $\log_{\hat{E}}$ is the formal logarithm of $\hat{E}$ and ψ is the $\mathbb{Q}_p$-linear trace operator for the Frobenius lift φ characterized by $\varphi(T) = (T+1)^p - 1$. Conversely, a power series in $\mathfrak{m}_T$ satisfying the relation (2) gives an admissible L_p-norm compatible system. In other words, the Λ-module $Z_{\hat{E}}$ is isomorphic to the module of admissible L_p-norm compatible systems.

In Kobayashi (2013), this theorem is proved in a similar way as in Washington's book (Washington 1997, Theorem 13.38), but it is much easier to use the theory of norm fields by Fontaine-Wintenberger as in the classical case. Kazuto Ota (2014) also pointed out that Theorem 1 can be easily generalized to higher dimensional formal groups of arbitrary finite height over absolutely unramified local rings by using the result of Knospe (1995). We recall the result of Knospe-Ota.

First we specify the setting and fix some notation. Let K be the fractional field of the ring of Witt vectors $W = W(k)$ of the finite field k of $q = p^{\nu}$ elements. We denote the Frobenius of W by σ. We fix a uniformizer $\pi \in \mathbb{Z}_p$ and a Frobenius lift $\varphi \in W[[T]]$ such that

$$\varphi(T) \equiv \pi T \mod \deg 2, \quad \varphi(T) \equiv T^p \mod pW[[T]],$$

and consider the associated Lubin-Tate formal group $\mathcal{F}_\pi$ of height 1 over $\mathbb{Z}_p$. We fix a system of π-power torsion points $(\varpi_n)_n$ such that $[\pi]\varpi_{n+1} = \varpi_n$ where $[\pi]$ is the multiplication by π of $\mathcal{F}_\pi$ and ϖ_1 is a non-zero π-torsion point. Let K_∞ be the field obtained from K by adjoining all π-power torsion points of $\mathcal{F}_\pi$ and K_n the field of π^n-torsion points over K. (So now K_∞ is not a $\mathbb{Z}_p$-extension but a $\mathbb{Z}_p^\times$-extension. However, the whole story does not essentially change.) We denote the maximal ideal of the integer ring $\mathcal{O}_{K_n}$ by $\mathfrak{m}_n$ and the maximal ideal of $W[[T]]$ by $\mathfrak{m}_T$. Let ψ be the unique σ^{-1}-semilinear map on $W[[T]]$ satisfying

$$\psi \circ \varphi = p, \quad \varphi \circ \psi(f)(T) = \sum_{[\pi]\varpi=0} f(T \oplus_{\mathcal{F}_\pi} \varpi)$$

where the sum runs through all π-torsion points and $\oplus_{\mathcal{F}_\pi}$ is the addition of $\mathcal{F}_\pi$.

Now we consider a d-dimensional formal group $\mathcal{G} = \operatorname{Spf} W[[X_1, \dots, X_d]]$ over W of finite height h. For simplicity, we use the multi-index notation and write the variables $(X_i)_i$ simply by X. (We always use the letter T as a one-variable parameter.) We let

$$\mathcal{P}_X = \{f(X) \in K[[X]] \mid df \in \hat{\Omega}^1_{W[[X]]/W},\ f(0) \in pW\}$$

and $\overline{\mathcal{P}}_X = \mathcal{P}_X/(pW[[X]])$. Similarly, we define one-variable $\mathcal{P}_T$ and $\overline{\mathcal{P}}_T$. Let F be the Frobenius on $\overline{\mathcal{P}}_X$, namely, the σ-semilinear ring morphism defined by $X \mapsto X^p$, and let V be the Verschiebung on $\overline{\mathcal{P}}_X$ defined as the σ^{-1}-semilinear map $\sum_n a_n X^n \mapsto \sum_n p\sigma^{-1}(a_{np})X^n$. The action of φ and ψ are extended on $\mathcal{P}_X$ and they are lifts of F and V on $\overline{\mathcal{P}}_X$.

In our setting, the Dieudonné module $M_\mathcal{G}$ of the special fiber of $\mathcal{G}$ may be defined explicitly as follows. For a closed form $\omega \in \hat{\Omega}^1_{W[[X]]/W}$, we define the primitive function F_ω of ω by the unique power series in $\mathcal{P}_X$ such that $dF_\omega = \omega$ and $F_\omega(0) = 0$. We put

$$Z^1_p(\mathcal{G}) := \{\omega \in \hat{\Omega}^1_{W[[X]]/W} \mid d\omega = 0,\ F_\omega(X \oplus_\mathcal{G} Y) - F_\omega(X) - F_\omega(Y) \in pW[[X]]\},$$

$$B^1_p(\mathcal{G}) := \{df \in \hat{\Omega}^1_{W[[X]]/W} \mid f(X) \in pW[[X]]\}.$$

Here $\oplus_\mathcal{G}$ is the addition of $\mathcal{G}$. We also consider the space of invariant differentials

$$L_\mathcal{G} := \{\omega \in \hat{\Omega}^1_{W[[X]]/W} \mid F_\omega(X \oplus_\mathcal{G} Y) = F_\omega(X) + F_\omega(Y)\}.$$

Then we define the Dieudonné module $M_\mathcal{G}$ by

$$M_\mathcal{G} := Z^1_p(\mathcal{G})/B^1_p(\mathcal{G}),$$

which depends only on the special fiber $\overline{\mathcal{G}}$. The action of φ and ψ induces the Frobenius F and the Verschiebung V on $M_{\mathcal{G}}$, and $M_{\mathcal{G}}$ is a free W-module of rank h. The Hodge filtration of $M_{\mathcal{G}}$ is given by

$$\mathrm{Fil}^i M_{\mathcal{G}} = \begin{cases} M_{\mathcal{G}} & (i < 1), \\ VL_{\mathcal{G}} & (i = 1) \\ \{0\} & (i > 1). \end{cases}$$

(We use the classical normalization of the Dieudonné module as in Fontaine (1977) and Perrin-Riou (1990). The reader may consult Katz (1979) for various normalizations, especially §5.5. Our module is $\mathbf{D}_p(G/W(k))$ there. The map $p^{-1}F$ defines the isomorphism from $M_{\mathcal{G}} \otimes \mathbb{Q}_p$ to the "usual" one by Grothendieck and Mazur-Messing (Katz 1979, (5.5.7)).) Let L be a finite extension of $\mathbb{Q}_p$. A splitting of the Hodge filtration of $M_{\mathcal{G},L} := M_{\mathcal{G}} \otimes_{\mathbb{Z}_p} L$ is an L-vector subspace $N \subset M_{\mathcal{G},L}$ such that $M_{\mathcal{G},L} = N + \mathrm{Fil}^1 M_{\mathcal{G},L}$ and $N \cap \mathrm{Fil}^1 M_{\mathcal{G},L} = \{0\}$.

We fix a basis $\omega_1, \ldots, \omega_d$ of $L_{\mathcal{G}}$ and denote their primitive functions by $\ell_1, \ldots, \ell_d$. Then the formal logarithm $\ell_{\mathcal{G}}$ is given by

$$\ell_{\mathcal{G}} : \mathcal{G} \to \mathbb{G}_a^{\oplus d}, \qquad X \mapsto (\ell_1(X), \ldots, \ell_d(X)).$$

We consider a monic polynomial $Q(t) = \sum_{n=0}^{m} a_n t^n \in \mathbb{Z}_p[t]$ satisfying two conditions:

- $Q(V)\omega_i = 0$ for all $i = 1, \ldots, d$ in $M_{\mathcal{G}}$.
- All roots have p-adic absolute values strictly greater than $|p|_p = 1/p$.

For example, the characteristic polynomial $\det(t^{\nu} I - V^{\nu} \mid M_{\mathcal{G}} \otimes K)$ as a K-vector space satisfies these conditions. If $\mathcal{G}$ is the basechange of a formal group $\mathcal{G}_0$ defined over $\mathbb{Z}_p$ and ω_i's are invariant differentials of $\mathcal{G}_0$ over $\mathbb{Z}_p$, we may take a smaller polynomial $\det(tI - V \mid M_{\mathcal{G}_0} \otimes \mathbb{Q}_p)$ as $Q(V)$.

Definition 2. An element $(P_n)_n \in \prod_{n=1}^{\infty} \mathcal{G}^{(n-1)}(\mathfrak{m}_n)$ is called a Q-norm system if it satisfies

$$\mathrm{tr}_{n+m/n}\, \ell_{\mathcal{G}}^{(n+m-1)}(P_{n+m}) + a_{m-1} \mathrm{tr}_{n+m-1/n}\, \ell_{\mathcal{G}}^{(n+m-2)}(P_{n+m-1}) + \cdots + a_0 \ell_{\mathcal{G}}^{(n-1)}(P_n) = 0$$

where we put $\mathcal{G}^{(n)} := \mathcal{G}^{\sigma^{-n}}$, $\ell^{(n)} := \ell_{\mathcal{G}}^{\sigma^{-n}}$ and $\mathrm{tr}_{n+m/n} := \mathrm{tr}_{K_{n+m}/K_n}$ is the usual trace of fields. We denote the set of Q-norm systems by $\varprojlim_Q \mathcal{G}^{(n-1)}(\mathfrak{m}_n)$. Furthermore, if the p-th power Frobenius map $\mathcal{G}^{(n)}(\mathcal{O}_{\mathbb{C}_p}/p) \to \mathcal{G}^{(n-1)}(\mathcal{O}_{\mathbb{C}_p}/p)$ sends P_{n+1} mod p to P_n mod p for all n, the system is called admissible.[1] We also say $(P_n)_n \in \mathbb{Q}_p \otimes$

[1]The definition of the admissibility is different from that of Kobayashi (2013), but they are equivalent by Theorem 2.

$\prod_{n=1}^{\infty} \mathcal{G}^{(n-1)}(\mathfrak{m}_n)$ is admissible if $(mP_n)_n \in \prod_{n=1}^{\infty} \mathcal{G}^{(n-1)}(\mathfrak{m}_n)$ is admissible for some non-zero integer m.

An important example of an admissible norm system is a system of Heegner points of higher order over ring class fields of an imaginary quadratic field with a split prime p (so locally over p it defines a system in a height one Lubin-Tate tower).

We say that a system $(P_n)_n \in \mathbb{Q}_p \otimes \prod_{n=1}^{\infty} \mathcal{G}^{(n-1)}(\mathfrak{m}_n)$ is interpolated by an element $f \in \mathcal{G}(\mathfrak{m}_T)$ if

$$f^{(n-1)}(\varpi_n) = P_n$$

for all $n \geq 1$.

Theorem 2 (Knospe 1995; Ota 2014). *Let $(P_n)_n \in \mathbb{Q}_p \otimes \prod_{n=1}^{\infty} \mathcal{G}^{(n)}(\mathfrak{m}_n)$ be a Q-norm system. Then $(P_n)_n$ is interpolated by $f \in \mathcal{G}(\mathfrak{m}_T)$ if and only if $(P_n)_n$ is admissible. Such f satisfies $Q(\psi)(\ell_{\mathcal{G}} \circ f) = 0$. Conversely, a power series in $\mathfrak{m}_n$ satisfying this relation gives an admissible Q-norm system.*

For $p > 2$, the theorem holds without $\otimes\mathbb{Q}_p$.

As in the classical case, the proof of Theorem 2 is obtained by combining the injectivity of $\varprojlim_Q \mathcal{G}^{(n-1)}(\mathcal{O}_{K_n}) \to \varprojlim_Q \mathcal{G}^{(n-1)}(\mathcal{O}_{K_n}/p)$ and the following two isomorphisms. One is the isomorphism by the theory of norm fields $\mathcal{G}(k[[T]]) \cong \varprojlim_n \mathcal{G}^{(n-1)}(\mathcal{O}_{K_n}/p)$ where the transition map is the p-th power Frobenius. The other is an isomorphism obtained by the Perrin-Riou lift (Perrin-Riou 1990, §4.1; Kobayashi 2013, §3.2) between $\mathcal{G}(k[[T]])$ and $Z_{\mathcal{G}}$, the $\mathbb{Z}_p[[\Gamma]]$-submodule of $\mathcal{G}(\mathfrak{m}_T)$ consisting of elements f such that $Q(\psi)(\ell_{\mathcal{G}} \circ f) = 0$. We briefly recall the Perrin-Riou lift. By Fontaine-Honda, we have a canonical isomorphism of $\mathbb{Z}_p$-modules

$$\mathcal{G}(k[[T]]) = \mathrm{Hom}_{W[F,V]}(M_{\mathcal{G}}, \overline{\mathcal{P}}_T). \tag{3}$$

Perrin-Riou (1990, §4.1) (see also Ota (2014) for our Lubin-Tate case) constructed the canonical lifting

$$\mathrm{Hom}_{W[F,V]}(M_{\mathcal{G}}, \overline{\mathcal{P}}_T) \longrightarrow \mathrm{Hom}_W(M_{\mathcal{G}}, \mathcal{P}_T), \quad x \longmapsto \hat{x} \tag{4}$$

characterized by

$$\psi(\varphi(\hat{x}(m)) - \hat{x}(Fm)) = 0$$

for all $m \in M_{\mathcal{G}}$, or in other words, $\psi(\hat{x}(m)) = \hat{x}(Vm)$ for all m. She also showed that this condition is equivalent to

$$P(\psi^{\nu}/p^{\nu})(\hat{x}(m)) = 0$$

for all $m \in M_{\mathcal{G}}$ where P is the characteristic polynomial

$$P(t) = \det_W(I - tF^{\nu} \mid M_{\mathcal{G}}) \in \mathbb{Z}_p[t].$$

By Theorem 2, the morphism

$$\mathcal{G}(k[[X]]) \otimes \mathbb{Q}_p \to \mathbb{Q}_p \otimes \prod_{n=1}^{\infty} \mathcal{G}^{(n-1)}(\mathfrak{m}_n), \quad x \mapsto (\omega \mapsto \hat{x}^{(n-1)}(\omega)(\varpi_n)) \tag{5}$$

is an isomorphism onto the set of admissible Q-norm systems. Here we identify $\mathcal{G}^{(n-1)}(\mathfrak{m}_n) \otimes \mathbb{Q}_p$ with its tangent space $\mathrm{Hom}_W(L_{\mathcal{G}^{(n-1)}}, K_n)$. Since the Perrin-Riou lift does not depend on Q, the set of admissible Q-norm systems is independent of the choice of Q. Therefore, we call the set of admissible Q-norm systems just the set of admissible norm systems for $\mathcal{G}$.

Following Perrin-Riou, we consider the morphism of $\mathbb{Z}_p$-modules

$$\delta_{\mathcal{G}} : \mathcal{G}(k[[X]]) \to \mathrm{Hom}_W(M^{(1)}, W[[T]]^{\psi=0})$$

defined by

$$x \longmapsto p^{-1}(\varphi(\hat{x}(m)) - \hat{x}(Fm)).$$

Here $M^{(1)} = M \otimes_{\sigma} W$ is the twist of M by σ. We write $\delta_{\mathcal{G}}$ simply by δ if there is no fear of confusion. We put $u = \pi/p$. Then the kernel and the cokernel of $\delta_{\mathcal{G}}$ are given by

$$\mathrm{Ker}\ \delta_{\mathcal{G}} \cong \mathrm{Hom}_W(M^{(1)}, W)^{V=u} \log_{\mathcal{F}_\pi}(T)$$

via (3) and (4), and

$$\mathrm{Coker}\ \delta_{\mathcal{G}} \cong [\mathrm{Hom}_W(M^{(1)}, W)/(V-u)\mathrm{Hom}_W(M^{(1)}, W)](1). \tag{6}$$

Here by definition, $x \in \mathrm{Hom}_W(M^{(1)}, W)^{V=u}$ if and only if $\sigma^{-1}x(V^{-1}(m)) = ux(m)$ for all $m \in M^{(1)}$, or equivalently, $x(Fm) = \pi\sigma x(m)$ for all m. The isomorphism (6) for the cokernel is given by looking at the first derivative, and the last symbol (1) in (6) is the twist by the Lubin-Tate character κ_π for $\mathcal{F}_\pi$. Since $W[[T]]^{\psi=0}$ is a free $W[[\mathrm{Gal}(K_\infty/K)]]$-module of rank 1 by the natural action of $\mathrm{Gal}(K_\infty/K)$ via κ_π, we have the following.

Proposition 1. *Suppose that* $\mathrm{Ker}\ \delta_{\mathcal{G}} = \mathrm{Coker}\ \delta_{\mathcal{G}} = 0$ *(e.g. $\mathcal{G}$ is biconnected), then $\mathcal{G}(k[[X]])$ has a W-module structure and it is a free* $W[[\mathrm{Gal}(K_\infty/K)]]$*-module of rank h. Hence so does the module of admissible Q-norm systems of $\mathcal{G}$.*

So far, we saw the theory of the universal norm subgroup and its variant only for formal groups. Here we add a few comments on the case of Galois representations. The concept of the universal norm subgroup is generalized in terms of Bloch-Kato's local Selmer groups, and investigated in Perrin-Riou (2000) mainly for crystalline representations and Berger (2005) for de Rham representations. A generalization of Q-norm compatible systems for crystalline representations is obtained from discussions in §2.3 and §2.4 of Perrin-Riou (1994) (especially from Proposition 2.4.2), which is the key of the construction of Perrin-Riou's big exponential map. Then the Perrin-Riou map is generalized for de Rham representations by Colmez (1998) and for de Rham (φ, Γ)-modules by Nakamura (2014).

3 Norm Construction of the p-Adic Height Pairings on Abelian Varieties at Non-ordinary Primes

The p-adic height pairing on abelian varieties is constructed by P. Schneider (1982) at ordinary primes by using the universal norm subgroup. Zarhin (1990) gave a different construction including the non-ordinary case. His observation is that the p-adic height pairing is constructed depending on a choice of a splitting of the Hodge filtration of the p-adic de Rham cohomology as $\mathbb{Q}_p$-vector spaces (not as K-vector spaces). In the good ordinary case, if we take the splitting obtained by the unit root subspace of the Dieudonné module, the resulting height pairing coincides with the pairing by Schneider. In this section, we use $\mathcal{G}(k[[T]])$ or equivalently, admissible norm systems to construct the p-adic height pairing at non-ordinary primes instead of the universal norm subgroup. We define a certain class of $\mathbb{Q}_p$-vector subspaces of the Dieudonné module, called adequate. We associate an adequate subspace to a $\mathbb{Z}_p[[\Gamma]]$-submodule of $\mathcal{G}(k[[T]])$ that plays a similar role with the splitting of the Hodge filtration in Zarhin's construction, and our height pairing depends on the choice of the adequate subspace. We show a correspondence between adequate subspaces and splittings of the Hodge filtration, and compare our height with Zarhin's. The correspondence is essentially given as the p-adic interpolation (smoothing) factor appeared in the Perrin-Riou map and this also suggests that the norm construction is more suitable for investigating integral properties.

3.1 Submodules Corresponding to Splittings of the Hodge Filtration

In this subsection, we construct a class of $\mathbb{Z}_p[[\Gamma]]$-submodules of $\mathcal{G}(k[[T]])$ that corresponds to splittings of the Hodge filtration as $\mathbb{Q}_p$-vector spaces. As pointed out in Nekovář (1993, §2.9), we require only the splitting as $\mathbb{Q}_p$-vector spaces but not as K-vector spaces. Our correspondence between $\mathbb{Z}_p[[\Gamma]]$-submodules of $\mathcal{G}(k[[T]])$ and splittings of the Hodge filtrations does not respect the structure of W-modules.

We follow the setting and notations in Sect. 2 if not otherwise specified. We denote by $\mathbf{t}$ the usual trace $K \to \mathbb{Q}_p$ and we also use the same symbol for maps induced by $\mathbf{t}$ by abuse of notation. For example, it induces canonical isomorphisms of $\mathbb{Z}_p$-modules

$$\mathbf{t}: \mathrm{Hom}_W(M_{\mathcal{G}}^{(1)}, W[[T]]^{\psi=0}) \cong \mathrm{Hom}_{\mathbb{Z}_p}(M_{\mathcal{G}}, \mathbb{Z}_p[[T]]^{\psi=0}),$$

$$\mathbf{t}: \mathrm{Hom}_W(M_{\mathcal{G}}^{(1)}, K) \cong \mathrm{Hom}_{\mathbb{Z}_p}(M_{\mathcal{G}}, \mathbb{Q}_p).$$

Let L be a finite extension of $\mathbb{Q}_p$ in $\mathbb{C}_p$. We often abbreviate $X \otimes_{\mathbb{Z}_p} L$ by X_L for a $\mathbb{Z}_p$-module X.

Definition 3. For an L-vector subspace $N \subset M_{\mathcal{G},L}$, we define $Z_{\mathcal{G},N}(k[[T]])$ as the inverse image by $\mathbf{t} \circ \delta_{\mathcal{G}}$ in $\mathcal{G}(k[[T]])_L$ of $\mathrm{Hom}_L(M_{\mathcal{G},L}/N, \mathbb{Z}_p[[T]]_L^{\psi=0})$, namely,

$$Z_{\mathcal{G},N}(k[[T]]) := \{x \in \mathcal{G}(k[[T]])_L \mid \mathbf{t} \circ (\varphi\hat{x}(\omega) - \hat{x}(F\omega)) = 0 \ \text{ for all } \omega \in N\}.$$

Remark 1. If N is a $K \otimes L$-submodule of $M_{\mathcal{G},L}$, the above condition is equivalent to $\varphi\hat{x}(\omega) = \hat{x}(F\omega)$ for all $\omega \in N$ since the pairing $K \times K \to \mathbb{Q}_p, (x, y) \mapsto \mathbf{t}(xy)$ is non-degenerate. Therefore this definition coincides with that in Perrin-Riou (1990) and Kobayashi (2013).

By the morphism (5), the module $Z_{\mathcal{G},N}(k[[T]])$ corresponds to a Λ-submodule of rank $[K : \mathbb{Q}_p]\dim_L(M_{\mathcal{G},L}/N)$ of admissible norm systems for $\mathcal{G}$. It seems not easy to describe $Z_{\mathcal{G},N}(k[[T]])$ purely in terms of norm relations for a general N but at the end of this subsection we give such descriptions for special types of N, which is sufficient for our application.

The map

$$\mathbb{Z}_p[[T]]^{\psi=0} \to \mathbb{Z}_p, \quad f \mapsto \mathrm{tr}_{1/0} f(\varpi_1)$$

induces an isomorphism

$$\mathrm{Hom}_L(M_{\mathcal{G},L}^{(1)}, \mathbb{Z}_p[[T]]_L^{\psi=0})_{/(\gamma-1)} \cong \mathrm{Hom}_L(M_{\mathcal{G},L}, L).$$

Here, for simplicity, we write $X/(\gamma-1)X$ as $X_{/(\gamma-1)}$ for a $\mathbb{Z}_p[[\Gamma]]$-module X. Combined with $\mathbf{t} \circ \delta_{\mathcal{G}}$, this induces a map $\tilde{\delta}_{\mathcal{G}} : \mathcal{G}(k[[T]])_{L/(\gamma-1)} \to \mathrm{Hom}_L(M_{\mathcal{G},L}, L)$,

$$\omega \longmapsto p^{-1}\mathbf{t} \circ \mathrm{tr}_{1/0}[(\varphi\hat{x}(\omega) - \hat{x}(F\omega))(\varpi_1)] = p^{-1}\mathbf{t} \circ \hat{x}((F-1)\omega)(0). \quad (7)$$

Here we used the following easy lemma.

Lemma 1. *Let $f(T)$ be an element of $\mathcal{P}_T$. Then*

$$\psi(f^{\sigma})(\varpi_n) = \mathrm{tr}_{n+1/n}(f(\varpi_{n+1}))$$

for $n \geq 1$ *and*

$$\psi(f^{\sigma})(0) = \mathrm{tr}_{1/0}(f(\varpi_1)) + f(0).$$

Proposition 2. *The morphism $\tilde{\delta}_{\mathcal{G}}$ is an isomorphism. In particular, the restriction of $\tilde{\delta}_{\mathcal{G}}$ induces an isomorphism*

$$\tilde{\delta}_{\mathcal{G},N}(= \tilde{\delta}_N) : \ Z_{\mathcal{G},N}(k[[T]])_{/(\gamma-1)} \cong \mathrm{Hom}_L(M_{\mathcal{G},L}/N, L). \tag{8}$$

Proof. The proof follows from the description of the kernel and the cokernel of $\delta_{\mathcal{G}}$.

Corollary 1. *The morphism*

$$\mathcal{G}(k[[T]])_{L/(\gamma-1)} \to \mathrm{Hom}_W(M_{\mathcal{G}}, K)_L, \quad x \mapsto (\omega \mapsto \hat{x}(\omega)(0))$$

is an isomorphism.

Proof. This follows from Proposition 2 and (7). Note that $F-1$ is invertible since $\mathcal{G}$ is connected. See also Claim 2 in p.256 of Knospe (1995).

For $i \geq 1$, we consider the composition

$$\Sigma_{\mathcal{G},i} : \ \mathcal{G}(k[[T]]) \otimes \mathbb{Q}_p \ \to \ \mathbb{Q}_p \otimes \varprojlim_{Q} \mathcal{G}^{(n-1)}(\mathfrak{m}_n) \ \to \ \mathcal{G}^{(i-1)}(\mathfrak{m}_i) \otimes \mathbb{Q}_p$$

with (5). We also put $\Sigma_{\mathcal{G}} := \mathrm{Tr}_{1/0}\Sigma_{\mathcal{G},1}$. We omit the index $\mathcal{G}$ in $\Sigma_{\mathcal{G},i}$ and $\Sigma_{\mathcal{G}}$ if there is no fear of confusion. We sometimes identify an element of $\mathcal{G}(k[[T]]) \otimes \mathbb{Q}_p$ with the admissible norm system by (5), and then $\Sigma_{\mathcal{G},i}$ is just the projection map $(P_n)_n \mapsto P_i$ and $\Sigma_{\mathcal{G}}$ is $(P_n)_n \mapsto \mathrm{Tr}_{1/0} P_1$.

Proposition 3. *The morphism $\Sigma_{\mathcal{G},i}$ is surjective. If $p > 2$, it is surjective without the tensor $\otimes\mathbb{Q}_p$.*

Proof. First consider the case $p > 2$. Suppose that $P_i \in \mathcal{G}^{(i-1)}(\mathfrak{m}_i)$ is given. Then $\overline{P}_i := P_i \mod p \in \mathcal{G}^{(i-1)}(\mathcal{O}_{K_i}/p)$ can be extended to a family

$$(\overline{P}_n)_n \in \varprojlim_{x \mapsto x^p} \mathcal{G}^{(n-1)}(\mathcal{O}_{K_n}/p) = \mathcal{G}(k[[T]]).$$

Hence by Perrin-Riou's lift, we have an admissible norm system $(Q_n)_n$ such that $Q_i \equiv P_i \mod p$. By Nakayama's lemma, we have the conclusion. The case $p = 2$ is similar. (Note that in general the Perrin-Riou lift (hence Σ_i) is well-defined only after the tensor $\mathbb{Q}_p$ when $p = 2$. cf. Kobayashi (2013, Proposition 3.1).)

By the logarithm map, $\mathcal{G}(W) \otimes \mathbb{Q}_p$ is naturally identified with $\mathrm{Hom}_W(\mathrm{Fil}^1 M_{\mathcal{G}}^{(1)}, K) = \mathrm{Hom}_W(VL_{\mathcal{G}}^{(1)}, K)$. Then Σ is regarded as the morphism of L-vector spaces

$$\mathcal{G}(k[[T]])_L \to \mathrm{Hom}_W(VL_{\mathcal{G}}^{(1)}, K)_L, \quad x \longmapsto (V\omega \mapsto \mathrm{tr}_{1/0}\hat{x}(\omega)(\varpi_1)).$$

This map is extended as

$$\Sigma^e : \mathcal{G}(k[[T]])_L \to \mathrm{Hom}_W(M_{\mathcal{G}}^{(1)}, K)_L, \quad x \longmapsto (\omega \mapsto p^{-1}\mathrm{tr}_{1/0}\hat{x}(F\omega)(\varpi_1)).$$

Proposition 4. *Suppose that the eigenvalues of V on $M_{\mathcal{G}}$ as $\mathbb{Q}_p$-vector spaces are different from* 1. *Then Σ^e is surjective. In particular, Σ is also surjective.*

Proof. By Lemma 1, we have

$$\mathbf{t} \circ \mathrm{tr}_{1/0}\hat{x}(\omega)(\varpi_1) = \mathbf{t} \circ \hat{x}((V-1)\omega)(0). \tag{9}$$

Since $V-1$ is invertible on $M_{\mathcal{G}}$ by our assumption, the assertion follows from Corollary 1.

The map $\Sigma(= \Sigma_{\mathcal{G}})$ induces an L-linear map $\Sigma_N(= \Sigma_{\mathcal{G},N}) : Z_{\mathcal{G},N}(k[[T]]) \to \mathcal{G}(pW)_L$, which factors through

$$\tilde{\Sigma}_N(= \tilde{\Sigma}_{\mathcal{G},N}) : \ Z_{\mathcal{G},N}(k[[T]])_{/(\gamma-1)} \longrightarrow \mathcal{G}(pW)_L.$$

Similarly, Σ^e induces the morphism

$$\Sigma_N^e : Z_{\mathcal{G},N}(k[[T]]) \to \mathrm{Hom}_W(M_{\mathcal{G}}^{(1)}, K)_L$$

and

$$\tilde{\Sigma}_N^e : Z_{\mathcal{G},N}(k[[T]])_{/(\gamma-1)} \longrightarrow \mathrm{Hom}_W(M_{\mathcal{G}}^{(1)}, K)_L.$$

Suppose that the Verschiebung V on $M_{\mathcal{G}}$ as a $\mathbb{Q}_p$-linear map does not have 1 as an eigenvalue. Then by Propositions 2 and 4, we have

$$\mathbf{t} \circ \tilde{\Sigma}^e : \mathcal{G}(k[[T]])_{/(\gamma-1)} \cong \mathrm{Hom}_{\mathbb{Z}_p}(M_{\mathcal{G}}, \mathbb{Q}_p)$$

and hence the $\mathbb{Q}_p$-linear isomorphism

$$\Phi = (\mathbf{t} \circ \tilde{\Sigma}^e \circ \tilde{\delta}_{\mathcal{G}}^{-1})^{\vee} : M_{\mathcal{G}} \xrightarrow{\ \cong\ } M_{\mathcal{G}}$$

where $\vee$ is the dual as $\mathbb{Q}_p$-vector spaces.

By (7) and (9), we have the following.

Proposition 5. *The $\mathbb{Q}_p$-linear map Φ is given by*

$$M_{\mathcal{G}} \longrightarrow M_{\mathcal{G}}, \qquad \omega \longmapsto (p-F)(F-1)^{-1}\omega.$$

Note that Φ is the interpolation factor of Perrin-Riou's theory. (cf. Perrin-Riou (1995) and Colmez (1998, Théorèm 3 and Remarque 5 (iii)). It looks slightly different because of our classical normalization of Dieudonné modules.) The map Φ is extended on $M_{\mathcal{G},L}$ linearly.

Corollary 2. *Assume that* 1 *is not an eigenvalue of* V *on* $M_{\mathcal{G}} \otimes \mathbb{Q}_p$ *as a* $\mathbb{Q}_p$*-vector space. Let* N *be an* $L[F]$*-submodule of* $M_{\mathcal{G},L}$*. Then* $\Phi(N) = N$*.*

We determine $\mathbb{Q}_p$-vector subspaces that are the image by Φ of complementary subspaces to the Hodge filtration of $M_{\mathcal{G},L}$.

Definition 4. We say an L-vector subspace $N \subset M_{\mathcal{G},L}$ is *adequate* if it satisfies the following two conditions.

1. $\dim_L (M_{\mathcal{G},L}/N) = [K : \mathbb{Q}_p] \dim \mathcal{G}$.
2. Σ_N is surjective.

Since $\dim_L Z_{\mathcal{G},N}(k[[T]])_{/(\gamma-1)} = \dim_L(M_{\mathcal{G},L}/N)$ by Proposition 2, the adequateness is equivalent to requiring that

$$\mathbf{t} \circ \tilde{\Sigma}_N : Z_{\mathcal{G},N}(k[[T]])_{/(\gamma-1)} \longrightarrow \mathrm{Hom}_L(\mathrm{Fil}^1 M_{\mathcal{G}}, L)$$

is an isomorphism of L-vector spaces. Therefore if N is adequate, $\tilde{\Sigma}_N^{-1}$ and $\tilde{\Sigma}$ induce an L-linear map

$$\mathrm{Hom}_L(\mathrm{Fil}^1 M_{\mathcal{G},L}^{(1)}, L) \cong Z_{\mathcal{G},N}(k[[T]])_{/(\gamma-1)} \to \mathcal{G}(k[[T]])_{/(\gamma-1)} \to \mathrm{Hom}_L(M_{\mathcal{G},L}, L),$$

which is the splitting of the Hodge filtration by $\Phi^{-1}(N)$. Conversely, a splitting of the Hodge filtration defines an adequate subspace by reversing the argument. Hence we have

Proposition 6. *Assume that* 1 *is not an eigenvalue of* V *on* $M_{\mathcal{G}} \otimes \mathbb{Q}_p$ *as a* $\mathbb{Q}_p$*-vector space. Then* Φ *defines a one-to-one correspondence between splittings of Hodge filtrations as* L*-vector spaces and adequate* L*-vector subspaces of* $M_{\mathcal{G},L}$*.*

This correspondence does not respect the canonical splitting by Wintenberger (1984) (cf. Kobayashi 2013, Remark 4.7) but it respects splittings as $L[F]$-modules by Corollary 2.

Now we study admissible norm systems coming from a particular type of N. Let e be an idempotent of $\mathrm{End}(\mathcal{G}/W) \otimes L$. Let $\mathfrak{q}$ be an L-vector subspace of the polynomial ring $L[t]$ and consider an L-vector subspace N of $M_{\mathcal{G},L}$ satisfying

$$eN = \mathfrak{q}\, e\mathrm{Fil}^1 M_{\mathcal{G},L} := \{f(V)\omega \mid f \in \mathfrak{q},\ \omega \in e\mathrm{Fil}^1 M_{\mathcal{G},L}\}. \qquad (10)$$

For a polynomial $f(t) = a_k t^k + a_{k-1}t^{k-1} + \cdots + a_0 \in L[t]$, we define a map

$$f(\mathrm{tr}) : \prod_{n=1}^{\infty} \mathcal{G}^{(n-1)}(\mathfrak{m}_n) \longrightarrow \prod_{n=1}^{\infty} K_n^{\oplus d}, \quad (P_n)_n \longmapsto (q_n)_n$$

where

$$q_n := a_k \mathrm{tr}_{n+k/n}\, \ell_{\mathcal{G}}^{(n+k-1)}(P_{n+k}) + a_{k-1}\mathrm{tr}_{n+k-1/n}\, \ell_{\mathcal{G}}^{(n+k-2)}(P_{n+k-1}) + \cdots + a_0 \ell_{\mathcal{G}}^{(n-1)}(P_n).$$

By definition, the kernel of $f(\mathrm{tr})$ is the set of f-norm systems.

Proposition 7. *Suppose that an L-vector subspace $N \subset M_{\mathcal{G},L}$ satisfies (10) and let $(P_n)_n \in L \otimes \prod_{n=1}^{\infty} \mathcal{G}^{(n-1)}(\mathfrak{m}_n)$ be an admissible norm system. Then $(eP_n)_n$ is an admissible norm system in the image of $Z_{\mathcal{G},N}(k[[T]])$ by (5) if and only if $(eP_n)_n$ is in the kernel of*

$$\mathrm{pr} \circ f(\mathrm{tr}) : \; L \otimes \prod_{n=1}^{\infty} \mathcal{G}^{(n-1)}(\mathfrak{m}_n) \longrightarrow L \otimes \prod_{n=2}^{\infty} (K_n/K_{n-1})^{\oplus d}$$

for all $f \in \mathfrak{q}$, namely,

$$eq_n = a_k\, \mathrm{tr}_{n+k/n}\, \ell_{\mathcal{G}}^{(n+k-1)}(eP_{n+k}) + \cdots + a_0\, \ell_{\mathcal{G}}^{(n-1)}(eP_n) \;\in\; K_{n-1}^{\oplus d}.$$

Here pr *is the natural projection.*

Proof. Since $(P_n)_n$ is an admissible norm system, it is the image of $x \in \mathcal{G}(k[[T]])_L$ by (5). By definition we have $ex \in Z_{\mathcal{G},N}(k[[T]])$ if and only if

$$(\varphi \circ \psi - p) f(\psi)\widehat{ex}(\omega)(T) = \varphi \hat{x}(\eta)(T) - \hat{x}(F\eta)(T) = 0$$

where $\eta = ef(V)V\omega \in eN$ for any $f \in \mathfrak{q}$ and $V\omega \in VL_{\mathcal{G}} = \mathrm{Fil}^1 M_{\mathcal{G}}$. (Note that $\widehat{ex} = \hat{x} \circ e$ and $\psi \circ \hat{x} = \hat{x} \circ V$, and eN is a K-vector space by (10).) If this holds, then $\mathrm{tr}_{n/n-1} eq_n = peq_n$ for $n \geq 2$ by Lemma 1. In particular, $eq_n \in K_{n-1}^{\oplus d}$. Conversely, if $(eP_n)_n$ is in the kernel of $\mathrm{pr} \circ f(\mathrm{tr})$, then

$$(\varphi \circ \psi - p) f(\psi)\widehat{ex}(\omega)(T)|_{T=\varpi_n} = 0$$

for $V\omega \in \mathrm{Fil}^1 M_{\mathcal{G}}$ and $n \geq 2$. Hence

$$(\varphi \circ \psi - p) f(\psi)\widehat{ex}(\omega)(T) = c \log_{\mathcal{F}_\pi} T$$

for some constant c. By applying ψ, we have $c = 0$.

Now we assume that $eN = \mathfrak{q}\, e\mathrm{Fil}^1 M_{\mathcal{G},L}$ is an F-invariant subspace of $M_{\mathcal{G},L}$. Let $Q \in L[t]$ be such that $Q(V)\mathrm{Fil}^1 M_{\mathcal{G},L} = \{0\}$ and all roots of Q have p-adic absolute values strictly greater than $|p|_p = 1/p$. Then eN is written in the form $eN = f\, e\mathrm{Fil}^1 M_{\mathcal{G},L}$ with $f | Q$.

Proposition 8. *Let N be a L-vector subspace of $M_{\mathcal{G},L}$ such that $eN = f\, e\mathrm{Fil}^1 M_{\mathcal{G},L}$ and eN is F-invariant. Let $(P_n)_n$ be an admissible norm system for $\mathcal{G}(k[[T]])_L$. Then $(eP_n)_n$ is an admissible norm system for $Z_{\mathcal{G},N}(k[[T]])$ if and only if $(eP_n)_n$ is an f-norm system.*

Proof. Suppose that $(P_n)_n$ is the image of $x \in \mathcal{G}(k[[T]])_L$ by (5). We take an element g of the ideal $(f) \subset \mathbb{Z}_p[t]$. Then $e\, g\mathrm{Fil}^1 M_{\mathcal{G},L} \subset eN$ by our assumption, and as in the proof of Proposition 7, we suppose that

$$(\varphi \circ \psi - p)g(\psi)\widehat{ex}(\omega)(T) = 0$$

for $\omega \in L_{\mathcal{G}}$. By putting $T = \varpi_n$ for $n \geq 0$, we have $g(\psi)\widehat{ex}(\omega)(\varpi_n) \in K_{n-1}^{\oplus d}$ for $n \geq 2$ and

$$g(\psi)\widehat{ex}(\omega)(\varpi_1) = g(\psi)\widehat{ex}(\omega)(0)$$

by Lemma 1. Then by taking $g = t^m f$ $(m = 0, \dots, n)$, we have

$$f(\psi)\widehat{ex}(\omega)(\varpi_n) = f(\psi)\widehat{ex}(\omega)(0).$$

Hence we have $eq_n \in K^{\oplus d}$ and it is constant for varying n. However, eq_n is an H-norm system for the polynomial $H = Q/f$ with $H(p) \neq 0$. Thus $eq_n = 0$. The converse is proven similarly as in the proof of Proposition 7.

Example 1. Let $\hat{E}$ be the formal group of an elliptic curve over $\mathbb{Q}_p$ with good supersingular reduction. First we let $N = L_{\hat{E}} \otimes \mathbb{Q}_p$. Then $N = (V - a_p)\mathrm{Fil}^1 M_{\hat{E},\mathbb{Q}_p}$, and by Proposition 7, an admissible norm system $(P_n)_n$ comes from $Z_{\hat{E},N}(k[[T]])$ if and only if

$$\mathrm{Tr}_{n+1/n} P_{n+1} - a_p P_n \in \hat{E}(\mathfrak{m}_{n-1}) \tag{11}$$

for $n \geq 2$. Then combined with $\mathrm{Tr}_{n+1/n-1} P_{n+1} - a_p \mathrm{Tr}_{n/n-1} P_n + pP_{n-1} = 0$, we have $\mathrm{Tr}_{n+1/n} P_{n+1} - a_p P_n + P_{n-1} = 0$. Conversely, this relation implies (11). Next we suppose $N = (V - \alpha)\mathrm{Fil}^1 M_{\hat{E},L} = L(F - \beta)\omega_{\hat{E}}$ where α, β are the roots of $t^2 - a_p t + p = 0$ and $L = \mathbb{Q}_p(\alpha)$. Then N is the α-eigenspace of the Frobenius. Hence Proposition 8 implies that an admissible norm system $(P_n)_n$ comes from $Z_{\hat{E},N}(k[[T]])$ if and only if $\mathrm{Tr}_{n+1/n} P_{n+1} = \alpha P_n$ for all $n \geq 1$.

3.2 The Zarhin-Nekovář Construction of p-Adic Height Pairings

In this subsection, we recall Zarhin's construction of the p-adic height pairing on abelian varieties, which is generalized to Galois representations by Nekovář.

Let F be a finite extension of $\mathbb{Q}$ in a fixed algebraic closure $\overline{\mathbb{Q}}$. We fix an embedding $\iota_p : \overline{\mathbb{Q}} \hookrightarrow \mathbb{C}_p$. We assume that F is unramified over p. Let L be a finite extension of $\mathbb{Q}_p$ (coefficient field) as before. Let $\log_p$ be the p-adic logarithm on $\mathbb{Z}_p^\times$ such that $\log_p p = 0$. We define the cyclotomic logarithm $\ell_{F,v}$ on $F_v^\times$ at a non-Archimedean place v by

$$\ell_{F,v}(x) = \begin{cases} -\log_p |x|_v = v(x)\log_p \mathbf{N}(v) & \text{if } v \nmid p \\ -\log_p \mathbf{N}_{F_v/\mathbb{Q}_p}(x) & \text{if } v \mid p \end{cases}$$

where $\mathbf{N}(v)$ is the number of elements of the residue field of F at v and we normalize as $v(\pi) = 1$ for a uniformizer π at v of F. We also denote $\ell_{F,v}$ by ℓ_v if there is no fear of confusion. We define the global cyclotomic logarithm ℓ_F by $\ell_F := \sum_v \ell_{F,v}$. Then $\ell_F(x) = 0$ for $x \in F^\times$.

Let A be an abelian variety over F and $A^\vee$ the dual abelian variety of A. For a finite place v of F, let $\mathcal{D}_0(A)(F_v)$ be the group of divisors algebraically equivalent to 0 defined over F_v and $Z_0(A)^0(F_v)$ the group of zero cycles of degree 0 defined over F_v. We denote by

$$(\mathcal{D}_0(A)(F_v) \times Z_0(A)^0(F_v))_e$$

the subgroup of $\mathcal{D}_0(A)(F_v) \times Z_0(A)^0(F_v)$ consisting of pairs with disjoint support.

Let N be an L-vector subspace of the Dieudonné module $M_A \otimes L$ complementary to the Hodge filtration. We recall the local p-adic height pairing

$$\langle\ ,\ \rangle_{p,v,N} : (\mathcal{D}_0(A)(F_v) \times Z_0(A)^0(F_v))_e \longrightarrow L$$

associated to N with the logarithm $\ell_v : F_v^\times \to \mathbb{Q}_p$.

First we recall the theta group. (See Moret-Bailly 1981; Mumford 1970.) Let $\mathcal{A}/\mathcal{O}_{F_v}$ be the Néron model of A/F_v and $\mathcal{A}^0$ the identity component. Let $\mathcal{A}^\vee$ be the Néron model of $A^\vee$. Then the rational equivalence class $\tilde{D}$ of $D \in \mathcal{D}_0(A)(F_v)$ defines a point in $A^\vee(F_v) = \mathcal{A}^\vee(\mathcal{O}_{F_v}) = \mathrm{Ext}^1_{fppf}(\mathcal{A}^0, \mathbb{G}_m)$. Hence we have an exact sequence

$$1 \longrightarrow \mathbb{G}_m \longrightarrow \mathcal{X}_{\tilde{D}} \longrightarrow \mathcal{A}^0 \longrightarrow 1 \tag{12}$$

as abelian fppf-sheaves on $\mathcal{O}_{F_v}$ (actually, it is exact as Zariski sheaves), and $\mathcal{X}_{\tilde{D}}$ is represented by a smooth separated commutative group scheme over $\mathcal{O}_{F_v}$.

Over Spec F_v, this exact sequence is isomorphic to

$$1 \longrightarrow \mathbb{G}_m \longrightarrow X_D \longrightarrow A \longrightarrow 1 \tag{13}$$

where X_D is given by $\mathbf{Spec}(\mathrm{Sym}\,\mathcal{O}_A(-D)) \setminus \{\text{the zero section}\}$, the line bundle associated to $\mathcal{O}_A(D)$ minus the zero section with group law from the primitivity (algebraically equivalent to zero). Hence attached to D, there is a section $s_D : A \setminus |D| \to X_D$ which is canonical up to a translation by an element of $\mathbb{G}_m$. Note that the isomorphism class of the extension (13) corresponds to the rational equivalence class of D. However, here, we choose the particular choice of the extension in the isomorphism class with the section s_D associated to the divisor D. We identify $\mathcal{X}_{\tilde{D}} \otimes F_v$ with X_D.

Suppose that A has good reduction at all places over p. We now define a morphism $\ell_{D,v}$ that makes the following diagram commutative.

$$\begin{array}{ccccccccc}
0 & \longrightarrow & \mathcal{O}_{F_v}^{\times}\hat{\otimes}L & \longrightarrow & \mathscr{X}_{\tilde{D}}(\mathcal{O}_{F_v})\hat{\otimes}L & \longrightarrow & \mathscr{A}^0(\mathcal{O}_{F_v})\hat{\otimes}L & \longrightarrow & 0 \\
 & & \downarrow & & \downarrow & & \| & & \\
0 & \longrightarrow & F_v^{\times}\hat{\otimes}L & \longrightarrow & X_D(F_v)\hat{\otimes}L & \longrightarrow & A(F_v)\hat{\otimes}L & \longrightarrow & 0 \\
 & & \downarrow{\scriptstyle \ell_v} & & \downarrow{\scriptstyle \ell_{D,v}} & & & & \\
 & & L & = & L. & & & &
\end{array}$$

(The rows are exact but vertical sequences are not.) When $v \nmid p$, the logarithm ℓ_v is unramified at v, namely, ℓ_v is trivial on $\mathcal{O}_{F_v}^{\times}$. Then ℓ_v is uniquely extended to $\ell_{D,v}$ so that the restriction to $\mathcal{X}_D(\mathcal{O}_{F_v})$ is trivial. Now assume that $v \mid p$. Let $M_{\mathcal{A}}$ be the Dieudonné module of the special fiber of $\mathcal{A}$ at v. We take an L-vector subspace N of $M_{\mathcal{A},L}$ such that $M_{\mathcal{A},L} = N \oplus \mathrm{Fil}^1 M_{\mathcal{A},L}$ as L-vector spaces, which defines a splitting as L-vector spaces of the exact sequence

$$0 \longrightarrow \mathrm{Fil}^1 M_{\mathscr{A},L} \longrightarrow M_{\mathscr{A},L} \longrightarrow M_{\mathscr{A},L}/\mathrm{Fil}^1 M_{\mathscr{A},L} \longrightarrow 0.$$

Since $M_{\mathcal{A},L}^{V=1} = \{0\}$, the exact sequence

$$0 \longrightarrow M_{\mathscr{A},L} \longrightarrow M_{\mathscr{X}_D,L} \longrightarrow M_{\mathbb{G}_m,L} \longrightarrow 0.$$

splits as F-modules by using the subspace $M_{\mathcal{X}_D,L}^{V=1} \otimes_{\mathbb{Q}_p} F_v \cong M_{\mathbb{G}_m,L}$. We put $N_D := N \oplus (M_{\mathcal{X}_D,L}^{V=1} \otimes F_v)$. Then we obtain an L-linear map

$$\mathrm{Fil}^1 M_{\mathcal{X}_D,L} \to M_{\mathcal{X}_D,L}/N_D \cong M_{\mathcal{A},L}/N \cong \mathrm{Fil}^1 M_{\mathcal{A},L}.$$

The pullback by this map defines a splitting as L-linear vector spaces

$$\begin{aligned} s_{D,v,N} : \ \mathcal{A}(F_v)\hat{\otimes}L = \mathrm{Hom}_L(\mathrm{Fil}^1 M_{\mathcal{A},L}, L) \ &\longrightarrow \ \mathrm{Hom}_L(\mathrm{Fil}^1 M_{\mathcal{X}_D,L}, L) \\ &= \mathcal{X}_D(F_v)\hat{\otimes}L. \end{aligned}$$

Here for $\mathcal{Y} = \mathcal{A}$ or $\mathcal{X}_D$, we identify

$$\mathrm{Hom}_{F_v\otimes L}(\mathrm{Fil}^1 M_{\mathcal{Y},L}^{(1)}, F_v \otimes L) = \mathrm{Hom}_L(\mathrm{Fil}^1 M_{\mathcal{Y},L}, L)$$

by the trace map $\mathbf{t} : F_v \to \mathbb{Q}_p$. We extend the logarithm ℓ_v to $\ell_{D,v}$ on $\mathcal{X}_D(F_v)\hat{\otimes}_{\mathbb{Z}_p} L$ so that it is trivial on the image of $s_{D,v,N}$.

The local p-adic height pairing at a finite place v is defined as

$$\langle D, \mathfrak{a} \rangle_{p,v,N} := \ell_{D,v}\left(\prod_i s_D(P_i)^{n_i}\right)$$

for a zero cycle $\mathfrak{a} = \sum_i n_i(P_i) \in Z_0(A)^0(\overline{F_v})$ prime to D fixed by $\mathrm{Gal}(\overline{F_v}/F_v)$. Here we are writing the group law of $\mathcal{X}_D$ multiplicatively. There is an ambiguity of a constant multiple for s_D but the pairing is well-defined since $\sum_i n_i = 0$.

Proposition 9. *The local p-adic height pairing*

$$\langle\ ,\ \rangle_{p,v,N} : (\mathcal{D}_0(A)(F_v) \times Z_0(A)^0(F_v))_e \longrightarrow L$$

has the following properties.

1. $\langle\ ,\ \rangle_{p,v,N}$ *is bilinear (whenever this makes sense).*
2. *If* $D = (h)$ *is principal, then*

$$\langle (h), \mathfrak{a}\rangle_{p,v,N} = \ell_v\Big(\prod_i h(P_i)^{n_i}\Big)$$

for a zero cycle $\mathfrak{a} = \sum n_i(P_i) \in Z_0(A)^0(\overline{F_v})$ *prime to* D *fixed by* $\mathrm{Gal}(\overline{F_v}/F_v)$.
3. *For any finite morphism* $\phi : A \to A$, *we have*

$$\langle \phi^* D, \mathfrak{a}\rangle_{p,v,N} = \langle D, \phi_* \mathfrak{a}\rangle_{p,v,N}.$$

4. *For any divisor* $D \in \mathcal{D}_0(A)(F_v)$ *and any point* $x_0 \in A(F_v) \setminus |D|$, *the morphism on* $A(F_v) \setminus |D|$

$$x \mapsto \langle D, (x) - (x_0)\rangle_{p,v,N}$$

is continuous for the v-adic topology on $A(F_v)$ *and the p-adic topology on* L.

If v is prime to p, the pairing is characterized by the above properties.

The global p-adic height pairing can be defined as the sum of local p-adic height pairings

$$A^\vee(F) \times A(F) \to L, \quad (a, b) \mapsto \sum_v \langle a', b'\rangle_{p,v,N}.$$

Here a' is a divisor that represents a and b' is a zero cycle $\sum n_i[b_i]$ of degree zero with $\sum n_i b_i = b$, and we choose a' and b' so that they have no point in common. The value $\sum_v \langle a', b'\rangle_{p,v,N}$ is independent of the choice a' and b'.

3.3 Norm Construction of p-Adic Height Pairings

In this subsection, we give a norm construction of the p-adic local height pairing associated to an adequate L-vector subspace N° of $M_{\hat{\mathcal{A}},L}$ for the formal group $\hat{\mathcal{A}}$ at $v|p$. Then we show that the pairing coincides with that of Zarhin-Nekovář associated to the splitting of the Hodge filtration by $c^{-1}(\Phi^{-1}(N^\circ))$ where $c : M_{\mathcal{A},L} \to M_{\hat{\mathcal{A}},L}$ is the natural map induced by the formal completion. Note that if $\mathcal{A}$ is ordinary at v, N° must be zero by the adequateness, and the splitting coincides with that by the unit root subspace.

We use the same notation as in the previous subsection. We assume that $v|p$ and put $K = F_v$ and $W = \mathcal{O}_{F_v}$ for simplicity. Let $\hat{\mathcal{A}}$ be the formal group of $\mathcal{A}$ over W. The formal completion induces an exact sequence of Dieudonné modules

$$0 \longrightarrow M^{\mathrm{unit}}_{\mathcal{A},L} \longrightarrow M_{\mathcal{A},L} \xrightarrow{\ c\ } M_{\hat{\mathcal{A}},L} \longrightarrow 0.$$

where $M^{\mathrm{unit}}_{\mathcal{A},L}$ is the unit root subspace of $M_{\mathcal{A},L}$ of the Frobenius F. (cf. Katz 1979, Corollary 5.7.7, 5.7.8). Let N° be an adequate L-vector subspace of $M_{\hat{\mathcal{A}},L}$. In the following we define a splitting

$$s^{\mathrm{norm}}_{D,v,N^\circ} : \hat{\mathcal{A}}(F_v)\hat{\otimes}L \longrightarrow \hat{\mathcal{X}}_D(F_v)\hat{\otimes}L.$$

Then the p-adic (local) height pairing is constructed just as Zarhin-Nekovář's except that we use the local section $s^{\mathrm{norm}}_{D,v,N^\circ}$ instead of $s_{D,v,N}$.

We consider L-linear maps

$$\tilde{\Sigma}_{\hat{\mathcal{A}},N^\circ} : Z_{\hat{\mathcal{A}},N^\circ}(k[[T]])_{/(\gamma-1)} \xrightarrow{\cong} \hat{\mathcal{A}}(pW)_L, \tag{14}$$

$$\tilde{\Sigma}_{\hat{\mathcal{X}}_D,N^\circ_D} : Z_{\hat{\mathcal{X}}_D,N^\circ_D}(k[[T]])_{/(\gamma-1)} \longrightarrow \hat{\mathcal{X}}_D(pW)_L. \tag{15}$$

Here we put $N^\circ_D := N^\circ \oplus (M^{V=1}_{\hat{\mathcal{X}}_D,L} \otimes K)$. Note that $\tilde{\Sigma}_{\hat{\mathcal{A}},N^\circ}$ is an isomorphism since N° is adequate. On the other hand, by Proposition 2, we have isomorphisms

$$\tilde{\delta}_{\hat{\mathcal{A}},N^\circ} : Z_{\hat{\mathcal{A}},N^\circ}(k[[T]])_{/(\gamma-1)} \cong \mathrm{Hom}_L(M_{\hat{\mathcal{A}},L}/N^\circ,\ L), \tag{16}$$

$$\tilde{\delta}_{\hat{\mathcal{X}}_D,N^\circ_D} : Z_{\hat{\mathcal{X}}_D,N^\circ_D}(k[[T]])_{/(\gamma-1)} \cong \mathrm{Hom}_L(M_{\hat{\mathcal{X}}_D,L}/N^\circ_D,\ L). \tag{17}$$

Since $M_{\hat{\mathcal{X}}_D,L} = M_{\hat{\mathcal{A}}} \oplus (M^{V=1}_{\hat{\mathcal{X}}_D,L} \otimes K)$ as L-vector spaces, isomorphisms (16) and (17) induce an isomorphism

$$Z_{\hat{\mathcal{A}},N^\circ}(k[[T]])_{/(\gamma-1)} \cong Z_{\hat{\mathcal{X}}_D,N^\circ_D}(k[[T]])_{/(\gamma-1)}. \tag{18}$$

Hence combined with (14) and (15), we have the desired section

$$s^{\mathrm{norm}}_{D,N^\circ} : \hat{\mathcal{A}}(F_v)\hat{\otimes}L \longrightarrow \hat{\mathcal{X}}_D(F_v)\hat{\otimes}L.$$

By construction, it is straightforward to see that $s^{\mathrm{norm}}_{D,N^\circ}$ coincides with $s_{D,N}$ for

$$N = c^{-1}(\Phi^{-1}(N^\circ)) = \Phi^{-1}(N^\circ) \oplus M^{\mathrm{unit}}_{\mathcal{A},L}$$

where we regard as $M_{\mathcal{A},L} = M^{\mathrm{unit}}_{\mathcal{A},L} \oplus M_{\hat{\mathcal{A}},L}$ by the splitting as F-modules. In particular, $s^{\mathrm{norm}}_{D,N^\circ}$ does not depend on the choice of the Lubin-Tate extension of $\mathbb{Q}_p$.

Now we describe $s^{\mathrm{norm}}_{D,N^\circ}$ in terms of admissible norm systems. Let N° be an adequate L-vector subspace of $M_{\hat{\mathcal{A}},L}$. For $x_0 \in \hat{\mathcal{A}}(W) \otimes L$, we can take an admissible norm system $(x_n)_n \in L \otimes \prod_{n=1}^{\infty} \hat{\mathcal{A}}^{(n-1)}(\mathfrak{m}_n)$ interpolated by an element of $Z_{\hat{\mathcal{A}},N^\circ}(k[[T]])$ with $\mathrm{Tr}_{1/0}\, x_1 = x_0$ by (14) (here we used the adequateness).

Lemma 2. *There exists an admissible norm system* $(\tilde{x}_n)_{n\geq 1} \in L \otimes \prod_{n=1}^{\infty} \hat{\mathcal{X}}_D^{(n-1)}(\mathfrak{m}_n)$ *that is a lift of* $(x_n)_n$ *for the exact sequence*

$$0 \longrightarrow \hat{\mathbb{G}}_m(\mathfrak{m}_n)_L \longrightarrow \hat{\mathcal{X}}_D^{(n-1)}(\mathfrak{m}_n)_L \longrightarrow \hat{\mathcal{A}}^{(n-1)}(\mathfrak{m}_n)_L \longrightarrow 0.$$

If $(x_n)_n$ *is a* Q*-norm system, then we can take* $(\tilde{x}_n)_n$ *so that it is also a* Q*-norm system.*

Proof. We take $x \in Z_{\hat{\mathcal{A}},N^\circ}(k[[T]])$ interpolating $(x_n)_n$. By (18), we can take a lift $\tilde{x} \in Z_{\hat{\mathcal{X}}_D,N_D^\circ}(k[[T]])$ of x. Then $\tilde{x}_n := \Sigma_{\hat{\mathcal{X}}_D,n,N_D^\circ}(\tilde{x})$ gives a desired system. Let $R \in \mathbb{Z}_p[t]$ be a monic polynomial such that $R(V)M_{\hat{\mathcal{A}}} = \{0\}$ and all roots of R have p-adic absolute values strictly greater than $|p|_p = 1/p$. Moreover, since $\hat{\mathcal{A}}$ is the formal group of an abelian variety, we may choose R so that $R(1) \neq 0$. Then by the definition of N_D°, we have $R(\psi)\hat{\tilde{x}}(\omega)(T) = c(\omega)\log_{\mathcal{F}_\pi}(T)$ for some constant $c(\omega)$ for all $\omega \in M_{\hat{\mathcal{X}}_D}$. Hence $(\tilde{x}_n)_n$ is an R-norm system. Suppose that (x_n) is a Q-norm system for $Q \in \mathbb{Z}_p[t]$. Then we may assume Q is a divisor of R and put $R = PQ$. We let $y_n = Q(\mathrm{tr})(\tilde{x}_n)$. Then $(y_n)_n$ is a P-norm system of $\hat{\mathbb{G}}_m$, but it must be also a norm compatible because it is obtained by the Perrin-Riou lift. However, since $P(1) \neq 0$, this system must be trivial.

It is straightforward to see that the local section $s^{\mathrm{norm}}_{D,w,N^\circ}$ can be given by

$$\hat{\mathcal{A}}(W)_L \longrightarrow \hat{\mathcal{X}}_D(W)_L, \qquad x \longmapsto \mathrm{Tr}_{1/0}\,\tilde{x}_1.$$

Here we take an admissible Q-norm system $(x_n)_n$ with $Q(1) \neq 0$ and $\mathrm{Tr}_{1/0} = x$, then $(\tilde{x}_n)$ is an admissible Q-norm lift of $(x_n)_n$.

We may also define a section

$$\hat{\mathcal{A}}(W)_L \longrightarrow \hat{\mathcal{X}}_D(W)_L/(\mathrm{Ker}\,\hat{\ell}_w \otimes L), \qquad x \longmapsto \mathrm{Tr}_{1/0}\,\tilde{x}_1$$

for any admissible lift $(\tilde{x}_n)$. The well-definedness is checked as follows. If $(\tilde{y}_n)_n$ is another admissible lift of (x_n), then $y_n - x_n$ defines an admissible norm system of $\hat{\mathbb{G}}_m$, which must be norm compatible. Hence $\mathrm{Tr}_{1/0}(\tilde{y}_1 - \tilde{x}_1)$ is a universal norm in $\hat{\mathbb{G}}_m(W)$, and therefore contained in $\mathrm{Ker}\,\hat{\ell}_w \otimes L$.

4 An Application for the p-Adic Gross-Zagier Formula

In this section we generalize the p-adic Gross-Zagier formula in Kobayashi (2013) to newforms for $\Gamma_0(N)$ of weight 2 with *arbitrary* Fourier coefficients (not necessarily in $\mathbb{Q}$). Most parts of the proof in Kobayashi (2013) also work in this case (or have been already proven in Kobayashi (2013)). The missing part is the theory of the p-adic height pairing on abelian varieties at non-ordinary primes developed in this paper.

Let $f = \sum_{n=1}^{\infty} a_n(f)q^n$ be a normalized eigen newform for $\Gamma_0(N)$ of weight 2. Let p be a prime number such that $(p, N) = 1$ and we fix a complex embedding $\iota_\infty : \overline{\mathbb{Q}} \hookrightarrow \mathbb{C}$ and a p-adic embedding $\iota_p : \overline{\mathbb{Q}} \hookrightarrow \mathbb{C}_p$. We fix real and purely imaginary periods $\Omega_f^{\pm}$ (e.g. Shimura periods) and a system of p-power roots of unity $(\zeta_{p^n})_n$ that is a generator of $\mathbb{Z}_p(1)$. Let α be a root of $t^2 - a_p(f)t + p = 0$ in $\mathbb{C}_p$ such that $|p/\alpha|_p < 1$ where $a_p(f)$ is regarded as an element of $\mathbb{C}_p$ by ι_∞ and ι_p. It is known that there is a one-variable p-adic analytic distribution $d\mu_{f,\alpha}$ of order < 1 satisfying that

$$\int_{\mathbb{Z}_p^\times} \chi(x)\, d\mu_{f,\alpha}(x) = \frac{\tau(\chi)}{\alpha^n} \frac{L(f, \overline{\chi}, 1)}{\Omega_f^{\chi(-1)}}$$

for a non-trivial Dirichlet character χ of conductor p^n with the Gauss sum $\tau(\chi)$, and

$$\int_{\mathbb{Z}_p^\times} d\mu_{f,\alpha} = \left(1 - \frac{1}{\alpha}\right)^2 \frac{L(f, 1)}{\Omega_f^+}.$$

(cf. Amice and Vélu 1975; Mazur et al. 1986; Višik 1976.) Here the L-values on the right hand side are algebraic and regarded as in $\mathbb{C}_p$ by ι_∞ and ι_p. Then the p-adic L-function $\mathscr{L}_p(f, \alpha, s)$ is defined as an analytic function of s on $\mathbb{Z}_p$ by

$$\mathscr{L}_p(f, \alpha, s) := \int_{\mathbb{Z}_p^\times} \langle x \rangle^{s-1}\, d\mu_{f,\alpha}(x),$$

where $\langle\ \rangle$ is the natural projection $\mathbb{Z}_p^\times \to 1 + 2p\mathbb{Z}_p$.

Let $\mathscr{K}$ be an imaginary quadratic field with discriminant $d_K = -\delta$ such that $(d_K, Np) = 1$. Let ε be the quadratic character attached to $\mathscr{K}/\mathbb{Q}$. We define the p-adic L-function of f over $\mathscr{K}$ by

$$\mathscr{L}_p(f/\mathscr{K}, \alpha, s) := \mathscr{L}_p(f, \alpha, s)\mathscr{L}_p(f \otimes \varepsilon, \varepsilon(p)\alpha, s) \frac{\Omega_f^+ \Omega_{f\otimes\varepsilon}^+ \delta^{\frac{1}{2}}}{\Omega_f}$$

where $f \otimes \varepsilon := \sum_{n=1}^{\infty} \varepsilon(n) a_n q^n$ is the twist of f by ε and

$$\Omega_f := \int_{X_0(N)(\mathbb{C})} \omega_f \wedge i\,\overline{\omega}_f$$

for $\omega_f = \sum a_n q^n dq/q = 2\pi i f(\tau) d\tau$ with $q = \exp(2\pi i \tau)$. If all rational primes dividing N split in $\mathscr{K}$, then $\mathscr{L}_p(f/\mathscr{K}, \alpha, 1) = 0$ since $L(f/\mathscr{K}, 1) = L(f, 1) L(f \otimes \varepsilon, 1) = 0$ by looking at the sign of the functional equation.

Let $X_0(N)$ be the modular curve over $\mathbb{Q}$ for $\Gamma_0(N)$ and let $J = J_0(N)$ be its Jacobian variety. There is an idempotent e_f in $\mathrm{End}_{\mathbb{Q}} J \otimes \mathbb{Q}_f$ such that

$$e_f(\Gamma(X_0(N), \Omega^1_{X_0(N)/\mathbb{Q}}) \otimes \mathbb{Q}_f) = \mathbb{Q}_f \omega_f.$$

Here $\mathbb{Q}_f$ is the field $\iota_\infty^{-1}(\mathbb{Q}(\{a_n(f)\}_n))$, and we let ω_f be defined over $\mathbb{Q}_f$ by ι_∞. Let L be the p-adic closure of $\iota_p(\mathbb{Q}_f)(\alpha)$ in $\mathbb{C}_p$. Let M_J be the Dieudonné module attached to the special fiber of J at p. By the Albanese map $X_0(N) \to J_0(N)$, $x \mapsto [x] - [\infty]$, the space of invariant holomorphic forms can be regarded as

$$L_J := \Gamma(X_0(N), \Omega^1_{X_0(N)/\mathbb{Q}}) \otimes_{\mathbb{Q}} \mathbb{Q}_p \subset M_J \otimes_{\mathbb{Z}_p} \mathbb{Q}_p$$

and $\mathrm{Fil}^1 M_J \otimes \mathbb{Q}_p = V L_J$. Then the Frobenius F acts on the space $e_f(M_J \otimes L)$ and the characteristic polynomial is $t^2 - a_p(f)t + p$. Let N_α be a subspace of $M_J \otimes L$ complementary to the Hodge filtration such that $e_f N_\alpha$ is the α-eigensubspace of F of $e_f(M_J \otimes L)$. Then we consider the p-adic height pairing on $J_0(N)(\mathscr{K})$ attached to $N_\alpha \otimes_{\mathbb{Q}} \mathscr{K}$ (with the cyclotomic logarithm $\ell_{\mathscr{K}}$)

$$\langle\, ,\, \rangle_{p,\mathscr{K},\alpha} : J_0(N)(\mathscr{K}) \times J_0(N)(\mathscr{K}) \longrightarrow L$$

where we identify $J_0(N)$ with its dual $J_0(N)^\vee$ by the principal polarization.

Let $\mathscr{H}$ be the Hilbert class field of $\mathscr{K}$. We assume that all rational prime numbers dividing N split in $\mathscr{K}$. Then there exists a Heegner point $z_{\mathscr{H}} \in X_0(N)(\mathscr{H})$ that represents a cyclic isogeny $\mathbb{C}/\mathcal{O}_{\mathscr{K}} \to \mathbb{C}/\mathscr{N}^{-1}$ of order N where $(N) = \mathscr{N}\mathscr{N}^*$ in $\mathcal{O}_{\mathscr{K}}$. (There exist several choices of Heegner points but we fix one since the p-adic height does not depend on the choice.) We also put $z_{\mathscr{H}} \in J_0(N)(\mathscr{H})$ by the negative of the Albanese map (see Remark 4.2 in Kobayashi (2013) for the sign convention) and put $z_{\mathscr{K},f} := e_f \mathrm{Tr}_{\mathscr{H}/\mathscr{K}} z_{\mathscr{H}} \in J_0(N)(\mathscr{K}) \otimes \mathbb{Q}_f$.

Then the p-adic Gross-Zagier formula for f is stated as follows.

Theorem 3. *Suppose that p splits in $\mathscr{K}$. Then*

$$\frac{d}{ds} \mathscr{L}_p(f/\mathscr{K}, \alpha, s)\,|_{s=1} = u^{-2} \left(1 - \frac{1}{\alpha}\right)^2 \left(1 - \frac{1}{\varepsilon(p)\alpha}\right)^2 \langle z_{\mathscr{K},f}, z_{\mathscr{K},f} \rangle_{p,\mathscr{K},\alpha}$$

where $u = \sharp\mathcal{O}_{\mathscr{K}}^\times/2$.

The formula is proven in Perrin-Riou (1987) for an ordinary p and in Kobayashi (2013) for a supersingular p for f with rational coefficients. If the p-adic height is a non-trivial function, we can also show the same formula for an inert prime p by a simple trick. (See the proof of Theorem 5.8 in Kobayashi (2013).) The strategy of the proof is as follows. We construct two p-adic modular forms F and G related to the p-adic height of the Heegner point and the derivative of the p-adic L-function respectively. We compare their Fourier coefficients by independent calculations and show that "the f-part" of F and G coincide, which gives the p-adic Gross-Zagier formula.

As in the complex case, the p-adic modular form F is defined as

$$F := \sum_{\sigma \in \mathrm{Gal}(\mathscr{H}/\mathscr{K})} \sum_{m=1}^{\infty} \langle z_{\mathscr{H}}, T_m z_{\mathscr{H}}^{\sigma} \rangle_{p,\mathscr{H},\alpha} q^m .$$

For G, we first construct the p-adic measure Φ on $\mathbb{Z}_p^\times$ with values in the space of p-adic modular forms by the p-adic convolution of the Eisenstein and theta measure. Then put

$$G(s) := \int_{\mathbb{Z}_p^\times} \langle x \rangle^{s-1} d\Phi$$

and $G = (d/ds)G(s)|_{s=1}$. "The f-part" of $G(s)$ is expected to be related to the p-adic L-function $\mathscr{L}_p(f/\mathscr{K}, \alpha, s)$. In fact, if p is ordinary, this is done by using Hida's ordinary projection which is considered as an analogue of Sturm's holomorphic projection and also enables us to take the f-part of $G(s)$ p-adically (essentially the Petersson inner product with f). In the non-ordinary case, the convergence condition arising from taking the f-part is not admissible and the direct analogue to the ordinary case does not work. (In fact, $\mathscr{L}_p(f/\mathscr{K}, \alpha, s)$ is in some sense a critical slope p-adic L-function since it cannot be characterized by the interpolation property because the corresponding power series has denominators of the logarithmic order.) In Kobayashi (2013), to overcome this difficulty, we introduced a (naive) two variable p-adic Eisenstein measure $\Phi^{(2)}$ on $\mathbb{Z}_p^\times \times \mathbb{Z}_p^\times$ whose restriction to the diagonal direction coincides with Φ (both variables are in the cyclotomic directions and "the f-part" of the integral of the function $\langle x \rangle^{s-1} \langle y \rangle^{t-1}$ with respect to $\Phi^{(2)}$ corresponds to the product $\mathscr{L}_p(f, \alpha, s)\mathscr{L}_p(f \otimes \varepsilon, \varepsilon(p)\alpha, t)$). This two-variable measure is constructed by using products of Eisenstein series (essentially, Kato's zeta element in the space of modular forms (Kato 2004)). Then the convergence condition related to taking the f-part of $\Phi^{(2)}$ on the vertical and horizontal directions is good enough and we can approximate the diagonal direction by using these directions to relate $G(s)$ to $\mathscr{L}_p(f/\mathscr{K}, \alpha, s)$. Calculations concerning G are written in Kobayashi (2013) for modular forms of weight 2 for $\Gamma_0(N)$ with general coefficients.

For the calculation of Fourier coefficients of F, we use the decomposition of the global height into the local heights. The local height at $v \nmid p$ is described geometrically as the intersection number on the integral model of $X_0(N)$. Hence the calculation is reduced to the classical case (Gross and Zagier 1986). For the local height at $v \mid p$, we show that it is essentially equal to zero. (We show that the "f-part" vanishes. [2]). The proof in Kobayashi (2013) also works in our case if we use the theory of the p-adic height developed in this paper. So we do not recall the details, but we just recall the setting to apply to our theory.

We now assume that p is non-ordinary, that is, $\iota_p(a_p(f))$ is not a p-adic unit. Hence there is no unit root part in $e_f(M_J \otimes_{\mathbb{Z}_p} L)$. Since we are interested in the f-part, we may assume that $(1 - e_f)N_\alpha$ is of the form $(M_J^{\text{unit}} \otimes L) \oplus N'$ with F-stable N'. Then N_α is F-stable and adequate by Corollary 2 and Proposition 6. We can also use the norm construction of the p-adic height pairing for $N_\alpha = \Phi(N_\alpha)$.

Let $\hat{J}$ be the formal group of the proper smooth model of $J_0(N)$ over $\mathbb{Z}_p$. Then by the Albanese map we can embed

$$\Gamma(X_0(N), \Omega^1_{X_0(N)/\mathbb{Q}}) \otimes \mathbb{Q}_p = L_{\hat{J}} \otimes \mathbb{Q}_p \subset M_{\hat{J}} \otimes \mathbb{Q}_p$$

and $\mathrm{Fil}^1 M_{\hat{J}} \otimes \mathbb{Q}_p = VL_{\hat{J}} \otimes \mathbb{Q}_p$. We also have $e_f(M_{\hat{J}} \otimes_{\mathbb{Z}_p} L) = e_f(M_J \otimes_{\mathbb{Z}_p} L)$.

Let $\mathscr{H}_n$ be the ring class field for the order $\mathbb{Z} + p^n \mathcal{O}_{\mathscr{K}}$. Let K be the completion of $\mathscr{H} = \mathscr{H}_0$ at a place w over p and let K_n be the completion of $\mathscr{H}_n$ at the place over w. (The place w is not necessary the place compatible with ι_p.) Since p splits in $\mathscr{K}$, the extension $K_n/\mathbb{Q}_p$ is abelian and $K_\infty = \cup_n K_n$ is a $\mathbb{Z}_p$-extension over K_1 obtained by a Lubin-Tate formal group over $\mathbb{Z}_p$ of height 1. We denote the integer ring of K by W and the residue field by k. By Proposition 5.1 of Kobayashi (2013), there is an admissible norm system $(c_n)_n \in e_f(L \otimes \prod_{n=1}^{\infty} \hat{J}(\mathfrak{m}_n))$ obtained from Heegner points such that

$$\mathrm{Tr}_{n+1/n} c_{n+1} - a_p(f) c_n + c_{n-1} = 0$$

for $n \geq 2$ in $\hat{J}(\mathfrak{m}_n)_L$ with $\mathrm{Tr}_{1/0}\, c_1 = z_{\mathscr{H}}$.[3] As in Example 1, this system is interpolated by an element of $Z_{N^\circ}(k[[T]]) \subset \hat{J}(k[[T]])_L$ with a $W \otimes_{\mathbb{Z}_p} L$-submodule $N^\circ \subset M_{\hat{J}/W} \otimes L$ such that $e_f N^\circ = e_f(L_{\hat{J}} \otimes W \otimes L)$. Now we put

$$c_{n,\alpha} := \mathrm{Tr}_{n+1/n} c_{n+1} - p c_n \otimes \alpha^{-1} \in \hat{J}(\mathfrak{m}_n)_L.$$

Then by Proposition 8, the system $(c_{n,\alpha})$ is an admissible norm system satisfying

$$\mathrm{Tr}_{n+1/n} c_{n+1,\alpha} = \alpha c_{n,\alpha}$$

---

[2] Since the local height is defined on divisors and not on divisor classes, we have to give an appropriate meaning of "the f-part".

[3] In Kobayashi (2013), we took c_n so that $\mathrm{Tr}_{1/0}\, c_1 = \alpha z_{\mathscr{H}}$. So c_n in this paper differs by α from that in Kobayashi (2013).

for $n \geq 1$ and interpolated by an element of $Z_{N_\alpha \otimes_{\mathbb{Q}_p} K}(k[[T]])$. Note that $e_f N_\alpha \otimes_{\mathbb{Q}_p} K$ is given as $(V-\alpha)e_f(\mathrm{Fil}^1 M_{\hat{J}} \otimes W \otimes L)$. By Lemma 2, we can take an admissible lift $(\tilde{c}_{n,\alpha}) \in L \otimes \prod_{n=1}^{\infty} \hat{\mathcal{X}}_D^{(n-1)}(\mathfrak{m}_n)$ satisfying $\mathrm{Tr}_{n+1/n}\tilde{c}_{n+1,\alpha} = \alpha\tilde{c}_{n,\alpha}$. Then we have

$$s^{\mathrm{norm}}_{D,w,N^\circ}(z_{\mathcal{H}}) = \mathrm{Tr}_{1/0}\ \tilde{c}_{1,\alpha} = \alpha^{-n+1}\mathrm{Tr}_{n/0}\ \tilde{c}_{n,\alpha}$$

for any n with $n \equiv 1 \bmod \nu$. Then the proof in Kobayashi (2013) works in our setting.

Acknowledgements The author thanks Kazuto Ota for discussion. This paper was written when the author was visiting l'Institute de Mathématiques de Jussieu in 2013, supported by Strategic Young Researcher Overseas Visits Program for Accelerating Brain Circulation by JSPS. He would like to thank sincerely Jan Nekovář for his hospitality. He is grateful to the referee for carefully reading the paper. The research is also partly supported by Inamori Foundation and Grant-in-Aid for Young Scientists (A) 25707001 by JSPS.

References

Amice, Y., Vélu, J.: Distributions p-adiques associées aux séries de Hecke. In: Journées Arithmétiques de Bordeaux (Conf., Univ. Bordeaux, Bordeaux, 1974), pp. 119–131; Asterisque, vol. 24–25. Soc. Math. France, Paris (1975)

Berger, L.: Représentations de de Rham et normes universelles. Bull. Soc. Math. France **133**(4), 601–618 (2005)

Coleman, R.: Division values in local fields. Invent. Math. **53**(2), 91–116 (1979)

Colmez, P.: Théorie d'Iwasawa des représentations de de Rham d'un corps local. Ann. Math. **148**(2), 485–571 (1998)

Fontaine, J-M.: Groupes p-divisibles sur les corps locaux. Astérisque, vol. 47–48, i+262pp. Société Mathématique de France, Paris (1977)

Gross, B., Zagier, D.: Heegner points and derivatives of L-series. Invent. Math. **84**(2), 225–320 (1986)

Kato, K.: Cohomologies p-adiques et applications arithmétiques. III (p-adic Hodge theory and values of zeta functions of modular forms). Astérisque **295** ix, 117–290 (2004)

Katz, N: Crystalline cohomology, Dieudonné modules, and Jacobi sums. In: Automorphic Forms, Representation Theory and Arithmetic (Bombay, 1979). Tata Institute of Fundamental Research Studies in Mathematics, vol. 10, pp. 165–246. Tata Institute of Fundamental Research, Bombay, (1981)

Knospe, H.: Iwasawa-theory of abelian varieties at primes of non-ordinary reduction. Manuscr. Math. **87**(2), 225–258 (1995)

Kobayashi, S.: Iwasawa theory for elliptic curves at supersingular primes. Invent. Math. **152**(1), 1–36 (2003)

Kobayashi, S.: The p-adic Gross-Zagier formula for elliptic curves at supersingular primes. Invent. Math. **191**(3), 527–629 (2013)

Mazur, B.: Rational points of abelian varieties with values in towers of number fields. Invent. Math. **18**, 183–266 (1972)

Mazur, B., Tate, J., Teitelbaum, J.: On p-adic analogues of the conjectures of Birch and Swinnerton-Dyer. Invent. Math. **84**(1), 1–48 (1986)

Moret-Bailly, L.: Familles de courbes et de variétés abéliennes sur $\mathbb{P}^1$, In: Séminaire sur les Pinceaux de Courbes de Genre au Moins Deux. Astérisque, vol. 86, vi+145pp. Société Mathématique de France, Paris (1981)

Mumford, D.: Abelian Variety. Tata Institute of Fundamental Research Studies in Mathematics, vol. 5. Oxford University Press, London (1970)

Nakamura, K.: Iwasawa theory of de Rham (ϕ, Γ)-modules over the Robba ring. Journal of the Institute of Mathematics of Jussieu. (13)(1), 65–118 (2014)

Nasybullin, A.G.: Elliptic curves with supersingular reduction over Γ-extensions. Uspehi Mat. Nauk 2 **32**(194), 221–222 (1977)

Nekovář, J.: On p-adic height pairings, In: Séminaire de Théorie des Nombres, Paris, 1990–91. Progress in Mathematics, vol. 108, pp. 127–202. Birkhäuser, Boston (1993)

Ota, K.: A generalization of the theory of Coleman power series. Tohoku Math. J. **66**(3), (2014)

Perrin-Riou, B.: Points de Heegner et dérivées de fonctions L p-adiques. Invent. Math. **89**(3), 455–510 (1987)

Perrin-Riou, B.: Théorie d'Iwasawa p-adique locale et globale. Invent. Math. **99**(2), 247–292 (1990)

Perrin-Riou, B.: Fonctions L p-adiques d'une courbe elliptique et points rationnels. Ann. Inst. Fourier (Grenoble) **43**(4), 945–995 (1993)

Perrin-Riou, B.: Théorie d'Iwasawa des représentations p-adiques sur un corps local. Invent. Math. **115**, 81–149 (1994)

Perrin-Riou, B.: Fonctions L p-adiques des représentations p-adiques. Astérisque, vol. 229, 198pp. (1995)

Perrin-Riou, B.: Représentations p-adiques et normes universelles. I. Le cas cristallin. J. Am. Math. Soc. **13**(3), 533–551 (2000)

Schneider, P.: p-adic height pairings I. Invent. Math. 69(3), 401–409 (1982)

Schneider, P.: Arithmetic of formal groups and applications I: universal norm subgroups. Invent. Math. **87**, 587–602 (1987)

Višik, M.: Non-Archimedean measures associated with Dirichlet series. Mat. Sb. (N.S.) **99**(2), 248–260; **141**, 296 (1976)

Washington, L.: Introduction to Cyclotomic Fields. Graduate Texts in Mathematics, vol. 83, 2nd edn. Springer, New York (1997)

Wintenberger, J-P.: Un scindage de la filtration de Hodge pour certaines variétés algébriques sur les corps locaux. Ann. Math. (2) **119**(3), 511–548 (1984)

Zarhin, Y.G.: p-adic heights on abelian varieties. In: Séminaire de Théorie des Nombres, Paris 1987–88. Progress in Mathematics, vol. 81, pp. 317–341. Birkhäuser, Boston (1990)

Iwasawa Modules Arising from Deformation Spaces of p-Divisible Formal Group Laws

Jan Kohlhaase

1 Introduction

Let p be a prime number, and let k be an algebraically closed field of characteristic p. Let $W = W(k)$ denote the ring of Witt vectors with coefficients in k, and let K denote the quotient field of W. We fix a p-divisible commutative formal group law G of height h over k and denote by $R := R_G^{\mathrm{def}}$ the universal deformation ring of G representing isomorphism classes of deformations of G to complete noetherian local W-algebras with residue class field k. Denote by $\mathbb{G}$ the universal deformation of G to R and by $\mathrm{Lie}(\mathbb{G})$ the Lie algebra of $\mathbb{G}$. For any integer m, the m-th tensor power $\mathrm{Lie}(\mathbb{G})^{\otimes m}$ of $\mathrm{Lie}(\mathbb{G})$ can be viewed as the space of global sections of a vector bundle on the universal deformation space $\mathrm{Spf}(R)$ which is equivariant for a natural action of the automorphism group $\Gamma := \mathrm{Aut}(G)$ of G.

If G is of dimension one, then the formal scheme $\mathrm{Spf}(R)$ is known as *the moduli space of Lubin-Tate*. It plays a crucial role in Harris' and Taylor's construction of the local Langlands correspondence for $\mathrm{GL}_h(\mathbb{Q}_p)$. Moreover, the Γ-representations $\mathrm{Lie}(\mathbb{G})^{\otimes m}$ and their cohomology figure prominently in stable homotopy theory (cf. the introduction to Devinatz and Hopkins (1995)). Still assuming G to be one dimensional, a detailed study of the Γ-representation R was given in Kohlhaase (2013). For $h = 2$ it led to the computation of the continuous Γ-cohomology of R, relying on the foundational work of Devinatz, Gross, Hopkins and Yu. The only prior analysis of p-adic representations stemming from equivariant vector bundles on deformation spaces of p-divisible formal groups concern the p-adic symmetric spaces of Drinfeld. These were studied extensively by Morita, Orlik, Schneider and Teitelbaum (cf. Orlik 2008; Schneider and Teitelbaum 2002 and our remarks at the end of Sect. 2).

J. Kohlhaase (✉)
Mathematisches Institut, Universität Münster, Münster, Germany
e-mail: kohlhaaj@math.uni-muenster.de

T. Bouganis and O. Venjakob (eds.), *Iwasawa Theory 2012*, Contributions in Mathematical and Computational Sciences 7, DOI 10.1007/978-3-642-55245-8_10,

The aim of the present article is to generalize and strengthen some of the results of Gross and Hopkins (1994) and of the author in Kohlhaase (2013). To this end, Sect. 1 and the first part of Sect. 2 give a survey of the theory of p-divisible commutative formal group laws. This includes the classification results of Dieudonné, Lazard and Manin, as well as the deformation theoretic results of Cartier, Lubin, Tate and Umemura. It follows from the work of Dieudonné and Manin that the group Γ is a compact Lie group over $\mathbb{Q}_p$ (cf. Corollary 1).

In the second part of Sect. 2, we prove that the action of Γ on $\mathrm{Lie}(\mathbb{G})^{\otimes m}$ extends to the Iwasawa algebra $\Lambda := W[\![\Gamma]\!]$ of Γ over W. This gives $\mathrm{Lie}(\mathbb{G})^{\otimes m}$ the structure of a pseudocompact module over Λ (cf. Corollary 2 and Theorem 5). In Sect. 3, we pass to the global rigid analytic sections $(\mathrm{Lie}(\mathbb{G})^{\otimes m})^{\mathrm{rig}}$ of our vector bundles and show that the action of Γ extends to a continuous action of the locally analytic distribution algebra $D(\Gamma)$ of Γ over K. As a consequence, the action of Γ on the strong continuous K-linear dual of $(\mathrm{Lie}(\mathbb{G})^{\otimes m})^{\mathrm{rig}}$ is locally analytic in the sense of Schneider and Teitelbaum (cf. Theorems 6 and 7).

We note that the continuity and the differentiability of the action of Γ on R^{rig} were first proven by Gross and Hopkins if G is of dimension one (cf. Gross and Hopkins 1994, Propositions 19.2 and 24.2). Using the structure theory of the algebra $D(\Gamma)$, we arrive at a more precise result for arbitrary m and G, avoiding the use of the period morphism. Our approach essentially relies on a basic lifting lemma for endomorphisms of G which is also at the heart of the strategy followed by Gross and Hopkins (cf. Lemma 1 and Proposition 1).

A major question that we have to leave open concerns the *coadmissibility* of the $D(\Gamma)$-modules $(\mathrm{Lie}(\mathbb{G})^{\otimes m})^{\mathrm{rig}}$ in the sense of Schneider and Teitelbaum (2003), section 6. Taking sections over suitable affinoid subdomains of $\mathrm{Spf}(R)^{\mathrm{rig}}$, it is related to the finiteness properties of the resulting Banach spaces as modules over certain Banach completions of $\Lambda \otimes_W K$. In Sect. 4, we assume G to be of dimension one and consider the restriction of $(\mathrm{Lie}(\mathbb{G})^{\otimes m})^{\mathrm{rig}}$ to an affinoid subdomain of $\mathrm{Spf}(R)^{\mathrm{rig}}$ over which the period morphism of Gross and Hopkins is an open immersion. By spelling out the action of the Lie algebra of Γ, we show that one naturally obtains a continuous module over a complete divided power enveloping algebra $\hat{U}_K^{\mathrm{dp}}(\mathring{\mathfrak{g}})$ constructed by Kostant (cf. Theorem 8). Here $\mathring{\mathfrak{g}}$ is a Chevalley order in the split form of the Lie algebra of Γ. If $h = 2$ and $m \geq -1$ then in fact $(\mathrm{Lie}(\mathbb{G})^{\otimes m})^{\mathrm{rig}}$ gives rise to a cyclic module over $\hat{U}_K^{\mathrm{dp}}(\mathring{\mathfrak{g}})$ (cf. Theorem 9). This result might indicate that $(\mathrm{Lie}(\mathbb{G})^{\otimes m})^{\mathrm{rig}}$ does not give rise to a coherent sheaf for the Fréchet-Stein structure of $D(\Gamma)$ considered in Schneider and Teitelbaum (2003), section 5 (cf. Remark 3).

Gross and Hopkins (1994) consider formal modules of dimension one and finite height over the valuation ring $\mathfrak{o}$ of an arbitrary non-Archimedean local field. The case of p-divisible formal groups corresponds to the case $\mathfrak{o} = \mathbb{Z}_p$. However, neither the deformation theory nor the theory of the period morphism have been worked out in detail for formal $\mathfrak{o}$-modules of dimension strictly greater than one. This is why we restrict to one dimensional formal groups in Sect. 4 and to p-divisible formal groups throughout.

Conventions and notation. If S is a commutative unital ring, if r is a positive integer, and if $X = (X_1, \dots, X_r)$ is a family of indeterminates, then we denote by $S[\![X]\!] = S[\![X_1, \dots, X_r]\!]$ the ring of formal power series in the variables $X_1, \dots, X_r$ over S. We write $f = f(X) = f(X_1, \dots, X_r)$ for an element $f \in S[\![X]\!]$. If $n = (n_1, \dots, n_r) \in \mathbb{N}^r$ is an r-tuple of non-negative integers then we set $|n| := n_1 + \dots + n_r$ and $X^n := X_1^{n_1} \cdots X_r^{n_r}$. If i and j are elements of a set then we denote by δ_{ij} the Kronecker symbol with value $1 \in S$ if $i = j$ and $0 \in S$ if $i \neq j$. If $\mathfrak{h}$ is a Lie algebra over S then we denote by $U(\mathfrak{h})$ the universal enveloping algebra of $\mathfrak{h}$ over S. Throughout the article, p will denote a fixed prime number.

2 Formal Group Laws

Let R be a commutative unital ring, and let d be a positive integer. A d-dimensional commutative *formal group law* (subsequently abbreviated to *formal group*) is a d-tuple $G = (G_1, \dots, G_d)$ of formal power series in $2d$ variables $G_i \in R[\![X, Y]\!] = R[\![X_1, \dots, X_d, Y_1, \dots, Y_d]\!]$, satisfying

(F1) $G_i(X, 0) = X_i$,
(F2) $G_i(X, Y) = G_i(Y, X)$, and
(F3) $G_i(G(X, Y), Z) = G_i(X, G(Y, Z))$

for all $1 \leq i \leq d$. It follows from the formal implicit function theorem (cf. Hazewinkel 1978, A.4.7) that for a given d-dimensional commutative formal group G there exists a unique d-tuple $\iota_G \in R[\![X]\!]^d$ of formal power series with trivial constant terms such that

$$G_i(X, \iota_G(X)) = 0 \quad \text{for all} \quad 1 \leq i \leq d$$

(cf. also Zink 1984, Korollar 1.5). Thus, if S is a commutative R-algebra, and if I is an ideal of S such that S is I-adically complete, then the set I^d becomes a commutative group with unit element $(0, \dots, 0)$ via

$$x +_G y := G(x, y) \quad \text{and} \quad -x := \iota_G(x).$$

Example 1. Let $R = \mathbb{Z}$ and $d = 1$. The formal group $\hat{\mathbb{G}}_a(X, Y) = X + Y$ is called the *one dimensional additive formal group*. We have $\iota_{\hat{\mathbb{G}}_a}(X) = -X$. The formal group $\hat{\mathbb{G}}_m(X, Y) = (1 + X)(1 + Y) - 1$ is called the *one dimensional multiplicative formal group*. We have $\iota_{\hat{\mathbb{G}}_m}(X) = \sum_{n=1}^{\infty}(-X)^n$.

Let G and H be formal groups over R of dimensions d and e, respectively. A *homomorphism from G to H* is an e-tuple $\varphi = (\varphi_1, \dots, \varphi_e)$ of power series $\varphi_i \in R[\![X]\!] = R[\![X_1, \dots, X_d]\!]$ in d-variables over R with trivial constant terms, satisfying

$$\varphi(G(X, Y)) = H(\varphi(X), \varphi(Y)).$$

If $\varphi : G \to G'$ and $\psi : G' \to G''$ are homomorphisms of formal groups then we define $\psi \circ \varphi$ through $(\psi \circ \varphi)(X) := \psi(\varphi(X))$. This is a homomorphism from G to G''. We let $\mathrm{End}(G)$ denote the set of *endomorphisms* of a d-dimensional commutative formal group G over R, i.e. of homomorphisms from G to G. It is a ring with unit $1_G = X = (X_1, \ldots, X_d)$, in which addition and multiplication are defined by $(\varphi +_G \psi)(X) := G(\varphi(X), \psi(X))$, $(-\varphi)(X) := \iota_G(\varphi(X))$ and $\psi \cdot \varphi := \psi \circ \varphi$. In particular, $\mathrm{End}(G)$ is a $\mathbb{Z}$-module. Given $m \in \mathbb{Z}$, we denote by $[m]_G \in R[\![X]\!]^d$ the corresponding endomorphism of G. We denote by $\mathrm{Aut}(G)$ the *automorphism group of* G, i.e. the group of units of the ring $\mathrm{End}(G)$.

Denoting by (X) the ideal of $R[\![X]\!]$ generated by $X_1, \ldots, X_d$, the free R-module

$$\mathrm{Lie}(G) := \mathrm{Hom}_R((X)/(X)^2, R)$$

of rank $d = \dim(G)$ is called the *Lie algebra of* G (or its *tangent space at* 1_G). It is an R-Lie algebra for the trivial Lie bracket. Non-commutative Lie algebras occur only for non-commutative formal groups (cf. Zink 1984, Kapitel I.7). An R-basis of $\mathrm{Lie}(G)$ is given by the linear forms $(\frac{\partial}{\partial X_i})_{1 \le i \le d}$ sending $f + (X)^2$ to $\frac{\partial f}{\partial X_i}(0)$. Here $\frac{\partial f}{\partial X_i}$ denotes the formal derivative of the power series f with respect to the variable X_i.

Any homomorphism $\varphi : G \to H$ of formal groups as above gives rise to an R-linear ring homomorphism $\varphi^* : R[\![Y_1, \ldots, Y_e]\!] \to R[\![X_1, \ldots, X_d]\!]$, determined by $\varphi^*(Y_i) = \varphi_i$ for all $1 \le i \le e$. It is called the *comorphism of* φ. It maps (Y) to (X), hence $(Y)^2$ to $(X)^2$, and therefore induces an R-linear map

$$\mathrm{Lie}(\varphi) : \mathrm{Lie}(G) \longrightarrow \mathrm{Lie}(H)$$

via $\mathrm{Lie}(\varphi)(\delta)(h + (Y)^2) := \delta(\varphi^*(h) + (X)^2)$. In the R-bases $(\frac{\partial}{\partial X_i})_i$ (resp. $(\frac{\partial}{\partial Y_j})_j$) of $\mathrm{Lie}(G)$ (resp. $\mathrm{Lie}(H)$), the map $\mathrm{Lie}(\varphi)$ is given by the *Jacobian matrix* $(\frac{\partial \varphi_i}{\partial X_j}(0))_{i,j} \in R^{e \times d}$ of φ. If $\varphi : G \to G'$ and $\psi : G' \to G''$ are homomorphisms of formal groups, then $(\psi \circ \varphi)^* = \varphi^* \circ \psi^*$ and $\mathrm{Lie}(\psi \circ \varphi) = \mathrm{Lie}(\psi) \circ \mathrm{Lie}(\varphi)$. If $H = G$ then one can use **(F1)** to show that the map $(\varphi \mapsto \mathrm{Lie}(\varphi)) : \mathrm{End}(G) \to \mathrm{End}_R(\mathrm{Lie}(G))$ is a homomorphism of rings. In particular, $\mathrm{Lie}(G)$ becomes a module over $\mathrm{End}(G)$ and we have $\mathrm{Lie}([m]_G) = m \cdot \mathrm{id}_{\mathrm{Lie}(G)}$ for any integer m.

If p is a prime number and if R is a complete noetherian local ring of residue characteristic p, then a homomorphism $\varphi : G \to H$ of formal groups is called an *isogeny* if the comorphism φ^* makes $R[\![X]\!]$ a finite free module over $R[\![Y]\!]$ (cf. Tate 1967, section 2.2). Of course, this can only happen if $d = e$. A formal group G over a complete noetherian local ring R with residue characteristic p is called *p-divisible*, if the homomorphism $[p]_G : G \to G$ is an isogeny. In this case the rank of $R[\![X]\!]$ over itself via $[p]_G^*$ is a power of p, say p^h (cf. Tate 1967, section 2.2; this result can also be deduced from Zink 1984, Satz 5.3). The integer $h =: \mathrm{ht}(G)$ is called the *height* of the p-divisible formal group G.

If $R = k$ is a perfect field of characteristic p, the necessary tools to effectively study the category of p-divisible commutative formal groups over k were first developed by Demazure (1986, Chapter III). His methods were later generalized by Cartier in order to describe commutative formal groups over arbitrary rings (cf. Lazard 1975, Chapters III & IV, or Zink 1984, Chapters III & IV).

Sticking to the case of a perfect field k of characteristic p, we denote by $W := W(k)$ the ring of Witt vectors over k. Let $\sigma = (x \mapsto x^p)$ denote the Frobenius automorphism of k, as well as its unique lift to a ring automorphism of W. Recall that a *σ^{-1}-crystal over k* is a pair (M, V), consisting of a finitely generated free W-module M and a map $V : M \to M$ which is σ^{-1}-linear, i.e. which is additive and satisfies

$$V(am) = \sigma^{-1}(a)V(m) \quad \text{for all} \quad a \in W,\ m \in M.$$

We shall be interested in those σ^{-1}-crystals (M, V) which satisfy the following two extra conditions (here **D** stands for Dieudonné):

(D1) $pM \subseteq V(M)$
(D2) V mod p is a nilpotent endomorphism of M/pM.

For the following fundamental result cf. Zink (1984), page 109.

Theorem 1 (Dieudonné). *If k is a perfect field of characteristic p then the category of p-divisible commutative formal groups over k is equivalent to the category of σ^{-1}-crystals over k, satisfying (D1) and (D2).* □

Let $W[F, V]$ be the non-commutative ring generated by two elements F and V over W subject to the relations

$$VF = FV = p, \quad Va = \sigma^{-1}(a)V \quad \text{and} \quad Fa = \sigma(a)F \quad \text{for all} \quad a \in W.$$

The equivalence of Theorem 1 associates with a p-divisible commutative formal group G its *(covariant) Cartier-Dieudonné module* M_G. This is a V-adically separated and complete module over $W[V, F]$ such that the action of V is injective. Since G is p-divisible, also the action of F is injective, and the underlying W-module of M_G is finitely generated and free. In particular, the pair (M_G, V) is a σ^{-1}-crystal over k, satisfying $pM_G = VFM_G \subseteq VM_G$, i.e. condition **(D1)**. Condition **(D2)** follows from the V-adic completeness of M_G. We also note that V and F give rise to a short exact sequence

$$0 \longrightarrow M_G/FM_G \xrightarrow{V} M_G/pM_G \longrightarrow M_G/VM_G \longrightarrow 0,$$

of k-vector spaces in which $\dim_k(M_G/pM_G) = \mathrm{ht}(G)$ and $\dim_k(M_G/VM_G) = \dim(G)$.

Conversely, if (M, V) is a σ^{-1}-crystal over k satisfying **(D1)**, then V is injective. In fact, **(D1)** implies that V becomes surjective (and hence bijective) over the quotient field K of W. Setting $F := V^{-1}p$, the W-module M becomes a module over $W[F, V]$ which is V-adically separated and complete if condition **(D2)** is satisfied.

Recall that a σ^{-1}*-isocrystal over* k is a pair (N, f) consisting of a finite dimensional K-vector space N and a σ^{-1}-linear bijection $f : N \to N$. If (M, V) is a σ^{-1}-crystal over k which satisfies **(D1)** then $(M \otimes_W K, V \otimes \mathrm{id}_K)$ is a σ^{-1}-isocrystal over k. The σ^{-1}-isocrystal which in this way is associated with the Cartier-Dieudonné module of a p-disivible commutative formal group G over k, classifies G up to isogeny (cf. Zink 1984, Satz 5.26 and the remarks on page 110; alternatively, consult Demazure 1986, Chapter IV.1).

Given integers r and s with $r > 0$, consider the σ^{-1}-isocrystal over k given by $(K[t]/(t^r - p^s), t \circ \sigma)$. Here $K[t]$ denotes the usual commutative polynomial ring in the variable t over K on which σ acts coefficientwise. If k is algebraically closed, we have the following fundamental classification result of Dieudonné and Manin (cf. Zink 1984, Satz 6.29; Demazure 1986, Chapter IV.4; Lazard 1975, Proposition VI.7.42).

Theorem 2 (Dieudonné-Manin). *If k is an algebraically closed field of characteristic p then the category of σ^{-1}-isocrystals over k is semisimple. The simple objects are given by the σ^{-1}-isocrystals $(K[t]/(t^r - p^s), t \circ \sigma)$, where r and s are relatively prime integers with $r > 0$.* □

To a pair (r, s) of integers as in Theorem 2 corresponds a particular p-divisible commutative formal group G_{rs} over k inside the isogeny class determined by the σ^{-1}-isocrystal $(K[t]/(t^r - p^s), t \circ \sigma)$. According to Lazard (1975), Proposition VI.7.42, the endomorphism ring of G_{rs} is isomorphic to the maximal order of the central division algebra of invariant $\frac{s}{r} + \mathbb{Z} \in \mathbb{Q}/\mathbb{Z}$ and dimension r^2 over $\mathbb{Q}_p$.

Corollary 1. *If G is a p-divisible commutative formal group over an algebraically closed field k of characteristic p then the endomorphism ring* End(G) *of G is an order in a finite dimensional semisimple $\mathbb{Q}_p$-algebra. Endowing* End(G) *with the p-adic topology and the automorphism group* Aut(G) *of G with the induced topology,* Aut(G) *is a compact Lie group over $\mathbb{Q}_p$.*

Proof. That End(G) is a p-adically separated and torsion free $\mathbb{Z}_p$-module can easily be proved directly, using that G is p-divisible. It also follows from the fact that the Cartier-Dieudonné module of G is free over W. According to Theorem 2 and the subsequent remarks there are central division algebras $D_1, \ldots, D_n$ over $\mathbb{Q}_p$ and natural numbers $m_1, \ldots, m_n$ such that

$$\mathrm{End}(G) \otimes_{\mathbb{Z}_p} \mathbb{Q}_p \simeq \mathrm{Mat}(m_1 \times m_1, D_1) \times \ldots \times \mathrm{Mat}(m_n \times m_n, D_n)$$

as $\mathbb{Q}_p$-algebras. Since End(G) is p-adically separated, it is bounded in End$(G) \otimes_{\mathbb{Z}_p} \mathbb{Q}_p$. Thus, it is a lattice in a finite dimensional $\mathbb{Q}_p$-vector space and must be finitely generated over $\mathbb{Z}_p$. This proves the first assertion. Endowing End(G) with the p-adic topology, it becomes a topological $\mathbb{Z}_p$-algebra and Aut(G) becomes a compact topological group for the subspace topology. By the above arguments, it is isomorphic to an open subgroup of $\prod_{i=1}^n \mathrm{GL}_{m_i}(D_i)$, hence naturally carries the structure of a Lie group over $\mathbb{Q}_p$.

3 Deformation Problems and Iwasawa Modules

We continue to denote by k a fixed algebraically closed field of characteristic p. We also fix a p-divisible commutative formal group G of dimension d over k. Denote by $W = W(k)$ the ring of Witt vectors of k and by $\mathcal{C}_k$ the category of complete noetherian commutative local W-algebras with residue class field k. Let R be an object of $\mathcal{C}_k$ and let $\mathfrak{m}$ be the maximal ideal of R. A *deformation of G to R* is a pair $(G', \rho_{G'})$, where G' is commutative formal group over R and $\rho_{G'} : G \to G' \bmod \mathfrak{m}$ is an isomorphism of formal groups over k. Two deformations $(G', \rho_{G'})$ and $(G'', \rho_{G''})$ of G to R are said to be isomorphic if there is an isomorphism $f : G' \to G''$ of formal groups over R such that the diagram

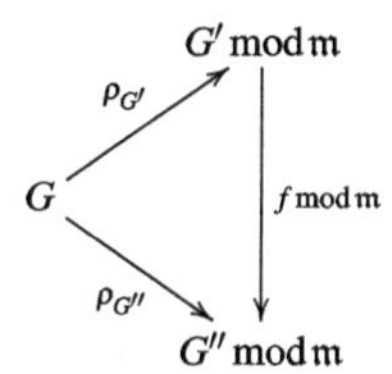

is commutative. Let Def_G denote the functor from $\mathcal{C}_k$ to the category *Sets* of sets which associates with an object R of $\mathcal{C}_k$ the set of isomorphism classes of deformations of G to R. If $\dim(G) = 1$, then the following theorem was first proved by Lubin and Tate (1966, Theorem 3.1), building on the work of Lazard. It was later generalized by Cartier und Umemura, independently (cf. Cartier 1968,1969; Umemura 1977).

Theorem 3. *The functor* $\mathrm{Def}_G : \mathcal{C}_k \to$ *Sets is representable, i.e. there is an object* R_G^{def} *of* $\mathcal{C}_k$ *and a deformation* $\mathbb{G}$ *of* G *to* R_G^{def} *with the following universal property. For any object* R *of* $\mathcal{C}_k$ *and any deformation* $(G', \rho_{G'})$ *of* G *to* R *there is a unique* W*-linear local ring homomorphism* $\varphi : R_G^{\mathrm{def}} \to R$ *and a unique isomorphism* $[\varphi] : \varphi_*(\mathbb{G}, \rho_{\mathbb{G}}) \simeq (G', \rho_{G'})$ *of deformations of* G *to* R.[1] *If* $h = \mathrm{ht}(G)$ *and* $d = \dim(G)$ *denote the height and the dimension of* G*, respectively, then the* W*-algebra* R_G^{def} *is non-canonically isomorphic to the power series ring* $W[\![u_1, \ldots, u_{(h-d)d}]\!]$ *in* $(h-d)d$ *variables over* W. □

It follows from the universal property of the deformation $(\mathbb{G}, \rho_{\mathbb{G}})$ that the automorphism group $\mathrm{Aut}(G)$ of G acts on the universal deformation ring R_G^{def} by W-linear local ring automorphisms. Indeed, given $\gamma \in \mathrm{Aut}(G)$, there is a unique W-linear local ring endomorphism γ of R_G^{def} and a unique isomorphism $[\gamma] : \gamma_*(\mathbb{G}, \rho_{\mathbb{G}}) \simeq (\mathbb{G}, \rho_{\mathbb{G}} \circ \gamma)$ of deformations of G to R_G^{def}. It follows from

[1]Here $\varphi_*(\mathbb{G}, \rho_{\mathbb{G}}) = (\varphi_*(\mathbb{G}), \rho_{\mathbb{G}})$, where $\varphi_*\mathbb{G}$ is obtained by applying φ to the coefficients of $\mathbb{G}$. Since φ induces an isomorphism between the residue class fields of R_G^{def} and R, we may identify $\mathbb{G} \bmod \mathfrak{m}_{R_G^{\mathrm{def}}}$ and $\varphi_*\mathbb{G} \bmod \mathfrak{m}$.

the uniqueness that the resulting map $\mathrm{Aut}(G) \to \mathrm{End}(R_G^{\mathrm{def}})$ factors through a homomorphism

$$\mathrm{Aut}(G) \longrightarrow \mathrm{Aut}(R_G^{\mathrm{def}})$$

of groups. It is this type of representation that we are concerned with in this article. To ease notation we shall denote by

$$R := R_G^{\mathrm{def}}$$

the universal deformation ring of our fixed p-divisible commutative formal group G over k. Let $\mathfrak{m}$ denote the maximal ideal of R. For any non-negative integer n we denote by $\mathbb{G}_n := \mathbb{G} \bmod \mathfrak{m}^{n+1}$ the reduction of the universal deformation $\mathbb{G}$ modulo the ideal $\mathfrak{m}^{n+1}$ of R. We have $G \simeq \mathbb{G}_0$ via $\rho_{\mathbb{G}}$.

Lemma 1. *If n is a non-negative integer then the ring homomorphism* $\mathrm{End}(\mathbb{G}_{n+1}) \to \mathrm{End}(\mathbb{G}_n)$, *induced by reduction modulo $\mathfrak{m}^{n+1}$, is injective.*

Proof. The formal group $\mathbb{G}_{n+1}$ is p-divisible because the comorphism $[p]^*_{\mathbb{G}_{n+1}}$ is finite and free. Indeed, it is so after reduction modulo $\mathfrak{m}$, and one can use Bourbaki (2006), III.2.1 Proposition 14 and III.5.3 Théorème 1, to conclude. Since the ideal $\mathfrak{m}^{n+1}(R/\mathfrak{m}^{n+2})$ of $R/\mathfrak{m}^{n+2}$ is nilpotent, the claim follows from the rigidity theorem in Zink (1984), Satz 5.30.

The preceding lemma allows us to regard all endomorphism rings $\mathrm{End}(\mathbb{G}_n)$ as subrings of $\mathrm{End}(\mathbb{G}_0)$. The main technical result of this section is the following assertion.

Proposition 1. *For any non-negative integer n the subring* $\mathrm{End}(\mathbb{G}_n)$ *of* $\mathrm{End}(\mathbb{G}_0)$ *contains* $p^n\,\mathrm{End}(\mathbb{G}_0)$.

Proof. We proceed by induction on n, the case $n = 0$ being trivial. Let $n \geq 1$ and assume the assertion to be true for $n - 1$. Set $R_n := R/\mathfrak{m}^{n+1}$. Let $\varphi \in p^{n-1}\,\mathrm{End}(\mathbb{G}_0) \subseteq \mathrm{End}(\mathbb{G}_{n-1})$ and choose a family $\tilde{\varphi} \in R_n[\![X]\!]^d$ of power series with trivial constant terms such that $\tilde{\varphi} \bmod \mathfrak{m}^n R_n = \varphi$. The d-tuple of power series $[p]_{\mathbb{G}_n} \circ \tilde{\varphi}$ is then a lift of $p\varphi$. We claim that it is an endomorphism of $\mathbb{G}_n$.

Note first that $[p]_{\mathbb{G}_n} \circ \tilde{\varphi}$ depends only on φ and not on the choice of a lift $\tilde{\varphi}$. Indeed, if $\tilde{\varphi}'$ is a second lift of φ with trivial constant terms, set $\psi := \tilde{\varphi}' - \tilde{\varphi}$. Setting $\chi := (\tilde{\varphi} + \psi) -_{\mathbb{G}_n} \tilde{\varphi}$, we have $\tilde{\varphi}' = \tilde{\varphi} +_{\mathbb{G}_n} \chi$. Further, the power series χ satisfies $\chi \bmod \mathfrak{m}^n = \varphi -_{\mathbb{G}_{n-1}} \varphi = 0$, hence has coefficients in $\mathfrak{m}^n R_n$. Since $p\mathfrak{m}^n \subseteq \mathfrak{m}^{n+1}$ and $(\mathfrak{m}^n)^m \subseteq \mathfrak{m}^{n+1}$ for any integer $m \geq 2$, we have $[p]_{\mathbb{G}_n} \circ \chi = 0$ and hence

$$[p]_{\mathbb{G}_n} \circ \tilde{\varphi}' = [p]_{\mathbb{G}_n}(\tilde{\varphi} +_{\mathbb{G}_n} \chi) = ([p]_{\mathbb{G}_n} \circ \tilde{\varphi}) +_{\mathbb{G}_n} ([p]_{\mathbb{G}_n} \circ \chi) = [p]_{\mathbb{G}_n} \circ \tilde{\varphi},$$

as desired.

If $\eta \in R_n[\![X]\!]^d$ is a family of power series with trivial constant terms, set $\delta_\eta := \delta_\eta(X,Y) := \eta(X +_{\mathbb{G}_n} Y) -_{\mathbb{G}_n} \eta(X) -_{\mathbb{G}_n} \eta(Y)$. Since $\tilde{\varphi}$ reduces to an endomorphism of $\mathbb{G}_{n-1}$, the power series $\delta_{\tilde{\varphi}}$ has coefficients in $\mathfrak{m}^n$. As above, this implies $[p]_{\mathbb{G}_n} \circ \delta_{\tilde{\varphi}} = 0$ and thus

$$\begin{aligned}\delta_{[p]_{\mathbb{G}_n} \circ \tilde{\varphi}} &= ([p]_{\mathbb{G}_n} \circ \tilde{\varphi})(X +_{\mathbb{G}_n} Y) -_{\mathbb{G}_n} ([p]_{\mathbb{G}_n} \circ \tilde{\varphi})(X) -_{\mathbb{G}_n} ([p]_{\mathbb{G}_n} \circ \tilde{\varphi})(Y) \\ &= [p]_{\mathbb{G}_n}(\delta_{\tilde{\varphi}}) = 0.\end{aligned}$$

As a consequence, $[p]_{\mathbb{G}_n} \circ \tilde{\varphi} \in \mathrm{End}(\mathbb{G}_n)$, and thus $p\varphi \in \mathrm{End}(\mathbb{G}_n)$. Since φ was arbitrary, we obtain the desired inclusion $p^n\,\mathrm{End}(\mathbb{G}_0) \subseteq \mathrm{End}(\mathbb{G}_n)$.

According to Corollary 1, the group $\mathrm{Aut}(G)$ is a profinite topological group. A basis of open neighborhoods of its identity is given by the subgroups $1 + p^n\,\mathrm{End}(G)$ with $n \geq 1$. If $\mathfrak{m}$ denotes the maximal ideal of the local ring R, the W-algebra R is a topological ring for the $\mathfrak{m}$-adic topology. We are now ready to prove the following result, a particular case of which was treated in Kohlhaase (2013), Proposition 3.1. The argument is borrowed from the proof of Gross and Hopkins (1994), Lemma 19.3. Let us put

$$\Gamma := \Gamma_0 := \mathrm{Aut}(G) \quad \text{and} \quad \Gamma_n := 1 + p^n\,\mathrm{End}(G) \quad \text{for} \quad n \geq 1.$$

Theorem 4. *The action of Γ on $R = R_G^{\mathrm{def}}$ is continuous in the sense that the map $((\gamma, f) \mapsto \gamma(f)) : \Gamma \times R \to R$ is a continuous map of topological spaces. Here $\Gamma \times R$ carries the product topology. If n is a non-negative integer then the induced action of Γ_n on $R/\mathfrak{m}^{n+1}$ is trivial.*

Proof. As in the proof of Kohlhaase (2013), Proposition 3.1, it suffices to prove the second statement. Let $\gamma \in \Gamma_n$ and consider the deformation $(\mathbb{G}_n, \rho_{\mathbb{G}} \circ \gamma)$ of G to $R_n = R/\mathfrak{m}^{n+1}$. Denote by $\mathrm{pr}_n : R \to R_n$ the natural projection and let γ_n denote the unique ring homomorphism $\gamma_n : R \to R_n$ for which there exists an isomorphism of deformations $[\gamma_n] : (\gamma_n)_*(\mathbb{G}, \rho_{\mathbb{G}}) \simeq (\mathbb{G}_n, \rho_{\mathbb{G}} \circ \gamma)$ (cf. Theorem 3). Note that also the ring homomorphism $\mathrm{pr}_n \circ \gamma : R \to R_n$ admits an isomorphism of deformations $(\mathrm{pr}_n \circ \gamma)_*(\mathbb{G}, \rho_{\mathbb{G}}) \simeq (\mathbb{G}_n, \rho_{\mathbb{G}} \circ \gamma)$, namely the reduction of $[\gamma]$ modulo $\mathfrak{m}^{n+1}$. By uniqueness, we must have $\gamma_n = \mathrm{pr}_n \circ \gamma$ and $[\gamma_n] = [\gamma] \mod \mathfrak{m}^{n+1}$.

Since the map $(\sigma \mapsto \rho_{\mathbb{G}} \circ \sigma \circ \rho_{\mathbb{G}}^{-1})$ is a ring isomorphism $\mathrm{End}(G) \to \mathrm{End}(\mathbb{G}_0)$, Proposition 1 shows that $\rho_{\mathbb{G}} \circ \gamma \circ \rho_{\mathbb{G}}^{-1} \in \mathrm{Aut}(\mathbb{G}_n)$ and therefore defines an isomorphism of deformations $(\mathrm{pr}_n)_*(\mathbb{G}, \rho_{\mathbb{G}}) = (\mathbb{G}_n, \rho_{\mathbb{G}}) \simeq (\mathbb{G}_n, \rho_{\mathbb{G}} \circ \gamma)$. By uniqueness again, we must have $\gamma_n = \mathrm{pr}_n \circ \gamma = \mathrm{pr}_n$. This implies that γ acts trivially on R_n and that $[\gamma] \mod \mathfrak{m}^{n+1} = \rho_{\mathbb{G}} \circ \gamma \circ \rho_{\mathbb{G}}^{-1}$.

If H is a profinite topological group then we denote by

$$\Lambda(H) := W[\![H]\!] := \varprojlim_{n \geq 1,\, N \trianglelefteq_o H} (W/p^n W)[H/N]$$

the *Iwasawa algebra* (or *completed group ring*) of H over W. The above projective limit runs over all positive integers n and over all open normal subgroups N of H. If n and n' are positive integers with $n' \leq n$, and if N and N' are two open normal subgroups of H with $N \subseteq N'$, then the transition map $(W/p^nW)[H/N] \to (W/p^{n'}W)[H/N']$ is the natural homomorphism of group rings induced by the surjective homomorphism $H/N \to H/N'$ of groups and the surjective ring homomorphism $W/p^nW \to W/p^{n'}W$. Endowing each ring $(W/p^nW)[H/N]$ with the discrete topology, $\Lambda(H)$ is a topological ring for the projective limit topology. It is a *pseudocompact ring* in the terminology of Brumer (1966), page 442, because each of the rings $(W/p^nW)[H/N]$ is Artinian. Recall that a complete Hausdorff topological $\Lambda(H)$-module M is called *pseudocompact*, if it admits a basis $(M_i)_{i\in I}$ of open neighborhoods of zero such that each M_i is a $\Lambda(H)$- submodule of M for which the $\Lambda(H)$-module M/M_i has finite length. For brevity, we will set

$$\Lambda := \Lambda(\mathrm{Aut}(G)).$$

Corollary 2. *The action of* $\mathrm{Aut}(G)$ *on* $R = R_G^{\mathrm{def}}$ *extends to an action of* Λ *and gives* R *the structure of a pseudocompact* Λ*-module.*

Proof. Since R is $\mathfrak{m}$-adically separated and complete, we may consider the natural isomorphism

$$R \simeq \varprojlim_{n\geq 0} R/\mathfrak{m}^{n+1}.$$

According to Theorem 4, the action of the group ring $W[\mathrm{Aut}(G)]$ on $R/\mathfrak{m}^{n+1}$ factors through $(W/p^{n+1}W)[\mathrm{Aut}(G)/(1 + p^n \mathrm{End}(G))]$ where $1 + p^n \mathrm{End}(G)$ is an open normal subgroup of $\mathrm{Aut}(G)$. Thus, $R/\mathfrak{m}^{n+1}$ can be viewed as a Λ-module via the natural ring homomorphism $\Lambda \to (W/p^{n+1}W)[\mathrm{Aut}(G)/(1 + p^n \mathrm{End}(G))]$. The transition maps in the above projective limit are then Λ-equivariant. This proves the first assertion.

As for the second assertion, the ideals $\mathfrak{m}^{n+1}$ of R are open and Λ-stable, being the kernels of the Λ-equivariant projections $R \to R/\mathfrak{m}^{n+1}$. They form a basis of open neighborhoods of zero of R, and the quotients $R/\mathfrak{m}^{n+1}$ are even of finite length over $W \subseteq \Lambda$.

Let $\mathrm{Lie}(\mathbb{G})$ denote the Lie algebra of the universal deformation $\mathbb{G}$ of G. This is a free module of rank $d = \dim(G)$ over R. Given $\gamma \in \mathrm{Aut}(G)$, we extend the ring automorphism $\gamma : R \to R$ to an automorphism $\gamma : R[\![X]\!] \to R[\![X]\!]$ by setting $\gamma(X_i) = X_i$ for all $1 \leq i \leq d$. It induces a homomorphism $\gamma : \mathrm{Lie}(\mathbb{G}) \to \mathrm{Lie}(\gamma_*\mathbb{G})$ of additive groups. We define $\tilde{\gamma} : \mathrm{Lie}(\mathbb{G}) \to \mathrm{Lie}(\mathbb{G})$ as the composite of the two additive maps

$$\mathrm{Lie}(\mathbb{G}) \xrightarrow{\gamma} \mathrm{Lie}(\gamma_*\mathbb{G}) \xrightarrow{\mathrm{Lie}([\gamma])} \mathrm{Lie}(\mathbb{G}),$$

with $[\gamma] : \gamma_*\mathbb{G} \to \mathbb{G}$ as above. Given a second element $\gamma' \in \mathrm{Aut}(G)$, we define $\gamma' : \mathrm{Lie}(\gamma_*\mathbb{G}) \to \mathrm{Lie}(\gamma'_*(\gamma_*\mathbb{G}))$ as before. Further, $\gamma'_*[\gamma] : \gamma'_*(\gamma_*\mathbb{G}) \to \gamma'_*\mathbb{G}$ denotes the homomorphism obtained by applying $\gamma' \in \mathrm{Aut}(R)$ to the coefficients of $[\gamma] \in R[\![X]\!]^d$. One readily checks that the diagram

$$\begin{array}{ccc} \mathrm{Lie}(\gamma_*\mathbb{G}) & \xrightarrow{\mathrm{Lie}([\gamma])} & \mathrm{Lie}(\mathbb{G}) \\ \downarrow{\gamma'} & & \downarrow{\gamma'} \\ \mathrm{Lie}(\gamma'_*(\gamma_*\mathbb{G})) & \xrightarrow[\mathrm{Lie}(\gamma'_*[\gamma])]{} & \mathrm{Lie}(\gamma'_*\mathbb{G}) \end{array}$$

is commutative. Further, the uniqueness assertion in Theorem 3 implies that $[\gamma'\gamma] = [\gamma'] \circ \gamma'_*[\gamma]$. Therefore,

$$\begin{aligned} (\gamma'\gamma)^\sim &= \mathrm{Lie}([\gamma'\gamma]) \circ (\gamma'\gamma) = \mathrm{Lie}([\gamma']) \circ \mathrm{Lie}(\gamma'_*[\gamma]) \circ \gamma' \circ \gamma \\ &= \mathrm{Lie}([\gamma']) \circ (\gamma' \circ \mathrm{Lie}([\gamma]) \circ (\gamma')^{-1}) \circ \gamma' \circ \gamma = \tilde{\gamma}' \circ \tilde{\gamma}. \end{aligned}$$

As a consequence, we obtain an action of $\mathrm{Aut}(G)$ on the additive group $\mathrm{Lie}(\mathbb{G})$ which is semilinear for the action on R in the sense that

$$\tilde{\gamma}(f \cdot \delta) = \gamma(f) \cdot \tilde{\gamma}(\delta) \quad \text{for all} \quad f \in R, \delta \in \mathrm{Lie}(\mathbb{G}).$$

To ease notation, we will again write $\gamma(\delta)$ for $\tilde{\gamma}(\delta)$.

Given a positive integer m we denote by $\mathrm{Lie}(\mathbb{G})^{\otimes m}$ the m-fold tensor product of $\mathrm{Lie}(\mathbb{G})$ over R with itself. This is a free R-module of rank d^m with a semilinear action of $\mathrm{Aut}(G)$ defined by

$$\gamma(\delta_1 \otimes \cdots \otimes \delta_m) := \gamma(\delta_1) \otimes \cdots \otimes \gamma(\delta_m).$$

We also set $\mathrm{Lie}(\mathbb{G})^{\otimes 0} := R$ and $\mathrm{Lie}(\mathbb{G})^{\otimes m} := \mathrm{Hom}_R(\mathrm{Lie}(\mathbb{G})^{\otimes(-m)}, R)$ if m is a negative integer. In the latter case $\mathrm{Lie}(\mathbb{G})^{\otimes m}$ is a free R-module of rank d^{-m} with a semilinear action of $\mathrm{Aut}(G)$ defined through

$$\gamma(\varphi)(\delta_1 \otimes \cdots \otimes \delta_{-m}) := \gamma(\varphi(\gamma^{-1}(\delta_1) \otimes \cdots \otimes \gamma^{-1}(\delta_{-m}))).$$

For any integer m we endow the R-module $\mathrm{Lie}(\mathbb{G})^{\otimes m}$ with the $\mathfrak{m}$-adic topology for which it is Hausdorff and complete. By the semilinearity of the $\mathrm{Aut}(G)$-action, the R-submodules $\mathfrak{m}^n \mathrm{Lie}(\mathbb{G})^{\otimes m}$ are $\mathrm{Aut}(G)$-stable for any non-negative integer n.

As an easy consequence of Proposition 1 and Theorem 4, we obtain the following result.

Theorem 5. *Let m and n be integers with $n \geq 0$. The action of* $\mathrm{Aut}(G)$ *on* $\mathrm{Lie}(\mathbb{G})^{\otimes m}$ *is continuous in the sense that the structure map* $\mathrm{Aut}(G) \times \mathrm{Lie}(\mathbb{G})^{\otimes m} \to \mathrm{Lie}(\mathbb{G})^{\otimes m}$ *of the action is continuous. Here the left hand side carries the product topology. The induced action of* $1 + p^{2n+1}\mathrm{End}(G)$ *on the quotient*

$\mathrm{Lie}(\mathbb{G})^{\otimes m}/\mathfrak{m}^{n+1}\,\mathrm{Lie}(\mathbb{G})^{\otimes m}$ *is trivial. In particular, the action of* $\mathrm{Aut}(G)$ *on* $\mathrm{Lie}(\mathbb{G})^{\otimes m}$ *extends to an action of* Λ *and gives* $\mathrm{Lie}(\mathbb{G})^{\otimes m}$ *the structure of a pseudocompact* Λ*-module.*

Proof. As in the proof of Theorem 4 and Corollary 2, it suffices to show that the action of $1 + p^{2n+1}\,\mathrm{End}(G)$ on $\mathrm{Lie}(\mathbb{G})^{\otimes m}/\mathfrak{m}^{n+1}\,\mathrm{Lie}(\mathbb{G})^{\otimes m}$ is trivial. By definition of the action and Theorem 4 we may assume $m = 1$. Setting $\mathbb{G}_n = \mathbb{G} \mod \mathfrak{m}^{n+1}$, as before, we have $\mathrm{Lie}(\mathbb{G})/\mathfrak{m}^{n+1}\,\mathrm{Lie}(\mathbb{G}) = \mathrm{Lie}(\mathbb{G}_n)$. Since $2n + 1 \geq n$, Theorem 4 and its proof show that the map $\gamma \mod \mathfrak{m}^{n+1} : \mathrm{Lie}(\mathbb{G}_n) \to \mathrm{Lie}(\mathbb{G}_n)$ is given by $\mathrm{Lie}(\rho_{\mathbb{G}} \circ \gamma \circ \rho_{\mathbb{G}}^{-1})$ where $\rho_{\mathbb{G}} \circ \gamma \circ \rho_{\mathbb{G}}^{-1}$ is contained in $1 + p^{2n+1}\,\mathrm{End}(\mathbb{G}_0) \subseteq 1 + p^{n+1}\,\mathrm{End}(\mathbb{G}_n)$ (cf. Proposition 1). Therefore, it suffices to show that the natural action of $1 + p^{n+1}\,\mathrm{End}(\mathbb{G}_n) \subset \mathrm{End}(\mathbb{G}_n)$ on $\mathrm{Lie}(\mathbb{G}_n)$ is trivial. However, if $\varphi \in \mathrm{End}(\mathbb{G}_n)$ and if $\delta \in \mathrm{Lie}(\mathbb{G}_n)$, then

$$\mathrm{Lie}(1 + p^{n+1}\varphi)(\delta) = \delta + p^{n+1}\,\mathrm{Lie}(\varphi)(\delta) = \delta,$$

because $p^{n+1} \in \mathfrak{m}^{n+1}$.

Before we continue, let us point out an important variant of the deformation problem considered above. It concerns the moduli problems considered by Rapoport and Zink (1996).

Let G be a fixed p-divisible group over the algebraically closed field k of characteristic p, i.e. an *fppf*-group scheme over $\mathrm{Spec}(k)$ for which multiplication by p is an epimorphism. Denoting by Nil_p the category of W-schemes on which p is locally nilpotent, let $\mathcal{M}_G : Nil_p \to Sets$ denote the set valued functor which associates to an object S of Nil_p the set of isomorphism classes of pairs $(G', \rho_{G'})$, where G' is a p-divisible group over S and $\rho_{G'} : G_{\overline{S}} \to G'_{\overline{S}}$ is a *quasi-isogeny* (cf. Rapoport and Zink 1996, Definition 2.8). Here $\overline{S}$ denotes the closed subscheme of S defined by the sheaf of ideals $p\mathcal{O}_S$. According to Rapoport and Zink (1996), Theorem 2.16, the functor $\mathcal{M}_G$ is represented by a formal scheme which is locally formally of finite type over $\mathrm{Spf}(W)$. If G is a p-divisible one dimensional commutative formal group law as in Sect. 1, then $\mathcal{M}_G$ is the disjoint union of open subschemes $\mathcal{M}^n_G$, $n \in \mathbb{Z}$, which are non-canonically isomorphic to $\mathrm{Spf}(R^{\mathrm{def}}_G)$ (cf. Rapoport and Zink 1996, Proposition 3.79). The reason is that any quasi-isogeny of height zero between one dimensional p-divisible formal group laws over k is an isomorphism.

One can generalize the moduli problem even further by considering deformations of p-divisible groups with additional structures such as polarizations or actions by maximal orders in finite dimensional semisimple $\mathbb{Q}_p$-algebras (cf. Rapoport and Zink 1996, Definition 3.21). The corresponding deformation functors are again representable, as was proven by Rapoport and Zink (1996, Theorem 3.25). An important example was studied by Drinfeld (cf. Rapoport and Zink 1996, 3.58). The generic fiber of the representing formal scheme is known as *Drinfeld's upper half space over* K. Instead of continuous representations of $\mathrm{Aut}(G)$ as in Theorem 4, it gives rise to an important class of *p-adic locally analytic representations* in the sense of Schneider and Teitelbaum. This particular class of representations was studied extensively by Morita, Orlik, Schneider and Teitelbaum (cf. Orlik 2008; Schneider and Teitelbaum 2002). It found aritheoremetic applications to the de Rham cohomology of varieties which are p-adically uniformized by Drinfeld's upper half space (cf. Kohlhaase and Schraen 2012). In the next section we shall see that the deformation spaces we consider here give rise to locally analytic representations, as well.

4 Rigidification and Local Analyticity

We keep the notation of the previous section and denote by k an algebraically closed field of characteristic p and by G a fixed commutative p-divisible formal group over k. Let h and d denote the height and the dimension of G, respectively. We denote by W the ring of Witt vectors of k and by K the quotient field of W. We let $R = R_G^{\mathrm{def}}$ denote the universal deformation ring of G (cf. Theorem 3).

According to Theorem 3, the rigidification $\mathrm{Spf}(R)^{\mathrm{rig}}$ of the formal scheme $\mathrm{Spf}(R)$ in the sense of Berthelot (cf. de Jong 1995, section 7) is isomorphic to the $(h-d)d$-dimensional rigid analytic open unit polydisc $\mathring{\mathbb{B}}_K^{(h-d)d}$ over K. We let

$$R^{\mathrm{rig}} := \mathcal{O}(\mathrm{Spf}(R)^{\mathrm{rig}})$$

denote the ring of global rigid analytic functions on $\mathrm{Spf}(R)^{\mathrm{rig}}$. Any isomorphism $R \simeq W[\![u]\!]$ of local W-algebras extends to an isomorphism

$$R^{\mathrm{rig}} \simeq \{ \sum_{\alpha \in \mathbb{N}^{(h-d)d}} c_\alpha u^\alpha \mid c_\alpha \in K \text{ and } \lim_{|\alpha| \to \infty} |c_\alpha| r^{|\alpha|} = 0 \text{ for all } 0 < r < 1\}$$

of K-algebras, where $|\cdot|$ denotes the p-adic absolute value on K. This allows us to view R^{rig} as a topological K-Fréchet algebra whose topology is defined by the family of norms $||\cdot||_\ell$, given by

$$||\sum_\alpha c_\alpha u^\alpha||_\ell := \sup_\alpha\{|c_\alpha| p^{-|\alpha|/\ell}\}$$

for any positive integer ℓ. Letting R_ℓ^{rig} denote the completion of R^{rig} with respect to the norm $||\cdot||_\ell$, the K-algebra R_ℓ^{rig} can be identified with the ring of rigid analytic functions on the affinoid subdomain

$$\mathbb{B}_\ell^{(h-d)d} := \{x \in \mathrm{Spf}(R)^{\mathrm{rig}} \mid |u_i(x)| \le p^{-1/\ell} \text{ for all } 1 \le i \le (h-d)d\}$$

of $\mathrm{Spf}(R)^{\mathrm{rig}}$. Further, $R^{\mathrm{rig}} \simeq \varprojlim_\ell R_\ell^{\mathrm{rig}}$ is the topological projective limit of the K-Banach algebras R_ℓ^{rig}. In fact, by a cofinality argument and Bosch et al. (1984),

6.1.3 Theorem 1, R^{rig} is the topological projective limit of the system of affinoid K-algebras corresponding to any nested admissible open affinoid covering of $\text{Spf}(R)^{\text{rig}}$.

By functoriality, the automorphism group $\Gamma = \text{Aut}(G)$ of G acts on $\text{Spf}(R)^{\text{rig}}$ by automorphisms of rigid analytic K-varieties. This gives rise to an action of Γ on R^{rig} by K-linear ring automorphisms. By the above cofinality argument, any of these automorphisms is continuous. The goal of this section is to show that the induced action on the strong topological K-linear dual of R^{rig} is *locally analytic* in the sense of Schneider and Teitelbaum (2002, page 451).

Fix an algebraic closure K^{alg} of K. According to de Jong (1995), Lemma 7.19, the maximal ideals of the ring $R_K := R \otimes_W K$ are in bijection with the points of $\text{Spf}(R)^{\text{rig}}$. It follows from Bosch et al. (1984), 7.1.1 Proposition 1, that the latter are in bijection with the $\text{Gal}(K^{\text{alg}}|K)$-orbits of

$$\mathring{\mathbb{B}}_K^{(h-d)d}(K^{\text{alg}}) := \{x \in (K^{\text{alg}})^{(h-d)d} \mid |x_i| < 1 \text{ for all } 1 \le i \le (h-d)d\}.$$

A point x representing one of these orbits corresponds to the kernel of the surjective K-linear ring homomorphism $R_K \to K(x) := K(x_1, \ldots, x_{(h-d)d}) \subseteq K^{\text{alg}}$, sending $f(u)$ to $f(x)$.

The following result constitutes the technical heart of this section. It is a straightforward generalization of Gross and Hopkins (1994), Lemma 19.3.

Proposition 2. *Let n and ℓ be integers with $n \ge 0$ and $\ell \ge 1$. If $\gamma \in \Gamma_n$ and if $f \in R^{\text{rig}}$ then $||\gamma(f) - f||_\ell \le p^{-n/\ell}||f||_\ell$.*

Proof. First assume $f = u_i$ for some $1 \le i \le (h-d)d$. If

$$\mathbb{B}_\ell^{(h-d)d}(K^{\text{alg}}) := \{x \in (K^{\text{alg}})^{(h-d)d} \mid |x_i| \le p^{-1/\ell} \text{ for all } 1 \le i \le (h-d)d\},$$

then $||g||_\ell = \sup\{|g(x)| \mid x \in \mathbb{B}_\ell^{(h-d)d}(K^{\text{alg}})\}$ for any $g \in R^{\text{rig}}$. Thus, we need to see that if $x \in \mathbb{B}_\ell^{(h-d)d}(K^{\text{alg}})$ and if $y := x\cdot\gamma = \gamma(u)(x)$, then $|x_i - y_i| \le p^{-(n+1)/\ell}$.

Denoting by W^{alg} the valuation ring of K^{alg}, consider the commutative diagram

$$\begin{array}{ccc} R & \xrightarrow{\gamma} & R \\ & \searrow \quad \swarrow & \\ & W^{\text{alg}} & \end{array}$$

of homomorphisms of W-algebras, in which the left and right oblique arrow is given by evaluation at y and x, respectively. Choosing $z \in W^{\text{alg}}$ with $|z| = p^{-1/\ell}$, we have $x_j \in zW^{\text{alg}}$ for any j. Further, $p \in zW^{\text{alg}}$ because $\ell \ge 1$. As a consequence, the right oblique arrow maps $\mathfrak{m}_R$ to zW^{alg}. Note that $\gamma(u_j) \in \mathfrak{m}_R$, so that we obtain $y_j = u_j(x \cdot \gamma) = \gamma(u_j)(x) \in zW^{\text{alg}}$, as well. Therefore, also the left oblique arrow maps $\mathfrak{m}_R$ to zW^{alg}. Now consider the induced diagram

$$\begin{array}{ccc} R/\mathfrak{m}_R^{n+1} & \xrightarrow{\quad\gamma\quad} & R/\mathfrak{m}_R^{n+1} \\ & \searrow \quad \swarrow & \\ & W^{\mathrm{alg}}/(z^{n+1}). & \end{array}$$

According to Theorem 4, the upper horizontal arrow is the identity. It follows that $x_i - y_i \in z^{n+1}W^{\mathrm{alg}}$, i.e. $|x_i - y_i| \leq p^{-(n+1)/\ell}$, as required. In particular, γ stabilizes $\mathbb{B}_\ell^{(h-d)d}(K^{\mathrm{alg}})$ and therefore is an isometry for the norm $||\cdot||_\ell$ on R^{rig}.

To prove the proposition, the continuity of γ allows us to assume $f = u^\alpha$ for some $\alpha \in \mathbb{N}^{(h-d)d}$. The assertion is trivial for $|\alpha| = 0$. If $|\alpha| > 0$ choose an index i with $\alpha_i > 0$. Define β through $\beta_j := \alpha_j$ if $j \neq i$ and $\beta_i := \alpha_i - 1$. If $x \in \mathbb{B}_\ell^{(h-d)d}(K^{\mathrm{alg}})$ and if $y = x \cdot \gamma$, then

$$\begin{aligned} |\gamma(u^\alpha)(x) - u^\alpha(x)| &= |y^\alpha - x^\alpha| = |y_i y^\beta - x_i x^\beta| \\ &\leq \max\{|y_i||y^\beta - x^\beta|, |y_i - x_i||x^\beta|\}. \end{aligned}$$

Here $|y_i||y^\beta - x^\beta| \leq p^{-1/\ell}||\gamma(u^\beta) - u^\beta||_\ell \leq p^{-(n+1)/\ell}||u^\beta||_\ell = p^{-n/\ell}||u^\alpha||_\ell$ by the induction hypothesis. Further, $|y_i - x_i||x^\beta| \leq p^{-(n+1)/\ell}p^{-|\beta|/\ell} = p^{-n/\ell}||u^\alpha||_\ell$, as seen above. Thus, we obtain $|\gamma(u^\alpha)(x) - u^\alpha(x)| \leq p^{-n/\ell}||u^\alpha||_\ell$ for all $x \in \mathbb{B}_\ell^{(h-d)d}(K^{\mathrm{alg}})$. This proves the proposition.

A topological group is a Lie group over $\mathbb{Q}_p$ if and only if it contains an open subgroup which is a *uniform pro-p group* (cf. Dixon et al. 2003, Definition 4.1 and Theorem 8.32). For the compact p-adic Lie group $\Gamma = \mathrm{Aut}(G)$ we have the following more precise result. We let

$$\varepsilon := 1 \quad \text{if} \quad p > 2 \quad \text{and} \quad \varepsilon := 2 \quad \text{if} \quad p = 2.$$

Lemma 2. *For any non-negative integer n we have $\Gamma_\varepsilon^{p^n} = \Gamma_{\varepsilon+n}$. The open subgroup $\Gamma_{\varepsilon+n}$ of Γ is a uniform pro-p group.*

Proof. As for the first assertion, the proofs of Dixon et al. (2003), Lemma 5.1 and Theorem 5.2, can be copied word by word on replacing $\mathrm{M}_d(\mathbb{Z}_p)$ by $\mathrm{End}(G)$ and $\mathrm{GL}_d(\mathbb{Z}_p)$ by $\mathrm{Aut}(G)$. Further, $\Gamma_{\varepsilon+n}$ is a powerful pro-p group by Dixon et al. (2003), Theorem 3.6 (i) and the remark after Definition 3.1. That it is uniform follows from Dixon et al. (2003), Theorem 3.6 (ii), and the first assertion.

Fix an integer $n \geq \varepsilon$. By Lemma 2 and Dixon et al. (2003), Theorem 3.6, the group Γ_n/Γ_{n+1} is a finite dimensional $\mathbb{F}_p$-vector space. Choosing elements $\gamma_1, \ldots, \gamma_r \in \Gamma_n$ whose images modulo Γ_{n+1} form an $\mathbb{F}_p$-basis of Γ_n/Γ_{n+1}, (Dixon et al. 2003), Theorem 4.9, shows that $(\gamma_1, \ldots, \gamma_r)$ is an *ordered basis* of Γ_n in the sense that the map $\mathbb{Z}_p^r \to \Gamma_n$, sending λ to $\gamma_1^{\lambda_1} \cdots \gamma_r^{\lambda_r}$, is a homeomorphism.

Set $b_i := \gamma_i - 1 \in \Lambda(\Gamma_n)$ and $b^\alpha := b_1^{\alpha_1} \cdots b_r^{\alpha_r}$ for any $\alpha \in \mathbb{N}^r$. By Dixon et al. (2003), Theorem 7.20, any element $\delta \in \Lambda(\Gamma_n)$ admits a unique expansion of the form

$$\lambda = \sum_{\alpha \in \mathbb{N}^r} d_\alpha b^\alpha \text{ with } d_\alpha \in W \text{ for all } \alpha \in \mathbb{N}^r.$$

For any $\ell \geq 1$ this allows us to define the K-norm $||\cdot||_\ell$ on the algebra $\Lambda(\Gamma_n)_K := \Lambda(\Gamma_n) \otimes_W K$ through

$$||\sum_\alpha d_\alpha b^\alpha||_\ell := \sup_\alpha \{|d_\alpha| p^{-\varepsilon|\alpha|/\ell}\}.$$

Remark 1. A more accurate notation would be the symbol $||\cdot||_\ell^{(n)}$ for the above norm on $\Lambda(\Gamma_n)_K$. It does generally not coincide with the restriction of $||\cdot||_\ell^{(m)}$ to $\Lambda(\Gamma_n)_K \subseteq \Lambda(\Gamma_m)_K$ if $n \geq m$. However, there is an explicit rescaling relation between the families of norms $(||\cdot||_\ell^{(n)})_\ell$ and $(||\cdot||_\ell^{(m)})_\ell$ on $\Lambda(\Gamma_n)_K$ (cf. Schmidt 2008, Proposition 6.2). Since we will never work with two different groups Γ_n and Γ_m at once, we decided to ease notation and use the somewhat ambiguous symbol $||\cdot||_\ell$.

By Schmidt (2008), Proposition 2.1 and Schneider and Teitelbaum (2003), Proposition 4.2, the norm $||\cdot||_\ell$ on $\Lambda(\Gamma_n)_K$ is submultiplicative whenever $\ell \geq 1$. As a consequence, the completion

$$\Lambda(\Gamma_n)_{K,\ell} = \{\sum_\alpha d_\alpha b^\alpha \mid d_\alpha \in K, \lim_{|\alpha|\to\infty} |d_\alpha| p^{-\varepsilon|\alpha|/\ell} = 0\}$$

of $\Lambda(\Gamma_n)_K$ with respect to $||\cdot||_\ell$ is a K-Banach algebra. The natural inclusions $\Lambda(\Gamma_n)_{K,\ell+1} \to \Lambda(\Gamma_n)_{K,\ell}$ endow the projective limit

$$D(\Gamma_n) := \varprojlim_\ell \Lambda(\Gamma_n)_{K,\ell}$$

with the structure of a K-Fréchet algebra. As is explained in Schneider and Teitelbaum (2003), section 4, a theorem of Amice allows us to identify it with the algebra of *K-valued locally analytic distributions on Γ_n*. Similarly, we denote by $D(\Gamma)$ the algebra of K-valued locally analytic distributions on Γ (cf. Schneider and Teitelbaum 2002, section 2).

Theorem 6. *For any integer $\ell \geq 1$ the action of Γ_ε on R^{rig} extends to R_ℓ^{rig} and makes R_ℓ^{rig} a topological Banach module over the K-Banach algebra $\Lambda(\Gamma_\varepsilon)_{K,\ell}$. The action of Γ on R^{rig} extends to a jointly continuous action of the K-Fréchet algebra $D(\Gamma)$. The action of Γ on the strong continuous K-linear dual $(R^{\mathrm{rig}})_b'$ of R^{rig} is locally analytic in the sense of Schneider and Teitelbaum (2002), page 451.*

Proof. First, we prove by induction on $|\alpha|$ that $||b^\alpha f||_\ell \leq ||b^\alpha||_\ell ||f||_\ell$ for any $f \in R^{\mathrm{rig}}$. This is clear if $|\alpha| = 0$. Otherwise, let i be the minimal index with $\alpha_i > 0$ and define β through $\beta_j = \alpha_j$ if $j \neq i$ and $\beta_i := \alpha_i - 1$. In this case, Proposition 2 and the induction hypothesis imply

$$\begin{aligned} ||b^\alpha f||_\ell &= ||(\gamma_i - 1)b^\beta f||_\ell \le p^{-\varepsilon/\ell}||b^\beta f||_\ell \\ &\le p^{-\varepsilon/\ell}p^{-\varepsilon|\beta|/\ell}||f||_\ell = ||b^\alpha||_\ell||f||_\ell, \end{aligned}$$

as required. This immediately gives $||\lambda \cdot f||_\ell \le ||\lambda||_\ell||f||_\ell$ for all $\lambda \in \Lambda(\Gamma_\varepsilon)_K$ and $f \in R_K$. Thus, the multiplication map $\Lambda(\Gamma_\varepsilon)_K \times R_K \to R_K$ is continuous, if $\Lambda(\Gamma_\varepsilon)_K$ and R_K are endowed with the respective $|| \cdot ||_\ell$-topologies, and if the left hand side carries the product topology. Since R_K is dense in R_ℓ^{rig}, we obtain a map $\Lambda(\Gamma_\varepsilon)_{K,\ell} \times R_\ell^{\mathrm{rig}} \to R_\ell^{\mathrm{rig}}$ by passing to completions. By continuity, it gives R_ℓ^{rig} the structure of a topological Banach module over $\Lambda(\Gamma_\varepsilon)_{K,\ell}$.

Passing to the projective limit, we obtain a continuous map $D(\Gamma_\varepsilon) \times R^{\mathrm{rig}} \to R^{\mathrm{rig}}$, giving R^{rig} the structure of a jointly continuous module over $D(\Gamma_\varepsilon)$. Since $D(\Gamma)$ is topologically isomorphic to the locally convex direct sum $\oplus_{\gamma\Gamma_\varepsilon \in \Gamma/\Gamma_\varepsilon}\gamma D(\Gamma_\varepsilon)$ (cf. Schneider and Teitelbaum 2002, page 447 bottom), R^{rig} is a jointly continuous module over $D(\Gamma)$.

It follows from Schneider (2002), Proposition 19.9 and the arguments proving the claim on page 98, that the K-Fréchet space R^{rig} is nuclear. Therefore, Schneider and Teitelbaum (2002), Corollary 3.4, implies that the locally convex K-vector space $(R^{\mathrm{rig}})'_b$ is of compact type and that the action of Γ obtained by dualizing is locally analytic.

Using Theorem 5, the preceding result can be generalized as follows. Fixing an integer m, the free R-module $\mathrm{Lie}(\mathbb{G})^{\otimes m}$ gives rise to a locally free coherent sheaf on $\mathrm{Spf}(R)$. For any positive integer ℓ we denote by $(\mathrm{Lie}(\mathbb{G})^{\otimes m})_\ell^{\mathrm{rig}}$ the sections of its rigidification over the affinoid subdomain $\mathbb{B}_\ell^{(h-d)d}$ of $\mathrm{Spf}(R)^{\mathrm{rig}}$. This is a free R_ℓ^{rig}-module for which the natural R_ℓ^{rig}-linear map

$$R_\ell^{\mathrm{rig}} \otimes_R \mathrm{Lie}(\mathbb{G})^{\otimes m} \longrightarrow (\mathrm{Lie}(\mathbb{G})^{\otimes m})_\ell^{\mathrm{rig}}$$

is bijective (cf. de Jong 1995, 7.1.11). We denote by $(\mathrm{Lie}(\mathbb{G})^{\otimes m})^{\mathrm{rig}}$ the space of global sections of the rigidification of $\mathrm{Lie}(\mathbb{G})^{\otimes m}$ over $\mathrm{Spf}(R)^{\mathrm{rig}}$. This is a free R^{rig}-module for which the natural R^{rig}-linear maps

$$R^{\mathrm{rig}} \otimes_R \mathrm{Lie}(\mathbb{G})^{\otimes m} \longrightarrow (\mathrm{Lie}(\mathbb{G})^{\otimes m})^{\mathrm{rig}} \longrightarrow \varprojlim_\ell (\mathrm{Lie}(\mathbb{G})^{\otimes m})_\ell^{\mathrm{rig}} \qquad (1)$$

are bijective. Further, $(\mathrm{Lie}(\mathbb{G})^{\otimes m})^{\mathrm{rig}} \simeq (\mathrm{Lie}(\mathbb{G})^{\mathrm{rig}})^{\otimes m}$, where the latter tensor products and dualities are with respect to R^{rig}.

By functoriality, the group $\Gamma = \mathrm{Aut}(G)$ acts on $(\mathrm{Lie}(\mathbb{G})^{\otimes m})^{\mathrm{rig}}$ in such a way that the left map in (1) becomes Γ-equivariant for the diagonal action on the left. In particular, it is semilinear for the action of Γ on R^{rig}. We endow $(\mathrm{Lie}(\mathbb{G})^{\otimes m})^{\mathrm{rig}}$ and $(\mathrm{Lie}(\mathbb{G})^{\otimes m})_\ell^{\mathrm{rig}}$ with the natural topologies of finitely generated modules over R^{rig} and R_ℓ^{rig}, respectively. This makes them a nuclear K-Fréchet space and a K-Banach space, respectively. The right map in (1) is then a topological isomorphism for the projective limit topology on the right. With the same cofinality argument as for R^{rig} one can show that any element of Γ acts on $(\mathrm{Lie}(\mathbb{G})^{\otimes m})^{\mathrm{rig}}$ through a continuous K-linear automorphism.

Theorem 7. *Let m be an integer. For any integer $\ell \geq 1$ the action of $\Gamma_{2\varepsilon-1}$ on $(\mathrm{Lie}(\mathbb{G})^{\otimes m})^{\mathrm{rig}}$ extends to $(\mathrm{Lie}(\mathbb{G})^{\otimes m})^{\mathrm{rig}}_{\ell}$ and makes $(\mathrm{Lie}(\mathbb{G})^{\otimes m})^{\mathrm{rig}}_{\ell}$ a topological Banach module over the K-Banach algebra $\Lambda(\Gamma_{2\varepsilon-1})_{K,\ell}$. The action of Γ on $(\mathrm{Lie}(\mathbb{G})^{\otimes m})^{\mathrm{rig}}$ extends to a jointly continuous action of the K-Fréchet algebra $D(\Gamma)$. The action of Γ on the strong continuous K-linear dual $[(\mathrm{Lie}(\mathbb{G})^{\otimes m})^{\mathrm{rig}}]'_b$ of $(\mathrm{Lie}(\mathbb{G})^{\otimes m})^{\mathrm{rig}}$ is locally analytic.*

Proof. Set $M^m_\ell := (\mathrm{Lie}(\mathbb{G})^{\otimes m})^{\mathrm{rig}}_\ell$. Any R-basis $(\delta_1, \ldots, \delta_s)$ of $\mathrm{Lie}(\mathbb{G})^{\otimes m}$ can be viewed as an R^{rig}_ℓ-basis of M^m_ℓ. Writing $M^m_\ell = \oplus_{i=1}^s R^{\mathrm{rig}}_\ell \delta_i$, the topology of M^m_ℓ is defined by the norm

$$||\sum_{i=1}^{s} f_i\delta_i||_\ell = \sup_i\{||f_i||_\ell\} \quad \text{if} \quad f_1, \ldots, f_s \in R^{\mathrm{rig}}_\ell.$$

We choose an ordered basis $(\gamma_1, \ldots, \gamma_r)$ of $\Gamma_{2\varepsilon-1}$ and let $b_i := \gamma_i - 1$ be as before. By induction on $|\alpha|$ we will first prove the fundamental estimate $||b^\alpha\delta||_\ell \leq ||b^\alpha||_\ell||\delta||_\ell$ for all $\alpha \in \mathbb{N}^r$ and $\delta \in (\mathrm{Lie}(\mathbb{G})^{\otimes m})^{\mathrm{rig}}$. As in the proof of Theorem 6 this is reduced to the case $|\alpha| = 1$, i.e. $b^\alpha = \gamma_i - 1$ for some $1 \leq i \leq r$. Further, we may assume $\delta = f\delta_j$ for some $f \in R^{\mathrm{rig}}$ and $1 \leq j \leq s$.

There are elements $r_1, \ldots, r_s \in R$ such that $\gamma_i(\delta_j) = \sum_{\nu=1}^s r_\nu\delta_\nu$. According to Theorem 5 we have $(\gamma_i - 1)(\delta_j) \in \mathfrak{m}^\varepsilon\, \mathrm{Lie}(\mathbb{G})^{\otimes m}$, i.e. $r_j - 1 \in \mathfrak{m}^\varepsilon$ and $r_\nu \in \mathfrak{m}^\varepsilon$ for $\nu \neq j$. We claim that $||r||_\ell \leq p^{-c/\ell}$ for any integer $c \geq 0$ and any element $r \in \mathfrak{m}^c$. Indeed, this is clear for $c = 0$. For general c, the ideal $\mathfrak{m}^c$ of R is generated by all elements of the form $p^a u^\beta$ with $a \in \mathbb{N}$, $\beta \in \mathbb{N}^{(h-d)d}$ and $a + |\beta| = c$. Since $\ell \geq 1$ we have $|p^a| = p^{-a} \leq p^{-a/\ell}$, and the claim follows from the multiplicativity of the norm $||\cdot||_\ell$ on R. Now

$$\begin{aligned} ||(\gamma_i - 1)(f\delta_j)||_\ell &\leq \max\{||(\gamma_i - 1)(f)\cdot\gamma_i(\delta_j)||_\ell, ||f\cdot(\gamma_i - 1)(\delta_j)||_\ell\} \\ &= \max\{||\sum_\nu(\gamma_i - 1)(f)r_\nu\delta_\nu||_\ell, ||f||_\ell||\delta_j - \sum_\nu r_\nu\delta_\nu||_\ell\}, \end{aligned}$$

where $||(\gamma_i-1)(f)\cdot r_\nu||_\ell \leq ||(\gamma_i-1)(f)||_\ell \leq p^{(2\varepsilon-1)/\ell}||f||_\ell$ by Proposition 2. Here $p^{(2\varepsilon-1)/\ell} \leq p^{-\varepsilon/\ell} = ||\gamma_i - 1||_\ell$. Moreover, $||r_j - 1||_\ell \leq p^{-\varepsilon/\ell}$ and $||r_\nu||_\ell \leq p^{-\varepsilon/\ell}$ if $\nu \neq j$ by the above claim. This finishes the proof of the fundamental estimate.

As an immediate consequence, we obtain that $||\lambda\cdot\delta||_\ell \leq ||\lambda||_\ell||\delta||_\ell$ for any $\lambda \in \Lambda(\Gamma_{2\varepsilon-1})_K$ and any $\delta \in \mathrm{Lie}(\mathbb{G})^{\otimes m} \otimes_W K$. The proof proceeds now as in Theorem 6.

According to Schneider and Teitelbaum (2003), Theorem 4.10, the projective system $(\Lambda(\Gamma_{2\varepsilon-1})_{K,\ell})_\ell$ of K-Banach algebras endow their projective limit $D(\Gamma_{2\varepsilon-1})$ with the structure of a *K-Fréchet-Stein algebra*. In the terminology of Schneider and Teitelbaum (2003), section 8, the family $((\mathrm{Lie}(\mathbb{G})^{\otimes m})^{\mathrm{rig}}_{\ell})_{\ell}$ is a *sheaf over* $(D(\Gamma_{2\varepsilon-1}), (\|\cdot\|_{\ell})_{\ell})$ with global sections $(\mathrm{Lie}(\mathbb{G})^{\otimes m})^{\mathrm{rig}}$ for any integer m. One of the main open questions in this setting is whether this sheaf is *coherent*, i.e. whether the $\Lambda(\Gamma_{2\varepsilon-1})_{K,\ell}$-modules $(\mathrm{Lie}(\mathbb{G})^{\otimes m})^{\mathrm{rig}}_{\ell}$ are finitely generated and whether the natural maps

$$\Lambda(\Gamma_{2\varepsilon-1})_{K,\ell} \otimes_{\Lambda(\Gamma_{2\varepsilon-1})_{K,\ell+1}} (\mathrm{Lie}(\mathbb{G})^{\otimes m})^{\mathrm{rig}}_{\ell+1} \longrightarrow (\mathrm{Lie}(\mathbb{G})^{\otimes m})^{\mathrm{rig}}_{\ell}$$

are always bijective. This would amount to the *admissibility* of the locally analytic Γ-representation $[(\mathrm{Lie}(\mathbb{G})^{\otimes m})^{\mathrm{rig}}]'_b$ in the sense of Schneider and Teitelbaum (2003), section 6. Nothing in this direction is known. In the next section, however, we will have a closer look at the case $\dim(G) = 1$ and $\ell = 1$. We will see that in order to obtain finitely generated objects, one might be forced to introduce yet another type of Banach algebras.

5 Non-commutative Divided Power Envelopes

In this final section we assume that our fixed p-divisible formal group G over the algebraically closed field k of characteristic p is of dimension one. If h denotes the height of G then the endomorphism ring of G is isomorphic to the maximal order $\mathfrak{o}_D$ of the central $\mathbb{Q}_p$-division algebra D of invariant $\frac{1}{h} + \mathbb{Z}$ (cf. Gross and Hopkins 1994, Proposition 13.10). In the following we will identify $\mathrm{End}(G)$ and $\mathfrak{o}_D$ (resp. $\mathrm{Aut}(G)$ and $\mathfrak{o}_D^*$). We will also exclude the trivial case $h = 1$. We continue to denote by $R = R_G^{\mathrm{def}}$ the universal deformation ring of G (cf. Theorem 3).

Consider the *period morphism* $\Phi : \mathrm{Spf}(R)^{\mathrm{rig}} \to \mathbb{P}_K^{h-1}$ of Gross and Hopkins, where $\mathbb{P}_K^{h-1}$ denotes the rigid analytic projective space of dimension $h-1$ over K (cf. Gross and Hopkins 1994, section 23). In projective coordinates Φ can be defined by $\Phi(x) = [\varphi_0(x) : \ldots : \varphi_{h-1}(x)]$ where $\varphi_0, \ldots, \varphi_{h-1} \in R^{\mathrm{rig}}$ are certain global rigid analytic functions on $\mathrm{Spf}(R)^{\mathrm{rig}}$ without any common zero. The power series expansions of the functions φ_i in suitable coordinates $u_1, \ldots, u_{h-1}$ can be written down explicitly by means of a closed formula of Yu (cf. Kohlhaase 2013, Proposition 1.5 and Remark 1.6). According to Gross and Hopkins (1994), Lemma 23.14, the function φ_0 does not have any zeroes on $\mathbb{B}_1^{h-1} \subset \mathrm{Spf}(R)^{\mathrm{rig}}$, hence is a unit in R_1^{rig}. We set

$$w_i := \frac{\varphi_i}{\varphi_0} \in R_1^{\mathrm{rig}} \quad \text{for} \quad 1 \le i \le h-1.$$

By Gross and Hopkins (1994), Lemma 23.14, any element $f \in R_1^{\mathrm{rig}}$ admits a unique expansion of the form $f = \sum_{\alpha \in \mathbb{N}^{h-1}} d_\alpha w^\alpha$ with $d_\alpha \in K$ and $\lim_{|\alpha| \to \infty} |d_\alpha| p^{-|\alpha|} = 0$. Further, Φ restricts to an isomorphism $\Phi : \mathbb{B}_1^{h-1} \to \Phi(\mathbb{B}_1^{h-1})$ (cf. Gross and Hopkins 1994, Corollary 23.15).

Denote by $\mathbb{Q}_{p^h}$ the unramified extension of degree h of $\mathbb{Q}_p$ and by $\mathbb{Z}_{p^h}$ its valuation ring. It was shown by Devinatz, Gross and Hopkins, that there exists an explicit closed embedding $\mathfrak{o}_D^* \hookrightarrow \mathrm{GL}_h(\mathbb{Q}_{p^h})$ of Lie groups over $\mathbb{Q}_p$ such that Φ is $\mathfrak{o}_D^*$-equivariant (cf. Kohlhaase 2013, Proposition 1.3 and Remark 1.4). Here $\mathfrak{o}_D^*$ acts on $\mathrm{Spf}(R)^{\mathrm{rig}}$ through the identification $\mathfrak{o}_D^* \simeq \mathrm{Aut}(G)$, and it acts by fractional linear transformations on $\mathbb{P}_K^{h-1}$ via the embedding $\mathfrak{o}_D^* \hookrightarrow \mathrm{GL}_h(\mathbb{Q}_{p^h})$.

The morphism Φ is constructed in such a way that $\Phi^*\mathcal{O}_{\mathbb{P}_K^{h-1}}(1) = \mathrm{Lie}(\mathbb{G})^{\mathrm{rig}}$. It follows from general properties of the inverse image functor that $\Phi^*\mathcal{O}_{\mathbb{P}_K^{h-1}}(m) = (\mathrm{Lie}(\mathbb{G})^{\otimes m})^{\mathrm{rig}}$ for any integer m. Restricting to $\mathbb{B}_1^{h-1}$, we obtain an $\mathfrak{o}_D^*$-equivariant and R_1^{rig}-linear isomorphism $(\mathrm{Lie}(\mathbb{G})^{\otimes m})_1^{\mathrm{rig}} \simeq R_1^{\mathrm{rig}} \cdot \varphi_0^m$ of free R_1^{rig}-modules of rank one.

We denote by $\mathfrak{d}$ the Lie algebra of the Lie group $\mathfrak{o}_D^*$ over $\mathbb{Q}_p$. It is isomorphic to the Lie algebra associated with the associative $\mathbb{Q}_p$-algebra D. According to Schneider and Teitelbaum (2002), page 450, the universal enveloping algebra $U_K(\mathfrak{d}) := U(\mathfrak{d} \otimes_{\mathbb{Q}_p} K)$ of $\mathfrak{d}$ over K embeds into the locally analytic distribution algebra $D(\Gamma_{2\varepsilon-1})$. Together with the natural map $D(\Gamma_{2\varepsilon-1}) \to \Lambda(\Gamma_{2\varepsilon-1})_{K,1}$, Theorem 7 allows us to view

$$M_1^m := (\mathrm{Lie}(\mathbb{G})^{\otimes m})_1^{\mathrm{rig}}$$

as a module over $U_K(\mathfrak{d}) \simeq U(\mathfrak{g} \otimes_{\mathbb{Q}_{p^h}} K) =: U_K(\mathfrak{g})$, where $\mathfrak{g} := \mathfrak{d} \otimes_{\mathbb{Q}_p} \mathbb{Q}_{p^h} \simeq \mathfrak{gl}_h$ as Lie algebras over $\mathbb{Q}_{p^h}$. Explicitly, the action of an element $\mathfrak{x} \in \mathfrak{g}$ on M_1^m is given by

$$\mathfrak{x}(\delta) = \frac{d}{dt}(\exp(t\mathfrak{x})(\delta))|_{t=0}.$$

Here $\exp : \mathfrak{g} \dashrightarrow \mathrm{GL}_h(\mathbb{Q}_{p^h})$ is the usual exponential map which is defined locally around zero in $\mathfrak{g}$. Further, a sufficiently small open subgroup of $\mathrm{GL}_h(\mathbb{Q}_{p^h})$ acts on M_1^m through the isomorphism $M_1^m \simeq \mathcal{O}_{\mathbb{P}_K^{h-1}}(m)(\Phi(\mathbb{B}_1^{h-1}))$. Writing an element $\mathfrak{x} \in \mathfrak{g}$ as a matrix $\mathfrak{x} = (a_{rs})_{0 \le r,s \le h-1}$ with coefficients $a_{rs} \in \mathbb{Q}_{p^h}$, fix indices $0 \le i,j \le h-1$ and denote by $\mathfrak{x}_{ij}$ the matrix with entry 1 at the place (i,j) and zero everywhere else. In the following we will formally put $w_0 := 1$.

Lemma 3. *Let i, j and m be integers with $0 \le i,j \le h-1$. If $f \in R_1^{\mathrm{rig}}$ then*

$$\mathfrak{x}_{ij}(f\varphi_0^m) = \begin{cases} w_i \frac{\partial f}{\partial w_j}\varphi_0^m, & \text{if } j \neq 0, \\ (mf - \sum_{\ell=1}^{h-1} w_\ell \frac{\partial f}{\partial w_\ell})\varphi_0^m, & \text{if } i = j = 0, \\ w_i(mf - \sum_{\ell=1}^{h-1} w_\ell \frac{\partial f}{\partial w_\ell})\varphi_0^m, & \text{if } i > j = 0. \end{cases}$$

Proof. If $i = j$ and if t is sufficiently close to zero in $\mathbb{Q}_{p^h}$ then $\exp(t\mathfrak{x}_{ii})$ is the diagonal matrix with entry $\exp(t)$ at the place (i,i) and 1 everywhere else on the diagonal. Recall that $\mathrm{GL}_h(\mathbb{Q}_{p^h})$ acts by fractional linear transformations on the projective coordinates $\varphi_0, \ldots, \varphi_{h-1}$ of $\mathbb{P}_K^{h-1}$. Thus, $\exp(t\mathfrak{x}_{ii})(w_\ell) = w_\ell$ if $\ell \neq i \neq 0$, $\exp(t\mathfrak{x}_{ii})(w_i) = \exp(t)w_i$ if $i \neq 0$, and $\exp(t\mathfrak{x}_{00})(w_\ell) = \frac{1}{\exp(t)}w_\ell$ for all $1 \leq \ell \leq h-1$.

If $i \neq j$ then $\exp(t\mathfrak{x}_{ij}) = 1 + t\mathfrak{x}_{ij}$ in $\mathrm{GL}_h(\mathbb{Q}_{p^h})$. Thus, $\exp(t\mathfrak{x}_{ij})(w_\ell) = w_\ell$ if $\ell \neq j \neq 0$, $\exp(t\mathfrak{x}_{ij})(w_j) = w_j + tw_i$ if $j \neq 0$, and $\exp(t\mathfrak{x}_{i0})(w_\ell) = w_\ell/(1+tw_i)$ for all $1 \leq \ell \leq h-1$. Writing $f = f(w_1, \ldots, w_{h-1})$ we have

$$\exp(t\mathfrak{x}_{ij})(f\varphi_0^m) = f(\exp(t\mathfrak{x}_{ij})(w_1), \ldots, \exp(t\mathfrak{x}_{ij})(w_{h-1})) \cdot \exp(t\mathfrak{x}_{ij})(\varphi_0)^m.$$

Here $\exp(t\mathfrak{x}_{ij})(\varphi_0) = \varphi_0$ if $j \neq 0$, $\exp(t\mathfrak{x}_{00})(\varphi_0) = \exp(t)\varphi_0$ and $\exp(t\mathfrak{x}_{i0})(\varphi_0) = \varphi_0 + t\varphi_i$ if $1 \leq i \leq h-1$. It is now an exercise in elementary calculus to derive the desired formulae.

Note that $(\mathrm{Lie}(\mathbb{G})^{\otimes m})^{\mathrm{rig}}$ is a $D(\Gamma_{2\varepsilon-1})$-stable K-subspace of M_1^m and hence is $\mathfrak{g}$-stable. If $m = 0$ then Lemma 3 shows that in order to describe the $\mathfrak{g}$-action in the coordinates $u_1, \ldots, u_{h-1}$, one essentially has to compute the functional matrix

$$F := (\frac{\partial u_i}{\partial w_j})_{1 \leq i,j \leq h-1}.$$

Proposition 3. *The matrix* $A := (\frac{\partial \varphi_i}{\partial u_j}\varphi_0 - \frac{\partial \varphi_0}{\partial u_j}\varphi_i)_{1 \leq i,j \leq h-1}$ *over* R^{rig} *is invertible over the localization* $R^{\mathrm{rig}}_{\varphi_0}$. *We have* $F = \varphi_0^2 A^{-1}$, *which is a matrix with entries in* $\varphi_0 R^{\mathrm{rig}}$. *Moreover, we have* $\sum_{j=1}^{h-1} \varphi_j \frac{\partial u_i}{\partial w_j} \in \varphi_0^2 R^{rig}$ *for any index* $1 \leq i \leq h-1$.

Proof. Let $B := (\frac{\partial \varphi_i}{\partial u_j})_{0 \leq i,j \leq h-1}$ with $\frac{\partial \varphi_i}{\partial u_0} := \varphi_i$. We have $B \in \mathrm{GL}_h(R^{\mathrm{rig}})$ by a result of Gross and Hopkins (1994, Corollary 21.17). Setting

$$N := \begin{pmatrix} 1 & 0 & \cdots & 0 \\ -\varphi_1 & \varphi_0 & & 0 \\ \vdots & & \ddots & \\ -\varphi_{h-1} & 0 & & \varphi_0 \end{pmatrix}, \text{ we have } NB = \begin{pmatrix} \varphi_0 & \frac{\partial \varphi_0}{\partial u_1} & \cdots & \frac{\partial \varphi_0}{\partial u_{h-1}} \\ 0 & & & \\ \vdots & & A & \\ 0 & & & \end{pmatrix}.$$

This already shows that A is invertible over $R^{\mathrm{rig}}_{\varphi_0}$. Denoting by $c_0, \ldots, c_{h-1}$ the columns of $B^{-1} = (c_{ij})_{i,j} \in \mathrm{GL}_h(R^{\mathrm{rig}})$, we obtain

$$(\varphi_0^{-1} \sum_{j=0}^{h-1} \varphi_j c_j, \varphi_0^{-1} c_1, \ldots, \varphi_0^{-1} c_{h-1}) = B^{-1}N^{-1} = \begin{pmatrix} \varphi_0^{-1} & * & \cdots & * \\ 0 & & & \\ \vdots & & A^{-1} & \\ 0 & & & \end{pmatrix}.$$

By the chain rain rule we have

$$\delta_{ij} = \frac{\partial w_i}{\partial w_j} = \sum_{\ell=1}^{h-1} \frac{\partial w_i}{\partial u_\ell} \cdot \frac{\partial u_\ell}{\partial w_j} = \sum_{\ell=1}^{h-1} \varphi_0^{-2} (\frac{\partial \varphi_i}{\partial u_\ell} \varphi_0 - \frac{\partial \varphi_0}{\partial u_\ell} \varphi_i) \frac{\partial u_\ell}{\partial w_j},$$

so that $F = \varphi_0^2 A^{-1}$. As seen above, the right hand side has entries in $\varphi_0 R^{\mathrm{rig}}$. Further, we have $\sum_{j=1}^{h-1} \varphi_j \frac{\partial u_i}{\partial w_j} = \sum_{j=1}^{h-1} \varphi_j \varphi_0 c_{ij} = -\varphi_0^2 c_{i0} \in \varphi_0^2 R^{\mathrm{rig}}$ for any index $1 \leq i \leq h-1$.

Together with Lemma 3, Proposition 3 shows that $\mathfrak{x}(u_i) \in R^{\mathrm{rig}}$ for any $\mathfrak{x} \in \mathfrak{g}$ and any $1 \leq i \leq h-1$, as was clear a priori. For $h = 2$, Lemma 3 and Proposition 3 reprove Gross and Hopkins (1994), formula (25.14).

Coming back to the $\mathfrak{g}$-module M_1^m for general m, consider the subalgebra $\mathfrak{sl}_h$ of $\mathfrak{g}$ over $\mathbb{Q}_{p^h}$. Let $\mathfrak{t}$ denote the Cartan subalgebra of diagonal matrices in $\mathfrak{sl}_h$, and let $\{\varepsilon_1, \ldots, \varepsilon_{h-1}\} \subset \mathfrak{t}^*$ denote the basis of the root system of $(\mathfrak{sl}_h, \mathfrak{t})$ given by $\varepsilon_i(\mathrm{diag}(t_0, \ldots, t_{h-1})) := t_{i-1} - t_i$. We let $\lambda_1 \in \mathfrak{t}^*$ denote the fundamental dominant weight defined by $\lambda_1 := \frac{1}{h} \sum_{i=1}^{h-1} (h-i)\varepsilon_i$. We have

$$\lambda_1(\mathrm{diag}(t_0, \ldots, t_{h-1})) = \frac{1}{h} \sum_{i=1}^{h-1} (h-i)(t_{i-1} - t_i) = \frac{1}{h}((h-1)t_0 - \sum_{i=1}^{h-1} t_i) = t_0$$

for any element $\mathrm{diag}(t_0, \ldots, t_{h-1}) \in \mathfrak{t} \subset \mathfrak{sl}_h$.

Proposition 4. *For any integer $m \geq 0$, the subspace $W := \sum_{|\alpha| \leq m} K \cdot w^\alpha \varphi_0^m$ of M_1^m is $\mathfrak{g}$-stable. The action of $\mathfrak{sl}_h$ on W is irreducible. More precisely, W is the irreducible $\mathfrak{sl}_h$-representation of highest weight $m \cdot \lambda_1$.*

Proof. It follows from Lemma 3 that W is stable under any element $\mathfrak{x}_{ij}$ with $j \neq 0$ or $i = j = 0$. If $1 \leq i \leq h-1$ and if n is a non-negative integer then

$$\mathfrak{x}_{i0}^n(w^\alpha \varphi_0^m) = [\prod_{\ell=0}^{n-1} (m - |\alpha| - \ell)] \cdot w^\alpha w_i^n \varphi_0^m,$$

as follows from Lemma 3 by induction. Therefore, $\mathfrak{x}_{i0}(w^\alpha \varphi_0^m) = 0$ if $|\alpha| = m$. If $|\alpha| < m$ then $\mathfrak{x}_{i0}(w^\alpha)$ has degree $|\alpha| + 1 \leq m$. This proves that W is $\mathfrak{g}$-stable.

The above formula also shows that W is generated by φ_0^m as an $\mathfrak{sl}_h$-representation. If $f\varphi_0^m \in W$ is non-zero, then Lemma 3 shows that $(\mathfrak{x}_{01}^{\alpha_1} \cdots \mathfrak{x}_{0(h-1)}^{\alpha_{h-1}})(f\varphi_0^m)$ is a non-zero scalar multiple of φ_0^m for a suitable multi-index α. Therefore, the $\mathfrak{sl}_h$-representation W is irreducible.

Finally, if $\mathfrak{x} = \mathrm{diag}(t_0, \ldots, t_{h-1}) \in \mathfrak{t}$ then $\mathfrak{x}(w^\alpha \varphi_0^m) = (t_0(m - |\alpha|) + \sum_{i=1}^{h-1} \alpha_i t_i) \cdot w^\alpha \varphi_0^m$ by Lemma 3. Here,

$$t_0(m-|\alpha|)+\sum_{i=1}^{h-1}\alpha_i t_i = t_0 m + \sum_{i=1}^{h-1}\alpha_i(t_i-t_0) = (m\cdot\lambda_1 - \sum_{i=1}^{h-1}\alpha_i\sum_{\ell=1}^{i}\varepsilon_\ell)(\mathfrak{x}).$$

This shows that $m\cdot\lambda_1$ is the highest weight of the $\mathfrak{sl}_h$-representation W.

Remark 2. The statement of Proposition 4 can be deduced from a stronger result of Gross and Hopkins. Namely, if $m=1$ then $\mathrm{Lie}(\mathbb{G})^{\mathrm{rig}}$ contains an h-dimensional algebraic representation of $\mathfrak{o}_D^*$ (cf. Gross and Hopkins 1994, Proposition 23.2). Under the restriction map $\mathrm{Lie}(\mathbb{G})^{\mathrm{rig}} \to \mathrm{Lie}(\mathbb{G})_1^{\mathrm{rig}}$, the derived representation of $\mathfrak{g}=\mathfrak{d}\otimes_{\mathbb{Q}_p}\mathbb{Q}_{p^h}$ maps isomorphically to the $\mathfrak{g}$-representation W above.

We will now see that the action of $\mathfrak{g}$ on M_1^m naturally extends to a certain divided power completion of the universal enveloping algebra $U_K(\mathfrak{g})$. Note that if i,j,r and s are indices between 0 and $h-1$, then $\mathfrak{x}_{ij}\cdot\mathfrak{x}_{rs}=\delta_{jr}\mathfrak{x}_{is}$ in $\mathfrak{g}\simeq\mathfrak{gl}_h$. Therefore,

$$[\mathfrak{x}_{ij},\mathfrak{x}_{rs}] = \delta_{jr}\mathfrak{x}_{is}-\delta_{is}\mathfrak{x}_{rj} = \begin{cases} 0, & \text{if } j\neq r \text{ and } i\neq s,\\ \mathfrak{x}_{is}, & \text{if } j=r \text{ and } i\neq s,\\ -\mathfrak{x}_{rj}, & \text{if } j\neq r \text{ and } i=s,\\ \mathfrak{x}_{ii}-\mathfrak{x}_{jj}, & \text{if } j=r \text{ and } i=s.\end{cases}$$

Setting $\mathfrak{x}'_{ij} := p^{\delta_{0i}-\delta_{0j}}\mathfrak{x}_{ij}$, one readily checks that the same relations hold on replacing $\mathfrak{x}_{ij}$ by $\mathfrak{x}'_{ij}$ and $\mathfrak{x}_{rs}$ by $\mathfrak{x}'_{rs}$ everywhere. It follows that the elements $\mathfrak{x}'_{ij}$ span a free $\mathbb{Z}_{p^h}$-Lie subalgebra of $\mathfrak{g}$ that we denote by $\mathring{\mathfrak{g}}$. Since $\mathrm{ad}(\mathfrak{x}'_{ij})^2=0$ if $i\neq j$, and since $(\varepsilon_{i+1}-\varepsilon_j)([\mathfrak{x}'_{ij},\mathfrak{x}'_{ij}])=2$ if $i<j$, it follows from Bourbaki (2006), VIII.12.7 Théorème 2 (iii), that the W-lattice $\mathring{\mathfrak{g}}$ of $\mathfrak{g}$ is the base extension from $\mathbb{Z}$ to W of a *Chevalley order* of $\mathfrak{g}$ in the sense of Bourbaki (2006), VIII.12.7 Définition 2.

For $0\leq i\leq h-1$ and $n\geq 0$ we set

$$\binom{\mathfrak{x}'_{ii}}{n} := \frac{\mathfrak{x}'_{ii}(\mathfrak{x}'_{ii}-1)\cdots(\mathfrak{x}'_{ii}-n+1)}{n!}\in U_K(\mathfrak{g}).$$

We let $\mathcal{U}$ denote the W-subalgebra of $U_K(\mathfrak{g})$ generated by the elements $(\mathfrak{x}'_{ij})^n/n!$ for $i\neq j$ and $n\geq 0$, as well as by the elements $\binom{\mathfrak{x}'_{ii}}{n}$ for $0\leq i\leq h-1$ and $n\geq 0$. It follows from Bourbaki (2006), VIII.7.12 Théorème 3, that $\mathcal{U}$ is a free W-module and that a W-basis of $\mathcal{U}$ is given by the elements

$$b_{\ell mn} := (\prod_{i<j}\frac{(\mathfrak{x}'_{ij})^{\ell_{ij}}}{\ell_{ij}!})\cdot(\prod_{i=0}^{h-1}\binom{\mathfrak{x}'_{ii}}{m_i})\cdot(\prod_{i>j}\frac{(\mathfrak{x}'_{ij})^{n_{ij}}}{n_{ij}!})$$

with $\ell=(\ell_{ij}), n=(n_{ij})\in\mathbb{N}^{h(h-1)/2}$ and $m=(m_i)\in\mathbb{N}^h$. Here the products of the $\mathfrak{x}'_{ij}$ for $i<j$ and $i>j$ have to be taken in a fixed but arbitrary ordering of the factors. For split semisimple Lie algebras these constructions and statements are due to Kostant (cf. Kostant 1966, Theorem 1, where $\mathcal{U}$ is denoted by B).

We denote by $\hat{\mathcal{U}}$ the p-adic completion of the ring $\mathcal{U}$ and set

$$\hat{U}_K^{\mathrm{dp}}(\mathring{\mathfrak{g}}) := \hat{\mathcal{U}} \otimes_W K.$$

According to the above freeness result, any element of $\hat{U}_K^{\mathrm{dp}}(\mathring{\mathfrak{g}})$ can be written uniquely in the form $\sum_{\ell,m,n} d_{\ell mn} b_{\ell mn}$ with coefficients $d_{\ell mn} \in K$ satisfying $d_{\ell mn} \to 0$ as $|\ell| + |m| + |n| \to \infty$. Therefore, $\hat{U}_K^{\mathrm{dp}}(\mathring{\mathfrak{g}})$ is a K-algebra containing $U_K(\mathfrak{g})$. We view it as a K-Banach algebra with unit ball $\hat{\mathcal{U}}$ and call it the *complete divided power enveloping algebra of* $\mathring{\mathfrak{g}}$.

Theorem 8. *For any integer m the action of $\mathfrak{g}$ on $(\mathrm{Lie}(\mathbb{G})^{\otimes m})_1^{\mathrm{rig}}$ extends to a continuous action of $\hat{U}_K^{\mathrm{dp}}(\mathring{\mathfrak{g}})$.*

Proof. The ring of continuous K-linear endomorphisms of $M_1^m = (\mathrm{Lie}(\mathbb{G})^{\otimes m})_1^{\mathrm{rig}}$ is a K-Banach algebra for the operator norm. Since the latter is submultiplicative, the set of endomorphisms with operator norm less than or equal to one is a p-adically separated and complete W-algebra. Therefore, it suffices to prove that any element of the form $(\mathfrak{x}'_{ij})^n/n!$, $i \neq j$, or $\binom{\mathfrak{x}'_{ii}}{n}$, $0 \leq i \leq h-1$, has operator norm less than or equal to one on M_1^m whenever $n \geq 0$. If $\alpha \in \mathbb{N}^{h-1}$ and $0 \leq i, j \leq h-1$ then

$$\mathfrak{x}_{ij}^n(w^\alpha \varphi_0^m) = \begin{cases} \alpha_j^n w^\alpha \varphi_0^m, & \text{if } i = j \neq 0, \\ (m - |\alpha|)^n w^\alpha \varphi_0^m, & \text{if } i = j = 0, \\ n!\binom{\alpha_j}{n} w^\alpha w_j^{-n} w_i^n \varphi_0^m, & \text{if } i \neq j \neq 0, \\ n!\binom{m-|\alpha|}{n} w^\alpha w_i^n \varphi_0^m, & \text{if } i \neq j = 0, \end{cases} \tag{2}$$

as follows from Lemma 3 by induction. Here the generalized binomial coefficients are defined by

$$\binom{x}{n} := \frac{x(x-1)\cdots(x-n+1)}{n!} \in \mathbb{Z}$$

for any integer x. Now $||(\sum_\alpha d_\alpha w^\alpha)\varphi_0^m||_1 = \sup_\alpha\{|d_\alpha| p^{-|\alpha|}\}$. Bearing in mind our convention $w_0 = 1$, we obtain the claim for $(\mathfrak{x}'_{ij})^n/n!$ if $i \neq j$. If $0 \leq i \leq h-1$ then we obtain

$$\binom{\mathfrak{x}'_{ii}}{n}(w^\alpha \varphi_0^m) = \begin{cases} \binom{\alpha_i}{n} w^\alpha \varphi_0^m, & \text{if } i \neq 0, \\ \binom{m-|\alpha|}{n} w^\alpha \varphi_0^m, & \text{if } i = 0. \end{cases}$$

This completes the proof.

Theorem 9. *Let m be an integer and set $c := w_1^{\max\{-1,m\}+1}\varphi_0^m$. The $U(\mathfrak{g})$-submodule $U(\mathfrak{g})\cdot c$ of $(\mathrm{Lie}(\mathbb{G})^{\otimes m})_1^{\mathrm{rig}}$ is dense. If $h = 2$ and $m \geq -1$ then $\hat{U}_K^{\mathrm{dp}}(\mathring{\mathfrak{g}})\cdot c = (\mathrm{Lie}(\mathbb{G})^{\otimes m})_1^{\mathrm{rig}}$.*

Proof. Equation (2) shows that $\mathfrak{x}_{01}^{\max\{-1,m\}+1}\mathfrak{x}_{10}^{\alpha_1}\cdots\mathfrak{x}_{(h-1)0}^{\alpha_{h-1}}\cdot c$ is a non-zero scalar multiple of $w^\alpha\varphi_0^m$. Thus, $K[w]\cdot\varphi_0^m \subset U_K(\mathfrak{g})\cdot c$, proving the first assertion.

If $h = 2$ and $m \geq -1$ let us be more precise. Setting $m' := \max\{-1,m\}+1$, $w := w_1$ and $\mathfrak{x} := \mathfrak{x}_{10}'$, we have $\mathfrak{x}^n\cdot c = (-1)^n n! p^{-n} w^{n+m'}\varphi_0^m$ for any $n \geq 0$ because $\binom{-1}{n} = (-1)^n$. If $f = \sum_{n\geq 0} d_n w^n \in R_1^{\mathrm{rig}}$ then $d_n p^n \to 0$ in K. Therefore, $\lambda := \sum_{n\geq 0} d_{n+m'}(-p)^n\frac{\mathfrak{x}^n}{n!}$ converges in $\hat{U}_K^{\mathrm{dp}}(\mathring{\mathfrak{g}})$ and we have $f\varphi_0^m - \lambda\cdot c = \sum_{n=0}^{m'-1} d_n w^n\varphi_0^m$. The latter is contained in $K[w]\cdot\varphi_0^m \subset U_K(\mathfrak{g})\cdot c$, as seen above.

Remark 3. By a result of Lazard, the image of $U_K(\mathfrak{g}) \simeq U_K(\mathfrak{d})$ in $\Lambda(\Gamma_{2\varepsilon-1})_{K,1}$ is dense (cf. Lazard 1965, Chapitre IV, Théorème 3.2.5). We state without proof that the completion of $U_K(\mathfrak{g})$ for the norm $||\cdot||_1$ embeds continuously into $\hat{U}_K^{\mathrm{dp}}(\mathring{\mathfrak{g}})$. However, a formal series like $\sum_{n\geq 0} p^n\frac{(\mathfrak{x}_{10}')^n}{n!} = \sum_{n\geq 0}\frac{\mathfrak{x}_{10}^n}{n!}$ does not converge in $\Lambda(\Gamma_{2\varepsilon-1})_{K,1}$. Therefore, one might have doubts whether M_1^m is still finitely generated over $\Lambda(\Gamma_{2\varepsilon-1})_{K,1}$.

References

Bosch, S., Güntzer, U., Remmert, R.: Non-Archimedean Analysis. Grundlehren der mathematischen Wissenschaften, vol. 261. Springer, Berlin/New York (1984)

Bourbaki, N.: Groupes et Algèbres de Lie, Chapitres 7 et 8, Springer, Berlin (2006)

Bourbaki, N.: Algèbre Commutative, Chapitres 1 à 4, Springer, Berlin (2006)

Brumer, A.: Pseudocompact algebras, profinite groups and class formations. J. Algebra **4**, 442–470 (1966)

Cartier, P.: Relèvements des groupes formels commutatifs. Séminaire Bourbaki **359**, 21e année (1968/1969)

de Jong, J.: Crystalline Dieudonné module theory via formal and rigid geometry. Publ. IHES **82**, 5–96 (1995)

Demazure, M.: Lectures on p-Divisible Groups. Lecture Notes in Mathematics, vol. 302. Springer, Berlin/Heidelberg (1986)

Devinatz, E.S., Hopkins, M.J.: The action of the Morava stabilizer group on the Lubin-Tate moduli space of lifts. Am. J. Math. **117**(3), 669–710 (1995)

Dixon, J.D., du Sautoy, M.P.F., Mann, A., Segal, D.: Analytic Pro-p Groups. Cambridge Studies in Advanced Mathematics, vol. 61, 2nd edn. Cambridge University Press, Cambridge/New York (2003)

Gross, B.H., Hopkins, M.J.: Equivariant vector bundles on the Lubin-Tate moduli space. Contemp. Math. **158**, 23–88 (1994)

Hazewinkel, M.: Formal Groups and Applications. Pure and Applied Mathematics, vol. 78. Academic, New York (1978)

Kohlhaase, J., Schraen, B.: Homological vanishing theorems for locally analytic representations. Math. Ann. **353**, 219–258 (2012)

Kohlhaase, J.: On the Iwasawa theory of the Lubin-Tate moduli space. Comput. Math. **149**, 793–839 (2013)

Kostant, B.: Groups over $\mathbb{Z}$. In: Algebraic Groups and Discontinuous Subgroups. Proceedings of Symposia in Pure Mathematics, Boulder, 1965, pp. 90–98. AMS, Providence (1966)
Lazard, M.: Groupes analytiques p-adiques. Publ. I.H.'E.S. **26**, 5–219 (1965)
Lazard, M.: Commutative Formal Groups. Lecture Notes in Mathematics, vol. 443. Springer, Berlin/New York (1975)
Lubin, J., Tate, J.: Formal moduli for one-parameter formal Lie groups, Bulletin de la S.M.F. tome **94**, 49–59 (1966)
Orlik, S.: Equivariant vector bundles on Drinfeld's upper half space. Invent. Math. **172**, 585–656 (2008)
Rapoport, M., Zink, Th.: Period Spaces for p-Divisible Groups. Annals of Mathematics Studies, vol. 141. Princeton University Press, Princeton (1996)
Schmidt, T.: Auslander regularity of p-adic distribution algebras. Represent. Theory **12**, 37–57 (2008)
Schneider, P.: Non-Archimedean Functional Analysis. Springer Monographs in Mathematics. Springer, Berlin/New York (2002)
Schneider, P., Teitelbaum, J.: Locally analytic distributions and p-adic representation theory, with applications to GL_2. J. AMS **15**, 443–468 (2002)
Schneider, P., Teitelbaum, J.: Algebras of p-adic distributions and admissible representations. Invent. Math. **153**, 145–196 (2003)
Schneider, P., Teitelbaum, J.: p-adic boundary values. In: Cohomologies p-adiques et applications aritheoremétiques I. Astérisque, vol. 278, pp. 51–125 (2002); Corrigendum in Cohomologies p-adiques et applications aritheoremétiques III. Astérisque, vol. 295, pp. 291–299 (2004)
Tate, J.: p-divisible Groups. In: Proceedings of Conference on Local Fields, Driebergen, 1966, pp. 158–183. Springer (1967)
Umemura, H.: Formal moduli for p-divisible formal groups. Nagoya Math. J. **42**, 1–7 (1977)
Zink, Th.: Cartiertheorie kommutativer formaler Gruppen. Teubner-Texte zur Mathematik, vol. 68. Teubner, Leipzig (1984)

The Structure of Selmer Groups of Elliptic Curves and Modular Symbols

Masato Kurihara

1 Introduction

1.1 Structure Theorem of Selmer Groups

Let E be an elliptic curve over $\mathbf{Q}$. Iwasawa theory, especially the main conjecture gives a formula on the order of the Tate Shafarevich group by using the p-adic L-function (cf. Schneider 1983). In this paper, as a sequel of Kurihara (2003b, 2012, 2014), we show that we can derive more information than the order, on the structure of the Selmer group and the Tate Shafarevich group from analytic quantities, in the setting of our paper, from modular symbols.

In this paper, we consider a prime number p such that

(i) p is a good ordinary prime > 2 for E,
(ii) The action of $G_{\mathbf{Q}}$ on the Tate module $T_p(E)$ is surjective where $G_{\mathbf{Q}}$ is the absolute Galois group of $\mathbf{Q}$,
(iii) The (algebraic) μ-invariant of $(E, \mathbf{Q}_\infty/\mathbf{Q})$ is zero where $\mathbf{Q}_\infty/\mathbf{Q}$ is the cyclotomic $\mathbf{Z}_p$-extension, namely the Selmer group $\mathrm{Sel}(E/\mathbf{Q}_\infty, E[p^\infty])$ (for the definition, see below) is a cofinitely generated $\mathbf{Z}_p$-module,
(iv) p does not divide the Tamagawa factor $\mathrm{Tam}(E) = \prod_{\ell:\mathrm{bad}}(E(\mathbf{Q}_\ell) : E^0(\mathbf{Q}_\ell))$, and p does not divide $\#E(\mathbf{F}_p)$ (namely not anomalous).

We note that the property (iii) is a conjecture of Greenberg since we are assuming (ii).

M. Kurihara (✉)
Department of Mathematics, Keio University, 3-14-1 Hiyoshi, Kohoku-ku, Yokohama, 223-8522, Japan
e-mail: kurihara@math.keio.ac.jp

T. Bouganis and O. Venjakob (eds.), *Iwasawa Theory 2012*, Contributions in Mathematical and Computational Sciences 7, DOI 10.1007/978-3-642-55245-8_11,

For a positive integer $N > 0$, we denote by $E[p^N]$ the Galois module of p^N-torsion points, and $E[p^\infty] = \bigcup_{N>0} E[p^N]$. For an algebraic extension $F/\mathbf{Q}$, $\mathrm{Sel}(E/F, E[p^N])$ is the classical Selmer group defined by

$$\mathrm{Sel}(E/F, E[p^N]) = \mathrm{Ker}(H^1(F, E[p^N]) \longrightarrow \prod_v H^1(F_v, E[p^N])/E(F_v) \otimes \mathbf{Z}/p^N),$$

so $\mathrm{Sel}(E/F, E[p^N])$ sits in an exact sequence

$$0 \longrightarrow E(F) \otimes \mathbf{Z}/p^N \longrightarrow \mathrm{Sel}(E/F, E[p^N]) \longrightarrow Ш(E/F)[p^N] \longrightarrow 0$$

where $Ш(E/F)$ is the Tate Shafarevich group over F. We define $\mathrm{Sel}(E/F, E[p^\infty]) = \varinjlim \mathrm{Sel}(E/F, E[p^N])$.

Let $\mathcal{P}^{(N)}$ be the set of prime numbers ℓ such that ℓ is a good reduction prime for E and $\ell \equiv 1 \pmod{p^N}$. For each ℓ, we fix a generator η_ℓ of $(\mathbf{Z}/\ell\mathbf{Z})^\times$ and define $\log_{\mathbf{F}_\ell}(a) \in \mathbf{Z}/(\ell-1)$ by $\eta_\ell^{\log_{\mathbf{F}_\ell}(a)} \equiv a \pmod{\ell}$.

Let $f(z) = \Sigma a_n e^{2\pi i n z}$ be the modular form corresponding to E. For a positive integer m and the cyclotomic field $\mathbf{Q}(\mu_m)$, we denote by $\sigma_a \in \mathrm{Gal}(\mathbf{Q}(\mu_m)/\mathbf{Q})$ the element such that $\sigma_a(\zeta) = \zeta^a$ for any $\zeta \in \mu_m$. We consider the modular element $\sum_{a=1,(a,m)=1}^m [\frac{a}{m}]\sigma_a \in \mathbf{C}[\mathrm{Gal}(\mathbf{Q}(\mu_m)/\mathbf{Q})]$ of Mazur and Tate (1987) where $[\frac{a}{m}] = 2\pi i \int_\infty^{a/m} f(z)dz$ is the usual modular symbol. We only consider the real part

$$\tilde{\theta}_{\mathbf{Q}(\mu_m)} = \sum_{\substack{a=1 \\ (a,m)=1}}^{m} \frac{\mathrm{Re}([\frac{a}{m}])}{\Omega_E^+} \sigma_a \in \mathbf{Q}[\mathrm{Gal}(\mathbf{Q}(\mu_m)/\mathbf{Q})] \tag{1}$$

where $\Omega_E^+ = \int_{E(\mathbf{R})} \omega_E$ is the Néron period. Suppose that m is a squarefree product of primes in $\mathcal{P}^{(N)}$. Since we are assuming the $G_{\mathbf{Q}}$-module $E[p]$ of p-torsion points is irreducible, we know $\tilde{\theta}_{\mathbf{Q}(\mu_m)} \in \mathbf{Z}_p[\mathrm{Gal}(\mathbf{Q}(\mu_m)/\mathbf{Q})]$ (cf. Stevens 1989). We consider the coefficient of $\tilde{\theta}_{\mathbf{Q}(\mu_m)}$ of "$\prod_{\ell|m}(\sigma_{\eta_\ell} - 1)$", more explicitly we define

$$\tilde{\delta}_m = \sum_{\substack{a=1 \\ (a,m)=1}}^{m} \frac{\mathrm{Re}([\frac{a}{m}])}{\Omega_E^+} (\prod_{\ell|m} \log_{\mathbf{F}_\ell}(a)) \in \mathbf{Z}/p^N \tag{2}$$

where $\log_{\mathbf{F}_\ell}(a)$ means the image of $\log_{\mathbf{F}_\ell}(a)$ under the canonical homomorphism $\mathbf{Z}/(\ell-1) \longrightarrow \mathbf{Z}/p^N$. Let $\mathrm{ord}_p : \mathbf{Z}/p^N \longrightarrow \{0, 1, \ldots, N-1, \infty\}$ be the p-adic valuation normalized as $\mathrm{ord}_p(p) = 1$ and $\mathrm{ord}_p(0) = \infty$. We note that $\mathrm{ord}_p(\tilde{\delta}_m)$ does not depend on the choices of η_ℓ for $\ell|m$. We define $\tilde{\delta}_1 = \theta_{\mathbf{Q}} = \mathrm{Re}([0])/\Omega_E^+ = L(E,1)/\Omega_E^+$.

For a squarefree product m of primes, we define $\epsilon(m)$ to be the number of prime divisors of m, namely $\epsilon(m) = r$ if $m = \ell_1 \cdot \ldots \cdot \ell_r$. Let $\mathcal{N}^{(N)}$ be the set of squarefree products of primes in $\mathcal{P}^{(N)}$. We suppose 1 is in $\mathcal{N}^{(N)}$. For each integer $i \geq 0$, we define the ideal $\Theta_i(\mathbf{Q})^{(N,\delta)}$ of $\mathbf{Z}/p^N$ to be the ideal generated by all $\tilde{\delta}_m$ such that $\epsilon(m) \leq i$ for all $m \in \mathcal{N}^{(N)}$;

$$\Theta_i(\mathbf{Q})^{(N,\delta)} = (\{\tilde{\delta}_m \mid \epsilon(m) \leq i \text{ and } m \in \mathcal{N}^{(N)}\}) \subset \mathbf{Z}/p^N. \tag{3}$$

We define $n_{i,N} \in \{0, 1, \ldots, N-1, \infty\}$ by $\Theta_i(\mathbf{Q})^{(N,\delta)} = p^{n_{i,N}}(\mathbf{Z}/p^N)$ (we define $n_{i,N} = \infty$ if $\Theta_i(\mathbf{Q})^{(N,\delta)} = 0$).

Theorem 1 (Kurihara 2014, Theorem B, Theorem 10.8 and (53)). *We assume that the main conjecture for $(E, \mathbf{Q}_\infty/\mathbf{Q})$ (see (9)) and the p-adic height pairing is non-degenerate.*

(1) *$n_{i,N}$ does not depend on N when N is sufficiently large (for example, when $N > 2\,\mathrm{ord}_p(\eta_0)$ where η_0 is the leading term of the p-adic L-function, see §10.7 in Kurihara (2014)). We put $n_i = n_{i,N}$ for $N \gg 0$. In other words, we define n_i by*

$$\varprojlim \Theta_i(\mathbf{Q})^{(N,\delta)} = p^{n_i}\mathbf{Z}_p \subset \mathbf{Z}_p.$$

We denote this ideal of $\mathbf{Z}_p$ by $\Theta_i(\mathbf{Q})^{(\delta)}$.

(2) *Consider the Pontrjagin dual* $\mathrm{Sel}(E/\mathbf{Q}, E[p^\infty])^\vee$ *of the Selmer group. Suppose that*

$$\mathrm{rank}_{\mathbf{Z}_p} \mathrm{Sel}(E/\mathbf{Q}, E[p^\infty])^\vee = r (\in \mathbf{Z}_{\geq 0}), \text{ and } \dim_{\mathbf{F}_p} \mathrm{Sel}(E/\mathbf{Q}, E[p])^\vee = a.$$

Then we have

$$\Theta_0(\mathbf{Q})^{(\delta)} = \ldots = \Theta_{r-1}(\mathbf{Q})^{(\delta)} = 0 \text{ and } \Theta_r(\mathbf{Q})^{(\delta)} \neq 0.$$

For any $i \geq r$, n_i is an even number, and

$$p^{n_r} = \#(\mathrm{Sel}(E/\mathbf{Q}, E[p^\infty])^\vee)_{\mathrm{tors}},$$

$$n_a = 0, \text{ and}$$

$$\mathrm{Sel}(E/\mathbf{Q}, E[p^\infty])^\vee \simeq \mathbf{Z}_p^{\oplus r} \oplus (\mathbf{Z}/p^{\frac{n_r - n_{r+2}}{2}})^{\oplus 2} \oplus (\mathbf{Z}/p^{\frac{n_{r+2} - n_{r+4}}{2}})^{\oplus 2} \oplus \ldots \oplus (\mathbf{Z}/p^{\frac{n_{a-2} - n_a}{2}})^{\oplus 2}$$

hold.

In particular, knowing $\Theta_i(\mathbf{Q})^{(\delta)}$ for all $i \geq 0$ completely determines the structure of $\mathrm{Sel}(E/\mathbf{Q}, E[p^\infty])^\vee$ as a $\mathbf{Z}_p$-module. Namely, the modular symbols determine the structure of the Selmer group under our assumptions.

1.2 Main Results

We define

$$\mathcal{P}_1^{(N)} = \{\ell \in \mathcal{P}^{(N)} \mid H^0(\mathbf{F}_\ell, E[p^N]) \simeq \mathbf{Z}/p^N\}.$$

This is an infinite set by Chebotarev density theorem since we are assuming (ii) (see Kurihara 2014, §5.8). We define $\mathcal{N}_1^{(N)}$ to be the set of squarefree products of primes in $\mathcal{P}_1^{(N)}$. Again, we suppose $1 \in \mathcal{N}_1^{(N)}$. We propose the following conjecture.

Conjecture 1. There is $m \in \mathcal{N}_1^{(N)}$ such that $\tilde{\delta}_m$ is a unit in $\mathbf{Z}/p^N$, namely

$$\mathrm{ord}_p(\tilde{\delta}_m) = 0.$$

Numerically, it is easy to compute $\tilde{\delta}_m$, so it is easy to check this conjecture.

Theorem 2 (Kurihara 2014, Theorem 10.8). *If we assume the main conjecture and the non-degeneracy of the p-adic height pairing, Conjecture 1 holds true.*

In fact, we obtain Conjecture 1, considering the case $i = s$ in Theorem 10.8 in Kurihara (2014) (see also Theorem 1 for $i = a$).

From now on, we do not assume the main conjecture (9) nor the non-degeneracy of the p-adic height pairing.

We define the Selmer group $\mathrm{Sel}(\mathbf{Z}[1/m], E[p^N])$ by

$$\mathrm{Sel}(\mathbf{Z}[1/m], E[p^N]) = \mathrm{Ker}(H^1(\mathbf{Q}, E[p^N]) \longrightarrow \prod_{v \nmid m} H^1(\mathbf{Q}_v, E[p^N])/E(\mathbf{Q}_v) \otimes \mathbf{Z}/p^N).$$

If all bad primes and p divide m, we know $\mathrm{Sel}(\mathbf{Z}[1/m], E[p^N])$ is equal to the étale cohomology group $H^1_{et}(\mathrm{Spec}\,\mathbf{Z}[1/m], E[p^N])$, which explains the notation "$\mathrm{Sel}(\mathbf{Z}[1/m], E[p^N])$". (We use $\mathrm{Sel}(\mathbf{Z}[1/m], E[p^N])$ for $m \in \mathcal{N}_1^{(N)}$ in this paper, but $E[p^N]$ is not an étale sheaf on $\mathrm{Spec}\,\mathbf{Z}[1/m]$ for such m.)

Let λ be the λ-invariant of $\mathrm{Sel}(E/\mathbf{Q}_\infty, E[p^\infty])^\vee$. We put $n_\lambda = \min\{n \in \mathbf{Z} \mid p^n - 1 \geq \lambda\}$ and $d_n = n_\lambda + Nn$ for $n \in \mathbf{Z}_{\geq 0}$. We define

$$\mathcal{P}_1^{(N,n)} = \{\ell \in \mathcal{P}_1^{(N)} \mid \ell \equiv 1 \pmod{p^{d_n}}\} \tag{4}$$

(then $\mathcal{P}_1^{(N,n)} \subset \mathcal{P}_1^{(N)}(\mathbf{Q}_{[n]})$ holds, see the end of Sect. 3.1 for this fact, and see Sect. 3.1 for the definition of the set $\mathcal{P}_1^{(N)}(\mathbf{Q}_{[n]})$). We denote by $\mathcal{N}_1^{(N,n)}$ the set of squarefree products of primes in $\mathcal{P}_1^{(N,n)}$.

In this paper, for any finite abelian p-extension $K/\mathbf{Q}$ in which all bad primes of E are unramified, we prove in Sect. 4 the following theorem for $\mathbf{Z}/p^N[\mathrm{Gal}(K/\mathbf{Q})]$-modules $\mathrm{Sel}(E/K, E[p^N])$ and $\mathrm{Sel}(O_K[1/m], E[p^N])$ (see Corollary 2 and Theorem 9). We simply state it in the case $K = \mathbf{Q}$ below. An essential ingredient in this paper is the Kolyvagin system of Gauss sum type. We construct Kolyvagin systems $\kappa_{m,\ell} \in \mathrm{Sel}(\mathbf{Z}[1/m\ell], E[p^N])$ for (m, ℓ) satisfying $\ell \in \mathcal{P}_1^{(N,\epsilon(m\ell)+1)}$ and $m\ell \in \mathcal{N}_1^{(N,\epsilon(m\ell)+1)}$ (see Sect. 3.4 and Proposition 2) by the method in Kurihara (2014). (We can construct these elements, using the half of the main conjecture proved by Kato (2004).) The essential difference between our Kolyvagin systems $\kappa_{m,\ell}$ of Gauss sum type and Kolyvagin systems in Mazur and Rubin (2004) is that our $\kappa_{m,\ell}$ is related to L-values. In particular, $\kappa_{m,\ell}$ satisfies a remarkable property $\phi_\ell(\kappa_{m,\ell}) = -\delta_{m\ell} t_{\ell,K}$ (see Proposition 2 (4)) though we do not explain the notation here.

Theorem 3. *Assume that* $\mathrm{ord}_p(\tilde{\delta}_m) = 0$ *for some* $m \in \mathcal{N}_1^{(N)}$.

(1) *The canonical homomorphism*

$$s_m : \mathrm{Sel}(E/\mathbf{Q}, E[p^N]) \longrightarrow \bigoplus_{\ell \mid m} E(\mathbf{Q}_\ell) \otimes \mathbf{Z}/p^N \simeq \bigoplus_{\ell \mid m} E(\mathbf{Q}_\ell) \otimes \mathbf{Z}/p^N \simeq (\mathbf{Z}/p^N)^{\epsilon(m)}$$

is injective.

(2) *Assume further that* $m \in \mathcal{N}_1^{(N,\epsilon(m)+1)}$ *and that* m *is admissible (for the definition of the notion "admissible", see the paragraph before Proposition 1). Then* $\mathrm{Sel}(\mathbf{Z}[1/m], E[p^N])$ *is a free* $\mathbf{Z}/p^N$*-module of rank* $\epsilon(m)$, *and* $\{\kappa_{\frac{m}{\ell},\ell}\}_{\ell \mid m}$ *is a basis of* $\mathrm{Sel}(\mathbf{Z}[1/m], E[p^N])$.

(3) *We define a matrix* $\mathcal{A}$ *as in* (27) *in Theorem 9, using* $\kappa_{\frac{m}{\ell},\ell}$. *Then* $\mathcal{A}$ *is a relation matrix of the Pontrjagin dual* $\mathrm{Sel}(E/\mathbf{Q}, E[p^N])^\vee$ *of the Selmer group; namely if* $f_A : (\mathbf{Z}/p^N)^{\epsilon(m)} \longrightarrow (\mathbf{Z}/p^N)^{\epsilon(m)}$ *is the homomorphism corresponding to the above matrix* $\mathcal{A}$, *then we have*

$$\mathrm{Coker}(f_A) \simeq \mathrm{Sel}(E/\mathbf{Q}, E[p^N])^\vee.$$

It is worth noting that we get nontrivial (moreover, linearly independent) elements in the Selmer groups.

The ideals $\Theta_i(\mathbf{Q})^{(\delta)}$ in Theorem 1 are not suitable for numerical computations because we have to compute *infinitely many* $\tilde{\delta}_m$. On the other hand, we can easily find m with $\mathrm{ord}_p(\tilde{\delta}_m) = 0$ numerically. Since s_m is injective, we can get information of the Selmer group from the image of s_m, which is an advantage of Theorem 3 and the next Theorem 4 (see also the comment in the end of Example 5 in Sect. 5.3).

We next consider the case $N = 1$, so $\mathrm{Sel}(E/\mathbf{Q}, E[p])$. Now we regard $\tilde{\delta}_m$ as an element of $\mathbf{F}_p$ for $m \in \mathcal{N}_1^{(1)}$. We say m is *δ-minimal* if $\tilde{\delta}_m \neq 0$ and $\tilde{\delta}_d = 0$ for all divisors d of m with $1 \leq d < m$. Our next conjecture claims that the structure(the dimension) of $\mathrm{Sel}(E/\mathbf{Q}, E[p])$ is determined by a δ-minimal m, therefore can be easily computed numerically.

Conjecture 2. If $m \in \mathcal{N}_1^{(1)}$ is δ-minimal, the canonical homomorphism

$$s_m : \mathrm{Sel}(E/\mathbf{Q}, E[p]) \longrightarrow \bigoplus_{\ell \mid m} E(\mathbf{Q}_\ell) \otimes \mathbf{Z}/p \simeq \bigoplus_{\ell \mid m} E(\mathbf{F}_\ell) \otimes \mathbf{Z}/p \simeq (\mathbf{Z}/p^N)^{\epsilon(m)}$$

is bijective. In particular, $\dim_{\mathbf{F}_p} \mathrm{Sel}(E/\mathbf{Q}, E[p]) = \epsilon(m)$.

If $m \in \mathcal{N}_1^{(1)}$ is δ-minimal, the above homomorphism $s_m : \mathrm{Sel}(E/\mathbf{Q}, E[p]) \longrightarrow (\mathbf{Z}/p^N)^{\epsilon(m)}$ is injective by Theorem 3 (1), so we know

$$\dim_{\mathbf{F}_p} \mathrm{Sel}(E/\mathbf{Q}, E[p]) \leq \epsilon(m).$$

Therefore, the problem is in showing the other inequality.

We note that the analogue of the above conjecture for ideal class groups does not hold (see Sect. 5.4). But we hope that Conjecture 2 holds for the Selmer groups of elliptic curves. We construct in Sect. 5 a modified version $\kappa_{m,\ell}^{q,q',z}$ of Kolyvagin systems of Gauss sum type for any (m, ℓ) with $m\ell \in \mathcal{N}_1^{(N)}$. (The Kolyvagin system $\kappa_{m,\ell}$ in Sect. 3 is defined for (m, ℓ) with $m\ell \in \mathcal{N}_1^{(N,\epsilon(m\ell)+1)}$, but $\kappa_{m,\ell}^{q,q',z}$ is defined for more general (m, ℓ), namely for (m, ℓ) with $m\ell \in \mathcal{N}_1^{(N)}$.) Using the modified Kolyvagin system $\kappa_{m,\ell}^{q,q',z}$, we prove the following.

Theorem 4. (1) *If $\epsilon(m) = 0$, 1, then Conjecture 2 is true.*
(2) *If there is $\ell \in \mathcal{P}^{(1)}$ which is δ-minimal (so $\epsilon(\ell) = 1$), then*

$$\mathrm{Sel}(E/\mathbf{Q}, E[p^\infty]) \simeq \mathbf{Q}_p/\mathbf{Z}_p.$$

Moreover, if there is $\ell \in \mathcal{P}_1^{(1)}$ which is δ-minimal and which satisfies $\ell \equiv 1 \pmod{p^{n_{\lambda'}+2}}$ where λ' is the analytic λ-invariant of $(E, \mathbf{Q}_\infty/\mathbf{Q})$, then the main conjecture (9) for $\mathrm{Sel}(E/\mathbf{Q}_\infty, E[p^\infty])$ holds true. In this case, $\mathrm{Sel}(E/\mathbf{Q}_\infty, E[p^\infty])^\vee$ is generated by one element as a $\mathbf{Z}_p[[\mathrm{Gal}(\mathbf{Q}_\infty/\mathbf{Q})]]$-module.

(3) *If $\epsilon(m) = 2$ and m is admissible, then Conjecture 2 is true.*
(4) *Suppose that $\epsilon(m) = 3$ and $m = \ell_1\ell_2\ell_3$. Assume that m is admissible and the natural maps $s_{\ell_i} : \mathrm{Sel}(E/\mathbf{Q}, E[p]) \longrightarrow E(\mathbf{F}_{\ell_i}) \otimes \mathbf{Z}/p$ are surjective both for $i = 1$ and $i = 2$. Then Conjecture 2 is true.*

In this way, we can determine the Selmer groups by finite numbers of computations in several cases. We give several numerical examples in Sect. 5.2.

Remark 1. Concerning the Fitting ideals and the annihilator ideals of some Selmer groups, we prove the following in this paper. Let $K/\mathbf{Q}$ be a finite abelian p-extension in which all bad primes of E are unramified. We take a finite set S of good reduction primes, which contains all ramifying primes in $K/\mathbf{Q}$ except p.

Let m be the product of primes in S. We prove that the initial Fitting ideal of the $R_K = \mathbf{Z}_p[\mathrm{Gal}(K/\mathbf{Q})]$-module $\mathrm{Sel}(O_K[1/m], E[p^\infty])^\vee$ is principal, and

$$\xi_{K,S} \in \mathrm{Fitt}_{0,R_K}(\mathrm{Sel}(O_K[1/m], E[p^\infty])^\vee)$$

where $\xi_{K,S}$ is an element of R_K which is explicitly constructed from modular symbols (see (17)). If the main conjecture (9) for $(E, \mathbf{Q}_\infty/\mathbf{Q})$ holds, the equality $\mathrm{Fitt}_{0,R_K}(\mathrm{Sel}(O_K[1/m], E[p^\infty])^\vee) = \xi_{K,S} R_K$ holds (see Remark 2). We prove the Iwasawa theoretical version in Theorem 6.

Let ϑ_K be the image of the p-adic L-function, which is also explicitly constructed from modular symbols. We show in Theorem 7

$$\vartheta_K \in \mathrm{Ann}_{R_K}(\mathrm{Sel}(O_K[1/m], E[p^\infty])^\vee).$$

Concerning the higher Fitting ideals (cf. Sect. 2.4), we show

$$\tilde{\delta}_m \in \mathrm{Fitt}_{\epsilon(m),\mathbf{Z}/p^N}(\mathrm{Sel}(E/\mathbf{Q}, E[p^N])^\vee)$$

where $\mathrm{Fitt}_{i,R}(M)$ is the i-th Fitting ideal of an R-module M. We prove a slightly generalized version for K which is in the cyclotomic $\mathbf{Z}_p$-extension $\mathbf{Q}_\infty$ of $\mathbf{Q}$ (see Theorem 8 and Corollary 1).

I would like to thank John Coates heartily for his helpful advice and for discussion with him, especially for the discussion in March 2013, which played an essential role in my producing this paper. I also thank heartily Kazuya Kato for his constant interest in the results of this paper. I also thank Kazuo Matsuno and Christian Wuthrich very much for their helping me to compute modular symbols.

2 Selmer Groups and p-Adic L-Functions

2.1 Modular Symbols and p-Adic L-Functions

Let E be an elliptic curve over $\mathbf{Q}$, and $f(z) = \Sigma a_n e^{2\pi i n z}$ the modular form corresponding to E. In this section, we assume that p is a prime number satisfying (i), (ii), (iii) in Sect. 1.1. We define $\mathcal{P}_{good} = \{\ell \mid \ell \text{ is a good reduction prime for } E\} \setminus \{p\}$. For any finite abelian extension $K/\mathbf{Q}$, we denote by K_∞/K the cyclotomic $\mathbf{Z}_p$-extension. For a real abelian field K of conductor m, we define $\tilde{\theta}_K$ to be the image of $\tilde{\theta}_{\mathbf{Q}(\mu_m)}$ in $\mathbf{Q}[\mathrm{Gal}(K/\mathbf{Q})]$ where $\tilde{\theta}_{\mathbf{Q}(\mu_m)}$ is defined in (1).

We write

$$R_K = \mathbf{Z}_p[\mathrm{Gal}(K/\mathbf{Q})] \text{ and } \Lambda_{K_\infty} = \mathbf{Z}_p[[\mathrm{Gal}(K_\infty/\mathbf{Q})]].$$

For any positive integer n, we simply write $R_{\mathbf{Q}(\mu_n)} = R_n$ in this subsection. For any positive integers d, c such that $d \mid c$, we define the norm map $\nu_{c,d} : R_d = \mathbf{Z}_p[\mathrm{Gal}(\mathbf{Q}(\mu_d)/\mathbf{Q})] \longrightarrow R_c = \mathbf{Z}_p[\mathrm{Gal}(\mathbf{Q}(\mu_c)/\mathbf{Q})]$ by $\sigma \mapsto \sum \tau$ where for $\sigma \in \mathrm{Gal}(\mathbf{Q}(\mu_d)/\mathbf{Q})$, τ runs over all elements of $\mathrm{Gal}(\mathbf{Q}(\mu_c)/\mathbf{Q})$ such that the restriction of τ to $\mathbf{Q}(\mu_d)$ is σ. Let m be a squarefree product of primes in $\mathcal{P}_{good}$, and n a positive integer. By our assumption (ii), we know $\tilde{\theta}_{\mathbf{Q}(\mu_{mp^n})} \in R_{mp^n}$ (cf. Stevens 1989). Let $\alpha \in \mathbf{Z}_p^\times$ be the unit root of $x^2 - a_p x + p = 0$ and put

$$\vartheta_{\mathbf{Q}(\mu_{mp^n})} = \alpha^{-n}(\tilde{\theta}_{\mathbf{Q}(\mu_{mp^n})} - \alpha^{-1}\nu_{mp^n,mp^{n-1}}(\tilde{\theta}_{\mathbf{Q}(\mu_{mp^{n-1}})})) \in R_{mp^n}$$

as usual. Then $\{\vartheta_{\mathbf{Q}(\mu_{mp^n})}\}_{n\geq 1}$ is a projective system (cf. Mazur and Tate 1987, the equation (4) on page 717) and we obtain an element $\vartheta_{\mathbf{Q}(\mu_{mp^\infty})} \in \Lambda_{\mathbf{Q}(\mu_{mp^\infty})}$, which is the p-adic L-function of Mazur and Swinnerton-Dyer.

We also use the notation $\Lambda_{np^\infty} = \Lambda_{\mathbf{Q}(\mu_{np^\infty})}$ for simplicity. Suppose that a prime ℓ does not divide mp, and $c_{m\ell,m} : \Lambda_{m\ell p^\infty} \longrightarrow \Lambda_{mp^\infty}$ is the natural restriction map. Then we know

$$c_{m\ell,m}(\vartheta_{\mathbf{Q}(\mu_{m\ell p^\infty})}) = (a_\ell - \sigma_\ell - \sigma_\ell^{-1})\vartheta_{\mathbf{Q}(\mu_{mp^\infty})} \tag{5}$$

(cf. Mazur and Tate 1987, the equation (1) on page 717).

We will construct a slightly modified element $\xi_{\mathbf{Q}(\mu_{mp^\infty})}$ in Λ_{mp^∞}. We put $P'_\ell(x) = x^2 - a_\ell x + \ell$. Let m be a squarefree product of $\mathcal{P}_{good}$. For any divisor d of m and a prime divisor ℓ of m/d, $\sigma_\ell \in \mathrm{Gal}(\mathbf{Q}(\mu_{dp^\infty})/\mathbf{Q}) = \varprojlim \mathrm{Gal}(\mathbf{Q}(\mu_{dp^n})/\mathbf{Q})$ is defined as the projective limit of $\sigma_\ell \in \mathrm{Gal}(\mathbf{Q}(\mu_{dp^n})/\mathbf{Q})$. We consider $P'_\ell(\sigma_\ell) \in \Lambda_{dp^\infty}$. Note that

$$-\sigma_\ell^{-1} = (-\sigma_\ell^{-1}P'_\ell(\sigma_\ell) - (a_\ell - \sigma_\ell - \sigma_\ell^{-1}))/(\ell - 1) \in \Lambda_{dp^\infty}. \tag{6}$$

We put $\alpha_{d,m} = (\prod_{\ell \mid \frac{m}{d}}(-\sigma_\ell^{-1}))\vartheta_{\mathbf{Q}(\mu_{dp^\infty})} \in \Lambda_{dp^\infty}$ and

$$\xi_{\mathbf{Q}(\mu_{mp^\infty})} = \sum_{d \mid m} \nu_{m,d}(\alpha_{d,m}) \in \Lambda_{mp^\infty}$$

where $\nu_{m,d} : \Lambda_{dp^\infty} \longrightarrow \Lambda_{mp^\infty}$ is the norm map defined similarly as above. (This modification $\xi_{\mathbf{Q}(\mu_{mp^\infty})}$ is done by the same spirit as Greither (2004) in which the Deligne-Ribet p-adic L-functions are treated.) Suppose that $\ell \in \mathcal{P}_{good}$ is prime to m. Then by the definition of $\xi_{\mathbf{Q}(\mu_{mp^\infty})}$ and (5) and (6), we have

$$\begin{aligned} c_{m\ell,m}(\xi_{\mathbf{Q}(\mu_{m\ell p^\infty})}) &= c_{m\ell,m}(\sum_{d \mid m} \nu_{m\ell,d}(\alpha_{d,m\ell}) + \sum_{d \mid m} \nu_{m\ell,d\ell}(\alpha_{d\ell,m\ell})) \\ &= (\ell - 1)\sum_{d \mid m} \nu_{m,d}(-\sigma_\ell^{-1}\alpha_{d,m}) + \sum_{d \mid m} \nu_{m,d}(c_{d\ell,d}(\alpha_{d\ell,m\ell})) \end{aligned}$$

$$
\begin{aligned}
&= (\ell - 1)\sum_{d|m} \nu_{m,d}(-\sigma_\ell^{-1}\alpha_{d,m}) + \sum_{d|m} \nu_{m,d}((a_\ell - \sigma_\ell - \sigma_\ell^{-1})\alpha_{d,m}) \\
&= (-\sigma_\ell^{-1} P_\ell'(\sigma_\ell)) \sum_{d|m} \nu_{m,d}(\alpha_{d,m}) \\
&= (-\sigma_\ell^{-1} P_\ell'(\sigma_\ell)) \xi_{\mathbf{Q}(\mu_{mp^\infty})}. \qquad (7)
\end{aligned}
$$

We denote by $\vartheta_{\mathbf{Q}(\mu_m)} \in R_{\mathbf{Q}(\mu_m)}$ the image of $\vartheta_{\mathbf{Q}(\mu_{mp^\infty})}$ under the natural map $\Lambda_{\mathbf{Q}(\mu_{mp^\infty})} \longrightarrow R_{\mathbf{Q}(\mu_m)}$. We have

$$
\vartheta_{\mathbf{Q}(\mu_m)} = (1 - \frac{\sigma_p}{\alpha})(1 - \frac{\sigma_p^{-1}}{\alpha})\tilde{\theta}_{\mathbf{Q}(\mu_m)}. \qquad (8)
$$

Since we are assuming $a_p \not\equiv 1 \pmod{p}$, we also have $\alpha \not\equiv 1 \pmod{p}$, so $(1 - \frac{\sigma_p}{\alpha})(1 - \frac{\sigma_p^{-1}}{\alpha})$ is a unit in $R_{\mathbf{Q}(m)}$ where $\mathbf{Q}(m)$ is the maximal p-subextension of $\mathbf{Q}$ in $\mathbf{Q}(\mu_m)$.

2.2 Selmer Groups

For any algebraic extension $F/\mathbf{Q}$, we denote by O_F the integral closure of $\mathbf{Z}$ in F. For a positive integer $m > 0$, we define a Selmer group $\mathrm{Sel}(O_F[1/m], E[p^\infty])$ by

$$
\begin{aligned}
\mathrm{Sel}(O_F[1/m], E[p^\infty]) = \mathrm{Ker}(&H^1(F, E[p^\infty]) \\
&\longrightarrow \prod_{v \nmid m} H^1(F_v, E[p^\infty])/E(F_v) \otimes \mathbf{Q}_p/\mathbf{Z}_p)
\end{aligned}
$$

where v runs over all primes of F which are prime to m. Similarly, for a positive integer N, we define $\mathrm{Sel}(O_F[1/m], E[p^N])$ by

$$
\begin{aligned}
\mathrm{Sel}(O_F[1/m], E[p^N]) = \mathrm{Ker}(&H^1(F, E[p^N]) \\
&\longrightarrow \prod_{v \nmid m} H^1(F_v, E[p^N])/E(F_v) \otimes \mathbf{Z}/p^N).
\end{aligned}
$$

In the case $m = 1$, we denote them by $\mathrm{Sel}(O_F, E[p^\infty])$, $\mathrm{Sel}(O_F, E[p^N])$, which are classical Selmer groups. We also use the notation $\mathrm{Sel}(E/F, E[p^\infty])$, $\mathrm{Sel}(E/F, E[p^N])$ for them, namely

$$
\mathrm{Sel}(E/F, E[p^\infty]) = \mathrm{Sel}(O_F, E[p^\infty]), \;\; \mathrm{Sel}(E/F, E[p^N]) = \mathrm{Sel}(O_F, E[p^N]).
$$

For a finite abelian extension $K/\mathbf{Q}$, we denote by K_∞/K the cyclotomic $\mathbf{Z}_p$-extension, and put $\Lambda_{K_\infty} = \mathbf{Z}_p[[\mathrm{Gal}(K_\infty/\mathbf{Q})]]$. The Pontrjagin dual $\mathrm{Sel}(O_{K_\infty}, E[p^\infty])^\vee$ is a torsion Λ_{K_∞}-module (Kato 2004, Theorem 17.4).

When the conductor of K is m, we define $\vartheta_{K_\infty} \in \Lambda_{K_\infty}$ to be the image of $\vartheta_{\mathbf{Q}(\mu_{mp^\infty})}$, and also $\xi_{K_\infty} \in \Lambda_{K_\infty}$ to be the image of $\xi_{\mathbf{Q}(\mu_{mp^\infty})}$. The Iwasawa main conjecture for $(E, \mathbf{Q}_\infty/\mathbf{Q})$ is the equality between the characteristic ideal of the Selmer group and the ideal generated by the p-adic L-function;

$$\mathrm{char}(\mathrm{Sel}(O_{\mathbf{Q}_\infty}, E[p^\infty])^\vee) = \vartheta_{\mathbf{Q}_\infty}\Lambda_{\mathbf{Q}_\infty}. \tag{9}$$

Since we are assuming the Galois action on the Tate module is surjective, we know $\vartheta_{\mathbf{Q}_\infty} \in \mathrm{char}(\mathrm{Sel}(O_{\mathbf{Q}_\infty}, E[p^\infty])^\vee)$ by Kato (2004, Theorem 17.4). Skinner and Urban (2014) proved the equality (9) under mild conditions. Namely, under our assumptions (i), (ii), they proved the main conjecture (9) if there is a bad prime ℓ which is ramified in $\mathbf{Q}(E[p])$ (Skinner and Urban 2014, Theorem 3.33).

More generally, let ψ be an even Dirichlet character and K be the abelian field corresponding to the kernel of ψ, namely K is the field such that ψ induces a faithful character of $\mathrm{Gal}(K/\mathbf{Q})$. We assume $K \cap \mathbf{Q}_\infty = \mathbf{Q}$. In this paper, for any finite abelian p-group G, any $\mathbf{Z}_p[G]$-module M and any character $\psi : G \longrightarrow \overline{\mathbf{Q}}_p^\times$, we define the ψ-quotient M_ψ by $M \otimes_{\mathbf{Z}_p[G]} O_\psi$ where $O_\psi = \mathbf{Z}_p[\mathrm{Image}\,\psi]$ which is regarded as a $\mathbf{Z}_p[G]$-module by $\sigma x = \psi(\sigma)x$ for any $\sigma \in G$ and $x \in O_\psi$. We consider $(\mathrm{Sel}(O_{K_\infty}, E[p^\infty])^\vee)_\psi$, which is a Λ_ψ-module where $\Lambda_\psi = (\Lambda_{K_\infty})_\psi = O_\psi[[\mathrm{Gal}(K_\infty/K)]]$. We denote the image of ϑ_{K_∞} in Λ_ψ by $\psi(\vartheta_{K_\infty})$. Then the main conjecture states

$$\mathrm{char}((\mathrm{Sel}(O_{K_\infty}, E[p^\infty])^\vee)_\psi) = \psi(\vartheta_{K_\infty})\Lambda_\psi. \tag{10}$$

We also note that $\psi(\vartheta_{K_\infty})\Lambda_\psi = \psi(\xi_{K_\infty})\Lambda_\psi$. By Kato (2004), we know $\psi(\vartheta_{K_\infty})$, $\psi(\xi_{K_\infty}) \in \mathrm{char}((\mathrm{Sel}(O_{K_\infty}, E[p^\infty])^\vee)_\psi)$.

Let $S \subset \mathcal{P}_{good}$ be a finite set of good primes, and $K/\mathbf{Q}$ be a finite abelian extension. We denote by $S_{\mathrm{ram}}(K)$ the subset of S which consists of all ramifying primes in K inside S. Recall that $P'_\ell(x) = x^2 - a_\ell x + \ell$. We define

$$\xi_{K_\infty,S} = \xi_{K_\infty} \prod_{\ell \in S \setminus S_{\mathrm{ram}}(K)} (-\sigma_\ell^{-1} P'_\ell(\sigma_\ell)).$$

So $\xi_{K_\infty,S} = \xi_{K_\infty}$ if S contains only ramifying primes in K. Suppose that S contains all ramifying primes in K and F is a subfield of K. We denote by $c_{K_\infty/F_\infty} : \Lambda_{K_\infty} \longrightarrow \Lambda_{F_\infty}$ the natural restriction map. Using (7) and the above definition of $\xi_{K_\infty,S}$, we have

$$c_{K_\infty/F_\infty}(\xi_{K_\infty,S}) = \xi_{F_\infty,S}. \tag{11}$$

For any positive integer m whose prime divisors are in $\mathcal{P}_{good}$, we have an exact sequence

$$0 \longrightarrow \mathrm{Sel}(O_{K_\infty}, E[p^\infty]) \longrightarrow \mathrm{Sel}(O_{K_\infty}[1/m], E[p^\infty]) \longrightarrow \bigoplus_{v|m} H^1(K_{\infty,v}, E[p^\infty]) \longrightarrow 0$$

because $E(K_{\infty,v}) \otimes \mathbf{Q}_p/\mathbf{Z}_p = 0$ (for the surjectivity of the third map, see Greenberg Lemma 4.6 in 1999). For a prime v of K_∞, let $K_{\infty,v,nr}/K_{\infty,v}$ be the maximal unramified extension, and $\Gamma_v = \mathrm{Gal}(K_{\infty,v,nr}/K_{\infty,v})$. Suppose v divides m. Since v is a good reduction prime, we have $H^1(K_{\infty,v}, E[p^\infty]) = \mathrm{Hom}_{Cont}(G_{K_{\infty,v,nr}}, E[p^\infty])^{\Gamma_v} = E[p^\infty](-1)^{\Gamma_v}$ where (-1) is the Tate twist. By the Weil pairing, the Pontrjagin dual of $E[p^\infty](-1)$ is the Tate module $T_p(E)$. Therefore, taking the Pontrjagin dual of the above exact sequence, we have an exact sequence

$$0 \longrightarrow \bigoplus_{v|m} T_p(E)_{\Gamma_v} \longrightarrow \mathrm{Sel}(O_{K_\infty}[1/m], E[p^\infty])^\vee \longrightarrow \mathrm{Sel}(O_{K_\infty}, E[p^\infty])^\vee \longrightarrow 0. \tag{12}$$

Note that $T_p(E)_{\Gamma_v}$ is free over $\mathbf{Z}_p$ because Γ_v is profinite of order prime to p.

Let $K/\mathbf{Q}$ be a finite abelian p-extension in which all bad primes of E are unramified. Suppose that S is a finite subset of $\mathcal{P}_{good}$ such that S contains all ramifying primes in $K/\mathbf{Q}$ except p. Let m be a squarefree product of all primes in S.

Theorem 5 (Greenberg). *$\mathrm{Sel}(O_{K_\infty}[1/m], E[p^\infty])^\vee$ is of projective dimension ≤ 1 as a Λ_{K_∞}-module.*

This is proved by Greenberg in 2011, Theorem 1 (the condition (iv) in Sect. 1.1 in this paper is not needed here, see also Proposition 3.3.1 in Greenberg (2011)). For more general p-adic representations, this is proved in Kurihara (2014, Proposition 2.18). We will give a sketch of the proof because some results in the proof will be used later.

Since we can take some finite abelian extension $K'/\mathbf{Q}$ such that $K_\infty = K'_\infty$ and $K' \cap \mathbf{Q}_\infty = \mathbf{Q}$, we may assume that $K \cap \mathbf{Q}_\infty = \mathbf{Q}$ and p is unramified in K. Since we are assuming that $E[p]$ is an irreducible $G_\mathbf{Q}$-module, we know that $\mathrm{Sel}(O_{K_\infty}, E[p^\infty])^\vee$ has no nontrivial finite $\mathbf{Z}_p[[\mathrm{Gal}(K_\infty/K)]]$-submodule by Greenberg (1999, Propositions 4.14, 4.15). We also assumed that the μ-invariant of $\mathrm{Sel}(O_{\mathbf{Q}_\infty}, E[p^\infty])^\vee$ is zero, which implies the vanishing of the μ-invariant of $\mathrm{Sel}(O_{K_\infty}, E[p^\infty])^\vee$ by Hachimori and Matsuno (1999). Therefore, $\mathrm{Sel}(O_{K_\infty}, E[p^\infty])^\vee$ is a free $\mathbf{Z}_p$-module of finite rank. By the exact sequence (12), $\mathrm{Sel}(O_{K_\infty}[1/m], E[p^\infty])^\vee$ is also a free $\mathbf{Z}_p$-module of finite rank.

Put $G = \mathrm{Gal}(K/\mathbf{Q})$. By the definition of the Selmer group and our assumption that all primes dividing m are good reduction primes, we have $\mathrm{Sel}(O_{K_\infty}[1/m], E[p^\infty])^G = \mathrm{Sel}(O_{\mathbf{Q}_\infty}[1/m], E[p^\infty])$. Since we assumed that the μ-invariant is zero, $\mathrm{Sel}(O_{\mathbf{Q}_\infty}[1/m], E[p^\infty])$ is divisible. This shows that the corestriction map $\mathrm{Sel}(O_{K_\infty}[1/m], E[p^\infty]) \longrightarrow \mathrm{Sel}(O_{\mathbf{Q}_\infty}[1/m], E[p^\infty])$ is surjective. Therefore, $\hat{H}^0(G, \mathrm{Sel}(O_{K_\infty}[1/m], E[p^\infty])) = 0$.

Next we will show that $H^1(G, \mathrm{Sel}(O_{K_\infty}[1/m], E[p^\infty])) = 0$. Let N_E be the conductor of E and put $m' = mpN_E$. We know $\mathrm{Sel}(O_{K_\infty}[1/m'], E[p^\infty])$ is equal to the étale cohomology group $H^1_{et}(\mathrm{Spec}\, O_{K_\infty}[1/m'], E[p^\infty])$. We have an exact sequence

$$0 \longrightarrow \mathrm{Sel}(O_{K_\infty}[1/m], E[p^\infty]) \longrightarrow \mathrm{Sel}(O_{K_\infty}[1/m'], E[p^\infty]) \longrightarrow \bigoplus_{v \mid \frac{m'}{m}} H^2_v(K_{\infty,v}) \longrightarrow 0 \tag{13}$$

where $H^2_v(K_{\infty,v}) = H^1(K_{\infty,v}, E[p^\infty])/(E(K_{\infty,v}) \otimes \mathbf{Q}_p/\mathbf{Z}_p)$, and the surjectivity of the third map follows from Greenberg Lemma 4.6 in Greenberg (1999). Let $E[p^\infty]^0$ be the kernel of $E[p^\infty] = E(\overline{\mathbf{Q}})[p^\infty] \longrightarrow E(\overline{\mathbf{F}}_p)[p^\infty]$ and $E[p^\infty]_{et} = E[p^\infty]/E[p^\infty]^0$. For a prime v of K_∞ above p, we denote by $K_{\infty,v,nr}$ the maximal unramified extension of $K_{\infty,v}$, and put $\Gamma_v = \mathrm{Gal}(K_{\infty,v,nr}/K_{\infty,v})$. We know the isomorphism $H^2_v(K_{\infty,v}) \xrightarrow{\simeq} H^1(K_{\infty,v,nr}, E[p^\infty]_{et})^{\Gamma_v}$ by Greenberg (1989, §2). If v is a prime of K_∞ not above p, we know $H^2_v(K_{\infty,v}) = H^1(K_{\infty,v}, E[p^\infty])$. Therefore, we get an isomorphism

$$\Big(\bigoplus_{v \mid \frac{m'}{m}} H^2_v(K_{\infty,v})\Big)^G = \bigoplus_{u \mid \frac{m'}{m}} H^2_u(\mathbf{Q}_{\infty,v})$$

where v (resp. u) runs over all primes of K_∞ (resp. $\mathbf{Q}_\infty$) above $m'/m = pN_E$. Thus, $\mathrm{Sel}(O_{K_\infty}[1/m'], E[p^\infty])^G \longrightarrow \bigoplus_{v \mid \frac{m'}{m}} H^2_v(K_{\infty,v})^G$ is surjective. On the other hand, we have $H^2_{et}(\mathrm{Spec}\, O_{K_\infty}[1/m'], E[p^\infty]) = 0$ (see Greenberg 1989, Propositions 3, 4). This implies that

$$H^1(G, H^1_{et}(\mathrm{Spec}\, O_{K_\infty}[1/m'], E[p^\infty])) = H^1(G, \mathrm{Sel}(O_{K_\infty}[1/m'], E[p^\infty])) = 0.$$

Taking the cohomology of the exact sequence (13), we get

$$H^1(G, \mathrm{Sel}(O_{K_\infty}[1/m], E[p^\infty])) = 0. \tag{14}$$

Therefore, $\mathrm{Sel}(O_{K_\infty}[1/m], E[p^\infty])$ is cohomologically trivial as a G-module by Serre (1968, Chap. IX, Théorème 8). This implies that $\mathrm{Sel}(O_{K_\infty}[1/m], E[p^\infty])^\vee$ is also cohomologically trivial. Since $\mathrm{Sel}(O_{K_\infty}[1/m], E[p^\infty])^\vee$ has no nontrivial finite submodule, the projective dimension of $\mathrm{Sel}(O_{K_\infty}[1/m], E[p^\infty])^\vee$ as a Λ_{K_∞}-module is ≤ 1 see for example, Popescu (2009, Proposition 2.3).

Theorem 6. *Let $K/\mathbf{Q}$ be a finite abelian p-extension in which all bad primes of E are unramified. We take a finite set S of good reduction primes which contains all ramifying primes in $K/\mathbf{Q}$ except p. Let m be the product of primes in S. Then*

(1) *$\xi_{K_\infty,S}$ is in the initial Fitting ideal* $\mathrm{Fitt}_{0,\Lambda_{K_\infty}}(\mathrm{Sel}(O_{K_\infty}[1/m], E[p^\infty])^\vee)$.
(2) *We have*

$$\mathrm{Fitt}_{0,\Lambda_{K_\infty}}(\mathrm{Sel}(O_{K_\infty}[1/m], E[p^\infty])^\vee) = \xi_{K_\infty,S}\Lambda_{K_\infty}$$

if and only if the main conjecture (9) for $(E, \mathbf{Q}_\infty/\mathbf{Q})$ holds.

Proof. As we explained in the proof of Theorem 5, we may assume that $K \cap \mathbf{Q}_\infty = \mathbf{Q}$. We recall that $\mathrm{Sel}(O_{K_\infty}[1/m], E[p^\infty])^\vee$ is a free $\mathbf{Z}_p$-module of finite rank under our assumptions.

(1) Let $\psi : \mathrm{Gal}(K/\mathbf{Q}) \longrightarrow \overline{\mathbf{Q}}_p^\times$ be a character of $\mathrm{Gal}(K/\mathbf{Q})$, not necessarily faithful. We study the Fitting ideal of the ψ-quotient $(\mathrm{Sel}(O_{K_\infty}[1/m], E[p^\infty])^\vee)_\psi = \mathrm{Sel}(O_{K_\infty}[1/m], E[p^\infty])^\vee \otimes_{\mathbf{Z}_p[\mathrm{Gal}(K/\mathbf{Q})]} O_\psi$. We denote by F the subfield of K corresponding to the kernel of ψ. We regard ψ as a faithful character of $\mathrm{Gal}(F/\mathbf{Q})$. Since $\mathrm{Sel}(O_{K_\infty}[1/m], E[p^\infty])^{\mathrm{Gal}(K/F)} = \mathrm{Sel}(O_{F_\infty}[1/m], E[p^\infty])$, we have

$$(\mathrm{Sel}(O_{K_\infty}[1/m], E[p^\infty])^\vee)_\psi = (\mathrm{Sel}(O_{F_\infty}[1/m], E[p^\infty])^\vee)_\psi$$

where the right hand side is defined to be $\mathrm{Sel}(O_{F_\infty}[1/m], E[p^\infty])^\vee \otimes_{\mathbf{Z}_p[\mathrm{Gal}(F/\mathbf{Q})]} O_\psi$.

We put $\Lambda_\psi = (\Lambda_{F_\infty})_\psi$. The group homomorphism ψ induces the ring homomorphism $\Lambda_{F_\infty} \longrightarrow \Lambda_\psi$ which we also denote by ψ. The composition with $c_{K_\infty/F_\infty} : \Lambda_{K_\infty} \longrightarrow \Lambda_{F_\infty}$ and the above ring homomorphism ψ is also denoted by $\psi : \Lambda_{K_\infty} \longrightarrow \Lambda_\psi$. Note that $F/\mathbf{Q}$ is a cyclic extension of degree a power of p. We denote by F' the subfield of F such that $[F : F'] = p$. We put $N_0 = N_{\mathrm{Gal}(F/F')} = \Sigma_{\sigma \in \mathrm{Gal}(F/F')}\sigma$. If we put $[F : \mathbf{Q}] = p^c$ and take a generator γ of $\mathrm{Gal}(F/\mathbf{Q})$, $N_0 = \Sigma_{i=0}^{p-1}\gamma^{p^{c-1}i}$ is a cyclotomic polynomial and $O_\psi = \mathbf{Z}_p[\mu_{p^c}] \simeq \mathbf{Z}_p[\mathrm{Gal}(F/\mathbf{Q})]/N_0$. For any $\mathbf{Z}_p[\mathrm{Gal}(F/\mathbf{Q})]$-module M, we define $M^\psi = \mathrm{Ker}(N_0 : M \longrightarrow M)$. Then the Pontrjagin dual of M^ψ is $(M^\psi)^\vee = (M^\vee)/N_0 = (M^\vee)_\psi$. By the same method as the proof of (14), we have $H^1(\mathrm{Gal}(F/F'), \mathrm{Sel}(O_{F_\infty}[1/m], E[p^\infty])) = 0$. Therefore, $\sigma - 1 : \mathrm{Sel}(O_{F_\infty}[1/m], E[p^\infty]) \longrightarrow \mathrm{Sel}(O_{F_\infty}[1/m], E[p^\infty])^\psi$ is surjective where $\sigma = \gamma^{p^{c-1}}$ is a generator of $\mathrm{Gal}(F/F')$. Therefore, taking the dual, we know that there is an injective homomorphism from $(\mathrm{Sel}(O_{F_\infty}[1/m], E[p^\infty])^\vee)_\psi$ to $\mathrm{Sel}(O_{F_\infty}[1/m], E[p^\infty])^\vee$ which is a free $\mathbf{Z}_p$-module. Therefore, $(\mathrm{Sel}(O_{F_\infty}[1/m], E[p^\infty])^\vee)_\psi$ contains no nontrivial finite Λ_ψ-submodule. This shows that

$$\mathrm{Fitt}_{0,\Lambda_\psi}((\mathrm{Sel}(O_{F_\infty}[1/m], E[p^\infty])^\vee)_\psi) = \mathrm{char}((\mathrm{Sel}(O_{F_\infty}[1/m], E[p^\infty])^\vee)_\psi).$$

Consider the ψ-quotient of the exact sequence (12);

$$(\bigoplus_{v|m} T_p(E)_{\Gamma_v})_\psi \longrightarrow (\mathrm{Sel}(O_{F_\infty}[1/m], E[p^\infty])^\vee)_\psi$$
$$\longrightarrow (\mathrm{Sel}(O_{F_\infty}, E[p^\infty])^\vee)_\psi \longrightarrow 0$$

where v runs over all primes of F_∞ above m. Since $\mathrm{Ext}^1_{\mathbf{Z}_p[\mathrm{Gal}(F/\mathbf{Q})]}(O_\psi, \mathrm{Sel}(O_{F_\infty}, E[p^\infty])) = \hat{H}^0(\mathrm{Gal}(F/\mathbf{Q}), \mathrm{Sel}(O_{F_\infty}, E[p^\infty]))$ is finite, the first map of the above exact sequence has finite kernel.

Suppose that ℓ is a prime divisor of m. If ℓ is unramified in F, we have

$$\mathrm{Fitt}_{0,\Lambda_\psi}((\bigoplus_{v|\ell} T_p(E)_{\Gamma_v})_\psi) = P'_\ell(\sigma_\ell)\Lambda_\psi$$

where $P'_\ell(x) = x^2 - a_\ell x + \ell$. If ℓ is ramified in F, $\psi(\ell) = 0$ and $(\bigoplus_{v|\ell} T_p(E)_{\Gamma_v})_\psi$ is finite. Therefore, we have

$$\mathrm{char}((\bigoplus_{v|m} T_p(E)_{\Gamma_v})_\psi) = (\prod_{\ell \in S \setminus S_{\mathrm{ram}}(F)} P'_\ell(\sigma_\ell))\Lambda_\psi.$$

Using the above exact sequence and Kato's theorem $\psi(\xi_{F_\infty}) \in \mathrm{char}((\mathrm{Sel}(O_{F_\infty}, E[p^\infty])^\vee)_\psi)$, we have

$$\mathrm{char}((\mathrm{Sel}(O_{F_\infty}[1/m], E[p^\infty])^\vee)_\psi) \supset \psi(\xi_{F_\infty})(\prod_{\ell \in S \setminus S_{\mathrm{ram}}(F)} P'_\ell(\sigma_\ell))\Lambda_\psi.$$

Since $\xi_{F_\infty}(\prod_{\ell \in S \setminus S_{\mathrm{ram}}(F)} P'_\ell(\sigma_\ell)) = \xi_{F_\infty,S}$ modulo unit and $c_{K_\infty/F_\infty}(\xi_{K_\infty,S}) = \xi_{F_\infty,S}$ by (11), we obtain

$$\psi(\xi_{K_\infty,S}) \in \mathrm{Fitt}_{0,(\Lambda_{K_\infty})_\psi}((\mathrm{Sel}(O_{K_\infty}[1/m], E[p^\infty])^\vee)_\psi) \tag{15}$$

for any character ψ of $\mathrm{Gal}(K/\mathbf{Q})$. Since the μ-invariant of $\mathrm{Sel}(O_{K_\infty}[1/m], E[p^\infty])^\vee$ is zero as we explained above, (15) implies

$$\xi_{K_\infty,S} \in \mathrm{Fitt}_{0,\Lambda_{K_\infty}}(\mathrm{Sel}(O_{K_\infty}[1/m], E[p^\infty])^\vee)$$

(see Lemma 4.1 in Kurihara 2003a, for example).

(2) We use the same notation ψ, F, etc. as above. At first, we assume (9). Then the algebraic λ-invariant of $\mathrm{Sel}(E/F_\infty, E[p^\infty])^\vee$ equals the analytic λ-invariant by Hachimori and Matsuno (1999) and Matsuno (2000), so the main conjecture $\mathrm{char}((\mathrm{Sel}(O_{F_\infty}, E[p^\infty])^\vee)_\psi) = \psi(\xi_{F_\infty})\Lambda_\psi$ also holds. Therefore, we have

$$\begin{aligned}\mathrm{char}((\mathrm{Sel}(O_{F_\infty}[1/m], E[p^\infty])^\vee)_\psi) &= \psi(\xi_{F_\infty})(\prod_{\ell \in S \setminus S_{\mathrm{ram}}(F)} P'_\ell(\sigma_\ell))\Lambda_\psi \\ &= \psi(\xi_{F_\infty,S})\Lambda_\psi = \psi(\xi_{K_\infty,S})\Lambda_\psi\end{aligned}$$

and

$$\mathrm{Fitt}_{0,(\Lambda_{K_\infty})_\psi}((\mathrm{Sel}(O_{K_\infty}[1/m], E[p^\infty])^\vee)_\psi) = \psi(\xi_{K_\infty,S}\Lambda_{K_\infty})\Lambda_\psi.$$

It follows from Kurihara (2003a, Corollary 4.2) that

$$\mathrm{Fitt}_{0,\Lambda_{K_\infty}}(\mathrm{Sel}(O_{K_\infty}[1/m], E[p^\infty])^\vee) = \xi_{K_\infty,S}\Lambda_{K_\infty}.$$

On the other hand, if we assume the above equality, taking the $\mathrm{Gal}(K/\mathbf{Q})$-invariant part of $\mathrm{Sel}(O_{K_\infty}[1/m], E[p^\infty])$, we get

$$\mathrm{Fitt}_{0,\Lambda_{\mathbf{Q}_\infty}}(\mathrm{Sel}(O_{\mathbf{Q}_\infty}[1/m], E[p^\infty])^\vee) = \xi_{\mathbf{Q}_\infty,S}\Lambda_{\mathbf{Q}_\infty,S},$$

which implies (9).

2.3 An Analogue of Stickelberger's Theorem

Let $K/\mathbf{Q}$ be a finite abelian p-extension. When the conductor of K is m, we define $\vartheta_K \in R_K = \mathbf{Z}_p[\mathrm{Gal}(K/\mathbf{Q})]$ to be the image of $\vartheta_{\mathbf{Q}(\mu_{mp^\infty})} \in \Lambda_{\mathbf{Q}(\mu_{mp^\infty})}$. Therefore, if m is prime to p, ϑ_K is the image of $\vartheta_{\mathbf{Q}(\mu_m)} = (1-\frac{\sigma_p}{\alpha})(1-\frac{\sigma_p^{-1}}{\alpha})\tilde{\theta}_{\mathbf{Q}(\mu_m)}$ by (8). If $m = m'p^n$ for some m' which is prime to p and for some $n \geq 2$, ϑ_K is the image of $\vartheta_{\mathbf{Q}(\mu_{m'p^n})} = \alpha^{-n}(\tilde{\theta}_{\mathbf{Q}(\mu_{m'p^n})} - \alpha^{-1}\nu_{m'p^n,m'p^{n-1}}(\tilde{\theta}_{\mathbf{Q}(\mu_{m'p^{n-1}})}))$.

For any positive integer n, we denote by $\mathbf{Q}(n)$ the maximal p-subextension of $\mathbf{Q}$ in $\mathbf{Q}(\mu_n)$.

Theorem 7. *For any finite abelian p-extension K in which all bad primes of E are unramified, ϑ_K annihilates $\mathrm{Sel}(O_K, E[p^\infty])^\vee$, namely we have*

$$\vartheta_K \,\mathrm{Sel}(O_K, E[p^\infty])^\vee = 0.$$

Proof. We may assume $K = \mathbf{Q}(mp^n)$ for some squarefree product m of primes in $\mathcal{P}_{good}$ and for some $n \in \mathbf{Z}_{\geq 0}$. By Theorem 6 (1), taking S to be the set of all prime divisors of m, we have $\xi_{K_\infty} \in \mathrm{Fitt}_{0,\Lambda_{K_\infty}}(\mathrm{Sel}(O_{K_\infty}[1/m], E[p^\infty])^\vee)$, which implies $\xi_{K_\infty}\,\mathrm{Sel}(O_{K_\infty}, E[p^\infty])^\vee = 0$. Let $\xi_K \in R_K = \mathbf{Z}_p[\mathrm{Gal}(K/\mathbf{Q})]$ be the image of ξ_{K_∞}. Since the natural map $\mathrm{Sel}(O_K, E[p^\infty]) \longrightarrow \mathrm{Sel}(O_{K_\infty}, E[p^\infty])$ is injective, we have $\xi_K\,\mathrm{Sel}(O_K, E[p^\infty])^\vee = 0$.

By the definitions of $\xi_{\mathbf{Q}(\mu_{mp^\infty})}$, $\xi_{\mathbf{Q}(mp^n)}$, $\vartheta_{\mathbf{Q}(mp^n)}$, we can write

$$\xi_K = \xi_{\mathbf{Q}(mp^n)} = \vartheta_{\mathbf{Q}(mp^n)} + \sum_{d|m, d \neq m} \lambda_d \nu_{m,d}(\vartheta_{\mathbf{Q}(dp^n)}) \tag{16}$$

for some $\lambda_d \in R_{\mathbf{Q}(mp^n)}$ where $\nu_{m,d} : R_{\mathbf{Q}(dp^n)} \longrightarrow R_{\mathbf{Q}(mp^n)}$ is the norm map defined similarly as in Sect. 2.1. We will prove this theorem by induction on m. Since $d < m$, we have $\vartheta_{\mathbf{Q}(dp^n)} \in \mathrm{Ann}_{R_{\mathbf{Q}(dp^n)}}(\mathrm{Sel}(O_{\mathbf{Q}(dp^n)}, E[p^\infty])^\vee)$ by the hypothesis of the induction. This implies that $\nu_{m,d}(\vartheta_{\mathbf{Q}(dp^n)})$ annihilates $\mathrm{Sel}(O_{\mathbf{Q}(mp^n)}, E[p^\infty])^\vee$. Since ξ_K is in $\mathrm{Ann}_{R_K}(\mathrm{Sel}(O_K, E[p^\infty])^\vee)$, the above equation implies that ϑ_K is in $\mathrm{Ann}_{R_K}(\mathrm{Sel}(O_K, E[p^\infty])^\vee)$.

Remark 2. Let K, S, m be as in Theorem 6. Under our assumptions, the control theorem works completely;

$$\mathrm{Sel}(O_K[1/m], E[p^\infty]) \xrightarrow{\simeq} \mathrm{Sel}(O_{K_\infty}[1/m], E[p^\infty])^{\mathrm{Gal}(K_\infty/K)}.$$

Therefore, Theorem 6 (1) implies that $\mathrm{Fitt}_{0,R_K}(\mathrm{Sel}(O_K[1/m], E[p^\infty])^\vee)$ is principal and

$$\xi_{K,S} \in \mathrm{Fitt}_{0,R_K}(\mathrm{Sel}(O_K[1/m], E[p^\infty])^\vee) \tag{17}$$

where $\xi_{K,S}$ is the image of $\xi_{K_\infty,S}$ in R_K.

Theorem 6 (2) implies that if we assume the main conjecture (9), we have

$$\mathrm{Fitt}_{0,R_K}(\mathrm{Sel}(O_K[1/m], E[p^\infty])^\vee) = \xi_{K,S} R_K\ . \tag{18}$$

2.4 Higher Fitting Ideals

For a commutative ring R and a finitely presented R-module M with n generators, let A be an $n \times m$ relation matrix of M. For an integer $i \geq 0$, $\mathrm{Fitt}_{i,R}(M)$ is defined to be the ideal of R generated by all $(n-i) \times (n-i)$ minors of A (cf. Northcott 1976; this ideal $\mathrm{Fitt}_{i,R}(M)$ does not depend on the choice of a relation matrix A).

Suppose that $K/\mathbf{Q}$ is a finite extension such that K is in the cyclotomic $\mathbf{Z}_p$-extension $\mathbf{Q}_\infty$ of $\mathbf{Q}$, and that m is a squarefree product of primes in $\mathcal{P}^{(N)}$. We define $K(m)$ by $K(m) = \mathbf{Q}(m)K$.

We put $\mathcal{G}_\ell = \mathrm{Gal}(\mathbf{Q}(\ell)/\mathbf{Q})$ and $\mathcal{G}_m = \mathrm{Gal}(\mathbf{Q}(m)/\mathbf{Q}) = \prod_{\ell|m} \mathcal{G}_\ell$. We have $\mathrm{Gal}(K(m)/K) = \mathcal{G}_m$. We put $n_\ell = \mathrm{ord}_p(\ell - 1)$. Suppose that $m = \ell_1 \cdot \ldots \cdot \ell_r$. We take a generator τ_{ℓ_i} of $\mathcal{G}_{\ell_i}$ and put $S_i = \tau_{\ell_i} - 1 \in R_{K(m)}$. We write n_i for n_{ℓ_i}. We identify $R_{K(m)}$ with

$$R_K[\mathcal{G}_m] = R_K[S_1, \ldots, S_r]/((1+S_1)^{p^{n_1}} - 1, \ldots, (1+S_r)^{p^{n_r}} - 1).$$

We consider $\vartheta_{K(m)} \in R_{K(m)}$ and write

$$\vartheta_{K(m)} = \sum_{i_1,\ldots,i_r \geq 0} a^{(m)}_{i_1,\ldots,i_r} S_1^{i_1} \cdot \ldots \cdot S_r^{i_r}$$

where $a^{(m)}_{i_1,\ldots,i_r} \in R_K$. Put $n_0 = \min\{n_1, \ldots, n_r\}$. For $s \in \mathbf{Z}_{>0}$, we define c_s to be the maximal positive integer c such that

$$T^{-1}((1+T)^{p^{n_0}} - 1) \in p^c \mathbf{Z}_p[T] + T^{s+1}\mathbf{Z}_p[T].$$

For example, $c_1 = n_0, \ldots, c_{p-2} = n_0$, $c_{p-1} = n_0 - 1, \ldots, c_{p^2-1} = n_0 - 2$. If $i_1, \ldots, i_r \leq s$, $a^{(m)}_{i_1,\ldots,i_r} \bmod p^{c_s}$ is well-defined (it does not depend on the choice of $a^{(m)}_{i_1,\ldots,i_r}$).

Theorem 8. *Let K be an intermediate field of the cyclotomic $\mathbf{Z}_p$-extension $\mathbf{Q}_\infty/\mathbf{Q}$ with $[K:\mathbf{Q}] < \infty$. Let c_s be the integer defined above for $s \in \mathbf{Z}_{>0}$ and m. Assume that $i_1, \ldots, i_r \leq s$ and $i_1 + \ldots + i_r \leq i$. Then we have*

$$a^{(m)}_{i_1,\ldots,i_r} \in \mathrm{Fitt}_{i,R_K/p^{c_s}}(\mathrm{Sel}(E/K, E[p^{c_s}])^\vee).$$

For $m = \ell_1 \cdot \ldots \cdot \ell_r$, we denote $(-1)^r$ times the coefficient of $S_1 \cdot \ldots \cdot S_r$ in $\vartheta_{K(m)}$ by δ_m. If ℓ_i splits completely in K for all $i = 1, \ldots, r$, we can write

$$\vartheta_{K(m)} \equiv \delta_m \prod_{i=1}^{r} (1 - \tau_{\ell_i}) = (-1)^r \delta_m S_1 \cdot \ldots \cdot S_r \pmod{p^N, S_1^2, \ldots, S_r^2} \qquad (19)$$

(see Kurihara 2014, §21). Taking $s = 1$ and $i = r$ in Theorem 8, we get

Corollary 1. *Let $K/\mathbf{Q}$ be a finite extension such that $K \subset \mathbf{Q}_\infty$. We have*

$$\delta_m \in \mathrm{Fitt}_{r,R_K/p^N}(\mathrm{Sel}(E/K, E[p^N])^\vee)$$

where $m = \ell_1 \cdot \ldots \cdot \ell_r$.

Proof of Theorem 8. We may assume $K = \mathbf{Q}(p^n)$ for some $n \geq 0$, so $K(m) = \mathbf{Q}(mp^n)$. First of all, we consider the image $\xi_{K(m)} \in R_{K(m)}$ of $\xi_{K(m)_\infty}$. Since $\mathrm{Sel}(E/K(m), E[p^\infty]) \longrightarrow \mathrm{Sel}(E/K(m)_\infty, E[p^\infty])$ is injective, $\xi_{K(m)}$ is in $\mathrm{Fitt}_{0,R_{K(m)}}(\mathrm{Sel}(E/K(m), E[p^\infty])^\vee)$ by Theorem 6 (1). We write

$$\xi_{K(m)} = \sum_{i_1,\ldots,i_r \geq 0} \alpha^{(m)}_{i_1,\ldots,i_r} S_1^{i_1} \cdot \ldots \cdot S_r^{i_r}$$

where $\alpha^{(m)}_{i_1,\ldots,i_r} \in R_K$. Assume that $i_1, \ldots, i_r \leq s$ and $i_1 + \ldots + i_r \leq i$. Then by Lemma 4.2 in Kurihara (2014) we have

$$\alpha^{(m)}_{i_1,\ldots,i_r} \in \mathrm{Fitt}_{i,R_K/p^{c_s}}(\mathrm{Sel}(E/K, E[p^{c_s}])^\vee).$$

On the other hand, since $K(m) = \mathbf{Q}(mp^n)$ for some $n \geq 0$, we have

$$\xi_{K(m)} = \vartheta_{K(m)} + \sum_{d|m, d \neq m} \lambda_d \nu_{m,d}(\vartheta_{\mathbf{Q}(dp^n)})$$

for some $\lambda_d \in R_{K(m)}$ by (16). This implies that the images of $\xi_{K(m)}$ and $\vartheta_{K(m)}$ under the canonical homomorphism

$$R_{K(m)} = R_K[S_1, \ldots, S_r]/I \longrightarrow R_K[[S_1, \ldots, S_r]]/J$$

coincide where $I = ((1+S_1)^{p^{n_1}} - 1, \ldots, (1+S_1)^{p^{n_r}} - 1)$ and $J = (S_1^{-1}(1+S_1)^{p^{n_1}} - 1, \ldots, S_r^{-1}(1+S_1)^{p^{n_r}} - 1, S_1^{s+1}, \ldots, S_r^{s+1})$. Therefore, $\alpha^{(m)}_{i_1,\ldots,i_r} \equiv a^{(m)}_{i_1,\ldots,i_r}$ mod p^{c_s} for $i_1,\ldots,i_r \leq s$. It follows that $a^{(m)}_{i_1,\ldots,i_r} \in \mathrm{Fitt}_{i,R_K/p^{c_s}}(\mathrm{Sel}(E/K, E[p^{c_s}])^\vee)$. This completes the proof of Theorem 8.

3 Review of Kolyvagin Systems of Gauss Sum Type for Elliptic Curves

In this section, we recall the results in Kurihara (2014) on Euler systems and Kolyvagin systems of Gauss sum type in the case of elliptic curves. From this section we assume all the assumptions (i), (ii), (iii), (iv) in Sect. 1.1.

3.1 Some Definitions

Recall that in Sect. 2 we defined $\mathcal{P}_{good}$ by $\mathcal{P}_{good} = \{\ell \mid \ell$ is a good reduction prime for $E\} \setminus \{p\}$, and $\mathcal{P}^{(N)}$ by

$$\mathcal{P}^{(N)} = \{\ell \in \mathcal{P}_{good} \mid \ell \equiv 1 \pmod{p^N}\}$$

for a positive integer $N > 0$. If ℓ is in $\mathcal{P}_{good}$, the absolute Galois group $G_{\mathbf{F}_\ell}$ acts on the group $E[p^N]$ of p^N-torsion points, so we consider $H^i(\mathbf{F}_\ell, E[p^N])$. We define

$$\mathcal{P}_0^{(N)} = \{\ell \in \mathcal{P}^{(N)} \mid H^0(\mathbf{F}_\ell, E[p^N]) \text{ contains an element of order } p^N\},$$
$$(\mathcal{P}_0')^{(N)} = \{\ell \in \mathcal{P}^{(N)} \mid H^0(\mathbf{F}_\ell, E[p^N]) = E[p^N]\}, \text{ and}$$
$$\mathcal{P}_1^{(N)} = \{\ell \in \mathcal{P}^{(N)} \mid H^0(\mathbf{F}_\ell, E[p^N]) \simeq \mathbf{Z}/p^N\}.$$

So $\mathcal{P}_0^{(N)} \supset (\mathcal{P}_0')^{(N)}$, $\mathcal{P}_0^{(N)} \supset \mathcal{P}_1^{(N)}$, and $(\mathcal{P}_0')^{(N)} \cap \mathcal{P}_1^{(N)} = \emptyset$. Suppose that ℓ is in $\mathcal{P}_1^{(N)}$. Then, since $\ell \equiv 1 \pmod{p^N}$, we have an exact sequence $0 \longrightarrow \mathbf{Z}/p^N \longrightarrow E[p^N] \longrightarrow \mathbf{Z}/p^N \longrightarrow 0$ of $G_{\mathbf{F}_\ell}$-modules where $G_{\mathbf{F}_\ell}$ acts on $\mathbf{Z}/p^N$ trivially. So the action of the Frobenius Frob_ℓ at ℓ on $E[p^N]$ can be written as $\begin{pmatrix} 1 & 1 \\ 0 & 1 \end{pmatrix}$ for a suitable basis of $E[p^N]$. Therefore, $H^1(\mathbf{F}_\ell, E[p^N])$ is also isomorphic to $\mathbf{Z}/p^N$ for $\ell \in \mathcal{P}_1^{(N)}$.

Let $t \in E[p^N]$ be an element of order p^N. We define

$$\mathcal{P}_{0,t}^{(N)} = \{\ell \in \mathcal{P}^{(N)} \mid t \in H^0(\mathbf{F}_\ell, E[p^N])\},$$
$$\mathcal{P}_{1,t}^{(N)} = \{\ell \in \mathcal{P}^{(N)} \mid H^0(\mathbf{F}_\ell, E[p^N]) = (\mathbf{Z}/p^N)t\}.$$

So, $\mathcal{P}_0^{(N)} = \bigcup_t \mathcal{P}_{0,t}^{(N)}$ and $\mathcal{P}_1^{(N)} = \bigcup_t \mathcal{P}_{1,t}^{(N)}$ where t runs over all elements of order p^N. Since we assumed that the Galois action on the Tate module is surjective, both $(\mathcal{P}_0')^{(N)}$ and $\mathcal{P}_{1,t}^{(N)}$ are infinite by Chebotarev density theorem (Kurihara 2014, §5.8).

We define $\mathcal{K}_{(p)}$ to be the set of number fields K such that $K/\mathbf{Q}$ is a finite abelian p-extension in which all bad primes of E are unramified. Suppose that K is in $\mathcal{K}_{(p)}$. We define

$$(\mathcal{P}_0')^{(N)}(K) = \{\ell \in (\mathcal{P}_0')^{(N)} \mid \ell \text{ splits completely in } K\},$$
$$\mathcal{P}_1^{(N)}(K) = \{\ell \in \mathcal{P}_1^{(N)} \mid \ell \text{ splits completely in } K\}.$$

Again by Chebotarev density theorem, both $(\mathcal{P}_0')^{(N)}(K)$ and $\mathcal{P}_1^{(N)}(K)$ are infinite (see Kurihara 2014, §5.8).

Suppose $\ell \in \mathcal{P}_{good}$. For a prime v above ℓ, we know $H^1(K_v, E[p^N])/(E(K_v) \otimes \mathbf{Z}/p^N) = H^0(\kappa(v), E[p^N](-1))$ where $\kappa(v)$ is the residue field of v. We put

$$\mathcal{H}_\ell^2(K) = \bigoplus_{v|\ell} H^0(\kappa(v), E[p^N](-1)). \tag{20}$$

If ℓ is in $(\mathcal{P}_0')^{(N)}(K)$ (resp. $\mathcal{P}_1^{(N)}(K)$), $\mathcal{H}_\ell^2(K)$ is a free R_K/p^N-module of rank 2 (resp. rank 1) where $R_K = \mathbf{Z}_p[\mathrm{Gal}(K/\mathbf{Q})]$ as before.

From now on, for a prime $\ell \in \mathcal{P}_0^{(N)}$, we fix a prime $\ell_{\overline{\mathbf{Q}}}$ of an algebraic closure $\overline{\mathbf{Q}}$ above ℓ. For any algebraic number field F, we denote the prime of F below $\ell_{\overline{\mathbf{Q}}}$ by ℓ_F, so when we consider finite extensions F_1/k, F_2/k such that $F_1 \subset F_2$, the primes ℓ_{F_2}, ℓ_{F_1} satisfy $\ell_{F_2}|\ell_{F_1}$.

We take a primitive p^n-th root of unity ζ_{p^n} such that $(\zeta_{p^n})_{n\geq 1} \in \mathbf{Z}_p(1) = \varprojlim \mu_{p^n}$, and fix it.

In the following, for each ℓ in $\mathcal{P}_0^{(N)}(K)$, we take $t_\ell \in H^0(\mathbf{F}_\ell, E[p^N])$ and fix it. We define

$$t_{\ell,K} = (t_\ell \otimes \zeta_{p^N}^{\otimes(-1)}, 0, \ldots, 0) \in \mathcal{H}_\ell^2(K) \tag{21}$$

where the right hand side is the element whose ℓ_K-component is $t_\ell \otimes \zeta_{p^N}^{\otimes(-1)}$ and other components are zero.

Suppose that K is in $\mathcal{K}_{(p)}$. Let K_∞/K be the cyclotomic $\mathbf{Z}_p$-extension, and K_n be the n-th layer. Since $\mathrm{Sel}(O_{K_\infty}, E[p^\infty])^\vee$ is a finitely generated $\mathbf{Z}_p$-module, the corestriction map $\mathrm{Sel}(O_{K_m}, E[p^N]) \longrightarrow \mathrm{Sel}(O_K, E[p^N])$ is the zero map if m is sufficiently large. We take the minimal $m > 0$ satisfying this property, and put $K_{[1]} = K_m$. We define inductively $K_{[n]}$ by $K_{[n]} = (K_{[n-1]})_{[1]}$ where we applied the above definition to $K_{[n-1]}$ instead of K.

We can compute how large $K_{[n]}$ is. Let λ be the λ-invariant of $\mathrm{Sel}(O_{K_\infty}, E[p^\infty])^\vee$. We take $a \in \mathbf{Z}_{\geq 0}$ such that $p^{a+1} - p^a \geq \lambda$. Suppose that $K = K'_m$ (m-th layer of K'_∞/K') for some K' such that p is unramified in K'. The corestriction map $\mathrm{Sel}(O_{K'_{a+1}}, E[p]) \longrightarrow \mathrm{Sel}(O_{K'_a}, E[p])$ is the zero map. Therefore, $\mathrm{Sel}(O_{K'_{a+N}}, E[p^N]) \longrightarrow \mathrm{Sel}(O_{K'_a}, E[p^N])$ is the zero map. Put $a' = \max(a - m, 0)$. Then $\mathrm{Sel}(O_{K_{a'+N}}, E[p^N]) \longrightarrow \mathrm{Sel}(O_{K_{a'}}, E[p^N])$ is the zero map. Therefore, we have $K_{[1]} \subset K_{a'+N}$. Also we know $K_{[n]} \subset K_{a'+nN}$.

Let n_λ, d_n be the numbers defined just before (4) in Sect. 1.2. Then we can show that if $\ell \in \mathcal{P}_1^{(N)}$ satisfies $\ell \equiv 1 \pmod{p^{d_n}}$, ℓ is in $\mathcal{P}_1^{(N)}(\mathbf{Q}_{[n]})$ by the same method as above.

3.2 Euler Systems of Gauss Sum Type for Elliptic Curves

For the Euler systems of Gauss sums, see Kolyvagin (1990) and Rubin (1991), and for Euler systems of Gauss sum type, see Kurihara (2012, 2014). We use the following lemma which is the global duality theorem (see Theorem 2.3.4 in Mazur and Rubin 2004).

Lemma 1. *Suppose that m is a product of primes in $\mathcal{P}_{good}$. We have an exact sequence*

$$0 \longrightarrow \mathrm{Sel}(O_K, E[p^N]) \longrightarrow \mathrm{Sel}(O_K[1/m], E[p^N]) \longrightarrow \bigoplus_{\ell | m} \mathcal{H}^2_\ell(K) \longrightarrow \mathrm{Sel}(O_K, E[p^N])^\vee.$$

We remark that we can take m such that the last map is surjective in our case (see Lemma 3 below).

Let K be a number field in $\mathcal{K}_{(p)}$ and $\ell \in \mathcal{P}_0^{(N)}(K_{[1]})$. We apply the above lemma to $K_{[1]}$ and obtain an exact sequence

$$\mathrm{Sel}(O_{K_{[1]}}[1/\ell], E[p^N]) \xrightarrow{\partial_\ell} \mathcal{H}^2_\ell(K_{[1]}) \xrightarrow{w_\ell} \mathrm{Sel}(O_{K_{[1]}}, E[p^N])^\vee.$$

Consider $\vartheta_{K_{[1]}} t_{\ell,K_{[1]}} \in \mathcal{H}_\ell^2(K_{[1]})$. By Theorem 7 we know $w_\ell(\vartheta_{K_{[1]}} t_{\ell,K_{[1]}}) = \vartheta_{K_{[1]}} w_\ell(t_{\ell,K_{[1]}}) = 0$. Therefore, there is an element $g \in \mathrm{Sel}(O_{K_{[1]}}[1/\ell], E[p^N])$ such that $\partial_\ell(g) = \vartheta_{K_{[1]}} t_{\ell,K_{[1]}}$. We define

$$g_{\ell,t_\ell}^{(K)} = \mathrm{Cor}_{K_{[1]}/K}(g) \in \mathrm{Sel}(O_K[1/\ell], E[p^N]). \tag{22}$$

This element $g_{\ell,t_\ell}^{(K)}$ does not depend on the choice of $g \in \mathrm{Sel}(O_{K_{[1]}}[1/\ell], E[p^N])$ (Kurihara 2014, §6.10). We write g_ℓ instead of $g_{\ell,t_\ell}^{(K)}$ when no confusion arises.

Remark 3. To define g_ℓ, we used in Kurihara (2014) the p-adic L-function θ_{K_∞} whose Euler factor at ℓ is $1 - \frac{a_\ell}{\ell}\sigma_\ell^{-1} + \frac{1}{\ell}\sigma_\ell^{-2}$. The element θ_{K_∞} can be constructed from ϑ_{K_∞} by the same method as when we constructed ξ_{K_∞} in Sect. 2.1. In the above definition (22), we used ϑ_K (namely ϑ_{K_∞}) instead of θ_{K_∞}.

3.3 *Kolyvagin Derivatives of Gauss Sum Type*

Let ℓ be a prime in $\mathcal{P}_{good}$. We define ∂_ℓ as a natural homomorphism

$$\partial_\ell : H^1(K, E[p^N]) \longrightarrow \mathcal{H}_\ell^2(K) = \bigoplus_{v|\ell} H^0(\kappa(v), E[p^N](-1))$$

where we used $H^1(K_v, E[p^N])/(E(K_v) \otimes \mathbf{Z}/p^N) = H^0(\kappa(v), E[p^N](-1))$.

Next, we assume $\ell \in \mathcal{P}_1^{(N)}(K)$. We denote by $\mathbf{Q}_\ell(\ell)$ the maximal p-subextension of $\mathbf{Q}_\ell$ inside $\mathbf{Q}_\ell(\mu_\ell)$. Put $\mathcal{G}_\ell = \mathrm{Gal}(\mathbf{Q}_\ell(\ell)/\mathbf{Q}_\ell)$. By Kummer theory, $\mathcal{G}_\ell$ is isomorphic to $\mu_{p^{n_\ell}}$ where $n_\ell = \mathrm{ord}_p(\ell - 1)$. We denote by τ_ℓ the corresponding element of $\mathcal{G}_\ell$ to $\zeta_{p^{n_\ell}}$ that is the primitive p^{n_ℓ}-th root of unity we fixed.

We consider the natural homomorphism $H^1(\mathbf{Q}_\ell, E[p^N]) \longrightarrow H^1(\mathbf{Q}_\ell(\ell), E[p^N])$ and denote the kernel by $H_{tr}^1(\mathbf{Q}_\ell, E[p^N])$. Let $\mathbf{Q}_{\ell,nr}$ be the maximal unramified extension of $\mathbf{Q}_\ell$. We identify $H^1(\mathbf{F}_\ell, E[p^N])$ with $H^1(\mathrm{Gal}(\mathbf{Q}_{\ell,nr}/\mathbf{Q}_\ell), E[p^N])$, and regard it as a subgroup of $H^1(\mathbf{Q}_\ell, E[p^N])$. Then both $H^1(\mathbf{F}_\ell, E[p^N])$ and $H_{tr}^1(\mathbf{Q}_\ell, E[p^N])$ are isomorphic to $\mathbf{Z}/p^N$, and we have decomposition

$$H^1(\mathbf{Q}_\ell, E[p^N]) = H^1(\mathbf{F}_\ell, E[p^N]) \oplus H_{tr}^1(\mathbf{Q}_\ell, E[p^N])$$

as an abelian group. We also note that $H^1(\mathbf{F}_\ell, E[p^N])$ coincides with the image of the Kummer map and is isomorphic to $E(\mathbf{Q}_\ell) \otimes \mathbf{Z}/p^N$. We consider the homomorphism

$$\phi' : H^1(\mathbf{Q}_\ell, E[p^N]) \longrightarrow H^1(\mathbf{F}_\ell, E[p^N]) \tag{23}$$

which is obtained from the above decomposition.

Note that $H^1(\mathbf{F}_\ell, E[p^N]) = E[p^N]/(\mathrm{Frob}_\ell - 1)$ where Frob_ℓ is the Frobenius at ℓ. Since ℓ is in $\mathcal{P}_1^{(N)}$, $\mathrm{Frob}_\ell^{-1} - 1 : E[p^N]/(\mathrm{Frob}_\ell - 1) \longrightarrow E[p^N]^{\mathrm{Frob}_\ell = 1} = H^0(\mathbf{F}_\ell, E[p^N])$ is an isomorphism. We define $\phi'' : H^1(\mathbf{Q}_\ell, E[p^N]) \longrightarrow$

$H^0(\mathbf{F}_\ell, E[p^N])$ as the composition of ϕ' and $H^1(\mathbf{F}_\ell, E[p^N]) \xrightarrow{\mathrm{Frob}_\ell^{-1}-1} H^0(\mathbf{F}_\ell, E[p^N])$. We define

$$\phi_\ell : H^1(K, E[p^N]) \longrightarrow \mathcal{H}^2_\ell(K)(1)$$

as the composition of the natural homomorphism $H^1(K, E[p^N]) \longrightarrow \bigoplus_{v|\ell} H^1(K_v, E[p^N])$ and ϕ'' for K_v. Using the primitive p^N-th root of unity ζ_{p^N} we fixed, we regard ϕ_ℓ as a homomorphism

$$\phi_\ell : H^1(K, E[p^N]) \longrightarrow \mathcal{H}^2_\ell(K).$$

For a prime $\ell \in \mathcal{P}_1^{(N)}(K)$, we put $\mathcal{G}_\ell = \mathrm{Gal}(\mathbf{Q}(\ell)/\mathbf{Q})$. We identify $\mathcal{G}_\ell$ with $\mathrm{Gal}(\mathbf{Q}_\ell(\ell)/\mathbf{Q}_\ell)$. Recall that we defined n_ℓ by $p^{n_\ell} = [\mathbf{Q}(\ell) : \mathbf{Q}]$, and we took a generator τ_ℓ of $\mathcal{G}_\ell$ above. We define

$$N_\ell = \sum_{i=0}^{p^{n_\ell}-1} \tau_\ell^i \in \mathbf{Z}[\mathcal{G}_\ell], \ D_\ell = \sum_{i=0}^{p^{n_\ell}-1} i\tau_\ell^i \in \mathbf{Z}[\mathcal{G}_\ell]$$

as usual.

We define $\mathcal{N}_1^{(N)}(K)$ to be the set of squarefree products of primes in $\mathcal{P}_1^{(N)}(K)$. We suppose $1 \in \mathcal{N}_1^{(N)}(K)$. For $m \in \mathcal{N}_1^{(N)}(K)$, we put $\mathcal{G}_m = \mathrm{Gal}(\mathbf{Q}(m)/\mathbf{Q})$, $N_m = \Pi_{\ell|m} N_\ell \in \mathbf{Z}[\mathcal{G}_m]$, and $D_m = \Pi_{\ell|m} D_\ell \in \mathbf{Z}[\mathcal{G}_m]$. Assume that ℓ is in $(\mathcal{P}_0')^{(N)}(K(m)_{[1]})$ and consider $g_{\ell,t_\ell}^{K(m)} \in \mathrm{Sel}(O_{K(m)}[1/\ell], E[p^N])$. We can check that $D_m g_{\ell,t_\ell}^{K(m)}$ is in $\mathrm{Sel}(O_{K(m)}[1/m\ell], E[p^N])^{\mathcal{G}_m}$. Using the fact that $\mathrm{Sel}(O_K[1/m\ell], E[p^N]) \xrightarrow{\simeq} \mathrm{Sel}(O_{K(m)}[1/m\ell], E[p^N])^{\mathcal{G}_m}$ is bijective by Lemma 2 below (cf. also Kurihara 2014, Lemma 7.6), we define

$$\kappa_{m,\ell} = \kappa_{m,\ell,t_\ell}^{(K)} \in \mathrm{Sel}(O_K[1/m\ell], E[p^N]) \tag{24}$$

to be the unique element whose image in $\mathrm{Sel}(O_{K(m)}[1/m\ell], E[p^N])$ is $D_m g_{\ell,t_\ell}^{(K(m))}$.

The following lemma will be also used in the next section.

Lemma 2. *Suppose that* $K, L \in \mathcal{K}_{(p)}$ *and* $K \subset L$. *For any* $m \in \mathbf{Z}_{>0}$, *the restriction map* $\mathrm{Sel}(O_K[1/m], E[p^N]) \xrightarrow{\simeq} \mathrm{Sel}(O_L[1/m], E[p^N])^{\mathrm{Gal}(L/K)}$ *is bijective.*

Proof. Let N_E be the conductor of E, $m' = mpN_E$, and m'' the product of primes which divide pN_E and which do not divide m. Put $G = \mathrm{Gal}(L/K)$. We have a commutative diagram of exact sequences

$$\begin{array}{ccccccc}
0 \longrightarrow & \mathrm{Sel}(O_K[1/m], E[p^N]) & \longrightarrow & \mathrm{Sel}(O_K[1/m'], E[p^N]) & \longrightarrow & \bigoplus_{v|m''} H^2_{K,v} \\
& \downarrow{\scriptstyle \alpha_1} & & \downarrow{\scriptstyle \alpha_2} & & \downarrow{\scriptstyle \alpha_3} \\
0 \longrightarrow & \mathrm{Sel}(O_L[1/m], E[p^N])^G & \longrightarrow & \mathrm{Sel}(O_L[1/m'], E[p^N])^G & \longrightarrow & (\bigoplus_{w|m''} H^2_{L,w})^G
\end{array}$$

where $H^2_{K,v} = H^1(K_v, E[p^N])/(E(K_v) \otimes \mathbf{Z}/p^N)$ and $H^2_{L,w} = H^1(L_w, E[p^N])/(E(L_w) \otimes \mathbf{Z}/p^N)$. Since $\mathrm{Sel}(O_L[1/m'], E[p^N]) = H^1_{et}(\mathrm{Spec}\, O_L[1/m'], E[p^N])$ and $H^0(L, E[p^N]) = 0$, α_2 is bijective. Suppose that v divides m'' and w is above v. When v divides N_E, since v is unramified in L and p is prime to $\mathrm{Tam}(E)$, $H^2_{K,v} \longrightarrow H^2_{L,w}$ is injective (Greenberg 1999, §3). When v is above p, $H^2_{K,v} \longrightarrow H^2_{L,w}$ is injective because $a_p \not\equiv 1 \pmod p$ (Greenberg 1999, §3). Hence α_3 is injective. Therefore, α_1 is bijective.

In Kurihara (2012), if m has a factorization $m = \ell_1 \cdot \ldots \cdot \ell_r$ such that $\ell_{i+1} \in \mathcal{P}_1^{(N)}(K(\ell_1 \cdot \ldots \cdot \ell_i))$ for all $i = 1,\ldots,r-1$, we called m *well-ordered*. But the word "well-ordered" might cause confusion, so we call m *admissible* in this paper if m satisfies the above condition. Note that we do not impose the condition $\ell_1 < \ldots < \ell_r$ in the above definition, and that m is admissible if there is one factorization as above. We sometimes call the set of prime divisors of m *admissible* if m is admissible.

Suppose that $m = \ell_1 \cdot \ldots \cdot \ell_r$. We define $\delta_m \in R_K/p^N$ by

$$\vartheta_{K(m)} \equiv \delta_m \prod_{i=1}^{r}(1-\tau_{\ell_i}) \pmod{p^N, (\tau_{\ell_1}-1)^2, \ldots, (\tau_{\ell_r}-1)^2} \tag{25}$$

(see Kurihara 2014, (21)).

We simply write $\kappa_{m,\ell}$ for $\kappa^{(K)}_{m,\ell,t_\ell}$. We have the following Proposition (Kurihara 2014, Propositions 7.7, 7.15 and Lemma 7.9).

Proposition 1. *Suppose that m is in $\mathcal{N}_1^{(N)}(K)$, and $\ell \in (\mathcal{P}'_0)^{(N)}(K(m)_{[1]})$. We take n_0 sufficiently large such that every prime of K_{n_0} dividing m is inert in K_∞/K_{n_0}. We further assume that $\ell \in (\mathcal{P}'_0)^{(N)}(K_{n_0+N})$. Then*

(0) $\kappa_{m,\ell} \in \mathrm{Sel}(O_K[1/m\ell], E[p^N])$.
(1) $\partial_r(\kappa_{m,\ell}) = \phi_r(\kappa_{\frac{m}{r},\ell})$ *for any prime divisor r of m.*
(2) $\partial_\ell(\kappa_{m,\ell}) = \delta_m t_{\ell,K}$.
(3) *Assume further that m is admissible. Then $\phi_r(\kappa_{m,\ell}) = 0$ for any prime divisor r of m.*

3.4 Construction of Kolyvagin Systems of Gauss Sum Type

In the previous subsection we constructed $\kappa_{m,\ell}$ for $m \in \mathcal{N}_1^{(N)}(K)$ and a prime $\ell \in (\mathcal{P}'_0)^{(N)}(K)$ satisfying some properties. In this subsection we construct $\kappa_{m,\ell}$ for $\ell \in \mathcal{P}_1^{(N)}(K)$ satisfying some properties (see Proposition 2). The property (4) in Proposition 2 is a beautiful property of our Kolyvagin systems of Gauss sum type, which is unique for Kolyvagin systems of Gauss sum type.

For a squarefree product m of primes, we define $\epsilon(m)$ to be the number of prime divisors of m, namely $\epsilon(m) = r$ if $m = \ell_1 \cdot \ldots \cdot \ell_r$.

For any prime number ℓ, we write $\mathcal{H}^2_\ell(K) = \bigoplus_{v|\ell} H^1(K_v, E[p^N])/(E(K_v) \otimes \mathbf{Z}/p^N)$, and consider the natural map

$$w_K : \bigoplus_\ell \mathcal{H}^2_\ell(K) \longrightarrow \mathrm{Sel}(O_K, E[p^N])^\vee$$

which is obtained by taking the dual of $\mathrm{Sel}(O_K, E[p^N]) \longrightarrow \bigoplus_v E(K_v) \otimes \mathbf{Z}/p^N$. We also consider the natural map

$$\partial_K : H^1(K, E[p^N]) \longrightarrow \bigoplus_\ell \mathcal{H}^2_\ell(K).$$

We use the following lemma which was proved in Kurihara (2014, Proposition 5.14 and Lemma 7.4 (2)).

Lemma 3. *Suppose that $K \in \mathcal{K}_{(p)}$ and $r_1,\ldots,r_s$ are s distinct primes in $\mathcal{P}_1^{(N)}(K)$. Assume that for each $i = 1,\ldots,s$, $\sigma_i \in \mathcal{H}^2_{r_i}(K)$ is given, and also $x \in \mathrm{Sel}(O_K, E[p^N])^\vee$ is given. Let K'/K be an extension such that $K' \in \mathcal{K}_{(p)}$. Then there are infinitely many $\ell \in \mathcal{P}_0^{(N)}(K)$ such that $w_K(t_{\ell,K}) = x$. We take such a prime ℓ and fix it. Then there are infinitely many $\ell' \in (\mathcal{P}'_0)^{(N)}(K')$ which satisfy the following properties:*

(i) $w_K(t_{\ell',K}) = w_K(t_{\ell,K}) = x$.
(ii) *There is an element $z \in \mathrm{Sel}(O_K[1/\ell\ell'], E[p^N])$ such that $\partial_K(z) = t_{\ell',K} - t_{\ell,K}$ and $\phi_{r_i}(z) = \sigma_i$ for each $i = 1,\ldots,s$.*

Assume that $m\ell$ is in $\mathcal{N}_1^{(N)}(K_{[\epsilon(m\ell)]})$. By Lemma 3 we can take $\ell' \in (\mathcal{P}'_0)^{(N)}$ satisfying the following properties:

(i) $\ell' \in (\mathcal{P}'_0)^{(N)}(K_{[\epsilon(m\ell)]}(m)_{[1]}K_{n_0+N})$ where n_0 is as in Proposition 1.
(ii) $w_{K_{[\epsilon(m\ell)]}}(t_{\ell',K_{[\epsilon(m\ell)]}}) = w_{K_{[\epsilon(m\ell)]}}(t_{\ell,K_{[\epsilon(m\ell)]}})$.
(iii) Let $\phi_r^{(K_{[\epsilon(m\ell)]})} : H^1(K_{[\epsilon(m\ell)]}, E[p^N]) \longrightarrow \mathcal{H}^2_r(K_{[\epsilon(m\ell)]})$ be the map ϕ_r for $K_{[\epsilon(m\ell)]}$. There is an element b' in $\mathrm{Sel}(O_{K_{[\epsilon(m\ell)]}}[1/\ell\ell'], E[p^N])$ such that

$$\partial_{K_{[\epsilon(m\ell)]}}(b') = t_{\ell',K_{[\epsilon(m\ell)]}} - t_{\ell,K_{[\epsilon(m\ell)]}}$$

and $\phi_r^{K_{[\epsilon(m\ell)]}}(b') = 0$ for all r dividing m.

We have already defined $\kappa_{m,\ell'}$ in the previous subsection. We put $b = \mathrm{Cor}_{K_{[\epsilon(m\ell)]}/K}(b')$ and define

$$\kappa_{m,\ell} = \kappa_{m,\ell'} - \delta_m b. \tag{26}$$

Then this element does not depend on the choice of ℓ' and b' (see Kurihara 2014, §7.10). In Kurihara (2014), we took b' which does not necessarily satisfy

$\phi_r^{K_{[\epsilon(m\ell)]}}(b') = 0$ in the definition of $\kappa_{m,\ell}$. But we adopted the above definition here because it is simpler and there is no loss of generality.

The next proposition was proved in Kurihara (2014, Propositions 7.13, 7.15, 7.16).

Proposition 2. *Suppose that $m\ell$ is in $\mathcal{N}_1^{(N)}(K_{[\epsilon(m\ell)]})$. Then*

(0) $\kappa_{m,\ell} \in \mathrm{Sel}(O_K[1/m\ell], E[p^N])$.
(1) $\partial_r(\kappa_{m,\ell}) = \phi_r(\kappa_{\frac{m}{r},\ell})$ *for any prime divisor r of m.*
(2) $\partial_\ell(\kappa_{m,\ell}) = \delta_m t_{\ell,K}$.
(3) *Assume further that m is admissible. Then $\phi_r(\kappa_{m,\ell}) = 0$ for any prime divisor r of m.*
(4) *Assume further that $m\ell$ is admissible, and $m\ell$ is in $\mathcal{N}_1^{(N)}(K_{[\epsilon(m\ell)+1]})$. Then we have*

$$\phi_\ell(\kappa_{m,\ell}) = -\delta_{m\ell} t_{\ell,K}.$$

4 Relations of Selmer Groups

In this section, we prove a generalized version of Theorem 3.

4.1 Injectivity Theorem

Suppose that K is in $\mathcal{K}_{(p)}$ and that m is in $\mathcal{N}_1^{(N)}(K)$. For a prime divisor r of m, we denote by

$$w_r : \mathcal{H}_r^2(K) \longrightarrow \mathrm{Sel}(O_K, E[p^N])^\vee$$

the homomorphism which is the dual of $\mathrm{Sel}(O_K, E[p^N]) \longrightarrow \bigoplus_{v|r} E(K_v) \otimes \mathbf{Z}/p^N$. Recall that $\mathcal{H}_r^2(K)$ is a free R_K/p^N-module of rank 1, generated by $t_{r,K}$.

Proposition 3. *We assume that δ_m is a unit of R_K/p^N for some $m \in \mathcal{N}_1^{(N)}(K)$. Then the natural homomorphism $\oplus_{r|m} w_r : \bigoplus_{r|m} \mathcal{H}_r^2(K) \longrightarrow \mathrm{Sel}(O_K, E[p^N])^\vee$ is surjective.*

Remark 4. We note that δ_m is numerically computable, in principle.

Proof (Proof of Proposition 3). Let x be an arbitrary element in $\mathrm{Sel}(O_K, E[p^N])^\vee$. Let $w_r : \mathcal{H}_r^2(K) \longrightarrow \mathrm{Sel}(O_K, E[p^N])^\vee$ be the natural homomorphism for each $r \mid m$. We will prove that x is in the submodule generated by all $w_r(t_{r,K})$ for $r \mid m$. Using Lemma 3, we can take a prime $\ell \in (\mathcal{P}'_0)^{(N)}(K(m)_{[1]}K_{n_0+N})$ such that $w_\ell(t_{\ell,K}) = x$ and ℓ is prime to m. We consider the Kolyvagin derivative $\kappa_{m,\ell}$ which was defined in (24). Consider the exact sequence

$$\mathrm{Sel}(O_K[1/m\ell], E[p^N]) \xrightarrow{\partial} \bigoplus_{\ell'|m\ell} \mathcal{H}^2_{\ell'}(K) \xrightarrow{w_K} \mathrm{Sel}(O_K, E[p^N])^\vee$$

(see Lemma 1) where $\partial = (\oplus\partial_{\ell'})_{\ell'|m\ell}$ and $w_K((z_{\ell'})_{\ell'|m\ell}) = \sum_{\ell'|m\ell} w_{\ell'}(z_{\ell'})$. For each $r \mid m$ we define $\lambda_r \in R_K/p^N$ by $\partial_r(\kappa_{m,\ell}) = \lambda_r t_{r,K} \in \mathcal{H}^2_r(K)$. The above exact sequence and Proposition 1 (2) imply that

$$\delta_m x + \sum_{r|m} \lambda_r w_r(t_{r,K}) = 0$$

in $\mathrm{Sel}(O_K, E[p^N])^\vee$. Since we assumed that δ_m is a unit, x is in the submodule generated by all $w_r(t_{r,K})$'s. This completes the proof of Proposition 3.

For a prime $\ell \in \mathcal{P}_1^{(N)}(K)$, we define

$$\mathcal{H}^1_{\ell,f}(K) = \bigoplus_{v|\ell} E(\kappa(v)) \otimes \mathbf{Z}/p^N.$$

Since $\kappa(v) = \mathbf{F}_\ell$, $E(\kappa(v)) \otimes \mathbf{Z}/p^N$ is isomorphic to $\mathbf{Z}/p^N$ and $\mathcal{H}^1_{\ell,f}(K)$ is a free R_K/p^N-module of rank 1.

Corollary 2. *Suppose that $m = \ell_1 \cdot \ldots \cdot \ell_a$ is in $\mathcal{N}_1^{(N)}(K)$. We assume that δ_m is a unit of R_K/p^N. Then the natural homomorphism*

$$s_m : \mathrm{Sel}(O_K, E[p^N]) \longrightarrow \bigoplus_{i=1}^{a} \mathcal{H}^1_{\ell_i,f}(K)$$

is injective.

Proof. This is obtained by taking the dual of the statement in Proposition 3.

4.2 Relation Matrices

Theorem 9. *Suppose that $m = \ell_1 \cdot \ldots \cdot \ell_a$ is in $\mathcal{N}_1^{(N)}(K_{[a+1]})$. We assume that m is admissible and that δ_m is a unit of R_K/p^N. Then*

(1) $\mathrm{Sel}(O_K[1/m], E[p^N])$ *is a free R_K/p^N-module of rank a.*
(2) $\{\kappa_{\frac{m}{\ell_i},\ell_i}\}_{1\le i\le a}$ *is a basis of* $\mathrm{Sel}(O_K[1/m], E[p^N])$.

(3) *The matrix*

$$\mathcal{A} = \begin{pmatrix} \delta_{\frac{m}{\ell_1}} & \phi_{\ell_1}(\kappa_{\frac{m}{\ell_1\ell_2},\ell_2}) & \dots & \phi_{\ell_1}(\kappa_{\frac{m}{\ell_1\ell_a},\ell_a}) \\ \phi_{\ell_2}(\kappa_{\frac{m}{\ell_1\ell_2},\ell_1}) & \delta_{\frac{m}{\ell_2}} & \dots & \phi_{\ell_2}(\kappa_{\frac{m}{\ell_2\ell_a},\ell_a}) \\ \cdot & \cdot & & \\ \cdot & \cdot & & \\ \phi_{\ell_a}(\kappa_{\frac{m}{\ell_1\ell_a},\ell_1}) & \phi_{\ell_a}(\kappa_{\frac{m}{\ell_2\ell_a},\ell_2}) & \dots & \delta_{\frac{m}{\ell_a}} \end{pmatrix} \tag{27}$$

is a relation matrix of $\mathrm{Sel}(E/K, E[p^N])^\vee$.

In particular, if $a = 2$, the above matrix is $\mathcal{A} = \begin{pmatrix} \delta_{\ell_2} & \phi_{\ell_1}(g_{\ell_2}) \\ \phi_{\ell_2}(g_{\ell_1}) & \delta_{\ell_1} \end{pmatrix}$. This is described in Remark 10.6 in Kurihara (2012) in the case of ideal class groups.

Proof (Proof of Theorem 9). (1) By Proposition 3, $\bigoplus_{i=1}^a \mathcal{H}_{\ell_i}^2(K) \longrightarrow \mathrm{Sel}(O_K, E[p^N])^\vee$ is surjective. Therefore, by Lemma 1 we have an exact sequence

$$0 \longrightarrow \mathrm{Sel}(O_K, E[p^N]) \longrightarrow \mathrm{Sel}(O_K[1/m], E[p^N]) \xrightarrow{\partial} \bigoplus_{i=1}^a \mathcal{H}_{\ell_i}^2(K) \longrightarrow \mathrm{Sel}(O_K, E[p^N])^\vee \longrightarrow 0. \tag{28}$$

It follows that $\#\,\mathrm{Sel}(O_K[1/m], E[p^N]) = \#\bigoplus_{i=1}^a \mathcal{H}_{\ell_i}^2(K) = \#(R_K/p^N)^a$.

Let m_{R_K} be the maximal ideal of R_K. By Lemma 2, $\mathrm{Sel}(\mathbf{Z}[1/m], E[p^N]) \xrightarrow{\simeq} \mathrm{Sel}(O_K[1/m], E[p^N])^{\mathrm{Gal}(K/\mathbf{Q})}$ is bijective. Since $H^0(\mathbf{Q}, E[p^\infty]) = 0$, the kernel of the multiplication by p on $\mathrm{Sel}(\mathbf{Z}[1/m], E[p^N])$ is $\mathrm{Sel}(\mathbf{Z}[1/m], E[p])$. Therefore, we have an isomorphism $\mathrm{Sel}(O_K[1/m], E[p^N])^\vee \otimes_{R_K} R_K/m_{R_K} \simeq \mathrm{Sel}(\mathbf{Z}[1/m], E[p])^\vee$. From the exact sequence

$$0 \longrightarrow \mathrm{Sel}(\mathbf{Z}, E[p]) \longrightarrow \mathrm{Sel}(\mathbf{Z}[1/m], E[p]) \longrightarrow \bigoplus_{i=1}^a \mathcal{H}_{\ell_i}^2(\mathbf{Q}) \longrightarrow \mathrm{Sel}(\mathbf{Z}, E[p])^\vee \longrightarrow 0,$$

and $\mathcal{H}_{\ell_i}^2(\mathbf{Q}) = H^0(\mathbf{F}_{\ell_i}, E[p]) \simeq \mathbf{F}_p$, we know that $\mathrm{Sel}(\mathbf{Z}[1/m], E[p])$ is generated by a elements. Therefore, by Nakayama's lemma, $\mathrm{Sel}(O_K[1/m], E[p^N])^\vee$ is generated by a elements. Since $\#\,\mathrm{Sel}(O_K[1/m], E[p^N])^\vee = \#(R_K/p^N)^a$, $\mathrm{Sel}(O_K[1/m], E[p^N])^\vee$ is a free R_K/p^N-module of rank a. This shows that $\mathrm{Sel}(O_K[1/m], E[p^N])$ is also a free R_K/p^N-module of rank a because R_K/p^N is a Gorenstein ring.

(2) We identify $\bigoplus_{i=1}^a \mathcal{H}_{\ell_i}^2(K)$ with $(R_K/p^N)^a$, using a basis $\{t_{\ell_i,K}\}_{1\le i\le a}$. Consider $\phi_{\ell_i} : \mathrm{Sel}(O_K[1/m], E[p^N]) \longrightarrow \mathcal{H}_{\ell_i}^2(K)$ and the direct sum of ϕ_{ℓ_i}, which we denote by Φ;

$$\Phi = \oplus_{i=1}^{a} \phi_{\ell_i} : \mathrm{Sel}(O_K[1/m], E[p^N]) \longrightarrow \bigoplus_{i=1}^{a} \mathcal{H}_{\ell_i}^2(K) \simeq (R_K/p^N)^a.$$

Recall that $\kappa_{\frac{m}{\ell_i},\ell_i}$ is an element of $\mathrm{Sel}(O_K[1/m], E[p^N])$ (Proposition 2 (0)). By Proposition 2 (3), (4), we have

$$\Phi(\kappa_{\frac{m}{\ell_i},\ell_i}) = -\delta_m e_i$$

for each i where $\{e_i\}_{1\le i\le a}$ is the standard basis of the free module $(R_K/p^N)^a$. Since we are assuming that δ_m is a unit, Φ is surjective. Since both the target and the source are free modules of the same rank, Φ is bijective. This implies Theorem 9 (2).

(3) Using the exact sequence (28) and the isomorphism Φ, we have an exact sequence

$$(R_n/p^N)^a \xrightarrow{\partial\circ\Phi^{-1}} \bigoplus_{1\le i\le a} \mathcal{H}_{\ell_i}^2(K_n) \xrightarrow{r} \mathrm{Sel}(O_K, E[p^N])^\vee \longrightarrow 0.$$

We take a basis $\{-\delta_m e_i\}_{1\le i\le a}$ of $(R_n/p^N)^a$ and a basis $\{t_{\ell_i,K}\}_{1\le i\le a}$ of $\bigoplus_{1\le i\le a} \mathcal{H}_{\ell_i}^2(K_n)$. Then the (i, j)-component of the matrix corresponding to $\partial \circ \Phi^{-1}$ is $\partial_{\ell_i}(\kappa_{\frac{m}{\ell_j},\ell_j})$. If $i = j$, this is $\delta_{\frac{m}{\ell_i}}$ by Proposition 2 (2). If $i \neq j$, we have $\partial_{\ell_i}(\kappa_{\frac{m}{\ell_j},\ell_j}) = \phi_{\ell_i}(\kappa_{\frac{m}{\ell_i\ell_j},\ell_j})$ by Proposition 2 (1). This completes the proof of Theorem 9.

Remark 5. Suppose that ℓ is in $\mathcal{P}_1^{(N)}(K)$. We define

$$\Phi_\ell' : H^1(K, E[p^N]) \longrightarrow \mathcal{H}_{\ell,f}^1(K)$$

as the composition of the natural map $H^1(K, E[p^N]) \longrightarrow \bigoplus_{v|\ell} H^1(K_v, E[p^N])$ and $\phi' : H^1(K_v, E[p^N]) \longrightarrow H^1(\kappa(v), E[p^N]) = E(\kappa(v)) \otimes \mathbf{Z}/p^N$ in (23). For $m \in \mathcal{N}_1^{(N)}(K)$, we define

$$\Phi_m' : H^1(K, E[p^N]) \longrightarrow \bigoplus_{\ell|m} \mathcal{H}_{\ell,f}^1(K)$$

as the direct sum of Φ_ℓ' for $\ell \mid m$. By definition, the restriction of Φ_m' to $\mathcal{S} = \mathrm{Sel}(E/K, E[p^N])$ coincides with the canonical map s_m;

$$(\Phi_m')_{|\mathcal{S}} = s_m : \mathrm{Sel}(E/K, E[p^N]) \longrightarrow \bigoplus_{\ell|m} \mathcal{H}_{\ell,f}^1(K)\,. \tag{29}$$

Since $\mathcal{H}^1_{\ell,f}(K)$ and $\mathcal{H}^2_\ell(K)$ are Pontrjagin dual each other, we can take the dual basis $t^*_{\ell,K}$ of $\mathcal{H}^1_{\ell,f}(K)$ as an R_K/p^N-module from the basis $t_{\ell,K}$ of $\mathcal{H}^2_\ell(K)$. Under the assumptions of Theorem 9, using the basis $\{t^*_{\ell_i,K}\}_{1\leq i\leq a}$ of $\bigoplus_{i=1}^a \mathcal{H}^1_{\ell,f}(K)$, $\{t_{\ell_i,K}\}_{1\leq i\leq a}$ of $\bigoplus_{i=1}^a \mathcal{H}^2_{\ell_i}(K)$ and the isomorphism Φ'_m, we have an exact sequence $\bigoplus_{\ell|m} \mathcal{H}^1_{\ell,f}(K) \xrightarrow{f} \bigoplus_{i=1}^a \mathcal{H}^2_{\ell_i}(K) \longrightarrow \mathrm{Sel}(E/K, E[p^N])^\vee \longrightarrow 0$. Then the matrix corresponding to f is an organizing matrix in the sense of Mazur and Rubin (2005) (cf. Kurihara 2014, §10).

5 Modified Kolyvagin Systems and Numerical Examples

5.1 Modified Kolyvagin Systems of Gauss Sum Type

In Sect. 3.4 we constructed Kolyvagin systems $\kappa_{m,\ell}$ for (m,ℓ) such that $m\ell \in \mathcal{N}_1^{(N)}(K_{[\epsilon(m\ell)+1]})$. But the condition $\ell \in \mathcal{P}_1^{(N)}(K_{[\epsilon(m\ell)+1]})$ is too strict, and it is not suitable for numerical computation. In this subsection, we define a modified version of Kolyvagin systems of Gauss sum type for (m,ℓ) such that $m\ell \in \mathcal{N}_1^{(N)}(K)$.

Suppose that K is in $\mathcal{K}_{(p)}$. For each $\ell \in \mathcal{P}_1^{(N)}(K)$, we fix $t_\ell \in H^0(\mathbf{F}_\ell, E[p^N])$ of order p^N, and consider $t_{\ell,K} \in \mathcal{H}^2_\ell(K)$, whose ℓ_K-component is $t_\ell \otimes \zeta_{p^N}^{\otimes(-1)}$ and other components are zero. Using $t_{\ell,K}$, we regard ∂_ℓ and ϕ_ℓ as homomorphisms $\partial_\ell : H^1(K, E[p^N]) \longrightarrow R_K/p^N$ and $\phi_\ell : H^1(K, E[p^N]) \longrightarrow R_K/p^N$.

We will define an element $\kappa_{m,\ell}^{q,q',z}$ in $\mathrm{Sel}(O_K[1/m\ell], E[p^N])$ for (m,ℓ) such that $m\ell \in \mathcal{N}_1(K)$ (and for some primes q, q' and some z in $\mathrm{Sel}(O_K[1/qq'], E[p^N])$). Consider (m,ℓ) such that ℓ is a prime and $m\ell \in \mathcal{N}_1(K)$. We take n_0 sufficiently large such that every prime of K_{n_0} dividing $m\ell$ is inert in K_∞/K_{n_0}. Then by Proposition 1 (1), for any $q \in (\mathcal{P}'_0)^{(N)}(K(m\ell)_{[1]}K_{n_0+N})$, $\kappa_{m\ell,q} \in \mathrm{Sel}(O_K[1/m\ell q], E[p^N])$ satisfies

$$\partial_r(\kappa_{m\ell,q}) = \phi_r(\kappa_{\frac{m\ell}{r},q})$$

for all r dividing $m\ell$. By Lemma 3, we can take $q, q' \in (\mathcal{P}'_0)^{(N)}(K(m\ell)_{[1]}K_{n_0+N})$ satisfying

- $w_K(t_{q,K}) = w_K(t_{q',K})$, and
- there is $z \in H^1_f(O_K[1/qq'], E[p^N])$ such that $\partial_K(z) = t_{q,K} - t_{q',K}$, $\phi_\ell(z) = 1$ and $\phi_r(z) = 0$ for any r dividing m.

For any $m \in \mathcal{N}_1(K)$, let δ_m be the element defined in (25). We define

$$\kappa_{m,\ell}^{q,q',z} = \kappa_{m\ell,q} - \kappa_{m\ell,q'} - \delta_{m\ell} z\,. \tag{30}$$

By Proposition 1 (2), we have $\kappa_{m,\ell}^{q,q',z} \in \mathrm{Sel}(O_K[1/m\ell], E[p^N])$.

Proposition 4. (0) $\kappa_{m,\ell}^{q,q',z}$ *is in* $\mathrm{Sel}(O_K[1/m\ell], E[p^N])$.

(1) *The element* $\kappa_{m,\ell}^{q,q',z}$ *satisfies* $\partial_r(\kappa_{m,\ell}^{q,q',z}) = \phi_r(\kappa_{\frac{m}{r},\ell}^{q,q',z})$ *for any prime divisor* r *of* m.

(2) *We further assume that* $m\ell$ *is admissible in the sense of the paragraph before Proposition 1. Then we have* $\phi_r(\kappa_{m,\ell}^{q,q',z}) = 0$ *for any prime divisor* r *of* m.

(3) *Under the same assumptions as* (2), $\phi_\ell(\kappa_{m,\ell}^{q,q',z}) = -\delta_{m\ell}$ *holds.*

Proof. (1) Using the definition of $\kappa_{m,\ell}^{q,q',z}$ and Proposition 1 (1), we have $\partial_r(\kappa_{m,\ell}^{q,q',z}) = \partial_r(\kappa_{m\ell,q} - \kappa_{m\ell,q'}) = \phi_r(\kappa_{\frac{m\ell}{r},q} - \kappa_{\frac{m\ell}{r},q'})$. Next, we use the definition of $\kappa_{\frac{m}{r},\ell}^{q,q',z}$ and $\phi_r(z) = 0$ to get $\phi_r(\kappa_{\frac{m\ell}{r},q} - \kappa_{\frac{m\ell}{r},q'}) = \phi_r(\kappa_{\frac{m}{r},\ell}^{q,q',z} + \delta_{\frac{m\ell}{r}} z) = \phi_r(\kappa_{\frac{m}{r},\ell}^{q,q',z})$. These computations imply (1).

(2) We have $\phi_r(\kappa_{m\ell,q}) = \phi_r(\kappa_{m\ell,q'}) = 0$ by Proposition 1 (3). This together with $\phi_r(z) = 0$ implies $\phi_r(\kappa_{m,\ell}^{q,q',z}) = \phi_r(\kappa_{m\ell,q} - \kappa_{m\ell,q'} - \delta_{m\ell} z) = 0$.

(3) We again use Proposition 1 (3) to get $\phi_\ell(\kappa_{m\ell,q}) = \phi_\ell(\kappa_{m\ell,q'}) = 0$. Since $\phi_\ell(z) = 1$, we have $\phi_\ell(\kappa_{m,\ell}^{q,q',z}) = \phi_\ell(\kappa_{m\ell,q} - \kappa_{m\ell,q'} - \delta_{m\ell} z) = -\delta_{m\ell}$. This completes the proof of Proposition 4.

5.2 *Proof of Theorem 4*

In this subsection we take $K = \mathbf{Q}$. For $m \in \mathcal{N}^{(N)} = \mathcal{N}^{(N)}(\mathbf{Q})$, we consider $\delta_m \in \mathbf{Z}/p^N$, which is defined from $\vartheta_{\mathbf{Q}(m)}$ by (25). We define $\tilde{\delta}_m \in \mathbf{Z}/p^N$ by

$$\tilde{\theta}_{\mathbf{Q}(m)} \equiv \tilde{\delta}_m \prod_{i=1}^{r} (\tau_{\ell_i} - 1) \pmod{p^N, (\tau_{\ell_1} - 1)^2, \ldots, (\tau_{\ell_r} - 1)^2} \tag{31}$$

where $m = \ell_1 \cdot \ldots \cdot \ell_r$. By (8), $\tilde{\theta}_{\mathbf{Q}(m)} = u\vartheta_{\mathbf{Q}(m)}$ for some unit $u \in R_{\mathbf{Q}(m)}^{\times}$. This together with (25) and (31) implies that

$$\mathrm{ord}_p(\tilde{\delta}_m) = \mathrm{ord}_p(\delta_m). \tag{32}$$

We take a generator $\eta_\ell \in (\mathbf{Z}/\ell\mathbf{Z})^\times$ such that the image of $\sigma_{\eta_\ell} \in \mathrm{Gal}(\mathbf{Q}(\mu_\ell)/\mathbf{Q}) \simeq (\mathbf{Z}/\ell)^\times$ in $\mathrm{Gal}(\mathbf{Q}(\ell)/\mathbf{Q}) \simeq (\mathbf{Z}/\ell)^\times \otimes \mathbf{Z}_p$ is τ_ℓ which is the generator we took. Then, using (31) and (1), we can easily check that Eq. (2) in Sect. 1.1 holds.

In the rest of this subsection, we take $N = 1$. We simply write $\mathcal{P}_1$ for $\mathcal{P}_1^{(1)}$, so

$$\mathcal{P}_1 = \{\ell \in \mathcal{P}_{good} \mid \ell \equiv 1 \pmod{p} \text{ and } E(\mathbf{F}_\ell) \simeq \mathbf{Z}/p\}.$$

The set of squarefree products of primes in $\mathcal{P}_1$ is denoted by $\mathcal{N}_1$.

We first prove the following lemma which is related to the functional equation of an elliptic curve.

Lemma 4. *Let ϵ be the root number of E. Suppose that $m \in \mathcal{N}_1$ is δ-minimal (for the definition of δ-minimalness, see the paragraph before Conjecture 2). Then we have $\epsilon = (-1)^{\epsilon(m)}$.*

Proof. By the functional equation (1.6.2) in Mazur and Tate (1987) and the above definition of $\tilde{\delta}_m$, we have $\epsilon(-1)^{\epsilon(m)}\tilde{\delta}_m \equiv \tilde{\delta}_m \pmod{p}$. This implies the conclusion.

For each $\ell \in \mathcal{P}_1$, we fix a generator $t_\ell \in \mathcal{H}^2_\ell(\mathbf{Q}) = H^0(\mathbf{F}_\ell, E[p](-1)) \simeq \mathbf{Z}/p = \mathbf{F}_p$, and regard ϕ_ℓ as a map $\phi_\ell : H^1(\mathbf{Q}, E[p]) \longrightarrow \mathbf{F}_p$. Note that the restriction of ϕ_ℓ to $\mathrm{Sel}(E/\mathbf{Q}, E[p])$ is the zero map if and only if the natural map $s_\ell : \mathrm{Sel}(E/\mathbf{Q}, E[p]) \longrightarrow E(\mathbf{F}_\ell) \otimes \mathbf{Z}/p \simeq \mathbf{F}_p$ is the zero map.

(I) Proof of Theorem 4 (1), (2).

Suppose that $\epsilon(m) = 0$, namely $m = 1$. Then $\delta_1 = \theta_{\mathbf{Q}} \bmod p = L(E,1)/\Omega_E^+ \bmod p$. If $\delta_1 \neq 0$, $\mathrm{Sel}(E/\mathbf{Q}, E[p]) = 0$ and s_1 is trivially bijective. Suppose next $\epsilon(m) = 1$, so $m = \ell \in \mathcal{P}_1$. It is sufficient to prove the next two propositions.

Proposition 5. *Assume that $\ell \in \mathcal{P}_1$ is δ-minimal. Then $\mathrm{Sel}(E/\mathbf{Q}, E[p])$ is 1-dimensional over $\mathbf{F}_p$, and $s_\ell : \mathrm{Sel}(E/\mathbf{Z}, E[p]) \longrightarrow \mathbf{F}_p$ is bijective. Moreover, the Selmer group $\mathrm{Sel}(E/\mathbf{Q}, E[p^\infty])^\vee$ with respect to the p-power torsion points $E[p^\infty]$ is a free $\mathbf{Z}_p$-module of rank 1, namely $\mathrm{Sel}(E/\mathbf{Q}, E[p^\infty])^\vee \simeq \mathbf{Z}_p$.*

Proof. We first assume $\mathrm{Sel}(E/\mathbf{Q}, E[p]) = 0$ and will obtain the contradiction. We consider $\kappa_{1,\ell}^{q,q',z} = \kappa_{\ell,q} - \kappa_{\ell,q'} - \delta_\ell z$, which was defined in (30). By Proposition 1 (1), we know $\partial_\ell(\kappa_{1,\ell}^{q,q',z}) = \phi_\ell(g_q - g_{q'})$. Consider the exact sequence (see Lemma 1)

$$0 \longrightarrow \mathrm{Sel}(E/\mathbf{Q}, E[p]) \longrightarrow \mathrm{Sel}(\mathbf{Z}[1/r], E[p]) \longrightarrow \mathcal{H}^2_r(\mathbf{Q})$$

for any $r \in \mathcal{P}_1$ where $\mathrm{Sel}(\mathbf{Z}[1/r], E[p]) \longrightarrow \mathcal{H}^2_r(\mathbf{Q}) \simeq \mathbf{F}_p$ is nothing but ∂_r. Since we assumed $\mathrm{Sel}(E/\mathbf{Q}, E[p]) = 0$, $\mathrm{Sel}(\mathbf{Z}[1/r], E[p]) \longrightarrow \mathcal{H}^2_r(\mathbf{Q}) \simeq \mathbf{F}_p$ is injective for any $r \in \mathcal{P}_1$. So $\partial_q(g_q) = \delta_1 = 0$ implies that $g_q = 0$. By the same method, we have $g_{q'} = 0$. Therefore, $\partial_\ell(\kappa_{1,\ell}^{q,q',z}) = \phi_\ell(g_q - g_{q'}) = 0$, which implies that $\kappa_{1,\ell}^{q,q',z} \in \mathrm{Sel}(E/\mathbf{Q}, E[p])$.

But Proposition 4 (3) tells us that $\phi_\ell(\kappa_{1,\ell}^{q,q',z}) = -\delta_\ell \neq 0$. Therefore, $\kappa_{1,\ell}^{q,q',z} \neq 0$, which contradicts our assumption $\mathrm{Sel}(E/\mathbf{Q}, E[p]) = 0$. Thus we get $\mathrm{Sel}(E/\mathbf{Q}, E[p]) \neq 0$.

On the other hand, by Corollary 2 we know that $s_\ell : \mathrm{Sel}(E/\mathbf{Q}, E[p]) \longrightarrow \mathbf{F}_p$ is injective, therefore bijective.

By Lemma 4, the root number ϵ is -1. This shows that $\mathrm{Sel}(E/\mathbf{Q}, E[p^\infty])^\vee$ has positive $\mathbf{Z}_p$-rank by the parity conjecture proved by Nekovář (2001, 2006). Therefore, we finally have $\mathrm{Sel}(E/\mathbf{Q}, E[p^\infty])^\vee \simeq \mathbf{Z}_p$, which completes the proof of Proposition 5.

If we assume a slightly stronger condition on ℓ, we also obtain the main conjecture. Let $\lambda' = \lambda^{an}$ be the analytic λ-invariant of the p-adic L-function $\vartheta_{\mathbf{Q}_\infty}$. We put $n_{\lambda'} = \min\{n \in \mathbf{Z} \mid p^n - 1 \geq \lambda'\}$.

Proposition 6. *Suppose that there is $\ell \in \mathcal{P}_1$ such that*

$$\ell \equiv 1 \ (mod\ p^{n_{\lambda'}+2}) \ and\ \tilde{\delta}_\ell \neq 0.$$

Then the main conjecture for $(E, \mathbf{Q}_\infty/\mathbf{Q})$ is true and $\mathrm{Sel}(E/\mathbf{Q}_\infty, E[p^\infty])^\vee$ *is generated by one element as a $\Lambda_{\mathbf{Q}_\infty}$-module.*

Proof. We use our Euler system $g_\ell^{(K)}$ in Sect. 3.2 instead of $\kappa_{1,\ell}^{q,q',z}$ which was used in the proof of Proposition 5. Let λ be the algebraic λ-invariant, namely the rank of $\mathrm{Sel}(E/\mathbf{Q}_\infty, E[p^\infty])^\vee$. Then $\lambda \leq \lambda'$ and $\vartheta_{\mathbf{Q}_\infty} \in \mathrm{char}(\mathrm{Sel}(O_{\mathbf{Q}_\infty}, E[p^\infty])^\vee)$ by Kato's theorem.

Put $K = \mathbf{Q}_{n_{\lambda'}}$ and $f = p^{n_{\lambda'}}$. Consider the group ring $R_K/p = \mathbf{F}_p[\mathrm{Gal}(K/\mathbf{Q})]$. We identify a generator γ of $\mathrm{Gal}(K/\mathbf{Q})$ with $1 + \mathrm{t}$, and identify R_K/p with $\mathbf{F}_p[[\mathrm{t}]]/(\mathrm{t}^f)$. The norm $N_{\mathrm{Gal}(K/\mathbf{Q})} = \Sigma_{i=0}^{f-1}\gamma^i$ is t^{f-1} by this identification, so our assumption $\lambda' \leq f - 1$ implies that the corestriction map $\mathrm{Sel}(E/K, E[p]) \longrightarrow \mathrm{Sel}(E/\mathbf{Q}, E[p])$ is the zero map because $\lambda \leq \lambda'$. Therefore, we have $\mathbf{Q}_{[1]} \subset K$. Since $p^{n_{\lambda'}+1} - p^{n_{\lambda'}} > p^{n_{\lambda'}} - 1 \geq \lambda' \geq \lambda$, the corestriction map $\mathrm{Sel}(E/\mathbf{Q}_{n_{\lambda'}+1}, E[p]) \longrightarrow \mathrm{Sel}(E/\mathbf{Q}_{n_{\lambda'}}, E[p]) = \mathrm{Sel}(E/K, E[p])$ is also the zero map. This shows that $\mathbf{Q}_{[2]} \subset \mathbf{Q}_{n_{\lambda'}+1}$.

Our assumption $\ell \equiv 1 \pmod{p^{n_{\lambda'}+2}}$ implies that ℓ splits completely in $\mathbf{Q}_{n_{\lambda'}+1}$, so we have $\ell \in \mathcal{P}_1(\mathbf{Q}_{[2]}) = \mathcal{P}_1(K_{[1]})$. Therefore, we can define

$$g_\ell^{(K)} \in \mathrm{Sel}(O_K[1/\ell], E[p])$$

in Sect. 3.2. Since $\ell \in \mathcal{P}_1(\mathbf{Q}_{[2]})$, we also have

$$\phi_\ell(g_\ell^{(\mathbf{Q})}) = -\delta_\ell^{(\mathbf{Q})} = -\delta_\ell$$

by Proposition 2 (4). It follows from our assumption $\delta_\ell \neq 0$ that $g_\ell^{(\mathbf{Q})} \neq 0$. Since $\mathrm{Cor}_{K/\mathbf{Q}}(g_\ell^{(K)}) = g_\ell^{(\mathbf{Q})}$ and the natural map $i : \mathrm{Sel}(\mathbf{Z}[1/\ell], E[p]) \longrightarrow \mathrm{Sel}(O_K[1/\ell], E[p])$ is injective, we get

$$i(g_\ell^{(\mathbf{Q})}) = N_{\mathrm{Gal}(K/\mathbf{Q})}g_\ell^{(K)} = \mathrm{t}^{f-1}g_\ell^{(K)} \neq 0.$$

Consider $\partial_\ell : \mathrm{Sel}(O_K[1/\ell], E[p]) \longrightarrow R_K/p$. By definition, we have $\partial_\ell(g_\ell^{(K)}) = u\mathrm{t}^{\lambda'}$ for some unit u of R_K/p. This shows that $\partial_\ell(\mathrm{t}^{f-\lambda'}g_\ell^{(K)}) = 0$, which implies that $\mathrm{t}^{f-\lambda'}g_\ell^{(K)} \in \mathrm{Sel}(E/K, E[p])$. The fact $\mathrm{t}^{f-1}g_\ell^{(K)} \neq 0$ implies the submodule generated by $\mathrm{t}^{f-\lambda'}g_\ell^{(K)}$ is isomorphic to $R_K/(p, \mathrm{t}^{\lambda'})$ as an R_K-module. Namely, we have

$$\mathrm{Sel}(E/K, E[p]) \supset \langle \mathrm{t}^{f-\lambda'}g_\ell^{(K)} \rangle \simeq R_K/(p, \mathrm{t}^{\lambda'}).$$

This implies that $\lambda = \lambda'$, and $\mathrm{Sel}(E/K, E[p]) \simeq R_K/(p, \mathrm{t}^{\lambda})$. Therefore, we have $\mathrm{Sel}(E/\mathbf{Q}_\infty, E[p])^\vee \simeq \Lambda_{\mathbf{Q}_\infty}/(p, \vartheta_{\mathbf{Q}_\infty})$. This together with Kato's theorem we mentioned implies that $\mathrm{Sel}(E/\mathbf{Q}_\infty, E[p^\infty])^\vee \simeq \Lambda_{\mathbf{Q}_\infty}/(\vartheta_{\mathbf{Q}_\infty})$.

(II) Proof of Theorem 4 (3).

Suppose that $m = \ell_1\ell_2 \in \mathcal{N}_1$ and m is δ-minimal. As in the proof of Proposition 5, we assume $\mathrm{Sel}(E/\mathbf{Q}, E[p]) = 0$ and will get the contradiction. We consider $\kappa_{\ell_1,\ell_2}^{q,q',z}$ defined in (30). Consider the exact sequence (see Lemma 1)

$$0 \longrightarrow \mathrm{Sel}(E/\mathbf{Q}, E[p]) \longrightarrow \mathrm{Sel}(\mathbf{Z}[1/\ell_1\ell_2 qq'], E[p]) \xrightarrow{\partial} \bigoplus_{v\in\{\ell_1,\ell_2,q,q'\}} \mathcal{H}_v^2(\mathbf{Q}).$$

By the same method as the proof of Proposition 5, $g_q = g_{q'} = 0$. Therefore, $\partial_{\ell_1}(\kappa_{\ell_1,q} - \kappa_{\ell_1,q'}) = \phi_{\ell_1}(g_q - g_{q'}) = 0$ by Proposition 1 (1). We have $\partial_q(\kappa_{\ell_1,q}) = \delta_{\ell_1} = 0$, $\partial_q(\kappa_{\ell_1,q'}) = 0$, $\partial_{q'}(\kappa_{\ell_1,q}) = 0$, $\partial_{q'}(\kappa_{\ell_1,q'}) = \delta_{\ell_1} = 0$. Therefore, $\partial(\kappa_{\ell_1,q} - \kappa_{\ell_1,q'}) = 0$. This together with $\mathrm{Sel}(E/\mathbf{Q}, E[p]) = 0$ shows that $\kappa_{\ell_1,q} - \kappa_{\ell_1,q'} = 0$. Therefore, using Proposition 1 (1), we have

$$\partial_{\ell_2}(\kappa_{\ell_1,\ell_2}^{q,q',z}) = \partial_{\ell_2}(\kappa_{m,q} - \kappa_{m,q'}) = \phi_{\ell_2}(\kappa_{\ell_1,q} - \kappa_{\ell_1,q'}) = 0.$$

By the same method as the above proof of $\kappa_{\ell_1,q} - \kappa_{\ell_1,q'} = 0$, we get $\kappa_{1,\ell_2}^{q,q',z} = 0$. This implies that $\partial_{\ell_1}(\kappa_{\ell_1,\ell_2}^{q,q',z}) = \phi_{\ell_1}(\kappa_{1,\ell_2}^{q,q',z}) = 0$ by Proposition 4 (1). It follows that $\partial(\kappa_{\ell_1,\ell_2}^{q,q',z}) = 0$, which implies $\kappa_{\ell_1,\ell_2}^{q,q',z} \in \mathrm{Sel}(E/\mathbf{Q}, E[p])$. But this is a contradiction because we assumed $\mathrm{Sel}(E/\mathbf{Q}, E[p]) = 0$ and

$$\phi_{\ell_2}(\kappa_{\ell_1,\ell_2}^{q,q',z}) = -\delta_m \neq 0$$

by Proposition 4 (3). Thus, we get $\mathrm{Sel}(E/\mathbf{Q}, E[p]) \neq 0$.

Now the root number is 1 by Lemma 4, therefore, by the parity conjecture proved by Nekovář (2001, 2006), we obtain $\dim_{\mathbf{F}_p} \mathrm{Sel}(E/\mathbf{Q}, E[p]) \geq 2$. On the other hand, by Corollary 2 we know that $s_m : \mathrm{Sel}(E/\mathbf{Q}, E[p]) \longrightarrow (\mathbf{F}_p)^{\oplus 2}$ is injective. Therefore, the injectivity of s_m implies the bijectivity of s_m. This completes the proof of Theorem 4 (3).

We give a simple corollary.

Corollary 3. *Suppose that there is $m \in \mathcal{N}_1$ such that m is δ-minimal and $\epsilon(m) = 2$. We further assume that the analytic λ-invariant λ' is 2. Then the main conjecture for $(E, \mathbf{Q}_\infty/\mathbf{Q})$ holds.*

Proof. Put $\mathrm{t} = \gamma - 1$ and identify $\Lambda_{\mathbf{Q}_\infty}/p$ with $\mathbf{F}_p[[\mathrm{t}]]$. Let $\mathcal{A}$ be the relation matrix of $S = \mathrm{Sel}(E/\mathbf{Q}_\infty, E[p^\infty])^\vee$. Since $S/(p, \mathrm{t}) = \mathrm{Sel}(E/\mathbf{Q}, E[p])^\vee \simeq \mathbf{F}_p \oplus \mathbf{F}_p$, t^2 divides $\det \mathcal{A}$ mod p. Therefore, the algebraic λ-invariant is also 2. This implies the main conjecture because $\det \mathcal{A}$ divides $\vartheta_{\mathbf{Q}_\infty}$ in $\Lambda_{\mathbf{Q}_\infty}$ (Kato 2004).

(III) Proof of Theorem 4 (4).

Lemma 5. *Suppose that ℓ, ℓ_1, ℓ_2 are distinct primes in $\mathcal{P}_1$ satisfying $\delta_\ell = \delta_{\ell\ell_1} = \delta_{\ell\ell_2} = 0$. Assume also that $s_\ell : \mathrm{Sel}(E/\mathbf{Q}, E[p]) \longrightarrow \mathbf{F}_p$ is bijective, and that $\ell\ell_1$, $\ell\ell_2$ are both admissible. We take q, q' such that they satisfy the conditions when we defined $\kappa^{q,q',z}_{\ell_1\ell_2,\ell}$. Then we have*

(1) $\mathrm{Sel}(E/\mathbf{Q}, E[p]) = \mathrm{Sel}(\mathbf{Z}[1/\ell], E[p])$,
(2) $\kappa^{q,q',z}_{\ell_1,\ell} = 0$, $\kappa^{q,q',z}_{\ell_2,\ell} = 0$, *and*
(3) $\kappa^{q,q',z}_{\ell_1\ell_2,\ell} \in \mathrm{Sel}(E/\mathbf{Q}, E[p])$.

Proof. (1) Since s_ℓ is bijective, taking the dual, we get the bijectivity of $\mathcal{H}^2_\ell(\mathbf{Q}) \longrightarrow \mathrm{Sel}(E/\mathbf{Q}, E[p])^\vee = \mathrm{Sel}(\mathbf{Z}, E[p])^\vee$. By the exact sequence

$$0 \longrightarrow \mathrm{Sel}(\mathbf{Z}, E[p]) \longrightarrow \mathrm{Sel}(\mathbf{Z}[1/\ell], E[p]) \xrightarrow{\partial_\ell} \mathcal{H}^2_\ell(\mathbf{Q}) \longrightarrow \mathrm{Sel}(\mathbf{Z}, E[p])^\vee \longrightarrow 0$$

in Lemma 1, we get $\mathrm{Sel}(E/\mathbf{Q}, E[p]) = \mathrm{Sel}(\mathbf{Z}, E[p]) = \mathrm{Sel}(\mathbf{Z}[1/\ell], E[p])$.

(2) We first note that the bijectivity of $s_\ell : \mathrm{Sel}(E/\mathbf{Q}, E[p]) \longrightarrow \mathbf{F}_p$ implies the bijectivity of $\phi_\ell : \mathrm{Sel}(E/\mathbf{Q}, E[p]) \longrightarrow \mathbf{F}_p$. Since $\partial_q(\kappa_{\ell,q}) = \delta_\ell = 0$, $\kappa_{\ell,q} \in \mathrm{Sel}(\mathbf{Z}[1/\ell], E[p]) = \mathrm{Sel}(E/\mathbf{Q}, E[p])$ where we used the property (1) which we have just proved. Proposition 1 (3) implies $\phi_\ell(\kappa_{\ell,q}) = 0$, which implies $\kappa_{\ell,q} = 0$ by the bijectivity of ϕ_ℓ. By the same method, we have $\kappa_{\ell,q'} = 0$. Therefore, we have

$$\kappa^{q,q',z}_{1,\ell} = \kappa_{\ell,q} - \kappa_{\ell,q'} - \delta_\ell z = 0.$$

Therefore, Proposition 4 (1) implies $\partial_{\ell_1}(\kappa^{q,q',z}_{\ell_1,\ell}) = \phi_{\ell_1}(\kappa^{q,q',z}_{1,\ell}) = 0$. This implies $\kappa^{q,q',z}_{\ell_1,\ell} \in \mathrm{Sel}(\mathbf{Z}[1/\ell], E[p]) = \mathrm{Sel}(E/\mathbf{Q}, E[p])$. Using Proposition 4 (3), we have

$$\phi_\ell(\kappa^{q,q',z}_{\ell_1,\ell}) = -\delta_{\ell\ell_1} = 0,$$

which implies $\kappa^{q,q',z}_{\ell_1,\ell} = 0$ by the bijectivity of ϕ_ℓ. The same proof works for $\kappa^{q,q',z}_{\ell_2,\ell}$.

(3) It follows from Proposition 4 (1) and Lemma 5 (2) that $\partial_{\ell_i}(\kappa^{q,q',z}_{\ell_1\ell_2,\ell}) = \phi_{\ell_i}(\kappa^{q,q',z}_{\frac{\ell_1\ell_2}{\ell_i},\ell}) = 0$ for each $i = 1, 2$. This implies $\kappa^{q,q',z}_{\ell_1\ell_2,\ell} \in \mathrm{Sel}(\mathbf{Z}[1/\ell], E[p])$. Using $\mathrm{Sel}(\mathbf{Z}[1/\ell], E[p]) = \mathrm{Sel}(E/\mathbf{Q}, E[p])$ which we proved in (1), we get the conclusion. This completes the proof of Lemma 5.

We next prove Theorem 4 (4). Assume that $m = \ell_1\ell_2\ell_3 \in \mathcal{N}_1$, m is δ-minimal, m is admissible, and $s_{\ell_i} : \mathrm{Sel}(E/\mathbf{Q}, E[p]) \longrightarrow \mathbf{F}_p$ is surjective for each $i = 1, 2$.

We assume $\dim_{\mathbf{F}_p} \mathrm{Sel}(E/\mathbf{Q}, E[p]) = 1$ and will get the contradiction. By this assumption, $s_{\ell_i} : \mathrm{Sel}(E/\mathbf{Q}, E[p]) \longrightarrow \mathbf{F}_p$ for each $i = 1, 2$ is bijective. This implies that $\phi_{\ell_i} : \mathrm{Sel}(E/\mathbf{Q}, E[p]) \longrightarrow \mathbf{F}_p$ for each $i = 1, 2$ is also bijective. By Lemma 5 (3) we get $\kappa^{q,q',z}_{\ell_2\ell_3,\ell_1} \in \mathrm{Sel}(E/\mathbf{Q}, E[p])$, taking q, q' satisfying the conditions when we defined this element. By Proposition 4 (3), we have $\phi_{\ell_1}(\kappa^{q,q',z}_{\ell_2\ell_3,\ell_1}) = -\delta_m \neq 0$, which implies $\kappa^{q,q',z}_{\ell_2\ell_3,\ell_1} \neq 0$. But by Proposition 4 (2), we have $\phi_{\ell_2}(\kappa^{q,q',z}_{\ell_2\ell_3,\ell_1}) = 0$. This contradicts the bijectivity of ϕ_{ℓ_2}. Therefore, we obtain $\dim_{\mathbf{F}_p} \mathrm{Sel}(E/\mathbf{Q}, E[p]) > 1$.

By Lemma 4 and our assumption that m is δ-minimal, we know that the root number ϵ is -1. This shows that $\dim_{\mathbf{F}_p} \mathrm{Sel}(E/\mathbf{Q}, E[p]) \geq 3$ by the parity conjecture proved by Nekovář (2001, 2006). On the other hand, Corollary 2 implies that $\dim_{\mathbf{F}_p} \mathrm{Sel}(E/\mathbf{Q}, E[p]) \leq 3$ and $s_m : \mathrm{Sel}(E/\mathbf{Q}, E[p]) \longrightarrow \mathbf{F}_p^{\oplus 3}$ is injective. Therefore, the above map s_m is bijective. This completes the proof of Theorem 4 (4).

5.3 Numerical Examples

In this section, we give several numerical examples.

Let $E = X_0(11)^{(d)}$ be the quadratic twist of $X_0(11)$ by d, namely $dy^2 = x^3 - 4x^2 - 160x - 1264$. We take $p = 3$. Then if $d \equiv 1 \pmod p$, p is a good ordinary prime which is not anomalous (namely $a_p(= a_3)$ for E satisfies $a_p \not\equiv 1 \pmod p$), and $p = 3$ does not divide $\mathrm{Tam}(E)$, and the Galois representation on $T_3(E)$ is surjective. In the following examples, we checked $\mu' = 0$ where μ' is the analytic μ-invariant. Then this implies that the algebraic μ-invariant is also zero (Kato 2004, Theorem 17.4 (3)) under our assumptions. In the computations of $\tilde{\delta}_m$ below, we have to fix a generator of $\mathrm{Gal}(\mathbf{Q}(\ell)/\mathbf{Q}) \simeq (\mathbf{Z}/\ell\mathbf{Z})^\times$ for a prime ℓ. We always take the least primitive root η_ℓ of $(\mathbf{Z}/\ell\mathbf{Z})^\times$. We compute $\tilde{\delta}_m$ using the formula in (2).

1. $d = 13$. We take $N = 1$. Since $\tilde{\delta}_7 = 20 \not\equiv 0 \pmod 3$, we know that $\mathrm{Fitt}_{1,\mathbf{F}_3}(\mathrm{Sel}(E/\mathbf{Q}, E[3])^\vee) = \mathbf{F}_3$ by Theorem 8, so $\mathrm{Sel}(E/\mathbf{Q}, E[3])$ is generated by one element.

   The root number is $\epsilon = (\frac{13}{11}) = -1$, so $L(E, 1) = 0$. We compute $\mathcal{P}_1 = \{7, 31, 73, \ldots\}$. Therefore, $\tilde{\delta}_7 \not\equiv 0 \pmod 3$ implies $\mathrm{Sel}(E/\mathbf{Q}, E[3]) \simeq \mathbf{F}_3$ and

   $$\mathrm{Sel}(E/\mathbf{Q}, E[3^\infty])^\vee \simeq \mathbf{Z}_3$$

   by Proposition 5. Also, it is easily computed that $\lambda' = 1$ in this case. This implies that $\mathrm{Sel}(E/\mathbf{Q}_\infty, E[3^\infty])^\vee \simeq \mathbf{Z}_3$, so the main conjecture also holds.

   We can find a point $P = (7045/36, -574201/216)$ of infinite order on the minimal Weierstrass model $y^2 + y = x^3 - x^2 - 1746x - 50295$ of $E = X_0(11)^{(13)}$. Therefore, we know $Ш(E/\mathbf{Q})[3^\infty] = 0$. We can easily check that $E(\mathbf{F}_7)$ is cyclic of order 6, and that the image of the point P in $E(\mathbf{F}_7)/3E(\mathbf{F}_7)$ is non-zero. So we also checked numerically that $s_7 : \mathrm{Sel}(E/\mathbf{Q}, E[3]) \longrightarrow E(\mathbf{F}_7)/3E(\mathbf{F}_7)$ is bijective as Proposition 5 claims.

2. $d = 40$. We know $\epsilon = (\frac{40}{11}) = -1$. We take $N = 1$. We can compute $\mathcal{P}_1 = \{7, 67, 73, \dots\}$, and $\tilde{\delta}_7 = -40 \not\equiv 0 \pmod 3$. This implies that $\mathrm{Sel}(E/\mathbf{Q}, E[3]) \simeq \mathbf{F}_3$ and $\mathrm{Sel}(E/\mathbf{Q}, E[3^\infty])^\vee \simeq \mathbf{Z}_3$ by Proposition 5.

In this case, we know $\lambda' = 7$. Therefore, $n_{\lambda'} = 2$. We can check $5347 \in \mathcal{P}_1$ (where $5347 \equiv 1 \pmod{3^5}$) and $\tilde{\delta}_{5347} = -412820 \not\equiv 0 \pmod 3$. Therefore, the main conjecture holds by Proposition 6. In this case, we can check that the p-adic L-function $\vartheta_{\mathbf{Q}_\infty}$ is divisible by $(1+t)^3 - 1$, so we have

$$\mathrm{rank}_{\mathbf{Z}_3} \mathrm{Sel}(E/\mathbf{Q}_1, E[3^\infty])^\vee = 3$$

where $\mathbf{Q}_1$ is the first layer of $\mathbf{Q}_\infty/\mathbf{Q}$.

In the following, for a prime $\ell \in \mathcal{P}$, we take a generator τ_ℓ of $\mathrm{Gal}(\mathbf{Q}(\ell)/\mathbf{Q}) \simeq (\mathbf{Z}/\ell\mathbf{Z})^\times$ and put $S = \tau_\ell - 1$. We write $\vartheta_{\mathbf{Q}(\ell)} = \Sigma a_i^{(\ell)} S^i$ where $a_i^{(\ell)} \in \mathbf{Z}_p$. Note that $\tilde{\delta}_\ell = a_1^{(\ell)}$.

3. $d = 157$. We know $\epsilon = (\frac{157}{11}) = 1$ and $L(E,1)/\Omega_E^+ = 45$. We take $N = 1$. We compute $a_2^{(37)} = -14065/2 \not\equiv 0 \pmod 3$. Since $37 \equiv 1 \pmod{3^2}$, $c_2 = 2-1 = 1$ and $a_2^{(37)}$ is in $\mathrm{Fitt}_{2,\mathbf{F}_3}(\mathrm{Sel}(E/\mathbf{Q}, E[3])^\vee)$ by Theorem 8, which implies that $\mathrm{Fitt}_{2,\mathbf{F}_3}(\mathrm{Sel}(E/\mathbf{Q}, E[3])) = \mathbf{F}_3$. Therefore, $\mathrm{Sel}(E/\mathbf{Q}, E[3])$ is generated by at most two elements.

We compute $\mathcal{P}_1 = \{7, 67, 73, 127, \dots\}$. Since $127 \equiv 1 \pmod 7$, 7×127 is admissible. We compute $\tilde{\delta}_{7\times 127} = 83165 \not\equiv 0 \pmod 3$. Therefore, 7×127 is δ-minimal. It follows from Theorem 4 (3) that $\mathrm{Sel}(E/\mathbf{Q}, E[3]) \simeq \mathbf{F}_3 \oplus \mathbf{F}_3$. In this example, we can check $\lambda' = 2$, so Corollary 3 together with the above computation implies the main conjecture. Since $L(E,1)/\Omega_E^+ = 45 \neq 0$, $\mathrm{rank}\, E(\mathbf{Q}) = 0$ by Kato, which implies $\mathrm{Sel}(E/\mathbf{Q}, E[3^\infty]) = Ш(E/\mathbf{Q})[3^\infty]$. Since $45 \in \mathrm{Fitt}_{0,\mathbf{Z}_3}(\mathrm{Sel}(E/\mathbf{Q}, E[3^\infty])^\vee)$, we have $\# Ш(E/\mathbf{Q})[3^\infty] \le 9$, and

$$Ш(E/\mathbf{Q})[3^\infty] \simeq \mathbf{Z}/3\mathbf{Z} \oplus \mathbf{Z}/3\mathbf{Z}.$$

4. $d = 265$. In this case, $\epsilon = (\frac{265}{11}) = 1$ and $L(E,1) = 0$. We take $N = 1$. As in Example (3), we compute $a_2^{(37)} = 16985 \not\equiv 0 \pmod 3$, which implies that $\mathrm{Sel}(E/\mathbf{Q}, E[3])$ is generated by at most two elements as above. We compute $\mathcal{P}_1 = \{7, 13, 31, 67, 103, 109, 127, \dots\}$. For an admissible pair $\{7, 127\}$, we have $\tilde{\delta}_{7\times 127} = -138880 \not\equiv 0 \pmod 3$. Therefore, 7×127 is δ-minimal and $\mathrm{Sel}(E/\mathbf{Q}, E[3]) \simeq \mathbf{F}_3 \oplus \mathbf{F}_3$ by Theorem 4 (3). Since $\lambda' = 2$ in this case, by Corollary 3 we know that the main conjecture holds.

Since $L(E,1) = 0$, we know $\mathrm{rank}\,\mathrm{Sel}(E/\mathbf{Q}, E[3^\infty])^\vee > 0$ by the main conjecture. This implies that

$$\mathrm{Sel}(E/\mathbf{Q}, E[3^\infty])^\vee \simeq \mathbf{Z}_3 \oplus \mathbf{Z}_3.$$

Now E has a minimal Weierstrass model $y^2 + y = x^3 - x^2 - 725658x - 430708782$. We can find rational points $P = (2403, 108146)$ and $Q = (5901, -448036)$ on this curve. We can also easily check that $E(\mathbf{F}_7)$ is cyclic group of order 6 and $E(\mathbf{F}_{31})$ is cyclic of order 39. The image of P in $E(\mathbf{F}_7)/3E(\mathbf{F}_7) \simeq \mathbf{Z}/3\mathbf{Z}$ is 0 (the identity element), and the image of Q in $E(\mathbf{F}_7)/3E(\mathbf{F}_7) \simeq \mathbf{Z}/3\mathbf{Z}$ is of order 3. On the other hand, the images of P and Q in $E(\mathbf{F}_{31})/3E(\mathbf{F}_{31}) \simeq \mathbf{Z}/3\mathbf{Z}$ do not vanish and coincide. This shows that P and Q are linearly independent over $\mathbf{Z}_3$. Therefore,

$$\operatorname{rank} E(\mathbf{Q}) = 2 \text{ and } Ш(E/\mathbf{Q})[3^\infty] = 0.$$

In the above argument we considered the images of $E(\mathbf{Q})$ in $E(\mathbf{F}_7)/3E(\mathbf{F}_7)$ and $E(\mathbf{F}_{31})/3E(\mathbf{F}_{31})$. What we explained above implies that the natural map $s_{7\times 31} : E(\mathbf{Q})/3E(\mathbf{Q}) \longrightarrow E(\mathbf{F}_7)/3E(\mathbf{F}_7) \oplus E(\mathbf{F}_{31})/3E(\mathbf{F}_{31})$ is bijective. In this example, $\tilde{\delta}_{7\times 31} = -15290 \not\equiv 0 \pmod 3$, so Conjecture 2 holds for $m = 7 \times 31$.

5. $d = 853$. We know $\epsilon = (\frac{853}{11}) = -1$. Take $N = 1$ at first. For $\ell = 271$, we have $a_3^{(271)} = 900852395/2 \not\equiv 0 \pmod 3$, which implies that $\dim_{\mathbf{F}_3} \mathrm{Sel}(E/\mathbf{Q}, E[3]) \leq 3$. We compute $\mathcal{P}_1 = \{7, 13, 67, 103, 109, \ldots, 463, \ldots\}$. We can find a rational point $P = (1194979057/51984, 40988136480065/11852352)$ on the minimal Weierstrass equation $y^2 + y = x^3 - x^2 - 7518626x - 14370149745$ of $E = X_0(11)^{(853)}$. We know that $E(\mathbf{F}_7)$ is cyclic of order 6, and $E(\mathbf{F}_{13})$ is cyclic of order 18. Both of the images of P in $E(\mathbf{F}_7)/3E(\mathbf{F}_7)$ and $E(\mathbf{F}_{13})/3E(\mathbf{F}_{13})$ are of order 3. Therefore, $s_\ell : \mathrm{Sel}(E/\mathbf{Q}, E[3]) \longrightarrow E(\mathbf{F}_\ell)/3E(\mathbf{F}_\ell)$ is surjective for each $\ell = 7, 13$. Since $13 = -1 \in (\mathbf{F}_7^\times)^3$, $463 = 1 \in (\mathbf{F}_7^\times)^3$ and $463 = 8 \in (\mathbf{F}_{13}^\times)^3$, $\{7, 13, 463\}$ is admissible. We can compute $\tilde{\delta}_{7\times 13\times 463} = -8676400 \not\equiv 0 \pmod 3$, and can check that $m = 7 \times 13 \times 463$ is δ-minimal. By Theorem 4 (4), we have

$$\mathrm{Sel}(E/\mathbf{Q}, E[3]) \simeq \mathbf{F}_3 \oplus \mathbf{F}_3 \oplus \mathbf{F}_3. \tag{33}$$

We have a rational point P of infinite order, so the rank of $E(\mathbf{Q})$ is ≥ 1. Take $N = 3$ and consider $\ell = 271$. Since $\tilde{\delta}_{271} = a_1^{(271)} = 35325 \equiv 9 \pmod{27}$, 9 is in $\mathrm{Fitt}_{1,\mathbf{Z}/p^3\mathbf{Z}}(\mathrm{Sel}(E/\mathbf{Q}, E[3^3])^\vee)$ by Corollary 1. This implies that $\operatorname{rank} E(\mathbf{Q}) = 1$ and $\#Ш(E/\mathbf{Q})[3^\infty] \leq 9$. This together with (33) implies that

$$Ш(E/\mathbf{Q})[3^\infty] \simeq \mathbf{Z}/3\mathbf{Z} \oplus \mathbf{Z}/3\mathbf{Z}. \tag{34}$$

Note that if we used only Theorem 1 and these computations, we could not get (33) nor (34) because we could not determine $\Theta_1(\mathbf{Q})^{(\delta)}$ by finite numbers of computations. We need Theorem 4 to obtain (33) and (34).

6. For positive integers d which are conductors of even Dirichlet characters (so $d = 4m$ or $d = 4m + 1$ for some m) satisfying $1 \leq d \leq 1{,}000$, $d \equiv 1 \pmod 3$, and

$d \not\equiv 0 \pmod{11}$, we computed $\mathrm{Sel}(E/\mathbf{Q}, E[3])$. Then $\dim \mathrm{Sel}(E/\mathbf{Q}, E[3]) = 0, 1, 2, 3$, and the case of dimension $= 3$ occurs only for $d = 853$ in Example (5).

7. We also considered negative twists. Take $d = -2963$. In this case, we know $L(E, 1) \neq 0$ and $L(E, 1)/\Omega_E^+ = 81$. We know from the main conjecture that the order of the 3-component of $Ш(E/\mathbf{Q})$ is 81, but *the main conjecture does not tell the structure of this group*. Take $N = 1$ and $\ell = 19$. Then we compute $a_2^{(19)} = 2753/2 \not\equiv 0 \pmod 3$ (we have $\vartheta_{\mathbf{Q}(19)} \equiv -432S + (2753/2)S^2 \bmod (9, S^3)$). Since $c_2 = 1$, this shows that $a_2^{(19)}$ is in $\mathrm{Fitt}_{2,\mathbf{F}_3}(\mathrm{Sel}(E/\mathbf{Q}, E[3])^\vee)$ by Theorem 8. Therefore, we have $\mathrm{Fitt}_{2,\mathbf{F}_3}(\mathrm{Sel}(E/\mathbf{Q}, E[3])) = \mathbf{F}_3$, which implies that $\mathrm{Sel}(E/\mathbf{Q}, E[3]) \simeq (\mathbf{F}_3)^{\oplus 2}$. This denies the possibility of $Ш(E/\mathbf{Q})[3^\infty] \simeq (\mathbf{Z}/3\mathbf{Z})^{\oplus 4}$, and we have

$$Ш(E/\mathbf{Q})[3^\infty] \simeq \mathbf{Z}/9\mathbf{Z} \oplus \mathbf{Z}/9\mathbf{Z}.$$

8. Let E be the curve $y^2 + xy + y = x^3 + x^2 - 15x + 16$ which is 563A1 in Cremona's book (Cremona 1992). We take $p = 3$. Since $a_3 = -1$, $\mathrm{Tam}(E) = 1$, $\mu = 0$ and the Galois representation on $T_3(E)$ is surjective, all the conditions we assumed are satisfied. We know $\epsilon = 1$ and $L(E, 1) = 0$. Take $N = 1$. We compute $\mathcal{P}_1 = \{13, 61, 103, 109, 127, 139, \ldots\}$. For admissible pairs $\{13, 103\}$, $\{13, 109\}$, we compute $\tilde{\delta}_{13\times 103} = -6{,}819 \equiv 0 \pmod 3$ and $\tilde{\delta}_{13\times 109} = -242 \not\equiv 0 \pmod 3$. From the latter, we know that

$$s_{13\times 109} : \mathrm{Sel}(E/\mathbf{Q}, E[3]) \xrightarrow{\simeq} (\mathbf{F}_3)^{\oplus 2}$$

is bijective by Theorem 4 (3). Since $\lambda' = 2$, the main conjecture also holds by Corollary 3. We know $L(E, 1) = 0$, so $\mathrm{Sel}(E/\mathbf{Q}, E[3^\infty]) \simeq (\mathbf{Z}_3)^{\oplus 2}$.

Numerically, we can find rational points $P = (2, -2)$ and $Q = (-4, 7)$ on this elliptic curve. We can check that $E(\mathbf{F}_{13})$ is cyclic of order 12, $E(\mathbf{F}_{103})$ is cyclic of order 84, and $E(\mathbf{F}_{109})$ is cyclic of order 102. The points P and Q have the same image and do not vanish in $E(\mathbf{F}_{13})/3E(\mathbf{F}_{13})$, but the image of P in $E(\mathbf{F}_{109})/3E(\mathbf{F}_{109})$ is zero, and the image of Q in $E(\mathbf{F}_{109})/3E(\mathbf{F}_{109})$ is non-zero. This shows that P and Q are linearly independent over $\mathbf{Z}_3$, and $s_{13\times 109}$ is certainly bijective. Since all the elements in $\mathrm{Sel}(E/\mathbf{Q}, E[3^\infty])$ come from the points, we have $Ш(E/\mathbf{Q})[3^\infty] = 0$. On the other hand, the image of P in $E(\mathbf{F}_{103})/3E(\mathbf{F}_{103})$ coincides with the image of Q, so $s_{13\times 103}$ is not bijective. This is an example for which $\tilde{\delta}_{13\times 103} \equiv 0 \pmod 3$ and $s_{13\times 103}$ is not bijective.

9. Let E be the elliptic curve $y^2 + xy + y = x^3 + x^2 - 10x + 6$ which has conductor 18097. We take $p = 3$. We know $a_3 = -1$, $\mathrm{Tam}(E) = 1$, $\mu = 0$ and the Galois representation on $T_3(E)$ is surjective, so all the conditions we assumed are satisfied. In this case, $\epsilon = -1$ and $L(E, 1) = 0$. Take $N = 1$. We compute $\mathcal{P}_1 = \{7, 19, 31, 43, 79, \ldots, 601, \ldots\}$. We know $\{7, 43, 601\}$ is admissible. We have $\tilde{\delta}_{7\times 43\times 601} = -2424748 \not\equiv 0 \pmod 3$, and $7 \times 43 \times 601$ is δ-minimal. We thank K. Matsuno heartily for his computing this value for us. The group $E(\mathbf{F}_7)$ is cyclic of order 9 and $E(\mathbf{F}_{43})$ is cyclic of order 42. The point $(0, 2)$

is on this elliptic curve, and has non-zero image both in $E(\mathbf{F}_7)/3E(\mathbf{F}_7)$ and $E(\mathbf{F}_{43})/3E(\mathbf{F}_{43})$. So both s_7 and s_{43} are surjective, and we can apply Theorem 4 (4) to get

$$s_{7\times 43\times 601} : \mathrm{Sel}(E/\mathbf{Q}, E[3]) \xrightarrow{\simeq} (\mathbf{F}_3)^{\oplus 3}$$

is bijective.

Numerically, we can find three rational points $P = (0,2)$, $Q = (2,-1)$, $R = (3,2)$ on this elliptic curve, and easily check that the restriction of $s_{7\times 43\times 601}$ to the subgroup generated by P, Q, R in $\mathrm{Sel}(E/\mathbf{Q}, E[3])$ is surjective. Therefore, we have checked numerically that $s_{7\times 43\times 601}$ is bijective. This also implies that $\operatorname{rank} E(\mathbf{Q}) = 3$ since $E(\mathbf{Q})_{\mathrm{tors}} = 0$. Therefore, all the elements of $\mathrm{Sel}(E/\mathbf{Q}, E[3^\infty])$ come from the rational points, and we have $\text{Ш}(E/\mathbf{Q})[3^\infty] = 0$.

5.4 A Remark on Ideal Class Groups

We consider the classical Stickelberger element

$$\tilde{\theta}^{St}_{\mathbf{Q}(\mu_m)} = \sum_{\substack{a=1 \\ (a,m)=1}}^{m} \left(\frac{1}{2} - \frac{a}{m}\right)\sigma_a^{-1} \in \mathbf{Q}[\mathrm{Gal}(\mathbf{Q}(\mu_m)/\mathbf{Q})]$$

(cf. (1)). Let $K = \mathbf{Q}(\sqrt{-d})$ be an imaginary quadratic field with conductor d, and χ be the corresponding quadratic character. Let m be a squarefree product whose prime divisors ℓ split in K and satisfy $\ell \equiv 1 \pmod p$. Using the above classical Stickelberger element, we define $\tilde{\delta}^{St}_{m,K}$ by

$$\tilde{\delta}^{St}_{m,K} = -\sum_{\substack{a=1 \\ (a,md)=1}}^{md} \frac{a}{md}\chi(a)\left(\prod_{\ell|m} \log_{\mathbf{F}_\ell}(a)\right)$$

(cf. (2)). We denote by Cl_K the class group of K, and define the notion "δ^{St}_K-minimalness" analogously. We consider the analogue of Conjecture 2 for $\tilde{\delta}^{St}_{m,K}$ and $\dim_{\mathbf{F}_p}(Cl_K/p)$. Namely, we ask whether $\dim_{\mathbf{F}_p}(Cl_K/p) = \epsilon(m)$ for a δ^{St}_K-minimal m. Then the analogue does not hold. For example, take $K = \mathbf{Q}(\sqrt{-23})$ and $p = 3$. We know $Cl_K \simeq \mathbf{Z}/3\mathbf{Z}$. Put $\ell_1 = 151$ and $\ell_2 = 211$. We compute $\tilde{\delta}^{St}_{\ell_1,K} = -270 \equiv 0 \pmod 3$, $\tilde{\delta}^{St}_{\ell_2,K} = -1272 \equiv 0 \pmod 3$, and $\tilde{\delta}^{St}_{\ell_1\cdot\ell_2,K} = -415012 \equiv 2 \pmod 3$. This means that $\ell_1 \cdot \ell_2$ is δ^{St}_K-minimal. But, of course, we know $\dim_{\mathbf{F}_p}(Cl_K/p) = 1 < 2 = \epsilon(\ell_1 \cdot \ell_2)$.

References

Cremona, J.E.: Algorithms for Modular Elliptic Curves. Cambridge University Press, Cambridge/ New York (1992)
Greenberg, R.: Iwasawa theory for p-adic representations. In: Algebraic Number Theory – In Honor of K. Iwasawa. Advanced Studies in Pure Math, vol. 17, pp. 97–137. Academic/ Kinokuniya, Boston/Tokyo (1989)
Greenberg, R.: Iwasawa theory for elliptic curves. In: Arithmetic Theory of Elliptic Curves, Cetraro, 1997, Springer Lecture Notes in Mathematics, vol. 1716, pp. 51–144. Springer, Berlin/New York (1999)
Greenberg, R.: Iwasawa Theory, Projective Modules, and Modular Representation. Memoirs of the AMS, vol. 992. American Mathematical Society, Providence (2011)
Greither, C.: Computing Fitting ideals of Iwasawa modules. Math. Z. **246**, 733–767 (2004)
Hachimori, Y., Matsuno, K.: An analogue of Kida's formula for the Selmer groups of elliptic curves. J. Algebra. Geom. **8**, 581–601 (1999)
Kato, K.: p-adic Hodge theory and values of zeta functions of modular forms (Cohomologies p-adiques et applications arithmétiques III). Astérisque **295**, 117–290 (2004)
Kolyvagin, V.A.: Euler systems. In: The Grothendieck Festschrift, vol. 2, pp. 435–483. Birkhäuser, Boston (1990)
Kurihara, M.: Iwasawa theory and Fitting ideals. J. Reine Angew. Math. **561**, 39–86 (2003a)
Kurihara, M.: On the structure of ideal class groups of CM-fields. Doc. Math. **Extra Volume Kato**, 539–563 (2003b)
Kurihara, M.: Refined Iwasawa theory and Kolyvagin systems of Gauss sum type. Proc. Lond. Math. Soc. **104**, 728–769 (2012)
Kurihara, M.: Refined Iwasawa theory for p-adic representations and the structure of Selmer groups. Münster J. Math. **7**, 149–223 (2014)
Matsuno K.: An analogue of Kida's formula for the p-adic L-functions of modular elliptic curves. J. Number Theory **84**, 80–92 (2000)
Mazur, B., Rubin, K.: Kolyvagin Systems. Memoirs of the AMS, vol. 168, p. 799. American Mathematical Society, Providence (2004)
Mazur, B., Rubin, K.: Organizing the arithmetic of elliptic curves. Adv. Math. **198**, 504–546 (2005)
Mazur, B., Tate, J.: Refined conjectures of the "Birch and Swinnerton-Dyer type". Duke Math. J. **54**, 711–750 (1987)
Nekovář, J.: On the parity of ranks of Selmer groups. II. C. R. Acad. Sci. Paris Sér. I Math. **332**, 99–104 (2001)
Nekovář, J.: Selmer Complexes. Astérisque, vol. 310. Société Mathématique de France, Paris (2006)
Northcott, D.G.: Finite Free Resolutions. Cambridge University Press, Cambridge/New York (1976)
Popescu, C.D.: On the Coates-Sinnott conjecture. Mathematische Nachrichten **282**, 1370–1390 (2009)
Rubin, K.: Kolyvagin's system of Gauss sums. In: van der Geer, G., et al. (ed.) Arithmetic Algebraic Geometry. Progress in Mathematics, vol. 89, pp. 309–324. Birkhäuser, Boston (1991)
Schneider, P.: Iwasawa L-functions of varieties over algebraic number fields, a first approach. Invent. Math. **71**, 251–293 (1983)
Serre, J.-P.: Corps Locaux, troisième édition. Hermann, Paris (1968)
Skinner, C., Urban, E.: The Iwasawa main conjecture for GL_2. Invent. Math. **195**, 1–277 (2014)
Stevens, G.: Stickelberger elements and modular parametrizations of elliptic curves. Invent. Math. **98**, 75–106 (1989)

P-Adic Integration on Ray Class Groups and Non-ordinary p-Adic L-Functions

David Loeffler

1 Introduction

P-adic L-functions attached to various classes of automorphic forms over number fields are an important object of study in Iwasawa theory. In most cases, these p-adic L-functions are constructed by interpolating the algebraic parts of critical values of classical (complex-analytic) L-functions of twists of the automorphic form by finite-order Hecke characters of the number field.

When the underlying automorphic form is ordinary at p, or when the number field is $\mathbf{Q}$, this p-adic interpolation is very well-understood. However, the case of non-ordinary automorphic forms for number fields $K \neq \mathbf{Q}$ has received comparatively little study. This paper grew out of the author's attempts to understand the work of Kim (2011), who considered the L-functions of weight 2 elliptic modular forms over imaginary quadratic fields.

In this paper, we develop a systematic theory of finite-order distributions and p-adic interpolation on ray class groups of number fields. Developing such a theory is a somewhat delicate issue, and appears to have been the subject of several mistakes in the literature to date; we hope this paper will go some way towards resolving the confusion. We then apply this theory to the study of p-adic L-functions for automorphic representations of GL_2 over a general number field, extending the work of Shai Haran (1987) in the ordinary case. Largely for simplicity we assume the automorphic representations have trivial central character and lowest cohomological infinity-type (as in the case of representations attached to modular elliptic curves over K).

Research supported by a Royal Society University Research Fellowship.

D. Loeffler (✉)
Mathematics Institute, University of Warwick, Coventry CV4 7AL, UK
e-mail: d.a.loeffler@warwick.ac.uk

T. Bouganis and O. Venjakob (eds.), *Iwasawa Theory 2012*, Contributions in Mathematical and Computational Sciences 7, DOI 10.1007/978-3-642-55245-8_12,

Applying our theory to the case of K imaginary quadratic with p split in K, we obtain proofs of two conjectures. Firstly, we prove a special case of a conjecture advanced of the author and Zerbes (Loeffler and Zerbes 2011), confirming the existence of two "extra" p-adic L-functions whose existence was predicted on the basis of general conjectures on Euler systems. Secondly, we consider the case when the Hecke eigenvalues at both primes above p are 0 (the "most supersingular" case); in this case Kim has predicted a decomposition of the L-functions analogous to the "signed L-functions" of Pollack (2003) for modular forms. Using the additional information arising from our two extra L-functions, we prove Kim's conjecture, constructing four bounded L-functions depending on a choice of sign at each of the two primes above p.

2 Integration on p-Adic Groups

In this section, we shall recall and slightly generalize some results concerning the space of locally analytic distributions on an abelian p-adic analytic group (the dual space of locally analytic functions); in particular, we are interested in when it is possible to uniquely interpolate a linear functional on locally constant functions by a locally analytic distribution satisfying some growth property.

In this section we fix a prime p and a coefficient field E, which will be a complete discretely-valued subfield of $\mathbf{C}_p$, endowed with the valuation v_p such that $v_p(p) = 1$.

2.1 *One-Variable Theory*

We first recall the well-known theory of finite-order functions and distributions on the group $\mathbf{Z}_p$.

For $r \in \mathbf{R}_{\geq 0}$, we define the space of *order r functions* $C^r(\mathbf{Z}_p, E)$ as the space of functions $\mathbf{Z}_p \to E$ admitting a local Taylor expansion of degree $\lfloor r \rfloor$ at every point with error term $o(z^r)$, cf. Colmez (2010, §I.5). We write $D^r(\mathbf{Z}_p, E)$, the space of *order r distributions*, for the dual of $C^r(\mathbf{Z}_p, E)$ (the space of linear functionals $C^r(\mathbf{Z}_p, E) \to E$ which are continuous with respect to the Banach space topology of $C^r(\mathbf{Z}_p, E)$; see op. cit. for the definition of this topology).

We may regard the space $C^{\text{la}}(\mathbf{Z}_p, E)$ of locally analytic E-valued functions on $\mathbf{Z}_p$ as a dense subspace of $C^r(\mathbf{Z}_p, E)$ for any r, so dually all of the spaces $D^r(\mathbf{Z}_p, E)$ may be regarded as subspaces of the E-algebra $D^{\text{la}}(\mathbf{Z}_p, E)$ of *locally analytic distributions*, the dual of $C^{\text{la}}(\mathbf{Z}_p, E)$. It is well known that $D^{\text{la}}(\mathbf{Z}_p, E)$

can be interpreted as the algebra of functions on a rigid-analytic space over E (isomorphic to the open unit disc), whose points parametrize continuous characters of $\mathbf{Z}_p$.

It is clear that the above definitions extend naturally if $\mathbf{Z}_p$ is replaced by any abelian p-adic analytic group of dimension 1, since any such group has a finite-index open subgroup H isomorphic to $\mathbf{Z}_p$; we say a function, distribution, etc. has a given property if its pullback under the map $\mathbf{Z}_p \cong a \cdot H \subseteq G$ has it for every $a \in G/H$.

We define a *locally constant distribution* to be a linear functional on the space $LC(G, E)$ of locally constant functions on $\mathbf{Z}_p$. This is clearly equivalent to the data of a finitely-additive E-valued function on open subsets of G, i.e. a "distribution" in the sense[1] of some textbooks such as Washington (1997).

Definition 1. Let G be an abelian p-adic analytic group of dimension 1, and let μ be an E-valued locally constant distribution on G. We say μ has *growth bounded by r* if there exists $C \in \mathbf{R}$ such that

$$\inf_{a \in G} v_p\, \mu\left(1_{a+p^m H}\right) \geq C - rm$$

for all $m \geq 0$, where H is some choice of subgroup of G isomorphic to $\mathbf{Z}_p$.

Remark 1. The definition is independent of the choice of H, although the constant C may not be so.

Theorem 1 (Amice–Vélu, Vishik). *Let G be an abelian p-adic Lie group of dimension 1.*

(i) If $\mu \in D^r(G, E)$, then the restriction of μ to $LC(G, E)$ has growth bounded by r.
(ii) Conversely, if $r < 1$ and μ is a locally constant distribution on G with growth bounded by r, there exists a unique element $\tilde{\mu} \in D^r(G, E)$ whose restriction to $LC(G, E)$ is μ.

Proof. This is a special case of Colmez (2010, Theorème II.3.2).

Remark 2. In the case $r \geq 1$, one can check that any locally constant distribution μ with growth bounded by r can be extended to an element $\tilde{\mu} \in D^r(G, E)$, but this is only uniquely defined modulo $\ell_0 D^{r-1}(G, E)$, where ℓ_0 is the order 1 distribution whose action on C^1 functions is $f \mapsto f'(0)$.

[1]To avoid confusion we shall not use the term "distribution" alone in this paper, but rather in company with an adjectival phrase, such as "locally analytic", and generally an "X distribution" shall mean "a continuous linear functional on the space of X functions" (with respect to some topology to be understood from the context).

2.2 The Case of Several Variables

We now consider the case of functions of several variables. First we consider the group $G = \mathbf{Z}_p^d$, for some integer $d \geq 1$. Let r be a fixed real number; for simplicity, we shall assume that $r < 1$.

Definition 2. Let $d \geq 1$ and let $f : G \to E$ be any continuous function. For $m \geq 0$ define the quantity $\varepsilon_m(f)$ by

$$\varepsilon_m(f) = \inf_{x \in G, y \in p^m G} v_p\left[f(x+y) - f(x)\right].$$

We say f has *order* r if $\varepsilon_m(f) - rm \to \infty$ as $m \to \infty$.

The space $C^r(G, E)$ of functions with this property is evidently an E-Banach space with the valuation

$$v_{C^r}(f) = \inf\left(\inf_{x \in G} v_p(f(x)), \inf_{m \geq 0}(\varepsilon_m(f) - rm)\right).$$

Remark 3. For $d = 1$ this reduces to the definition of the valuation denoted by v'_{C^r} in Colmez (2010); in op. cit. the notation v_{C^r} is used for another slightly different valuation on $C^r(\mathbf{Z}_p, E)$, which is not equal to v'_{C^r} but induces the same topology.

If we define, for any $f \in C^r(G, E)$, a sequence of functions $(f_m)_{m \geq 0}$ by

$$f_m = \sum_{j_1=0}^{p^m-1} \cdots \sum_{j_d=0}^{p^m-1} f(j_1, \dots, j_d) 1_{(j_1, \dots, j_d) + p^m G},$$

then it is clear from the definition of the valuation v_{C^r} that we have $f_m \to f(x)$ in the topology of $C^r(G, E)$, so in particular the space $LC(G, E)$ of locally constant E-valued functions on G is dense in $C^r(G, E)$.

We define a space $D^r(G, E)$ as the dual of $C^r(G, E)$, as before.

As in the case $d = 1$ all of these constructions are clearly local on G and thus extend[2] to abelian groups having an open subgroup isomorphic to $\mathbf{Z}_p^d$. Since any abelian p-adic analytic group has this property for some $d \geq 0$, we obtain well-defined spaces $C^r(G, E)$ and dually $D^r(G, E)$ for any such group G.

Remark 4. This includes the case where G is finite, so $d = 0$; in this case we clearly have $C^r(G, E) = C^{\mathrm{la}}(G, E) = \mathrm{Maps}(G, E)$ and $D^r(G, E) = D^{\mathrm{la}}(G, E) = E[G]$ for any $r \geq 0$.

---

[2] A little care is required here since for $k \geq 1$ pullback along the inclusion $\mathbf{Z}_p \xrightarrow{\times p^k} \mathbf{Z}_p$ is a continuous map $C^r(\mathbf{Z}_p, E) \to C^r(\mathbf{Z}_p, E)$ but *not* an isometry if $r \neq 0$. So for a general G the space $C^r(G, E)$ has a canonical topology, but not a canonical valuation inducing this topology.

We now consider the problem of interpolating locally constant distributions by order r distributions.

Definition 3. Let μ be a locally constant distribution on G. We say that μ has *growth bounded by* r if there is a constant C and an open subgroup H of G isomorphic to $\mathbf{Z}_p^d$ such that

$$\inf_{a \in G} v_p \mu\left(1_{a+p^m H}\right) \geq C - rm$$

for all $m \geq 0$.

The natural analogue of the theorem of Amice–Vélu–Vishik is the following.

Theorem 2. *Let G be an abelian p-adic analytic group, $r \in [0, 1)$, and μ a locally constant distribution on G with growth bounded by r. Then there is a unique element $\tilde{\mu} \in D^r(G, E)$ whose restriction to $LC(\mathbf{Z}_p, E)$ is μ.*

Proof. The proof of this result in the general case is very similar to the case $d = 1$, so we shall give only a brief sketch. Clearly one may assume $G = H = \mathbf{Z}_p^d$. One first shows that the space of locally constant functions on $\mathbf{Z}_p^d$ has a "wavelet basis" consisting of the indicator functions of the sets

$$S(i_1, \dots, i_d) = (i_1 + p^{\ell(i_1)}\mathbf{Z}_p) \times \dots \times (i_d + p^{\ell(i_d)}\mathbf{Z}_p)$$

for $i_1, \dots, i_d \geq 0$, where $\ell(n)$ is as defined in Colmez (2010, §1.3). One then checks that the functions

$$e_{(i_1,\dots,i_d),r} := p^{\sup_j \lfloor r\ell(i_j) \rfloor} 1_{S(i_1,\dots,i_d)}$$

form a Banach basis of $C^r(\mathbf{Z}_p^d, E)$, so a linear functional on locally constant functions taking bounded values on the $e_{(i_1,\dots,i_d),r}$ extends uniquely to the whole of $C^r(\mathbf{Z}_p^d, E)$.

Remark 5. The case when $G = \mathcal{O}_L$, for some finite extension $L/\mathbf{Q}_p$, has been studied independently by De Ieso (2012), who shows the following more general result: a locally polynomial distribution of degree k which has growth of order r in a sense generalizing Definition 3 admits a unique extension to an element of $D^r(G, E)$ if (and only if) $r < k + 1$. This reduces to the theorem above for $k = 0$.

2.3 *Groups with a Quasi-factorization*

The constructions we have used for C^r and D^r seem to be essentially the only sensible approaches to defining such spaces which are functorial with respect to arbitrary automorphisms of G. However, if one assumes a little more structure on the group G then there are other possibilities, which allow a greater range of locally constant distributions to be p-adically interpolated.

Definition 4. Let G be an abelian p-adic analytic group, and let $\mathcal{G}$ be the Lie algebra of G. A *quasi-factorization* of G is a decomposition $\mathcal{G} = \bigoplus \mathcal{H}_i$ of $\mathcal{G}$ as a direct sum of subspaces.

If $(\mathcal{H}_i)$ is a quasi-factorization of G, we can clearly choose (non-uniquely) closed subgroups $H_i \subseteq G$ such that $\mathcal{H}_i$ is the Lie algebra of H_i and $\prod H_i$ is an open subgroup of G. We shall say that (H_i) are subgroups compatible with the quasi-factorization $(\mathcal{H}_i)$.

Definition 5. Let G be an abelian p-adic analytic group with a quasi-factorization $\mathcal{H} = (\mathcal{H}_1, \dots, \mathcal{H}_g)$, and let $(H_1, \dots, H_g)$ be subgroups compatible with $\mathcal{H}$. Let $f : G \to E$ be continuous. Define

$$\varepsilon_{m_1,\dots,m_g}(f) = \inf_{\substack{x \in G \\ y \in p^{m_1} H_1 \times \dots \times p^{m_g} H_g}} v_p \left[f(x+y) - f(x) \right].$$

We say f has *order* $(r_1, \dots, r_g)$, where $r_i \in [0, 1)$, if we have

$$\varepsilon_{m_1,\dots,m_g}(f) - (r_1 m_1 + \dots + r_g m_g) \to \infty \tag{1}$$

as $(m_1, \dots, m_g) \to \infty$ (with respect to the filter of cofinite subsets of $\mathbf{N}^g$, i.e. for any N there are finitely many g-tuples $(m_1, \dots, m_g)$ such that the above expression is $\le N$)

We define $C^{(r_1,\dots,r_g)}(G, E)$ to be the space of functions of order $(r_1, \dots, r_g)$, equipped with the valuation given by

$$v_{(r_1,\dots,r_g)}(f) = \inf \left[\inf_{x \in G} v_p(f(x)), \inf_{(m_1,\dots,m_n)} \left(\varepsilon_{m_1,\dots,m_g}(f) - (r_1 m_1 + \dots + r_g m_g) \right) \right],$$

which makes it into an E-Banach space. We write $D^{(r_1,\dots,r_g)}(G, E)$ for its dual. It is clear that as topological vector spaces these do not depend on the choice of the (H_i).

Remark 6. One can check that if $G = H_1 \times \dots \times H_g$, then $C^{(r_1,\dots,r_g)}(G, E)$ is isomorphic to the completed tensor product

$$\widehat{\bigotimes_{1 \le i \le g}} C^{r_i}(H_i, E).$$

The spaces $C^{(r_1,\dots,r_g)}(G, E)$ are clearly invariant under automorphisms of G preserving the quasi-factorization $\mathcal{H}$.

Definition 6. Let G be an abelian p-adic analytic group with a quasi-factorization $\mathcal{H} = (\mathcal{H}_1, \dots, \mathcal{H}_g)$, and let $(H_1, \dots, H_g)$ be subgroups compatible with $\mathcal{H}$. Let μ be a locally constant distribution on G.

We say μ has *growth bounded by* $(r_1, \dots, r_d)$ if the following condition holds: there is a constant C such that for all $m_1, \dots, m_g \in \mathbf{Z}_{\ge 0}$, we have

$$\inf_{a \in G} v_p \mu \left(1_{a+(p^{m_1} H_1) \times \cdots \times (p^{m_g} H_g)}\right) \geq C - (r_1 m_1 + \cdots + r_g m_g).$$

It is clear that whether or not μ has growth bounded by (r_i) does not depend on the choice of (H_i), although the constant C will so depend.

Theorem 3. *Suppose μ is a locally constant distribution with growth bounded by $(r_1, \ldots, r_g)$, where $r_i \in [0, 1)$. Then there is a unique extension of μ to an element*

$$\tilde{\mu} \in D^{(r_1, \ldots, r_g)}(G, E).$$

Proof. As usual it suffices to assume that $G = H_1 \times \cdots \times H_g$. Then the isomorphism

$$C^{(r_1, \ldots, r_g)}(G, E) \cong \widehat{\bigotimes_{1 \leq i \leq g}} C^{r_i}(H_i, E)$$

gives an explicit Banach space basis of $C^{(r_1, \ldots, r_g)}(G, E)$, and the result is now clear.

Remark 7. The special case of Theorem 3 where $G = \mathbf{Z}_p^2$, with its natural quasi-factorization, is studied in §7.1 of Kim (2011), which was the inspiration for many of the results of this paper.

3 Distributions on Ray Class Groups

We now consider the application of the above machinery to global settings. We give a fairly general formulation, in the hope that this theory will be useful in contexts other than those we consider in the sections below; sadly, this comes at the cost of somewhat cumbersome notation.

Let K be a number field. As usual, we define a "modulus" of K to be a finite formal product $\prod_{i=1}^{N} v_i^{n_i}$ where each v_i is either a finite prime of K or a real place of K, and $n_i \in \mathbf{Z}_{\geq 0}$, with $n_i \leq 1$ if v_i is infinite. We define a "pseudo-modulus" to be a similar formal product but with some of the n_i at finite places allowed to be ∞. If $\mathfrak{F}$ is a pseudo-modulus, then the ray class group of K modulo $\mathfrak{F}$ is defined, and we denote this group by $G_{\mathfrak{F}}$.

Let $\mathfrak{F}$ be a pseudo-modulus of K, and let Σ be the finite set of primes $\mathfrak{p}$ of K such that $\mathfrak{p}^\infty \mid \mathfrak{F}$.

Let F be a finite extension of either $\mathbf{Q}$ or of $\mathbf{Q}_p$ for some prime p, and V a finite-dimensional F-vector space.

Definition 7. A *growth parameter* is a family $\delta = (\delta_{\mathfrak{p}})_{\mathfrak{p} \in \Sigma}$ of non-zero elements of $\mathcal{O}_F$ indexed by the primes in Σ.

Definition 8. A *ray class distribution* modulo $\mathfrak{F}$ with growth parameter δ and values in V to be the data of, for each modulus $\mathfrak{f}$ dividing $\mathfrak{F}$, an element

$$\Phi_{\mathfrak{f}} \in F[G_{\mathfrak{f}}] \otimes_F V,$$

such that:

1. We have $\mathrm{Norm}_{\mathfrak{f}}^{\mathfrak{f}'} \Phi_{\mathfrak{f}'} = \Phi_{\mathfrak{f}}$ for all pairs of moduli $(\mathfrak{f}, \mathfrak{f}')$ with $\mathfrak{f} \mid \mathfrak{f}'$ and $\mathfrak{f}' \mid \mathfrak{F}$.
2. There exists an $\mathcal{O}_F$-lattice $\Lambda \subseteq V$ such that for every $\mathfrak{f} \mid \mathfrak{F}$, we have $\Phi_{\mathfrak{f}} \in \mathcal{O}_F[G_{\mathfrak{f}}] \otimes_{\mathcal{O}_F} \delta_{\mathfrak{f}}^{-1} \Lambda$, where $\delta_{\mathfrak{f}} = \prod_{\mathfrak{p} \in \Sigma} \delta_{\mathfrak{p}}^{-v_{\mathfrak{p}}(\mathfrak{f})}$.

We shall see in subsequent sections that systems of elements of this kind appear in several contexts as steps in the construction of p-adic L-functions attached to automorphic forms.

Since the natural maps $G_{\mathfrak{F}} \to G_{\mathfrak{f}}$ for $\mathfrak{f} \mid \mathfrak{F}$ are surjective, and $G_{\mathfrak{F}}$ is equal to the inverse limit of the finite groups $G_{\mathfrak{f}}$ over moduli $\mathfrak{f} \mid \mathfrak{F}$, condition (1) implies that we may interpret a ray class distribution as a locally constant distribution on $G_{\mathfrak{F}}$ with values in V. Condition (2) can then be interpreted as a growth condition on this locally constant distribution. Our goal is to investigate how this interacts with the growth conditions studied in Sect. 2 above.

If the coefficient field F is a number field, then choosing a prime $\mathfrak{q}$ of F allows us to interpret V-valued ray class distributions as $V_{\mathfrak{q}}$-valued ray class distributions, where $V_{\mathfrak{q}} = F_{\mathfrak{q}} \otimes_F V$. So we shall assume henceforth that $F = E$ is a finite extension of $\mathbf{Q}_p$ for some rational prime p, as in Sect. 2.

Let us write $\mathfrak{F} = \mathfrak{g} \cdot \prod_{\mathfrak{p} \in \Sigma} \mathfrak{p}^{\infty}$, for some modulus $\mathfrak{g}$ coprime to Σ. Then we have an exact sequence

$$0 \to \overline{\mathcal{E}}_{\mathfrak{g}} \to \prod_{\mathfrak{p} \in \Sigma} \mathcal{O}_{K,\mathfrak{p}}^{\times} \to G_{\mathfrak{F}} \to G_{\mathfrak{g}} \to 0$$

where $\overline{\mathcal{E}}_{\mathfrak{g}}$ is the closure of the image in $\prod_{\mathfrak{p} \in \Sigma} \mathcal{O}_{K,\mathfrak{p}}^{\times}$ of the group of units of $\mathcal{O}_K$ congruent[3] to 1 modulo $\mathfrak{g}$. Since $G_{\mathfrak{g}}$ is finite, we see that if Σ is a subset of the primes above p, then $G_{\mathfrak{F}}$ is a p-adic analytic group; and if

$$\mathfrak{f} = \prod_{\mathfrak{p} \in \Sigma} \mathfrak{p}^{n_{\mathfrak{p}}}$$

for some integers $n_{\mathfrak{p}}$, the kernel of the natural surjection $G_{\mathfrak{F}} \to G_{\mathfrak{f}\mathfrak{g}}$ is the image $H_{\mathfrak{f}}$ of the subgroup

$$U_{\mathfrak{f}} := \prod_{\mathfrak{p} \in \Sigma} (1 + \mathfrak{p}^{n_{\mathfrak{p}}} \mathcal{O}_{K,\mathfrak{p}})^{\times} \subseteq \prod_{\mathfrak{p} \in \Sigma} \mathcal{O}_{K,\mathfrak{p}}^{\times}.$$

The subgroups $H_{\mathfrak{f}}$ for $\mathfrak{f} \mid \Sigma^{\infty}$ form a basis of neighbourhoods of 1 in $G_{\mathfrak{F}}$, so we may regard a ray class distribution as a locally constant distribution on $G_{\mathfrak{F}}$ satisfying

[3] We adopt the usual convention in class field theory that if $\mathfrak{g}$ is divisible by a real infinite place v, "congruent to 1 modulo v" means that the image of the unit concerned under the corresponding real embedding of K should be positive.

a boundedness condition relative to the subgroups $H_{\mathfrak{f}}$. Our task is to investigate how this interacts with the boundedness conditions considered in Sect. 2 above.

3.1 Criteria for Distributions of Order r

The following theorem gives a sufficient condition for a ray class distribution to define a finite-order distribution on the group $G_{\mathfrak{F}}$ in the sense of Sect. 2.2, when Σ is a subset of the of primes above p (so that $G_{\mathfrak{F}}$ is a p-adic analytic group). We continue to assume that the coefficient field E is a finite extension of $\mathbf{Q}_p$, and (as before) we write v_p for the valuation on E such that $v_p(p) = 1$. We shall assume for simplicity that the coefficient space V is simply E; the general case follows immediately from this.

Theorem 4. *Let Θ be a ray class distribution modulo $\mathfrak{F}$. Suppose Θ has growth parameter $(\alpha_{\mathfrak{p}})_{\mathfrak{p}\in\Sigma}$, and define*

$$r = \sum_{\mathfrak{p}\in\Sigma} e_{\mathfrak{p}} v_p(\alpha_{\mathfrak{p}}),$$

where $e_{\mathfrak{p}}$ is the absolute ramification index of the prime $\mathfrak{p}$. Then $\Theta_{\mathfrak{F}}$, viewed as a locally constant distribution on $G_{\mathfrak{F}}$, has growth bounded by r in the sense of Definition 3.

In particular, if $r < 1$, there is a unique element $\tilde{\Theta} \in D^r(G_{\mathfrak{F}}, E)$ extending Θ.

Proof. Let $\mathfrak{p} \in \Sigma$. We can choose some integer $c \geq 0$ such that the p-adic logarithm converges on $U_{\mathfrak{p},k} = (1 + \mathfrak{p}^m \mathcal{O}_{\mathfrak{p}})^\times \subseteq \mathcal{O}_{\mathfrak{p}}^\times$ for all $k \geq c$, and identifies $U_{\mathfrak{p},k}$ with the additive group $\mathfrak{p}^k \mathcal{O}_{\mathfrak{p}}$. In particular, $U_{\mathfrak{p},c}$ is isomorphic to $\mathbf{Z}_p^{[K_{\mathfrak{p}}:\mathbf{Q}_p]}$, and for $m \geq 0$ we have

$$U_{\mathfrak{p},c}^{p^m} \cong p^m \mathfrak{p}_i^c \mathcal{O}_{\mathfrak{p}} = \mathfrak{p}^{c+e_{\mathfrak{p}}m} \mathcal{O}_{\mathfrak{p}} \cong U_{\mathfrak{p},c+e_{\mathfrak{p}}m}.$$

Now let $\mathfrak{p}_1, \dots, \mathfrak{p}_g$ be the primes in Σ and choose a constant c_i for each. Then the subgroup $U = U_{\mathfrak{p}_1,c_1} \times \dots \times U_{\mathfrak{p}_g,c_g}$ is an open subgroup of $(\mathcal{O}_K \otimes \mathbf{Z}_p)^\times = \prod_{i=1}^g \mathcal{O}_{\mathfrak{p}_i}^\times$, and $U \cong \mathbf{Z}_p^{[K:\mathbf{Q}]}$. By increasing the c_i if necessary, we may assume that the image of U in $G_{\mathfrak{F}}$ is torsion-free and hence isomorphic to $\mathbf{Z}_p^h$ for some $h \leq [K : \mathbf{Q}]$.

Let H be the image of U. Then the image of U^{p^m} is H^{p^m}, clearly; so a locally constant distribution on $G_{\mathfrak{F}}$ has order r if and only if there is C such that $v_p \mu(1_{x \cdot H^{p^m}}) \geq C - rm$ (this is just Definition 3, but with the group law on $G_{\mathfrak{F}}$ written multiplicatively).

However, we have

$$U^{p^m} = \prod_{i=1}^{g} U_{\mathfrak{p}_i, c_i + e_{\mathfrak{p}_i} m} = \prod_{i=1}^{g} (1 + \mathfrak{p}_i^{c_i + e_{\mathfrak{p}_i} m} \mathcal{O}_{\mathfrak{p}_i})^\times.$$

Thus $G_{\mathfrak{F}}/H^{p^m}$ is the ray class group of modulus

$$\mathfrak{f}_m := \mathfrak{g} \cdot \prod_{i=1}^{g} \mathfrak{p}_i^{c_i + e_{\mathfrak{p}_i} m},$$

where $\mathfrak{g}$ is (as above) the modulus such that $\mathfrak{F} = \mathfrak{g} \cdot \prod_{i=1}^{g} \mathfrak{p}_i^\infty$. So, if Θ is a ray class distribution with growth parameter $(\alpha_\mathfrak{p})_{\mathfrak{p} \in \Sigma}$, then we know that the valuation of $\Theta(X)$ where X is any coset of H^{p^m} is bounded below by

$$C - \sum_{i=1}^{g} v_p(\alpha_{\mathfrak{p}_i}) \cdot \operatorname{ord}_{\mathfrak{p}_i}(\mathfrak{f}_m) = C' - m \sum_{i=1}^{g} e_{\mathfrak{p}_i} v_p(\alpha_{\mathfrak{p}_i}) = C' - rm$$

for some constants C, C', where r is as defined in the statement of the theorem. So a ray class distribution with growth parameter $(\alpha_\mathfrak{p})_{\mathfrak{p} \in \Sigma}$ defines a locally constant distribution on the p-adic analytic group $G_{\mathfrak{F}}$ whose growth is bounded by r, as required.

Note that this argument does not depend on the dimension of $G_{\mathfrak{F}}$ (which is useful, since this dimension depends on whether Leopoldt's conjecture holds for K). This result is essentially the best possible (at least using the present methods) when there is a unique prime of K above p, or when K is totally real and Leopoldt's conjecture holds (as we show in the next section). However, for other fields K finer statements are possible using the theory of quasi-factorizations developed in Sect. 2.3, as we shall see below.

3.2 A Converse Result for Totally Real Fields

Let us now suppose K is totally real. We also suppose that Leopoldt's conjecture holds for K, so the image of $\mathcal{E}$ in $\prod_{\mathfrak{p} \mid p} \mathcal{O}_\mathfrak{p}^\times$ has rank $n - 1$. We shall take $\mathfrak{F} = p^\infty \mathfrak{g}$ for some modulus $\mathfrak{g}$ coprime to p, so $\Sigma = \Sigma_p$ is the set of all primes dividing p. Leopoldt's conjecture implies that $G_{p^\infty \mathfrak{g}}$ is a p-adic analytic group of dimension 1; thus, in particular, the notions of finite-order functions and distributions on $G_{p^\infty \mathfrak{g}}$ are just the standard ones.

We shall prove the following theorem:

Theorem 5. *Let $\underline{\alpha} = (\alpha_\mathfrak{p})_{\mathfrak{p} \in \Sigma_p}$ be elements of E, and let $h \in \mathbf{R}_{\geq 0}$. Then the implication*

Every ray class distribution modulo $p^{\infty}\mathfrak{g}$ of growth parameter $\underline{\alpha}$ is a locally constant distribution on $G_{p^\infty \mathfrak{g}}$ with growth bounded by h

is true if and only if the inequality

$$\sum_{\mathfrak{p}\in\Sigma_p} e_{\mathfrak{p}} v_p(\alpha_{\mathfrak{p}}) \leq h$$

holds.

We retain the notation of the proof of Theorem 4, so $\mathfrak{p}_1, \ldots, \mathfrak{p}_g$ are the primes above p, and for each i, c_i is a constant such that for all $k \geq c_i$ the logarithm map identifies $U_{\mathfrak{p}_i,k}$ with the additive group $(\mathfrak{p}_i)^m$, and the image of $U = U_{\mathfrak{p}_1,c_1} \times \ldots U_{\mathfrak{p}_g,c_g}$ in G is torsion-free and hence isomorphic to $\mathbf{Z}_p$.

Enlarging the c_i further if necessary, we may assume that there is an integer w independent of i such that for each i the image of $\mathfrak{p}_i^{c_i}$ under the trace map $Tr_{K_{\mathfrak{p}_i}/\mathbf{Q}_p}$ is $p^w\mathbf{Z}_p$. Let $d_i = [K_{\mathfrak{p}_i} : \mathbf{Q}_p]$.

Proposition 1. *For each i there exists a basis $b_{i1}, \ldots, b_{id_i}$ of $\mathfrak{p}_i^{c_i}$ as a $\mathbf{Z}_p$-module such that*

$$\mathrm{Tr}_{K_{\mathfrak{p}_i}/\mathbf{Q}_p}(b_{ij}) = p^w$$

for each j.

Proof. Elementary linear algebra.

Proposition 2. *The isomorphism*

$$U \cong \mathbf{Z}_p^n$$

(where $n = [K : \mathbf{Q}] = \sum_{i=1}^{g} d_i$) given by the bases $b_{i1}, \ldots, b_{id_i}$ for $1 \leq i \leq g$ identifies the closure of $\mathcal{E} \cap U$ with the submodule

$$\Delta = \{(x_1, \ldots, x_d) \in \mathbf{Z}_p^d : \sum_{j=1}^{d} x_j = 0\}.$$

Proof. With respect to the basis $b_1, \ldots, b_{e_i f_i}$ of $\mathfrak{p}_i^{c_i}$, the trace map is given by $(x_1, \ldots, x_{d_i}) \mapsto p^w \sum_j x_j$, so our isomorphism

$$U \xrightarrow{\log} \prod_{i=1}^{g} \mathfrak{p}_i^{c_i} \xrightarrow{\cong} \mathbf{Z}_p^n$$

identifies Δ with the elements of U whose norm down to $\mathbf{Q}$ is 1. However, since U is torsion-free, any $u \in U \cap \mathcal{E}$ satisfies $\mathrm{Norm}_{K/\mathbf{Q}}(u) = 1$, so Δ contains the closure of $\mathcal{E} \cap U$, which we write as Δ'.

By Leopoldt's conjecture (which we are assuming), Δ' has $\mathbf{Z}_p$-rank $(n-1)$, the same as Δ; so the quotient $\frac{\Delta}{\Delta'}$ is finite. However, we obviously have

$$\frac{U}{\Delta'} \cong \mathbf{Z}_p \oplus \frac{\Delta}{\Delta'};$$

and $\frac{U}{\Delta'} = U(p^\infty, \mathfrak{p}_1^{c_1} \dots \mathfrak{p}_g^{c_g})$ is torsion-free, so we must have $\Delta = \Delta'$.

Proposition 3. *Let H be the image of U in $G_{p^\infty\mathfrak{g}}$, and let $H_m = H^{p^m}$ as above. Suppose $m \ge 1$. Then if $r_1, \dots, r_g$ are integers such that the image in G of $U_{\mathfrak{p}_1,r_1} \times \dots \times U_{\mathfrak{p}_g,r_g}$ is contained in H_m, we must have $r_i \ge c_i + e_i m$ for all i.*

Proof. If the image of $U_{\mathfrak{p}_1,r_1} \times \dots \times U_{\mathfrak{p}_g,r_g}$ is contained in H_m, then the same must also be true with $(r_1, \dots, r_g)$ replaced by $(r_1', \dots, r_g')$ where $r_i' = \sup(r_i, c_i)$. So we may assume without loss of generality that $r_i \ge c_i$ for all i. Now the result is clear from the above description of the image of the global units.

This result clearly implies Theorem 5.

Remark 8. Curiously, the results above seem to conflict with some statements asserted without proof in Panchishkin (1994) (and quoted by some other subsequent works). The group studied in op. cit. corresponds in our notation to $G_{\mathfrak{g}p^\infty}$, where $\mathfrak{g}$ is the product of the infinite places, and K is assumed totally real. (Thus $G_{\mathfrak{g}p^\infty}$ corresponds via class field theory to the Galois group of the maximal abelian extension of K unramified outside p and the infinite places; it is denoted by Gal_p in op. cit.)

In Definition 4.2 of op. cit., a locally constant distribution μ on $G_{\mathfrak{g}p^\infty}$ is defined to be 1-*admissible* if

$$\sup_a \left|\mu\left(1_{a+(\mathfrak{m})}\right)\right| = o\left(|\mathfrak{m}|_p^{-1}\right).$$

Paragraph 4.3 of op. cit. then claims that a 1-admissible measure extends uniquely to an element of $D^{\mathrm{la}}(G_{\mathfrak{g}p^\infty}, E)$ which, regarded as a rigid-analytic function on the space $\mathfrak{X}_p$ parametrizing characters of $G_{\mathfrak{g}p^\infty}$, has growth $o(\log(1+X))$.

This is consistent with Theorems 4 and 5 above if there is only one prime above p. However, if there are multiple primes above p, it is not so clear how $|\mathfrak{m}|$ is to be defined for general ideals $\mathfrak{m} \mid p^\infty$ of K, and the uniqueness assertion of Conjecture 6.2 of op. cit. (which is asserted to be a consequence of the theory of h-admissible measures) contradicts Theorem 5 of the present paper.

3.3 Imaginary Quadratic Fields

Let K be an imaginary quadratic field. If p is inert or ramified in K then there is only one prime $\mathfrak{p}$ of K above p, and hence there is no canonical direct sum decomposition of $\mathrm{Lie}\, G_{p^\infty} \cong K_{\mathfrak{p}}$. On the other hand, we can find one when p splits:

Proposition 4. *If K is imaginary quadratic and $p = \mathfrak{p}\overline{\mathfrak{p}}$ is split, then the images of* $\mathrm{Lie}\, \mathcal{O}_{K,\mathfrak{p}}^\times$ *and* $\mathrm{Lie}\, \mathcal{O}_{K,\overline{\mathfrak{p}}}^\times$ *in* $\mathrm{Lie}(G_{p^\infty})$ *form a quasi-factorization.*

Proposition 5. *Let μ be a ray class distribution modulo $p^\infty\mathfrak{g}$ with growth parameter $(\alpha_\mathfrak{p}, \alpha_{\overline{\mathfrak{p}}})$. Then μ has growth bounded by $(v_p(\alpha_\mathfrak{p}), v_p(\alpha_{\overline{\mathfrak{p}}}))$ in the sense of Definition 6.*

Proof. Clear by construction.

Combining the above with Theorem 3 we have the following:

Theorem 6. *Let K be an imaginary quadratic field in which $p = \mathfrak{p}\overline{\mathfrak{p}}$ is split, and let Θ be a ray class distribution modulo $p^\infty\mathfrak{g}$ with values in V and growth parameter $(\alpha_\mathfrak{p}, \alpha_{\overline{\mathfrak{p}}})$. Let $r = v_p(\alpha_\mathfrak{p})$ and $s = v_p(\alpha_{\overline{\mathfrak{p}}})$. If $r, s < 1$, then there is a unique distribution*

$$\tilde{\Theta} \in D^{(r,s)}(G_{p^\infty}, E) \otimes_E V$$

such that the restriction of $\tilde{\Theta}$ to $LC(G_{p^\infty}, E) \otimes_E V$ is Θ. If we have $r + s < 1$, then $\tilde{\Theta}$ lies in $D^{r+s}(G_{p^\infty}, E) \otimes_E V$ and agrees with the element constructed in Theorem 4.

3.4 Fields of Higher Degree

When K is a non-totally-real extension of degree >2, one can sometimes find interesting quasi-factorizations of G_{p^∞} by considering subsets of the primes above p. If we identify G_{p^∞} with the Galois group of the maximal abelian extension $K(p^\infty)/K$ unramified outside p, we seek to write $K(p^\infty)$ as a compositum of smaller extensions, each ramified at as few primes as possible. We give a few examples of the behaviour that occurs when p is totally split.

Example 1 (Mixed signature cubic fields). Suppose K is a non-totally-real cubic field, and p is totally split in K, say $p = \mathfrak{p}_1\mathfrak{p}_2\mathfrak{p}_3$. Then G_{p^∞} has dimension 2 (since Leopoldt's conjecture is trivially true in this case), and the images of the groups $\mathcal{O}_{K,\mathfrak{p}_i}^\times$ in G_{p^∞} give three pairwise-disjoint one-dimensional subspaces of $\mathrm{Lie}\, G_{p^\infty}$, so any two of these form a quasi-factorization. One therefore obtains three quasi-factorizations of G_{p^∞}.

One checks that with respect to the quasi-factorization given by $\mathfrak{p}_1$ and $\mathfrak{p}_2$, a ray class distribution of parameter $(\alpha_{\mathfrak{p}_1}, \alpha_{\mathfrak{p}_2}, \alpha_{\mathfrak{p}_3})$ has growth bounded by (r, s) if and only if $v_p(\alpha_{\mathfrak{p}_1}) + v_p(\alpha_{\mathfrak{p}_3}) \le r$ and $v_p(\alpha_{\mathfrak{p}_2}) + v_p(\alpha_{\mathfrak{p}_3}) \le s$. For instance, if all α_i are equal and their common valuation h is $< \frac{1}{2}$, then can construct a locally analytic distribution (indeed three of them), while Theorem 4 would only apply if $h < \frac{1}{3}$. For the cases $\frac{1}{3} \le h < \frac{1}{2}$, it is not immediately obvious whether the interpolations provided by the three quasi-factorizations are equal or not, but we shall see in the next section that this is indeed the case.

Example 2 (Quartic fields). Suppose K is a totally imaginary quartic field, and p is totally split in K. As in the previous example, $\mathcal{O}_K^\times$ has rank 1 and thus Leopoldt's conjecture is automatic, so G_{p^∞} has rank 3. We can obtain a quasi-factorization of

$G_{p\infty}$ by considering the images of the Lie algebras of $\mathcal{O}_{K,\mathfrak{p}_i}^\times$ for $1 \le i \le 3$; then (much as in the previous case) we find that the condition for a ray class distribution to have growth bounded by (r, s, t) is that

$$v_p(\alpha_1) + v_p(\alpha_4) \le r,$$
$$v_p(\alpha_2) + v_p(\alpha_4) \le s,$$
$$v_p(\alpha_3) + v_p(\alpha_4) \le t.$$

A slightly more intricate class of quasi-factorizations can be obtained as follows. We can choose a basis for Lie $\mathcal{O}_{K,\mathfrak{p}_i}^\times$ for each i such that the Lie algebra of the subgroup $\overline{\mathcal{E}}_K$ corresponds to the subspace Δ of $\mathbf{Q}_p^4$ spanned by $(1, 1, 1, 1)$. Then choosing a quasi-factorization amounts to choosing a basis of the 3-dimensional space of linear functionals on $\mathbf{Q}_p^4$ which vanish on Δ. One such basis is given by the functionals mapping (x_1, x_2, x_3, x_4) to $\{x_1 - x_4, x_2 - x_4, x_3 - x_4\}$, and this gives the quasi-factorization we have already seen. However, we can also consider the functionals $\{x_1 - x_2, x_2 - x_3, x_3 - x_4\}$. One checks then that the condition for a ray class distribution to have growth bounded by (r, s, t) is

$$v_p(\alpha_1) + v_p(\alpha_2) \le r,$$
$$v_p(\alpha_2) + v_p(\alpha_3) \le s,$$
$$v_p(\alpha_3) + v_p(\alpha_4) \le t.$$

One can generalize the last example to a much wider range of number fields under suitable assumptions on the position of Lie $\overline{\mathcal{E}}_K$ inside Lie$(\mathbf{Z}_p \otimes \mathcal{O}_K)^\times$ (which amounts to assuming a strong form of Leopoldt's conjecture, such as that formulated in Calegari and Mazur (2009)).

3.5 A Representation-Theoretic Perspective

For simplicity let us suppose that K has class number 1 and $\mathfrak{g} = (1)$, so $G := G_{p\infty} = U/\overline{\Delta}$ where $U = \prod_{i=1}^{g} \mathcal{O}_{\mathfrak{p}_i}^\times$ and $\Delta = \mathcal{O}_K^\times$. Pick real numbers $r_1, \dots, r_g$ with $r_i \in [0, 1) \cap v_p(E^\times)$.

We consider the inclusion

$$LC(U, E)^\Delta \hookrightarrow C^{(r_1,\dots,r_g)}(U, E)^\Delta,$$

where we give U the obvious quasi-factorization. We have, of course, $LC(U, E)^\Delta = LC(U/\overline{\Delta}, E) = LC(G, E)$. The following proposition is immediate from the definitions:

Proposition 6. *Let μ be an element of $D^{(r_1,\dots,r_g)}(U, E)$. Then the restriction of μ to $LC(U, E)^{\Delta}$ is a ray class distribution of parameter $(\alpha_1, \dots, \alpha_g)$, where the α_i are any elements with $v_p(\alpha_i) = r_i$.*

Conversely, any ray class distribution of parameter $(\alpha_1, \dots, \alpha_g)$ defines a linear functional on $LC(U, E)^{\Delta}$ which is continuous with respect to the subspace topology given by the inclusion into $C^{(r_1,\dots,r_g)}(U, E)$.

However, even though $LC(U, E)$ is dense in $C^{(r_1,\dots,r_g)}(U, E)$, it does not necessarily follow that $LC(U, E)^{\Delta}$ is dense in $C^{(r_1,\dots,r_g)}(U, E)^{\Delta}$; the completion of the invariants may be smaller than the invariants of the completion.

Example 3. Consider the case where U is $\mathbf{Z}_p^2$ with its natural quasi-factorization, and $\overline{\Delta}$ is the subgroup $\{(x, y) : x + y = 0\}$. Then a continuous function $f : U \to E$ is Δ-invariant if and only if there is a function $h \in C^0(\mathbf{Z}_p, E)$ such that $f(x, y) = h(x+y)$. So the coefficients in the Mahler–Amice expansion $f(x, y) = \sum_{m,n} a_{m,n}\binom{x}{m}\binom{y}{n}$ are given by $a_{m,n} = b_{m+n}$, where $h(x) = \sum_{n\geq 0} b_n\binom{x}{n}$. Since the functions

$$p^{\lfloor r_1\ell(m)+r_2\ell(n)\rfloor}\binom{x}{m}\binom{y}{n}$$

are a Banach basis of $C^{(r_1,r_2)}(U, E)$, we see that f is in this space if and only if

$$\lim_{m,n\to\infty} v_p(b_{m+n}) - r_1\ell(m) - r_2\ell(n) = \infty.$$

The supremum of $r_1\ell(m) + r_2\ell(n)$ over pairs (m, n) with $m + n = k$ is $(r_1 + r_2)\ell(k) + O(1)$ as $k \to \infty$, so this is equivalent to

$$\lim_{k\to\infty} v_p(b_k) - (r_1 + r_2)\ell(k) = \infty,$$

which is precisely the condition that h is C^r where $r = r_1 + r_2$. Moreover, this gives a Banach space isomorphism between $C^r(\mathbf{Z}_p, E)$ and $C^{(r_1,r_2)}(U, E)^{\Delta}$ preserving the subspaces of locally constant functions, so if $r_1 + r_2 \geq 1$ we see that $LC(U, E)^{\Delta}$ is not dense in $C^{(r_1,r_2)}(U, E)^{\Delta}$.

Remark 9. In Sect. 3.4, we gave criteria for locally constant functions on $U/\overline{\Delta}$ to be dense in a Banach space whose topology is coarser than that of $C^{(r_1,\dots,r_g)}(U, E)^{\Delta}$. This in particular implies that such functions are dense in $C^{(r_1,\dots,r_g)}(U, E)^{\Delta}$ in these cases, and hence that the multiple possible choices of auxilliary quasi-factorizations considered in the examples above do all give the same distribution on G when they apply.

4 Construction of p-Adic L-Functions

We now give the motivating example of a ray class distribution: the distribution constructed from the Mazur–Tate elements of an automorphic representation of GL_2/K.

4.1 Mazur–Tate Elements

Let K be a number field, and Π an automorphic representation of GL_2/K. We make the following simplifying assumptions:

1. Π is cohomological in trivial weight, i.e. the $(\mathfrak{g}, K_\infty)$-cohomology of Π does not vanish, where K_∞ is a maximal compact connected subgroup of $\mathrm{GL}_2(K \otimes \mathbf{R})$;
2. The central character of Π is trivial.

Conditions (1) and (2) are satisfied, for instance, if Π is the base-change to K of the automorphic representation of $\mathrm{GL}_2/\mathbf{Q}$ attached to a modular form of weight 2 and trivial nebentypus. It follows from (1) and (2) that there exists a finite extension $F/\mathbf{Q}$ inside $\mathbf{C}$ such that the $\mathrm{GL}_2(\mathbf{A}_K^\infty)$-representation Π^∞ can be defined over F, and in particular the Hecke eigenvalues $a_\mathfrak{l}(\Pi)$ for each prime $\mathfrak{l}$ of K are in $\mathcal{O}_F$.

Theorem 7 (Haran). *Under the above assumptions, there exists a finitely-generated $\mathcal{O}_F$-submodule Λ_Π of $\mathbf{C}$, and for each ideal $\mathfrak{f}$ of K coprime to the conductor of Π an element*

$$\Theta_\mathfrak{f}(\Pi) \in \mathbf{Z}[G_\mathfrak{f}] \otimes_\mathbf{Z} \Lambda_\Pi,$$

where $G_\mathfrak{f}$ is the ray class group modulo $\mathfrak{f}$, such that the following relations hold:

(i) (Special values) If ω is a primitive ray class character of conductor $\mathfrak{f}$, then

$$\omega\left(\Theta_\mathfrak{f}(\Pi)\right) = \frac{L_\mathfrak{f}(\Pi, \omega, 1)}{\tau(\omega) \cdot |\mathfrak{f}|^{1/2} \cdot (4\pi)^{[K:\mathbf{Q}]}}$$

where $L_\mathfrak{f}(\Pi, \omega, s)$ denotes the L-function of Π twisted by ω, without the Euler factors at ∞ or at primes dividing $\mathfrak{f}$, and $\tau(\omega)$ is the Gauss sum (normalized so that $|\tau(\omega)| = 1$);

(ii) (Norm relation) For $\mathfrak{l}$ a prime not dividing the conductor of Π, we have

$$\mathrm{Norm}_\mathfrak{f}^{\mathfrak{f}\mathfrak{l}}\, \Theta_{\mathfrak{f}\mathfrak{l}}(\Pi) = \begin{cases} \left(a_\mathfrak{l}(\Pi) - \sigma_\mathfrak{l} - \sigma_\mathfrak{l}^{-1}\right) \Theta_\mathfrak{f}(\Pi) & \text{if } \mathfrak{l} \nmid \mathfrak{f}, \\ a_\mathfrak{l}(\Pi)\Theta_\mathfrak{f}(\Pi) - \Theta_{\mathfrak{f}/\mathfrak{l}}(\Pi) & \text{if } \mathfrak{l} \mid \mathfrak{f}. \end{cases}$$

where $\sigma_\mathfrak{l}$ denotes the class of $\mathfrak{l}$ in $G_\mathfrak{f}$.

Proof. See Haran (1987). (Note that in the main text of op. cit. it is assumed that K is totally imaginary, but this assumption can easily be removed, as noted in section 7.)

Remark 10. In fact one can obtain essentially the same result under far weaker assumptions: it suffices to suppose that Π is cuspidal, cohomological (of any weight), and critical. However, the necessary generalizations of Shai Haran's arguments are time-consuming, so we shall not consider this more general setting here. For the case when K is totally real, but Π has arbitrary critical weight and central character, see Dimitrov (2013).

We refer to the elements $\Theta_{\mathfrak{f}}(\Pi) \in \mathbf{Z}[G_{\mathfrak{f}}] \otimes_{\mathbf{Z}} \Lambda_\Pi$ as *Mazur–Tate elements*, since they are closely analogous to the group ring elements considered in Mazur and Tate (1987). The coefficient module Λ_Π is in fact rather small, as the following result (an analogue of the Manin–Drinfeld theorem) shows:

Proposition 7. *The $\mathcal{O}_F$-submodule $\Lambda_\Pi \subseteq \mathbf{C}$ may be taken to have rank 1.*

Proof. We know that Λ_Π is finitely-generated, so it suffices to show that $F \otimes_{\mathcal{O}_F} \Lambda_\Pi$ is 1-dimensional.

To prove this, we must delve a little into the details of Shai Haran's construction. The elements $\Theta_{\mathfrak{f}}(\Pi)$ for varying Π are all obtained from a "universal Mazur–Tate element" in the module $\mathbf{Z}[G_{\mathfrak{f}}] \otimes H_r^{\mathrm{BM}}(Y, \mathbf{Z})$, where Y is a locally symmetric space for $\mathrm{GL}_2(\mathbf{A}_K)$ and H_r^{BM} is Borel–Moore homology (the homological analogue of compactly supported cohomology). Here $r = r_1 + r_2$, where r_1 and r_2 are the numbers of real and complex places of K. The element $\Theta_{\mathfrak{f}}(\Pi)$ is obtained by integrating the universal Mazur–Tate element against a class in $H^r_{\mathrm{par}}(Y, \mathbf{C})$ arising from Π. However, it follows from the results of Harder (1987) that there is a unique Hecke-invariant section of the projection map

$$H^r_c(Y, \mathbf{C}) \to H^r_{\mathrm{par}}(Y, \mathbf{C}),$$

and the Π-isotypical component of $H^r_c(Y, \mathbf{C})$ is 1-dimensional and descends to $H^r_c(Y, F)$. Thus, after renormalizing by a single (probably transcendental) complex constant depending on Π, the Mazur–Tate elements for Π can be obtained as values of the perfect pairing

$$H_r^{\mathrm{BM}}(Y, F) \times H^r_c(Y, F) \to F,$$

so we deduce that $F \otimes_{\mathcal{O}_F} \Lambda_\Pi$ has dimension 1 as claimed.

4.2 Stabilization

We now introduce "Σ-stabilized" versions of the $\Theta_{\mathfrak{f}}(\Pi)$, again following Haran (1987) closely. Let $\mathfrak{F}$ be a pseudo-modulus of K, and Σ the set of primes $\mathfrak{p}$ such that $\mathfrak{p}^\infty \mid \mathfrak{F}$, as before. We assume that none of the primes $\mathfrak{p} \in \Sigma$ divide the conductor of Π.

Definition 9. An *Σ-refinement* of Π is the data of, for each $\mathfrak{p} \in \Sigma$, a root $\alpha_\mathfrak{p} \in \overline{F}$ of the polynomial

$$P_\mathfrak{p}(\Pi) = X^2 - a_\mathfrak{p}(\Pi)X + N_{K/\mathbf{Q}}(\mathfrak{p}) \in F[X]$$

(the local L-factor of Π at $\mathfrak{p}$).

We write $\beta_\mathfrak{p}$ for the complementary root to $\alpha_\mathfrak{p}$. By enlarging F if necessary, we may assume that the $\alpha_\mathfrak{p}$ and $\beta_\mathfrak{p}$ all lie in F.

We introduce a formal operator $R_\mathfrak{p}$ on the $\Theta_\mathfrak{f}(\Pi)$'s by $R_\mathfrak{p} \cdot \Theta_\mathfrak{f}(\Pi) = \Theta_{\mathfrak{f}/\mathfrak{p}}(\Pi)$ whenever $\mathfrak{p} \mid \mathfrak{f}$. Then it is easy to see that the $R_\mathfrak{p}$ for different $\mathfrak{p} \mid \mathfrak{f}$ commute with each other, so we can define

$$\Theta_\mathfrak{f}^\Sigma(\Pi) = \alpha_\mathfrak{f}^{-1}\left(\prod_{\mathfrak{p}\in\Sigma}(1 - \alpha_\mathfrak{p}^{-1}R_\mathfrak{p})\right)\Theta_\mathfrak{f}(\Pi),$$

whenever $\mathfrak{f}$ is divisible by all $\mathfrak{p} \in \Sigma$, where $\alpha_\mathfrak{f}$ stands for the product

$$\alpha_\mathfrak{f} := \prod_{\mathfrak{p}\in\Sigma} \alpha_\mathfrak{p}^{-v_\mathfrak{p}(\mathfrak{f})}.$$

Then one checks that we have

$$\mathrm{Norm}_\mathfrak{f}^{\mathfrak{f}\mathfrak{p}}\left(\Theta_{\mathfrak{f}\mathfrak{p}}^\Sigma(\Pi)\right) = \Theta_\mathfrak{f}^\Sigma(\Pi)$$

for any $\mathfrak{p}$ such that $\mathfrak{p} \mid \mathfrak{f}$ and $\mathfrak{p} \in \Sigma$. We can therefore extend the definition of $\Theta_\mathfrak{f}^\Sigma(\Pi)$ uniquely to all $\mathfrak{f} \mid \mathfrak{F}_\infty$ in such a way that this formula still holds.

Corollary 1. *Let $\mathfrak{F}$ be a pseudo-modulus of K and write $\mathfrak{F} = \mathfrak{g}\mathfrak{F}_\infty$, where $\mathfrak{F} = \prod_{\mathfrak{p}\in\Sigma}\mathfrak{p}^\infty$ and $\mathfrak{g}$ is coprime to Σ. Then the elements*

$$\{\Theta_{\mathfrak{f}\mathfrak{g}}^\Sigma(\Pi) : \mathfrak{f} \mid \mathfrak{F}_\infty\}$$

defined above form a ray class distribution modulo $\mathfrak{F}$, with values in $F \otimes_{\mathcal{O}_F} \Lambda_\Pi$ and growth parameter $(\alpha_\mathfrak{p})_{\mathfrak{p}\in\Sigma}$.

4.3 Consequences for p-Adic L-Functions

We now summarize the results on p-adic L-functions that can be obtained by applying our theory to the ray class distributions constructed in Sect. 4.2. As before, let F be a number field, and E be the completion of F at some choice of prime above p; and let v_p denote the p-adic valuation on E normalized so that $v_p(p) = 1$.

Theorem 8. *Let K be an arbitrary number field and Σ a subset of the primes of K above p, and let $\mathfrak{F}$ be the pseudo-modulus $\mathfrak{g} \cdot \prod_{\mathfrak{p}\in\Sigma}\mathfrak{p}^\infty$, where $\mathfrak{g}$ is a modulus coprime* *to Σ. Let Π be an automorphic representation of GL_2/K with Hecke eigenvalues in F, satisfying the hypotheses (1) and (2) of Sect. 4.1, and unramified at the primes in Σ. Let $\underline{\alpha} = (\alpha_{\mathfrak{p}})_{\mathfrak{p}\in\Sigma}$ be a Σ-refinement of Π defined over F. If we have*

$$h = \sum_{\mathfrak{p}\in\Sigma} e_{\mathfrak{p}} v_p(\alpha_{\mathfrak{p}}) < 1,$$

then there is a unique distribution $L_{p,\underline{\alpha}}(\Pi) \in D^h(G_{\mathfrak{F}}, E) \otimes_{\mathcal{O}_F} \Lambda_\Pi$ such that for all finite-order characters ω of $G_{\mathfrak{F}}$ whose conductor is divisible by $\mathfrak{g}$, we have

$$L_{p,\underline{\alpha}}(\Pi)(\omega) = \left(\prod_{\mathfrak{p}\in\Sigma} \alpha_{\mathfrak{p}}^{-\mathrm{ord}_{\mathfrak{p}}\mathfrak{f}}\right) \cdot \left(\prod_{\mathfrak{p}\in\Sigma,\mathfrak{p}\nmid\mathfrak{f}} (1-\alpha_{\mathfrak{p}}^{-1}\omega(\mathfrak{p}))(1-\alpha_{\mathfrak{p}}^{-1}\omega(\mathfrak{p})^{-1})\right) \cdot \frac{L_{\mathfrak{f}}(\Pi,\omega,1)}{\tau(\omega)\cdot|\mathfrak{f}|\cdot(4\pi)^{[K:\mathbf{Q}]}}, \tag{2}$$

where $\mathfrak{f}$ is the conductor of ω.

Proof. This is immediate from Theorem 7 and Corollary 1 except for the formula for $L_{p,\underline{\alpha}}(\Pi)(\omega)$ when $\mathfrak{f}$ is not divisible by all primes in Σ; the latter follows via a short explicit calculation from part (ii) of Theorem 7.

If K is imaginary quadratic and p is split, we obtain a slightly finer statement using the quasi-factorization above:

Theorem 9. *Let K be an imaginary quadratic field in which $p = \mathfrak{p}\overline{\mathfrak{p}}$ splits, and let Π be an automorphic representation satisfying the hypotheses (1) and (2) of Sect. 4.1 with Hecke eigenvalues in F, and unramified at $\mathfrak{p}$ and $\overline{\mathfrak{p}}$. Let $\underline{\alpha} = (\alpha_{\mathfrak{p}}, \alpha_{\overline{\mathfrak{p}}})$ be a p-refinement of Π defined over F.*

If we have

$$r = v_p(\alpha_{\mathfrak{p}}) < 1 \quad \text{and} \quad s = v_p(\alpha_{\overline{\mathfrak{p}}}) < 1,$$

then there is a unique distribution $L_{p,\underline{\alpha}}(\Pi) \in D^{(r,s)}(G_{p\infty}, E) \otimes_{\mathcal{O}_F} \Lambda_\Pi$ satisfying the interpolating property (2).

Note that every Π will admit at least one p-refinement satisfying the hypotheses of Theorem 9, and indeed if Π is non-ordinary at $\mathfrak{p}$ and $\overline{\mathfrak{p}}$ then all four p-refinements do so; but at most two of the four p-refinements, and in many cases none at all, will satisfy the hypotheses of Theorem 8.

Remark 11. Applying Theorem 9 to the base-change of a classical modular form, we obtain Conjecture 6.13 of our earlier paper (Loeffler and Zerbes 2011) for all weight 2 modular forms whose central character is trivial.

Remark 12. Examples of ray class distributions appear to arise in more or less any context where p-adic interpolation of L-values of non-ordinary automorphic representations over general number fields is considered. For instance, the works of Ash–Ginzburg (1994) on p-adic L-functions for GL_{2n} over a number field, and of

Januszewski on Rankin–Selberg convolutions on $\mathrm{GL}_n \times \mathrm{GL}_{n-1}$ over a number field (Januszewski 2011), can both be viewed as giving rise to a ray class distribution (with some parameter depending on the Hecke eigenvalues of the automorphic representations involved). Thus the theory developed in this paper allows these works to be extended to non-ordinary automorphic representations of (sufficiently small) non-zero slope.

5 The $a_p = 0$ Case

We now concentrate on a special case of Theorem 9, where K is imaginary quadratic with p split and $a_{\mathfrak{p}}(\Pi) = a_{\overline{\mathfrak{p}}}(\Pi) = 0$ (as in the case of the automorphic representation attached to a modular elliptic curve with good supersingular reduction at the primes above p, assuming $p \geq 5$). Then the two Hecke polynomials at $\mathfrak{p}$ and $\overline{\mathfrak{p}}$ are the same, and we write $\{\alpha, \beta\}$ for their common set of roots, which are both square roots of $-p$ and in particular have normalized p-adic valuation $\frac{1}{2}$. Thus Theorem 9 furnishes four p-adic L-functions.

Remark 13. None of these four L-functions can be obtained using Theorem 8. Moreover, the restriction of any of the four to a "cyclotomic line" in the space of characters of G_{p^∞} (i.e. a coset of the subgroup of characters factoring through the norm map) is a distribution of order 1, and thus not determined uniquely by its interpolating property. Thus, paradoxically, it is easier to interpolate L-values at a larger set of twists than a smaller one.

We know that the module Λ_Π in which our Mazur–Tate elements are valued has rank 1 over $\mathcal{O}_F$. We may thus choose an $\mathcal{O}_E$-basis Ω_Π of $\mathcal{O}_E \otimes_{\mathcal{O}_F} \Lambda_\Pi$. Let us write

$$\mu_{\alpha,\alpha} = \frac{L_{p,(\alpha,\alpha)}(\Pi)}{\Omega_\Pi} \in D^{(1/2,1/2)}(G_{p^\infty}, E)$$

and similarly for $\mu_{\alpha,\beta}, \mu_{\beta,\alpha}, \mu_{\beta,\beta}$.

Proposition 8. *The distributions*

$$\begin{aligned}
\mu_{+,+} &= \mu_{\alpha,\alpha} + \mu_{\alpha,\beta} + \mu_{\beta,\alpha} + \mu_{\beta,\beta} \\
\mu_{+,-} &= \mu_{\alpha,\alpha} - \mu_{\alpha,\beta} + \mu_{\beta,\alpha} - \mu_{\beta,\beta} \\
\mu_{-,+} &= \mu_{\alpha,\alpha} + \mu_{\alpha,\beta} - \mu_{\beta,\alpha} - \mu_{\beta,\beta} \\
\mu_{-,-} &= \mu_{\alpha,\alpha} - \mu_{\alpha,\beta} - \mu_{\beta,\alpha} + \mu_{\beta,\beta}
\end{aligned}$$

all lie in $D^{(1/2,1/2)}(G_{p^\infty}, E)$, and have the property that if ω is a finite-order character of conductor $\mathfrak{p}^{n_\mathfrak{p}} \overline{\mathfrak{p}}^{n_{\overline{\mathfrak{p}}}}$, with both $n_\mathfrak{p}, n_{\overline{\mathfrak{p}}} > 0$, then $\mu_{*,\circ}$ vanishes at ω unless $* = (-1)^{n_\mathfrak{p}}$ and $\circ = (-1)^{n_{\overline{\mathfrak{p}}}}$.

Proof. Clear from the interpolating property of the μ's.

Thus each of the distributions $\mu_{*,\circ}$, for $*, \circ \in \{+,-\}$, vanishes at three-quarters of the finite-order characters of the p-adic Lie group G_{p^∞}.

We now interpret this statement in terms of divisibility by half-logarithms. Let $\log_{\mathfrak{p}}^{+}$ be Pollack's half-logarithm on $\mathcal{O}_{K,\mathfrak{p}}^{\times} \cong \mathbf{Z}_p^{\times}$ (cf. Pollack 2003) which is a distribution of order 1/2 with the property that $\log_{\mathfrak{p}}^{+}(\chi)$ vanishes at all characters χ of $\mathbf{Z}_p^{\times}$ of conductor p^n with n an odd integer (and no others). Similarly, let $\log_{\mathfrak{p}}^{-}$ be the half-logarithm which vanishes at characters of conductor p^n with n positive and even. We define $\log_{\mathfrak{p}}^{+}$ and $\log_{\mathfrak{p}}^{-}$ as distributions on G_{p^∞} by pushforward, and similarly $\log_{\bar{\mathfrak{p}}}^{\pm}$.

Proposition 9. *For $*, \circ \in \{+,-\}$, the distribution $\mu_{*,\circ}$ is divisible in $D^{\mathrm{la}}(G_{p^\infty})$ by $\log_{\mathfrak{p}}^{*} \log_{\bar{\mathfrak{p}}}^{\circ}$.*

Proof. We begin by introducing distributions

$$\begin{aligned}
\mu_{+,\alpha} &= \mu_{\alpha,\alpha} + \mu_{\beta,\alpha}\\
\mu_{+,\beta} &= \mu_{\alpha,\beta} + \mu_{\beta,\beta}\\
\mu_{-,\alpha} &= \mu_{\alpha,\alpha} - \mu_{\beta,\alpha}\\
\mu_{-,\beta} &= \mu_{\alpha,\beta} - \mu_{\beta,\beta}.
\end{aligned}$$

We claim that $\mu_{+,\alpha}$ and $\mu_{+,\beta}$ are divisible by $\log_{\mathfrak{p}}^{+}$, and $\mu_{-,\alpha}$ and $\mu_{-,\beta}$ by $\log_{\mathfrak{p}}^{-}$. We prove only the first of these four statements, since the proofs of the other three are virtually identical. If χ is any finite-order character such that $n_{\mathfrak{p}}(\chi)$ (the power of $\mathfrak{p}$ dividing the conductor of χ) is positive and odd, then $\mu_{+,\alpha}$ vanishes at χ by construction.

Since $\mu_{+,\alpha}$ is in $D^{(1/2,1/2)}(G_{p^\infty}, E)$, it follows that it must vanish at any character of G whose pullback to $\mathcal{O}_{K,\mathfrak{p}}^{\times}$ is of finite order with odd p-power conductor. Hence $\mu_{+,\alpha}$ is divisible by $\log_{\mathfrak{p}}^{+}$ in $D^{(1/2,1/2)}(G_{p^\infty}, E)$ (and the quotient clearly has order $(0, 1/2)$).

We now apply the same argument in the other variable, to show that $\mu_{\alpha,+} = \mu_{\alpha,\alpha} + \mu_{\alpha,\beta}$ is divisible by $\log_{\bar{\mathfrak{p}}}^{+}$, etc. This shows that $\mu_{+,+}$ is divisible by both $\log_{\mathfrak{p}}^{+}$ and $\log_{\bar{\mathfrak{p}}}^{+}$. These two distributions are coprime (if we regard them as rigid-analytic functions on the character space of G_{p^∞}, the intersection of their zero loci is a subvariety of codimension 2) so we are done.

Corollary 2. *There exist four bounded measures*

$$L_p^{+,+}, L_p^{+,-}, L_p^{-,+}, L_p^{-,-} \in \Lambda_F(G_{p^\infty})$$

such that

$$\frac{L_{p,(\alpha,\alpha)}(\Pi)}{\Omega_\Pi} = \frac{1}{4}\Big(L_p^{+,+} \log_{\mathfrak{p}}^{+} \log_{\bar{\mathfrak{p}}}^{+} + L_p^{+,-} \log_{\mathfrak{p}}^{+} \log_{\bar{\mathfrak{p}}}^{-} + L_p^{-,+} \log_{\mathfrak{p}}^{-} \log_{\bar{\mathfrak{p}}}^{+} + L_p^{-,-} \log_{\mathfrak{p}}^{-} \log_{\bar{\mathfrak{p}}}^{-}\Big)$$

and similarly for the other three unbounded L-functions.

Remark 14. If Π is the base-change to K of a weight 2 modular form, then $\mu_{\alpha,\alpha}$ and $\mu_{\beta,\beta}$ coincide with the distributions constructed in Kim (2011), up to minor differences in the local interpolation factors. Hence the above corollary proves the conjecture formulated in op. cit. However, it does not appear to be possible to construct the signed distributions $L_p^{+,+}$ etc. solely from $\mu_{\alpha,\alpha}$ and $\mu_{\beta,\beta}$; it seems to be necessary to use $\mu_{\alpha\beta}$ and $\mu_{\beta\alpha}$ as well, and these last two are apparently not amenable to construction via Rankin–Selberg convolution techniques as in op. cit.

Acknowledgements I would like to thank B.D. Kim for sending me an early draft of his paper (Kim 2011), together with many informative remarks on its contents. I am also grateful to Matthew Emerton, Günter Harder, Mladen Dimitrov and Sarah Zerbes for interesting and helpful conversations on this topic; and to the anonymous referee for several valuable comments and corrections.

References

Ash, A., Ginzburg, D.: p-adic L-functions for GL($2n$). Invent. Math. **116**(1–3), 27–73 (1994)

Calegari, F., Mazur, B.: Nearly ordinary Galois deformations over arbitrary number fields. J. Inst. Math. Jussieu **8**(1), 99–177 (2009)

Colmez, P.: Fonctions d'une variable p-adique. Astérisque **330**, 13–59 (2010)

De Ieso, M.: Espaces de fonctions de classe C^r sur O_F. Indagationes Mathematicae **24**(3), 530–556 (2013)

Dimitrov, M.: Automorphic symbols, p-adic L-functions and ordinary cohomology of Hilbert modular varieties. Amer. J. Math. **135**(4), 1117–1155 (2013)

Haran, S.M.J.: p-adic L-functions for modular forms. Compos. Math. **62**(1), 31–46 (1987)

Harder, G.: Eisenstein cohomology of arithmetic groups. The case GL_2. Invent. Math. **89**(1), 37–118 (1987)

Januszewski, F.: Modular symbols for reductive groups and p-adic Rankin-Selberg convolutions over number fields. J. Reine Angew. Math. **653**, 1–45 (2011)

Kim, B.D.: Two-variable p-adic L-functions of modular forms for non-ordinary primes. J. Num. Theo. (2011, preprint)

Loeffler, D., Zerbes, S.L.: Iwasawa theory and p-adic L-functions for $\mathbb{Z}_p^2$-extensions. Int. J. Num. Theo. (2011, preprint)

Mazur, B., Tate, J.: Refined conjectures of the "Birch and Swinnerton-Dyer type". Duke Math. J. **54**(2), 711–750 (1987)

Panchishkin, A.A.: Motives over totally real fields and p-adic L-functions. Ann. Inst. Fourier (Grenoble) **44**(4), 989–1023 (1994)

Pollack, R.: On the p-adic L-function of a modular form at a supersingular prime. Duke Math. J. **118**(3), 523–558 (2003)

Washington, L.C.: Introduction to Cyclotomic Fields. Graduate Texts in Mathematics, vol. 83, 2nd edn. Springer, New York (1997)

On Equivariant Characteristic Ideals of Real Classes

Thong Nguyen Quang Do

1 Introduction

Fix an odd prime p and a Galois extension F/k of totally real number fields, with group $G = \mathrm{Gal}(F/k)$. Let $F_\infty = \bigcup_{n \geq 0} F_n$ the cyclotomic $\mathbb{Z}_p$-extension of F, $\Gamma = \mathrm{Gal}(F_\infty/F)$, $G_\infty = \mathrm{Gal}(F_\infty/k)$, $\Lambda = \mathbb{Z}_p[[\Gamma]]$, $\mathbb{A} = \mathbb{Z}_p[[G_\infty]]$. Take $S_f = S_p \cup Ram(F/k)$, $S = S_\infty \cup S_f$, where S_∞ (resp. S_p) is the set of Archimedean primes (resp. primes above p) of k and $Ram(F/k)$ is the set of places of k which ramify in F/k. By abuse of language, we also denote by S the set of primes above S in any extension of k. Let $G_S(F_n)$ be the Galois group over F_n of the maximal algebraic S-ramified (i.e. unramified outside S) extension of F_n. Since p is odd and S contains $S_p \cup S_\infty$, it is known that $cd_p(G_S(F_n)) \leq 2$ (see e.g. Neukirch et al. 2008, Propos 8.3.18). Furthermore, the continuous cohomology groups $H^i_S(F_n, \mathbb{Z}_p(m)) := H^i(G_S(F_n), \mathbb{Z}_p(m))$, $m \in \mathbb{Z}$, coincide with the étale cohomology groups $H^i_{\text{ét}}(\mathcal{O}_{F_n}[1/S], \mathbb{Z}_p(m))$. We are interested in the $\mathbb{A}$-modules $H^i_{Iw,S}(F_\infty, \mathbb{Z}_p(m)) := \varprojlim_{\text{cores}} H^i_S(F_n, \mathbb{Z}_p(m))$ ($H^i_{Iw}(.)$ for short if there is no ambiguity on S), to which can be attached invariants containing important arithmetical information.

For instance, if G is *abelian* and the μ-invariant associated to F_∞/F vanishes, it can be shown that, for any $m \equiv 0 \pmod 2$, the initial $\mathbb{A}$ – Fitting ideal of $H^2_{Iw}(F_\infty, \mathbb{Z}_p(m))$ is given by the formula:

$$Fit_{\mathbb{A}}(H^2_{Iw}(F_\infty, \mathbb{Z}_p(m))) = tw_m(IG_\infty . \theta_S) \tag{1}$$

T. Nguyen Quang Do (✉)
UFRST, F-25030, Besancon, France
e-mail: thong_nguyenquangdo@yahoo.fr

T. Bouganis and O. Venjakob (eds.), *Iwasawa Theory 2012*, Contributions in Mathematical and Computational Sciences 7, DOI 10.1007/978-3-642-55245-8_13,

where tw_m denotes the automorphism of the total ring of fractions $Q\mathbb{A}$ induced by the mth-Iwasawa twist $\sigma \mapsto \kappa_{cyc}^m \sigma^{-1}$, κ_{cyc} is the cyclotomic character, IG_∞ is the augmentation ideal of $\mathbb{Z}_p[[G_\infty]]$ and $\theta_S \in Q\ \mathbb{A}$ is the Deligne-Ribet pseudo-measure associated with G_∞ (Nguyen Quang Do 2005, thms 3.1.2, 3.3.3). Note that for $m < 0$, we implicitly assume the validity of the mth-twist of Leopoldt's conjecture; see Proposition 4 below.

The identity (1) was shown by using the (abelian) Equivariant Main Conjecture (EMC for short) of Iwasawa theory in the formulation of Ritter and Weiss (2002, theorem 11). Here "equivariant" means that the Galois action of G is taken into account. One salient feature of formula (1) is that it allows a proof by descent of the p-part of the Coates-Sinnott conjecture, as well as a weak form of the p-part of Brumer's conjecture (Nguyen Quang Do 2005, thms 4.3, 5.2). On the same subject, let us report the approach of Burns and Greither (2003b, theorems 3.1, 5.1 and corollary 2) using the Iwasawa theory of complexes initiated by Kato and subsequently extended by Nekovar (2006). See also Greither and Popescu (2013).

After the recent proof of the non commutative EMC (under the hypothesis that $\mu = 0$, see Kakde 2013; Ritter and Weiss 2011), no doubt that the above results could be extended to the case where G is non abelian (this has been done recently by Nickel 2013). But in this article, we are mainly interested in the odd twists, $m \equiv 1$ (mod 2), which are not a priori covered by the EMC. Since there are so many "main conjectures" floating around, a short explanation is in order. Let us come back to the classical Main Conjecture, or Wiles' theorem (WMC for short, see Wiles 1990) and take $G = (1)$ for simplicity. The Galois group $\mathfrak{X}_\infty$ over F_∞ of the maximal abelian (p)-ramified pro-p-extension of F_∞ is a Λ-torsion module because F is totally real, and the Λ-characteristic series of $\mathfrak{X}_\infty$ is precisely related by the WMC to the Deligne-Ribet pseudo-measure attached to the maximal pro-p-quotient of $G_S^{ab}(F)$ (where $S = S_\infty \cup S_p$). If we allow – just for this discussion – the base field F to be a CM-field, then $(tor_\Lambda \mathfrak{X}_\infty)^+ = \mathfrak{X}_\infty^+$ by the weak Leopoldt conjecture, and it is related by Spiegelung (reflection theorems) to the "minus part" X_∞^- of the Galois group over F_∞ of the maximal abelian unramified pro-p-extension of F_∞, so that we know the Λ-characteristic series of X_∞^- in terms of L_p-functions. However our intended study of odd twists actually concerns the "plus part" X_∞^+, at least because if $\zeta_p \in F$, then $X'_\infty(m-1)$ is contained in $H^2_{Iw}(F_\infty, \mathbb{Z}_p(m))$ (see Proposition 4 below), where X'_∞ is obtained from X_∞ by adding the condition that all (p)-primes (hence all finite primes, because at a non (p)-prime, the local cyclotomic $\mathbb{Z}_p$-extension coincides with the local maximal non ramified pro-p-extension) must be totally split. A related problem is Greenberg's celebrated conjecture – a "reasonable" generalization of Vandiver's – which asserts the finiteness of X_∞^+ (or equivalently of X'^+_∞). Not much is known on the plus parts. Let us recall some results in the case where $k = \mathbb{Q}$, F is totally real and G is abelian:

- It is well known that the WMC is equivalent to the so-called "Gras type" equality (because it implies the Gras conjecture):

$$char_\Lambda X_\infty = char_\Lambda(\overline{U}_\infty / \overline{C}_\infty) \tag{2}$$

where $\overline{U}_\infty$ (resp. $\overline{C}_\infty$) denotes the inverse limit (w.r.t. norms) of the p-completions $\overline{U}_n$ (resp. $\overline{C}_n$) of the groups of units (resp. circular units) along the cyclotomic tower of F. But the right hand side Λ-characteristic series is not known. Besides, this Gras type equality comes from the multiplicativity of $char_\Lambda(.)$ in the exact sequence of class-field theory relative to inertia:

$$0 \to \overline{U}_\infty/\overline{C}_\infty \to \mathfrak{U}_\infty/\overline{C}_\infty \to \mathfrak{X}_\infty \to X_\infty \to 0,$$

where $\mathfrak{U}_\infty$ is the semi-local analogue of $\overline{U}_\infty$. But when passing from Λ to $\mathbb{A}$, no straightforward equivariant generalization (of $\mathbb{A}$-characteristic series of modules and their multiplicativity) is known.

- Galois annihilators of X'_∞ have been explicitly computed in Nguyen Quang Do and Nicolas (2011) and Solomon (2010), as well as Fitting ideals in the semi-simple case in Nguyen Quang Do and Nicolas (2011).

In this context, the main result of this paper will be an *explicit equivariant generalization of the Gras type equality* (2) *for all odd twists* (see Theorem 2 below).

The proof will proceed in essentially two stages:

- Use a "limit theorem" in the style of Burns and Greither (2003a, thm 6.1; this is also an EMC, but we don't call it so for fear of overload), but expressed in the framework of the Iwasawa theory of perfect torsion complexes as in Witte (2006), to relate the $\mathbb{A}$-determinant of $H^2_{Iw}(F_\infty, \mathbb{Z}_p(m))$, m odd, to that of a suitable quotient of $H^1_{Iw}(F_\infty, \mathbb{Z}_p(m))$
- Use an axiomatization of a method originally introduced by Kraft and Schoof for real quadratic fields (Kraft and Schoof 1995) to compute explicitly the latter determinant (which will actually be an $\mathbb{A}$-initial Fitting ideal).

Summary of Iwasawa theoretic notations

F a number field
$F_\infty = \bigcup_{n\geq 0} F_n$ the cyclotomic $\mathbb{Z}_p$-extension of F
$\Gamma_n = \mathrm{Gal}(F_n/F)$, $\Gamma = \mathrm{Gal}(F_\infty/F)$, $\Lambda = \mathbb{Z}_p[[\Gamma]]$
If moreover $F/\mathbb{Q}$ is Galois
$G_n = \mathrm{Gal}(F_n/\mathbb{Q})$, $G_\infty = \mathrm{Gal}(F_\infty/\mathbb{Q})$, $\mathbb{A} = \mathbb{Z}_p[[G_\infty]]$
A_n (resp. A'_n) = the p-group of ideal classes (resp. (p)-classes) of F_n
X_∞ (resp. X'_∞) = $\varprojlim A_n$ (resp. A'_n) w.r.t. norms
$\mathfrak{X}_\infty$ = the Galois group over F_∞ of the maximal abelian pro-p-extension of F_∞ which is unramified outside p
$\bar{U}_n$ = the p-completion of the group of units of F_n
$\bar{U}_\infty = \varprojlim \bar{U}_n$ w.r.t. norms
If moreover $F/\mathbb{Q}$ is abelian:
$\bar{C}_n$ = the p-completion of the group of circular units of F_n
$\bar{C}_\infty = \varprojlim \bar{C}_n$ w.r.t. norms

2 Generators and Relations, and Fitting Ideals

We study in this section an axiomatic method for describing certain (initial) Fitting ideals by generators and relations. It was first introduced by Kraft and Schoof (1995) for quadratic real fields, and subsequently applied by Solomon (2010) to the cyclotomic field $\mathbb{Q}(\zeta_p)^+$.

2.1 General Case

Since the process is purely algebraic, we can relax here the assumptions on k and F imposed in the introduction. So *F/k will just be a Galois extension with group G*, and the usual Galois picture will be:

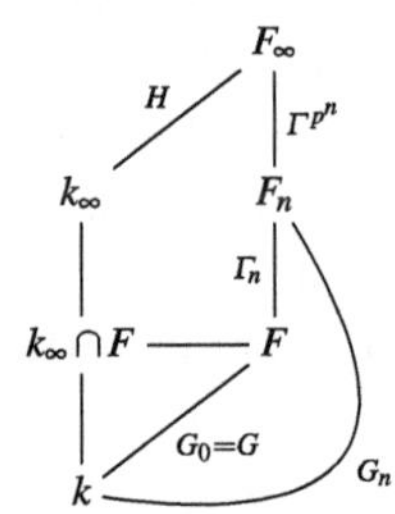

Note that $\mathbb{A} = \Lambda[H]$ is equal to $\Lambda[G]$ if and only if $k_\infty \cap F = k$. Put $R_n = \mathbb{Z}_p[G_n]$, $R_{n,a} = \mathbb{Z}/p^a\,[G_n]$.

Hypothesis: We are given a projective (w.r.t. norms) system of R_n-modules V_n which are $\mathbb{Z}_p$-free of finite type, as well as a projective subsystem $W_n \subset V_n$ such that each W_n is R_n-free of rank 1. In other words, there is a norm coherent system $\eta = (\eta_n)$, $\eta_n \in V_n$, such that $W_n = R_n . \eta_n$ for all $n \geq 0$. Without any originality, a pair $(V_n, W_n)_{n \geq 0}$ as above will be called *admissible*.

At the time being, we don't worry about the existence of such systems $(W_n, V_n)_n$. Arithmetical examples will be given later in Sects. 2 and 3.

Goal: denoting by B_n the quotient V_n/W_n, describe the module $t\,B_n$ (where $t(.)$ means $\mathbb{Z}_p$-torsion) by generators and relations.

In the sequel, for two left modules X, Y, the Galois action on $\mathrm{Hom}(X, Y)$ will always be defined by ${}^{\sigma}f(x) = \sigma(f(\sigma^{-1}x))$. The Pontryagin dual of X will be denoted by X^*.

Proposition 1. *For any $n \geq 0$,*

$$(t\,B_n)^* \simeq R_n/\mathcal{D}_n, \text{ where } \mathcal{D}_n = \{\sum_{\sigma \in G_n} f(\sigma^{-1}\eta_n).\sigma \;;\; f \in Hom(V_n, \mathbb{Z}_p)\}.$$

Proof. We follow essentially the argument of Kraft and Schoof (1995, thm 2.4). For any $a \geq 0$, the snake lemma applied to the p^a-th power map yields an exact sequence:

$$0 \to B_n[p^a] \to W_n/p^a \to V_n/p^a \to B_n/p^a \to 0$$

(the injectivity on the left is due to the fact that V_n is $\mathbb{Z}_p$-free).

Applying the functor $\mathrm{Hom}_{G_n}(.\,, R_{n,a})$, we get another exact sequence of $R_{n,a}$-modules:

$$\to \mathrm{Hom}_{G_n}\big(V_n, R_{n,a}\big) \to \mathrm{Hom}_{G_n}\big(W_n, R_{n,a}\big) \to \mathrm{Hom}_{G_n}\big(B_n[p^a], R_{n,a}\big) \to$$

But $R_{n,a}$ is a Gorenstein ring, which means that $R^*_{n,a}$ is a free $R_{n,a}$-module of rank 1, hence, for any finite $R_{n,a}$-module M, the canonical isomorphism $\mathrm{Hom}_{R_n}\big(M, R^*_{n,a}\big) \tilde{\to} M^*$ gives rise to an $R_{n,a}$-isomorphism $\mathrm{Hom}_{R_n}\big(M, R_{n,a}\big) \tilde{\to} M^*$, $f \mapsto \psi \circ f$, where ψ is a chosen $R_{n,a}$-generator $R_{n,a} \to \mathbb{Q}_p/\mathbb{Z}_p$. It follows that the functor $\mathrm{Hom}_{R_n}\big(.\,, R_{n,a}\big)$ is exact and we have

$$\to \mathrm{Hom}_{R_n}\big(V_n, R_{n,a}\big) \to \mathrm{Hom}_{R_n}\big(W_n, R_{n,a}\big) \to B_n^*/p^a \to 0$$

Taking $\varprojlim_a$, we derive an exact sequence:

$$\to \mathrm{Hom}_{R_n}(V_n, R_n) \xrightarrow{res} \mathrm{Hom}_{R_n}(W_n, R_n) \to (tB_n)^* \to 0\,.$$

But $W_n = R_n.\eta_n$, hence an element $f \in \mathrm{Hom}_{R_n}(W_n, R_n)$ is determined by the value $f(\eta_n)$, i.e. $\mathrm{Hom}_{R_n}(W_n, R_n) \tilde{\to} R_n$, $f \mapsto f(\eta_n)$, and the above exact sequence becomes: $\mathrm{Hom}_{R_n}(V_n, R_n) \to R_n \to (tB_n)^* \to 0$.

In other words, $(tB_n)^* \simeq R_n/\{f(\eta_n)\,;\, f \in \mathrm{Hom}_{G_n}(V_n, R_n)\}$. The canonical isomorphism $\mathrm{Hom}_{\mathbb{Z}_p}(M, \mathbb{Z}_p) \tilde{\to} \mathrm{Hom}_{R_n}(M, R_n)$, $f \mapsto F$ such that $F(m) = \sum_{\sigma \in G_n} f(\sigma^{-1}m).\sigma$, $\forall\, m \in M$, gives the desired result.

Corollary 1. $Ann_{R_n}\big((tB_n)^*\big) = Fit_{R_n}\big((tB_n)^*\big) = \mathfrak{D}_n$

Proof. The Proposition 1 shows at the same time that $(tB_n)^*$ is R_n-cyclic, hence its R_n-annihilator and R_n-Fitting ideal coincide, and that $Fit_{R_n}\big((tB_n)^*\big) = \mathfrak{D}_n$.

Definition 1. In the sequel, we shall denote by $\mathfrak{D}(F_\infty)$ (or $\mathfrak{D}_\infty$ for short) the projective limit $\varprojlim \mathfrak{D}_n$ w.r.t. norms (the detailed transition maps can be found e.g. in Solomon 2010, proposition 1). Of course $\mathfrak{D}_\infty$ depends on the admissible pair $(V_n, W_n)_{n \geq 0}$. Corollary 1 shows that $Ann_{\mathbb{A}}(\varprojlim (tB_n)^*) = Fit_{\mathbb{A}}(\varprojlim (tB_n)^*) = \mathfrak{D}_\infty$.

2.2 A Kummerian Description

Let $E = F(\zeta_p)$ and $\Delta = \mathrm{Gal}(E/F)$. Fix a norm coherent system $\boldsymbol{\zeta} = \left(\zeta_{p^n}\right)_{n\geq 0}$ of generators of the groups μ_{p^n} of p^n-th roots of unity. Attach to E_∞/E the following objects: U_n (resp. U'_n) = group of units (resp. (p)-units) of E_n

$$\overline{U}_n = \overline{U}_n \otimes \mathbb{Z}_p.$$

$\mathfrak{X}_n$ = Galois group over E_n of the maximal abelian (p)-ramified pro-p-extension of E_n. At the level of E_n, we have the Kummer pairing:

$$H^1\big(G_S(E_n), \mu_{p^n}\big)\times\mathfrak{X}_n/p^n \to \mu_{p^n},\ (x,\rho) \mapsto \left(x^{\frac{1}{p^n}}\right)^{\rho-1}.$$

Let $\langle .,.\rangle_n \;:\; \mathrm{Hom}\big(\mathfrak{X}_n, \mathbb{Z}/p^n\big)\times\mathfrak{X}_n/p^n \;\to\; \mathbb{Z}/p^n$ be the pairing defined by $\left(x^{\frac{1}{p^n}}\right)^{\rho-1} = \zeta_{p^n}^{\langle x,\rho\rangle_n}$. Take V_n^E to be $\overline{U}_n/tors$, or $\overline{U}'_n/tors$, or more generally a free $\mathbb{Z}_p$-module such that $V_n^E/p^n \hookrightarrow \mathrm{Hom}\,\big(G_S(E_n), \mu_{p^n}\big)$. In this subsection, we relax the condition of Galois freeness on W_n^E, which we assume only to be cyclic: $W_n^E = \mathbb{Z}[J_n].\,\eta_n$, where $J_n = \mathrm{Gal}(E_n/k)$.

Define

$$\begin{aligned}
\overline{\mathfrak{D}}_n(E) &= \{\sum_{\sigma\in J_n} f(\sigma^{-1}\eta_n).\sigma; f \in (V_n/p^n)^*\}\\
&= \{\sum_{\sigma\in J_n} \langle\sigma^{-1}.\,\eta_n, \rho\rangle_n\sigma \;;\; \rho \in \mathfrak{X}_n/p^n\}\\
&= \{\sum_{\sigma\in J_n} \langle\eta_n, \sigma.\,\rho\rangle_n\, \kappa_{cyc}^{-1}(\sigma)\sigma \;;\; \rho \in \mathfrak{X}_n/p^n\}
\end{aligned}$$

(the second equality comes from the Kummer pairing just recalled; the third from a functorial property of this pairing). Recall that

$$\mathfrak{D}_n(E) := \{\sum_{\sigma\in J_n} f(\sigma^{-1}\eta_n).\sigma \;;\; f \in \mathrm{Hom}(V_n^E, \mathbb{Z}_p)\}\,.$$

Proposition 2. *With the pair $(V_n^E, W_n^E)_{n\geq 0}$ chosen as above*

$$\mathfrak{D}(E_\infty) := \varprojlim \mathfrak{D}_n(E) = \varprojlim \overline{\mathfrak{D}}_n(E)$$

is cyclic as a $\mathbb{Z}_p[[Gal(E_\infty/\mathbb{Q})]]$-module.

Proof. Since Hom $(V_n^E, \mathbb{Z}_p) \simeq \left(V_n^E \otimes \mathbb{Q}_p/\mathbb{Z}_p\right)^*$, the equality $\varprojlim \mathcal{D}_n(E) = \varprojlim \overline{\mathcal{D}}_n(E)$ is straightforward. Notice next that in the sum $\sum_{\sigma \in J_n} \langle \eta_n, \sigma.\rho \rangle_n \, \kappa_{cyc}^{-1}(\sigma)\sigma$, the value of the pairing $\langle \eta_n, \sigma.\rho \rangle_n$ does not depend on ρ, only on the image of ρ in the cyclic group $\mathrm{Gal}\left(E_n(\eta_n^{\frac{1}{p^n}})/E_n\right)$. Choosing a generator τ_n of this group, we see that the sum above is a multiple of $\sum_{\sigma \in J_n} \langle \eta_n, \sigma.\tau_n \rangle_n \, \kappa_{cyc}^{-1}(\sigma)\sigma$. By considering the extension of E_∞ obtained by adjoining all p^n-th roots of all the $\eta_n' s$ and their conjugates under the action of $\mathrm{Gal}(E_\infty/\mathbb{Q})$, we get the desired cyclicity result.

This proposition will be applied in Sect. 4.4.2 to cut out $\mathcal{D}(E_\infty)$ by characters of $\mathrm{Gal}(E/F)$.

As usually happens, the "new" object $\mathcal{D}_\infty$ appears afterwards to be not so new. Some previous occurrences must be recorded:

- For $k = \mathbb{Q}$ and $E = \mathbb{Q}(\zeta_p)$ and for a special choice of $\boldsymbol{\eta}$, $\mathcal{D}_\infty$ plays an important role in Ihara's theory of universal power series for Jacobi sums (see Ihara et al. 1987). In Nguyen Quang Do and Nicolas (2011, § 4.1), $\mathcal{D}_\infty$ is interpreted as a certain module of "p-adic Gauss sums", and both Nguyen Quang Do and Nicolas (2011) and Solomon (2010) show that $(\mathcal{D}_\infty)^\#$ is the $\mathbb{A}$-initial Fitting ideal of $X_\infty'^{+}$, where the sign $(\,.\,)^\#$ means inverting Galois action.
- In exactly the same setting, $\mathcal{D}_\infty$ appears on the Galois side of Sharifi's conjecture on the two-variable L_p-function of Mazur-Kitagawa (Sharifi 2011, propos. 6.2). Note that a weak form of this conjecture has been recently proved for all fields $\mathbb{Q}(\zeta_{p^n})$ by Fukaya and Kato (2012).
- For k totally real and E/k abelian, containing ζ_p, and for a special choice of $\boldsymbol{\eta}$, $(\mathcal{D}_\infty)^\#$ is shown in Nguyen Quang Do and Nicolas (2011) and Solomon (2010) to annihilate $X_\infty'^{+}$. Moreover, in the semi-simple case $(p \nmid |G|)$, thms 3.4.2 and 5.3.2 of Nguyen Quang Do and Nicolas (2011) assert that for any odd character ψ of G, the ψ-part $((\mathcal{D}_\infty)^\#)^\psi$ is isomorphic to the Fitting ideal of $(X_\infty')^{\psi^*}$, where $\psi^* = \omega\psi^{-1}$ denotes the "mirror" of ψ, ω being the Teichmüller character.
- For $k/\mathbb{Q}$ abelian and $F = k(\zeta_p)$, $\mathcal{D}_\infty$ appears prominently in the explicit reciprocity law of Coleman, generalized by Perrin-Riou (Flach 2011, § 1.3; Perrin-Riou 1994, thm 4.3.2). Let us just recall the starting point in Perrin-Riou (1994): denoting $\bigoplus_{v|p} H^1(E_{n,v}, \mathbb{Z}_p(m))$ by $H^1_{s\ell}(E_n, \mathbb{Z}_p(m))$ and $\varprojlim H^1_{s\ell}(E_n, \mathbb{Z}_p(m))$ by $H^1_{Iw,s\ell}(E_\infty, \mathbb{Z}_p(m))$, Perrin-Riou constructs a "cup-product" at the infinite level $H^1_{Iw,s\ell}(E_\infty, \mathbb{Z}_p(1)) \times H^1_{Iw,s\ell}(E_\infty, \mathbb{Z}_p) \overset{\cup}{\rightarrow} (\mathcal{O}_E \otimes \mathbb{Z}_p)[[G_\infty]]$ such that, writing π_n for the natural projection $(\mathcal{O}_F \otimes \mathbb{Z}_p)[[G_\infty]] \rightarrow \mathbb{Z}_p[G_n]$, $\pi_n(\mathbf{x} \cup \mathbf{y}) = \sum_{\sigma \in G_n} (\sigma^{-1} x_n \cup y_n).\sigma$ (recall that the cup-product at finite levels does not commute with corestriction).

3 Semi-simple Example

In this section, to illustrate the "Gras type" approach via the WMC, we intend to study a (particular) semi-simple case, which the reader can skip if pressed for time. The following hypotheses will be assumed: Let $F/\mathbb{Q}$ be a *totally real abelian* extension of conductor f, such that p does not divide the order of $G = \mathrm{Gal}(F/\mathbb{Q})$. Let U_F (resp. U'_F) be the group of units (resp. (p)-units) of F. The group $\mathrm{Cyc}(F)$ of F is the subgroup of F^* generated by -1 and all the elements $N_{\mathbb{Q}(\zeta_f)/F\cap\mathbb{Q}(\zeta_f)}(1-\zeta_f^a)$, $(a,f)=1$. The group C_F (resp. C'_F) of circular units (resp. circular (p)-units) in Sinnott's sense is defined as $C_F = U_F \cap Cyc(F)$ (resp. $C'_F = U'_F \cap Cyc(F)$). Write U_n, C_n, etc. for U_{F_n}, C_{F_n}, etc. and $\overline{(.)}$ for the p-completion. We take $V_n = \overline{U}'_n$ (which is $\mathbb{Z}_p$-free of rank $[F_n:\mathbb{Q}]$) and look for candidates for the W'_ns. Since $p\not|\,|G|$, any (p)-place of F is totally ramified in F_∞. Let us denote by s the number of (p)-places of F (hence also of F_∞). We keep the Galois notations of the beginning of Sect. 1.

Lemma 1. *For any $n \geq 1$, the $\mathbb{Z}_p[G_n]$-module $\overline{C}'_n$ is free (necessarily of rank 1) if and only if $s=1$.*

Proof. Let us first consider only the Λ-module structure. The cohomology groups $H^i(\Gamma_n, \overline{C}'_n)$ are computed in general in Nguyen Quang Do and Lescop (2006). In our situation, $\overline{C}'_n$ is Γ_n-cohomologically trivial if and only if $s=1$ (Nguyen Quang Do and Lescop 2006, propos. 2.8). Since Γ_n is a p-group, $\mathbb{Z}_p[\Gamma_n]$ is a local algebra, and for a $\mathbb{Z}_p[\Gamma_n]$-module without torsion, cohomological triviality is equivalent to $\mathbb{Z}_p[\Gamma_n]$-projectivity, hence to $\mathbb{Z}_p[\Gamma_n]$-freeness here. To pass from Γ_n to G_n, just notice that $H^i(G_n, \overline{C}'_n)^{\Delta} \overset{\text{res}}{\simeq} H^i(\Gamma_n, \overline{C}'_n)$ because $p\not|\,|\Delta|$, hence $s=1$ if and only if $\overline{C}'_n$ is $\mathbb{Z}_p[G_n]$-projective (of $\mathbb{Z}_p$-rank equal to $|G_n|$). By decomposing $\overline{C}'_n$ into χ-parts, $\overline{C}'_n = \bigoplus_{\chi\in\hat{G}} (\overline{C}'_n)^{\chi}$, we see that $s=1$ if and only if each $(\overline{C}'_n)^{\chi}$ is free over the local algebra $\mathbb{Z}_p[\chi][\Gamma_n]$, if and only if $\overline{C}'_n$ is $\mathbb{Z}_p[G_n]$-free (necessarily of rank 1).

Summarizing, if $s=1$, one can take $V_n = \overline{U}'_n$, $W_n = \overline{C}'_n$ in order to apply Proposition 1 to $tB_n = B_n = \overline{U}'_n/\overline{C}'_n$. Note that in the semi-simple case with $s=1$, $\overline{U}'_n/\overline{C}'_n \simeq \overline{U}_n/\overline{C}_n$ for any $n \geq 1$ (Nguyen Quang Do and Lescop 2006, lemma 2.7) and also $X'_\infty \simeq X_\infty$ in the notations of the introduction (Nguyen Quang Do and Lescop 2006, lemma 1.5). Recall that X_∞ (resp. X'_∞) is the unramified (resp. totally split at all finite places) Iwasawa module above F_∞. We can now determine the $\mathbb{A}$-Fitting ideal of X_∞ in our particular case:

Proposition 3. *Let $F/\mathbb{Q}$ be a totally real abelian extension, such that $p\not|\,|G|$ and $s=1$. Then $\mathfrak{D}(F_\infty)^{\#} = Fit_{\mathbb{A}}(X_\infty)$, where $(.)^{\#}$ means inverting the Galois action.*

This is a particular case ($s = 1$) of thm 5.3.2 of Nguyen Quang Do and Nicolas (2011). Typical examples are $F = \mathbb{Q}(\zeta_p)^+$, or $F = \mathbb{Q}(\sqrt{d})$, d a square free positive integer such that $\left(\frac{d}{p}\right) \neq 1$.

Proof. Since $B_n = \overline{U}_n/\overline{C}_n$ is finite, Proposition 1 shows that $\mathcal{D}(F_\infty) = \mathcal{D}_\infty = Fit_{\mathbb{A}}(\varprojlim(B_n^*))$. It remains to describe $\varprojlim(B_n^*)$. Denote $\overline{U}_\infty = \varprojlim \overline{U}_n$, $\overline{C}_\infty = \varprojlim \overline{C}_n$, $Y_\infty = \overline{U}_\infty/\overline{C}_\infty$, $X_\infty^0 =$ the maximal finite submodule of X_∞. With $s = 1$, we have the following codescent exact sequences (Nguyen Quang Do and Lescop 2006, proposition 4.7):

$$0 \longrightarrow (Y_\infty)_{\Gamma_n} \longrightarrow B_n \longrightarrow (X_\infty^0)^{\Gamma_n} \longrightarrow 0\,.$$

Taking duals and $\varprojlim$ we get an exact sequence of $\mathbb{A}$-modules:

$0 \longrightarrow (X_\infty^0)^* \longrightarrow \varprojlim(B_n^*) \longrightarrow \alpha(Y_\infty) \longrightarrow 0$, where $\alpha(.)$ denotes the Iwasawa adjoint (with additional action by G). In particular, $(X_\infty^0)^* = (\varprojlim B_n^*)^0$. Let us take Fitting ideals in this exact sequence. A well known lemma (generally attributed to Cornacchia; see e.g. Nguyen Quang Do and Nicolas 2011, lemma 3.4.2) states that for any torsion $\Lambda[\chi]$-module M, $Fit_{\Lambda[\chi]}(M) = Fit_{\Lambda[\chi]}(M^0).\,char_{\Lambda[\chi]}(M)$ where $char_{\Lambda[\chi]}(.)$ denotes the characteristic ideal. In the semi-simple situation, we can put the χ-parts together to get: $Fit_{\mathbb{A}}(\varprojlim B_n^*) = Fit_{\mathbb{A}}(X_\infty^0).\,char_{\mathbb{A}}(\alpha(Y_\infty))$ (with an obvious definition of $char_{\mathbb{A}}(.)$ here). Let $(M)^\#$ be the module M with inverted Galois action. It is classically known that $\alpha(M)$ is pseudo-isomorphic to $M^\#$, and that $Fit_{\mathbb{A}}(M^*) = Fit_{\mathbb{A}}(M^\#)$ since the p-Sylow subgroups of the G_n''s are cyclic (Mazur and Wiles 1984, appendix). Hence $\mathcal{D}_\infty = Fit_{\mathbb{A}}(\varprojlim B_n^*) = Fit_{\mathbb{A}}(X_\infty^0)^\#.\,char_{\mathbb{A}}(Y_\infty)^\#$. As $char_{\mathbb{A}}(Y_\infty) = char_{\mathbb{A}}(X_\infty)$ in the "Gras type" formulation of the WMC, we get: $(\mathcal{D}_\infty)^\# = Fit_{\mathbb{A}}(X_\infty^0).\,char_{\mathbb{A}}(X_\infty) = Fit_{\mathbb{A}}(X_\infty)$.

Remarks. 1. In spite of the presence of the algebra $\mathbb{A}$, Proposition 3 is not a genuine equivariant result. In particular, the definition of the characteristic ideal $char_{\mathbb{A}}(.)$ cannot be generalized to the non semi-simple case.

2. The ideal $\mathcal{D}(F_\infty)$ can easily be made explicit in kummerian terms using Sect. 2.2. It suffices to start from the base field $E = F(\zeta_p)$ and then use (co) descent from E to F, or from E_∞ to F_∞, which works smoothly because p does not divide $[E : F]$.

3. It is well known that $\overline{U}'_\infty = H^1_{I_w,S_p}(F_\infty, \mathbb{Z}_p(1))$ in full generality. In the situation of Proposition 3, we have also $X_\infty = X'_\infty = H^2_{I_w,S_p}(F_\infty, \mathbb{Z}_p(1))$ (for details, see Proposition 4 below). We can also consider the modules $H^2_{I_w,S_p}(F_\infty, \mathbb{Z}_p(m))$, m odd, as in the introduction and describe explicitly their $\mathbb{A}$-Fitting ideals by taking Tate twists above E_∞, and then doing (co)descent as in 2) (for details, Sect. 4.4.2 below).

4 Equivariant Study of the Non Semi-simple Case

In this case, as we noticed before, two major difficulties are encountered right from the start: the notion of characteristic ideals of torsion modules (with appropriate multiplicative properties) is no longer available; neither is the "Gras type" formulation of the WMC. The solution to both problems will come from the equivariant Iwasawa theory of complexes. Among many existing formulations, that of M. Witte (2006) seems the best suited to our purpose. Let us recall the minimal amount of definitions and results that we need, referring to Witte (2006) for further details or greater generality.

4.1 Perfect Torsion Complexes and Characteristic Ideals

(see Witte 2006, §1) Let R be a commutative noetherian ring and $Q(R)$ its total ring of fractions. A complex $C^{\cdot}$ of R-modules is called a *torsion* complex if $Q(R)\underset{R}{\otimes} C^{\cdot}$ is acyclic; *perfect* if $C^{\cdot}$ is quasi-isomorphic to a bounded complex of projective R-modules. To a perfect complex $C^{\cdot}$ one can attach its Knudsen-Mumford determinant, denoted $det_R\, C^{\cdot}$. If moreover $C^{\cdot}$ is torsion, the natural isomorphism $Q(R) \simeq Q(R)\underset{R}{\otimes} det_R\, C^{\cdot}$ allows to see $det_R\, C^{\cdot}$ as an invertible fractional ideal of R. Recall that these invertible fractional ideals form a group $\mathfrak{J}(R)$, which is isomorphic to $Q(R)^{\times}/R^{\times}$ if the ring R is semi-local.

Definition 2. The characteristic ideal of a perfect torsion complex $C^{\cdot}$ is $char_R(C^{\cdot}) = (det_R\, C^{\cdot})^{-1} \in \mathfrak{J}(R)$.

Examples. If R is a noetherian and normal domain and M is a torsion module which is perfect (considered as a complex concentrated in degree 0), i.e. of finite projective dimension, then $char_R(M)$ coincides with the "content" of M in the sense of Bourbaki. If $R = \Lambda = \mathbb{Z}_p[[T]]$, then $char_{\Lambda}(M)$ is the usual characteristic ideal. This justifies the name of equivariant characteristic ideal for $char_{\mathbb{A}}(M)$.

Many functorial properties of $char_R(.)$ are gathered in Witte (2006, proposition 1.5). We are particularly interested in the following:

If $C^{\cdot}$ is a perfect torsion complex of R-modules such that the cohomology modules of $C^{\cdot}$ are themselves perfect, then $char_R(C^{\cdot}) = \prod_{n\in\mathbb{Z}} (char_R\, H^n C^{\cdot})^{(-1)^n}$.

4.2 Iwasawa Cohomology Complexes

For the rest of the paper, unless otherwise specified, $F/\mathbb{Q}$ will be an *abelian* number field (not necessarily totally real). With the notations and conventions of the beginning of Sect. 1, let us extract from Witte (2006, § 3) some perfect complexes and cohomology modules for our use (the situation in Witte (2006) is more general):

The Iwasawa complex of $\mathbb{Z}_p(m)$ relative to S is the cochain complex of continuous étale cohomology $\mathbf{R}\Gamma_{Iw}(F_\infty, \mathbb{Z}_p(m))$ as constructed by U. Jannsen. This is a perfect $\mathbb{A}$-complex whose cohomology modules are $H^i_{Iw}(F_\infty, \mathbb{Z}_p(m)) = H^i_{Iw,S}(F_\infty, \mathbb{Z}_p(m)) = \varprojlim_{\text{cores}} H^i_S(F_n, \mathbb{Z}_p(m))$ for $i = 1, 2$, zero otherwise.

Let us gather in an overall proposition many known properties of these cohomology groups. Our main reference will be Kolster et al. (1996, sections 1 and 2) (in which $S = S_p$, but the proofs remain valid for any finite set S containing S_p). Let us fix again some notations:

$E = F(\zeta_p)$, $E_\infty = F(\zeta_{p^\infty})$, $\Gamma^\times = \mathrm{Gal}(E_\infty/F)$

r_1 (resp. r_2) = number of real (resp. complex) places of F

$\overline{U}_S(E) = U_S(E) \otimes \mathbb{Z}_p$ = the p-adic completion of the S-units of E

$\overline{U}_S(E_\infty) = \varprojlim \overline{U}_S(E_n)$ w.r.t. norms, $X'(E_\infty)$ = the totally split Iwasawa module above E_∞.

Proposition 4. *(i) For any $m \in \mathbb{Z}$, $\mathrm{rank}_{\mathbb{Z}_p} H^1_S(F, \mathbb{Z}_p(m)) = d_m + \delta_m$, where $d_m = r_1 + r_2$ (resp. r_2) if m is odd (resp. even), and $\delta_m = \mathrm{rank}_{\mathbb{Z}_p}$ $(X'(E_\infty)(m-1)^{\Gamma^\times})$.*

(ii) For any $m \in \mathbb{Z}$, $m \neq 0$, $\mathrm{tor}_{\mathbb{Z}_p} H^1_S(F, \mathbb{Z}_p(m)) = H^0_S(F, \mathbb{Q}_p/\mathbb{Z}_p(m))$. In particular, if m is odd and F totally real, $H^1_S(F, \mathbb{Z}_p(m))$ is $\mathbb{Z}_p$-free.

(iii) For any $m \in \mathbb{Z}$, $m \neq 1$, there is a natural codescent exact sequence:

$$0 \to \left(\overline{U}_S(E_\infty)(m-1)/\Lambda\text{-tors}\right)_{\Gamma^\times} \to H^1_S(F, \mathbb{Z}_p(m))/\mathbb{Z}_p\text{-tors} \to X'(E_\infty)(m-1)^{\Gamma^\times} \to 0$$

(iv) For any $m \in \mathbb{Z}$, $m \neq 1$, the Poitou-Tate sequence for H^2_S can be written as: $0 \to X'(E_\infty)(m-1)_{\Gamma^\times} \to H^2_S(F, \mathbb{Z}_p(m)) \to \bigoplus_{v \in S_f} H^2(F_v, \mathbb{Z}_p(m)) \simeq \bigoplus_{v \in S_f} H^0(F_v, \mathbb{Q}_p/\mathbb{Z}_p(1-m))^* \to H^0_S(F, \mathbb{Q}_p/\mathbb{Z}_p(1-m))^* \to 0$

For $m = 1$, the leftmost term must be replaced by $A_S(F)$, the p-part of the S_f-class group of F.

Remarks. It is conjectured that $X'(E_\infty)(m-1)^{\Gamma^\times}$ is finite (i.e. $\delta_m = 0$) for all $m \in \mathbb{Z}$ (see e.g. Kolster et al. 1996, p. 637). These are the so-called *mth-twists of Leopoldt's conjecture*. The case $m = 0$ (resp. 1) corresponds to Leopoldt's (resp. Gross') conjecture. For $m \geq 2$, $\delta_m = 0$ because $K_{2m-2}\mathcal{O}_F$ is finite. Recall that we implicitly suppose that $\delta_m = 0$ for all $m \in \mathbb{Z}$.

At the infinite level, we have the following

Lemma 2. *The $\mathbb{A}$-modules $H^i_{Iw}(F_\infty, \mathbb{Z}_p(m))$, $i = 1, 2$, are perfect.*

Proof. Let $\mathbb{Q}_\infty$ be the cyclotomic $\mathbb{Z}_p$-extension of $\mathbb{Q}$ and $\Lambda = \mathbb{Z}_p[[\mathrm{Gal}(\mathbb{Q}_\infty/\mathbb{Q})]]$. Since Λ is a regular noetherian ring, every Λ-noetherian module is perfect.

For the cohomology modules over $\mathbb{A}$, we just have to use the following quasi-isomorphism established e.g. by Fukaya and Kato (2006) and Nekovář (2006) etc.: the natural ring homomorphism $\Lambda \to \mathbb{A}$ induces a quasi-isomorphism $\mathbb{A} \underset{\Lambda}{\otimes} \mathbf{R}\,\Gamma_{Iw}(\mathbb{Q}_\infty, \mathbb{Z}_p(m)) \longrightarrow \mathbf{R}\,\Gamma_{Iw}(F_\infty, \mathbb{Z}_p(m))$.

4.3 The Limit Theorem

As we explained in the introduction, the "limit theorem" of Burns and Greither (2003a, thm 6.1) is actually an EMC, expressed in the language of the Iwasawa theory of complexes, which encapsulates equivariant generalizations of both the WMC (for the minus part of X_∞) and its formulation "à la Gras" (for the plus part). It is ultimately derived from the WMC, but only after some hard work. We recall here its presentation by M. Witte (only for the characters κ_{cyc}^m; the characters in Witte (2006) are more general). Our main reference will be Witte (2006, sections 6 and 7). The "limit theorem" will relate a "zeta-element" and a "special cyclotomic element" constructed as follows:

Let $f = N\,p^d$, $p \nmid N$, the conductor of F. Introduce $L_\infty = \mathbb{Q}(\zeta_{Np^\infty}) \supset F_\infty$. At the nth-levels, define the Stickelberg elements:

$$Stick_n = \sum_{\substack{0<a<fp^n \\ (a,fp)=1}} \left(\frac{a}{fp^n} - \frac{1}{2}\right)\Big(\prod_{\ell \text{ prime}} Frob_\ell^{-v_\ell(a)}\Big) \in \mathbb{Q}_p[\mathrm{Gal}(L_n/\mathbb{Q})]\,,$$

and at the infinite level: $Stick_\infty = \varprojlim Stick_n \in Q\big(\mathbb{Z}_p[[\mathrm{Gal}(L_\infty/\mathbb{Q})]]\big)$.

Denote by $pr_\pm$ the projectors onto the (± 1)-eigenspaces of complex conjugation and define the *zeta-element* $\mathcal{L}_S(F_\infty, \kappa_{cyc}^{1-m})$ as the image by the natural map $Q\,\mathbb{Z}_p[[(\mathrm{Gal}(L_\infty/\mathbb{Q}))]] \longrightarrow Q\,\mathbb{A} = Q\,\mathbb{Z}_p[[G_\infty]]$ of the element $Tw_{m-1}(pr_+ - pr_- Stick_\infty)$, where Tw_a is the Tate twist induced by $\sigma \mapsto \kappa_{cyc}^a(\sigma).\sigma$.

Example (Witte 2006, proposition 6.3). Let $F_\infty = \mathbb{Q}_\infty$ and $S = S_p$ and write ω for the Teichmüller character. Then the image by the natural map $Q\,\mathbb{Z}_p[[\Gamma]] \to \mathbb{Q}_p$ of $\mathcal{L}_S(\mathbb{Q}_\infty, \kappa_{cyc}^{1-m})$ is $L_p(1-m, \omega^m.\,rec^{-1})$, where $rec\colon \mathrm{Gal}(\mathbb{Q}(\zeta_N)/\mathbb{Q}) \to (\mathbb{Z}/N\mathbb{Z})^\times$ is the isomorphism induced by $Frob_\ell^{-1} \mapsto \ell$, and $L_p(.)$ is the Kubota-Leopoldt p-adic L-function.

Define now the *special cyclotomic element* $\eta(F_\infty, \kappa_{cyc}^m)$ as the image by the composite map $H^1_{Iw}(L_\infty, \mathbb{Z}_p(1)) = \overline{U}_S(L_\infty) \overset{Tw_{m-1}}{\longrightarrow} H^1_{Iw}(L_\infty, \mathbb{Z}_p(m)) = \overline{U}_S(L_\infty)(m-1) \overset{nat}{\longrightarrow} H^1_{Iw}(F_\infty, \mathbb{Z}_p(m))$ of the element $pr_+((1-\zeta_{Np^n})_{n\geq 0})$. The element $\eta(F_\infty, \kappa_{cyc}^m)$ allows to modify the Iwasawa complex $\mathbf{R}\Gamma_{Iw}(F_\infty, \mathbb{Z}_p(m))$ to get a perfect torsion complex. Precisely, there exists a unique morphism $\mathbb{A}\eta(F_\infty, \kappa_{cyc}^m)[-1] \to \mathbf{R}\Gamma_{Iw}(F_\infty, \mathbb{Z}_p(m))$ in the derived category which induces the natural inclusion on cohomology, hence a unique (up to quasi-isomorphism) complex denoted by $\mathbf{R}\,\Gamma_{Iw}/\eta(F_\infty,\kappa_{cyc}^m)$ which takes place in the following distinguished triangle

$$\mathbb{A}\,\eta(F_\infty,\kappa_{cyc}^m)[-1] \to \mathbf{R}\ \Gamma_{Iw}(F_\infty,\mathbb{Z}_p(m)) \to \mathbf{R}\,\Gamma_{Iw}/\eta(F_\infty,\kappa_{cyc}^m)\,.$$

Lemma 3 (Witte 2006, lemma 7.2).

(i) $\mathbf{R}\,\Gamma_{Iw}/\eta(F_\infty,\kappa_{cyc}^m)$ *is a perfect torsion complex of* $\mathbb{A}$*-modules*
(ii) *If* F_∞ *is totally real and* m *is odd, then* $\mathbb{A}\eta(F_\infty,\kappa_{cyc}^m)$ *is a free* $\mathbb{A}$*-module of rank 1.*

We can now state the "limit theorem" (Burns and Greither 2003a, theorem 6.1; Witte 2006, theorem 7.4).

Theorem 1. *For an abelian number field* $F/\mathbb{Q}$*, with* $S = S_\infty \cup S_p \cup Ram(F/\mathbb{Q})$*, for any* $m \in \mathbb{Z}$*:*

(i) *At any prime* $\mathcal{P}$ *of codimension 1 of* $\mathbb{A}$*, containing* p*, the localized complex* $\big(\mathbf{R}\,\Gamma_{Iw}/\eta(F_\infty,\kappa_{cyc}^m)\big)_{\mathcal{P}}$ *is acyclic ("vanishing of the* μ*-invariant").*
(ii) $\mathcal{L}_S(F_\infty,\kappa_{cyc}^{1-m})$ *generates the* $\mathbb{A}$*-characteristic ideal of* $\mathbf{R}\,\Gamma_{Iw}/\eta(F_\infty,\kappa_{cyc}^m)$*.*

To stress the difference between the (equivariant) limit theorem above and results obtained character by character (such as in Huber and Kings 2003), let us cite the following comparison lemma (Witte 2006, lemma 7.6):

Lemma 4. *Suppose for simplification that* F *is linearly disjoint from* $\mathbb{Q}_\infty$ *and* $p^2 \nmid f$*. Let* $\phi\ :\ \Omega = \mathcal{O}_p[[G_\infty]] \to \tilde{\Omega} = \prod_{\chi\in\hat{G}} \mathcal{O}_p[[Gal(\mathbb{Q}_\infty/\mathbb{Q})]]$ *be the normalisation of* Ω *in* $Q\,\Omega$*, where* $\mathcal{O}_p$ *is the ring obtained from* $\mathbb{Z}_p$ *by adding all the values of all the characters of* G*. Then*

$$char_{\tilde{\Omega}}\big(\mathbf{L}\,\phi_*\ \mathbf{R}\ \Gamma_{Iw}/\eta(F_\infty,\mathcal{O}_p(m)\big) = L_S(F_\infty,\kappa_{cyc}^{1-m}\big).\,\tilde{\Omega}$$

4.4 The Determinant of $H^2_{Iw}(F_\infty,\mathbb{Z}_p(m))$

From now on, F_∞ is *totally real* and $m \in \mathbb{Z}$ is *odd*. The mth - twisted Leopoldt conjectures for the totally real field F are known to hold for all odd $m \neq 1$: for $m > 1$, see Proposition 4(i); for $m < 0$, see (Schneider 1979, § 5), Corollary 5. To state and prove our main result, we shall proceed in several steps.

Because the cohomology $\mathbb{A}$-modules $H^i_{Iw}(.)$ are perfect (Lemma 2), it follows from Theorem 1(ii) and the last property cited in section 4-1 that

$$\begin{aligned} &char_{\mathbb{A}}\big(H^1_{Iw}(F_\infty,\mathbb{Z}_p(m))/\mathbb{A}\eta(F_\infty,\kappa_{cyc}^m)\big)^{-1}.\,char_{\mathbb{A}}\ H^2_{Iw}(F_\infty,\mathbb{Z}_p(m))\\ &= (\mathcal{L}_S(F_\infty,\kappa_{cyc}^{1-m}))\\ &= (1)\,, \end{aligned}$$

so that it remains only to determine the first equivariant characteristic series, appealing to the algebraic results of Sect. 1.

4.4.1

The point is to choose the admissible pair $(V_n, W_n)_{n\geq 0}$ attached to F_∞/F. Fix an odd integer $m \neq 1$ and for any $n \geq 0$, write $\eta_n^{(m)}$ for the image of $\eta(F_\infty, \kappa_{cyc}^m)$ by the natural map $H^1_{Iw}(F_\infty, \mathbb{Z}_p(m)) \to H^1_S(F_n, \mathbb{Z}_p(m))$. Then $\eta_n^{(m)}$ is $\mathbb{Z}_p[G_n]$-free: this comes from Lemma 3(ii) and the codescent exact sequence of Proposition 4(i); for a direct argument, see Beliaeva and Belliard (2012, thm. 3.4). Note that highly non trivial ingredients are needed for both proofs: Bloch-Kato's reciprocity law for the first, twisted Leopoldt's conjecture for the second. We shall take $V_n = H^1_S(F_n, \mathbb{Z}_p(m))$ and $W_n = \mathbb{Z}_p[G_n].\,\eta_n^{(m)}$. Adding a superscript $(.)^{(m)}$ to the notations of Sect. 1, let $\mathcal{D}^{(m)}(F_\infty) = \mathrm{Fit}_{\mathbb{A}}\,(\varprojlim B_n^*)$, which we must relate to the desired $\mathbb{A}$-determinant.

Remark. The reason for choosing a twist $m \neq 1$ is that codescent on S-units (corresponding to $m = 1$) is notoriously not smooth, especially in case of p-decomposition. We shall reintegrate $m = 1$ in Theorem 2 by using a "twisting trick".

Since the p-Sylow subgroups of the G_n's are no longer necessarily cyclic, the argument on Fitting ideals used in the semi-simple case (Proposition 3) no longer works. We must consequently change the definition of $V_n = H^1_S(F_n, \mathbb{Z}_p(m))$, replacing it by $V_n' = \text{image of } \big(H^1_{Iw}(F_\infty, \mathbb{Z}_p(m)) \to H^1_S(F_n, \mathbb{Z}_p(m))\big)$. We don't change $W_n = \mathbb{Z}_p[G_n].\,\eta_n^{(m)}$ and we define $B_n' = V_n'/W_n$. By Proposition 4(iii), $\varprojlim V_n' = \varprojlim V_n = H^1_{Iw}(F_\infty, \mathbb{Z}_p(m))$, hence $\varprojlim B_n' = \varprojlim B_n := B_\infty = H^1_{Iw}(F_\infty, \mathbb{Z}_p(m))/\mathbb{A}\eta(F_\infty, \kappa_{cyc}^m)$, whereas $\varprojlim B_n'^* = \alpha(\varprojlim B_n') = \alpha(B_\infty)$, $\alpha(.)$ denoting the Iwasawa adjoint, with additional action by $H = \mathrm{Gal}(F_\infty/k_\infty)$. Note that this adjoint module is naturally isomorphic to $\mathrm{Ext}^1_{\mathbb{A}}(B_\infty, \mathbb{A})$ (see e.g. Nguyen Quang Do 2005, § 3) over $\mathbb{A}$. Over Λ, we know that $\alpha(B_\infty)$ and $(B_\infty)^\#$ are pseudo-isomorphic, hence the existence of an exact sequence (non canonical) of Λ-modules $0 \to \alpha(B_\infty) \to (B_\infty)^\# \to \Phi \to 0$, where Φ is a finite abelian p-group. Since H acts on the first two terms, it also acts on the third, i.e. the above sequence is indeed exact over $\mathbb{A}$.

4.4.2

Since $\varprojlim V_n' = \varprojlim V_n$, the same reasoning exactly as in Proposition 1 shows that $\mathrm{Fit}_{\mathbb{A}}(\varprojlim B_n'^*) = \mathrm{Fit}_{\mathbb{A}}(\alpha(B_\infty)) = \mathcal{D}^{(m)}(F_\infty)$ (the same $\mathcal{D}^{(m)}(F_\infty)$ as before). In particular, this Fitting ideal is principal according to Proposition 1 and an obvious descent from $E_\infty = F_\infty(\mu_{p^\infty})$ to F_∞. It remains to compare the two principal ideals $I = \mathrm{Fit}_{\mathbb{A}}(\alpha(B_\infty))$ and $J = char_{\mathbb{A}}(B_\infty)^{\#}$ by using localization:

Lemma 5. *(Burns and Greither 2003a, lemma 6.1) Let R be a Cohen-Macaulay ring of dimension 2 and let I, J be two invertible fractional ideals of R. Then $I = J$ if and only if $I_{\mathcal{P}} = J_{\mathcal{P}}$ for all height one prime ideals $\mathcal{P}$ of R.*

Cutting out if necessary by the characters of the non-p-part of H, we can suppose that our ring $\mathbb{A}$ is as in Lemma 5 and proceed to localization:

- At a height one prime $\mathcal{P}$ not containing p, $\mathbb{A}_{\mathcal{P}}$ is a discrete valuation ring in which p is invertible. It follows that Fit (.) and det $(.)^{-1}$ coincide over $\mathbb{A}_{\mathcal{P}}$, and $\Phi_{\mathcal{P}} = (0)$, hence $I_{\mathcal{P}} = J_{\mathcal{P}}$.
- At the unique height one prime $\mathcal{P}$ of $\mathbb{A}$ containing p, the vanishing of the μ-invariant (Theorem 1(i)) means that $(B_\infty)^{\#}_{\mathcal{P}}$ vanishes, hence also $\alpha(B_\infty)_{\mathcal{P}}$.

We have thus shown that $char_{\mathbb{A}}(B_\infty)^{\#} = \mathrm{Fit}_{\mathbb{A}}(\alpha(B_\infty)) = \mathcal{D}^{(m)}(F_\infty)$. We can now state and prove our main result:

Theorem 2. *Let $F/\mathbb{Q}$ be a totally real abelian number field, $E = F(\zeta_p)$, $\mathbb{A} = \mathbb{Z}_p[[Gal(F_\infty/\mathbb{Q})]]$, $\mathbb{B} = \mathbb{Z}_p[[Gal(E_\infty/\mathbb{Q})]]$. Then, for any odd $m \in \mathbb{Z}$, we have:*

1. $\mathrm{char}_{\mathbb{B}}(H^2_{Iw}(E_\infty, \mathbb{Z}_p(m))) = \mathcal{D}^{(1)}(E_\infty)^{\#}(m-1) = tw_{m-1}\mathcal{D}^{(1)}(E_\infty)$, *where tw_{m-1} is the Iwasawa twist induced by $\sigma \mapsto \kappa_{cyc}^{m-1}(\sigma)\sigma^{-1}$.*
2. $\mathrm{char}_{\mathbb{A}}(H^2_{Iw}(F_\infty, \mathbb{Z}_p(m)) = (e_{m-1}\mathcal{D}^{(1)}(E_\infty))^{\#}$, *where e_{m-1} is the idempotent of Δ associated to the power ω^{m-1} of a generator ω of $\hat{\Delta}$.*

Proof. For any odd $m \neq 1$, we have just seen that

$$\mathrm{char}_{\mathbb{A}}(H^2_{Iw}(F_\infty, \mathbb{Z}_p(m)))^{\#} = \mathcal{D}^{(m)}(F_\infty) = \nu\mathcal{D}^{(m)}(E_\infty),$$

where ν denotes the norm map of Δ. But $\mathcal{D}^{(m)}(E_\infty) = \mathcal{D}^{(1)}(E_\infty)(m-1)$, so that $\nu\mathcal{D}^{(m)}(E_\infty) = e_{m-1}\mathcal{D}^{(1)}(E_\infty)$, because Δ is of order prime to p. Besides, denoting by π the natural projection $\mathbb{B} \to \mathbb{A}$, we know that $\mathbf{L}\pi_* \mathbf{R}\,\Gamma_{Iw}(E_\infty, \mathbb{Z}_p(m))$ is naturally quasi-isomorphic to $\mathbf{R}\,\Gamma_{Iw}(F_\infty, \mathbb{Z}_p(m))$ (Witte 2006, propos. 3.6 (ii)). Hence, by definition of the determinant, $\mathrm{char}_{\mathbb{A}}(H^2_{Iw}(F_\infty, \mathbb{Z}_p(m))) = \nu\,(char_{\mathbb{B}}\, H^2_{Iw}(E_\infty, \mathbb{Z}_p(m))) = e_{m-1}(char_{\mathbb{B}}\, H^2_{Iw}(E_\infty, \mathbb{Z}_p(1)))$. Since the idempotent e_{m-1} depends only on the residue class of $(m-1) \bmod (p-1)$, we can conclude that $\mathrm{char}_{\mathbb{B}}\, H^2_{Iw}(E_\infty, \mathbb{Z}_p(1)) = \mathcal{D}^{(1)}(E_\infty)^{\#}$ and $\mathrm{char}_{\mathbb{B}}\, H^2_{Iw}(E_\infty, \mathbb{Z}_p(m)) = \mathcal{D}^{(1)}(E_\infty)^{\#}(m-1) = tw_{m-1}\mathcal{D}^{(1)}(E_\infty)$ for all odd $m \in \mathbb{Z}$.

Note that in the semi-simple case, Theorem 2 above contains thm. 5.3.2 of Nguyen Quang Do and Nicolas (2011).

4.5 The Fitting Ideal of $H^2_{Iw}(F_\infty, \mathbb{Z}_p(m))$

The next natural step would be to perform (co)descent on Theorem 2. Because $cd_p\, G_S(F) \leq 2$, the map $H^2_{Iw}(F_\infty, \mathbb{Z}_p(m))_{\Gamma^{p^n}} \to H^2_S(F_n, \mathbb{Z}_p(m))$ is an isomorphism, but the latter module needs no longer be perfect (or, equivalently, cohomologically trivial) over $\mathbb{Z}_p[G_n]$. Actually, in the exact sequence of Proposition 4(iv), our knowledge of the cohomology of the codescent module $X'(E_\infty)(m-1)_{\Gamma^{p^n}}$ is … less than perfect. A way to turn the difficulty would be to replace determinants by Fitting ideals, which are compatible with codescent. But for this we need an equivariant analogue of Cornacchia's lemma which was used in the proof of Proposition 3. To this end, we would like to add a technical condition which the reader would rightly find too brutal if it were imposed *ex abrupto* without any explanation. Hence the following preliminaries:

4.5.1

Let us recall briefly the setting of Cornacchia's lemma: for a noetherian Λ-module M, denote by M^0 its maximal finite submodule and write $\tilde{M} = M/M^0$. Since the global projective dimension of Λ is equal to 2, $pd_\Lambda \tilde{M} \leq 1$ because $\tilde{M}^0 = (0)$ (Auslander-Buchsbaum), hence $\mathrm{Fit}_\Lambda(M) = \mathrm{Fit}_\Lambda(M^0).\mathrm{Fit}_\Lambda(\tilde{M})$. If moreover M is Λ-torsion, $\mathrm{Fit}_\Lambda(\tilde{M}) = \mathrm{char}_\Lambda(\tilde{M}) = \mathrm{char}_\Lambda(M)$. The difficulty when passing from Λ to $\mathbb{A}$ is that the Auslander-Buchbaum result is no longer available. We shall use instead a weak substitute due to Greither. Recall that $\mathbb{A} = \Lambda[H]$, where $H = \mathrm{Gal}(F_\infty/k_\infty)$.

Lemma 6 (Greither 2000, propos. 2.4). *If an $\mathbb{A}$-noetherian torsion module N is cohomologically trivial over H and has no non-trivial finite submodule, then $pd_{\mathbb{A}}(N) \leq 1$.*

In the sequel, we shall work over $E_\infty = F(\zeta_{p^\infty})$, fix an odd $m \in \mathbb{Z}$, $m \neq 1$, and consider the module $M := H^2_{Iw}(E_\infty, \mathbb{Z}_p(m))$. According to Proposition 4 and after taking $\varprojlim$ with respect to corestriction maps, we have an exact sequence:

$$0 \to X'(E_\infty)(m-1) \to M \to W_m(E_\infty) \to 0\,,$$

where $W_m(E_\infty)$ is defined tautologically and will be given an explicit description below. This shows in particular that $M^0 = X'(E_\infty)^0(m-1)$. We want to get hold of $\tilde{M} = M/M^0$. Applying the snake lemma to the commutative diagram:

$$\begin{array}{ccccccccc}
0 & \longrightarrow & X'(E_\infty)(m-1)^0 & \longrightarrow & M & \longrightarrow & \tilde{M} & \longrightarrow & 0 \\
 & & \downarrow & & \| & & \downarrow & & \\
0 & \longrightarrow & X'(E_\infty)(m-1) & \longrightarrow & M & \longrightarrow & W_m(E_\infty) & \longrightarrow & 0\,,
\end{array}$$

we get an exact sequence $0 \to \widetilde{X'(E_\infty)}(m-1) \to \tilde{M} \to W_m(E_\infty) \to 0$. We intend to apply Greither's lemma to the rightmost module $W_m(E_\infty)$, which is the inverse limit of the kernels $W_m(E_n) = Ker\Big(\underset{v\in S_f}{\oplus} H^2(E_{n,v}, \mathbb{Z}_p(m)) \simeq \underset{v\in S_f}{\oplus} H^0(E_{n,v}, \mathbb{Q}_p/\mathbb{Z}_p(1-m))^* \to H^0(E_n, \mathbb{Q}_p/\mathbb{Z}_p(1-m))^*\Big)$. The limit of the semi-local modules is $\underset{v\in S_f}{\oplus} H^2_{Iw}(E_{\infty,v}, \mathbb{Z}_p(m))$, which is perfect (for the same reason as in Lemma 2), hence H-cohomologically trivial. As for the limit of the global $H^0(.)^*$'s, note that the transition maps reduce to p^a-power maps between twisted roots of unity; since $H^i(H, \underleftarrow{\lim}) = \underleftarrow{\lim}\, H^i(H, .)$ and H is finite, we readily get the H-cohomological triviality of the limit.

Summarizing, Lemma 6 applies and $W_m(E_\infty)$ has projective dimension ≤ 1. Going down to F_∞, we get an exact sequence $0 \to e_{m-1}\tilde{X}'(E_\infty) \to H^2_{Iw}(F_\infty, \mathbb{Z}_p(m)) \to W_m(F_\infty) \to 0$, with $pd_{\mathbb{A}} W_m(F_\infty) \leq 1$. Since m is odd, the leftmost module lives in the "plus" part. Let $E_\infty^{(m-1)}$ be the subfield of E_∞ cut out by the character ω^{m-1} (notations of Theorem 2).

Assume Greenberg's conjecture for $E_\infty^{(m-1)}/E^{(m-1)}$, which is equivalent to the vanishing of $e_{m-1}\tilde{X}'(E_\infty)$ and yields an exact sequence:

$$0 \to e_{m-1}X'(E_\infty)^0 \to H^2_{Iw}(F_\infty, \mathbb{Z}_p(m)) \to W_m(F_\infty) = \widetilde{H^2_{Iw}(F_\infty, \mathbb{Z}_p(m))} \to 0 \tag{3}$$

Since $pd_{\mathbb{A}}\, W_m(F_\infty) \leq 1$, the Fitting ideal behaves multiplicatively and we have: $\mathrm{Fit}_{\mathbb{A}}\, H^2_{Iw}(F_\infty, \mathbb{Z}_p(m)) = \mathrm{Fit}_{\mathbb{A}}\big(e_{m-1}X'(E_\infty)^0\big).\mathrm{Fit}_{\mathbb{A}}\, W_m(F_\infty)$. Moreover, $\mathrm{Fit}_{\mathbb{A}}\, W_m(F_\infty) = \mathrm{char}_{\mathbb{A}}\, W_m(F_\infty) = \mathrm{char}_{\mathbb{A}}\, \widetilde{H^2_{Iw}(F_\infty, \mathbb{Z}_p(m))} = \mathrm{char}_{\mathbb{A}}\, H^2_{Iw}(F_\infty, \mathbb{Z}_p(m))$ (the last equality is obtained by the already used localization argument). Finally:

Theorem 3. *For any odd $m \in \mathbb{Z}$, $m \neq 1$, assume Greenberg's conjecture for $E_\infty^{(m-1)}/E^{(m-1)}$. Then:*

1. At infinite level,

$$\mathrm{Fit}_{\mathbb{A}}\, H^2_{Iw}(F_\infty, \mathbb{Z}_p(m)) = \mathrm{Fit}_{\mathbb{A}}(e_{m-1}\, X'(E_\infty)^0).\,(e_{m-1}\, \mathfrak{D}^{(1)}(E_\infty))^{\#}\,.$$

2. At any finite level $n \geq 0$,

$$\mathrm{Fit}_{\mathbb{Z}_p[G_n]}\, H^2_S(F_n, \mathbb{Z}_p(m)) = \mathrm{Fit}_{\mathbb{Z}_p[G_n]}\, (e_{m-1}\, X'(E_\infty)^0)_{\Gamma^{p^n}}.\,(e_{m-1}\, \mathfrak{D}_n(E))^{\#}\,.$$

Remark. The perfectness of the two last terms in the exact sequence (3) implies that of the first term, hence the existence of $\mathrm{char}_{\mathbb{A}}(e_{m-1}\, X'(E_\infty)^0)$. But the usual localization argument shows that this ideal is (1), and we recover a particular case of Theorem 2(ii).

4.5.2

Recall that the module $\mathcal{D}_n(E)$ was defined in Sect. 2.2 and its quotient mod p^n was described explicitly in kummerian terms. The interest of Theorem 3(ii) lies in its comparison with already known results on refinements (for m odd) of the Coates-Sinnott conjecture on Galois annihilators of the modules $H^2_S(F_n, \mathbb{Z}_p(m)) \simeq K_{2m-2}(\mathcal{O}_{F_n}[1/S], \mathbb{Z}_p(m))$ (by Quillen-Lichtenbaum's conjecture, now a theorem – more precisely a consequence of the so called Milnor-Bloch-Kato conjecture, proved by Voevodsky and others; see e.g. Weibel (2009) and the references therein). Note that for m odd, the usual formulation of Coates-Sinnott gives no information other than "zero kills everybody". A refined conjecture was formulated by Snaith (2006) (resp. Nickel 2011) in terms of leading terms (rather than values) of Artin L-functions at negative integers for abelian (resp. general) Galois extensions of number fields, and shown to be a consequence of the equivariant Tamagawa number conjecture (ETNC) for the Tate motives attached to these extensions. Let us return to the situation of the introduction, where F/k is an abelian extension with group G, k is totally real and F is CM. We need to recall quickly the construction of a "canonical fractional ideal" by Snaith and Nickel. We follow the presentation of Nickel (2011), adapting it to our situation:

- For the algebraic part, fix $m \geq 2$ and let $\mathcal{H}_{1-m}(F) = \bigoplus_{S_\infty} (2\pi i)^{m-1}\mathbb{Z}$, with action of complex conjugation (diagonally on S_∞ and on $(2\pi i)^{m-1}$). The Borel regulator $\rho_{1-m} : K_{2m-1}(\mathcal{O}_F) \to \mathcal{H}^+_{1-m}(F) \otimes \mathbb{Q}$ yields the existence of a $\mathbb{Q}[G]$-isomorphism $\phi_{1-m} : \mathcal{H}_{1-m}(F)^+ \otimes \mathbb{Q} \xrightarrow{\sim} K_{2m-1}(\mathcal{O}_F) \otimes \mathbb{Q}$. This allows, by applying the Quillen-Lichtenbaum conjecture (now a theorem), to construct a G-equivariant embedding (we fix m and drop the index) $\phi : \mathcal{H}_{1-m}(F)^+ \otimes \mathbb{Z}_p \hookrightarrow H^1_S(F, \mathbb{Z}_p(m))$ (note that this $H^1_S(.)$ does not depend on $S \supset S_p$, by the localization exact sequence in étale cohomology, see Soulé 1979, III 3). The algebraic part of the canonical ideal will be $\mathrm{Fit}_{\mathbb{Z}_p[G]}((coker\,\phi)^*)$.
- For the analytic part, consider the ring $R(G)$ of virtual characters of G with values in $\overline{\mathbb{Q}}$ and let L be a finite Galois extension of $\mathbb{Q}$ such that each representation of G can be realized over L; let also V_χ be an $L[G]$-module with character χ. We can form regulator maps:

$$R_{\phi_{1-m}} : R(G) \longrightarrow \mathbb{C}^*$$

$$\chi \mapsto det\Big(\rho_{1-m}.\phi_{1-m} \mid \mathrm{Hom}_G(V_{\check{\chi}}, \mathcal{H}^+_{1-m}(F) \otimes \mathbb{C}\Big)$$

(where $\check{\chi}$ denotes the contragredient character).

Define then a function $\mathcal{A}^S_{\phi_{1-m}} : R(G) \to \mathbb{C}^*$, $\chi \mapsto R_{\phi_{1-m}}(\chi)/L^*_S(1-m,\chi)$, where $L_S(s,\chi)$ is the S-truncated Artin L-function and $L^*_S(1-m,\chi)$ is the leading term of this function at $(1-m)$. Gross' higher analogue of Stark's conjecture states that $\mathcal{A}^S_{\phi_{1-m}}(\chi^\sigma) = \mathcal{A}^S_{\phi_{1-m}}(\chi)^\sigma$ for all $\sigma \in Aut\,(\mathbb{C})$. Assuming this conjecture and choosing an identification $\mathbb{C} \simeq \mathbb{C}_p$, we can now define the "p-adic canonical ideal" (we drop the index in ϕ_{1-m}).

Definition 3. $\mathfrak{K}^S_{1-m}(p) = (\mathrm{Fit}_{\mathbb{Z}_p[G]}(coker\,\phi)^*.\,(\mathcal{A}^S_\phi)^{-1}))^\#$.

NB: this is actually an "intermediate" ideal in the construction of Snaith-Nickel, but it is all we need.

Theorem 4 (Snaith 2006 and Nickel 2011). *Suppose that Gross' conjecture, and also the ETNC for the pair* $(\mathbb{Q}(1-m)_F, \mathbb{Z}[\frac{1}{2}][G])$, $m \geq 2$, *hold for the abelian extension* F/k, *where* k *is totally real and* F *is* CM. *Then* $\mathrm{Fit}_{\mathbb{Z}_p[G]} H^2_S(F, \mathbb{Z}_p(m)) = \mathfrak{K}^S_{1-m}(p)$.

Proof. See Nickel (2011), end of the proof of thm 4.1, as well as remark, p. 14.

Remarks and forecasting:

1. For even $m \geq 2$, further calculations show that Theorem 4 contains the p-part of the usual Coates-Sinnott conjecture.
2. If $k = \mathbb{Q}$, Gross' conjecture and the ETNC hold true, and Theorem 4 becomes unconditional. Its comparison with Theorem 3 could give an analytic meaning to the parasite modules $(e_{m-1}X'(E_\infty)^0)_{\Gamma^{p^n}}$. "Numerical" information could also be obtained by computing the orders of the groups $K_{2m-2}(\mathcal{O}_{F_n}, \mathbb{Z}_p(m))$. For example, in the semi-simple case, where Γ_n intervenes in place of G_n, this order was computed by Martin (2011) in terms of values of L_p-functions at *positive* integers.
3. Instead of leading terms of Artin L-functions, one could also appeal to derivatives as in Burns et al. (2012). A natural (but resting only on thin air) query would be: is there any conceptual link with Theorem 2, knowing that $\mathcal{D}_n(E)$ can be interpreted as a module of "p-adic Gauss sums" (Nguyen Quang Do and Nicolas 2011, § 4.1)?
4. A natural expectation would be the extension of Theorem 2 to the relative abelian case ($k \neq \mathbb{Q}$). But then a serious obstacle is the absence of special elements (at least non conjecturally). Partial progress has been made by Nicolas (2013), starting from the idea of replacing special elements by L_p-functions over k; this is a natural idea since, when $k = \mathbb{Q}$, special elements and L_p-functions are "equivalent" by Coleman's theory.
5. Finally, one would of course wish to deal with the non abelian case, in view of the non commutative EMC recently proved by Kakde (2013) and Ritter and Weiss (2011). But a non commutative analogue of the "limit theorem" is missing.

Acknowledgements It is a pleasure for the author to thank Dr. Malte Witte for many useful discussions on his "limit theorem", as well as the referee for his careful reading of the manuscript.

References

Beliaeva, T., Belliard, J.-R.: Indices isotypiques des éléments cyclotomiques. Tokyo J. Math. **35**(1), 139–164 (2012)

Burns, D., Greither, C.: On the equivariant Tamagawa number conjecture for Tate motives. Invent. Math. **153**(2), 303–359 (2003a)

Burns, D., Greither, C.: Equivariant Weierstrass preparation and values of L-functions at negative integers. Doc. Math. **Extra Vol.**, 157–185 (2003b). (electronic) Kazuya Kato's Fiftieth Birthday

Burns, D., de Jeu, R., Gangl, H.: On special elements in higher algebraic K-theory and the Lichtenbaum-Gross conjecture. Adv. Math. **230**(3), 1502–1529 (2012)

Flach, M.: On the cyclotomic main conjecture for the prime 2. J. Reine Angew. Math. **661**, 1–36 (2011)

Fukaya, T., Kato, K.: A formulation of conjectures on p-adic zeta functions in noncommutative Iwasawa theory. In: Proceedings of the St. Petersburg Mathematical Society, Vol. XII. American Mathematical Society, Series 2, vol. 219, pp. 1–85. American mathematical society, Providence (2006)

Fukaya, T., Kato, K.: On conjectures of Sharifi. In: Iwasawa 2012 Conference, Heidelberg, 30 July–3 Aug 2012

Greither, C.: Some cases of Brumer's conjecture for abelian CM extensions of totally real fields. Math. Z. **233**(3), 515–534 (2000)

Greither, C., Popescu, C: An equivariant main conjecture in Iwasawa theory and applications. J. Algebr. Geom. (2013, to appear). arXiv 1103.3069v1

Huber, A., Kings, G.: Bloch-Kato conjecture and main conjecture of Iwasawa theory for Dirichlet characters. Duke Math. J. **119**(3), 393–464 (2003)

Ihara, Y., Kaneko, M., Yukinari, A.: On some properties of the universal power series for Jacobi sums. In: Galois Representations and Arithmetic Algebraic Geometry (Kyoto, 1985/Tokyo, 1986). Advanced Studies in Pure Mathematics, vol. 12, pp. 65–86. North-Holland, Amsterdam (1987)

Kakde, M.: The main conjecture of Iwasawa theory for totally real fields. Invent. Math. (2013, to appear). arXiv 1008.0142v3

Kolster, M., Nguyen Quang Do, T., Fleckinger, V.: Twisted S-units, p-adic class number formulas, and the Lichtenbaum conjectures. Duke Math. J. **84**(3), 679–717 (1996)

Kraft, J.S., Schoof, R.: Computing Iwasawa modules of real quadratic number fields. Compos. Math. **97**(1–2), 135–155 (1995). Special issue in honour of Frans Oort

Martin, A.: Théorie d'Iwasawa des noyaux sauvages étales et sommes de Gauß, thèse de l'Université de Franche-Comté (2011) (Ph.D. thesis, Université de Franche-Comté 2011)

Mazur, B., Wiles, A.: Class fields of abelian extensions of **Q**. Invent. Math. **76**(2), 179–330 (1984)

Nekovář, J.: Selmer Complexes. Astérisque, vol. 310, pp. viii+559. Société mathématique de France, Paris (2006)

Neukirch, J., Schmidt, A, Wingberg, K.: Cohomology of Number Fields. Grundlehren der Mathematischen Wissenschaften (Fundamental Principles of Mathematical Sciences), vol. 323, 2nd edn. Springer, Berlin (2008)

Nickel, A.: Leading terms of Artin L-series at negative integers and annihilation of higher K-groups. Math. Proc. Camb. Philos. Soc. **151**(1), 1–22 (2011)

Nickel, A.: Equivariant Iwasawa theory and non-abelian Stark-type conjectures. Proc. Lond. Math. Soc. **106**(6), 1223–1247 (2013)

Nicolas, V.: Conjecture équivariante des nombres de Tamagawa pour les motifs de Tate dans le cas abélien relatif, thèse de l'Université de Franche-Comté (2013) (Ph.D. thesis, Université de Franche-Comté 2013)

Nguyen Quang Do, T.: Conjecture Principale Équivariante, idéaux de Fitting et annulateurs en théorie d'Iwasawa. J. Théor. Nombres Bordeaux **17**(2), 643–668 (2005)

Nguyen Quang Do, T., Lescop, M.: Iwasawa descent and co-descent for units modulo circular units. Pure Appl. Math. Q. **2**(2, part 2), 465–496 (2006). With an appendix by J.-R. Belliard

Nguyen Quang Do, T., Nicolas, V.: Nombres de Weil, sommes de Gauss et annulateurs galoisiens. Am. J. Math. **133**(6), 1533–1571 (2011)

Perrin-Riou, B.: Théorie d'Iwasawa des représentations p-adiques sur un corps local. Invent. Math. **115**(1), 81–161 (1994). With an appendix by Jean-Marc Fontaine

Ritter, J., Weiss, A.: Toward equivariant Iwasawa theory. Manuscripta Math. **109**(2), 131–146 (2002)

Ritter, J., Weiss, A.: On the "main conjecture" of equivariant Iwasawa theory. J. Am. Math. Soc. **24**(4), 1015–1050 (2011)

Schneider, P.: Über gewisse Galoiscohomologiegruppen. Math. Z. **168**, 181–205 (1979)

Sharifi, R.: A reciprocity map and the two-variable p-adic L-function. Ann. Math. (2) **173**(1), 251–300 (2011)

Snaith, V.P.: Stark's conjecture and new Stickelberger phenomena. Can. J. Math. **58**(2), 419–448 (2006)

Solomon, D.: Some new maps and ideals in classical Iwasawa theory with applications (2010). arXiv 0905.4336v3

Soulé, v: K-théorie des anneaux d'entiers de corps de nombres et cohomologie étale. Invent. Math. **55**(3), 251–295 (1979)

Weibel, C.: The norm residue isomorphism theorem. J. Topol. **2**(2), 346–372 (2009)

Wiles, A.: The Iwasawa conjecture for totally real fields. Ann. Math. (2) **131**(3), 493–540 (1990)

Witte, M.: On the equivariant main conjecture of Iwasawa theory. Acta Arith. **122**(3), 275–296 (2006)

Nearly Overconvergent Modular Forms

Eric Urban

1 Introduction

The purpose of this paper is to define and give the basic properties of nearly overconvergent (elliptic) modular forms. Nearly holomorphic forms were introduced by G. Shimura in the 1970s for proving algebraicity results for special values of L-functions (Shimura 1976). He defined the notion of algebraicity of those by evaluating them at CM-points. After introducing a sheaf-theoretic definition, it is also possible to give an algebraic and even integral structure on the space of nearly holomorphic forms, allowing to study congruences between them. This naturally leads to the notion of nearly overconvergent forms. For that matter, we can think that nearly overconvergent forms are to overconvergent forms what nearly holomorphic forms are to classical holomorphic modular forms. The notion of nearly overconvergent forms came to the author when working on his joint project with C. Skinner (see Urban (2013) for an account of this work in preparation (Skinner and Urban in preparation)) and appears as a natural way to study certain p-adic families of nearly holomorphic forms and its application to Bloch-Kato type conjectures. In the aforementioned work where the case of unitary groups[1] is considered, the notion is not absolutely necessary but it is clearly in the background of our construction and keeping it in mind makes the strategy more transparent.

An important feature of nearly overconvergent forms is that its space is equipped with an action of the Atkin operator U_p and that this action is completely continuous. This allows to have a spectral decomposition and to study p-adic families of nearly overconvergent forms. Another remarkable fact which is not really surprising

[1]It will be clear to the reader that it could be generalized to any Shimura variety of PEL type.

E. Urban (✉)
Department of Mathematics, Columbia University, 2990 Broadway, New York, NY 10027, USA
e-mail: urban@math.columbia.edu; urban@math.jussieu.fr

T. Bouganis and O. Venjakob (eds.), *Iwasawa Theory 2012*, Contributions in Mathematical and Computational Sciences 7, DOI 10.1007/978-3-642-55245-8_14,

but useful is that this space embeds naturally in the space of p-adic forms. In particular, this allows us to define the p-adic q-expansion of these forms. All the tools and differential operators that are used in the classical theory are also available here thanks to the sheaf theoretic definition and the Gauss-Manin connection. In particular, we can define the Maass-Shimura differential operator for families and the overconvergent projection which is a generalization in this context of the holomorphic projection of Shimura. Our theory is easily generalisable to general Shimura varieties of PEL type. To make this notion more appealing, I decided to include an illustration (which is not considered in Skinner and Urban in preparation) of its potential application to the construction of p-adic L-functions in the non-ordinary case. In the works of Hida (1985, 1988) on the construction of 3-variable Rankin-Selberg p-adic L-functions attached to ordinary families, the fact that the ordinary idempotent is the p-adic equivalent notion of the holomorphic projector makes it play a crucial role in the construction of Hida's p-adic measure. Here the spectral theory of the U_p-operator on the space of nearly overconvergent forms and the overconvergent projection play that important role.

We now review quickly the content of the different sections. In the Sect. 2, we recall the notion of nearly holomorphic forms and give its sheaf-theoretic definition. This allows to give an algebraic and integral version for nearly holomorphic forms and define their polynomial q-expansions as well as an arithmetic version of the classical differential operators of this theory. We also check that Shimura's rationality of nearly holomorphic forms is equivalent to ours. In Sect. 3, we introduce the space of nearly overconvergent forms and we prove it embeds in the space of p-adic forms. We also study the spectral theory of U_p on them and give a q-expansion principle. Then we define the differential operators in families and the overconvergent projection. In the last section we apply the tools introduced before to make the construction of the Rankin-Selberg p-adic L-functions on the product of two eigencurve of tame level 1. When restricted to the ordinary locus, this p-adic L-function is nothing else but the 3-variable p-adic L-function of Hida.

After the basic material of this work was obtained, I learned from M. Harris that he had also given a sheaf theoretic definition[2] using the theory of jets which is valid for general Shimura varieties of Shimura's nearly holomorphic forms in Harris (1985, 1986) and the fact his definition is equivalent to Shimura's has been verified by his former student Mark Nappari in his Thesis (Nappari 1992). It should be easy to see that our description is equivalent to his in the PEL case. However, Harris did not study nor introduce the nearly overconvergent version. I would like also to mention that some authors have introduced an ad hoc definition of nearly overconvergent forms as polynomials in E_2 with overconvergent forms as coefficients. However this definition cannot be generalized to other groups and is not convenient for the spectral theory of the U_p-operator (or the slope decomposition). Finally we recently learned from V. Rotger that H. Darmon and V. Rotger have independently introduced a definition similar to ours in Darmon and Rotger (to

[2]His definition follows a suggestion of P. Deligne.

appear) using the work Coleman et al. (1995). The reader will see that our work is independent of loc. cit. and recover the result of Coleman et al. (1995) has a by-product and can therefore be generalized to any Shimura variety of PEL type.

This text grew out from the handwritten lecture notes of a graduate course the author gave at Columbia University during the Spring 2012. After, the work Skinner and Urban (in preparation) was presented at various conferences[3] including the Iwasawa 2012 conference held in Heidelberg, several colleagues suggested him to write up an account in the $GL(2)$-case of this notion of nearly overconvergent forms. This text and in particular the application to the p-adic Rankin-Selberg L-function would not have existed without their suggestions.

Notations. Throughout this paper p is a fixed prime. Let $\overline{\mathbf{Q}}$ and $\overline{\mathbf{Q}}_p$ be, respectively, algebraic closures of $\mathbf{Q}$ and $\mathbf{Q}_p$ and let $\mathbf{C}$ be the field of complex numbers. We fix embeddings $\iota_\infty : \overline{\mathbf{Q}} \hookrightarrow \mathbf{C}$ and $\iota_p : \overline{\mathbf{Q}} \hookrightarrow \overline{\mathbf{Q}}_p$. Throughout we implicitly view $\overline{\mathbf{Q}}$ as a subfield of $\mathbf{C}$ and $\overline{\mathbf{Q}}_p$ via the embeddings ι_∞ and ι_p. We fix an identification $\overline{\mathbf{Q}}_p \cong \mathbf{C}$ compatible with the embeddings ι_p and ι_∞. For any rigid analytic space $\mathfrak{X}$ over a p-adic number field, we denote respectively by $A(\mathfrak{X}), A^b(\mathfrak{X}), A^0(\mathfrak{X})$ and $F(\mathfrak{X})$ the rings of analytic function, of bounded analytic function, of analytic functions bounded by 1 and of meromorphic functions on $\mathfrak{X}$.

2 Nearly Holomorphic Modular Forms

2.1 Classical Definition

In this paragraph, for the purpose to set notations we recall[4] some classical definitions and operations on modular forms.

2.1.1 Definition

We recall that the subgroup of 2×2 matrices of positive determinant $GL_2(\mathbf{R})^+$ acts on on the Poincaré upper half plane

$$\mathfrak{h} := \{\tau = x + iy \in \mathbf{C} \mid y > 0\}$$

[3]The first half of this note was also presented in my lecture given at H. Hida's 60th birthday conference.

[4]Those facts are mainly due to Shimura

by the usual formula

$$\gamma.\tau = \frac{a\tau + b}{c\tau + d} \text{ for } \gamma = \left(\begin{smallmatrix} a & b \\ c & d \end{smallmatrix}\right) \in GL_2(\mathbf{R})^+ \text{ and } \tau \in \mathfrak{h}$$

Let f be a complex valued function defined on $\mathfrak{h}$. For any integer $k \geq 0$ and $\gamma \in GL_2(\mathbf{R})^+$, we set

$$f|_k\gamma(\tau) := det(\gamma)^{k/2}(c\tau + d)^{-k} f(\gamma.\tau)$$

Let $r \geq 0$ be another integer. We say that f is a nearly holomorphic modular form of weight k and order $\leq r$ for an arithmetic group $\Gamma \subset SL_2(\mathbf{R})$ if f satisfies the following properties:

(a) f is C^∞ on $\mathfrak{h}$,
(b) $f|_k\gamma = f$ for all $\gamma \in \Gamma$,
(c) There are holomorphic functions $f_0, \ldots, f_r$ on $\mathfrak{h}$ such that

$$f(\tau) = f_0(\tau) + \frac{1}{y} f_1(\tau) + \cdots + \frac{1}{y^r} f_r(\tau)$$

(d) f has a finite limit at the cusps.

If f is a C^∞ function on $\mathfrak{h}$, we set following Shimura's $\epsilon.f$ the function of $\mathfrak{h}$ defined by

$$(\epsilon.f)(\tau) := 8\pi i y^2 \frac{\partial f}{\partial \bar{\tau}}(\tau) \tag{1}$$

It is easy to check that the condition (c) can be replaced by the condition

(c') $\epsilon^{r+1}.f = 0$

Moreover if f is nearly holomorphic of weight k and order $\leq r$, then $\epsilon.f$ is nearly holomorphic of weight $k-2$ and order $\leq r-1$. We denote by $\mathcal{N}_k^r(\Gamma, \mathbf{C})$ the space of nearly holomorphic form of weight k, order $\leq r$ and level Γ. For $r = 0$, this is the space of holomorphic modular form of weight k and level Γ and we will sometimes use the standard notation $\mathcal{M}_k(\Gamma, \mathbf{C})$ instead of $\mathcal{N}_k^0(\Gamma, \mathbf{C})$.

2.1.2 The Maass-Shimura Differential Operator

We recall the differential operator δ_k on the space of nearly holomorphic forms of weight k defined by

$$\delta_k.f := \frac{1}{2i\pi} y^{-k} \frac{\partial}{\partial \tau}(y^k f) = \frac{1}{2i\pi}(\frac{\partial f}{\partial \tau} + \frac{k}{2iy} f)$$

An easy computation shows that $\delta_k.f$ is of weight $k+2$ and its degree of near holomorphy is increased by one. For a positive integer s, let δ_k^s be the differential operator defined by

$$\delta_k^s := \delta_{k+2s-2} \circ \cdots \circ \delta_k$$

The following easy lemma is due to Shimura and is proved by induction on r. We will generalize it to nearly overconvergent forms in the next section.

Lemma 1. *Let $f \in \mathcal{N}_k^r(\Gamma, \mathbf{C})$. Assume that $k > 2r$. Then there exist $g_0, \ldots, g_r$ with $g_i \in \mathcal{M}_{k-2i}(\Gamma, \mathbf{C})$ such that*

$$f = g_0 + \delta_{k-2} g_1 + \cdots + \delta_{k-2r}^r g_r$$

It is easy to check, the forms g_i's are unique. When the hypothesis of this lemma are satisfied, Shimura defines the holomorphic projection $\mathcal{H}(f)$ of f as the holomorphic form defined by

$$\mathcal{H}(f) := g_0$$

Remark 1. The conclusion of this lemma is wrong if the assumption $k > 2r$ is not satisfied. The most important example is given by the Eisenstein series E_2 of weight 2 and level 1.

2.2 *Sheaf Theoretic Definition*

2.2.1 Sheaves of Differential Forms

Let $Y = Y_\Gamma := \Gamma \backslash \mathcal{H}$ and $X = X_\Gamma := \Gamma \backslash (\mathfrak{h} \sqcup \mathbf{P}^1(\mathbf{Q}))$ be respectively the open modular curve and complete modular curve of level Γ. Let $\mathcal{E}$ be the universal elliptic curve over Y_Γ and let $\mathbf{p} : \bar{\mathcal{E}} \to X_\Gamma$ be its Kuga-Sato compactification over X_Γ. We consider the sheaf of invariant relative differential forms with logarithmic poles along $\partial\bar{\mathcal{E}} = \bar{\mathcal{E}} \backslash \mathcal{E}$ which is a normal crossing divisor of $\bar{\mathcal{E}}$.

$$\omega := \mathbf{p}_* \Omega^1_{\bar{\mathcal{E}}/X}(log(\partial\bar{\mathcal{E}}))$$

It is a locally free sheaf of rank one in the holomorphic topos of X. We also consider

$$\mathcal{H}^1_{dR} := R^1\mathbf{p}_* \Omega^\bullet_{\bar{\mathcal{E}}/X}(log(\partial\bar{\mathcal{E}}))$$

the sheaf of relative degree one de Rham cohomology of $\bar{\mathcal{E}}$ over X with logarithmic poles along $\partial\bar{\mathcal{E}}$. The Hodge filtration induces the exact sequence

$$0 \to \omega \to \mathcal{H}^1_{dR} \to \omega^\vee \to 0 \tag{2}$$

and in the C^∞-topos, this exact sequence splits to give the Hodge decomposition:

$$\mathcal{H}^1_{dR} = \omega \oplus \overline{\omega}$$

2.2.2 Complex Uniformization

More concretely, let π be the projection $\mathfrak{h} \to Y_\Gamma$ and $\pi^*\mathcal{E}$ be the pull back of $\mathcal{E}$ by π. We have

$$\pi^*\mathcal{E} = (\mathbf{C} \times \mathfrak{h})/\mathbf{Z}^2$$

where the action of $\mathbf{Z}^2$ on $\mathbf{C} \times \mathfrak{h}$ is defines by $(z,\tau).(a,b) = (z + a + b\tau, \tau)$ for $(z,\tau) \in \mathbf{C} \times \mathfrak{h}$ and $(a,b) \in \mathbf{Z}^2$. The fiber $\mathcal{E}_\tau$ of $\pi^*\mathcal{E}$ at $\tau \in \mathfrak{h}$ can be identified with $\mathbf{C}/L_\tau$ with $L_\tau = \mathbf{Z} + \tau\mathbf{Z} \subset \mathbf{C}$. We have

$$\pi^*\omega = \mathcal{O}_{\mathfrak{h}} dz$$

with $\mathcal{O}_{\mathfrak{h}}$ the sheaf of holomorphic function on $\mathfrak{h}$. Note also that

$$\mathcal{E} = \Gamma\backslash\mathbf{C} \times \mathfrak{h}/\mathbf{Z}^2$$

with the action of Γ on $\mathbf{C} \times \mathfrak{h}/\mathbf{Z}^2$ given by $\gamma.(z,\tau) = ((c\tau + d)^{-1}z, \gamma.\tau)$. We therefore have

$$\gamma^* dz = (c\tau + d)^{-1} dz$$

From this relation and the condition at the cusps, it is easy and well known to see that

$$H^0(X_\Gamma, \omega^{\otimes^k}) \cong \mathcal{M}_k(\Gamma, \mathbf{C})$$

Let $\mathcal{C}^\infty_{\mathfrak{h}}$ the sheaf of C^∞ functions on $\mathfrak{h}$. Then the Hodge decomposition of $\pi^*\mathcal{H}^1_{dR}$ reads

$$\pi^*\mathcal{H}^1_{dR} \otimes \mathcal{C}^\infty_{\mathfrak{h}} = \mathcal{C}^\infty_{\mathfrak{h}} dz \oplus \mathcal{C}^\infty_{\mathfrak{h}} d\overline{z}$$

On the other hand, by the Riemann-Hilbert correspondence, we have

$$\pi^*\mathcal{H}^1_{dR} = \pi^* R^1\mathbf{p}_*\mathbf{Z} \otimes \mathcal{O}_{\mathfrak{h}} = Hom(R_1\mathbf{p}_*\mathbf{Z}, \mathcal{O}_{\mathfrak{h}}) = \mathcal{O}_{\mathfrak{h}}\alpha \oplus \mathcal{O}_{\mathfrak{h}}\beta$$

where α, β is the basis of horizontal sections inducing on $H_1(\mathcal{E}_\tau, \mathbf{Z}) = L_\tau$ the linear forms $\alpha(a + b\tau) = a$ and $\beta(a + b\tau) = b$ so that we have

$$dz = \alpha + \tau\beta \text{ and } d\overline{z} = \alpha + \overline{\tau}\beta$$

From the action of Γ on the differential form dz, it is then easy to see that

$$\gamma^* \cdot \begin{pmatrix} dz \\ \beta \end{pmatrix} = \begin{pmatrix} (c\tau + d)^{-1} & 0 \\ -c & (c\tau + d) \end{pmatrix} \begin{pmatrix} dz \\ \beta \end{pmatrix}$$

We define the holomorphic sheaf of X_Γ:

$$\mathcal{H}_k^r := \omega^{\otimes^{k-r}} \otimes Sym^r(\mathcal{H}_{dR}^1)$$

Then we have the following proposition.

Proposition 1. *The Hodge decomposition induces a canonical isomorphism*

$$H^0(X_\Gamma, \mathcal{H}_k^r) \cong \mathcal{N}_k^r(\Gamma, \mathbf{C})$$

Proof. Let $\eta \in H^0(X_\Gamma, \mathcal{H}_k^r)$. Then $\pi^*\eta(\tau) = \sum_{l=0}^r f_l(\tau) dz^{\otimes^{k-l}} \beta^{\otimes^l}$ where the f_l's are holomorphic functions on $\mathfrak{h}$. Since we have $\beta = \frac{1}{2iy}(dz - d\overline{z})$, we deduce:

$$\pi^*\eta(\tau) = \sum_{l=0}^r \frac{f_l(\tau)}{(2iy)^l} \sum_{i=0}^l (-1)^i \binom{l}{i} dz^{\otimes^{k-i}} d\overline{z}^{\otimes^i}$$

The projection of $\pi^*\eta$ on the $(k,0)$-component of $H^0(\mathfrak{h}, \pi^*\mathcal{H}_k^r)$ is therefore given by $f(\tau) dz^{\otimes^k}$ with

$$f(\tau) = \sum_{l=0}^r \frac{f_l(\tau)}{(2iy)^l}$$

It is clearly a nearly holomorphic form. It is useful to remark that the projection on the $(k,0)$-component is injective.[5] Conversely, if $f(\tau) = \sum_{l=0}^r \frac{f_l(\tau)}{y^l}$ is a nearly holomorphic form of weight k and order $\leq r$, using the injectivity of the projection onto the $(k,0)$-component, it is straightforward to see that $\sum_{l=0}^r (2i)^l f_l(\tau) dz^{\otimes^{k-l}} \beta^{\otimes^l}$ is invariant by Γ and defines an element of $H^0(X_\Gamma, \mathcal{H}_k^r)$ projecting onto $f(\tau) dz^{\otimes^k}$ via the Hodge decomposition. □

The quotient by the first step of the de Rham filtration of $\mathcal{H}_k^r$ induces by Poincaré duality the following canonical exact sequence

$$0 \to \omega^{\otimes^k} \to \mathcal{H}_k^r \to \mathcal{H}_{k-2}^{r-1} \to 0 \tag{3}$$

The map $\mathcal{H}_k^r \to \mathcal{H}_{k-2}^{r-1}$ induces the morphism ϵ of (1).

[5] We will see a similar fact in the p-adic case. See Proposition 6.

2.3 Rational and Integral Structures

2.3.1 Rational and Integral Nearly Holomorphic Forms

Let N be a positive integer and let us assume $\Gamma = \Gamma_1(N)$ with $N \geq 3$. Then $X_\Gamma = X_1(N)$ is defined over $\mathbf{Z}[\frac{1}{N}]$ as well as ω, $\mathcal{H}^1_{dR}$ and $\mathcal{H}^r_k$. Recall that $Y_\Gamma = Y_1(N)$ classifies the isomorphism classes of pairs $(E, \alpha_N)_{/S}$ where $E_{/S}$ is an elliptic scheme over an $\mathbf{Z}[\frac{1}{N}]$-schemes S and α_N is a $\Gamma_1(N)$-level structure for E (i.e. an injection of group scheme: $\mu_{N/S} \hookrightarrow E[N]_{/S}$). Moreover the generalized universal elliptic curve is defined over $X_1(N)$ and we can define the sheaves ω, $\mathcal{H}^1_{dR}$ and $\mathcal{H}^r_k$ over $X_1(N)_{/\mathbf{Z}[\frac{1}{N}]}$ as in the previous paragraph. The exact sequence (3) is also defined over $\mathbf{Q}$. For any $\mathbf{Z}[\frac{1}{N}]$-algebra A, we define $\mathcal{M}_k(N, A)$ and $\mathcal{N}^r_k(N, A)$ respectively as the global section of $\omega^{\otimes k}_{/A}$ and $\mathcal{H}^r_{k/A}$. This gives integral and rational definitions of the space of nearly holomorphic modular forms.

2.3.2 Nearly Holomorphic Forms as Functors

It follows from the exact sequence (2) that $\mathcal{H}^1_{dR}$ is locally free over $Y_1(N)_{\mathbf{Z}[1/N]}$. A similar result would hold for general Shimura varieties of PEL type. In that case the torsion-freeness result from the basic properties of relative de Rham cohomology (for example see Katz 1970). We may therefore consider $\mathcal{T}$ the B-torsor over $(Y_1(N))_{Zar}$ of isomorphisms $\psi_U : \mathcal{H}^1_{dR/U} \cong \mathcal{O}_U \oplus \mathcal{O}_U$ inducing an isomorphism $\psi^1_U : \omega_{/U} \cong \mathcal{O}_U \oplus \{0\}$ such that the isomorphism $(\mathcal{H}^1_{dR}/\omega)_{/U} \cong \mathcal{O}_U$ induced by ψ_U is dual to ψ^1_U for all Zariski open subset $U \subset Y_\Gamma$.

For any ring R, we denote by $R[X]_r$ the R-module of polynomial in X of degree $\leq r$. Let B the Borel subgroup of SL_2 of upper triangular matrices. Then we consider the representation ρ^r_k of $B(R)$ on $R[X]_r$ defined by

$$\rho^r_k((\begin{smallmatrix} a & b \\ 0 & a^{-1} \end{smallmatrix})).P(X) = a^k P(a^{-2}X + ba^{-1})$$

Then over $Y_1(N)$, for any $\mathbf{Z}[\frac{1}{N}]$-algebra, we have

$$\mathcal{T} \times^B A[X]_r(k) \cong \mathcal{H}^r_{k/A}$$

This can be checked easily for principal rings and this implies the isomorphism above since the formation of both left and right hand sides commute to base change.

Concretely, we can see it the following way. For an elliptic scheme $E_{/R}$ we consider a basis (ω, ω') of $\mathcal{H}^1_{dR}(E/R)$ such that ω is a basis of $\omega_{E/R} = H^0(E, \Omega^1_{E/R})$ and $\langle \omega, \omega' \rangle_{dR} = 1$. Then $f \in \mathcal{N}^r_k(N, A)$ can be seen as a functorial rule assigning to any A-algebra R and a quadruplet $(E, \alpha_N, \omega, \omega')_{/R}$ a polynomial

$$f(E, \alpha_N, \omega, \omega')(X) = \sum_{l=0}^{r} b_l X^l \in R[X]_r$$

defined such that the pull-back of f to $\mathcal{H}^1_{dR}(E/R)$ is $\sum_{l=0}^{r} b_l \omega^{\otimes^{k-l}} \otimes \omega'^{\otimes^l}$. For any $a \in R^\times, b \in R$, we have

$$f(E, \alpha_N, a\omega, a^{-1}\omega' + b\omega)(X) = a^{-k} f(E, \alpha_N, \omega, \omega')(a^2 X - ab)$$

The condition that f is finite at the cusps is expressed in terms of q-expansion as usual. It will be defined in the next paragraph.

Proposition 2. *Let $f \in \mathcal{N}_k^r(N, A)$ and $\epsilon.f \in \mathcal{N}_{k-2}^{r-1}(N, A)$ the image of f by the projection* (3). *Then for any quadruplet $(E, \alpha_N, \omega, \omega')_{/R}$, we have*

$$(\epsilon.f)(E, \alpha_N, \omega, \omega')(X) = \frac{d}{dX} f(E, \alpha_N, \omega, \omega')(X)$$

Proof. For any ring R, we have the exact sequence:

$$0 \to R[X]_0(k) \to R[X]_r(k) \to R[X]_{r-1}(k-2)$$

where the right hand side map is given by $P(X) \mapsto P'(X)$. It is clearly $B(R)$-equivariant. By taking the contracted product of this exact sequence with $\mathcal{T}$, we obtain the exact sequence of sheaves (3) which implies our claim. Notice that the map of sheaves inducing ϵ is surjective only if $r!$ is invertible in A. □

2.3.3 Polynomial q-Expansions

We consider *Tate*(q) the Tate curve over $\mathbf{Z}[\frac{1}{N}]((q))$ with its canonical invariant differential form ω_{can} and canonical $\Gamma_1(N)$ level structure $\alpha_{N,can}$. We have the Gauss-Manin connection:

$$\nabla : \mathcal{H}^1_{dR}(Tate(q)/\mathbf{Z}[\frac{1}{N}]((q))) \to \mathcal{H}^1_{dR}(Tate(q)/\mathbf{Z}[\frac{1}{N}]((q))) \otimes \Omega^1_{\mathbf{Z}[\frac{1}{N}]((q)))/\mathbf{Z}[\frac{1}{N}]}$$

and let

$$u_{can} := \nabla(q\frac{d}{dq})(\omega_{can})$$

Then (ω_{can}, u_{can}) form a basis of $\mathcal{H}^1_{dR}(Tate(q)/\mathbf{Z}[\frac{1}{N}]((q)))$ and u_{can} is horizontal; moreover $\langle \omega_{can}, u_{can} \rangle_{dR} = 1$ (see for instance the appendix 2 of Katz 1972). For any $\mathbf{Z}[\frac{1}{N}]$-algebra A and $f \in \mathcal{N}_k^r(N, A)$, we consider

$$f(q, X) := f(Tate(q)_{/A((q))}, \alpha_{can}, \omega_{can}, u_{can})(X) \in A[[q]][X]_r.$$

We call it the polynomial q-expansion of the nearly holomorphic form f.

Remark 2. We can think of the variable X as $\frac{-1}{4\pi y}$.

2.3.4 The Nearly Holomorphic Form E_2

It is well-known that the Eisenstein series of weight 2 and level 1 is a nearly holomorphic form of order 1. Its given by

$$E_2(\tau) = -\frac{1}{24} + \frac{1}{8\pi y} + \sum_{n=1}^{\infty} \sigma_n^1 q^n$$

where σ_n^1 is the sum of the positive divisor of n and $q = e^{2i\pi\tau}$.

We can define E_2 as a functorial rule. Let R be a ring with $\frac{1}{6} \in R$ and E be an elliptic curve over R. Recall (see for instance Katz 1972, Appendix 1) that any basis $\omega \in \omega_{E/R}$ defines a Weierstrass equation for E:

$$Y^2 = 4X^3 - g_2 X - g_3$$

such that $\omega = \frac{dX}{Y}$ and $\eta = X\frac{dX}{Y}$ form a R-basis of $H^1_{dR}(E/R)$. Moreover if we replace ω by $\lambda\omega$, η is replaced by $\lambda^{-1}\eta$. Therefore $\omega \otimes \eta \in \omega_{E/R} \otimes H^1_{dR}(E/R)$ is independent of the choice of ω. It therefore defines a section of $\mathcal{H}_2^1$ over $Y_{/\mathbf{Z}[\frac{1}{6}]}$. If (ω, ω') is a basis of $H^1_{dR}(E/R)$ such that $\langle \omega, \omega' \rangle_{dR} = 1$ then we can put

$$\tilde{E}_2(E, \omega, \omega') = \langle \eta, \omega' \rangle_{dR} + X \langle \omega, \eta \rangle_{dR}$$

where $\langle \cdot, \cdot \rangle_{dR}$ stands for the Poincaré pairing on $H^1_{dR}(E/R)$.

Its polynomial q-expansion is given by

$$\tilde{E}_2(q, X) = \tilde{E}_2(Tate(q), \omega_{can}, u_{can})(X) = \frac{P(q)}{12} + X$$

because $u_{can} = -\frac{P(q)}{12}\omega_{can} + \eta_{can}$ and $\langle \omega_{can}, \eta_{can} \rangle_{dR} = 1$ and where $P(q)$ is defined in Katz (1972) by

$$P(q) = 1 - 24 \sum_{n=1}^{\infty} \sigma_n^1 q^n$$

Since the q-expansion of $\tilde{E}_2$ is finite at $q = 0$, $\tilde{E}_2$ defines a section of $\mathcal{H}_2^1$ over X. From the q-expansion, it is easy to see that $\tilde{E}_2 = -2E_2$. We clearly have $\epsilon.\tilde{E}_2 = 1$ (i.e. is the constant modular form of weight 0 taking the value 1).

Remark 3. We can see easily that multiplying by $\tilde{E}_2$ is useful to get a splitting of $\mathcal{H}_k^1$:

$$0 \to \omega^{\otimes k} \to \mathcal{H}_k^1 \to \omega^{\otimes k-2} \to 0$$

In particular, that shows that nearly holomorphic forms are polynomial in E_2 with holomorphic forms as coefficients. As mentioned in the introduction, this provides a way to give an ad hoc definition of nearly overconvergent forms.

2.4 Differential Operators

Recall that the Gauss-Manin connection

$$\nabla : \mathcal{H}^1_{dR} \to \mathcal{H}^1_{dR} \otimes \Omega_{X_1(N)/\mathbf{Z}[\frac{1}{N}]}(log(Cusp))$$

induces the Kodaira-Spence isomorphism $\omega^{\otimes^2} \cong \Omega_{X_1(N)/\mathbf{Z}[\frac{1}{N}]}(log(Cusp))$ and a connection

$$\nabla : Sym^k(\mathcal{H}^1_{dR}) \to Sym^k(\mathcal{H}^1_{dR}) \otimes \Omega_{X_1(N)/\mathbf{Z}[\frac{1}{N}]}(log(Cusp)) \cong Sym^k(\mathcal{H}^1_{dR}) \otimes \omega^{\otimes^2}$$

The Hodge filtration of $Sym^k(\mathcal{H}^1_{dR})$ is given by $Fil^0 \supset \cdots \supset Fil^k \supset Fil^{k+1} = \{0\}$ with

$$Fil^{k-r} = \mathcal{H}^r_k = \omega^{\otimes^{k-r}} \otimes Sym^r(\mathcal{H}^1_{dR}) \text{ for } 0 \leq r \leq k$$

Since ∇ satisfies Griffith transversality, when $k \geq r \geq 0$, it sends Fil^{k-r} into $Fil^{k-r-1} \otimes \Omega_{X_1(N)/\mathbf{Z}[\frac{1}{N}]}(log(Cusp))$. We therefore get the sheaf theoretic version of the Maass-Shimura operator[6]:

$$\delta_k : \mathcal{H}^r_k \to \mathcal{H}^{r+1}_{k+2} \quad \text{for} \quad 0 \leq r \leq k$$

We still denote δ_k the corresponding operator

$$\delta_k : \mathcal{N}^r_k(N, A) \to \mathcal{N}^{r+1}_{k+2}(N, A)$$

The following proposition gives the effect of δ_k on the polynomial q-expansion.

Proposition 3. *Let $f \in \mathcal{N}^r_k(N, A)$. Then the polynomial q-expansion of $\delta_k f$ is given by:*

$$(\delta_k f)(q, X) = X^k D(X^{-k} f(q, X))$$

---

[6]We leave it as an exercise to check that this operator corresponds to the classical Maass-Shimura operator via the isomorphism of Proposition 1.

where D is the differential operator on $A[[q]][X]$ given by $D = q\frac{\partial}{\partial q} - X^2\frac{\partial}{\partial X}$. In other words, if $f(q,X) = \sum_{i=0}^{r} f_i(q)X^i$, we have

$$(\delta_k f)(q,X) = \sum_{i=0}^{r} q\frac{d}{dq} f_i(q)X^i + (k-i)f_i(q)X^{i+1}$$

Proof. The pull-back of f to $\phi^*\mathcal{H}_k^r$ with ϕ : Spec $(A((q))) \to X_1(N)_{/A}$ corresponds to $(Tate(q), \alpha_{N,can})_{/A((q))}$ is given by

$$\phi^* f = \sum_{i=0}^{r} f_i(q)\omega_{can}^{\otimes k-i} \otimes u_{can}^{\otimes i}$$

Since ∇ induces $\frac{dq}{q} \cong \omega_{can}^2$, $\nabla(q\frac{d}{dq})(\omega_{can}) = u_{can}$ and $\nabla(q\frac{d}{dq})(u_{can}) = 0$, we have:

$$\nabla(q\frac{d}{dq})(\phi^* f) = \sum_{i=0}^{r} \frac{d}{dq} f_i(q)\omega_{can}^{\otimes k-i+2} \otimes u_{can}^{\otimes i} + (k-i)f_i(q)\omega_{can}^{\otimes k-i-1} \otimes u_{can}^{\otimes i+1}$$

This implies our claim by the definition of the polynomial q-expansion. □

Remark 4. We can rewrite the formula for δ_k in the following way:

$$\delta_k = D + kX$$

Using the relation $[D, X] = -X^2$, we easily show by induction that

$$\delta_k^r = \sum_{j=0}^{r} \binom{r}{j} \frac{\Gamma(k+r)}{\Gamma(k+r-j)} . X^j D^{r-j} \tag{4}$$

Notice in particular that for $s \leq r$ and h holomorphic, we have

$$(\epsilon_{k+2r}^s \circ \delta_k^r)h = \sum_{j=s}^{r} \binom{r}{j} \frac{\Gamma(k+r)\Gamma(j+1)}{\Gamma(k+r-j)\Gamma(j+1-s)} . X^{j-s}(q\frac{d}{dq})^{r-j} h \tag{5}$$

2.5 Hecke Operators

Let $R_1(N)$ be the abstract Hecke algebra attached the pair $(\Gamma_1(N), \Delta_1(N))$ by Shimura (1971, chapt. 3). This algebra is generated over $\mathbf{Z}$ by the operators T_n for n running in the set of natural integers. If ℓ is a prime dividing N, the operator T_ℓ is sometimes called U_ℓ. These operators act on the space of nearly holomorphic forms by the usual standard formulas and preserve the weight and degree of nearly holomorphy. Moreover the Hecke operators respect the rationality. After an appropriate normalization, it can be seen that integrality is preserved as well.

For any ring $A \subset \mathbf{C}$, we then denote by $h_k^r(N, A) \subset End_{\mathbf{C}}(\mathcal{N}_k^r(N, \mathbf{C}))$ the subalgebra generated over A by the image of the T_n's. If $\mathbf{Z}[\frac{1}{N}] \subset A \subset B$, then the above remark shows that we have

$$h_k^r(N, A) \otimes_A B \cong h_k^r(N, B)$$

An easy computation shows that, for each integer n and $f \in \mathcal{N}_k^r(N, \mathbf{C})$ we have:

$$(\delta_k.f)|_{k+2}T_n = n.\delta_k(f|_k T_n)$$
$$\epsilon.(f|_k T_n) = n.(\epsilon.f)|_{k-2}T_n$$

Remark 5. From these formulae, we see that if f is a holomorphic eigenform of weight k. Then $\delta_k^r f$ is a nearly holomorphic eigenform of weight $k + 2r$. Moreover the system of Hecke eigenvalues of $\delta_k^r f$ is different than the one of any holomorphic Hecke eigenform of weight $k + 2r$ if $r > 0$.

2.6 Rationality and CM-Points

2.6.1 Evaluation at CM-Points

We review quickly the rationality notion introduced by Shimura. For $K \subset \overline{\mathbf{Q}}$ an imaginary quadratic field and $\tau \in \mathfrak{h} \cap \mathcal{K}$, the elliptic curve E_τ has complex multiplication by $\mathcal{K}$ and is therefore defined over $\mathcal{K}^{ab} \subset \overline{\mathbf{Q}}$ the maximal abelian extension of $\mathcal{K}$ by the theory of Complex Multiplication. We then denote by ω_τ an invariant Kähler differential of E_τ defined over $\mathcal{K}^{ab}$ and we denote by Ω_τ the corresponding CM period defined by

$$\omega_\tau = \Omega_\tau dz$$

Then $(\omega_\tau, \overline{\omega}_\tau)$ forms a basis of $H^1_{dR}(E_\tau/\mathcal{K}^{ab})$. Let E be a number field and $f \in \mathcal{N}_k^r(N, E)$. Let $\alpha_{N,\tau}$ the $\Gamma_1(N)$-level structure of E_τ induced by $\frac{1}{N}\mathbf{Z}/\mathbf{Z} \subset \mathbf{C}/L_\tau$. Then the polynomial $f(E_\tau, \alpha_{N,\tau}, \omega_\tau, \overline{\omega}_\tau)(X)$ belongs to $E\mathcal{K}^{ab}[X]$. Since we have $\overline{\omega}_\tau = \overline{\Omega}_\tau d\overline{z}$, we deduce that $f(E_\tau, \alpha_{N,\tau}, \omega_\tau, \overline{\omega}_\tau)(0)$ is the left hand side of (6) and that we therefore have

$$\frac{f(\tau)}{\Omega_\tau^k} \in E\mathcal{K}^{ab} \tag{6}$$

According to Shimura, a nearly holomorphic form is defined as rational if and only if it satisfies (6) for any imaginary quadratic field $\mathcal{K}$ and almost all $\tau \in \mathfrak{h} \cap \mathcal{K}$. It can be easily seen his definition is equivalent to our sheaf theoretic definition.

Proposition 4. *Let* $f \in \mathcal{N}_k^r(N, \mathbf{C})$ *and* E *be a number field such that for any imaginary quadratic field* $\mathcal{K} \subset \overline{\mathbf{Q}}$ *and almost all* $\tau \in \mathfrak{h} \cap \mathcal{K}$*, we have*

$$\frac{f(\tau)}{\Omega_\tau^k} \in E.\mathcal{K}^{ab},$$

then, $f \in \mathcal{N}_k^r(N, E\mathbf{Q}^{ab})$.

Proof. We just give a sketch under the assumption $k > 2r$ since the general case can be deduced after multiplying f by E_2. Thanks to a Galois descent argument, we may assume E contains the eigenvalues of all Hecke operators acting on $\mathcal{N}_k^r(N, \mathbf{C})$. By Lemma 1, we can decompose f *as*

$$f = f_0 + \delta_{k-2} f_1 + \cdots + \delta_{k-2r}^r f_r$$

with f_i holomorphic of weight $k - 2i$. Now we remark that if T is a Hecke operator defined over E, then $f|_k T$ satisfies (6). This follows easily from the definition of the action of Hecke operators using isogenies. Moreover from Remark 5, the system of Hecke eigenvalues of nearly holomorphic forms $\delta^i h$ and $\delta^{i'} h'$ are distinct for any two holomorphic forms h and h' when $i \neq i'$; we deduce that $\delta^i f_i$ satisfy (6). We may assume therefore $f = \delta_{k-2r}^r g$ for an holomorphic form g of weight $k - 2r$. In fact using a similar argument, we may even assume g is an eigenform. Then $g = \lambda.g_0$ with g_0 defined over E. Since $\delta_{k-2r}^r g_0$ is defined over E, we deduce from Sect. 2.6.1 that $\delta_{k-2r}^r g_0$ satisfies (6) and therefore $\lambda \in E\mathcal{K}^{ab}$ and $f \in \mathcal{N}_k^r(N, E\mathcal{K}^{ab})$. Since this can be done for any $\mathcal{K}$ the result follows. □

3 Nearly Overconvergent Forms

In this section, we introduce our definition of nearly overconvergent modular forms and show they are p-adic modular forms of a special type. We use the spectral theory of the Atkin U_p-operator on them and we define p-adic families of such forms. We study also the effect of differential operators on them and define an analogue of the holomorphic projection. These tools are useful to study certain p-adic families of modular forms and also to study p-adic L-functions.

3.1 Katz p-Adic Modular Forms

3.1.1 Definition

We fix p a prime. Let X_{rig} be the generic fiber in the sense of rigid geometry of the formal completion of $X_{/\mathbf{Z}_p}$ along its special fiber. Let $A \in H^0(X_{/\mathcal{F}_p}, \omega^{\otimes p-1})$ be the Hasse invariant and let $\tilde{A}^q$ be a lifting of A^q to characteristic 0 for q sufficiently large. For $\rho \in p^{\mathbf{Q}} \cap [p^{-1/p+1}, 1]$, we write $X^{\geq \rho}$ for the rigid affinoid subspace of X_{rig} defined as the set of $x \in X_{rig}$ satisfying $|\tilde{A}^q(x)|_p \geq \rho^q$. For $\rho = 1$ we get the ordinary locus of X_{rig} and we denote it X_{ord}. The space of p-adic modular forms of weight k is defined as

$$M_k^{p-adic}(N) := H^0(X_{\mathrm{ord}}, \omega^{\otimes^k})$$

The space of overconvergent forms of weight k is the subspace of p-adic forms which are defined on some strict neighborhood of X_{ord} so:

$$M_k^{\dagger}(N) := \varinjlim_{\rho<1} H^0(X^{\geq \rho}, \omega^{\otimes^k})$$

3.1.2 Frobenius and Θ

Let $\varphi : X_{\mathrm{ord}} \to X_{\mathrm{ord}}$ the lifting of Frobenius induced on Y_{ord} by $(E, \alpha_N) \mapsto (E^{(\varphi)}, \alpha_N^{(\varphi)})$ where $E^{(\varphi)} := E/E[p]^{\circ}$ and $\alpha_N^{(\varphi)}$ is the composition of α_N and the Frobenius isogeny $E \to E^{(\varphi)}$. We get a φ^*-linear morphism obtained as the composite

$$\Phi : \mathcal{H}_{dR}^1 \to \varphi^* \mathcal{H}_{dR}^1 = \mathcal{H}_{dR}^1(\mathcal{E}^{\varphi}/X_{\mathrm{ord}}) \to \mathcal{H}_{dR}^1$$

This morphism stabilizes the Hodge filtration of $\mathcal{H}_{dR}^1$ and we know by Dwork that there is a unique Φ-stable splitting, called the unit root splitting:

$$\mathcal{H}_{dR/X_{\mathrm{ord}}}^1 = \omega_{/X_{\mathrm{ord}}} \oplus \mathcal{U}_{/X_{\mathrm{ord}}}$$

such that Φ is invertible on $\mathcal{U}$ and $\mathcal{U}$ is a free sheaf of rank 1 generated by its subsheaf of horizontal sections for the Gauss-Manin connection. This unit root splitting induces a splitting of $\mathcal{H}_{k/X_{\mathrm{ord}}}^r$ and therefore of a canonical projection

$$\mathcal{H}_{k/X_{\mathrm{ord}}}^r \to \omega_{/X_{\mathrm{ord}}}^{\otimes k} \tag{7}$$

We now recall the definition of the Theta operator on the space of p-adic modular forms of weight k. At the level of sheaves, Θ is defined as the composite of the following maps of sheaves over the ordinary locus:

$$\omega_{/X_{\mathrm{ord}}}^{\otimes k} \stackrel{\delta_k}{\to} \mathcal{H}_{k+2/X_{\mathrm{ord}}}^1 \to \omega_{/X_{\mathrm{ord}}}^{\otimes k+2}$$

where he second arrow is the one given by (7) for $r = 1$. This defines

$$\Theta : M_k^{p-adic}(N) \to M_{k+2}^{p-adic}(N)$$

It follows from Propositions 3 and 6 below that on the level of q-expansion, we have:

$$\Theta(f)(q) = q\frac{d}{dq}f(q)$$

for all $f \in M_k^{p-adic}(N)$. The following proposition will be useful in the next paragraph.

Proposition 5. *For any $\rho < 1$ and any Zariski open $V \subset X^{\geq\rho}$, the unit root splitting on $V_{ord} := U \cap X_{ord}$ does not extend to a splitting of the Hodge filtration of $\mathcal{H}^1_{dR}$ over any finite cover of V.*

Proof. We show it by contradiction. Let us assume that this splitting extends to some finite cover S of V for some $\rho < 1$.

$$\mathcal{H}^1_{dR}(\mathcal{E}/S) = \omega_{\mathcal{E}/S} \oplus \mathcal{U}(\mathcal{E}/S).$$

Let $S_{\mathrm{ord}} = S \times_{V_{\mathrm{ord}}} V$. Since $\mathcal{U}(\mathcal{E}/S) \otimes \mathcal{O}_{S_{\mathrm{ord}}}$ is stable by Φ so is $\mathcal{U}(\mathcal{E}/S)$. Since V is a strict neighborhood of V_{ord}, we can find a finite extension L of $\mathbf{Q}_p$ and $x \in S(L)\backslash S_{\mathrm{ord}}(L)$. Then, we will obtain a splitting $H^1_{dR}(\mathcal{E}_x/L) = \omega_{\mathcal{E}_x/L} \oplus \mathcal{U}(\mathcal{E}_x/L)$ with Φ_x inducing a semi-linear invertible endomorphism of $\mathcal{U}(\mathcal{E}_x/L)$. Let k_L be the residue field of L. By the results of Berthelot and Ogus (1978, chapt. 7), the pair $(H^1_{dR}(\mathcal{E}_x/L), \Phi_x)$ is isomorphic to $(H^1_{crys}(\mathcal{E}_{x,0}/W(k_L)) \otimes L, F^* \otimes id_L)$ where $H^1_{crys}(\mathcal{E}_{x,0}/W(k_L))$ stands for the crystalline cohomology of the special fiber $\mathcal{E}_{x,0/k_L}$ of $\mathcal{E}_x$ over the ring of Witt vectors of k_L and where $\Phi_x = F^* \otimes id_L$ where F^* is the "crystalline" Frobenius induced by the Frobenius isogeny $\mathcal{E}_{x,0} \to \mathcal{E}_{x,0}^{(p)}$ in characteristic p. Since it has a splitting of the form $H^1_{crys}(\mathcal{E}_{x,0}/O_L) = Fil^1 \oplus U$ with F^* inversible on U, $\mathcal{E}_{x,0}$ has to be ordinary which is a contradiction since $x \notin S_{\mathrm{ord}}(L)$. □

3.2 Nearly Overconvergent Forms as p-Adic Modular Forms

3.2.1 Definition of Nearly Overconvergent Forms

For each ρ, $H^0(X^{\geq\rho}, \mathcal{H}_k^r)$ is naturally a $\mathbf{Q}_p$-Banach space for the Supremum norm $|\cdot|_\rho$ and if $\rho' < \rho < 1$, the map $H^0(X^{\geq\rho'}, \mathcal{H}_k^r) \hookrightarrow H^0(X^{\geq\rho}, \mathcal{H}_k^r)$ is completely continuous. We define the space of nearly overconvergent forms of weight k and order $\leq r$ by

$$\mathcal{N}_k^{r,\dagger}(N) := \lim_{\substack{\rightarrow \\ \rho<1}} \mathcal{N}_k^{r,\rho}(N)$$

with $\mathcal{N}_k^{r,\rho}(N) := H^0(X^{\geq\rho}, \mathcal{H}_k^r)$. We can define the operators δ_k and ϵ on nearly overconvergent forms since they are defined at the level of sheaves. Moreover we can define the polynomial q-expansion of a nearly overconvergent form and it is straightforward to check that the action of δ_k and ϵ on this q-expansion is the same as for nearly holomorphic forms.

Remark 6. For any nearly overconvergent form f of weight k and order at most r, we can easily show that there exist overconvergent forms $g_0, \ldots, g_r$ such that

$$f = g_0 + g_1.E_2 + \cdots + g_r.E_2^r$$

where for $i = 0, \ldots, r$, g_i is of weight $k - 2i$. This could be used as an ad hoc definition of nearly overconvergent forms but it would be uneasy to show this space is stable by the action of Hecke operators and that it has a slope decomposition as we will see in the next sections. Moreover, this definition would not be suited for generalization to higher rank reductive groups.

3.2.2 Embedding Into the Space of p-Adic Modular Forms

We now want to consider nearly overconvergent modular forms as p-adic modular forms. Using (7), we get a map

$$H^0(X^{\geq \rho}, \mathcal{H}_k^r) \to H^0(X_{\mathrm{ord}}, \mathcal{H}_k^r) \to H^0(X_{\mathrm{ord}}, \omega^{\otimes k}) \tag{8}$$

We have the following

Proposition 6. *The maps* (8) *induces a canonical injection*

$$\mathcal{N}_k^{r,\dagger}(N) \hookrightarrow M_k^{p-adic}(N) \tag{9}$$

fitting in the commutative diagram:

$$\begin{array}{ccc} \mathcal{N}_k^{r,\dagger}(N) & \hookrightarrow & M_k^{p-adic}(N) \\ \downarrow & & \downarrow \\ \mathbf{Q}_p[[q]][X]_r & \longrightarrow & \mathbf{Q}_p[[q]] \end{array}$$

where the bottom map is induced by evaluating $X = 0$.

Proof. The fact that the diagram commutes follows from the fact that u_{can} belongs to the fiber of the unit root sheaf $\mathcal{U}$ at the $\mathbf{Z}_p((q))$-point defining $Tate(q)$. Indeed, it is explained in the appendix 2 of Katz (1972) that u_{can} is fixed by Frobenius. We are left with proving the injectivity. We consider f in the kernel of this map. Let $U \subset X^{\geq r} \cap Y_{rig}$ be an irreducible affinoid. It is the generic fiber in the sense of Raynaud of an affine formal scheme $Spf(R)$ with R a p-adically complete domain. Let $\mathcal{E}_{/R}$ the universal elliptic curve over R. Let us choose a basis (ω, ω') of $\mathcal{H}_{dR}^1(\mathcal{E}/R)$ as in the previous section and such that $\langle \omega, \omega' \rangle_{dR} = 1$. Let $h \in R$ be a lifting of the Hasse invariant of $\mathcal{E} \times_R Spec(R/pR)$ and let $S := \widehat{R[1/h]}$ where the hat here stands for

p-adic completion. Then $\mathcal{E}/S$ has ordinary reduction and the unit root splitting over S defines a basis (ω, u) of $\mathcal{H}^1_{dR}(\mathcal{E}/S)$. We must have

$$u = \omega' + \lambda.\omega$$

with $\lambda \in S$. But U is a strict neighborhood of $U_{\text{ord}} = U \cap X_{\text{ord}}$ which is the generic fiber of $Spf(S)$, we know by the previous proposition that λ is not algebraic over R. Let $Q(X) := f(\mathcal{E}_{/R}, \alpha_N, \omega, \omega')(X) \in R[X]_r$. We want to show that $Q(X) = 0$. By assumption, we know that $Q(\lambda) = f(\mathcal{E}_{/S}, \alpha_N, \omega, \omega' + \lambda\omega)(0) = f(\mathcal{E}_{/S}, \alpha_N, \omega, u)(0) = 0$. Since λ is not algebraic over R, this is possible only if $Q(X) = 0$. Since this can be done for any pair (ω, ω') we conclude that $f \equiv 0$. □

If f is a nearly overconvergent form, the p-adic q-expansion of f is by definition the q-expansion of the image of f in the space of p-adic forms. The following corollary can be thought of as a *polynomial q-expansion principle* for the degree of near overconvergence.

Corollary 1. *Let $f \in \mathcal{N}_k^{r,\dagger}(N)$. If $f(q, X)$ is of degree r then there is no $g \in \mathcal{N}_k^{r-1,\dagger}(N)$ having the same p-adic q-expansions.*

Proof. We prove this by contradiction. Let us assume that such a g exists. Let $h = f - g$. Since $f(q, X)$ is of degree r and $g \in \mathcal{N}_k^{r-1,\dagger}(N)$, $h(q, X)$ is still of degree r and therefore h is non-zero. However, by assumption $h(q, 0) = 0$. This implies that $h = 0$ by the diagram of the previous proposition, which is a contradiction. □

3.2.3 E_2, Θ and Overconvergence

In the following two corollaries, we recover the main results of Coleman et al. (1995) using the polynomial q-expansion principle. It can be easily generalized to modular forms for other Shimura varieties.

Corollary 2. *The p-adic modular form E_2 is not overconvergent.*

Proof. By the Corollary 1, this is immediate since the polynomial q expansion of E_2 is of degree 1. □

Corollary 3. *If f is overconvergent of weight k and $k \neq 0$, then $\Theta.f$ is not overconvergent.*

Proof. It follows from Propositions 3 and 6, that $\Theta.f$ is the image of the nearly overconvergent form $\delta_k.f$ in the space of p-adic modular forms by the map (9). Moreover

$$(\delta_k.f)(q, X) = q\frac{d}{dq}f + kXf(q)$$

It is therefore of degree 1 since $k \neq 0$ and the result follows from Corollary 1. □

3.2.4 The Overconvergent Projection

We give a p-adic version of the holomorphic projector.

Lemma 2. *Let f be a nearly overconvergent form of weight k and order $\leq r$ such that $k > 2r$. Then for each $i = 0, \cdots, r$, there exists a unique overconvergent form g_i of weight $k - 2i$ such that*

$$f = \sum_{i=1}^{r} \delta_{k-2i}^{i} \cdot g_i$$

Proof. This is special case of the proof of Proposition 9 below. □

We define the overconvergent projection $\mathcal{H}^{\dagger}(f)$ of f by:

$$\mathcal{H}^{\dagger}(f) := g_0$$

It is an overconvergent version of the holomorphic projection since if f is holomorphic, then we clearly have:

$$\mathcal{H}^{\dagger}(f) = \mathcal{H}(f)$$

which means that $\mathcal{H}^{\dagger}(f)$ is holomorphic.

Remark 7. Let $f \in \mathcal{N}_k^{r,\dagger}(N, \mathbf{Q}_p)$ and $g \in \mathcal{N}_l^{s,\dagger}(N, \mathbf{Q}_p)$ such that $k + l > 2s + 2r$. Then the following holds $\mathcal{H}^{\dagger}(f\delta_l^m g) = (-1)^m \mathcal{H}^{\dagger}(g\delta_k^m f)$. One can also show that when a Hecke equivariant p-adic Petersson inner product is defined then δ and ϵ are very close to be adjoint operators. This implies a formula of the type $\langle f, g\rangle_{p-adic} = \langle f, \mathcal{H}^{\dagger}(g)\rangle_{p-adic}$ when f is overconvergent. We hope to come back to this in a future paper.

3.2.5 Action of the Atkin-Hecke Operator U_p

If $\rho > p^{-1/p+1}$, it follows from the theory of the canonical subgroup (Katz-Lubin) that we can extend canonically φ on X_{ord} into

$$\varphi : X^{\geq \rho} \to X^{\geq \rho^p}.$$

Let $\mathcal{E}/X^{\geq \rho^p}$ be the generalized universal elliptic curve over $X^{\geq \rho^p}$ and let $\mathcal{E}^{(\varphi)}/X^{\geq \rho}$ be its pullback by φ. We have degree p isogeny

$$\mathcal{E} \stackrel{F_\varphi}{\to} \mathcal{E}^{(\varphi)}$$

over $X^{\geq \rho}$ and we denote $V_\varphi : \mathcal{E}^{(\varphi)} \to \mathcal{E}$ the dual isogeny. On the level of sheaves, the operator U_p is defined as the composition of the following maps.

$$\mathcal{H}^r_{k/X^{\geq\rho}} \xrightarrow{V^*_\varphi} \mathcal{H}^{r,(\varphi)}_k{}_{/X^{\geq\rho}} = \mathcal{H}^r_{k/X^{\geq\rho^p}} \otimes_{\varphi^*} \mathcal{O}_{/X^{\geq\rho}} \overset{Id\otimes\frac{1}{p}\cdot Tr}{\to} \mathcal{H}^r_{k/X^{\geq\rho^p}} \overset{j}{\to} \mathcal{H}^r_{k/X^{\geq\rho}}$$

where j is induced by the completely continuous inclusion $\mathcal{O}_{X^{\geq\rho^p}} \to \mathcal{O}_{X^{\geq\rho}}$ defined by the restriction of analytic function on $X^{\geq\rho^p}$ to $X^{\geq\rho}$ and Tr is induced by the trace of the degree p map $\varphi^* : \mathcal{O}_{X^{\geq\rho^p}} \to \mathcal{O}_{X^{\geq\rho}}$. Since j is completely continuous, U_p induces a completely continuous endomorphism of $\mathcal{N}^{r,\rho}_k(N, \mathbf{Q}_p)$. The following proposition is easy to prove.

Proposition 7. *Let $f \in \mathcal{N}^{r,\dagger}_k(N, \mathbf{Q}_p)$. Let us write its polynomial q-expansion as:*

$$f(q, X) = \sum_{n=0}^{\infty} a(n, f)(X)q^n$$

Then we have:

(i) $(f|U_p)(q, X) = \sum_{n=0}^{\infty} a(np, f)(pX)q^n$
(ii) $\epsilon(f|U_p) = p.(\epsilon f)|U_p$
(iii) $(\delta_k f)|U_p = p\delta_k(f|U_p)$

Proof. (i) follows from a standard computation and (ii) and (iii) follow from (i) and the effect of δ_k and ϵ on the polynomial q-expansion explained in Sect. 2. □

Let $\hat{\mathcal{N}}^{\infty,\rho}_k(N, \mathbf{Q}_p)$ be the p-adic completion of

$$\mathcal{N}^{\infty,\rho}_k(N, \mathbf{Q}_p) := \bigcup_{r\geq 0} \mathcal{N}^{r,\rho}_k(N, \mathbf{Q}_p).$$

Then we have:

Corollary 4. *The action of U_p on $\hat{\mathcal{N}}^{\infty,\rho}_k(N, \mathbf{Q}_p)$ is completely continuous.*

Proof. It follows easily from the lemma below for the sequence $M_i = \mathcal{N}^{i,\rho}_k(N)$ and the relation (ii) of the previous proposition. □

Lemma 3. *Let M_i be an increasing sequence of Banach modules over a p-adic Banach algebra A. Let u be an endomorphism on $M := \bigcup_i M_i$ such that*

(i) u induces a completely continuous endomorphism on each of the M_i's.
(ii) Let α_i be the norm of the operator on the Hausdorff quotient of M_i/M_{i-1} induced by u. Then the sequence α_i converges to 0.

Then u induces a completely continuous operator on the p-adic completion of M.

Proof. This is an easy exercise which is left to the reader. □

Remark 8. We can give a sheaf theoretic definition of $\hat{\mathcal{N}}^{\infty,\rho}_k$. Let $\mathcal{A}$ be the ring of analytic functions defined over $\mathbf{Q}_p$ on the closed unit disc. It is isomorphic to the power series in X with $\mathbf{Q}_p$-coefficient converging to 0. We denote it $\mathcal{A}_k$ if one equips it with the representation of the standard Iwahori subgroup of $SL_2(\mathbf{Z}_p)$ defined by:

$$((\begin{smallmatrix} a & b \\ c & d \end{smallmatrix}).f)(X) = (a + cX)^k f(\frac{b + dX}{a + cX})$$

If we restrict this representation to the Borel subgroup we get a representation that contains $\mathbf{Z}_p[X]_r(k)$ for all r. In fact, it is the p-adic completion of $\mathbf{Z}_p[X](k) = \bigcup_r \mathbf{Z}_p[X]_r(k)$. It is not difficult to see, one can define a sheaf of Banach spaces on $X^{\geq \rho}$ (in the sense of Andreatta et al. to appear) by considering the contracted product

$$\mathcal{H}_k^\infty := \mathcal{T} \times^B \mathcal{A}_k$$

Then, we easily see that

$$\hat{\mathcal{N}}_k^{\infty,\rho} = H^0(X^{\geq \rho}, \mathcal{H}_k^\infty)$$

3.2.6 Slopes of Nearly Overconvergent Forms

Let $f \in \mathcal{N}_k^{r,\dagger}(N, \overline{\mathbf{Q}}_p)$ which is an eigenform for U_p for the eigenvalue α. If $\lambda \neq 0$, we say f is of finite slope $\alpha = v_p(\lambda)$. The following proposition compares the slope and the degree of near overconvergence and extends the classicity result of Coleman to nearly overconvergent forms.

Proposition 8. *Let $f \in \mathcal{N}_k^{\infty,\dagger}(N, \overline{\mathbf{Q}}_p)$, then the following properties hold*

(i) If f is of slope α, then its degree of overconvergence r satisfies $r \leq \alpha$,
(ii) If f is of degree r and slope $\alpha < k - 1 - r$, then f is a classical nearly holomorphic form.

Proof. The part (i) is easy. If r is the degree of near overconvergence of f, then $g = \epsilon^r.f$ is a non trivial overconvergent form and by Proposition 7 its slope is $\alpha - r$. Since a slope has to be non-negative, (i) follows. The part (ii) is a straightforward generalization of the result for $r = 0$ which is a theorem of Coleman (1996). We may assume that r is the exact degree of near overconvergence. By the point (i), we therefore have $\alpha \geq r$. From the assumption, we deduce $k - 1 - r \geq r$. Therefore $k > 2r$ and we may apply Lemma 2:

$$f = \sum_{i=o}^{r} \delta_{k-2i}^i g_i$$

with g_i overconvergent of weight $k - 2i$ for each i. By uniqueness of the g_i's, we see easily that the $\delta_{k-2i} g_i$ are eigenforms for U_p with the same eigenvalue as f. So for each i, g_i is of slope $\alpha - i < k - 1 - r - i \leq (k - 2i) - 1$. Therefore it is classical by the theorem of Coleman. This implies f is classical nearly holomorphic. □

3.3 Families of Nearly Overconvergent Forms

3.3.1 Weight Space

Let $\mathfrak{X}$ be the rigid analytic space over $\mathbf{Q}_p$ such that for any p-adic field $L \subset \overline{\mathbf{Q}}_p$ $\mathfrak{X}(L) = Hom_{cont}(\mathbf{Z}_p^\times, L^\times)$. Any integer $k \in \mathbf{Z}$ can be seen as the point $[k] \in \mathfrak{X}(\mathbf{Q}_p)$ defined as $[k](x) = x^k$ for all $x \in \mathbf{Z}_p^\times$. Recall we have the decomposition $\mathbf{Z}_p = \Delta \times 1 + q\mathbf{Z}_p$ where $\Delta \subset \mathbf{Z}_p^\times$ is the subgroup of roots of unity contained in $\mathbf{Z}_p^\times$ and $q = p$ if p odd and $q = 4$ if $p = 2$. We can decompose $\mathfrak{X}$ as a disjoint union

$$\mathfrak{X} = \bigsqcup_{\psi \in \hat{\Delta}} B_\psi$$

where $\hat{\Delta}$ is the set of characters of Δ and B_ψ is identified to the open unit disc of center 1 in $\overline{\mathbf{Q}}_p$. If $\kappa \in \mathfrak{X}(L)$ then it correspond to $u_\kappa \in B_\psi(L)$ if $\kappa|_\Delta = \psi$ and $\kappa(1+q) = u_\kappa$.

3.3.2 Families

Let $\mathfrak{U} \subset \mathfrak{X}$ be an affinoid subdomain. It is known from the works of Coleman-Mazur (1998) and Pilloni (to appear), that there exist $\rho_{\mathfrak{U}} \in [0,1) \cap p^{\mathbf{Q}}$ such that for all $\rho \geq \rho_{\mathfrak{U}}$, there exists an orthonomalizable Banach space $\mathcal{M}_{\mathfrak{U}}^\rho$ over $A(\mathfrak{U})$ such that for all $\kappa \in \mathfrak{U}(\overline{\mathbf{Q}}_p)$, we have

$$\mathcal{M}_{\mathfrak{U}}^\rho \otimes_\kappa \overline{\mathbf{Q}}_p \cong H^0(X^{\geq\rho}, \omega_\kappa)$$

We consider the sheaf[7] $\Omega_{\mathfrak{U}}$ over $\mathfrak{U} \times X^{\geq\rho}$ associated to the $A(\mathfrak{U} \times X^{\geq\rho})$-module $\mathcal{M}_{\mathfrak{U}}^\rho$ and we put

$$\mathcal{H}_{\mathfrak{U}}^r := \Omega_{\mathfrak{U}} \otimes \mathcal{H}_0^r$$

For any weight $k \in \mathbf{Z}$ such that $[k] \in \mathfrak{U}(\mathbf{Q}_p)$, we recover $\mathcal{H}_k^r$ by the *pull-back*

$$([k] \times id_{X^{\geq}\rho})^* \mathcal{H}_{\mathfrak{U}}^r \cong \mathcal{H}_k^r \tag{10}$$

For general weight $\kappa \in \mathfrak{U}(L)$, we define the sheaf $\mathcal{H}_\kappa^r = \omega_\kappa \otimes \mathcal{H}_0^r$ where ω_κ is the invertible sheaf defined in Pilloni (to appear, §3). We define the space of nearly overconvergent forms of weight κ by

[7] In Pilloni (to appear), this sheaf is constructed in a purely geometric way and the existence of $\mathcal{M}_{\mathfrak{U}}^\rho$ is deduced from it.

$$\mathcal{N}_{\kappa}^{r,\rho}(N,L) := H^0(X^{\geq\rho}, \mathcal{H}^r_{\kappa/L})$$

and the space of $\mathfrak{U}$-families of nearly overconvergent forms:

$$\mathcal{N}_{\mathfrak{U}}^{r,\rho}(N) := H^0(X^{\geq\rho}, \mathcal{H}^r_{\mathfrak{U}})$$

We also define $\mathcal{N}_{\kappa}^{r,\dagger}(N)$ and $\mathcal{N}_{\mathfrak{U}}^{r,\dagger}(N)$ the spaces we obtain by taking the inductive limit over ρ. The space $\mathcal{N}_{\mathfrak{U}}^{r,\rho}(N)$ is a Banach module over $A(\mathfrak{U})$ and for any weight $\kappa \in \mathfrak{U}(L)$, we have

$$\mathcal{N}_{\mathfrak{U}}^{r,\rho}(N) \otimes_{\kappa} L = \mathcal{N}_{\kappa}^{r,\rho}(N,L)$$

This follows easily from (10) and the fact that $X^{\geq\rho}$ is an affinoid.

As in the previous section, one can define an action of U_p on these spaces and show it is completely continuous. For any integer r, any affinoid $\mathfrak{U} \subset \mathfrak{X}$ and $\rho \geq \rho_{\mathfrak{U}}$, we may consider the Fredholm determinant

$$P^r_{\mathfrak{U}}(X) = P^r_{\mathfrak{U}}(\kappa, X) := det(1 - X.U_p|\mathcal{N}_{\mathfrak{U}}^{r,\rho}) \in A(\mathfrak{U})[[X]]$$

because one can show as in Pilloni (to appear) that $\mathcal{N}_{\mathfrak{U}}^{r,\rho}(N)$ is $A(\mathfrak{U})$-projective. A standard argument shows this Fredholm determinant is independent of ρ. For $\mathfrak{U} = \{\kappa\}$, we just write $P^r_{\kappa}(X)$. If $r = 0$, we omit r from the notation.

For any integer m, we consider the map $[m] : \mathfrak{X} \to \mathfrak{X}$ defined by $\kappa \mapsto \kappa.[m]$ and we denote $\mathfrak{U}[m]$, the image of $\mathfrak{U}$ by this map. We easily see from its algebraic definition, that the operator ϵ can be defined in families and induces a short exact sequence:

$$0 \to \mathcal{M}^{\rho}_{\mathfrak{U}} \to \mathcal{N}^{r,\rho}_{\mathfrak{U}} \to \mathcal{N}^{r-1,\rho}_{\mathfrak{U}[-2]} \to 0$$

From Proposition 7 and the above exact sequence one easily sees by induction on r that

$$P^r_{\mathfrak{U}}(\kappa, X) = \prod_{i=0}^{r} P_{\mathfrak{U}[-2i]}(\kappa.[-2i], p^i.X)$$

Let us define $\mathcal{N}_{\mathfrak{U}}^{\hat{\infty},\rho}(N)$ as the p-adic completion of $\mathcal{N}_{\mathfrak{U}}^{\infty,\rho}(N) := \bigcup_{r\geq 0} \mathcal{N}_{\mathfrak{U}}^{r,\rho}(N)$. Then it follows from Lemma 3 again that U_p acts completely continuously on it with the Fredholm determinant given by the converging product

$$P^{\infty}_{\mathfrak{U}}(\kappa, X) = \prod_{i=0}^{\infty} P_{\mathfrak{U}[-2i]}(\kappa.[2i], p^i X)$$

Definition 1. Let $\mathfrak{U} \subset \mathfrak{X}$ be an affinoid subdomain and $Q(X) \in A(\mathfrak{U})[X]$ be a polynomial of degree d such that $Q(0) = 1$. The pair $(Q, \mathfrak{U})$ is said admissible for nearly overconvergent forms (reps. for overconvergent forms) if there is a factorization

$$P_{\mathfrak{U}}^{\infty}(X) = Q(X)R(X) \quad (resp. \quad P_{\mathfrak{U}}(X) = Q(X)R(X)\,)$$

with P and Q relatively prime and $Q^*(0) \in A(\mathfrak{U})^{\times}$ with $Q^*(X) := X^d Q(1/X)$.

If $(Q, \mathfrak{U})$ is admissible for nearly overconvergent forms, it results from Coleman-Riesz-Serre theory (Coleman 1997) that there is a unique U_p-stable decomposition

$$\mathcal{N}_{\mathfrak{U}}^{\infty,\rho} = \mathcal{N}_{Q,\mathfrak{U}} \oplus \mathcal{S}_{Q,\mathfrak{U}}$$

such that $\mathcal{N}_{Q,\mathfrak{U}}$ is projective of finite rank over $A(\mathfrak{U})$ with

(i) $det(1 - U_p X | \mathcal{N}_{Q,\mathfrak{U}}) = P(X)$
(ii) $Q^*(U_p)$ is invertible on $\mathcal{S}_{Q,\mathfrak{U}}$

It is worth noticing also that the projector $e_{Q,\mathfrak{U}}$ of $\mathcal{N}_{\mathfrak{U}}^{\infty,\rho}$ onto $\mathcal{N}_{Q,\mathfrak{U}}$ can be expressed as $S(U_p)$ for some entire power series $S(X) \in XA(\mathfrak{U})X$. If we have two admissible pairs $(Q, \mathfrak{U})$ and $(Q', \mathfrak{U}')$, we write $(Q, \mathfrak{U}) < (Q', \mathfrak{U}')$ if $\mathfrak{U} \subset \mathfrak{U}'$ and if Q divides the image $Q'|_{\mathfrak{U}}(X)$ of $Q'(X)$ by the canonical map $A(\mathfrak{U}')[X] \to A(\mathfrak{U})[X]$. When this happens, we easily see from the properties of the Riesz decomposition that

$$e_{Q,\mathfrak{U}} \circ (e_{Q',\mathfrak{U}'} \otimes_{A(\mathfrak{U}')} 1_{A(\mathfrak{U})}) = e_{Q,\mathfrak{U}} \tag{11}$$

We have dropped ρ from the notation in $\mathcal{N}_{Q,\mathfrak{U}}$ since this space is clearly independent of ρ by a standard argument. We define $\mathcal{N}_{\mathfrak{U}}^{fs}$ as the inductive limit of the $\mathcal{N}_{Q,\mathfrak{U}}$ over the Q's. Since $\mathcal{N}_{Q,\mathfrak{U}}$ is of finite rank and U_p-stable it is easy to see that there exists r such that

$$\mathcal{N}_{Q,\mathfrak{U}} \subset \mathcal{N}_{\mathfrak{U}}^{r,\dagger}$$

Remark 9. If $\alpha_{Q,\mathfrak{U}}$ is the maximal[8] valuation taken by the values of the analytic function $Q^*(0) \in A(\mathfrak{U})$ on $\mathfrak{U}$, then one can easily see that $r \leq \alpha_{Q,\mathfrak{U}}$ by the point (i) of Proposition 8.

3.3.3 Families of q-Expansions and Polynomial q-Expansions

By evaluating at the Tate object we have defined in Sect. 2, we can define the polynomial q-expansion of an element $F \in \mathcal{N}_{\mathfrak{U}}^{r,\rho}(N)$ that we write

[8] This maximum is $< \infty$ since $Q^*(0) \in A(\mathfrak{U})^{\times}$.

$F(q,X) \in A(\mathfrak{U})[X]_r[[q]]$. The evaluation $F_\kappa(q,X)$ at κ of $F(q,X)$ is the polynomial q-expansion of the nearly overconvergent form of weight κ obtained by specializing F at κ. We also denote $F_\kappa(q) = F_\kappa(q,0)$ the p-adic q-expansion of the specialization of F at κ. In what follow, we show that when the slope is bounded a family of q-expansion of nearly overconvergent forms is equivalent to a family of nearly overconvergent forms.

Let $F(q) \in A(\mathfrak{U})[[q]]$ and $\Sigma \subset \mathfrak{U}(\overline{\mathbf{Q}}_p)$ a Zariski dense subset of points. We say that $F(q)$ is a Σ-family of q-expansions of nearly overconvergent form of type $(Q,\mathfrak{U})$ if for all but finitely many $\kappa \in \Sigma$ the evaluation $F_\kappa(q)$ of $F(q)$ at κ is the p-adic q-expansion of a nearly overconvergent form of weight κ and type Q_κ (i.e. is annihilated by $Q_\kappa^*(U_p)$). Let $\mathcal{N}^{\Sigma}_{Q,\mathfrak{U}}$ be the $A(\mathfrak{U})$-module of families of q-expansion of nearly overconvergent forms of type $(Q,\mathfrak{U})$. Similarly, we can define $\mathcal{N}^{\Sigma,pol}_{Q,\mathfrak{U}} \subset A(\mathfrak{U})[X][[q]]$ the subspace of polynomial q-expansion satisfying a smiler property for specialization at points in Σ with an obvious map:

$$\mathcal{N}^{\Sigma,pol}_{Q,\mathfrak{U}} \to \mathcal{N}^{\Sigma}_{Q,\mathfrak{U}}$$

given by the evaluation X at 0. Then we have:

Lemma 4. *The q-expansion map and polynomial expansion maps induce the isomorphisms*

$$\mathcal{N}_{Q,\mathfrak{U}} \cong \mathcal{N}^{\Sigma,pol}_{Q,\mathfrak{U}} \cong \mathcal{N}^{\Sigma}_{Q,\mathfrak{U}}.$$

Proof. From Proposition 6, it suffices to show that the q-expansion map induces:

$$\mathcal{N}_{Q,\mathfrak{U}} \cong \mathcal{N}^{\Sigma}_{Q,\mathfrak{U}}.$$

The argument to prove this is well-known but we don't know a reference for it. We therefore sketch it below. Notice first that for any $\kappa_0 \in \mathfrak{U}(\overline{\mathbf{Q}}_p)$ the evaluation map at κ_0 induces an injective map:

$$\mathcal{N}^{\Sigma}_{Q,\mathfrak{U}} \otimes_{\kappa_0} \overline{\mathbf{Q}}_p \hookrightarrow \overline{\mathbf{Q}}_p[[q]] \tag{12}$$

Indeed if $F \in \mathcal{N}^{\Sigma}_{Q,\mathfrak{U}}$ is such that $F_{\kappa_0}(q) = 0$ then if $\varpi_{\kappa_0} \in A(\mathfrak{U})$ is a generator of the ideal of the elements of $A(\mathfrak{U})$ vanishing at κ_0, we have $F(q) = \varpi_{\kappa_0}.G(g)$ for some $G \in A(\mathfrak{U})[[q]]$. Clearly for any $\kappa \in \Sigma\backslash\{\kappa_0\}$, we have $G_\kappa(q) = \frac{1}{\varpi_{\kappa_0}(\kappa)} F_\kappa(q)$ is the q-expansion of a nearly over convergent for of weight κ and type Q_κ. Therefore $G \in \mathcal{N}^{\Sigma}_{Q,\mathfrak{U}}$ and our first claim is proved. Now let $\kappa \in \Sigma$. We have the following commutative diagram:

$$\begin{array}{ccc} \mathcal{N}_{Q,\mathfrak{U}} \otimes_\kappa \overline{\mathbf{Q}}_p & \xrightarrow{(3)} & \mathcal{N}^{\Sigma}_{Q,\mathfrak{U}} \otimes_\kappa \overline{\mathbf{Q}}_p \\ {\scriptstyle (1)}\downarrow & & \downarrow{\scriptstyle (2)} \\ \mathcal{N}^{\Sigma}_{Q_\kappa,\kappa} & \xrightarrow{(4)} & \overline{\mathbf{Q}}_p[[q]] \end{array}$$

Since (2) and (4) are injectives and (1) is an isomorphism, we deduce (3) is injective. Now since the image of (2) is included in the image of (4) and (1) is surjective, we deduce that (3) is an isomorphism of finite vector spaces. Since $\mathcal{N}^{\Sigma}_{Q,\mathfrak{U}}$ is torsion free over $A(\mathfrak{U})$ such that $\mathcal{N}^{\Sigma}_{Q,\mathfrak{U}} \otimes_{\kappa} \overline{\mathbf{Q}}_p$ has bounded dimension when κ runs in Σ, a standard argument shows that $\mathcal{N}^{\Sigma}_{Q,\mathfrak{U}}$ is of finite type over $A(\mathfrak{U})$ (see for instance Wiles 1988, §1.2). Notice that the injectivity of (3) below is true for all κ and therefore we deduce that the map

$$\mathcal{N}_{Q,\mathfrak{U}} \to \mathcal{N}^{\Sigma}_{Q,\mathfrak{U}} \tag{13}$$

is injective with a torsion cokernel of finite type. We want now to prove the surjectivity. Let $F(q) \in \mathcal{N}^{\Sigma}_{Q,\mathfrak{U}}$ and let $I_F \subset A(\mathfrak{U})$ be the ideal of element a such that $a.F(q)$ is in the image of (13) and let a_F be a generator of I_F. Let $G \in \mathcal{N}_{Q,\mathfrak{U}}$ whose image is $a_F.F(q)$. For any κ_0 such that $a_F(\kappa_0) = 0$, we get that $G_{\kappa_0} = 0$. By the isomorphism (1), G is therefore divisible by ϖ_{κ_0} and thus $\frac{a_F}{\varpi_{\kappa_0}} \in I_F$ which contradicts the fact that a_F is a generator of I_F. Therefore a_F does not vanish on $\mathfrak{U}$ and F is in the image of (13). This proves the surjectivity we have claimed. □

More generally, for any $\mathbf{Q}_p$-Banach space M, we can define $\mathcal{N}^{\Sigma}_{Q,\mathfrak{U}}(M)$ the subspace of elements $F \in A(\mathfrak{U})\hat{\otimes}M[[q]]$ such that for almost all $\kappa \in \Sigma$, the evaluation F_κ at κ of $F(q)$ is the q-expansion of an element of $\mathcal{N}_{Q_\kappa,\kappa}(M) = \mathcal{N}_{Q_\kappa,\kappa} \otimes M$. Similarly, one defines $\mathcal{N}^{\Sigma,pol}_{Q,\mathfrak{U}}(M)$. Then it is easy to deduce the following:

Corollary 5. *We have the isomorphisms:*

$$\mathcal{N}_{Q,\mathfrak{U}}\hat{\otimes}M \cong \mathcal{N}^{\Sigma,pol}_{Q,\mathfrak{U}}(M) \cong \mathcal{N}^{\Sigma}_{Q,\mathfrak{U}}(M).$$

Proof. Left to the reader. □

3.3.4 Maass-Shimura Operator and Overconvergent Projection in p-Adic Families

The formula for the action of the Maass-Shimura operators on the q-expansion suggests it behaves well in families. We explain this here using the lemma (13). This could be avoided but it would take more time than we want to devote to this here. We explain this in a remark below.

We defined the analytic function $Log(\kappa)$ on $\mathfrak{X}$ by the formula:

$$Log(\kappa) = \frac{log_p(\kappa(1+q)^t)}{log_p((1+q)^t)}$$

where log_p is the p-adic logarithm defined by the usual Taylor expansion $log_p(x) = -\sum_{n=1}^{\infty} \frac{(1-x)^n}{n}$ for all $x \in \mathbf{C}_p$ such that $|x-1|_p \leq p^{-1}$ and t is an integer greater than $1/v_p(\kappa(1+q))$. Of course, from the definition we have $Log([k]) = k$. Moreover Log is clearly an analytic function on $\mathfrak{X}$.

If $F(q,X) = \sum_{n=0}^{\infty} a_n(X,F)q^n \in A(\mathfrak{U})[X]_r[[q]]$, we define

$$\delta F(q,X) := D.F(q,X) + Log(\kappa)XF(q,X)$$

where

$$D = q\frac{\partial}{\partial q} - X^2\frac{\partial}{\partial X}.$$

If $F(q,X) \in \mathcal{N}_{Q,\mathfrak{U}}^{\Sigma,pol}$ for a Zariski-dense set of classical weight Σ, it is clear that $\delta.F(q,X) \in \mathcal{N}_{\tilde{Q},\mathfrak{U}}^{\tilde{\Sigma},pol}$ with $\tilde{\Sigma} = \Sigma[2]$ and $\tilde{Q}(\kappa,X) = Q(\kappa.[-2],pX)$. We therefore thanks to Lemma 4 deduce we have a map

$$\delta : \mathcal{N}_{\mathfrak{U}}^{\dagger,r,\mathrm{fs}} \to \mathcal{N}_{\mathfrak{U}[2]}^{\dagger,r+1,\mathrm{fs}}$$

Remark also, it is straightforward to see that the effect of the operator ϵ on the polynomial q-expansion of families is the partial differentiation with respect to X:

$$(\epsilon.F)(q,X) = \frac{\partial}{\partial X}F(q,X) \quad \forall F \in \mathcal{N}_{\mathfrak{U}}^{\dagger,r,\mathrm{fs}}$$

Remark 10. Like in Remark 8, we can give a sheaf theoretic definition of $\hat{\mathcal{N}}_{\mathfrak{U}}^{\infty,\rho}$. For simplicity, let us assume that all the p-adic characters in $\mathfrak{U}$ are analytic on $\mathbf{Z}_p$. Let $\mathcal{A}_{\mathfrak{U}} := A(\mathfrak{U})\hat{\otimes}\mathcal{A}$. Elements in $\mathcal{A}_{\mathfrak{U}}$, can be seen as rigid analytic functions on $\mathfrak{U} \times \mathbf{Z}_p$. It is equipped with the representation of the standard Iwahori subgroup of $SL_2(\mathbf{Z}_p)$ defined by:

$$(\left(\begin{smallmatrix} a & b \\ c & d \end{smallmatrix}\right).f)(\kappa,X) = \kappa(a+cX)f(\frac{b+dX}{a+cX})$$

Again as in Remark 8, one can define but now using the technics of Pilloni (to appear) a sheaf of Banach spaces on $X^{\geq\rho} \times \mathfrak{U}$

$$\mathcal{H}_{\mathfrak{U}}^{\infty} := \mathcal{T} \times^B \mathcal{A}_{\mathfrak{U}} \cong \omega_{\mathfrak{U}} \otimes \mathcal{H}_0^{\infty}$$

and show that we have:

$$\hat{\mathcal{N}}_{\mathfrak{U}}^{\infty,\rho} = H^0(X^{\geq\rho} \times \mathfrak{U}, \mathcal{H}_{\mathfrak{U}}^{\infty})$$

Since $\mathcal{A}_{\mathfrak{U}}$ is a representation of the Lie algebra of sl_2, it would be possible to define a connection using the BGG formalism like in the algebraic case (see for instance Tilouine 2011, §3.2)

$$\mathcal{H}_{\mathfrak{U}}^{\infty} \to \mathcal{H}_{\mathfrak{U}[2]}^{\infty}$$

One would then obtain the Maass-Shimura operator in family without the finite slope condition:

$$\delta : \hat{\mathcal{N}}_{\mathfrak{U}}^{\rho,\infty} \to \hat{\mathcal{N}}_{\mathfrak{U}[2]}^{\rho,\infty}$$

It is then easy to verify that $\delta(\mathcal{N}_{\mathfrak{U}}^{\rho,r}) \subset \mathcal{N}_{\mathfrak{U}[2]}^{\rho,r+1}$. We leave the details of this construction for another occasion or to the interested reader.

Finally, we want to mention that Robert Harron and Liang Xiao (2013) have also given a geometric construction of this operator in family using a splitting of the Hodge filtration and showing the definition is independent of the chosen splitting. The above sketched construction can be done without such a choice but it probably boils down to a similar argument.

Now we have the following proposition.

Proposition 9. *Let $\mathfrak{U} \subset \mathfrak{X}$ be an open affinoid subdomain and $F \in \mathcal{N}_{\mathfrak{U}}^{\dagger,r,fs}$, then for each $i \in \{0,\dots,r\}$ there exists*

$$G_i \in \frac{1}{\prod_{j=2}^{2r}(Log(\kappa) - j)} \mathcal{N}_{\mathfrak{U}[-2i]}^{\dagger,0,fs}$$

such that

$$F = G_0 + \delta.G_1 + \cdots + \delta^r.G_r$$

Moreover, this decomposition is unique.

If $\mathfrak{U} = \{\kappa\}$ such that $Log(\kappa) \notin \{2,3,\dots 2r\}$, the result holds as well.

Proof. It is sufficient to prove this when $\mathfrak{U}$ is open since we can obtain the general result after specialization. We prove this by induction on r. Notice that for $G \in \mathcal{N}_{\mathfrak{U}[-2r]}^{\dagger,0,fs}$, we have by (4)

$$\epsilon^r.\delta^r.G = r!.\prod_{i=1}^{r}(Log(\kappa) - r - i).G$$

since the left hand and right hand sides coincide after evaluation at classical weights bigger than $2r$. We put

$$G_r := \frac{1}{r!.\prod_{i=1}^{r}(Log(\kappa) - r - i)} \epsilon^r F$$

then $G_r \in \frac{1}{\prod_{i=1}^{r}(Log(\kappa)-r-i)} \mathcal{N}_{\mathfrak{U}[-2r]}^{\dagger,0,fs}$ and $F - \delta^r.G_r$ is by construction of degree of nearly overconvergent less or equal to $r-1$. We conclude by induction. □

Then we define

$$\mathcal{H}^{\dagger}(F) := G_0$$

This is the overconvergent (or holomorphic) projection in family since it clearly coincides with the holomorphic projection for nearly holomorphic forms of weight $k > 2r$.

Lemma 5. *For any nearly overconvergent family of finite slope $F \in \mathcal{N}_{\mathfrak{U}}^{\dagger,r,\mathrm{fs}}$, and Heke operator T, we have*

$$\mathcal{H}^{\dagger}(T.F) = T.\mathcal{H}^{\dagger}(F).$$

In particular, for any admissible pair $(Q, \mathfrak{U})$ and we have

$$e_{Q,\mathfrak{U}}(\mathcal{H}^{\dagger}(F)) = \mathcal{H}^{\dagger}(e_{Q,\mathfrak{U}}(F)).$$

Proof. It follows easily from the relation $\delta^j(T(n).F) = n^j T(n).\delta^j(F)$ and the uniqueness of the G_i's in the decomposition of Proposition 4. □

4 Application to Rankin-Selberg p-Adic L-Functions

Let $\mathcal{E}$ be the eigencurve of level 1 constructed by Coleman and Mazur (1998). In this section, we give the main lines of a construction of a p-adic L-function on $\mathcal{E}\times\mathcal{E}\times\mathfrak{X}$. The general case of arbitrary tame level can be done exactly the same way. The restriction of our p-adic L-function to the ordinary part of the eigencurve, gives Hida's p-adic L-function constructed in Hida (1988, 1993). Our method follows closely Hida's construction for ordinary families of eigenforms. We are able to treat the general case using the framework of nearly overconvergent forms. We will omit the details of computation that are similar to Hida's construction and will focus on how we get rid of the ordinary assumptions. We don't pretend to any originality here. We just want to give an illustration of the theory of nearly overconvergent forms to the construction of p-adic L-functions in the non-ordinary case.

4.1 Review on Rankin-Selberg L-Function for Elliptic Modular Forms

We recall the definition and integral representations of the Rankin-Selberg L-function of two elliptic modular forms and its critical values. Let f and g be two elliptic normalized newforms of weights k and l with $k > l$ and nebentypus ψ and ξ respectively of level M. We denote their Fourier expansion by:

$$f(z) = \sum_{n=1}^{\infty} a_n q^n \text{ and } g(z) = \sum_{n=1}^{\infty} b_n q^n$$

Shimura is probably the first one to study in Shimura (1976) the algebraicities of the critical values of

$$D_M(s,f,g) := L(\psi\xi, k+l-2s-2)(\sum_{n=1}^{\infty} a_n b_n n^{-s})$$

More precisely he proved that for every integer $m \in \{0,\ldots,k-l-1\}$, then

$$\frac{D(l+m,f,g)}{\pi^{l+2m+1}\langle f,f\rangle_M} \in \overline{\mathbf{Q}}$$

Here $\langle f,f\rangle_M$ is the Petersson inner product of f with itself. Recall it is defined by the formula

$$\langle f,g\rangle_M = \int_{\Gamma_1(M)\backslash\mathfrak{h}} \overline{f(\tau)}g(\tau)y^{k-2}dxdy$$

When $0 \le 2m < k-l$, the essential ingredient in the proof of Shimura was to establish a formula of the type

$$D_M(l+m,f,g) = \langle f, g\delta^m_{k-l-2m}E\rangle_M = \langle f, \mathcal{H}(g\delta^m_{k-l-2m}E)\rangle_M$$

where E is a suitable holomorphic Eisenstein series of weight $k-l-2m$.

When f and g vary in Hida families and m is also allowed to vary p-adically, Hida has constructed a 3-variable p-adic L-function interpolating a suitable p-normalization these numbers. We now recall the precise formula that is used to interpolate these special values in Hida (1988). We first need some standard notations. For any integer M, we put $\tau_M = \left(\begin{smallmatrix} 0 & -1 \\ M & 0 \end{smallmatrix}\right)$ and for any modular form h, we denote by h^ρ the form defined by $h^\rho(\tau) = \overline{h(-\bar{\tau})}$ for $\tau \in \mathfrak{h}$. For any Dirichlet character χ of level M and any integer $j \ge 2$ such that $\chi(-1) = (-1)^j$, we denote by $E_j(\chi)$ the Eisenstein series of level M, nebentypus χ and weight j whose q-expansion is given by

$$E_j(\chi)(\tau) = \frac{L(1-m,\chi)}{2} + \sum_{n=1}^{\infty}(\sum_{\substack{d|n \\ (d,M)=1}} \chi(d)d^{j-1})q^n$$

Proposition 10 (Hida 1988, Thm 6.6). *Let L be an integer such that f and g are of level Lp^β, then we have*

$$D_{Lp}(l+m,f,g) = t.\pi^{l+2m+1}\langle f^\rho|_k\tau_{Lp^\beta}, \mathcal{H}(g|_l\tau_{Lp^\beta}\delta^m_{k-l-2m}(E_{k-l-2m,Lp}(\psi\xi)\rangle_{Lp^\beta}$$

with

$$t = \frac{2^{k+l+2m}(Lp^{\beta})^{\frac{k-l}{2}-m-1} i^{l-k}}{m!(l+m-1)!}.$$

and $m \in \mathbf{Z}$ *with* $0 \leq m < (k-l)/2$.

4.2 The p-Adic Petersson Inner Product

For simplicity, we assume the tame level is 1. Fix and admissible pair $(R, \mathfrak{V})$ for overconvergent forms and let $\mathcal{M}_{R,\mathfrak{V}}$ the corresponding associated space of $\mathfrak{V}$-families of overconvergent forms. Let $\mathbf{T}_{R,\mathfrak{V}}$ the Hecke algebra acting on $\mathcal{M}_{R,\mathfrak{V}}$. A standard argument using the q-expansion principle shows that the pairing

$$\mathcal{M}_{R,\mathfrak{V}} \otimes_{A(\mathfrak{U})} \mathbf{T}_{R,\mathfrak{V}} \to A(\mathfrak{V})$$

given by

$$(T, f) := a(1, f|T)$$

is a perfect duality. Since the level[9] is 1, we also know that $\mathbf{T}_{R,\mathfrak{V}}$ is reduced. Therefore, the trace map induces a non-degenerate pairing on $\mathbf{T}_{Q,\mathfrak{U}}$ with ideal discriminant $\mathfrak{d}_{R,\mathfrak{V}} \subset A(\mathfrak{V})$ whose set of zeros is the set of weight where the map $\mathcal{E}_{Q,\mathfrak{U}} \to \mathfrak{U}$ is ramified. In particular, we have a canonical isomorphism:

$$\mathcal{M}_{R,\mathfrak{V}} \otimes F(\mathfrak{V}) \cong T_{R,\mathfrak{V}} \otimes F(\mathfrak{V}) \tag{14}$$

4.2.1

From this, we deduce a Hecke-equivariant pairing

$$(-,-)_{R,\mathfrak{V}} : \mathcal{M}_{R,\mathfrak{V}} \otimes_{A(\mathfrak{V})} \otimes \mathcal{M}_{R,\mathfrak{V}} \longrightarrow F(\mathfrak{V})$$

Let now $\mathcal{F}$ be a Galois extension of $F(\mathfrak{V})$ the field of fraction of $A(\mathfrak{V})$. We assume that for each irreducible component $\mathcal{C}$ of $\mathcal{E}_{R,\mathfrak{V}}$, $\mathcal{F}$ contains the function field $F(\mathcal{C})$ of $\mathcal{C}$. For each irreducible component $\mathcal{C}$, we define the corresponding idempotent $1_{\mathcal{C}} \in T_{R,\mathfrak{V}} \otimes \mathcal{F}$ and we write $F_{\mathcal{C}}$ for the element defined by

---

[9] When the level is not 1, one uses the theory of primitive forms which described the maximal semi-simple direct factor of $T_{R,\mathfrak{V}} \otimes F(\mathfrak{V})$

$$F_{\mathcal{C}} := \sum_{n=1}^{\infty} \lambda_{\mathcal{C}}(T_n) q^n \in \mathcal{F}[[q]]$$

where $\lambda_{\mathcal{C}}$ is the character of the Hecke algebra defined by $T.1_{\mathcal{C}} \otimes id_{\mathcal{F}} = 1_{\mathcal{C}} \otimes \lambda_{\mathcal{C}}(T)$. If we denote by $(-,-)_{\mathcal{F}}$ the scalar extension of $(-,-)_{R,\mathfrak{V}}$ to $\mathcal{F}$ then the Hecke invariance of the inner product implies that

$$a(1, 1_{\mathcal{C}}.G) = (F_{\mathcal{C}}, G)_{\mathcal{F}}$$

is the coefficient of $F_{\mathcal{C}}$ when one writes G as a linear decomposition of the eigen families $F_{\mathcal{C}}$'s.

Remark 11. This construction can be easily extended to the space $\mathcal{N}_{R,\mathfrak{V}}$ for any admissible psi $(R, \mathfrak{V})$ for nearly overconvergent forms.

4.3 The Nearly Overconvergent Eisenstein Family

We consider the Eisenstein family $E(q) \in A^0(\mathfrak{X})[[q]]$ such that for each weight $\kappa \in \mathfrak{X}(\overline{\mathbf{Q}}_p)$, its evaluation at κ is given by

$$E(\kappa, q) = \sum_{n=1}^{\infty} a(n, E, \kappa) q^n := \sum_{\substack{n=1 \\ (n,p)=1}}^{\infty} \sum_{d|n} \langle d \rangle_{\kappa} . d^{-1} q^n.$$

where for any $m \in \mathbf{Z}_p^{\times}$, $\langle m \rangle_{\kappa} \in A(\mathfrak{X})$ stands for the analytic function of $\mathfrak{X}$ defined by

$$\kappa \mapsto \kappa(m)$$

In particular, when $\kappa = [k].\psi$ with ψ a ramified finite order character of $\mathbf{Z}_p^{\times}$, then $E(\kappa, q)$ is the q-expansion of

$$E_k^{(p)}(\psi)(\tau) := E_k(\psi)(\tau) - E_k(\psi)(p\tau).$$

We define the nearly overconvergent Eisenstein family $\Theta.E \in A^0(\mathfrak{X} \times \mathfrak{X})[[q]]$ by

$$\Theta.E(\kappa, \kappa') := \sum_{\substack{n=1 \\ (n,p)=1}}^{\infty} \langle n \rangle_{\kappa} a(n, E, \kappa') q^n$$

Lemma 6. *If $\kappa = [r]$ and $\kappa' = [k]\psi$, the evaluation at (κ, κ') of $\Theta.E$ is*

$$\Theta.E(\kappa, \kappa') = \Theta^r . E_k^{(p)}(\psi)(q).$$

It is the p-adic q-expansion of the nearly holomorphic Eisenstein series $\delta_k^r E_k^{(p)}(\psi)$.

Proof. The first part is obvious and the second part follows from the formula (4) and the canonical diagram of Proposition 6. □

4.4 Construction of the Rankin-Selberg L-Function on $\mathcal{E} \times \mathcal{E} \times \mathfrak{X}$

4.4.1 Some Preparation

Let $(Q, \mathfrak{U})$ be an admissible pair for overconvergent forms of tame level 1 and let $T_{Q,\mathfrak{U}}$ be the corresponding Hecke algebra over $A(\mathfrak{U})$. By definition it is the ring of analytic function on the affinoid subdomain $\mathcal{E}_{Q,\mathfrak{U}}$ sitting over the affinoid subdomain $Z_{Q,\mathfrak{U}}$ associated to $(Q, \mathfrak{U})$ of the spectral curve of the U_p-operator. Recall that

$$Z_{Q,\mathfrak{U}} = Max(A(\mathfrak{U})[X]/Q^*(X)) \subset Z_{U_p} \subset \mathbf{A}^1_{\mathbf{r}ig} \times \mathfrak{U}$$

where Z_{U_p} is the spectral curve attache to U_p (i.e. the set of points $(\alpha, \kappa) \in \mathbf{A}^1_{\mathbf{r}ig} \times \mathfrak{U}$ such that $P^0_\kappa(\alpha) = 0$) and

$$T_{Q,\mathfrak{U}} = A(\mathcal{E}_{Q,\mathfrak{U}}) \text{ with } \mathcal{E}_{Q,\mathfrak{U}} = \mathcal{E} \otimes_{Z_{U_p}} Z_{Q,\mathfrak{U}}$$

The universal family of overconvergent modular eigenforms of type $(Q, \mathfrak{U})$ is given by

$$G_{Q,\mathfrak{U}} := \sum_{n=1}^{\infty} T(n)q^n \in T_{Q,\mathfrak{U}}[[q]]$$

Tautologically, for any point $x \in \mathcal{E}_{Q,\mathfrak{U}}$ of weight $\kappa_x \in \mathfrak{U}$, the evaluation $G_{Q,\mathfrak{U}}(x)$ at x of $G_{Q,\mathfrak{U}}$ is the overconvergent normalized eigenform f_x of weight κ_x associated to x.

Let

$$G^E_{Q,\mathfrak{U}} := G_{Q,\mathfrak{U}}.\Theta.E \in T_{Q,\mathfrak{U}} \otimes A^b(\mathfrak{X} \times \mathfrak{X})[[q]] = A^b(\mathcal{E}_{Q,\mathfrak{U}} \times \mathfrak{X} \times \mathfrak{X})[[q]]$$

The Fourier coefficients of this Fourier expansions are analytic functions on $\mathcal{E}_{Q,\mathfrak{U}} \times \mathfrak{X} \times \mathfrak{X}$. Let now $(R, \mathfrak{V})$ be an admissible pair for nearly overconvergent forms of tame level 1. Then we consider

$$G^E_{Q,\mathfrak{U},R,\mathfrak{V}}(q) \in A^b(\mathfrak{V} \times \mathcal{E}_{Q,\mathfrak{U}} \times \mathfrak{X})[[q]]$$

defined by

$$G^E_{Q,\mathfrak{U},R,\mathfrak{V}}(\kappa, y, \nu)(q) := e_{R,\mathfrak{V}}.G^E_{Q,\mathfrak{U}}(y, \nu, \kappa\kappa_y^{-1}\nu^{-2})(q)$$

and where $e_{R,\mathfrak{V}} = S(U_p)$ for some $S \in X.A(\mathfrak{V})[[X]]$ is the projector of $\mathcal{N}_{\mathfrak{V}}^{\infty,fs}$ onto $\mathcal{N}_{R,\mathfrak{V}}$.

Proposition 11. *With the notation above $G^E_{Q,\mathfrak{U},R,\mathfrak{V}}(q)$ is the q-expansion of an element of $G^E_{Q,\mathfrak{U},R,\mathfrak{V}} \in \mathcal{N}_{R,\mathfrak{V}} \otimes_{\mathbf{Q}_p} A^b(\mathcal{E}_{Q,\mathfrak{U}} \times \mathfrak{X})$. Moreover if we have $(Q,\mathfrak{U}) < (Q',\mathfrak{U}')$ and $(R,\mathfrak{V}) < (R'\mathfrak{V}')$, then $G^E_{Q,\mathfrak{U},R,\mathfrak{V}}$ is the image of $G^E_{Q',\mathfrak{U}',R',\mathfrak{V}'}$ by the natural map*

$$\mathcal{N}_{R',\mathfrak{V}'} \otimes_{\mathbf{Q}_p} A^b(\mathcal{E}_{Q',\mathfrak{U}'} \times \mathfrak{X}) \to \mathcal{N}_{R,\mathfrak{V}} \otimes_{\mathbf{Q}_p} A^b(\mathcal{E}_{Q,\mathfrak{U}} \times \mathfrak{X})$$

Proof. By Corollary 5, with $(Q,\mathfrak{U})$ replaced by $(R,\mathfrak{V})$ and with $M = A^b(\mathcal{E}_{Q,\mathfrak{U}} \times \mathfrak{X})$, it is sufficient to show that the specialization at a Zariski dense set of arithmetic points of $([k],x,[r]) \in \mathfrak{V} \times \mathcal{E}_{Q,\mathfrak{U}} \times \mathfrak{X}$ is the q-expansion of a nearly holomorphic form of weight k annihilated by $R^*_{[k]}(U_p)$. It is sufficient to choose the triplet $([k],x,[r])$ such that $\kappa_x = [l]$ with $l \in \mathbf{Z}_{\geq 2}$, $r \geq 0$ such that $k-l-2r \geq 0$ since such triplets form a Zariski dense set of $\mathfrak{V} \times \mathcal{E}_{Q,\mathfrak{U}} \times \mathfrak{X}$. The evaluation at such a triplet is easily seen to be the p-adic q-expansion of $e_{R,k}(g_x.\Theta^r.E_{k-l-2r})$. By definition of $e_{R,k}$ it follows that this form belongs to $\mathcal{N}_{R,k}$. The second part of the proposition is a trivial consequence of (11). □

4.4.2 A 3-Variable p-Adic Meromorphic Function

Let $Z_{R,\mathfrak{V}} \subset A^1_{rig} \times \mathfrak{V}$ the affinoid of the spectral curve $Z^\infty_{U_p}$ attached to $P^\infty_{\mathfrak{V}}$. This affinoid is a priori not contained in the spectral curve attached to U_p but the eigencurve is still sitting over it since $Z_{U_p} \subset Z^\infty_{U_p}$. We can therefore consider

$$\mathcal{E}_{R,\mathfrak{V}} = \mathcal{E}_N \times_{Z^\infty_{U-p}} Z_{R,\mathfrak{V}}$$

and the $\mathcal{E}_{R,\mathfrak{V}}$'s form an admissible covering of $\mathcal{E}_N$ when the $(R,\mathfrak{V})$ vary.

Let $\mathcal{C} \subset \mathcal{E}_{R,\mathfrak{V}}$ be an irreducible component. Then we set

$$D_{\mathcal{C},Q,\mathfrak{U}} := a(1, 1_{\mathcal{C}}.\mathcal{H}^\dagger(G^E_{Q,\mathfrak{U},R,\mathfrak{V}})) \in F(\mathcal{C}) \otimes F(\mathcal{E}_{Q,\mathfrak{U}} \times \mathfrak{X})$$

Remark 12. If $H_{\mathcal{C}} \subset A(\mathcal{C})$ is a denominator of $1_{\mathcal{C}}$, then the poles of $D_{\mathcal{C},Q,\mathfrak{U}}$ come from the zeros of $H_{\mathcal{C}}$ and the poles of the overconvergent projector. Therefore we have:

$$H_{\mathcal{C}}.\prod_{i=2}^{2r_{R,\mathfrak{V}}} (Log(\kappa) - i).D_{\mathcal{C},Q,\mathfrak{U}} \in A(\mathcal{C} \times \mathcal{E}_{Q,\mathfrak{U}} \times \mathfrak{X}) \quad (15)$$

We denote by $D_{R,\mathfrak{V},Q,\mathfrak{U}}$ the unique element of

$$F(\mathcal{E}_{R,\mathfrak{V}} \times \mathcal{E}_{Q,\mathfrak{U}} \times \mathfrak{X}) = \prod_{\substack{\mathcal{C} \subset \mathcal{E}_{R,\mathfrak{V}} \\ irreducible}} F(\mathcal{C}) \otimes F(\mathcal{E}_{Q,\mathfrak{U}} \times \mathfrak{X})$$

restricting to $D_{\mathcal{C},Q,\mathfrak{U}}$ on $\mathcal{C} \times \mathcal{E}_{Q,\mathfrak{U}} \times \mathfrak{X}$ for each irreducible component $\mathcal{C}$ of $\mathcal{E}_{R,\mathfrak{V}}$. It can be constructed as the image of $\mathcal{H}^\dagger(G^E_{Q,\mathfrak{U},R,\mathfrak{V}})$ in $T_{R,\mathfrak{V}} \otimes F(\mathfrak{V}) \otimes F(\mathcal{E}_{Q,\mathfrak{U}} \times \mathfrak{X}) = F(\mathcal{E}_{R,\mathfrak{V}} \times \mathcal{E}_{Q,\mathfrak{U}} \times \mathfrak{X})$ by the map (14) tensored by $F(\mathcal{E}_{Q,\mathfrak{U}} \times \mathfrak{X})$

We have the following result:

Lemma 7. *There exist a meromorphic function $\mathcal{D}$ on $\mathcal{E} \times \mathcal{E} \times \mathfrak{X}$ whose restriction to $F(\mathcal{E}_{R,\mathfrak{V}} \times \mathcal{E}_{Q,\mathfrak{U}} \times \mathfrak{X})$ gives $D_{R,\mathfrak{V},Q,\mathfrak{U}}$ for any quadruplet $(R, \mathfrak{V}, Q, \mathfrak{U})$.*

Proof. If we have pairs $(R, \mathfrak{V}), (Q, \mathfrak{U}), (R', \mathfrak{V}'), (Q', \mathfrak{U})$ with $(R, \mathfrak{V}) < (R', \mathfrak{V}')$ and $(Q, \mathfrak{U}) < (Q', \mathfrak{U}')$ then we have by (11) $e_{R',\mathfrak{V}'}|_{\mathfrak{V}} \circ e_{R,\mathfrak{V}} = e_{R,\mathfrak{V}}$ and $e_{Q',\mathfrak{U}'}|_{\mathfrak{U}} \circ e_{Q,\mathfrak{U}} = e_{Q,\mathfrak{U}}$. Since the overconvergent projection is Hecke equivariant, we deduce that

$$e_{R,\mathfrak{V}} \otimes e_{Q,\mathfrak{V}}.(D_{R',\mathfrak{V}',Q',\mathfrak{U}'}|_{\mathfrak{V}\times\mathfrak{U}}) = D_{R,\mathfrak{V},Q,\mathfrak{U}}$$

and that the $D_{R,\mathfrak{V},Q,\mathfrak{U}}$'s glue to define a memormorphic function $\mathcal{D} \in F(\mathcal{E} \times \mathcal{E} \times \mathfrak{X})$.

4.4.3 The Interpolation Property

For $x \in \mathcal{E}(\overline{\mathbf{Q}}_p)$, we denote θ_x the corresponding character of the Hecke algebra. If x is attached to a classical form, we denote by f_x the eigenform attached to x. By definition,

$$\iota_p \circ \iota_\infty^{-1}(f_x(q)) = \sum_{n=1}^{\infty} \theta_x(T_n) q^n$$

We denote by k_x its weight, ψ_x its nebentypus and p^{m_x} its minimal level with m_x a positive integer. We will always assume $k_x \geq 2$ and that p^{m_x} is the conductor of ψ_x. We consider the complex number $W(f_x)$ defined by

$$h_x := f_x^\rho|\tau_{p^{m_x}} = W(f_x) f_x$$

It is a complex number of norm 1 called the root number of f_x.

If $\mathcal{E}$ is smooth at x, then there is only one irreducible component containing it and if $\mathcal{C}$ is the irreducible component of an affinoid $\mathcal{E}_{R,\mathfrak{V}}$ containing x, then

$$H_{\mathcal{C}}(\kappa_x) \neq 0 \tag{16}$$

In that case, we can define the specialization $1_x \in T_{R_{\kappa_x},\kappa_x}$ of $1_{\mathcal{C}}$ at κ_x and it satisfies:

$$T_n.1_x = \theta_x(T_n).1_x \quad \forall n$$

In general, $T_{R_{\kappa_x},\kappa_x}$ is not semi-simple so 1_x is not necessarily the (generalized) θ_x-eigenspace projector. But if the projection map $\mathcal{E} \to \mathfrak{X}$ is étale at x then it is. We know it is the case when x is non-critical; recall that x is said non-critical if

$$v_p(\theta_x(U_p)) < k_x - 1.$$

If all the slopes of R_{κ_x} are strictly less that $k_x - 1$ (something that we can assume after shrinking $\mathcal{E}_{R,\mathfrak{V}}$), the image of $1_{\mathcal{C}}$ into the Hecke algebra acting on the space of forms $\mathcal{M}_{R_{\kappa_x},\kappa_x}$ is the projector 1_{f_x} attached to the new form f_x. Moreover we can show by the same computations as Hida (1985, sect. 4) that

$$a(1, 1_{f_x}.g) = a(p, f_x)^{m_x-n}.p^{(n-m_x)(k_x/2-1)} \frac{\langle f_x^\rho | \tau_{p^n}, g\rangle_{p^n}}{\langle h_x, f_x\rangle_{p^{m_x}}} \tag{17}$$

for any $g \in \mathcal{M}_{R_{\kappa_x},\kappa_x}$ of level p^n with $n \geq m_x$.

For any $\nu \in \mathfrak{X}(\overline{\mathbf{Q}}_p)$, we write $k_\nu := \log(\nu)$. We say ν is arithmetic if $k_\nu \in \mathbf{Z}$ and we denote ψ_ν the finite order character such that $\nu = [k_\nu].\psi_\nu$.

We have the following theorem.

Theorem 1. *Let L be a finite extension of $\overline{\mathbf{Q}}_p$ and $(x, y) \in \mathcal{E} \times \mathcal{E}(L)$ and any arithmetic $\nu = [r].\psi_\nu$ such that κ_x, κ_y are arithmetic and satisfy the following*

(i) $k_x - k_y > r \geq 0$,
(ii) x is classical and non-critical,
(iii) y is classical,
(iv) The level of f_y^ρ equals the level of $f_y^\rho|\psi_\nu$
(v) ψ_x and ψ_y are ramified.

Then we have

$$\mathcal{D}(x, y, \nu) = (-1)^{k_y} W(f_x) W(f_y^\rho) a(p, f_x)^{m_x-m_y} p^{m_x(1-\frac{k_x}{2})+m_y(r+\frac{k_y}{2})} \tag{18}$$
$$\times \Gamma(k_y + r)\Gamma(k_x + 1) \frac{D_{p^n}(f_x, f_y^\rho|\psi_\nu, k_y + r)}{(2i\pi)^{k_x+k_y+2r+1}\pi^{1-k_x}\langle f_x, f_x\rangle_{p^{m_x}}}$$

Proof. This computation follows closely those of Hida (1993). We treat the case $k_x - k_y > 2r \geq 0$. The case $k_x - k_y > r \geq (k_x - k_y)/2$ can be treated similarly (see for instance Hida 1993) and is obtained using the functional equation for the nearly holomorphic Eisenstein series. We also assume ψ_ν is trivial to lighten the notations.

By our hypothesis, we can choose a quadruplet $(R, \mathfrak{V}, Q, \mathfrak{U})$ such that

(a) $(x, y) \in \mathcal{E}_{R,\mathfrak{V}} \times \mathcal{E}_{Q,\mathfrak{U}}(L)$
(b) The eigenvalues of $R_{\kappa_x}^*(X) \in L[X]$ are of valuation smaller than $k_x - 1$

Then, $\mathcal{D}(x, y, [r]) = D_{R,\mathfrak{V},Q,\mathfrak{U}}(x, y, [r])$. By the condition (b), we know that $T_{R_{\kappa_x},\kappa_x} = T_{R,\mathfrak{V}} \otimes_{\kappa_x} L$ is semi-simple and therefore the map $\mathcal{E}_{R,\mathfrak{V}} \to \mathfrak{V}$ is étale at x. In particular, $\mathcal{E}_{R,\mathfrak{V}}$ is smooth at x and x belongs to only one irreducible component $\mathcal{C}$ of $\mathcal{E}_{R,\mathfrak{V}}$. By construction, we have:

$$\begin{aligned} D_{R,\mathfrak{V},Q,\mathfrak{U}}(x, y, [r]) &= a(1, 1_x \circ \mathcal{H}^\dagger(e_{R_{\kappa_x},\kappa_x}(G_{Q,\mathfrak{U}}(y)\Theta E([r], \kappa_x\kappa_y^{-1}[2r]^{-1})))) \\ &= a(1, 1_{f_x}(\mathcal{H}^\dagger(e_{R_{\kappa_x},\kappa_x}(f_y\delta^r_{k_x-k_y-2r}E^{(p)}_{k_x-k_y-2r})))) \\ &= a(1, 1_{f_x} \circ e_{R_{\kappa_x},\kappa_x}(\mathcal{H}^\dagger(g))) \end{aligned}$$

with $g = f_y\delta^r_{k_x-k_y-2r}E^{(p)}_{k_x-k_y-2r}(\psi_x\psi_y^{-1})$. Since g is nearly holomorphic of order $\leq r$ and weight $k_x > 2r$, we have $\mathcal{H}^\dagger(g) = \mathcal{H}(g)$ is holomorphic. Since $\mathcal{H}(g)$ is an holomorphic form of level p^n with $n = Max(m_x, m_y)$, we have

$$\begin{aligned} D_{R,\mathfrak{V},Q,\mathfrak{U}}(x, y, [r]) &= a(1, 1_{f_x} \circ e_{R_{\kappa_x},\kappa_x}\mathcal{H}(g)) \\ &= a(p, f_x)^{m_x-n}.p^{(n-m_x)(\frac{k_x}{2}-1)}.\frac{\langle f_x^\rho|\tau_{p^n}, \mathcal{H}(g)\rangle_{p^n}}{\langle h_x, f_x\rangle_{p^{m_x}}} \\ &= a(p, f_x)^{m_x-n}.p^{(n-m_x)(\frac{k_x}{2}-1)}.\frac{\langle f_x^\rho|\tau_{p^n}, g\rangle_{p^n}}{\langle h_x, f_x\rangle_{p^{m_x}}} \end{aligned} \tag{19}$$

As in Hida (1993), we now transform f_y a little:

$$\begin{aligned} f_y &= (-1)^{k_y}f_y|\tau_{p^n}|\tau_{p^n} = (-1)^{k_y}.p^{\frac{k_y}{2}(n-m_y)}.f_y|\tau_{p^{m_y}}[p^{n-m_y}]|\tau_{p^n} \\ &= (-1)^{k_y}W(f_y^\rho).p^{\frac{k_y}{2}(n-m_y)}.f_y^c|[p^{n-m_y}]|\tau_{p^n} \end{aligned}$$

By replacing this in the expression above we get for (19):

$$\begin{gathered} a(p, f_x)^{m_x-n}.(-1)^{k_y}W(f_y^\rho).p^{(n-m_x)(\frac{k_x}{2}-1)}.p^{\frac{k_y}{2}(n-m_y)}.\times \\ \frac{\langle f_x^\rho|\tau_{p^n}, f_y^c|[p^{n-m_y}]|\tau_{p^n}E^{(p)}_{k_x-k_y-2r}(\psi_x\psi_y^{-1})\rangle_{p^n}}{\langle h_x, f_x\rangle_{p^{m_x}}} = \\ a(p, f_x)^{m_x-n}.(-1)^{k_y}W(f_y^\rho).p^{n(k_y+r)+m_x(1-\frac{k_x}{2})-m_y\frac{k_y}{2}} \times \\ (k_x + k_y + r + 1)!r!.\frac{D_{p^n}(f_x, f_y^\rho|[p^{n-m_y}], k_y + r)}{\pi^{k_y+2r+1}2^{k_x+k_y+2r}i^{k_y-k_x}\langle h_x, f_x\rangle_{p^{m_x}}} \end{gathered} \tag{20}$$

Now using the fact that for p dividing M, we have:

$$D_M(f, g[p^m], s) = a(p, f)^m p^{-ms}D_M(f, g, s))$$

we deduce that (19) is equal to

$$a(p, f_x)^{m_x - m_y}.(-1)^{k_y} W(f_y^\rho).p^{m_x(1-\frac{k_x}{2})+m_y(r+\frac{k_y}{2})} \times$$
$$(k_y + r - 1)!r!.\frac{D_{p^n}(f_x, f_y^\rho|[p^{n-m_y}], k_y + r)}{\pi^{k_y+2r+1}2^{k_x+k_y+2r}i^{k_y-k_x}\langle h_x, f_x\rangle_{p^{m_x}}}$$

and the specialization formula stated in the theorem follows. □

Remark 13. (a) This result is still true if x is classical and critical if it is not θ-critical. The condition (iv) and (v) are not necessary and could be removed at the expanse to modify the formula by adding some Euler factors at p.

(b) From the construction, we see that this meromorphic function has possible poles along certain hypersurfaces of $\mathcal{E} \times \mathcal{E} \times \mathfrak{X}$ corresponding to intersections of the irreducible components of the first variable and also along certain hypersurfaces created by the overconvergent projection. This happens when the overconvergent form f_x is at the same time the specialization of a family of overconvergent forms and a family of positive order nearly overconvergent forms. It is easy to see that implies x is θ-critical. In the next section, we review the definition of a θ-critical point and compute the residue of $\mathfrak{D}$ when the weight map at this point is étale.

4.4.4 Residue at an étale θ-Critical Point

Let $x_1 \in \mathcal{E}(L)$ of classical weight $k_1 \geq 2$ and slope $k_1 - 1$. We say that x is θ-critical if there exist x_0 of weight $k_0 = 2 - k_1$ such that $f_{x_1} = \Theta^{k_1-1} f_{x_0}$. Here we denote f_{x_0} the ordinary form of weight $k_0 = 2 - k_1$ attached to x_0. We then write $x_1 = \theta(x_0)$. We have the following result.

Theorem 2. *Let x_0 and x_1 as above. Assume that $\kappa : \mathcal{E} \to \mathfrak{X}$ is étale at $x_1 = \theta(x_0)$ or equivalently that $\mathcal{E}$ is smooth at x_0. Then the order of the pole of $\mathfrak{D}(x, y, \nu)$ at x_1 is at most one and*

$$Res|_{x=x_1}(\mathfrak{D}(x, y, \nu)) = \frac{\prod_{j=0}^{k_1-2}(Log(\nu\kappa_y) - j)(Log(\nu) - j)}{(k_1 - 1)!}\mathfrak{D}(x_0, y, \nu[1 - k_1]) \tag{21}$$

for all $(y, \nu) \in \mathcal{E} \times \mathfrak{X}$

Proof. The fact that $\mathcal{E} \stackrel{\kappa}{\to} \mathfrak{X}$ is étale at x_1 is equivalent to $\mathcal{E}$ smooth at x_0 is well-known and follows from R. Coleman's work.

We choose $(R_0, \mathfrak{V}_0)$ such that $x_0 \in \mathcal{E}_{R_0,\mathfrak{V}_0}(L)$. Consider the pair $(R_1, \mathfrak{V}_1)$ with $R_1(\kappa, X) = R_0(\kappa[2 - 2k_1], p^{k_1-1}X)$ and $\mathfrak{V}_1 = \mathfrak{V}_0[2k_1 - 2]$.

For $i = 0, 1$, let $\mathcal{C}_i$ be the (unique) irreducible component of $\mathcal{E}_{R_i,\mathfrak{V}_i}$ containing x_i and let consider $F_i = F_{\mathcal{C}_i}$ the corresponding Coleman family. Let $G = G^E_{R_1,\mathfrak{V}_1,Q,\mathfrak{U}}$ for some admissible pair $(Q, \mathfrak{U})$.

Let $\mathcal{F}$ be an extension of $F(\mathfrak{V})$ as in Sect. 4.2.1. Then by definition of $D = D_{C_1,Q,\mathfrak{U}}$ and of the overconvergent projection, we have

$$G = D.F_1 + D'\delta^{k_1-1}F_0 + H \tag{22}$$

with some $D' \in \mathcal{F} \otimes A^b(\mathcal{E}_{Q,\mathfrak{U}} \times \mathfrak{X})$ and $H \in \mathcal{N}_{R_1,\mathfrak{V}_1} \otimes A^b(\mathcal{E}_{Q,\mathfrak{U}} \times \mathfrak{X})$ such that $(H, F_i)_{\mathcal{F}} = 0$ for $i = 0, 1$ where $(-, -)_{\mathcal{F}}$ is the p-adic Petersson inner product defined in Sect. 4.2.1. Notice that by our hypothesis, $F_1(x_1) = f_{x_1} = \delta^{k_1-1}F_0(x_0)$. Since G is regular at $[k_1]$ and F_1 and $\delta^{k_1-1}F_0$ are the only families of nearly overconvergent forms of finite slope specializing to f_{x_1}, this implies that $D + D'$ is regular at x_1. Therefore in particular the order of the pole of D at x_1 is the same as the order of the pole of D' at x_1 and

$$Res|_{x=x_1}(D(x, y, \nu)) = -Res|_{x=x_1}(D'(x, y, \nu)) \tag{23}$$

From (22), we have

$$\epsilon^{k_1-1}G(\kappa_x, y, \nu) = \prod_{j=0}^{k_1-2}(2 - Log(\kappa_x) + j)D'(x, y, \nu).F_0 + \epsilon^{k_1-1}H \tag{24}$$

Since the eigencurve is smooth at x_0, this implies that $\prod_{j=0}^{k_1-2}(2 - Log(\kappa_x) + j)D'$ is regular at $x = x_1$ and therefore the pole of D' at x_1 is at most simple. Moreover, we get

$$Res|_{x=x_1}(D'(x, y, \nu)) = \frac{(-1)^{k_1}}{(k_1-1)!}.a(1, 1_{\mathcal{C}_0}.\epsilon^{k_1-1}G([k_1], y, \nu)) \tag{25}$$

Now we want to evaluate $\epsilon^{k_1-1}G(\kappa, y, \nu)$. For any classical triplet $(\kappa, y, \nu) = ([k], y, [r])$ with $k - k_y > 2r \geq 0$ and $\psi = \psi_y\psi_{x_1}$, we deduce from the evaluation of (5) at $X = 0$ that

$$\begin{aligned}
\epsilon^{k_1-1}G(x, y, \nu)(q) &= \epsilon^{k_1\ 1}e_{R_1,\mathfrak{V}_1}f_y\Theta^r.E_{k-k_y-2r}(\psi)(q) \\
&= e_{R_0,\mathfrak{V}_0}\epsilon^{k_1-1}f_y\Theta^r.E_{k-k_y-2r}(\psi)(q) \\
&= e_{R_0,\mathfrak{V}_0}f_y\epsilon^{k_1-1}\Theta^r.E_{k-k_y-2r}(\psi)(q) \\
&= \frac{\Gamma(k-k_y-r)r!}{\Gamma(k-k_y-r-k_1+1)(r-k_1+1)!}e_{R_0,\mathfrak{V}_0}f_y\Theta^{r-k_1+1}E_{k-k_y-2r}(\psi)(q)
\end{aligned}$$

We deduce that

$$\epsilon^{k_1-1} G(\kappa, y, \nu) = \prod_{j=1}^{k_1-1} (Log(\kappa\kappa_y^{-1}\nu^{-1}) - j)(Log(\nu) - j + 1)$$
$$\times G^E_{R_0,\mathfrak{V}_0,Q,\mathfrak{U}}(\kappa[2-2k_1], y, \nu[1-k_1])$$

since the left and right hand sides of the above have the same evaluations on a Zariski dense set of point of $\mathfrak{X} \times \mathcal{E}_{Q,\mathfrak{U}} \times \mathfrak{X}$. Evaluating at $\kappa = [k_1]$ gives:

$$\epsilon^{k_1-1} G([k_1], y, \nu) = \prod_{j=0}^{k_1-2} (j - Log(\nu\kappa_y))(Log(\nu) - j)$$
$$\times G^E_{R_0,\mathfrak{V}_0,Q,\mathfrak{U}}([2-k_1], y, \nu[1-k_1]) \qquad (26)$$

Since

$$a(1, 1_{\mathcal{C}_0} G^E_{R_0,\mathfrak{V}_0,Q,\mathfrak{U}}([2-k_1], y, \nu)) = \mathcal{D}(x_0, y, \nu)$$

for $(y, \nu) \in \mathcal{E}_{Q,\mathfrak{U}} \times \mathfrak{X}$, the formula (21) follows from (23), (25) and (26). □

Remark 14. This residue formula has a flavor similar to the work of Bellaiche (2012) in which it is proved that the standard p-adic L-function attached to a θ-critical point is divisible by a similar product of p-adic Log's.

Remark 15. It is also possible to define a two variable Rankin-Selberg p-adic L-function interpolating the critical values $D_p(f_x, f_y, k_x - 1)$ by replacing $\Theta.E(\kappa, \kappa')$ by $E^{ord}(\kappa)$ in our construction where for $\kappa = [k].\psi_\kappa$ and $k \in \mathbf{Z}_{\geq 2}$ we have

$$E^{ord}(\kappa) = \frac{L(1-k,\psi)}{2} + \sum_{n=1}^{\infty}\Big(\sum_{\substack{d|n \\ (n,p)=1}} \kappa(d)d^{-1}\Big)q^n.$$

Since $E(\kappa)$ has a pole at $\kappa = [0]$, this two-variable p-adic L-function would have a pole along the hypersurface defined by $\kappa_x = \kappa_y$. It should be easy to compute the corresponding residue and obtain a formula similar to the one of Hida's Theorem 3 Hida (1993, p. 228).

Acknowledgements The author would like to thank Giovanni Rosso and Chris Skinner for interesting conversations during the preparation of this work and Vincent Pilloni for pointing out an error in a previous version of this text. He would like also to thank Pierre Colmez who encouraged him to write this note. He is also grateful to the organizers of the conference Iwasawa 2012 held in Heidelberg for their invitation and for giving the opportunity to publish this paper in the proceedings of this conference. This work was also lectured during the Postech winter school in January 2013. The author would like to thank the organizers of this workshop for their invitation. Finally the author would like to thank the Florence Gould Foundation for its support when he was a Member at the Institute for Advanced Studies and when some part of this work was conceived.

References

Andreatta, F., Iovita, A., Pilloni, V.: Families of siegel modular forms. Ann. Math. (to appear)

Bellaiche, J.: Critical p-adic L-functions. Invent. Math. **189**(1), 1–60 (2012)

Berthelot, P., Ogus, A.: Notes on Crystalline Cohomology. Princeton University Press/University of Tokyo Press, Princeton/Tokyo, vi+243pp. (1978)

Coleman, R.: Overconvergent and classical modular forms. Invent. Math. **124**, 215–241 (1996)

Coleman, R.: p-adic Banach spaces and families of modular forms. Invent. Math. **127**(3), 417–479 (1997)

Coleman, R., Mazur, B.: The eigencurve. In: Galois Representations in Arithmetic Algebraic Geometry (Durham, 1996). Volume 254 of London Mathematical Society Lecture Note Series, pp. 1–113. Cambridge University Press, Cambridge (1998)

Coleman, R., Gouvêa, F., Fernando Q., Jochnowitz, N.: E_2, Θ and overconvergence. Internat. Math. Res. Notices 1, 23-41 (1995)

Darmon, H., Rotger, V.: Diagonal cycles and Euler systems I: a p-adic Gross-Zagier formula. Ann. Sci. Éc. Norm. Supér. **47**(4), 779–832 (2014)

Harris, M.: Arithmetic vector bundles and automorphic forms on Shimura varieties. I. Invent. Math. **82**, 151–189 (1985)

Harris, M.: Arithmetic vector bundles and automorphic forms on Shimura varieties. II. Compos. Math. tome **60**, 323–378 (1986)

Hida, H.: A p-adic measure attached to the zeta functions associated with two elliptic modular forms. I. Invent. Math. **79**, 159–195 (1985)

Hida, H.: A p-adic measure attached to the zeta functions associated with two elliptic modular forms. II. Annales de l'Institut Fourier tome **38**(3), 1–83 (1988)

Hida, H.: Elementary Theory of L-Functions and Eisenstein Series. London Mathematical Society Student Texts, vol. 26. Cambridge University Press, Cambridge/New York (1993)

Katz, N.: Nilpotent connections and the monodromy theorem: applications of a result of Turrittin. Publications mathématiques de l'Institut des Hautes Études Scientifique **39**, 175–232 (1970)

Katz, N.: p-adic properties of modular schemes and modular forms. In: International Summer School on Modular Functions, Antwerp (1972)

Nappari, M.: Holomorphic forms canonically attached to nearly holomorphic automorphic forms. Thesis, Brandeis University (1992)

Pilloni, V.: Overconvergent modular forms. Ann. Inst. Fourier **63**(1), 219–239 (2013)

Pilloni, V.: Formes modulaires surconvergentes. Annales de l'Institut Fourier (to appear)

Shimura, G.: Introduction to the Arithmetic Theory of Automorphic Functions, Kanō Memorial Lectures, No. 1, Publications of the Mathematical Society of Japan, No. 11, xiv+267pp. Iwanami Shoten/Princeton University Press, Tokyo/Princeton (1971)

Shimura, G.: The special values of the zeta functions associated with cusp forms. Commun. Pure Appl. Math. **29**(6), 783–804 (1976)

Skinner, C.M., Urban, E.: p-adic families of nearly holomorphic forms and applications (in preparation)

Tilouine, J.: Companion forms and classicity in the $GL_2(\mathbf{Q})$-case. In: Number Theory. Ramanujan Mathematical Society Lecture Notes Series, vol. 15, pp. 119–141. Ramanujan Mathematical Society, Mysore (2011)

Urban, E.: On the ranks of Selmer groups of elliptic curves over $\mathbf{Q}$. In: Automorphic Representations and L-Functions. Proceedings of the International Colloquium held at the Tata Institute of Fundamental Research, Mumbai, pp. 651–680 (2013)

Wiles, A.: On ordinary λ-adic representations associated to modular forms. Invent. Math. **94**(3), 529–573 (1988)

Harron, R., Xiao, L.: Gauss-Manin connections for p-adic families of nearly overconvergent modular forms. arXiv:1308.1732

Noncommutative L-Functions for Varieties over Finite Fields

Malte Witte

1 Introduction

Let $\mathcal{F}$ be an ℓ-adic sheaf on a separated scheme X of finite type over a finite field $\mathbb{F}$ of characteristic different from ℓ. The L-function of $\mathcal{F}$ is defined as the product over all closed points x of X of the characteristic polynomials of the geometric Frobenius automorphism $\mathfrak{F}_\xi$ at a geometric point ξ over x acting on the stalk $\mathcal{F}_\xi$:

$$L(\mathcal{F}, T) = \prod_{x} \det(\mathrm{id} - \mathfrak{F}_\xi T^{\deg x} \colon \mathcal{F}_\xi)^{-1}.$$

The Grothendieck trace formula relates the L-function to the action of the geometric Frobenius $\mathfrak{F}_{\mathbb{F}}$ on the ℓ-adic cohomology groups with proper support over the base change

$$\overline{X} = X \times_{\operatorname{Spec} \mathbb{F}} \operatorname{Spec} \overline{\mathbb{F}}$$

of X to the algebraic closure $\overline{\mathbb{F}}$ of $\mathbb{F}$:

$$L(\mathcal{F}, T) = \prod_{i \in \mathbb{Z}} \det(\mathrm{id} - \mathfrak{F}_{\mathbb{F}} T \colon \mathrm{H}^i_c(\overline{X}, \mathcal{F}))^{(-1)^{i+1}}.$$

It was used by Grothendieck to establish the rationality and the functional equation of the zeta function of X, both of which are parts of the Weil conjectures.

The Grothendieck trace formula may also be viewed as an equality between two elements of the first K-group of the power series ring $\mathbb{Z}_\ell[[T]]$. Since the ring $\mathbb{Z}_\ell[[T]]$

M. Witte (✉)
Mathematisches Institut, Universität Heidelberg, Im Neuenheimer Feld 288, Heidelberg, Germany
e-mail: witte@mathi.uni-heidelberg.de

T. Bouganis and O. Venjakob (eds.), *Iwasawa Theory 2012*, Contributions in Mathematical and Computational Sciences 7, DOI 10.1007/978-3-642-55245-8_15,

is a semilocal commutative ring, $K_1(\mathbb{Z}_\ell[[T]])$ may be identified with the group of units $\mathbb{Z}_\ell[[T]]^\times$ via the map induced by the determinant. For each closed point x of X, the $\mathbb{Z}_\ell[[T]]$-automorphism $\mathrm{id} - \mathfrak{F}_\xi T$ on $\mathbb{Z}_\ell[[T]] \otimes_{\mathbb{Z}_\ell} \mathcal{F}_\xi$ defines a class in $K_1(\mathbb{Z}_\ell[[T]])$. The product of all these classes converges in the profinite topology induced on $K_1(\mathbb{Z}_\ell[[T]])$ by the isomorphism

$$K_1(\mathbb{Z}_\ell[[T]]) \cong \varprojlim_n K_1(\mathbb{Z}_\ell[[T]]/(\ell^n, T^n)).$$

The image of the limit under the determinant map agrees with the inverse of the L-function of $\mathcal{F}$. On the other hand, the $\mathbb{Z}_\ell[[T]]$-automorphisms

$$\mathbb{Z}_\ell[[T]] \otimes_{\mathbb{Z}_\ell} \mathrm{H}^i_c(\overline{X}, \mathcal{F}) \xrightarrow{\mathrm{id}-\mathfrak{F}_{\mathbb{F}}T} \mathbb{Z}_\ell[[T]] \otimes_{\mathbb{Z}_\ell} \mathrm{H}^i_c(\overline{X}, \mathcal{F})$$

also give rise to elements in the group $K_1(\mathbb{Z}_\ell[[T]])$. The Grothendieck trace formula may thus be translated into an equality between the alternating product of those elements and the class corresponding to the L-function.

In this article, we will show that in the above reformulation of the Grothendieck trace formula, one may replace $\mathbb{Z}_\ell$ by any adic $\mathbb{Z}_\ell$-algebra, i.e. a compact, semilocal $\mathbb{Z}_\ell$-algebra Λ whose Jacobson radical is finitely generated. These rings play an important role in noncommutative Iwasawa theory.

A central step in this reformation is the development of a convenient framework, in which one can put the K-theoretic machinery to use. This was accomplished in Witte (2008), using the notion of Waldhausen categories. For any adic ring Λ, we introduced in loc. cit. a Waldhausen category of perfect complexes of adic sheaves of Λ-modules on X. Furthermore, we presented an explicit construction of a Waldhausen exact functor $\mathrm{R}\Gamma_c(\overline{X}, \mathcal{F}^\bullet)$ that computes the cohomology with proper support for any perfect complex $\mathcal{F}^\bullet$.

By suitably adapting the classical construction, we define the L-function of such a complex $\mathcal{F}^\bullet$ as an element $L(\mathcal{F}^\bullet, T)$ of $K_1(\Lambda[[T]])$. The automorphism $\mathrm{id} - \mathfrak{F}_{\mathbb{F}}T$ on $\Lambda[[T]] \otimes_\Lambda \mathrm{R}\Gamma_c(\overline{X}, \mathcal{F}^\bullet)$ gives rise to another class in $K_1(\Lambda[[T]])$. Below, we shall prove the following theorem.

Theorem 1. *Let $\mathcal{F}^\bullet$ be a perfect complex of adic sheaves of Λ-modules on X. Then*

$$L(\mathcal{F}^\bullet, T) = \left[\Lambda[[T]] \otimes_\Lambda \mathrm{R}\Gamma_c(\overline{X}, \mathcal{F}^\bullet) \xrightarrow{\mathrm{id}-\mathfrak{F}_{\mathbb{F}}T} \Lambda[[T]] \otimes_\Lambda \mathrm{R}\Gamma_c(\overline{X}, \mathcal{F}^\bullet)\right]^{-1}$$

in $K_1(\Lambda[[T]])$.

The rough line of argumentation in the proof is as follows. As in the proof of the classical Grothendieck trace formula, one may reduce everything to the case of X being a smooth geometrically connected curve over the finite field $\mathbb{F}$. Moreover, one can replace Λ by $\mathbb{Z}_\ell[\mathrm{Gal}(L/K)]$, where L is a Galois extension of the function field K of X. By the classical Grothendieck trace formula, we know that our theorem is true if we further enlarge $\mathbb{Z}_\ell[\mathrm{Gal}(L/K)]$ to the maximal order M in a split semisimple algebra. The crucial step is then to show that

$$\mathrm{SK}_1(\mathbb{Z}_\ell[\mathrm{Gal}(L/K)][[T]]) = \ker \mathrm{K}_1(\mathbb{Z}_\ell[\mathrm{Gal}(L/K)][[T]]) \to \mathrm{K}_1(M[[T]])$$

vanishes in the limit as L tends to the separable closure of K. This is achieved as follows: By using results of Oliver (1988) we prove that

$$\mathrm{SK}_1(\mathcal{O}_F[G][[T]]) = \mathrm{SK}_1(\mathcal{O}_F[G])$$

for any finite group G and the valuation ring $\mathcal{O}_F$ of any finite extension F of $\mathbb{Q}_\ell$. (This fact also has some other interesting applications, see e.g. Chinburg et al. (2012).) The vanishing of

$$\varprojlim_L \mathrm{SK}_1(\mathcal{O}_F[\mathrm{Gal}(L/K)])$$

then follows by using an argument from Fukaya and Kato (2006).

Outline. Section 2 recalls briefly Waldhausen's construction of algebraic K-theory. In Sect. 3 we introduce a special Waldhausen category that computes the K-theory of an adic ring. A similar construction is then used in Sect. 4 to define the categories of perfect complexes of adic sheaves. In Sect. 5 we prove the abovementioned results for $\mathrm{SK}_1(\mathbb{Z}_\ell[G][[T]])$. In Sect. 6 we define the L-function of a perfect complex of adic sheaves. Section 7 contains the proof of the Grothendieck trace formula for these L-functions.

2 Waldhausen Categories

Waldhausen (1985) introduced a construction of algebraic K-theory that is both more transparent and more flexible than Quillen's original approach. He associates K-groups to any category of the following kind.

Definition 1. A *Waldhausen category* $\mathbf{W}$ is a category with a zero object $*$, together with two subcategories $\mathrm{co}(\mathbf{W})$ (*cofibrations*) and $\mathrm{w}(\mathbf{W})$ (*weak equivalences*) subject to the following set of axioms.

1. Any isomorphism in $\mathbf{W}$ is a morphism in $\mathrm{co}(\mathbf{W})$ and $\mathrm{w}(\mathbf{W})$.
2. For every object A in $\mathbf{W}$, the unique map $* \to A$ is in $\mathrm{co}(\mathbf{W})$.
3. If $A \to B$ is a map in $\mathrm{co}(\mathbf{W})$ and $A \to C$ is a map in $\mathbf{W}$, then the pushout $B \cup_A C$ exists and the canonical map $C \to B \cup_A C$ is in $\mathrm{co}(\mathbf{W})$.
4. If in the commutative diagram

$$\begin{array}{ccccc} B & \longleftarrow & A & \xrightarrow{f} & C \\ \downarrow & & \downarrow & & \downarrow \\ B' & \longleftarrow & A' & \xrightarrow[g]{} & C' \end{array}$$

the morphisms f and g are cofibrations and the downwards pointing arrows are weak equivalences, then the natural map $B \cup_A C \to B' \cup_{A'} C'$ is a weak equivalence.

We denote maps from A to B in co(**W**) by $A \rightarrowtail B$, those in w(**W**) by $A \xrightarrow{\sim} B$. If $C = B \cup_A *$ is a cokernel of the cofibration $A \rightarrowtail B$, then we denote the natural quotient map from B to C by $B \twoheadrightarrow C$. The sequence

$$A \rightarrowtail B \twoheadrightarrow C$$

is called *exact sequence* or *cofibre sequence*.

Definition 2. A functor between Waldhausen categories is called *(Waldhausen) exact* if it preserves cofibrations, weak equivalences, and pushouts along cofibrations.

If **W** is a Waldhausen category, then Waldhausen's S-construction yields a topological space $\mathbb{K}(\mathbf{W})$ and Waldhausen exact functors $F\colon \mathbf{W} \to \mathbf{W}'$ yield continuous maps $\mathbb{K}(F)\colon \mathbb{K}(\mathbf{W}) \to \mathbb{K}(\mathbf{W}')$ (Waldhausen 1985).

Definition 3. The *n-th* K-*group of* **W** is defined to be the n-th homotopy group of $\mathbb{K}(\mathbf{W})$:

$$\mathrm{K}_n(\mathbf{W}) = \pi_n(\mathbb{K}(\mathbf{W})).$$

Example 1.

1. Any exact category **E** may be viewed as a Waldhausen category by taking the admissible monomorphisms as cofibrations and isomorphisms as weak equivalences. Then the Waldhausen K-groups of **E** agree with the Quillen K-groups of **E** (Thomason and Trobaugh 1990, Theorem 1.11.2).
2. Let $\mathbf{Kom}^b(\mathbf{E})$ be the category of bounded complexes over the exact category **E** with degreewise admissible monomorphisms as cofibrations and quasi-isomorphisms (in the category of complexes of an ambient abelian category **A**) as weak equivalences. By the Gillet-Waldhausen theorem (Thomason and Trobaugh 1990, Theorem 1.11.7), the Waldhausen K-groups of $\mathbf{Kom}^b(\mathbf{E})$ also agree with the K-groups of **E**.
3. In fact, Thomason showed that if **W** is any sufficiently nice Waldhausen category of complexes and $F\colon \mathbf{W} \to \mathbf{Kom}^b(\mathbf{E})$ a Waldhausen exact functor that induces an equivalence of the derived categories of **W** and $\mathbf{Kom}^b(\mathbf{E})$, then F induces an isomorphism of the corresponding K-groups (Thomason and Trobaugh 1990, Theorem 1.9.8).

Remark 1. In the view of Example 1.(3) one might wonder wether it is possible to define a reasonable K-theory for triangulated categories. However, Schlichting (2002) shows that such a construction fails to exist.

The zeroth K-group of a Waldhausen category can be described fairly explicitly as follows.

Proposition 1. *Let* $\mathbf{W}$ *be a Waldhausen category. The group* $\mathrm{K}_0(\mathbf{W})$ *is the abelian group generated by the objects of* $\mathbf{W}$ *modulo the relations*

1. $[A] = [B]$ *if there exists a weak equivalence* $A \xrightarrow{\sim} B$,
2. $[B] = [A][C]$ *if there exists a cofibre sequence* $A \rightarrowtail B \twoheadrightarrow C$.

Proof. See Thomason and Trobaugh (1990, §1.5.6). □

There also exists a description of $\mathrm{K}_1(\mathbf{W})$ for general $\mathbf{W}$ as the kernel of a certain group homomorphism (Muro and Tonks 2007). We shall come back to this description later in a more specific situation.

3 The K-Theory of Adic Rings

All rings will be associative with unity, but not necessarily commutative. For any ring R, we let

$$\mathrm{Jac}(R) = \{x \in R | 1 - rx \text{ is invertible for any } r \in R\}$$

denote the *Jacobson radical* of R, i.e. the intersection of all maximal left ideals. It is the largest two-sided ideal I of R such that $1 + I \subset R^{\times}$ (Lam 1991, Chapter 2, §4). The ring R is called *semilocal* if $R/\mathrm{Jac}(R)$ is artinian.

Definition 4. A ring Λ is called an *adic ring* if it satisfies any of the following equivalent conditions:

1. Λ is compact, semilocal and the Jacobson radical is finitely generated.
2. For each integer $n \geq 1$, the ideal $\mathrm{Jac}(\Lambda)^n$ is of finite index in Λ and

$$\Lambda = \varprojlim_n \Lambda/\mathrm{Jac}(\Lambda)^n.$$

3. There exists a twosided ideal I such that for each integer $n \geq 1$, the ideal I^n is of finite index in Λ and

$$\Lambda = \varprojlim_n \Lambda/I^n.$$

Example 2. The following rings are adic rings:

1. Any finite ring,
2. $\mathbb{Z}_\ell$,
3. The group ring $\Lambda[G]$ for any finite group G and any adic ring Λ,

4. The power series ring $\Lambda[[T]]$ for any adic ring Λ and an indeterminate T that commutes with all elements of Λ,
5. The profinite group ring $\Lambda[[G]]$, when Λ is a adic $\mathbb{Z}_\ell$-algebra and G is a topologically finitely generated profinite group whose ℓ-Sylow subgroup has finite index in G.

Note that adic rings are not noetherian in general, the power series over $\mathbb{Z}_\ell$ in two noncommuting indeterminates being a counterexample.

We will now examine the K-theory of Λ.

Definition 5. Let R be any ring. A complex $M^\bullet$ of left R-modules is called *strictly perfect* if it is strictly bounded and for every n, the module M^n is finitely generated and projective. We let $\mathbf{SP}(R)$ denote the Waldhausen category of strictly perfect complexes, with quasi-isomorphisms as weak equivalences and injective complex morphisms as cofibrations.

Definition 6. Let R and S be two rings. We denote by $R^{\mathrm{op}}\text{-}\mathbf{SP}(S)$ the Waldhausen category of complexes of S-R-bimodules (with S acting from the left, R acting from the right) which are strictly perfect as complexes of S-modules. The weak equivalences are given by quasi-isomorphisms, the cofibrations are the injective complex morphisms.

By Example 1 we know that the Waldhausen K-theory of $\mathbf{SP}(R)$ coincides with the Quillen K-theory of R:

$$\mathrm{K}_n(\mathbf{SP}(R)) = \mathrm{K}_n(R).$$

For complexes $M^\bullet$ and $N^\bullet$ of right and left R-modules, respectively, we let

$$(M \otimes_R N)^\bullet$$

denote the total complex of the bicomplex $M^\bullet \otimes_R N^\bullet$. Any complex $M^\bullet$ in $R^{\mathrm{op}}\text{-}\mathbf{SP}(S)$ clearly gives rise to a Waldhausen exact functor

$$(M \otimes_R (-))^\bullet : \mathbf{SP}(R) \to \mathbf{SP}(S).$$

and hence, to homomorphisms $\mathrm{K}_n(R) \to \mathrm{K}_n(S)$.

Let now Λ be an adic ring. The first algebraic K-group of Λ has the following useful property.

Proposition 2 (Fukaya and Kato 2006, Prop. 1.5.3). *Let Λ be an adic ring. Then*

$$\mathrm{K}_1(\Lambda) = \varprojlim_{I \in \mathfrak{I}_\Lambda} \mathrm{K}_1(\Lambda/I)$$

In particular, $\mathrm{K}_1(\Lambda)$ *is a profinite group.*

It will be convenient to introduce another Waldhausen category that computes the K-theory of Λ.

Definition 7. Let R be any ring. A complex $M^\bullet$ of left R-modules is called *DG-flat* if every module M^n is flat (but not necessarily finitely generated) and for every acyclic complex $N^\bullet$ of right R-modules, the complex $(N \otimes_R M)^\bullet$ is acyclic.

Remark 2. The notion of *DG*-flatness goes back to Avramov and Foxby (1991). Every bounded above complex of flat modules is *DG*-flat, but a *DG*-flat complex does not need to be bounded above and not every unbounded complex of flat modules is *DG*-flat. We refer to the cited reference for examples. Unbounded complexes will appear quite naturally in later constructions. If one desires, one can avoid the homological algebra of unbounded complexes by using appropriate truncation operations, but the author feels that the concept of *DG*-flatness is the more elegant solution. See also Remark 5 in this regard.

We shall denote the lattice of open ideals of an adic ring Λ by $\mathfrak{I}_\Lambda$.

Definition 8. Let Λ be an adic ring. We denote by $\mathbf{PDG}^{\mathrm{cont}}(\Lambda)$ the following Waldhausen category. The objects of $\mathbf{PDG}^{\mathrm{cont}}(\Lambda)$ are inverse system $(P_I^\bullet)_{I\in\mathfrak{I}_\Lambda}$ satisfying the following conditions:

1. For each $I \in \mathfrak{I}_\Lambda$, $P_I^\bullet$ is a *DG*-flat complex of left Λ/I-modules and *perfect*, i.e. quasi-isomorphic to a complex in $\mathbf{SP}(\Lambda/I)$,
2. For each $I \subset J \in \mathfrak{I}_\Lambda$, the transition morphism of the system
$$\varphi_{IJ} : P_I^\bullet \to P_J^\bullet$$
induces an isomorphism
$$\Lambda/J \otimes_{\Lambda/I} P_I^\bullet \cong P_J^\bullet.$$

A morphism of inverse systems $(f_I : P_I^\bullet \to Q_I^\bullet)_{I\in\mathfrak{I}_\Lambda}$ in $\mathbf{PDG}^{\mathrm{cont}}(\Lambda)$ is a weak equivalence if every f_I is a quasi-isomorphism. It is a cofibration if every f_I is injective and the system $(\operatorname{coker} f_I)_{I\in\mathfrak{I}_\Lambda}$ is in $\mathbf{PDG}^{\mathrm{cont}}(\Lambda)$.

To see that $\mathbf{PDG}^{\mathrm{cont}}(\Lambda)$ is indeed a Waldhausen category one uses that the category $\mathbf{PDG}^{\mathrm{cont}}(\Lambda)$ is a full subcategory of the category of complexes over the abelian category of $\mathfrak{I}_\Lambda$-systems of Λ-modules which is closed under shifts and extensions; see Witte (2008, Prop. 5.4.5) for a detailed proof.

Proposition 3. *The Waldhausen exact functor*
$$\mathbf{SP}(\Lambda) \to \mathbf{PDG}^{\mathrm{cont}}(\Lambda), \qquad P^\bullet \to (\Lambda/I \otimes_\Lambda P^\bullet)_{I\in\mathfrak{I}_\Lambda}$$
identifies $\mathbf{SP}(\Lambda)$ *with a full Waldhausen subcategory of* $\mathbf{PDG}^{\mathrm{cont}}(\Lambda)$. *Moreover, it induces isomorphisms*
$$\mathrm{K}_n(\mathbf{SP}(\Lambda)) \cong \mathrm{K}_n(\mathbf{PDG}^{\mathrm{cont}}(\Lambda)).$$

Proof. The main step is to show that for every object $(Q_I^\bullet)_{I\in\mathfrak{I}_\Lambda}$ in $\mathbf{PDG}^{\mathrm{cont}}(\Lambda)$, the complex

$$\varprojlim_{I\in\mathfrak{I}_\Lambda} Q_I^\bullet$$

is a perfect complex of Λ-modules. This is proved using the argument of Fukaya and Kato (2006, Proposition 1.6.5). The assertion about the K-theory is then an easy consequence of the Waldhausen approximation theorem. We refer to Witte (2008, Proposition 5.2.5) for the details. □

We will now extend the definition of the tensor product to $\mathbf{PDG}^{\mathrm{cont}}(\Lambda)$.

Definition 9. For $(P_I^\bullet)_{I\in\mathfrak{I}_\Lambda} \in \mathbf{PDG}^{\mathrm{cont}}(\Lambda)$ and $M^\bullet \in \Lambda^{\mathrm{op}}\text{-}\mathbf{SP}(\Lambda')$ we set

$$\Psi_M\left((P_I^\bullet)_{I\in\mathfrak{I}_\Lambda}\right) = (\varprojlim_{J\in\mathfrak{I}_\Lambda} \Lambda'/I \otimes_{\Lambda'} (M\otimes_\Lambda P_J)^\bullet)_{I\in\mathfrak{I}_{\Lambda'}}$$

and obtain a Waldhausen exact functor

$$\Psi_M\colon \mathbf{PDG}^{\mathrm{cont}}(\Lambda) \to \mathbf{PDG}^{\mathrm{cont}}(\Lambda').$$

Note that the annihilator

$$A = \left\{x \in \Lambda \mid \quad \Lambda'/I \otimes_{\Lambda'} M^\bullet x = 0\right\}$$

is an open two-sided ideal of Λ and therefore, an element of $\mathfrak{I}_\Lambda$. If $J \in \mathfrak{I}_\Lambda$ is contained in this annihilator, then we have

$$\Lambda'/I \otimes_{\Lambda'} (M\otimes_\Lambda P_J)^\bullet \cong (\Lambda'/I \otimes_{\Lambda'} M \otimes_{\Lambda/A} \Lambda/A \otimes_{\Lambda/J} P_J)^\bullet \cong \Lambda'/I \otimes_{\Lambda'} (M\otimes_\Lambda P_A)^\bullet.$$

In particular, since $\mathfrak{I}_\Lambda$ is filtered,

$$\varprojlim_{J\in\mathfrak{I}_\Lambda} \Lambda'/I \otimes_{\Lambda'} (M\otimes_\Lambda P_J)^\bullet \cong \Lambda'/I \otimes_{\Lambda'} (M\otimes_\Lambda P_A)^\bullet.$$

With this description, it is clear that $\Psi_M\left((\mathcal{F}_I^\bullet)_{I\in\mathfrak{I}_\Lambda}\right)$ is indeed an object in the category $\mathbf{PDG}^{\mathrm{cont}}(X, \Lambda')$. We refer to Witte (2008, Prop. 5.5.4) for the easy proof that Ψ_M is Waldhausen exact.

Proposition 4. *Let $M^\bullet$ be a complex in $\Lambda^{\mathrm{op}}\text{-}\mathbf{SP}(\Lambda')$. Then the following diagram commutes.*

$$\begin{array}{ccc}
\mathrm{K}_n(\mathbf{SP}(\Lambda)) & \xrightarrow{\cong} & \mathrm{K}_n(\mathbf{PDG}^{\mathrm{cont}}(\Lambda)) \\
\Big\downarrow {\scriptstyle \mathrm{K}_n(M^\bullet\otimes_\Lambda(-))} & & \Big\downarrow {\scriptstyle \mathrm{K}_n(\Psi_M)} \\
\mathrm{K}_n(\mathbf{SP}(\Lambda')) & \xrightarrow{\cong} & \mathrm{K}_n(\mathbf{PDG}^{\mathrm{cont}}(\Lambda'))
\end{array}$$

Proof. Let $P^\bullet$ be a strictly perfect complex in $\mathbf{SP}(\Lambda)$. There exists a canonical isomorphism

$$\left(\Lambda'/I \otimes_{\Lambda'} (M \otimes_\Lambda P)^\bullet\right)_{I\in\mathfrak{I}_{\Lambda'}} \cong \big(\varprojlim_{J\in\mathfrak{I}_\Lambda} \Lambda'/I \otimes_{\Lambda'} (M \otimes_\Lambda \Lambda/J \otimes_\Lambda P)^\bullet\big)_{I\in\mathfrak{I}_{\Lambda'}}.$$

□

From Muro and Tonks (2007) we deduce the following properties of the group $\mathrm{K}_1(\Lambda)$.

Proposition 5. *The group* $\mathrm{K}_1(\Lambda)$ *is generated by the weak autoequivalences*

$$(f_I\colon P_I^\bullet \xrightarrow{\sim} P_I^\bullet)_{I\in\mathfrak{I}_\Lambda}$$

in $\mathbf{PDG}^{\mathrm{cont}}(\Lambda)$. *Moreover, we have the following relations:*

1. $[(f_I\colon P_I^\bullet \xrightarrow{\sim} P_I^\bullet)_{I\in\mathfrak{I}_\Lambda}] = [(g_I\colon P_I^\bullet \xrightarrow{\sim} P_I^\bullet)_{I\in\mathfrak{I}_\Lambda}][(h_I\colon P_I^\bullet \xrightarrow{\sim} P_I^\bullet)_{I\in\mathfrak{I}_\Lambda}]$ *if for each* $I \in \mathfrak{I}_\Lambda$, *one has* $f_I = g_I \circ h_I$,
2. $[(f_I\colon P_I^\bullet \xrightarrow{\sim} P_I^\bullet)_{I\in\mathfrak{I}_\Lambda}] = [(g_I\colon Q_I^\bullet \xrightarrow{\sim} Q_I^\bullet)_{I\in\mathfrak{I}_\Lambda}]$ *if for each* $I \in \mathfrak{I}_\Lambda$, *there exists a quasi-isomorphism* $a_I\colon P_I^\bullet \xrightarrow{\sim} Q_I^\bullet$ *such that the square*

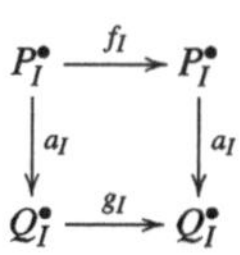

commutes up to homotopy,
3. $[(g_I\colon P_I'^\bullet \xrightarrow{\sim} P_I'^\bullet)_{I\in\mathfrak{I}_\Lambda}] = [(f_I\colon P_I^\bullet \xrightarrow{\sim} P_I^\bullet)_{I\in\mathfrak{I}_\Lambda}][(h_I\colon P_I''^\bullet \xrightarrow{\sim} P_I''^\bullet)_{I\in\mathfrak{I}_\Lambda}]$ *if for each* $I \in \mathfrak{I}_\Lambda$, *there exists an exact sequence* $P_I \rightarrowtail P_I' \twoheadrightarrow P_I''$ *such that the diagram*

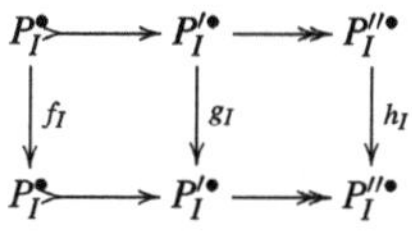

commutes in the strict sense.

Proof. The description of $\mathrm{K}_1(\mathbf{PDG}^{\mathrm{cont}}(\Lambda))$ as the kernel of

$$\mathfrak{D}_1\mathbf{PDG}^{\mathrm{cont}}(\Lambda) \xrightarrow{\partial} \mathfrak{D}_0\mathbf{PDG}^{\mathrm{cont}}(\Lambda)$$

given in Muro and Tonks (2007) shows that the weak autoequivalences are indeed elements of $\mathrm{K}_1(\mathbf{PDG}^{\mathrm{cont}}(\Lambda))$. Together with Proposition 2 this description also implies that relations (1) and (3) are satisfied. For relation (2), one can use

(Witte 2008, Lemma 3.1.6). Finally, the classical description of $\mathrm{K}_1(\Lambda)$ implies that $\mathrm{K}_1(\mathbf{PDG}^{\mathrm{cont}}(\Lambda))$ is already generated by isomorphisms of finitely generated, projective modules viewed as strictly perfect complexes concentrated in degree 0.

□

Remark 3. Despite the relatively explicit description of $\mathrm{K}_1(\mathbf{W})$ for a Waldhausen category $\mathbf{W}$ in Muro and Tonks (2007) it is not an easy task to deduce from it a presentation of $\mathrm{K}_1(\mathbf{W})$ as an abelian group. We refer to Muro and Tonks (2008) for a partial result in this direction.

In particular, one should not expect that the relations (1)–(3) describe the group $\mathrm{K}_1(\mathbf{PDG}^{\mathrm{cont}}(\Lambda))$ completely. However, they will suffice for the purpose of this paper.

4 Perfect Complexes of Adic Sheaves

We let $\mathbb{F}$ denote a finite field of characteristic p, with $q = p^{\nu}$ elements. Furthermore, we fix an algebraic closure $\overline{\mathbb{F}}$ of $\mathbb{F}$.

Write $\mathbf{Sch}_{\mathbb{F}}^{\mathrm{sep}}$ for the category of separated $\mathbb{F}$-schemes of finite type. For any scheme X in $\mathbf{Sch}_{\mathbb{F}}^{\mathrm{sep}}$ and any adic ring Λ we introduced in Witte (2008) a Waldhausen category $\mathbf{PDG}^{\mathrm{cont}}(X, \Lambda)$ of perfect complexes of adic sheaves on X. Below, we will recall the definition.

Definition 10. Let R be a finite ring and X be a scheme in $\mathbf{Sch}_{\mathbb{F}}^{\mathrm{sep}}$. A complex $\mathcal{F}^\bullet$ of étale sheaves of left R-modules on X is called *strictly perfect* if it is strictly bounded and each $\mathcal{F}^n$ is constructible and flat. A complex is called *perfect* if it is quasi-isomorphic to a strictly perfect complex. It is *DG-flat* if for each geometric point of X, the complex of stalks is *DG*-flat.

Definition 11. We will denote by $\mathbf{PDG}(X, R)$ the *category of DG-flat perfect complexes of R-modules* on X. It is a Waldhausen category with quasi-isomorphisms as weak equivalences and injective complex morphisms with cokernel in $\mathbf{PDG}(X, R)$ as cofibrations.

Definition 12. Let X be a scheme in $\mathbf{Sch}_{\mathbb{F}}^{\mathrm{sep}}$ and let Λ be an adic ring. The *category of perfect complexes of adic sheaves* $\mathbf{PDG}^{\mathrm{cont}}(X, \Lambda)$ is the following Waldhausen category. The objects of $\mathbf{PDG}^{\mathrm{cont}}(X, \Lambda)$ are inverse system $(\mathcal{F}_I^\bullet)_{I \in \mathfrak{I}_\Lambda}$ such that:

1. For each $I \in \mathfrak{I}_\Lambda$, $\mathcal{F}_I^\bullet$ is in $\mathbf{PDG}(X, \Lambda/I)$,
2. For each $I \subset J \in \mathfrak{I}_\Lambda$, the transition morphism

$$\varphi_{IJ} : \mathcal{F}_I^\bullet \to \mathcal{F}_J^\bullet$$

of the system induces an isomorphism

$$\Lambda/J \otimes_{\Lambda/I} \mathcal{F}_I^\bullet \cong \mathcal{F}_J^\bullet.$$

Weak equivalences and cofibrations are those morphisms of inverse systems that are weak equivalences or cofibrations for each $I \in \mathfrak{I}_\Lambda$, respectively.

We refer to Witte (2008, Cor. 4.1.4, Prop. 5.4.5) for the straightforward verification that $\mathbf{PDG}(X, \Lambda)$ and $\mathbf{PDG}^{\mathrm{cont}}(X, \Lambda)$ are indeed Waldhausen categories.

Remark 4. If Λ is a finite ring, the zero ideal is open and hence, an element in $\mathfrak{I}_\Lambda$. In particular, the following Waldhausen exact functors are mutually inverse equivalences for finite rings Λ:

$$\mathbf{PDG}^{\mathrm{cont}}(X, \Lambda) \to \mathbf{PDG}(X, \Lambda), \qquad (\mathcal{F}_I^\bullet)_{I \in \mathfrak{I}_\Lambda} \mapsto \mathcal{F}_{(0)}^\bullet,$$
$$\mathbf{PDG}(X, \Lambda) \to \mathbf{PDG}^{\mathrm{cont}}(X, \Lambda), \qquad \mathcal{F}^\bullet \mapsto (\Lambda/I \otimes_\Lambda \mathcal{F}^\bullet)_{I \in \mathfrak{I}_\Lambda}.$$

We use these functors to identify the two categories.

If $\Lambda = \mathbb{Z}_\ell$, then the subcategory of complexes concentrated in degree 0 of $\mathbf{PDG}^{\mathrm{cont}}(X, \mathbb{Z}_\ell)$ corresponds precisely to the exact category of those constructible ℓ-adic sheaves on X in the sense of Grothendieck (1977, Exp. VI, Def. 1.1.1, Exp. V, Def. 3.1.1) which are flat. In this sense, we recover the classical theory.

If $f: Y \to X$ is a morphism in $\mathbf{Sch}_{\mathbb{F}}^{\mathrm{sep}}$, we define a Waldhausen exact functor

$$f^*: \mathbf{PDG}^{\mathrm{cont}}(X, \Lambda) \to \mathbf{PDG}^{\mathrm{cont}}(Y, \Lambda), \qquad (\mathcal{F}_I^\bullet)_{I \in \mathfrak{I}_\Lambda} \mapsto (f^*\mathcal{F}_I^\bullet)_{I \in \mathfrak{I}_\Lambda}.$$

We will also need a Waldhausen exact functor that computes higher direct images with proper support. For the purposes of this article it suffices to use the following construction.

For any X in $\mathbf{Sch}_{\mathbb{F}}^{\mathrm{sep}}$ and any complex $\mathcal{G}^\bullet$ of abelian étale sheaves on X we let $\mathrm{G}_X^\bullet \mathcal{G}$ denote the Godement resolution of $\mathcal{G}^\bullet$ (Witte 2008, Def. 4.2.1). We note that $\mathrm{G}_X^\bullet \mathcal{G}$ is a complex of flabby sheaves quasi-isomorphic to $\mathcal{G}^\bullet$ (Witte 2008, Lemma 4.2.3, Prop. 4.26). Moreover, for any finite ring Λ the Godement resolution is a Waldhausen exact functor

$$\mathrm{G}_X^\bullet: \mathbf{PDG}(X, \Lambda) \to \mathbf{PDG}(X, \Lambda)$$

(Witte 2008, Cor. 4.2.8).

Remark 5. We point out that with the above definition, $\mathrm{G}_X^\bullet \mathcal{G}$ is in general not bounded above, even if $\mathcal{G}^\bullet$ is concentrated in degree 0 and that $\mathrm{G}_X^n \mathcal{G}$ is not a constructible sheaf of Λ-modules, even if this is true for all $\mathcal{G}^n$, $n \in \mathbb{Z}$. Since X is of finite Krull dimension, one can also work with a truncated version of the Godement resolution that preserves boundedness of complexes, but we do not know of any resolution with good functorial properties taking strictly perfect complexes to strictly perfect complexes.

By a theorem of Nagata (1963, Thm. 2) there exist a factorisation $f = p \circ j$ for any morphism $f: X \to Y$ in the category $\mathbf{Sch}_{\mathbb{F}}^{\mathrm{sep}}$ of separated schemes of finite type over $\mathbb{F}$ such that $j: X \hookrightarrow X'$ is an open immersion and $p: X' \to Y$ is a proper morphism.

Definition 13. Define a Waldhausen exact functor

$$\begin{aligned} \mathrm{R}\, f_!: \mathbf{PDG}^{\mathrm{cont}}(X, \Lambda) &\to \mathbf{PDG}^{\mathrm{cont}}(Y, \Lambda) \\ (\mathcal{F}_I^\bullet)_{I \in \mathfrak{J}_\Lambda} &\mapsto (p_* \, \mathrm{G}^\bullet_{X'} \, j_! \mathcal{F}_I)_{I \in \mathfrak{J}_\Lambda} \end{aligned}$$

For the verification that the definition of $\mathrm{R}\, f_!$ makes sense and produces a Waldhausen exact functor we refer to Witte (2008, Prop. 4.3.4, Prop. 4.3.8, Def. 5.4.13).

Obviously, this definition depends on the particular choice of the compactification $f = p \circ j$. However, all possible choices will induce the same homomorphisms

$$\mathrm{K}_n(\mathrm{R}\, f_!): \mathrm{K}_n(\mathbf{PDG}^{\mathrm{cont}}(X, \Lambda)) \to \mathrm{K}_n(\mathbf{PDG}^{\mathrm{cont}}(Y, \Lambda))$$

and this is all we need.

Remark 6. In Witte (2008, Section 4.5) we present a way to make the construction of $\mathrm{R}\, f_!$ independent of the choice of a particular compactification.

Proposition 6. *Let $f: X \to Y$ be a morphism in $\mathbf{Sch}^{\mathrm{sep}}_{\mathbb{F}}$.*

1. *$\mathrm{K}_n(\mathrm{R}\, f_!)$ is independent of the choice of a compactification $f = p \circ j$.*
2. *Let $\mathbb{F}'$ be a subfield of $\mathbb{F}$ and consider f as a morphism in $\mathbf{Sch}^{\mathrm{sep}}_{\mathbb{F}'}$. Then $\mathrm{K}_n(\mathrm{R}\, f_!)$ remains the same.*
3. *If $g: Y \to Z$ is another morphism in $\mathbf{Sch}^{\mathrm{sep}}_{\mathbb{F}}$, then*

$$\mathrm{K}_n(\mathrm{R}(g \circ f)_!) = \mathrm{K}_n(\mathrm{R}\, g_!) \circ \mathrm{K}_n(\mathrm{R}\, f_!)$$

4. *For any cartesian square*

$$\begin{array}{ccc} Y \times_X Z & \xrightarrow{f'} & Z \\ \downarrow{\scriptstyle g'} & & \downarrow{\scriptstyle g} \\ Y & \xrightarrow{f} & X \end{array}$$

in $\mathbf{Sch}^{\mathrm{sep}}_{\mathbb{F}}$ we have

$$\mathrm{K}_n(f^* \,\mathrm{R}\, g_!) = \mathrm{K}_n(\mathrm{R}\, g'_! \, f'^*)$$

Proof. All of this follows easily from Artin et al. (1972, Exposé XXVII). See also Witte (2008, Section 4.5). □

Definition 14. Let X be a scheme in $\mathbf{Sch}^{\mathrm{sep}}_{\mathbb{F}}$ and write $h: X \to \operatorname{Spec} \mathbb{F}$ for the structure map, $s: \operatorname{Spec} \overline{\mathbb{F}} \to \operatorname{Spec} \mathbb{F}$ for the map induced by the embedding into the algebraic closure. We define the Waldhausen exact functor

$$\mathrm{R}_{\mathbb{F}}\, \Gamma_c(\overline{X}, -)\colon \mathbf{PDG}^{\mathrm{cont}}(X, \Lambda) \to \mathbf{PDG}^{\mathrm{cont}}(\Lambda)$$

to be the composition of

$$\mathrm{R}\, h_!\colon \mathbf{PDG}^{\mathrm{cont}}(X, \Lambda) \to \mathbf{PDG}^{\mathrm{cont}}(\operatorname{Spec} \mathbb{F}, \Lambda)$$

with the section functor

$$\mathbf{PDG}^{\mathrm{cont}}(\operatorname{Spec} \mathbb{F}, \Lambda) \to \mathbf{PDG}^{\mathrm{cont}}(\Lambda), \qquad (\mathcal{F}_I^\bullet)_{I \in \mathfrak{I}_\Lambda} \to (\Gamma(\operatorname{Spec} \overline{\mathbb{F}}, s^*\mathcal{F}_I^\bullet))_{I \in \mathfrak{I}_\Lambda}.$$

Remark 7. If $\mathbb{F}'$ is a subfield of $\mathbb{F}$, then $\mathrm{R}_{\mathbb{F}}\, \Gamma_c(\overline{X}, -)$ and $\mathrm{R}_{\mathbb{F}'}\, \Gamma_c(\overline{X}, -)$ are in fact quasi-isomorphic and hence, they induce the same homomorphism of K-groups. Nevertheless, it will be convenient to distinguish between the two functors. We will omit the index if the base field is clear from the context.

The definition of Ψ_M extends to $\mathbf{PDG}^{\mathrm{cont}}(X, \Lambda)$.

Definition 15. For $(\mathcal{F}_I^\bullet)_{I \in \mathfrak{I}_\Lambda} \in \mathbf{PDG}^{\mathrm{cont}}(X, \Lambda)$ and $M^\bullet \in \Lambda^{\mathrm{op}}\text{-}\mathbf{SP}(\Lambda')$ we set

$$\Psi_M\left((\mathcal{F}_I^\bullet)_{I \in \mathfrak{I}_\Lambda}\right) = (\varprojlim_{J \in \mathfrak{I}_\Lambda} \Lambda'/I \otimes_{\Lambda'} (M \otimes_\Lambda \mathcal{F}_J)^\bullet)_{I \in \mathfrak{I}_{\Lambda'}}$$

and obtain a Waldhausen exact functor

$$\Psi_M\colon \mathbf{PDG}^{\mathrm{cont}}(X, \Lambda) \to \mathbf{PDG}^{\mathrm{cont}}(X, \Lambda').$$

Proposition 7. *Let $M^\bullet$ be a complex in Λ^{op}-$\mathbf{SP}(\Lambda')$. Then the natural morphism*

$$\Psi_M\, \mathrm{R}\, \Gamma_c(\overline{X}, -) \to \mathrm{R}\, \Gamma_c(\overline{X}, \Psi_M(-))$$

is a weak equivalence. In particular, the following diagram commutes.

$$\begin{array}{ccc}
\mathrm{K}_n(\mathbf{PDG}^{\mathrm{cont}}(X,\Lambda)) & \xrightarrow{\mathrm{K}_n \mathrm{R}\Gamma_c(\overline{X},-)} & \mathrm{K}_n(\mathbf{PDG}^{\mathrm{cont}}(\Lambda)) \\
\downarrow{\scriptstyle \mathrm{K}_n(\Psi_M)} & & \downarrow{\scriptstyle \mathrm{K}_n(\Psi_M)} \\
\mathrm{K}_n(\mathbf{PDG}^{\mathrm{cont}}(X,\Lambda')) & \xrightarrow{\mathrm{K}_n \mathrm{R}\Gamma_c(\overline{X},-)} & \mathrm{K}_n(\mathbf{PDG}^{\mathrm{cont}}(\Lambda'))
\end{array}$$

Proof. See Witte (2008, Proposition 5.5.7). □

Finally, we need the following result. Let X be a connected scheme and $f\colon Y \to X$ a finite *Galois covering* of X with *Galois group* G, i.e. f is finite étale and the degree of f is equal to the order of $G = \operatorname{Aut}_X(Y)$. We set

$$\mathbb{Z}_\ell[G]_X^\sharp = (\mathbb{Z}_\ell[G]/I \otimes_{\mathbb{Z}[G]} f_! f^*\mathbb{Z})_{I \in \mathfrak{I}_{\mathbb{Z}_\ell[G]}}.$$

Then $\mathbb{Z}_\ell[G]^\sharp_X$ is a system of locally constant sheaves which we may view as an object of $\mathbf{PDG}^{\text{cont}}(X, \mathbb{Z}_\ell[G])$ concentrated in degree 0. It corresponds to the continuous Galois module $\mathbb{Z}_\ell[G]$ on which the fundamental group of X acts contragrediently.

Lemma 1. *Let X be a connected scheme in $\mathbf{Sch}^{\text{sep}}_{\mathbb{F}}$. Let R be a finite $\mathbb{Z}_\ell$-algebra and let $\mathcal{F}^\bullet$ be a bounded complex of flat, locally constant, and constructible sheaves of R-modules. Then there exists a finite Galois covering Y of X with Galois group G and a complex $M^\bullet$ in $\mathbb{Z}_\ell[G]^{op}$-$\mathbf{SP}(R)$ such that*

$$\mathcal{F}^\bullet \cong \Psi_M(\mathbb{Z}_\ell[G]^\sharp_X)$$

in $\mathbf{PDG}(X, R)$.

Proof. Choose a large enough Galois covering $f: Y \to X$ such that $f^*\mathcal{F}^\bullet$ is a complex of constant sheaves and set $M^\bullet = \Gamma(Y, f^*\mathcal{F}^\bullet)$. This is in a natural way a complex in $\mathbb{Z}_\ell[G]^{op}$-$\mathbf{SP}(R)$ and

$$M^\bullet \otimes_{\mathbb{Z}_\ell[G]} \mathbb{Z}_\ell[G]^\sharp_X \cong \mathcal{F}^\bullet$$

See Witte (2008, Section 5.6) for further details. □

5 On $\mathrm{SK}_1(\mathbb{Z}_\ell[G][[T]])$

In this section, we will fix a finite extension F of $\mathbb{Q}_\ell$ and write $\mathcal{O}_F$ for the valuation ring of F. Let G be a finite group and let T denote an indeterminate that commutes with every element of $\mathcal{O}_F[G]$. In the end, we will only need the case $F = \mathbb{Q}_\ell$, but the more general results of this section may be interesting also for other applications.

Recall that there exists a finite extension F' of F such that $F'[G]$ is *split semisimiple*:

$$F'[G] \cong \prod_{k=1}^{r} \mathrm{End}^{\mathrm{op}}_{F'}(F'^{s_k})$$

for some integers $r, s_1, \dots, s_r$. Let M be a *maximal $\mathcal{O}_F$-order* in $F'[G]$, i.e. an $\mathcal{O}_F$-lattice in $F'[G]$ which is a subring and which is maximal with respect to this property. Then

$$M \cong \prod_{k=1}^{r} \mathrm{End}^{\mathrm{op}}_{\mathcal{O}_{F'}}(\mathcal{O}_{F'}^{s_k})$$

according to Oliver (1988, Theorem 1.9). In particular, the determinant map induces an isomorphism

$$\mathrm{K}_1(M) \cong \bigoplus_{k=1}^{r} \mathcal{O}_{F'}^{\times}.$$

If $\mathcal{O}_F[G] \subset M$, then Theorem 2.5 of loc. cit. implies

$$\mathrm{SK}_1(\mathcal{O}_F[G]) = \ker \mathrm{K}_1(\mathcal{O}_F[G]) \to \mathrm{K}_1(F[G]) = \ker \mathrm{K}_1(\mathcal{O}_F[G]) \to \mathrm{K}_1(M).$$

Analogously, we define a subgroup in $\mathrm{K}_1(\mathcal{O}_F[G][[T]])$.

Definition 16. Let G be a finite group and choose a finite extension F' of F such that $F'[G]$ is split semisimple. We set

$$\mathrm{SK}_1(\mathcal{O}_F[G][[T]]) = \ker \mathrm{K}_1(\mathcal{O}_F[G][[T]]) \to \mathrm{K}_1(M[[T]])$$

where M denotes a maximal $\mathcal{O}_F$-order in $F'[G]$ containing $\mathcal{O}_F[G]$.

Remark 8. Let $\overline{Q}$ be an algebraic closure of the fraction field of $\mathcal{O}_F[[T]]$. Then

$$\mathrm{SK}_1(\mathcal{O}_F[G][[T]]) = \ker \mathrm{K}_1(\mathcal{O}_F[G][[T]]) \to \mathrm{K}_1(\overline{Q}[G])$$

such that our definition agrees with the definition given in Chinburg and Pappas (2013).

Lemma 2. *For any finite group G,*

$$\mathrm{SK}_1(\mathcal{O}_F[G][[T]]) \cong \varprojlim_n \mathrm{SK}_1(\mathcal{O}_F[G \times \mathbb{Z}/(\ell^n)]).$$

Proof. Let F' and F'' be splitting fields for $\mathcal{O}_F[G]$ and $\mathcal{O}_F[G \times \mathbb{Z}/(\ell^n)]$, respectively and denote the corresponding maximal orders by M and M'. The commutativity of the diagram

$$\begin{array}{ccc}
\mathrm{K}_1(M[\mathbb{Z}/(\ell^n)]) & \longrightarrow & \mathrm{K}_1(M') \\
{\scriptstyle \cong}\big\downarrow{\scriptstyle \det} & & {\scriptstyle \cong}\big\downarrow{\scriptstyle \det} \\
\bigoplus_{k=1}^{r} \mathcal{O}_{F'}[\mathbb{Z}/(\ell^n)]^{\times} & \xrightarrow{\subset} & \bigoplus_{k=1}^{r'} \mathcal{O}_{F''}^{\times}
\end{array}$$

implies that

$$\mathrm{SK}_1(\mathcal{O}_F[G \times \mathbb{Z}/(\ell^n)]) = \ker \mathrm{K}_1(\mathcal{O}_F[G \times \mathbb{Z}/(\ell^n)]) \to \mathrm{K}_1(M[\mathbb{Z}/(\ell^n)]).$$

By Neukirch et al. (2000, Theorem 5.3.5) the choice of a topological generator $\gamma \in \mathbb{Z}_\ell$ induces an isomorphism

$$\mathcal{O}_F[[T]] \cong \varprojlim_n \mathcal{O}_F[\mathbb{Z}/(\ell^n)], \qquad T \mapsto \gamma - 1.$$

In particular, we have compatible isomorphisms

$$\mathrm{K}_1(\mathcal{O}_F[G][[T]]) \cong \varprojlim_n \mathrm{K}_1(\mathcal{O}_F[G \times \mathbb{Z}/(\ell^n)]), \quad \mathrm{K}_1(M[[T]]) \cong \varprojlim_n \mathrm{K}_1(M[\mathbb{Z}/(\ell^n)])$$

by Proposition 2. Hence, we obtain an isomorphism

$$\mathrm{SK}_1(\mathcal{O}_F[G][[T]]) \cong \varprojlim_n \mathrm{SK}_1(\mathcal{O}_F[G \times \mathbb{Z}/(\ell^n)])$$

as claimed. □

Proposition 8. *For any finite group G,*

$$\mathrm{SK}_1(\mathcal{O}_F[G][[T]]) \cong \mathrm{SK}_1(\mathcal{O}_F[G]).$$

Proof. By Lemma 2 it suffices to prove that the projection $\mathcal{O}_F[G \times \mathbb{Z}/(\ell^n)] \to \mathcal{O}_F[G]$ induces an isomorphism

$$\mathrm{SK}_1(\mathcal{O}_F[G \times \mathbb{Z}/(\ell^n)]) \cong \mathrm{SK}_1(\mathcal{O}_F[G]).$$

Let $g_1, \ldots, g_k$ be a system of representatives for the F-conjugacy classes of elements of order prime to ℓ in G. (Two elements g, h of order r prime to ℓ are called F-conjugated if $g^a = xhx^{-1}$ for some $x \in G$, $a \in \mathrm{Gal}(F(\zeta_r)/F) \subset (\mathbb{Z}/(r))^\times$.) Let r_i denote the order of g_i and set

$$N_i(G) = \left\{x \in G \quad \mid x g_i x^{-1} = g_i^a \text{ for some } a \in \mathrm{Gal}(F(\zeta_{r_i})/F)\right\},$$
$$Z_i(G) = \left\{x \in G \quad \mid x g_i x^{-1} = g_i\right\}.$$

Furthermore, let

$$\mathrm{H}_2^{ab}(Z_i(G), \mathbb{Z}) = \mathrm{Im}\left(\bigoplus_{\substack{H \subset Z_i(G) \\ \text{abelian}}} \mathrm{H}_2(H, \mathbb{Z}) \to \mathrm{H}_2(Z_i(G), \mathbb{Z}) \right)$$

denote the subgroup of the second homology group generated by elements induced up from abelian subgroups of $Z_i(G)$. According to Oliver (1988, Theorem 12.5) there exists an isomorphism

$$\mathrm{SK}_1(\mathcal{O}_F[G]) \cong \bigoplus_{i=1}^{k} \mathrm{H}_0(N_i(G)/Z_i(G), \mathrm{H}_2(Z_i(G), \mathbb{Z})/\,\mathrm{H}_2^{ab}(Z_i(G), \mathbb{Z}))_{(\ell)}.$$

Now, $(g_1, 0), \ldots, (g_k, 0)$ is a system of representatives for the F-conjugacy classes of elements of order prime to ℓ in $G \times \mathbb{Z}/(\ell^n)$ and

$$N_i(G \times \mathbb{Z}/(\ell^n)) = N_i(G) \times \mathbb{Z}/(\ell^n), \qquad Z_i(G \times \mathbb{Z}/(\ell^n)) = Z_i(G) \times \mathbb{Z}/(\ell^n).$$

By Oliver (1988, Prop. 8.12), we have

$$\begin{aligned}\mathrm{H}_2(Z_i(G) \times \mathbb{Z}/(\ell^n), \mathbb{Z})/\,\mathrm{H}_2^{ab}(Z_i(G) \times \mathbb{Z}/(\ell)^n, \mathbb{Z}) = \\ \mathrm{H}_2(Z_i(G), \mathbb{Z})/\,\mathrm{H}_2^{ab}(Z_i(G), \mathbb{Z}) \times \mathrm{H}_2(\mathbb{Z}/(\ell^n), \mathbb{Z})/\,\mathrm{H}_2^{ab}(\mathbb{Z}/(\ell^n), \mathbb{Z})\end{aligned}$$

and clearly,

$$\mathrm{H}_2(\mathbb{Z}/(\ell^n), \mathbb{Z}) = \mathrm{H}_2^{ab}(\mathbb{Z}/(\ell^n), \mathbb{Z}).$$

From this, the claim of the proposition follows. □

Remark 9. In the case that F is unramified over $\mathbb{Q}_\ell$ the above equality was independently observed by Chinburg, Pappas, and Taylor (Chinburg and Pappas 2013) (using a different approach) some years after the first version of this article had been made available.

The following proposition was proved in Fukaya and Kato (2006, Proposition 2.3.7) in the case of number fields and $F = \mathbb{Q}_\ell$.

Proposition 9. *Let Q be a function field of transcendence degree 1 over a finite field $\mathbb{F}$ and let ℓ be any prime. Then*

$$\varprojlim_{L} \mathrm{SK}_1(\mathcal{O}_F[\mathrm{Gal}(L/Q)]) = 0.$$

where L runs through the finite Galois extensions of Q in a fixed separable closure $\overline{Q}$ of Q.

Proof. Let F' be a totally ramified extension of F. Using Oliver (1988, Theorem 8.7.(i)) and the functoriality of the construction of the isomorphism in Oliver (1988, Theorem 12.5) one checks that the inclusion $\mathcal{O}_F \to \mathcal{O}_{F'}$ induces an isomorphism

$$\mathrm{SK}_1(\mathcal{O}_F[G]) \to \mathrm{SK}_1(\mathcal{O}_{F'}[G])$$

for any finite group G. Hence, we may assume that F is unramified over $\mathbb{Q}_\ell$. We then have a surjection

$$\mathrm{H}_2(G, \mathcal{O}_F[G_r]) \to \mathrm{SK}_1(\mathcal{O}_F[G])$$

with $\mathcal{O}_F[G_r]$ denoting the $\mathcal{O}_F$-module generated by the elements in G of order prime to ℓ (Oliver 1988, Theorem 12.10).

By the same argument as in the proof of Proposition 2.3.7 in Fukaya and Kato (2006), it now suffices to prove that

$$\mathrm{H}^2(\mathrm{Gal}(\overline{Q}/L), F/\mathcal{O}_F) = 0$$

for any finite extension L of Q.

If ℓ is different from the characteristic of $\mathbb{F}$, then the vanishing of this group can be deduced via the same argument as the analogous statement for number fields given in Schneider (1979, § 4, Satz 1): Let

$$L_\infty = \bigcup_n L(\zeta_{\ell^n}).$$

Then

$$\mathrm{H}^2(\mathrm{Gal}(\overline{Q}/L), F/\mathcal{O}_F) = \mathrm{H}^1(\mathrm{Gal}(L_\infty/L), \mathrm{H}^1(\mathrm{Gal}(\overline{Q}/L_\infty), F/\mathcal{O}_F))$$

and by Kummer theory,

$$\mathrm{H}^1(\mathrm{Gal}(\overline{Q}/L_\infty), F/\mathcal{O}_F) = L_\infty^\times \otimes_{\mathbb{Z}} F/\mathcal{O}_F(-1).$$

Now

$$\mathrm{H}^1(\mathrm{Gal}(L_\infty/L), L_\infty^\times \otimes_{\mathbb{Z}} F/\mathcal{O}_F(-1)) = \varinjlim_n \mathrm{H}^1(\mathrm{Gal}(L_\infty/L), L(\zeta_{\ell^n})^\times \otimes_{\mathbb{Z}} F/\mathcal{O}_F(-1))$$

by Neukirch et al. (2000, Proposition 1.5.1) and

$$\mathrm{H}^1(\mathrm{Gal}(L_\infty/L), L(\zeta_{\ell^n})^\times \otimes_{\mathbb{Z}} F/\mathcal{O}_F(-1)) = \left(L(\zeta_{\ell^n})^\times \otimes_{\mathbb{Z}} F/\mathcal{O}_F(-1)\right)_{\mathrm{Gal}(L_\infty/L)}$$

by loc. cit., Proposition 1.6.13. Since the latter group is a factor group of

$$\left(L(\zeta_{\ell^n})^\times \otimes_{\mathbb{Z}} F/\mathcal{O}_F(-1)\right)_{\mathrm{Gal}(L_\infty/L(\zeta_{\ell^n}))} = 0,$$

the claim is proved.

If ℓ is equal to the characteristic of $\mathbb{F}$, then the cohomological ℓ-dimension of $\mathrm{Gal}(\overline{Q}/L)$ is known to be 1 (Neukirch et al. 2000, Theorem 10.1.11.(iv)) and hence, the second cohomology group of $F/\mathcal{O}_F$ vanishes for trivial reasons. □

6 L-Functions

Consider an adic ring Λ and let $\Lambda[[T]]$ denote the ring of power series in the indeterminate T (where T is assumed to commute with every element of Λ). The ring $\Lambda[[T]]$ is still an adic ring whose Jacobson radical $\mathrm{Jac}(\Lambda[[T]])$ is generated by $\mathrm{Jac}(\Lambda)$ and T. In particular, we conclude from Proposition 2 that

$$\mathrm{K}_1(\Lambda[[T]]) = \varprojlim_n \mathrm{K}_1(\Lambda[[T]]/\,\mathrm{Jac}(\Lambda[[T]])^n)$$

is a profinite group.

Let $\mathbb{F}$ be a finite field. We write X_0 for the set of closed points of a scheme X in $\mathbf{Sch}_{\mathbb{F}}^{\mathrm{sep}}$. If $x \in X_0$ is a closed point, then

$$\overline{x} = x \times_{\operatorname{Spec}\mathbb{F}} \operatorname{Spec}\overline{\mathbb{F}}$$

consists of finitely many points, whose number is given by the degree $\deg(x)$ of x, i.e. the degree of the residue field $k(x)$ of x as a field extension of $\mathbb{F}$. Let

$$s_x\colon \overline{x} \to X$$

denote the structure map. For any complex

$$\mathcal{F}^\bullet = (\mathcal{F}_I^\bullet)_{I\in\mathfrak{I}_\Lambda}$$

in $\mathbf{PDG}^{\mathrm{cont}}(X,\Lambda)$, we write

$$\mathcal{F}_x^\bullet = (\Gamma(\overline{x}, s_x^*\mathcal{F}_I^\bullet))_{I\in\mathfrak{I}_\Lambda}.$$

This defines a Waldhausen exact functor

$$\mathbf{PDG}^{\mathrm{cont}}(X,\Lambda) \to \mathbf{PDG}^{\mathrm{cont}}(\Lambda), \qquad \mathcal{F}^\bullet \mapsto \mathcal{F}_x^\bullet.$$

Note that $\mathcal{F}_x^\bullet$ can also be written as the product over the stalks of $\mathcal{F}$ in the points of $\overline{x}$:

$$\mathcal{F}_x^\bullet \cong \prod_{\xi\in\overline{x}} ((\mathcal{F}_I^\bullet)_\xi)_{I\in\Lambda}.$$

The geometric Frobenius automorphism

$$\mathfrak{F}_{\mathbb{F}} \in \mathrm{Gal}(\overline{\mathbb{F}}/\mathbb{F})$$

operates on $\mathcal{F}_x^\bullet$ through its action on $\overline{x}$. Hence, it also operates on $\Psi_{\Lambda[[T]]}(\mathcal{F}_x^\bullet)$. Here,

$$\Psi_{\Lambda[[T]]}\colon \mathbf{PDG}^{\mathrm{cont}}(\Lambda) \to \mathbf{PDG}^{\mathrm{cont}}(\Lambda[[T]])$$

denotes the change of ring functor with respect to the $\Lambda[[T]]$-Λ-bimodule $\Lambda[[T]]$, as constructed in Definition 9. The morphism

$$\mathrm{id} - \mathfrak{F}_{\mathbb{F}} T \colon \Psi_{\Lambda[[T]]}(\mathcal{F}^{\bullet}_{\bar{x}}) \to \Psi_{\Lambda[[T]]}(\mathcal{F}^{\bullet}_{\bar{x}}).$$

is a natural isomorphism whose inverse is given by

$$\sum_{n=0}^{\infty} \mathfrak{F}_{\mathbb{F}}^{n} T^{n}.$$

Definition 17. The class

$$E_x(\mathcal{F}^{\bullet}, T) = [\Psi_{\Lambda[[T]]}(\mathcal{F}^{\bullet}_{\bar{x}}) \xrightarrow{\mathrm{id} - \mathfrak{F}_{\mathbb{F}} T} \Psi_{\Lambda[[T]]}(\mathcal{F}^{\bullet}_{\bar{x}})]^{-1}$$

in $\mathrm{K}_1(\Lambda[[T]])$ is called the *Euler factor* of $\mathcal{F}^{\bullet}$ at x.

One can easily verify that the Euler factor is multiplicative on exact sequences and that

$$E_x(\mathcal{F}^{\bullet}, T) = E_x(\mathcal{G}^{\bullet}, T)$$

if the complexes $\mathcal{F}^{\bullet}$ and $\mathcal{G}^{\bullet}$ are quasi-isomorphic. Hence, the above assignment extends to a homomorphism

$$E_x(-, T) \colon \mathrm{K}_0(\mathbf{PDG}^{\mathrm{cont}}(X, \Lambda)) \to \mathrm{K}_1(\Lambda[[T]]).$$

Lemma 3. *Let $\xi \in \bar{x}$ be a geometric point. Then*

$$E_x(\mathcal{F}^{\bullet}, T) = [\Psi_{\Lambda[[T]]}(\mathcal{F}^{\bullet}_{\xi}) \xrightarrow{\mathrm{id} - \mathfrak{F}_{k(x)} T^{\deg(x)}} \Psi_{\Lambda[[T]]}(\mathcal{F}^{\bullet}_{\xi})]^{-1}.$$

Proof. The Frobenius automorphism $\mathfrak{F}_{\mathbb{F}}$ induces isomorphisms $\mathcal{F}^{\bullet}_{\mathfrak{F}_{\mathbb{F}}^{k}\xi} \cong \mathcal{F}^{\bullet}_{\xi}$ for $k = 1, \dots, \deg(x)$. For $k = \deg(x)$ we have $\mathfrak{F}_{\mathbb{F}}^{k}\xi = \xi$ and the isomorphism is given by the operation of $\mathfrak{F}_{k(x)}$ on $\mathcal{F}^{\bullet}_{\xi}$. Hence, we may identify $\mathcal{F}^{\bullet}_{\bar{x}}$ with the complex $(\mathcal{F}^{\bullet}_{\xi})^{\deg(x)}$, on which the Frobenius $\mathfrak{F}_{\mathbb{F}}$ acts through the matrix

$$\begin{pmatrix} 0 & \dots\dots & 0 & \mathfrak{F}_{k(x)} \\ \mathrm{id} & 0 & \dots\dots & 0 \\ 0 & \mathrm{id} & 0 \ \dots & 0 \\ \vdots & \ddots & \ddots \ \ddots & \vdots \\ 0 & \dots & 0 \ \ \mathrm{id} & 0 \end{pmatrix}.$$

Let A be the automorphism of $\Psi_{\Lambda[[T]]}((\mathcal{F}^{\bullet}_{\xi})^{\deg(x)})$ given by the matrix

$$\begin{pmatrix} \mathrm{id} & 0 & \dots\dots\dots & & 0 \\ \mathrm{id}T & \mathrm{id} & 0 & \dots & 0 \\ \mathrm{id}T^2 & \mathrm{id}T & \mathrm{id} & 0 & \vdots \\ \vdots & \ddots & \ddots & \ddots & 0 \\ \mathrm{id}T^{\deg(x)-1} & \dots & \mathrm{id}T^2 & \mathrm{id}T & \mathrm{id} \end{pmatrix}$$

Then $A(\mathrm{id} - \mathfrak{F}_{\mathbb{F}}T)$ corresponds to the matrix

$$\begin{pmatrix} \mathrm{id} & 0 & \dots & 0 & -\mathfrak{F}_{k(x)}T \\ 0 & \mathrm{id} & 0 & \dots & -\mathfrak{F}_{k(x)}T^2 \\ \vdots & & \ddots & & \vdots \\ 0 & \dots & 0 & \mathrm{id} & -\mathfrak{F}_{k(x)}T^{\deg(x)-1} \\ 0 & \dots\dots & & 0 & \left(\mathrm{id} - \mathfrak{F}_{k(x)}T^{\deg(x)}\right) \end{pmatrix}$$

Moreover, we have $[A] = 1$ in $\mathrm{K}_1(\Lambda[[T]])$. Hence,

$$[\Psi_{\Lambda[[T]]}(\mathcal{F}^\bullet_x) \xrightarrow{\mathrm{id}-\mathfrak{F}_{\mathbb{F}}T} \Psi_{\Lambda[[T]]}(\mathcal{F}^\bullet_x)] = [\Psi_{\Lambda[[T]]}(\mathcal{F}^\bullet_\xi) \xrightarrow{\mathrm{id}-\mathfrak{F}_{k(x)}T^{\deg(x)}} \Psi_{\Lambda[[T]]}(\mathcal{F}^\bullet_\xi)]$$

as claimed. □

Proposition 10. *The infinite product*

$$\prod_{x\in X_0} E_x(\mathcal{F}^\bullet, T)$$

converges in the profinite topology of $\mathrm{K}_1(\Lambda[[T]])$.

Proof. For each integer m, there exist only finitely many closed points $x \in X_0$ with $\deg(x) < m$. If $\deg(x) \geq m$, then we conclude from Lemma 3 that the image of $E_x(\mathcal{F}^\bullet, T)$ in $\mathrm{K}_1(\Lambda[T]/(T^m))$ is 1. □

Definition 18. The *L-function* of the complex $\mathcal{F}^\bullet$ in $\mathbf{PDG}^{\mathrm{cont}}(X, \Lambda)$ is given by

$$L_{\mathbb{F}}(\mathcal{F}^\bullet, T) = \prod_{x\in X_0} E_x(\mathcal{F}^\bullet, T) \in \mathrm{K}_1(\Lambda[[T]])$$

Remark 10. If $\mathbb{F}'$ is a subfield of $\mathbb{F}$, then Lemma 3 implies that

$$L_{\mathbb{F}'}(\mathcal{F}^\bullet, T) = L_{\mathbb{F}}(\mathcal{F}^\bullet, T^{[\mathbb{F}:\mathbb{F}']}) \in \mathrm{K}_1(\Lambda[[T]]).$$

Remark 11. If Λ is commutative, the determinant induces an isomorphism

$$\det\colon \mathrm{K}_1(\Lambda[[T]]) \to \Lambda[[T]]^\times.$$

In particular, we see that the L-function agrees with the one defined in Deligne (1977, Fonction L mod ℓ^n) in the case of commutative adic rings.

7 The Grothendieck Trace Formula

In this section, we will prove the Grothendieck trace formula for our L-functions.

Definition 19. For a scheme X in $\mathbf{Sch}_{\mathbb{F}}^{\mathrm{sep}}$ and a complex $\mathcal{F}^\bullet$ in $\mathbf{PDG}^{\mathrm{cont}}(X, \Lambda)$ we let $\mathcal{L}_{\mathbb{F}}(\mathcal{F}^\bullet, T)$ denote the element

$$\left[\Psi_{\Lambda[[T]]}\left(\mathrm{R}_{\mathbb{F}}\, \Gamma_c(\overline{X}, \mathcal{F}^\bullet)\right) \xrightarrow{\mathrm{id}-\mathfrak{F}_{\mathbb{F}}T} \Psi_{\Lambda[[T]]}\left(\mathrm{R}_{\mathbb{F}}\, \Gamma_c(\overline{X}, \mathcal{F}^\bullet)\right)\right]^{-1}$$

in $\mathrm{K}_1(\Lambda[[T]])$.

Theorem 2 (Grothendieck trace formula). *Let $\mathbb{F}$ be a finite field of characteristic p and let Λ be an adic ring such that p is invertible in Λ. Then*

$$L_{\mathbb{F}}(\mathcal{F}^\bullet, T) = \mathcal{L}_{\mathbb{F}}(\mathcal{F}^\bullet, T)$$

for every scheme X in $\mathbf{Sch}_{\mathbb{F}}^{\mathrm{sep}}$ and every complex $\mathcal{F}^\bullet$ in $\mathbf{PDG}^{\mathrm{cont}}(X, \Lambda)$.

We proceed by a series of lemmas, following closely along the lines of Milne (1980, Chapter VI, §13).

Lemma 4. *Let U be an open subscheme of X with closed complement Z. Theorem 2 is true for X if it is true for U and Z.*

Proof. Write $j\colon U \hookrightarrow X$ and $i\colon Z \hookrightarrow X$ for the corresponding immersions,

$$u\colon U \to \operatorname{Spec} \mathbb{F}, \qquad x\colon X \to \operatorname{Spec} \mathbb{F}, \qquad z\colon Z \to \operatorname{Spec} \mathbb{F}$$

for the structure morphisms. Clearly,

$$L_{\mathbb{F}}(\mathcal{F}^\bullet, T) = L_{\mathbb{F}}(j^*\mathcal{F}^\bullet, T) L_{\mathbb{F}}(i^*\mathcal{F}^\bullet, T)$$

On the other hand, we have an exact sequence

$$\mathrm{R}\, x_! j_! j^* \mathcal{F}^\bullet \rightarrowtail \mathrm{R}\, x_! \mathcal{F}^\bullet \twoheadrightarrow \mathrm{R}\, x_! i_* i^* \mathcal{F}^\bullet$$

and (chains of) quasi-isomorphisms

$$\mathrm{R}\, u_! j^* \mathcal{F}^\bullet \simeq \mathrm{R}\, x_! j_! j^* \mathcal{F}^\bullet \qquad \mathrm{R}\, z_! i^* \mathcal{F}^\bullet \simeq \mathrm{R}\, x_! i_* i^* \mathcal{F}^\bullet.$$

Hence,

$$[\mathrm{R}\, x_! \mathcal{F}^\bullet] = [\mathrm{R}\, u_! j^* \mathcal{F}^\bullet][\mathrm{R}\, z_! i^* \mathcal{F}^\bullet] \tag{1}$$

in $\mathrm{K}_0\,\mathbf{PDG}^{\mathrm{cont}}(\operatorname{Spec}\mathbb{F},\Lambda)$ by Proposition 1. Using Proposition 5 one checks that the assignment

$$\mathcal{G}^\bullet \mapsto [\Psi_{\Lambda[[T]]}(\Gamma(\operatorname{Spec}\overline{\mathbb{F}}, s^*\mathcal{G}^\bullet)) \xrightarrow{\mathrm{id}-\mathfrak{F}_{\mathbb{F}}T} \Psi_{\Lambda[[T]]}(\Gamma(\operatorname{Spec}\overline{\mathbb{F}}, s^*\mathcal{G}^\bullet))]^{-1}$$

extends to a homomorphism

$$\mathrm{K}_0\,\mathbf{PDG}^{\mathrm{cont}}(\operatorname{Spec}\mathbb{F},\Lambda) \to K_1(\Lambda[[T]]),$$

which preserves relation (1). □

Lemma 5. *Let Λ' be a second adic ring and $M^\bullet$ be a complex in Λ^{op}-$\mathbf{SP}(\Lambda')$. For any $\mathcal{F}^\bullet$ in $\mathbf{PDG}^{\mathrm{cont}}(X,\Lambda)$ we have*

$$\begin{aligned}\mathrm{K}_1(\Psi_{M\otimes_\Lambda \Lambda[[T]]})(\mathcal{L}_{\mathbb{F}}(\mathcal{F}^\bullet,T)) &= \mathcal{L}_{\mathbb{F}}(\Psi_M\mathcal{F}^\bullet,T),\\ \mathrm{K}_1(\Psi_{M\otimes_\Lambda \Lambda[[T]]})(L_{\mathbb{F}}(\mathcal{F}^\bullet,T)) &= L_{\mathbb{F}}(\Psi_M\mathcal{F}^\bullet,T)\end{aligned}$$

in $\mathbf{PDG}^{\mathrm{cont}}(\Lambda'[[T]])$.

Proof. For $\mathcal{L}_{\mathbb{F}}(\mathcal{F}^\bullet,T)$, this follows from the definition and from Proposition 7. It is true for $\mathcal{L}_{\mathbb{F}}(\mathcal{F}^\bullet,T)$ because it is true for all $E_x(\mathcal{F}^\bullet,T)$, $x\in X_0$. □

Next, we prove that the formula is compatible with change of the base field.

Lemma 6. *Let $\mathbb{F}'$ be a subfield of $\mathbb{F}$ and X a scheme in $\mathbf{Sch}^{\mathrm{sep}}_{\mathbb{F}}$. Then*

$$\mathcal{L}_{\mathbb{F}'}(\mathcal{F}^\bullet,T) = \mathcal{L}_{\mathbb{F}}(\mathcal{F}^\bullet,T^{[\mathbb{F}:\mathbb{F}']}).$$

Proof. Let $r\colon \operatorname{Spec}\mathbb{F}\to\operatorname{Spec}\mathbb{F}'$ be the morphism induced by the inclusion $\mathbb{F}'\subset\mathbb{F}$ and write

$$h\colon X\times_{\operatorname{Spec}\mathbb{F}}\operatorname{Spec}\overline{\mathbb{F}}\to X, \qquad h'\colon X\times_{\operatorname{Spec}\mathbb{F}'}\operatorname{Spec}\overline{\mathbb{F}}\to X$$

$$s\colon\operatorname{Spec}\overline{\mathbb{F}}\to\operatorname{Spec}\mathbb{F}, \qquad s'\colon\operatorname{Spec}\overline{\mathbb{F}}\to\operatorname{Spec}\mathbb{F}'$$

for the corresponding structure morphisms. For any $\mathcal{F}^\bullet$ in $\mathbf{PDG}^{\mathrm{cont}}(X,\Lambda)$, the complexes $r_*\,\mathrm{R}\,h_!\mathcal{F}^\bullet$, $\mathrm{R}\,r_!\,\mathrm{R}\,h_!\mathcal{F}^\bullet$, and $\mathrm{R}\,h'_!\mathcal{F}^\bullet$ in $\mathbf{PDG}^{\mathrm{cont}}(\operatorname{Spec}\mathbb{F}',\Lambda)$ are quasi-isomorphic. Moreover, for any complex $\mathcal{G}^\bullet$ in $\mathbf{PDG}^{\mathrm{cont}}(\operatorname{Spec}\mathbb{F},\Lambda)$, the following diagram is commutative:

$$\begin{array}{ccc}\Gamma(\operatorname{Spec}\overline{\mathbb{F}},s^*r^*r_*\mathcal{G}^\bullet) & \xrightarrow{\cong} & \bigoplus_{k=1}^{[\mathbb{F}:\mathbb{F}']}\Gamma(\operatorname{Spec}\overline{\mathbb{F}},s^*\mathcal{G}^\bullet)\\ \Big\downarrow{\scriptstyle \mathfrak{F}_{\mathbb{F}'}} & & \Big\downarrow{\scriptstyle \begin{pmatrix}0 & \dots & 0 & \mathfrak{F}_{\mathbb{F}}\\ \mathrm{id} & 0 & \dots & 0\\ \vdots & \ddots & \ddots & \vdots\\ 0 & \dots & \mathrm{id} & 0\end{pmatrix}}\\ \Gamma(\operatorname{Spec}\overline{\mathbb{F}},s^*r^*r_*\mathcal{G}^\bullet) & \xrightarrow{\cong} & \bigoplus_{k=1}^{[\mathbb{F}:\mathbb{F}']}\Gamma(\operatorname{Spec}\overline{\mathbb{F}},s^*\mathcal{G}^\bullet)\end{array}$$

As in the proof of Lemma 3 one concludes

$$\left[\Psi_{\Lambda[[T]]}\big(\mathrm{R}_{\mathbb{F}'}\,\Gamma_c(\overline{X},\mathcal{F}^\bullet)\big)\xrightarrow{\mathrm{id}-\mathfrak{F}_{\mathbb{F}'}T}\Psi_{\Lambda[[T]]}\big(\mathrm{R}_{\mathbb{F}'}\,\Gamma_c(\overline{X},\mathcal{F}^\bullet)\big)\right]=$$

$$\left[\Psi_{\Lambda[[T]]}\big(\Gamma(\operatorname{Spec}\overline{\mathbb{F}},s^*r^*r_*\,\mathrm{R}\,h_!\mathcal{F}^\bullet)\big)\xrightarrow{\mathrm{id}-\mathfrak{F}_{\mathbb{F}'}T}\Psi_{\Lambda[[T]]}\big(\Gamma(\operatorname{Spec}\overline{\mathbb{F}},s^*r^*r_*\,\mathrm{R}\,h_!\mathcal{F}^\bullet)\big)\right]=$$

$$\left[\Psi_{\Lambda[[T]]}\big(\Gamma(\operatorname{Spec}\overline{\mathbb{F}},s^*\,\mathrm{R}\,h_!\mathcal{F}^\bullet)\big)\xrightarrow{\mathrm{id}-\mathfrak{F}_{\mathbb{F}}T^{[\mathbb{F}:\mathbb{F}']}}\Psi_{\Lambda[[T]]}\big(\Gamma(\operatorname{Spec}\overline{\mathbb{F}},s^*\,\mathrm{R}\,h_!\mathcal{F}^\bullet)\big)\right]=$$

$$\left[\Psi_{\Lambda[[T]]}\big(\mathrm{R}_{\mathbb{F}}\,\Gamma_c(\overline{X},\mathcal{F}^\bullet)\big)\xrightarrow{\mathrm{id}-\mathfrak{F}_{\mathbb{F}}T^{[\mathbb{F}:\mathbb{F}']}}\Psi_{\Lambda[[T]]}\big(\mathrm{R}_{\mathbb{F}}\,\Gamma_c(\overline{X},\mathcal{F}^\bullet)\big)\right].$$

□

Clearly, Theorem 2 is true for schemes of dimension 0. Next, we consider the case that X is a curve.

Lemma 7. *The formula in Theorem 2 is true for any smooth and geometrically connected curve X, $\Lambda = \mathbb{Z}_\ell[G]$, and $\mathcal{F}^\bullet = \mathbb{Z}_\ell[G]^\sharp_X$, where ℓ is a prime different from the characteristic of $\mathbb{F}$ and G is the Galois group of a finite Galois covering of X.*

Proof. Let Q be the function field of X and let F the function field of a finite Galois covering of X, i.e. F/Q is a finite Galois extension unramified in the closed points of X. Let d_F denote the element

$$d_F = L(\mathbb{Z}_\ell[\mathrm{Gal}(F/Q)]^\sharp_X, T)\mathcal{L}(\mathbb{Z}_\ell[\mathrm{Gal}(F/Q)]^\sharp_X, T)^{-1}$$

in $\mathrm{K}_1(\mathbb{Z}_\ell[\mathrm{Gal}(F/Q)][[T]])$.

Note that d_F does not change if we replace X by an open subscheme of X. Hence, by shrinking X appropriately, we may extend the definition of d_F to arbitrary finite Galois extensions F of Q. If $F \subset F'$ and F'/Q is Galois, then $d_{F'}$ is mapped onto d_F under the canonical homomorphism

$$\mathrm{K}_1(\mathbb{Z}_\ell[\mathrm{Gal}(F'/Q)][[T]]) \to \mathrm{K}_1(\mathbb{Z}_\ell[\mathrm{Gal}(F/Q)][[T]]).$$

To see this, choose X sufficiently small such that F'/Q is unramified in the closed points of X and apply Lemma 5.

Let L be a splitting field for $\mathbb{Q}_\ell[\mathrm{Gal}(F/Q)]$ and

$$M \subset L[\mathrm{Gal}(F/Q)]$$

a maximal $\mathbb{Z}_\ell$-order containing $\mathcal{O}_L[G]$. Recall that

$$M = \prod_{k=1}^{r} \mathrm{End}^{\mathrm{op}}_{\mathcal{O}_L}(P_k)$$

for some finitely generated free $\mathcal{O}_L$-modules P_k and some $r > 0$ (Oliver 1988, Theorem 1.9). We may thus consider P_k as an element in $\mathbb{Z}_\ell[G]^{\mathrm{op}}\text{-}\mathbf{SP}(\mathcal{O}_L)$. By the classical Grothendieck trace formula (Deligne 1977, Fonction L mod ℓ^n, Theorem 2.2.(a)) applied to the $\mathcal{O}_L$-sheaves $\Psi_{P_k}(\mathbb{Z}_\ell[\mathrm{Gal}(F/Q)]_X^\sharp)$ and by Lemma 5, the image of d_F under the homomorphism

$$\mathrm{K}_1(\mathbb{Z}_\ell[\mathrm{Gal}(F/Q)][[T]]) \to \mathrm{K}_1(M[[T]]) \xrightarrow{\cong} \bigoplus_{k=1}^{r} \mathcal{O}_L[[T]]^\times$$

is trivial; hence $d_F \in \mathrm{SK}_1(\mathbb{Z}_\ell[\mathrm{Gal}(F/Q)][[T]]) = \mathrm{SK}_1(\mathbb{Z}_\ell[\mathrm{Gal}(F/Q)])$. From Proposition 9 we conclude $d_F = 1$. □

Lemma 8. *The formula in Theorem 2 is true for any scheme X in $\mathbf{Sch}_{\mathbb{F}}^{\mathrm{sep}}$ of dimension less or equal 1, any adic ring Λ with $p \in \Lambda^\times$ and any complex $\mathcal{F}^\bullet$ in $\mathbf{PDG}^{\mathrm{cont}}(X, \Lambda)$.*

Proof. By Proposition 2 it suffices to consider finite rings Λ. The ℓ-Sylow subgroups of Λ are subrings of Λ and Λ is equal to their direct product. Since p is invertible, the p-Sylow subgroup is trivial. Hence, we may further assume that Λ is a $\mathbb{Z}_\ell$-algebra for $\ell \neq p$.

Shrinking X if necessary we may assume that X is smooth, irreducible curve and that $\mathcal{F}^\bullet$ is a strictly perfect complex of locally constant sheaves. By replacing $\mathbb{F}$ with its algebraic closure in the function field of X and using Lemma 6, we may assume that X is geometrically connected. By Lemmas 1 and 5 we have

$$\mathcal{L}(\mathcal{F}^\bullet, T) = \mathrm{K}_1(\Psi_{M \otimes_{\mathbb{Z}_\ell[G]} \mathbb{Z}_\ell[G][[T]]})(\mathcal{L}(\mathbb{Z}_\ell[G]_X^\sharp, T))$$

for a suitable Galois group G and a complex $M^\bullet$ in $\mathbb{Z}_\ell[G]^{\mathrm{op}}\text{-}\mathbf{SP}(\Lambda)$. Likewise,

$$L(\mathcal{F}^\bullet, T) = \mathrm{K}_1(\Psi_{M \otimes_{\mathbb{Z}_\ell[G]} \mathbb{Z}_\ell[G][[T]]})(L(\mathbb{Z}_\ell[G]_X^\sharp, T)).$$

Now the assertion follows from Lemma 7. □

We complete the proof of Theorem 2 by induction on the dimension d of X. By shrinking X if necessary we may assume that there exists a morphism $f: X \to Y$ such that Y and all fibres of f have dimension less than d. Then Proposition 6.(3) and the induction hypothesis imply

$$\mathcal{L}(\mathcal{F}^\bullet, T) = \mathcal{L}(\mathrm{R}\, f_! \mathcal{F}^\bullet, T) = L(\mathrm{R}\, f_! \mathcal{F}^\bullet, T).$$

Let now y be a closed point of Y. Write $f_y: X_y \to X$ for the fibre over y. Then

$$\begin{aligned} E_y(\mathrm{R}\, f_! \mathcal{F}^\bullet, T) &= \left[\Psi_{\Lambda[[T]]}\left(\mathrm{R}\,\Gamma_c(\overline{X}_y, f_y^* \mathcal{F}^\bullet)\right) \xrightarrow{\mathrm{id} - \mathfrak{F}_{\mathbb{F}} T} \Psi_{\Lambda[[T]]}\left(\mathrm{R}\,\Gamma_c(\overline{X}_y, f_y^* \mathcal{F}^\bullet)\right)\right]^{-1} \\ &= L(f_y^* \mathcal{F}^\bullet, T) \end{aligned}$$

by Proposition 6.(4) and the induction hypothesis. Since clearly

$$L(\mathcal{F}^\bullet, T) = \prod_{y \in Y_0} L(f_y^* \mathcal{F}^\bullet, T),$$

Theorem 2 follows.

Remark 12. The formula in Theorem 2 is also valid if Λ is a finite field of characteristic p, see Deligne (1977, Fonction L mod ℓ^n, Theorem 2.2.(b)). However, it does not extend to general adic $\mathbb{Z}_p$-algebras. We refer to loc. cit., §4.5 for a counterexample. More precisely, if Λ is a commutative $\mathbb{Z}_p$-algebra, then the difference between $L_{\mathbb{F}}(\mathcal{F}^\bullet, T)$ and $\mathcal{L}_{\mathbb{F}}(\mathcal{F}^\bullet, T)$ is given by a unit in

$$\Lambda\langle T\rangle = \varprojlim_{I \in \mathfrak{I}(\Lambda)} \Lambda/I[T]$$

(Emerton and Kisin 2001). In Witte (2013) we generalise this result to noncommutative $\mathbb{Z}_p$-algebras Λ.

Acknowledgements The author would like to thank Annette Huber and Alexander Schmidt for their encouragement and for valuable discussions. He also thanks the anonymous referee for various corrections.

References

Artin, M., Grothendieck, A., Verdier, J.L.: Théorie des topos et cohomologie étale des schémas (SGA 4–3). Lecture Notes in Mathematics, vol. 305. Springer, Berlin (1972)

Avramov, L.L., Foxby, H.-B.: Homological dimensions of unbounded complexes. J. Pure Appl. Algebra **71**, 129–155 (1991)

Chinburg, T., Pappas, G., Taylor, M.J.: K_1 of a p-adic group ring II. The determinantal kernel SK_1 (2013, preprint). arXiv:1303.5337v1

Chinburg, T., Pappas, G., Taylor, M.J.: Higher adeles and non-abelian Riemann-Roch (2012, preprint). arXiv:1204.4520v2

Deligne, P.: Cohomologie étale (SGA $4\frac{1}{2}$). Lecture Notes in Mathematics, vol. 569. Springer, Berlin (1977)

Emerton, M., Kisin, M.: Unit L-functions and a conjecture of Katz. Ann. Math. **153**(2), 329–354 (2001)

Fukaya,T., Kato, K.: A formulation of conjectures on p-adic zeta functions in non-commutative Iwasawa theory. In: Proceedings of the St. Petersburg Mathematical Society, vol. XII, American Mathematical Society, Series 2, vol. 219, pp. 1–85. American mathematical society, Providence (2006)

Grothendieck, A.: Cohomologie ℓ-adique et fonctions L (SGA 5). Lecture Notes in Mathematics, vol. 589. Springer, Berlin (1977)

Lam, T.Y.: A First Course in Noncommutative Rings. Graduate Texts in Mathematics, vol. 131. Springer, Berlin (1991)

Milne, J.S.: Etale Cohomology. Princeton Mathematical Series, vol. 33. Princeton University Press, Princeton (1980)

Muro, F., Tonks, A.: The 1-type of a Waldhausen K-theory spectrum. Adv. Math. **216**(1), 178–211 (2007)

Muro, F., Tonks, A.: On K_1 of a Waldhausen category. K-theory and noncommutative geometry. EMS Series of Congress Reports, pp. 91–116 (2008)

Nagata, M.: A generalization of the imbedding problem of an abstract variety in a complete variety. J. Math. Kyoto Univ. **3**, 89–102 (1963)

Neukirch, J., Schmidt, A., Wingberg, K.: Cohomology of Number Fields. Grundlehren der mathematischen Wissenschaften, vol. 323. Springer, Berlin/Heidelberg (2000)

Oliver, R.: Whitehead Groups of Finite Groups. London Mathematical Society lecture notes series, vol. 132. Cambridge University Press, Cambridge (1988)

Schlichting, M.: A note on K-theory and triangulated categories. Invent. Math. **150**(1), 111–116 (2002)

Schneider, P.: Über gewisse Galoiscohomologiegruppen. Math. Z. **260**, 181–205 (1979)

Thomason, R.W., Trobaugh, T.: Higher algebraic K-theory of schemes and derived categories. The Grothendieck Festschrift, vol. III, Progress in Mathematics, vol. 88, pp. 247–435. Birkhäuser, Boston (1990)

Waldhausen, F.: Algebraic K-theory of spaces. In: Algebraic and Geometric Topology (Berlin Heidelberg). Lecture Notes in Mathematics, vol. 1126, pp. 318–419. Springer, Berlin (1985)

Witte, M.: Noncommutative Iwasawa main conjectures for varieties over finite fields. Ph.D. thesis, Universität Leipzig (2008). Available at `http://www.dart-europe.eu/full.php?id=162794`

Witte, M.: Unit L-functions for étale sheaves of modules over noncommutative rings (2013, preprint). arXiv:1309.2493v1

On $\hat{\mathbb{Z}}$-Zeta Function

Zdzisław Wojtkowiak

1 Introduction

In this note we propose a definition of zeta function for $\mathbb{Q}$ which generalizes or rather incorporates all p-adic L-series of Kubota-Leopoldt for $\mathbb{Q}$. It also incorporates so called Soulé classes – which we call ℓ-adic polylogarithms evaluated at $\overrightarrow{10}$ – which are elements of $H^1_{\text{et}}(\mathbb{Q}; \mathbb{Z}_\ell(n))$.

This is very important because the Soulé class in $H^1_{\text{et}}(\mathbb{Q}; \mathbb{Z}_\ell(n))$ is an analog of the real number $\zeta(n)$. Both determine an extension of $\mathbb{Z}(0)$ by $\mathbb{Z}(n)$ in different realizations.

We shall define this new $\hat{\mathbb{Z}}$-zeta function as a measure more precisely as a cocycle on $G_{\mathbb{Q}}$ with values in measures. This point of view on p-adic zeta functions is indicated in de Shalit (1987, pages 15–16). This measure appears naturally when studying the representation of the absolute Galois group $G_{\mathbb{Q}}$ on $\pi_1^{\text{et}}(\mathbb{P}^1_{\bar{\mathbb{Q}}} \setminus \{0, 1, \infty\}, \overrightarrow{01})$.

During the meeting IWASAWA 2012 in Heidelberg Prof. John Coates mentioned the need of zeta functions defined on $\hat{\mathbb{Z}}$ and with values in $\hat{\mathbb{Z}}$, if the author of the present note understood correctly his remark. The measure we are studying, has perhaps some properties required by him. Therefore the author decided to write this note.

Another remark, which somebody made during his talk was that in the p-adic case one must work very hard to prove congruence relations and then construct a measure. In this approach one gets immediately measures which however depend on elements of Galois group.

Z. Wojtkowiak (✉)
Laboratoire Jean Alexandre Dieudonné, U.R.A. au C.N.R.S., No 168, Département de Mathématiques, Université de Nice-Sophia Antipolis, Parc Valrose – B.P.N° 71, 06108 Nice Cedex 2, France
e-mail: wojtkow@unice.fr; Zdzislaw.WOJTKOWIAK@unice.fr

T. Bouganis and O. Venjakob (eds.), *Iwasawa Theory 2012*, Contributions in Mathematical and Computational Sciences 7, DOI 10.1007/978-3-642-55245-8_16,

Some time ago the author has planned to write together with Hiroaki Nakamura a paper on adelic polylogarithms, which should generalize our work (Nakamura and Wojtkowiak 2002). Though we never even started such a paper, this project served also as a motivation to write the present note.

The action of G_K on $\pi_1^{\text{et}}(E_{\bar{K}}\backslash\{0\}, \overrightarrow{0})$, where E_K is an elliptic curve over K, leads also to measures (see Nakamura 1995, 2013). We do not know if one gets p-adic non-Archimedean zeta functions of E_K in this case. However if the elliptic curve has a complex multiplication then it seems clear that the obtained measure leads to p-adic non-Archimedean zeta functions of the elliptic curves with a complex multiplication. This was our impression when we looked at various papers on the subject, for example Yager (1986) or de Shalit (1987).

On the other side in Wojtkowiak, studying the action of $G_{\mathbb{Q}}$ on torsors of paths on $\mathbb{P}^1_{\bar{\mathbb{Q}}} \setminus \{0, 1, \infty\}$ we get measures on $(\mathbb{Z}_p)^r$, which should lead to p-adic non-Archimedean multi-zeta functions. We do not have generalizations of these measures to $\hat{\mathbb{Z}} \otimes \mathbb{Q}$-valued measures on $(\hat{\mathbb{Z}})^r$.

2 Formulation of Main Results

Let us define

$$\hat{\mathbb{Z}} := \varprojlim \mathbb{Z}/N\mathbb{Z},$$

where the structure maps are the projections

$$\mathbb{Z}/MN\mathbb{Z} \to \mathbb{Z}/N\mathbb{Z}.$$

We define the Iwasawa algebra

$$\hat{\mathbb{Z}}[[\hat{\mathbb{Z}}]] := \varprojlim \mathbb{Z}/N\mathbb{Z}[\mathbb{Z}/M\mathbb{Z}],$$

where the structure maps are

$$\mathbb{Z}/NN_1\mathbb{Z}[\mathbb{Z}/MM_1\mathbb{Z}] \to \mathbb{Z}/N\mathbb{Z}[\mathbb{Z}/M\mathbb{Z}].$$

We define an action of the group $\hat{\mathbb{Z}}^\times$ on $\hat{\mathbb{Z}}[[\hat{\mathbb{Z}}]]$. We define this action on finite levels. If $c \in \hat{\mathbb{Z}}^\times$ and $\sum_{i=0}^{N-1} a_i[i] \in \hat{\mathbb{Z}}[\mathbb{Z}/N\mathbb{Z}]$ then

$$c(\sum_{i=0}^{N-1} a_i[i]) = \sum_{i=0}^{N-1} ca_i[c^{-1}i], \tag{1}$$

where $[a]$ denotes the class of a modulo N.

Let $\chi : G_{\mathbb{Q}} \to \hat{\mathbb{Z}}^{\times}$ be the cyclotomic character. Composing χ with the action of $\hat{\mathbb{Z}}^{\times}$ on $\hat{\mathbb{Z}}[[\hat{\mathbb{Z}}]]$ defined above, we get an action of $G_{\mathbb{Q}}$ on $\hat{\mathbb{Z}}[[\hat{\mathbb{Z}}]]$. Let

$$Z^1(G_{\mathbb{Q}}, \hat{\mathbb{Z}}[[\hat{\mathbb{Z}}]])$$

be the set of cocycles on $G_{\mathbb{Q}}$ with values in the $G_{\mathbb{Q}}$-module $\hat{\mathbb{Z}}[[\hat{\mathbb{Z}}]]$.

Using the action of $G_{\mathbb{Q}}$ on the tower of coverings $\mathbb{P}^1_{\bar{\mathbb{Q}}} \setminus (\{0,\infty\} \cup \mu_N)$ of $\mathbb{P}^1_{\bar{\mathbb{Q}}} \setminus \{0,1,\infty\}$ we get an element belonging to $Z^1(G_{\mathbb{Q}}, \hat{\mathbb{Z}}[[\hat{\mathbb{Z}}]])$ and which we denote by

$$\hat{\zeta}_{\mathbb{Q}}\,.$$

In fact this element already appeared in Nakamura and Wojtkowiak (2002) and before in Ichimura and Sakaguchi (1987). In Nakamura and Wojtkowiak (2002) it is called adelic Kummer-Heisenberg values.

For any rational prime p there is a projection

$$pr_p : \hat{\mathbb{Z}}[[\hat{\mathbb{Z}}]] \to \mathbb{Z}_p[[\mathbb{Z}_p]]$$

compatible with the action of $G_{\mathbb{Q}}$ on $\hat{\mathbb{Z}}[[\hat{\mathbb{Z}}]]$ and $\mathbb{Z}_p[[\mathbb{Z}_p]]$. Composing $\hat{\zeta}_{\mathbb{Q}}$ with the projection pr_p we get an element which we denote

$$\zeta_p \in Z^1\big(G_{\mathbb{Q}}, \mathbb{Z}_p[[\mathbb{Z}_p]]\big).$$

In Wojtkowiak where we are dealing only with one fixed prime, the element ζ_p is denoted $K_1(\overrightarrow{10})$. The element ζ_p allows to recover p-adic L-functions of Kubota-Leopoldt. We recall definitions and results from Wojtkowiak.

Let $\omega : \mathbb{Z}_p^{\times} \to \mu_{p-1} \subset \mathbb{Z}_p^{\times}$ be the Teichmüller character. For $x \in \mathbb{Z}_p^{\times}$ we set

$$[x] := x\omega(x)^{-1}\,.$$

We denote by $\chi_p : G_{\mathbb{Q}} \to \mathbb{Z}_p^{\times}$ the cyclotomic character restricted to $\mathrm{Gal}(\mathbb{Q}(\mu_{p^\infty})/\mathbb{Q})$.

Definition 1. Let p be a rational prime. Let $0 \leq \beta < p-1$. We define the function

$$\mathcal{Z}_p^{\beta} : \mathbb{Z}_p \times G_{\mathbb{Q}} \to \mathbb{Z}_p$$

by the formula

$$\mathcal{Z}_p^{\beta}(1-s,\sigma) := \int_{\mathbb{Z}_p^{\times}} [x]^s x^{-1} \omega(x)^{\beta} d\zeta_p(\sigma)(x)\,.$$

Theorem 1 (see Wojtkowiak, Corollary 8.3). *Let p be a rational prime. Let β be even and $0 \leq \beta < p-1$. Let $\sigma \in G_{\mathbb{Q}}$ be such that $(\chi_p(\sigma))^{p-1} \neq 1$. Then*

$$\mathcal{Z}_p^{\beta}(1-s,\sigma) = \frac{1}{2}\big([\chi_p(\sigma)]^s \omega(\chi_p(\sigma))^{\beta} - 1\big) L_p(1-s,\omega^{\beta}),$$

where $L_p(1-s,\omega^{\beta})$ is the Kubota-Leopoldt p-adic L-function.

We recall that p-adic L-functions were first defined in Kubota and Leopoldt (1964). The definition we used in Wojtkowiak is that of Lang (1990).

Now we shall look at functions $\mathcal{Z}_p^{\beta}$ for β odd.

Definition 2. Let $s \in \mathbb{Z}_p$. Let $\sigma \in G_{\mathbb{Q}}$ and let $x \in \mathbb{Z}_p$. We define a $G_{\mathbb{Q}}$-module

$$\mathbb{Z}_p\langle s\rangle$$

defining the action of $G_{\mathbb{Q}}$ on $\mathbb{Z}_p$ by

$$\sigma(x) := [\chi_p(\sigma)]^s x.$$

Let $0 \leq \beta < p-1$. We define a $G_{\mathbb{Q}}$-module

$$\big(\omega(\chi_p)^{\beta}\big)$$

defining the action of $G_{\mathbb{Q}}$ on $\mathbb{Z}_p$ by

$$\sigma(x) := (\omega(\chi_p(\sigma)))^{\beta} x.$$

We denote by

$$\mathbb{Z}_p\langle s\rangle \otimes_{\mathbb{Z}_p} \big(\omega(\chi_p)^{\beta}\big)$$

the $\mathbb{Z}_p$-module $\mathbb{Z}_p$ equipped with the action of $G_{\mathbb{Q}}$ given by

$$\sigma(x) := [\chi_p(\sigma)]^s (\omega(\chi_p(\sigma)))^{\beta} x.$$

Observe that if k is an integer then

$$\mathbb{Z}_p(k) = \mathbb{Z}_p\langle k\rangle \otimes_{\mathbb{Z}_p} \big(\omega(\chi_p)^k\big).$$

In Theorem 1 we have seen that for β even the element $\hat{\zeta}_{\mathbb{Q}}$ gives non-Archimedean p-adic L-series. The next result shows that for β odd we get Soulé classes for the field $\mathbb{Q}$ (ℓ-adic Galois polylogarithms evaluated at $\overrightarrow{10}$).

Theorem 2. *Let p be a rational prime and let β be an integer such that $0 \le \beta < p-1$.*

(i) For a fixed $s \in \mathbb{Z}_p$, the function

$$G_{\mathbb{Q}} \ni \sigma \mapsto \mathcal{Z}_p^{\beta}(1-s,\sigma) \in \mathbb{Z}_p\langle s\rangle \otimes_{\mathbb{Z}_p} \left(\omega(\chi_p)^{\beta}\right)$$

on $G_{\mathbb{Q}}$ with values in a $G_{\mathbb{Q}}$-module $\mathbb{Z}_p\langle s\rangle \otimes_{\mathbb{Z}_p} \left(\omega(\chi_p)^{\beta}\right)$ is a cocycle;

(ii) Let β be odd and let k be a positive integer such that $k \equiv \beta$ modulo $p-1$. Then

$$\mathcal{Z}_p^{\beta}(1-k,\sigma) = (1-p^{k-1})(k-1)!l_k(\overrightarrow{10})(\sigma) = s_k(\sigma)\,,$$

where $l_k(\overrightarrow{10})$ is the p-adic Galois polylogarithm (see Wojtkowiak 2005) and s_k is the Soulé class (see Soulé 1987).

(The point (ii)) of the theorem is of course well known, see Ihara (1986, Corollary of Theorem 10) and Deligne (1989). The proof of the point (i) of the theorem is indicated in Sect. 3.)

3 Measures

In this section we define the element $\hat{\zeta}_{\mathbb{Q}}$. Let

$$V_N := \mathbb{P}^1_{\bar{\mathbb{Q}}} \setminus (\{0,\infty\} \cup \mu_N)$$

and let

$$f_N^{MN} : V_{MN} \to V_N$$

be given by $f_N^{MN}(\mathfrak{z}) = \mathfrak{z}^M$. We get a projective system of finite coverings of

$$V_1 = \mathbb{P}^1_{\bar{\mathbb{Q}}} \setminus \{0,1,\infty\}\,.$$

Applying the functor π_1^{et} we get a projective system of pro-finite groups

$$\{\pi_1^{\mathrm{et}}(V_N,\overrightarrow{01})\}_{N\in\mathbb{N}}\,. \tag{2}$$

We fix an embedding $\bar{\mathbb{Q}} \subset \mathbb{C}$. Hence we have a comparison homomorphism

$$\pi_1(V_N(\mathbb{C}),\overrightarrow{01}) \to \pi_1^{\mathrm{et}}(V_N,\overrightarrow{01})$$

and therefore the natural topological generators of $\pi_1^{\text{et}}(V_N, \overrightarrow{01})$. We denote by

$$x_N$$

(loop around 0) and by

$$y_{i,N}$$

(loop around ξ_N^i) for $0 \leq i < N$, the standard free generators of $\pi_1^{\text{et}}(V_N, \overrightarrow{01})$. We denote by

$$\Gamma^2 \pi_1^{\text{et}}(V_N, \overrightarrow{01})$$

the commutator subgroup of the group $\pi_1^{\text{et}}(V_N, \overrightarrow{01})$.

We have

$$(f_N^{MN})_*(x_{MN}) = (x_N)^M \tag{3}$$

and

$$(f_N^{MN})_*(y_{i,MN}) \equiv y_{i',N} \mod \Gamma^2 \pi_1^{et}(V_N, \overrightarrow{01}) \tag{4}$$

for $0 \leq i < MN$ and where $i' \equiv i$ modulo N and $0 \leq i' < N$.

Let p_N be the canonical path on V_N from $\overrightarrow{01}$ to $\frac{1}{N}\overrightarrow{10}$, the interval [0, 1]. For any $\sigma \in G_{\mathbb{Q}}$, the elements

$$\mathfrak{f}_{p_N}(\sigma) := p_N^{-1} \cdot \sigma(p_N) \in \pi_1^{\text{et}}(V_N, \overrightarrow{01})$$

form a coherent family of the projective system (2).

Here and later our convention of composing a path α from y to z with a path β from x to y will be that $\alpha \cdot \beta$ is defined as a path from x to z.

For $N \in \mathbb{N}$ and $0 \leq i < N$ we define elements

$$\alpha_i^N(\sigma) \in \hat{\mathbb{Z}}$$

by the congruence

$$\mathfrak{f}_{p_N}(\sigma) \equiv \prod_{i=0}^{N-1} (y_{i,N})^{\alpha_i^N(\sigma)} \mod \Gamma^2 \pi_1^{\text{et}}(V_N, \overrightarrow{01}) . \tag{5}$$

It follows immediately from the congruences (4) that the functions

$$\mathbb{Z}/N\mathbb{Z} \ni i \mapsto \alpha_i^N(\sigma) \in \hat{\mathbb{Z}} \tag{6}$$

form a distribution on $\hat{\mathbb{Z}}$ with values in $\hat{\mathbb{Z}}$, hence they form a measure. This measure we denote by

$$\hat{\zeta}_{\mathbb{Q}}(\sigma)\,.$$

The element $\hat{\zeta}_{\mathbb{Q}}(\sigma)$ belongs to $\hat{\mathbb{Z}}[[\hat{\mathbb{Z}}]]$. Hence we get a function

$$\hat{\zeta}_{\mathbb{Q}} : G_{\mathbb{Q}} \to \hat{\mathbb{Z}}[[\hat{\mathbb{Z}}]]\,.$$

Definition 3. The function

$$\hat{\zeta}_{\mathbb{Q}} : G_{\mathbb{Q}} \to \hat{\mathbb{Z}}[[\hat{\mathbb{Z}}]]$$

we call $\hat{\mathbb{Z}}$-zeta function of the field $\mathbb{Q}$.

We shall study properties of the function $\hat{\zeta}_{\mathbb{Q}}$. First however we state a result about the action of $G_{\mathbb{Q}}$ on $\pi_1^{\mathrm{et}}(V_N, \overrightarrow{01})$. Next we shall calculate explicitly the coefficients $\alpha_i^N(\sigma)$.

We recall that $\bar{\mathbb{Q}} \subset \mathbb{C}$. We set

$$\xi_m := e^{\frac{2\pi\sqrt{-1}}{m}}$$

for any $m \in \mathbb{N}$.

Below the suffix $i\chi(\sigma)$ at $y_{i\chi(\sigma),N}$ means the unique integer $0 \le r < N$ such that $i\chi(\sigma) \equiv r$ modulo N. The same remark applies to other situations when suffixes are in $\{0, 1, \ldots, N-1\}$.

Proposition 1. *Let $\sigma \in G_{\mathbb{Q}}$. We have*

$$\sigma(x_N) = (x_N)^{\chi(\sigma)}$$

and

$$\sigma(y_{i,N}) \equiv (y_{i\chi(\sigma),N})^{\chi(\sigma)} \mod \Gamma^2\pi_1^{\mathrm{et}}(V_N, \overrightarrow{01})$$

for $0 \le i < N$.

Proof. Let $\mathfrak{z}$ be a local parameter at 0 corresponding to $\overrightarrow{01}$. We apply $\sigma \cdot y_{i,N} \cdot \sigma^{-1}$ to the germ of $(1 - \xi_N^{-i\chi(\sigma)}\mathfrak{z})^{\frac{1}{m}}$ at $\overrightarrow{01}$. After applying σ^{-1} we get $(1 - \xi_N^{-i}\mathfrak{z})^{\frac{1}{m}}$. Then we apply $y_{i,N}$ and we get $\xi_m(1-\xi_N^{-i}\mathfrak{z})^{\frac{1}{m}}$. Hence finally after acting by σ we get $\xi_m^{\chi(\sigma)}(1-\xi_N^{-i\chi(\sigma)}\mathfrak{z})^{\frac{1}{m}}$. The effect is the same as applying $(y_{i\chi(\sigma),N})^{\chi(\sigma)}$ to $(1 - \xi_N^{-i\chi(\sigma)}\mathfrak{z})^{\frac{1}{m}}$. □

Proposition 2. *The coefficient*

$$\alpha_i^N(\sigma) = \left(\alpha_i^{N,m}(\sigma)\right)_{m\in\mathbb{N}} \in \varprojlim_m(\mathbb{Z}/m\mathbb{Z}) = \hat{\mathbb{Z}}$$

is given by the formula

$$\frac{\sigma\big((1-\xi_N^{-i\chi(\sigma^{-1})})^{\frac{1}{m}}\big)}{(1-\xi_N^{-i})^{\frac{1}{m}}}=\xi_m^{\alpha_i^{N,m}(\sigma)}$$

for $0<i<N$ *and by the formula*

$$\sigma(N^{-\frac{1}{m}})=\xi_m^{\alpha_0^{N,m}(\sigma)}N^{-\frac{1}{m}}$$

for $i=0$.

Proof. We act on the germ of $(1-\xi_N^{-i}\mathfrak{z})^{\frac{1}{m}}$ by the path $p_N^{-1}\cdot\sigma(p_N)$. Notice that the local parameter at 1 corresponding to $\frac{1}{N}\overrightarrow{10}$ is $s=N(1-\mathfrak{z})$. We get that

$$p_N^{-1}\cdot\sigma(p_N):(1-\xi_N^{-i}\mathfrak{z})^{\frac{1}{m}}\to\frac{\sigma\big((1-\xi_N^{-i\chi(\sigma^{-1})})^{\frac{1}{m}}\big)}{(1-\xi_N^{-i})^{\frac{1}{m}}}(1-\xi_N^{-i}\mathfrak{z})^{\frac{1}{m}}\,.$$

On the other side it follows from the congruences (5) that we get $\xi_m^{\alpha_i^{N,m}(\sigma)}(1-\xi_N^{-i}\mathfrak{z})^{\frac{1}{m}}$. Hence we get the formula for $0<i<N$.

To calculate the coefficient $\alpha_0^N(\sigma)$ we act on the germ of $(1-\mathfrak{z})^{\frac{1}{m}}$. Applying $p_N\cdot\sigma^{-1}$ to $(1-\mathfrak{z})^{\frac{1}{m}}$ we get $N^{-\frac{1}{m}}s^{\frac{1}{m}}$. Next applying $p_N^{-1}\cdot\sigma$ we get $\sigma(N^{-\frac{1}{m}})N^{\frac{1}{m}}(1-\mathfrak{z})^{\frac{1}{m}}$. □

We indicate the cocycle and the continuity properties. Let τ and σ belong to $G_{\mathbb{Q}}$. Then it follows from the equality

$$\mathfrak{f}_{p_N}(\tau\sigma)=\mathfrak{f}_{p_N}(\tau)\cdot\tau(\mathfrak{f}_{p_N}(\sigma))$$

(see for example the proof of Wojtkowiak 2004, Proposition 1.0.7) that

$$\alpha_i^N(\tau\sigma)=\alpha_i^N(\tau)+\chi(\tau)\alpha_{i\chi(\tau)^{-1}}^N(\sigma)\,. \tag{7}$$

The direct consequences of the equality (7) are the point (i) of Theorem 2 and the following result.

Proposition 3. *(i) The function*

$$\hat{\xi}_{\mathbb{Q}}:G_{\mathbb{Q}}\to\hat{\mathbb{Z}}[[\hat{\mathbb{Z}}]]$$

is a cocycle.

(ii) Let k be a positive integer. We have

$$\int_{\mathbb{Z}_p} x^{k-1}\omega(x)^\beta d\zeta_p(\tau\sigma) =$$

$$\int_{\mathbb{Z}_p} x^{k-1}\omega(x)^\beta d\zeta_p(\tau) + \chi_p(\tau)^k \omega(\chi_p(\tau))^\beta \int_{\mathbb{Z}_p} x^{k-1}\omega(x)^\beta d\zeta_p(\sigma).$$

Observe that in the point (ii) of the proposition we recover the cocycle property of p-adic Galois polylogarithms $l_k(\overrightarrow{10})$ (Soulé classes for the field $\mathbb{Q}$), see for example Wojtkowiak (2005, Corollary 11.0.12.).

The proof of the continuity of the function $\mathcal{Z}_p^\beta(s,\sigma)$ is also straightforward, so we formulate only the result.

Proposition 4. *The function of two variables*

$$\mathcal{Z}_p^\beta : \mathbb{Z}_p \times G_{\mathbb{Q}} \to \mathbb{Z}_p$$

is continuous.

Theorem 1 is the main reason that we call the cocycle $\hat{\zeta}_{\mathbb{Q}}$, the $\hat{\mathbb{Z}}$-zeta function of the field $\mathbb{Q}$. The paper Wojtkowiak is still not published. Hence we give a sketch of a proof of Theorem 1.

Proof of Theorem 1.

Proof. It follows from Nakamura and Wojtkowiak (2002, Proposition 3) that

$$l_k(\overrightarrow{10})(\sigma) = \frac{1}{(k-1)!}\int_{\mathbb{Z}_p} x^{k-1} d\zeta_p(\sigma)(x). \tag{8}$$

In Wojtkowiak (2009, Proposition 3.1) we have shown that

$$l_{2k}(\overrightarrow{10})(\sigma) = -\frac{B_{2k}}{2(2k)!}(\chi_p(\sigma)^{2k} - 1), \tag{9}$$

where B_{2k} is the $2k$th Bernoulli number. (The result is already stated in Ihara (1990) without proof.) It is an elementary property of the measure $\zeta_p(\sigma)$ that

$$\int_{\mathbb{Z}_p} x^{k-1} d\zeta_p(\sigma)(x) = \frac{1}{1-p^{k-1}}\int_{\mathbb{Z}_p^\times} x^{k-1} d\zeta_p(\sigma)(x). \tag{10}$$

It follows immediately from the equalities (8)–(10) that

$$\frac{2}{(\chi_p(\sigma)^{2k} - 1)}\int_{\mathbb{Z}_p^\times} x^{2k-1} d\zeta_p(\sigma)(x) = -(1-p^{2k-1})\frac{B_{2k}}{2k}.$$

Theorem 1 follows from the last equality. □

4 Relations with Iwasawa Theory

We recall that we have defined the action of $\mathbb{Z}_p^\times$ (which we identify with the group $\mathrm{Gal}(\mathbb{Q}(\mu_{p^\infty})/\mathbb{Q})$ via the cyclotomic character χ_p) on $\mathbb{Z}_p[[\mathbb{Z}_p]]$ via the formula (1). This action of $\mathbb{Z}_p^\times$ is the consequence of the equality

$$\mathfrak{f}_{p^N}(\tau\sigma) = \mathfrak{f}_{p^N}(\tau) \cdot \tau(\mathfrak{f}_{p^N}(\sigma)).$$

Let $\mathcal{M}_\infty$ be the maximal abelian pro-p extension of $\mathbb{Q}(\mu_{p^\infty})$, which is unramified outside p. In the Iwasawa theory the action of $\mathbb{Z}_p^\times = \mathrm{Gal}(\mathbb{Q}(\mu_{p^\infty})/\mathbb{Q})$ on $\mathrm{Gal}(\mathcal{M}_\infty/\mathbb{Q}(\mu_{p^\infty}))$ is given via inner automorphisms. If $\sigma \in \mathrm{Gal}(\mathcal{M}_\infty/\mathbb{Q}(\mu_{p^\infty}))$, $\tau \in \mathbb{Z}_p^\times$ and $\tilde{\tau}$ is a lifting of τ to $\mathrm{Gal}(\mathcal{M}_\infty/\mathbb{Q})$ then

$$\tau(\sigma) := \tilde{\tau}\sigma\tilde{\tau}^{-1}$$

(see Coates and Sujatha 2006, page 5).

It follows from Wojtkowiak (2004) that for any $\tau,\ \sigma \in G_{\mathbb{Q}}$ we have

$$\mathfrak{f}_{p^N}(\tau\sigma\tau^{-1}) = \mathfrak{f}_{p^N}(\tau) \cdot \tau(\mathfrak{f}_{p^N}(\sigma)) \cdot \big((\tau\sigma\tau^{-1})((\mathfrak{f}_{p^N}(\tau)^{-1})\big)$$

Hence comparing coefficients modulo $\Gamma^2\pi_1^{\text{ét}}(V_N, \overrightarrow{01})$ we get

$$\alpha_i^N(\tau\sigma\tau^{-1}) = \alpha_i^N(\tau) - \chi(\sigma)\alpha^N_{i\chi(\sigma)^{-1}}(\tau) + \chi(\tau)\alpha^N_{i\chi(\tau)^{-1}}(\sigma).$$

Now we assume that $N = p^n$ and $\sigma \in G_{\mathbb{Q}(\mu_{p^\infty})}$. Then we get

$$\alpha_i^{p^n}(\tau\sigma\tau^{-1}) = \chi_p(\tau)\alpha^{p^n}_{i\chi_p(\tau)^{-1}}(\sigma).$$

Hence for $\sigma \in G_{\mathbb{Q}(\mu_{p^\infty})}$, the Iwasawa action of $\mathbb{Z}_p^\times$ on measures $\zeta_p(\sigma)$ is given by the formula (1).

Let $n \geq m$. We set

$$K_{n,m} := \mathbb{Q}(\mu_{p^n})\big((1-\xi_{p^n}^i)^{\frac{1}{p^m}} \mid 0 < i < p^n\big)$$

and

$$\mathcal{K}_\infty := \bigcup_{n \geq m \geq 1} K_{n,m}.$$

It follows from Proposition 2 that the cocycle $\zeta_p : G_{\mathbb{Q}} \to \mathbb{Z}_p[[\mathbb{Z}_p]]$ factors through the quotient group

$$\mathrm{Gal}(\mathcal{K}_\infty/\mathbb{Q})$$

of $G_{\mathbb{Q}}$.

Observe that $\mathcal{K}_\infty$ is an abelian pro-p extension of $\mathbb{Q}(\mu_{p^\infty})$ which is unramified outside p. Moreover $\mathcal{K}_\infty$ is Galois over $\mathbb{Q}$. Hence we have a surjective morphism of $\mathbb{Z}_p[[\mathbb{Z}_p^\times]]$-modules

$$Gal\big(\mathcal{M}_\infty/\mathbb{Q}(\mu_{p^\infty})\big) \to Gal(\mathcal{K}_\infty/\mathbb{Q}(\mu_{p^\infty}))\,.$$

Proposition 5. *(i) The cocycle $\zeta_p : G_\mathbb{Q} \to \mathbb{Z}_p[[\mathbb{Z}_p]]$ vanishes on $G_{\mathcal{K}_\infty}$, hence it induces*

$$\zeta_p : Gal(\mathcal{K}_\infty/\mathbb{Q}) \to \mathbb{Z}_p[[\mathbb{Z}_p]]\,.$$

(ii) The restriction of ζ_p to $\mathrm{Gal}(\mathcal{K}_\infty/\mathbb{Q}(\mu_{p^\infty}))$ *is a morphism of $\mathbb{Z}_p[[\mathbb{Z}_p^\times]]$-modules.*
(iii) The cocycle ζ_p : $\mathrm{Gal}(\mathcal{K}_\infty/\mathbb{Q}) \to \mathbb{Z}_p[[\mathbb{Z}_p]]$ *is injective.*

Proof. The points (i) and (iii) follow from Proposition 2. The point (ii) was shown in the discussion before the proposition when we observe that the Iwasawa action on measures $\zeta_p(\sigma)$ for $\sigma \in G_{\mathbb{Q}(\mu_{p^\infty})}$ is given by the formula (1). □

Let $p > 2$. The action of $-1 \in \mathbb{Z}_p^\times$ decomposes any $\mathbb{Z}_p^\times$-module M on the direct sum

$$M = M^+ \oplus M^-\,,$$

where -1 acts on M^+ as the identity and on M^- as the multiplication by -1.

Proposition 6. *Let $p > 2$. The image of* $\mathrm{Gal}\big(\mathcal{K}_\infty/\mathbb{Q}(\mu_{p^\infty})\big)$ *in $\mathbb{Z}_p[[\mathbb{Z}_p]]$ by ζ_p is contained in $\mathbb{Z}_p[[\mathbb{Z}_p]]^-$.*

Proof. Let $\sigma \in G_{\mathbb{Q}(\mu_{p^\infty})}$. Observe that the involution induced by -1 maps the element $\sum_{i=0}^{p^n-1} \alpha_i^{p^n}(\sigma)[i]$ into $-\sum_{i=0}^{p^n-1} \alpha_{-i}^{p^n}(\sigma)[i]$. It follows from Proposition 2 that
$\sum_{i=0}^{p^n-1} \alpha_i^{p^n}(\sigma)[i] - \sum_{i=0}^{p^n-1} \alpha_{-i}^{p^n}(\sigma)[i] = 0$. Hence $\zeta_p(\sigma) \in \mathbb{Z}_p[[\mathbb{Z}_p]]^-$. □

Corollary 1. *The plus part of $Gal\big(\mathcal{K}_\infty/\mathbb{Q}(\mu_{p^\infty})\big)$ is 0.*

We denote by $(-1)_*$ the morphism induced by $-1 \in \mathbb{Z}_p^\times$. We denote by δ_0 the Dirac distribution on $\mathbb{Z}_p$ concentrated at 0.

We recall the definition of the Bernoulli measure $E_{1,c}$ on $\mathbb{Z}_p$ (see Lang 1990). Let $c \in \mathbb{Z}_p^\times$. The Bernoulli measure

$$E_{1,c} = \big(E_{1,c}^{(n)} : \mathbb{Z}/p^n\mathbb{Z} \to \mathbb{Q}_p\big)_{n\in\mathbb{N}}$$

on $\mathbb{Z}_p$ is defined by

$$E_{1,c}^{(n)}(i) = \frac{i}{p^n} - c\frac{\langle c^{-1}i\rangle}{p^n} + \frac{c-1}{2}$$

for $0 \leq i < p^n$, where $0 \leq \langle c^{-1}i \rangle < p^n$ and $\langle c^{-1}i \rangle \equiv c^{-1}i$ modulo p^n.

Proposition 7. *Let $p > 2$ and let $\sigma \in G_{\mathbb{Q}}$. We have*

$$\zeta_p(\sigma) + (-1)_* \zeta_p(\sigma) = E_{1,\chi_p(\sigma)} + \frac{1 - \chi_p(\sigma)}{2} \delta_0 .$$

Proof. It follows from Wojtkowiak, Lemma 4.1 (see also Nakamura and Wojtkowiak 2012, proof of Proposition 5.13) that $\alpha_i^{p^n}(\sigma) - \alpha_{-i}^{p^n}(\sigma) = E^{(n)}_{1,\chi_p(\sigma)}(i)$ for $0 < i < p^n$. On the other side $E^{(n)}_{1,\chi_p(\sigma)}(0) = \frac{\chi_p(\sigma)-1}{2}$ for any n. Hence the proposition follows. □

We were hoping that studying the representation of $G_{\mathbb{Q}}$ on the étale fundamental group $\pi_1^{\text{ét}}(\mathbb{P}^1_{\bar{\mathbb{Q}}} \setminus \{0, 1, \infty\}, \overrightarrow{01})_{pro-p}$ we can recover the Main Conjecture (see Coates and Sujatha 2006, Theorem 1.4.3). It seems however that it is not the case. In fact in Ichimura and Sakaguchi (1987) one can find more relations with the Main Conjecture that in the present paper.

On the other side the Main Conjecture should appear, manifest somewhere, when studying representations of $G_{\mathbb{Q}}$ on π_1.

Acknowledgements These research were started in January 2011 during our visit in Max-Planck-Institut für Mathematik in Bonn. We would like to thank very much MPI for support. We would like also thank to Professor C. Greither for invitation on the conference Iwasawa 2008 in Kloster Irsee. We acknowledge the financial help of the *Laboratoire de Dieudonné*, which allows us to participate in the meeting Iwasawa 2012 in Heidelberg.

References

Coates, J., Sujatha, R.: Cyclotomic Fields and Zeta Values. Springer Monographs in Mathematics. Springer, Berlin (2006)

Deligne, P.: Le groupe fondamental de la droite projective moins trois points. In: Galois Groups over Q. Mathematical Sciences Research Institute Publications, vol. 16, pp. 79–297. Springer, New York (1989)

de Shalit, E.: Iwasawa Theory of Elliptic Curves with Complex Multiplication. Perspective in Mathematics, vol. 3. Academic, Boston (1987)

Ichimura, H., Sakaguchi, K.: The non-vanishing of a certain Kummer character χ_m (after Soulé), and some related topics. In: Galois Representations and Arithmetic Algebraic Geometry. Advanced Studies in Pure Mathematics, vol. 12, pp. 53–64. Kinokuniya Co./North-Holland/Elsevier, Tokyo/Amsterdam/New York (1987)

Ihara, Y.: Profinite braid groups, Galois representations and complex multiplications. Ann. Math. **123**, 43–106 (1986)

Ihara, Y.: Braids, Galois groups, and some arithmetic functions. In: Proceedings of the International Congress of Mathematics, Kyoto, pp. 99–120. Springer (1990)

Kubota, T., Leopoldt, H.W.: Eine p-adische Theorie der Zetawerte, I. J. Reine und angew. Math. **214/215**, 328–339 (1964)

Lang, S.: Cyclotomic Fields I and II. Graduate Texts in Mathematics, vol. 121. Springer, New York (1990)

Nakamura, H.: On exterior Galois representations associated with open elliptic curves. J. Math. Sci. Univ. Tokyo **2**, 197–231 (1995)
Nakamura, H.: On arithmetic monodromy representations of Eisenstein type in fundamental groups of once punctured elliptic curves. Publ. RIMS Kyoto Univ. **49**(3), 413–496 (2013)
Nakamura, H., Wojtkowiak, Z.: On the explicit formulae for ℓ-adic polylogarithms. In: Arithmetic Fundamental Groups and Noncommutative Algebra. Proceedings of Symposia in Pure Mathematics (AMS), vol. 70, pp. 285–294. American Mathematical Society, Providence (2002)
Nakamura, H., Wojtkowiak, Z.: Tensor and homotopy criteria for functional equations of l-adic and classical iterated integrals. In: Non-abelian Fundamental Groups and Iwasawa Theory. London Mathematical Society Lecture Note Series, vol. 393, pp. 258–310. Cambridge University Press, Cambridge/New York (2012)
Soulé, C.: Éléments Cyclotomiques en K-Théorie. Astérisque **147/148**, 225–258 (1987)
Wojtkowiak, Z.: On ℓ-adic iterated integrals, I analog of Zagier conjecture. Nagoya Math. J. **176**, 113–158 (2004)
Wojtkowiak, Z.: On ℓ-adic iterated integrals, II functional equations and l-adic polylogarithms. Nagoya Math. J. **177**, 117–153 (2005)
Wojtkowiak, Z.: On ℓ-adic Galois periods, relations between coefficients of Galois representations on fundamental groups of a projective line minus a finite number of points. Publ. Math. de Besançon, Algèbra et Théorie des Nombres, 2007–2009, pp. 155–174, Février (2009)
Wojtkowiak, Z.: On ℓ-adic Galois L-functions (2014). arXiv:1403.2209 [math.NT]
Yager, R.I.: On two variable p-adic L-functions. Ann. Math. **115**, 411–449 (1986)

Erratum to: Modular Symbols in Iwasawa Theory

Takako Fukaya, Kazuya Kato, and Romyar Sharifi

Erratum to:
T. Bouganis and O. Venjakob (eds.), *Iwasawa Theory 2012*, Contributions in Mathematical and Computational Sciences 7, DOI 10.1007/978-3-642-55245-8_5,

---

The commutative diagram in Figure 1.10 was not the intended diagram. The correct diagram follows.

$$\begin{array}{ccccc} \mathcal{S} & \xrightarrow{z} & K & \xrightarrow{\mathrm{reg}} & \mathfrak{S} \\ {\scriptstyle \mathrm{mod}\, I}\downarrow & & \downarrow{\scriptstyle \infty} & & \downarrow{\scriptstyle \mathrm{mod}\, I} \\ P & \xrightarrow{\varpi} & Y & \xrightarrow{\xi'\Upsilon} & P. \end{array}$$

The reference citation on page 180 should be Section 2.7 and not Section 7 in the first paragraph.

The online version of the updated chapter can be found under
DOI 10.1007/978-3-642-55245-8_5

T. Fukaya • K. Kato
Department of Mathematics, University of Chicago, 5734 S. University Ave., Chicago, IL 60637, USA
e-mail: takako@math.uchicago.edu; kkato@math.uchicago.edu

R. Sharifi (✉)
Department of Mathematics, University of Arizona, 617 N. Santa Rita Ave., Tucson, AZ 85711, USA
e-mail: sharifi@math.arizona.edu

T. Bouganis and O. Venjakob (eds.), *Iwasawa Theory 2012*, Contributions in Mathematical and Computational Sciences 7, DOI 10.1007/978-3-642-55245-8_17,

MIX
Papier aus verantwortungsvollen Quellen
Paper from responsible sources
FSC® C105338

If you have any concerns about our products,
you can contact us on
ProductSafety@springernature.com

In case Publisher is established outside the EU,
the EU authorized representative is:
Springer Nature Customer Service Center GmbH
Europaplatz 3, 69115 Heidelberg, Germany

Printed by Libri Plureos GmbH
in Hamburg, Germany